T0140583

Lecture Notes in Electrical Engineering

Volume 422

Board of Series editors

Leopoldo Angrisani, Napoli, Italy
Marco Arteaga, Coyoacán, México
Samarjit Chakraborty, München, Germany
Jiming Chen, Hangzhou, P.R. China
Tan Kay Chen, Singapore, Singapore
Rüdiger Dillmann, Karlsruhe, Germany
Haibin Duan, Beijing, China
Gianluigi Ferrari, Parma, Italy
Manuel Ferre, Madrid, Spain
Sandra Hirche, München, Germany
Faryar Jabbari, Irvine, USA
Janusz Kacprzyk, Warsaw, Poland
Alaa Khamis, New Cairo City, Egypt
Torsten Kroeger, Stanford, USA
Tan Cher Ming, Singapore, Singapore
Wolfgang Minker, Ulm, Germany
Pradeep Misra, Dayton, USA
Sebastian Möller, Berlin, Germany
Subhas Mukhopadyay, Palmerston, New Zealand
Cun-Zheng Ning, Tempe, USA
Toyoaki Nishida, Sakyo-ku, Japan
Bijaya Ketan Panigrahi, New Delhi, India
Federica Pascucci, Roma, Italy
Tariq Samad, Minneapolis, USA
Gan Woon Seng, Nanyang Avenue, Singapore
Germano Veiga, Porto, Portugal
Haitao Wu, Beijing, China
Junjie James Zhang, Charlotte, USA

About this Series

"Lecture Notes in Electrical Engineering (LNEE)" is a book series which reports the latest research and developments in Electrical Engineering, namely:

- Communication, Networks, and Information Theory
- Computer Engineering
- Signal, Image, Speech and Information Processing
- Circuits and Systems
- Bioengineering

LNEE publishes authored monographs and contributed volumes which present cutting edge research information as well as new perspectives on classical fields, while maintaining Springer's high standards of academic excellence. Also considered for publication are lecture materials, proceedings, and other related materials of exceptionally high quality and interest. The subject matter should be original and timely, reporting the latest research and developments in all areas of electrical engineering.

The audience for the books in LNEE consists of advanced level students, researchers, and industry professionals working at the forefront of their fields. Much like Springer's other Lecture Notes series, LNEE will be distributed through Springer's print and electronic publishing channels.

More information about this series at http://www.springer.com/series/7818

Neil Y. Yen · Jason C. Hung
Editors

Frontier Computing

Theory, Technologies and Applications FC 2016

Editors
Neil Y. Yen
School of Computer Science and Engineering
University of Aizu
Aizu-Wakamatsu, Fukushima
Japan

Jason C. Hung
Department of Information Technology
Overseas Chinese University
Taichung
Taiwan

ISSN 1876-1100 ISSN 1876-1119 (electronic)
Lecture Notes in Electrical Engineering
ISBN 978-981-10-9811-6 ISBN 978-981-10-3187-8 (eBook)
DOI 10.1007/978-981-10-3187-8

© Springer Nature Singapore Pte Ltd. 2018
Softcover reprint of the hardcover 1st edition 2017
This work is subject to copyright. All rights are reserved by the Publisher, whether the whole or part of the material is concerned, specifically the rights of translation, reprinting, reuse of illustrations, recitation, broadcasting, reproduction on microfilms or in any other physical way, and transmission or information storage and retrieval, electronic adaptation, computer software, or by similar or dissimilar methodology now known or hereafter developed.
The use of general descriptive names, registered names, trademarks, service marks, etc. in this publication does not imply, even in the absence of a specific statement, that such names are exempt from the relevant protective laws and regulations and therefore free for general use.
The publisher, the authors and the editors are safe to assume that the advice and information in this book are believed to be true and accurate at the date of publication. Neither the publisher nor the authors or the editors give a warranty, express or implied, with respect to the material contained herein or for any errors or omissions that may have been made. The publisher remains neutral with regard to jurisdictional claims in published maps and institutional affiliations.

Printed on acid-free paper

This Springer imprint is published by Springer Nature
The registered company is Springer Nature Singapore Pte Ltd.
The registered company address is: 152 Beach Road, #21-01/04 Gateway East, Singapore 189721, Singapore

Message from General Chairs

The International Conference on Frontier Computing—Theory, Technologies, and Applications (FC) was first proposed in early 2010 on an IET executive meeting. This conference series aims at providing an open forum to reach a comprehensive understanding to the recent advances and emergence in information technology, science, and engineering, with the themes in the scope of Communication Network Technology and Applications, Communication Network Technology and Applications, Business Intelligence and Knowledge Management, Web Intelligence, and any related fields that prompt the development of information technology. This will be the fourth event of the series, in which fruitful results can be found in the digital library or conference proceedings of FC 2010 (Taichung, Taiwan), FC 2012 (Xining, China), FC 2013 (Gwangju, Korea), and FC 2015 (Bangkok, Thailand). Each event brings together the researchers worldwide to have excited and fruitful discussions as well as the future collaborations.

The papers accepted for inclusion in the conference proceeding primarily cover the topics: database and data mining, networking and communications, Web and Internet of things, embedded system, soft computing, social network analysis, security and privacy, optics communication, and ubiquitous/pervasive computing. Many papers have shown their great academic potential and value, and, in addition, indicate promising directions of research in the focused realm of this conference series. We believe that the presentations of these accepted papers will be more exciting than the papers themselves, and lead to creative and innovative applications. We hope that the attendees (and readers as well) will find these results useful and inspiring to your field of specialization and future research.

On behalf of the organizing committee, we would like to thank the members of the organizing and the program committees, the authors, and the speakers for their dedication and contributions that make this conference possible. We would like to thank and welcome all participants to the capital city of Japan—Tokyo. Tokyo is a city with a long and remarkable history. To get a picture of North-East Asia, this city will certainly be an entry. Though most of countries may share some similar characteristics, you will find that culture of Japan is very rich from different perspectives, such as art, religion, nomadic lifestyle, food, and music. Tokyo is a world-class and well-known city, with modern facilities and convenience traffic. We encourage the participants to take this chance to see and experience Japan, especially the remote counties and the nomadic lifestyle there. We also sincerely hope that all participants from overseas and from Japan enjoy the technical discussions at the conference, build a strong friendship, and establish ties for future collaborations.

We send our sincere appreciations to the authors for their valuable contributions and the other participants of this conference. The conference would not have been possible without their support. Thanks are also due to the many experts who contributed to making the event a success.

July 2016

Neil Y. Yen, University of Aizu, Japan
Jason C. Hung, Overseas Chinese University, Taiwan
FC 2016 Steering Committee Chairs

Organization Committee

General Chairs
Jen-Shiun Chiang, Tamkang University, Taiwan

Vice General Chairs
Han-Chieh Chao, National Ilan University, Taiwan
Young-Ae Jung, Sun Moon University, Korea
Qingguo Zhou, Lanzhou University, China

Steering Committee Chairs
Kuan-Ching Li, Providence University, Taiwan
Jason C. Hung, Overseas Chinese University, Taiwan
Neil Y. Yen, University of Aizu, Japan

International Advisory Board
Jinannong Cao, Hong Kong Polytechnic University, Hong Kong
Su-Ching Chen, University of Florida, USA
Fatos Xhafa, Technical University of Catalonia, Spain
Jianhua Ma, Hosei University, Japan
Runhe Huang, Hosei University, Japan
Qun Jin, Waseda University, Japan
Victor Leung, University of British Columbia, Canada
Qing Li, City University of Hong Kong, Hong Kong
Zheng Xu, Shanghai University, China
Jean-Luc Gaudiot, University of California-Irvine, USA
Mu-Yen Chen, National Taichung University of Science and Technology, Taiwan

Program Chairs
Hai Jiang, Arkansas State University, USA
Meng-Yen Hsieh, Providence University, Taiwan
Zhou Rui, Lanzhou University, China

Vice Program Chairs
Deqiang Han, Beijing University of Technology, China
Hsuan-Fu Wang, Chung Chou University of Science and Technology, Taiwan
Jun-Hong Shen, Asia University, Taiwan
Fang-Biau Ueng, National Chung Hsing University, Taiwan

Workshop Chairs
Wei-Chen Wu, Hsin Sheng College of Medical Care and Management, Taiwan
Kehan Zeng, University of Macau, Macau
You-Shyang Chen, Hwa Hsia University of Technology, Taiwan
Chengjiu Yin, Kyushu University, Japan

Special Session Chairs
Yu-Wei Chan, Chung Chou University of Science and Technology, Taiwan
Masaru Fukushi, Yamaguchi University, Japan
Yan Pei, University of Aizu, Japan

Tutorial Chairs
Yutaka Watanobe, University of Aizu, Japan
Yun Yang, Swinburne University of Technology, Melbourne, Australia
Beihong Jin, Professor, Chinese Academy of Sciences, China

Publicity Chairs
Vladimír Smejkal, Brno University of Technology, Czech Republic
Fei Wu, Zhejiang University, China
Francisco Isidro Massetto, Federal University of ABC, Brazil
Shing-Chern You, National Taipei University of Technology, Taiwan
Riz Sulaiman, Universiti Kebangsaan Malaysia, Malaysia
Wei Tsang Ooi, National University of Singapore, Singapore
Yusuke Manabe, Chiba Institute of Technology, Japan
Soumya Banerjee, Birla Institute of Technology, India
Tran Thien Phuc, Hochimin City University of Technology, Vietnam
Jindrich Kodl, Authorised Expert in Security of Information Systems, Cryptology and Informatics, Czech Republic
Poonphon Suesaowaluk, Assumption University of Thailand, Thailand
Jenn-Wei Lin, Fu Jen University, Taiwan
Jun Shen, University of Wollongong, Australia
Gerald Schaefer, Loughborough University, UK
Yishui Zhu, Chang'an University, China
Chun-Hong Huang, Lunghwa University of Science and Technology, Taiwan
Yung-Hui Chen, Lunghwa University of Science and Technology, Taiwan
Kuan-Chou Lai, National Taichung University of Education, Taiwan
Kuan-Cheng Lin, National Chung Hsing University, Taiwan

Keynote Speaker

Hideyuki Takagi
Professor
Chair of Department of Art and Infomration Design, School of Design
Chair of Human Science Course/Human Science International Course, Graduate School of Design
Kyushu University

Email: takagi@design.kyushu-u.ac.jp
URL: http://www.design.kyushu-u.ac.jp/~takagi

Fitness Landscape Information for Acceleration and Applications for Human Science

Abstract

We introduce two topics in evolutionary computation (EC) research. The first topic is to estimate fitness landscape information and use it to accelerate EC convergence. Some of these approaches include approximation of fitness landscape and estimate a rough point of the global optimum, estimation of unimodal or multimodal shape of the landscape, and estimation of the global optimum from the movement direction of individuals between generations. The estimated location of the global optimum can be used as an elite individual, and shape information of a fitness landscape can be used to change EC operational parameters adaptively. They result fast EC convergence. The second topic is a new type of EC applications, human science. Some of these researches include measuring mental dynamic range, obtaining unknown knowledge on auditory system, and awareness science as a support system of human awareness.

Biography

Professor Takagi had worked for the Central Research Laboratories of Matsushita Electric Industrial Co., Ltd. from April 1981 to March 1995. During these 14 years, he has researched on software engineering, speech recognition, neural networks, fuzzy systems, and genetic algorithms.

He was a Visiting Industrial Fellow of the University of California at Berkeley from October 1991 to September 1993, hosted by Prof. Lotfi A. Zadeh of Computer Science Division. Besides his visiting research, he played an educational role through research projects and technical meeting with students, and BISC special seminar on Soft Computing of Computer Science Division. He has worked for Kyushu Institute of Design since April 1995 as an associate professor. He belonged to Department of Acoustic Design in 1995–1999 and Department of Art and Information Design in 1999–2003. Kyushu Institute of Design and Kyushu University merged into one on October 1, 2003, and his affiliation name changed to Faculty of Design, Kyushu University. Now, he is a Professor of Kyushu University.

Professor Takagi is interested in computational intelligence such as neural networks, fuzzy systems, evolutionary computations, and other so-called soft computing technologies, especially cooperation of these technologies. Currently, his interest focuses on interactive evolutionary computation which aims the cooperation of human and evolutionary computation. He is also interested in signal processing and human-machine interface.

Contents

Part I
Frontier Computing Main Conference

Building the Profile of Web Events Based on Website Measurement

Zheng Xu, Junyu Xuan, Yiwei Zhu and Xiao Wei

Abstract Nowadays, Web makes it possible to study emergencies from web information due to its real-time, open, and dynamic features. After the emergence of a web event, there will be numerous websites publishing webpages to cover this web event. Measuring temporal features in evolution course of web events can help people timely know and understand which events are emergencies, so harms to the society caused by emergencies can be reduced. In this paper, website preference is formally defined and mined by three proposed strategies which are all explicitly or implicitly based on the three-level networks: website-level, webpage-level and keyword-level. An iterative algorithm is firstly introduced to calculate outbreak power of web events, and increased web pages of events, increased attributes of events, distribution of attributes in web pages and the relationships of attributes are embedded into this iterative algorithm as the variables. By means of prior knowledge, membership grade of web events belong to each type can be calculated, and then the type of web events can be discriminated. Experiments on real data set demonstrate the proposed algorithm is both efficient and effective, and it is capable of providing accurate results of discrimination.

Keywords Website preference · Web mining · Web events

Z. Xu (✉)
The Third Research Institute of the Ministry of Public Security, Shanghai, China
e-mail: xuzheng@shu.edu.cn

J. Xuan
Shanghai University, Shanghai, China

Y. Zhu
Zhejiang Business Technology Institute, Ningbo, China

X. Wei
Shanghai Institute of Technology, Shanghai, China

© Springer Nature Singapore Pte Ltd. 2018

N.Y. Yen and J.C. Hung (eds.), *Frontier Computing*, Lecture Notes
in Electrical Engineering 422, DOI 10.1007/978-981-10-3187-8_1

1 Introduction

Web event is what social Medias (i.e., BBS, blog, and news sites) discuss via cyber and influence on our real society. People can discuss web event in various forms, such as commenting news, posting and replying in forum, or recording and messaging in blog, etc. A web event could be a hot news story, like a scandal, or the cover of a social event in the real world on the web, like an earthquake. In general, the content of a web event is not stationary, but it will change with the evolution of this web event. This evolution may come from the development of a news story and the change of interests of its web followers. At a given time, a web event is generally composed by some sub-topics which focus on different aspects of this web event. These discussions, which describe lots web events, have an impact on the evolution of web event. In return, our society will be influenced by the information in web. So the detection and prediction of web events evolution is a meaningful work. To get this goal, we put our hands to measure and analyse evolution features of web events.

The identification of website preferences web events is of significant. The merits are: (1) instead of the most visited websites, the specified websites can be recommended to users and organizations who interest on the aspects of web events covered by these websites. Then, users can follow their interested aspects of web events by following the recommended websites; (2) identify the malicious websites which only publish malicious aspect of web events. If the slander information of a web event is only come from one website, it is possible that this website is spreading slander information. From above incidents, the tapping phone is one kind of social event happened in our society but mapped on the web. By the mapping, social events spread, evolve and mutate in the web along with interaction with real world. And we call such events as social events mapped on web. The latter incident is caused by message on web and impact on real world. In other words, this kind of event happened in virtual world but evolve with human interference. We call such events as web sentiment events. All of these two kinds of events are called web event. Some web events have much bad influence on society. To avoid these bad influences, it is necessary to monitor and predict the evaluative tendency of web events. Therefore, how to collect and organize web events in the intelligent and automatic way, and how to track and measure dynamic evolution of web events are becoming an important subject in the field of information processing.

In this paper, website preference is formally defined and mined by three proposed strategies which are all explicitly or implicitly based on the three-level networks: website-level, webpage-level and keyword-level. An iterative algorithm is firstly introduced to calculate outbreak power of web events, and increased web pages of events, increased attributes of events, distribution of attributes in web pages and the relationships of attributes are embedded into this iterative algorithm as the variables. By means of prior knowledge, membership grade of web events belong to each type can be calculated, and then the type of web events can be dis-criminated. Experiments on real data set demonstrate the proposed algorithm is both efficient and effective, and it is capable of providing accurate results of discrimination.

2 Related Work

The evolution is a basic feature of web events and is also a part of studies on Topic Detecting and Tracking (TDT) [1–3]. Traditional TDT involves detecting unknown events, gathering and segmenting information, detecting when the event first reported, detecting follow-up reports of events and tracking events' tendency. Generally, TDT technology attempts to detect unknown web events and make related news pages clustered. Although TDT tracks development of web events, it does not measure the dynamic evolution process of web events. So we cannot have a global and clear understanding of web events. Qi [4] suggests that a website should be evaluated from three aspects: usefulness, service quality and physical accessibility. The qualities of content and structure of websites will impact on their usage preferences which means the efficiency of using these websites. And the content is more important than structure in the long run [5]. The content and structure of website is evaluated to fit better the needs of visitors by reorganizing the documents [6]. There are also many extensions of LDA which have considered different aspects of documents. There are also some works trying to release the independent of documents and discovered topics by considering the citation relations between documents [7] and relations of topics. However, all these works are still based on 'bag-of-words' assumption and the relations of keywords within documents are ignored. Some researchers were ware of this gap. In our previous work, in order to detect and describe the real time urban emergency event, the 5W (What, Where, When, Who, and Why) model is proposed by Xu [8]. Xuan [9] proposed a framework to identify the different underlying levels of semantic uncertainty in terms of Web events, and then utilize these for Webpage recommendations. The basic idea is to consider a Web event as a system composed of different keywords, and the uncertainty of this keyword system is related to the uncertainty of the particular Web event. Liu [10] explored a Markov random field based method for discovering the core semantics of event. The method makes semantics collaborative computation for learning association relation distribution and makes information gradient computation for discovering k redundancy-free texts as the core semantics of event. A crowdsourcing based burst computation algorithm of an urban emergency event is developed in order to convey information about the event clearly and to help particular social groups or governments to process events effectively [11–15].

3 Iterative Method

In this paper, the communities of keyword level network are adopted to represent subtopics of a given web event. Since the each keyword is a semantic unit of a web event, the community of a number of keywords, which have relative close relation with each others, can be seen as a sub-topic of a web event. The most straightforward method to get the preferences of websites would be to detect the communities of keyword level network and then these detected communities could be seen as the different sub-topics of a web event. The preferences of websites could be computed as the membership degree on each community. The procedure of this method can be described as (Fig. 1),

(1) Construct keyword network of a number of webpages published on a number of websites;

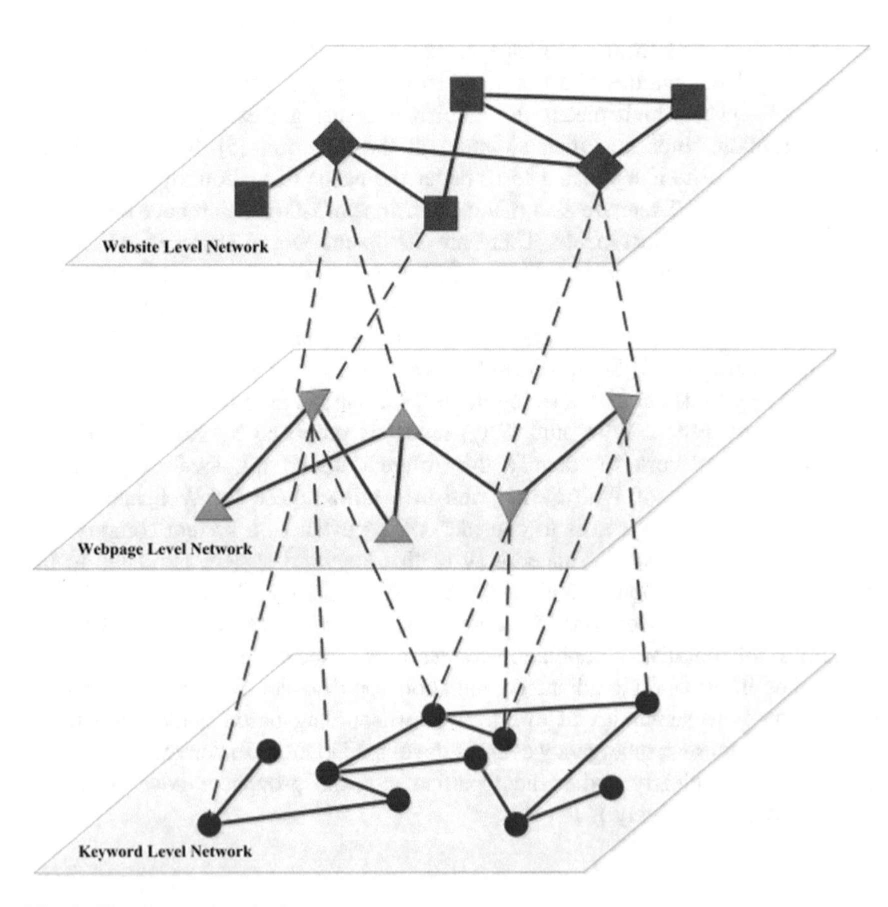

Fig. 1 The proposed method

(2) Do community detection on this keyword network;
(3) Compute the membership degree of each website on the detected communities.

The communities of keyword level network are only based on the keyword relations between each other, a horizontal relation in Fig. 1. This relation implies that the keywords, which have close association relations with each others, will be more likely to describe same sub-topic of a web event. Actually, the webpage level network will also influence the formation of communities at keyword level. When all the keywords are in the same webpage by the mapping relations between keywords and webpages, it is also possible that they are talking the same sub-topic of a web event. However, the relations in the keyword level network, ALNK, does not take the mapping relations into consideration, which only consider the statistical values of co-occurrence relations on all the webpages. For example, two keywords, ki and kj, have a small co-occurrence relation which means that they do not usually show in the webpages simultaneously. However, if two webpages which contain keywords, ki and kj, respectively and they are in the same community of webpage level network, keywords, ki and kj, are also have big probability to talk about same sub-topic of a web event. Similarly, the communities of webpage level are also influenced by the mapping relations between webpages and websites. Inspired by their inter-dependency and inter-limitation relations of websites, webpages and keywords, a iterative algorithm is proposed to optimize the formation of keyword communities/sub-topics.

4 Fuzzy Based Algorithm for Type Discrimination of Web Events

With the time changing, the emergent degree of web events changes is in dynamic change. One event in different segments has different emergent degree, so for a web event, it may go through three states: general state, hot state, and emergent state. Fewer domestic and foreign scholars study on emergent level classification of web events in different segments, so that the lack of a prior knowledge of type discrimination of web events in different segments. Therefore, we study the changes of features and emergent degree of web events in evolution course, and we can obtain the relationship between emergent degree and outbreak power, fluctuation power. Then by studying these relationships, we extract features of different emergent degree, establish evolution model of web events, and construct the membership model for type discrimination of web events as prior knowledge. Thereby to provide effective guidance for the type prediction of web event in later section.

For the result of algorithm, it describes the emergent degree of web events and it is called outbreak power. In this paper, "day" is the minimum time granularity. Source data of temporal features of web events is collected from different news sites daily, Algorithm 1 calculate the daily outbreak power of web events based on these source data, and then time series data of outbreak power of web events in a certain time interval are obtained, as shown in Fig. 2. The outbreak power of web events, which is calculated by Algorithm 1, combines the increased webpages, increased attributes of events, and distribution of attributes in webpages. The Algorithm 1 considers the physical attributes of web events, semantic content, and distribution of web events on web. So the outbreak power we get can comprehensively describe the evolution course of web events.

Herein, 100 web events were selected as the experimental object. And 60 web events among experimental object as training set to establish prior knowledge of web events, and the remaining 40 web events were as test set of type discrimination.

In experiment, we first trained 60 web events in training set, annotated the web events according to their emergent degree, so these 60 web events were labelled as emergent event, hot event or general event. By statistics on the training set, we calculated the membership frequency of temporal features belonging to each type when temporal features took different values, and combined with prior knowledge of our cognition on web events, we got the membership distribution of each temporal feature belonging to different types of web events. Here, Fig. 3 shows the membership distribution of average outbreak power belonging to different types of web events.

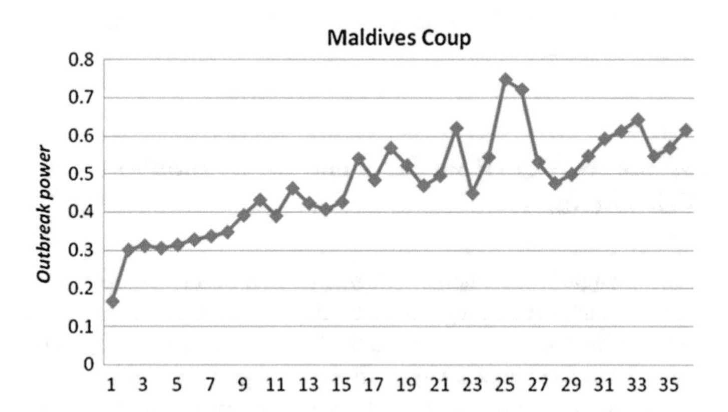

Fig. 2 The outbreak power of web events in a certain time

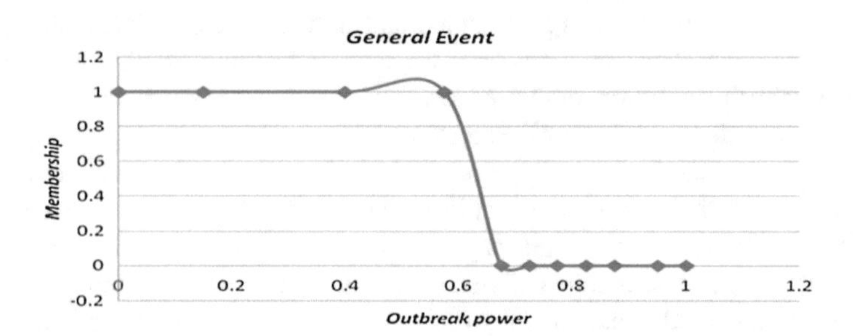

Fig. 3 The membership distribution of average outbreak power

5 Conclusions

In this paper, website preference is formally defined and mined by three proposed strategies which are all explicitly or implicitly based on the three-level networks: website-level, webpage-level and keyword-level. An iterative algorithm is firstly introduced to calculate outbreak power of web events, and increased web pages of events, increased attributes of events, distribution of attributes in web pages and the relationships of attributes are embedded into this iterative algorithm as the variables. By means of prior knowledge, membership grade of web events be-long to each type can be calculated, and then the type of web events can be dis-criminated. Experiments on real data set demonstrate the proposed algorithm is both efficient and effective, and it is capable of providing accurate results of discrimination.

Acknowledgements This work was supported in part by the National Science and Technology Major Project under Grant 2013ZX01033002-003, in part by the National High Technology Research and Development Program of China (863 Program) under Grant 2013AA014603, in part by the National Science Foundation of China under Grant 61300202, in part by the China Post-doctoral Science Foundation under Grant 2014M560085, and in part by the Science Foundation of Shanghai under Grant 13ZR1452900.

References

1. C. Yang, X. Shi, and C. Wei. Discovering Event Evolution Graphs from News Corpora. IEEE Trans. On Systems, Man and Cybernetics—Part A: 39(4):850–863, 2009.
2. Juha Makkonen. Investigation on event evolution in TDT. In Proceedings of the 2003 Conference of the North American Chapter of the Association for Computational Linguistics on Human Language, pp. 43–48, 2003.
3. J. Allan, G. Carbonell, G. Doddington, J. Yamron, and Y. Yang. Topic Detection and Tracking Pilot Study Final Report. In Proceedings of the Broadcast News Transcription and Understanding Workshop, 1998.
4. Shanshan Qi, Crystal Ip, Rosanna Leung, and Rob Law. 2010. A new framework on website evaluation. In E-Business and E-Government (ICEE), 2010 International Conference on. IEEE, 78–81.
5. Michael J Davern, Dov Te'eni, and Jae Yun Moon. 2000. Content versus structure in information environments: A longitudinal analysis of Website preferences. In Proceedings of the twenty first international conference on Information systems. Association for Information Systems, 564–570.
6. Barbara Poblete and Ricardo Baeza-Yates. 2006. A content and structure website mining model. In Proceedings of the 15th international conference on World Wide Web. ACM, 957–958.
7. Jonathan Chang and David M Blei. 2010. Hierarchical relational models for document networks. The Annals of Applied Statistics 4, 1 (2010), 124–150.
8. Z. Xu et al. Crowdsourcing based Description of Urban Emergency Events using Social Media Big Data. IEEE Transactions on Cloud Computing. doi:10.1109/TCC.2016.2517638.
9. J. Xuan, X. Luo, G. Zhang, J. Lu, and Z. Xu. Uncertainty Analysis for the Keyword System of Web Events. IEEE Transactions on Systems, Man, and Cybernetics: Systems. doi:10.1109/TSMC.2015.2470645.

10. Z. Xu et al. The Semantic Analysis of Knowledge Map for the Traffic Violations from the Surveillance Video Big Data. Computer Systems Science and Engineering, 30(5):403–410, 2015.
11. Z. Xu et al. Crowdsourcing based Social Media Data Analysis of Urban Emergency Events. Multimedia Tools And Applications, doi:10.1007/s11042-015-2731-1.
12. Z. Xu et al. Incremental building association link network. Computer systems science and engineering, 26(3):153–162, 2011.
13. X. Luo, Zheng Xu, J. Yu, and X. Chen. Building Association Link Network for Semantic Link on Web Resources. IEEE transactions on automation science and engineering, 2011, 8 (3), 482–494.
14. C. Hu, Zheng Xu, et al. Semantic Link Network based Model for Organizing Multimedia Big Data. IEEE Transactions on Emerging Topics in Computing, 2014, 2(3), 376–387.
15. Z. Xu et al. Knowle: a Semantic Link Network based System for Organizing Large Scale Online News Events. Future Generation Computer Systems, 2015, 43–44, 40–50.

Building Domain Keywords Using Cognitive Based Sentences Framework

Zheng Xu, Weidong Liu, Yiwei Zhu and Shunxiang Zhang

Abstract As the novel web social media emerges on the web, large scale unordered sentences are springing up in the forms: news headlines, microblogs, comments and so on. Domain keywords extraction is very important for information extraction, information retrieval, classification, clustering, topic detection and tracking, and so on. Although these massive sentences contain rich information, their loose semantic association and highly unordered semantic organization make web users extremely difficult to capture the rich information due to the lack of semantic coherence. Sentence ordering is a significant research area focusing on obtaining coherent sentence orders which could assist web user to easily understand these unordered sentences. TextRank is a common graph-based algorithm for keywords extraction. For TextRank, only edge weights are taken into account. We proposed a new text ranking formula that takes into account both edge and node weights of words, named F2N-Rank. The results show our model can obtain coherent sentence orders with higher accuracy in less iterations. The proposed sentence ordering model can be applied in automatic text organization and summarization.

Keywords Domain keywords · TextRank · Sentence ordering

Z. Xu (✉)
The Third Research Institute of the Ministry of Public Security, Shanghai, China
e-mail: xuzheng@shu.edu.cn

W. Liu
Shanghai University, Shanghai, China

Y. Zhu
Zhejiang Business Technology Institute, Ningbo, China

S. Zhang
Anhui University of Science and Technology, Huainan, China

© Springer Nature Singapore Pte Ltd. 2018

N.Y. Yen and J.C. Hung (eds.), *Frontier Computing*, Lecture Notes
in Electrical Engineering 422, DOI 10.1007/978-981-10-3187-8_2

1 Introduction

Domain keywords can serve as a highly condensed summary for a domain, and they can be used as labels for a domain. Domain keywords should be ordered by the "importance" of keywords. With the boom of microblogs, massive unordered sentences are emerging on the web as a main message passing form. Although these sentences contain much useful information, loose semantic association and unordered sentence organization make web users lost in the large scale data when they face these massive unordered sentences. Web users normally expect these sentences are well ordered according to their semantic coherence since coherent sentence orders can assist them to easily understanding the content of these sentences. However, such sentence ordering problem is burdensome computation even though the 10 sentence scale is small.

In the study of keywords extraction, supervised methods [1–3] always depend on the trained model and the domain it is trained on. And in unsupervised methods [4, 5], algorithms based on term frequency and based on graph are the most common methods. Algorithms based on term frequency such as TF, ATF, ATF*DF, ATF*DF are easy to realize but their precisions are not very high. Algorithms based on graph, such as TextRank [1], are more effective than algorithms based on term frequency for they take into account the relationships among words. To overcome the above limitations, we adopt markov random field as special case of association link network which have been widely used in many tasks from learning technologies to knowledge discovery. Compared with association link network, markov random field has stronger ability in representation and inference since it implies association relation distribution and can make inference on the distribution. More importantly, our markov random field incorporates three cognitive logical structures which respectively guide sentence ordering model to link different sentence conditioned on different cognitive structures. What is more, we develop sound cognitive mechanistic for fast sentence ordering such as decision making process on cognitive logical structure, keywords spreading process and sentence activation process working on markov random field for ordering sentences.

TextRank is a common graph-based algorithm for keywords extraction. For TextRank, only edge weights are taken into account. We proposed a new text ranking formula that takes into account both edge and node weights of words, named F2N-Rank. The results show our model can obtain coherent sentence orders with higher accuracy in less iterations. The proposed sentence ordering model can be applied in automatic text organization and summarization.

2 Problem Formulation

Markov random field is a basic undirected probabilistic graphic model with outstanding abilities in semantic representation and inference. Inspired the above outstanding abilities of Markov random field, we propose semantic Markov random field which is built by the limited number of association relations in power serials presentation represent and can inference the whole distribution of association relation in sentences to be order. In this paper, we can regard semantic Markov random field as a special case of association link network, which is constructed by association relations under different cognitive structures.

Sentence ordering task can be regarded as restrictive writing, since the content of sentences is known and the sentence order is unknown. The human beings' task is to order these sentences for well semantic coherence. Based on cognitive process of writing [6–8], writing can be characterized as a "journey of discovery". On this journey, association knowledge is continually activated and spreads to generate a coherent sentence order. Spreading activation model assumes that specific keywords distribute on semantic link network [9–13] and spreading activation process is a semantic processing on semantic link network, where keywords are continually spreading their influences into relevant keywords and these keywords can activate relevant sentence. Figure 1 shows sentence ordering process in restrictive writing, which spans on three memory modules: (1) long term memory, (2) short term memory and (3) working memory as follows.

(1) Long term memory contains the relatively stable entities and relations. These entities and relations are stored by semantic link network. Besides, some cognitive logical structures are stored in this module. Keywords can spread towards different directions on semantic link network under different cognitive logical structures.
(2) Short term memory contains cognitive logical structure schema and a word activation window with m activated keywords since Millers Law pointed out that the number of objects an average human can hold in short term memory is seven, plus or minus two. These keywords are activated from semantic link network by spreading activation process. Which keywords to be activated are conditioned on three different cognitive logical structures.

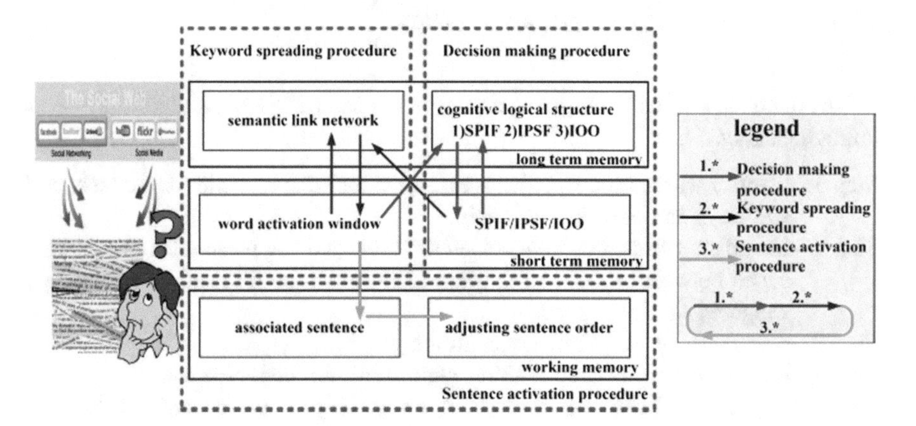

Fig. 1 The sentence ordering process

(3) Working memory contains a newly generated sentence closely associated with activated keywords and a sentence order to be adjusted by spreading activation process. The spreading activation process will link all the unordered sentences toward well semantic coherence.

3 The Proposed Algorithm

TextRank algorithm only focuses on the relationship among nodes, and node weights are not taken into account. Equation (1) integrates TextRank formula with the node weight $(F(V_i))$.

$$FS(V_i) = (1 - d) * \mathbf{F(V_i)} + d * \mathbf{F(V_i)} * \sum_{V_j \in In(V_i)} \frac{w_{ji}}{\sum_{V_k \in Out(V_j)} w_{jk}} FS(V_j) \quad (1)$$

There are several formulas can be used to calculate the value of $F(V_i)$, such as TF, ATF, ATF*DF. ATF*DF is the most suitable of the three formulas because it takes into account both term frequency and document frequency. However, the simple combination of ATF and DF does not account for their proportions. Here, the idea of F-measure is introduced for calculating $F(V_i)$. The formulas are given as followings:

$$F(V_i) = \frac{(1 + \beta^2) * ATF(V_i) * DF(V_i)}{\beta^2 * ATF(V_i) + DF(V_i)} \quad (\beta = 2) \quad (2)$$

$$ATF(V_i) = \frac{\sum_{|D|} \frac{n_{i,j}}{\sum_k n_{k,j}}}{|\{d : t_i \in d\}|} \quad (3)$$

$$DF(V_i) = \log \frac{|\{d : t_i \in d\}|}{|D|} \quad (4)$$

The main steps of extracting domain keywords using F2N-Rank algorithm are as followings (Fig. 2):

Step 1 Identify words (nouns, adjectives, and so on) that suitable for the task, and add them as nodes in the graph.

Step 2 Identify relations that connect such words, and use these relations to draw edges between nodes in the graph. Edges can be directed or undirected, weighted or unweighted.

Step 3 Calculate the weight of nodes in the graph.

Step 4 Iterate the graph-based ranking algorithm until convergence.

Step 5 Sort nodes based on their final score. Top T words are the domain keywords.

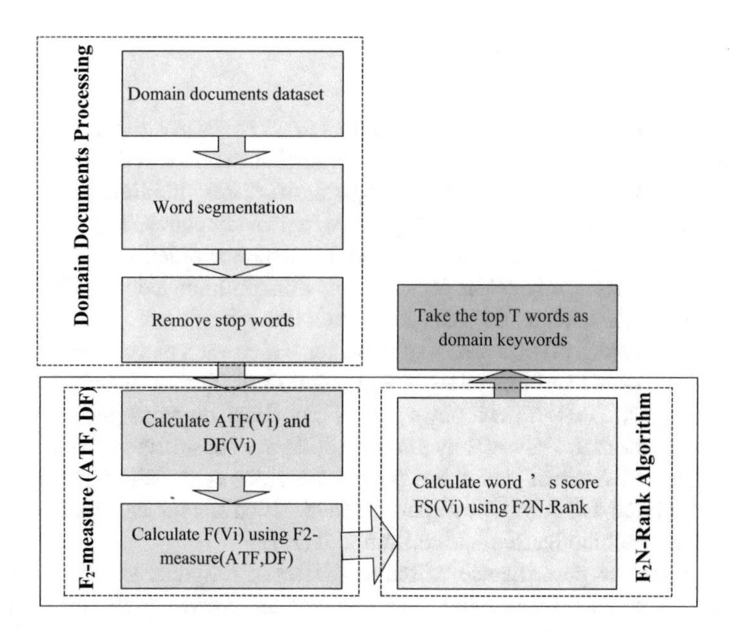

Fig. 2 The flow chat

Cognitive decision making is a core module of cognitive memory-inspired sentence ordering model (CM-SOM) since: (1) cognitive decision making module decides which cognitive logical structure and (2) different cognitive logical structures decides different sentence orders. As such, how to learn an optimal policy for cognitive logical structure decision making is an important issue for the cognitive decision learning model since such policy decides how to the logical structures shift during sentence ordering. To build the above ordering model and obtain a coherent sentence order, we propose the modules as follow.

(1) Cognitive logical structure learning module. This module proposes three logical structure and construct cognitive logical structure based markov random field. Different cognitive logical structures and corresponding markov random filed shift in sentence ordering procedure;
(2) Spreading and activation computation module. When a cognitive logical structure is selected, this module mainly makes keyword spreading and sentence activation based on makov random field, which is similar to the semantic association ability in human memory.
(3) Cognitive decision making learning module. This module learns the decision making policy for shifting cognitive logical structure and such policy can guide which cognitive logical structure to develop sentence before a new sentence is linked.

4 Experiment and Results

We collect 2 datasets which consist of Reuters news as dataset 1 and paper abstracts as dataset 2. Domain data is used to learn logical structure and construct semantic markov random fields under different logical structures; training data is used to learn the decision making policy of cogitative logical structures. Test data is used to test the sentence ordering results generated by sentence ordering model. For each dataset, we will randomly select 50% texts as domain knowledge; 25% as texts as training data; 25% texts as test data from each dataset.

Dataset 1 includes 60,000 pieces of news for which each piece of these news has average 15.67 words and these news are crawled from Reuters website from March 2009 to August 2009. These news are about three domains including health, environment, internet. Dataset 2 includes 50,000 paper abstracts for which each one has average 13.27 words and these papers are from Association for Computing Machinery-digital Library. These papers cover 10 different categories including data mining, machine learning, algorithm and so on.

To evaluate the performance of ranking Tibetan religious keywords, we conducted a performance measurement using precision. Now, we discuss the evaluation of three different ranking algorithms. We compared algorithms which are: F2N-Rank, TextRank and ATF*DF. Results are shown in Fig. 3 by measuring the precision for top N keywords. We can see that F2N-Rank clearly outperformed both TextRank and ATF*DF. For F2N-Rank, TextRank and ATF*DF, the average precision are 78.6, 62.2 and 49.2%. The improvement over TextRank is around 16% in average precision and 29% over ATF*DF. Using F2N-Rank for domain keywords extraction has showed better results.

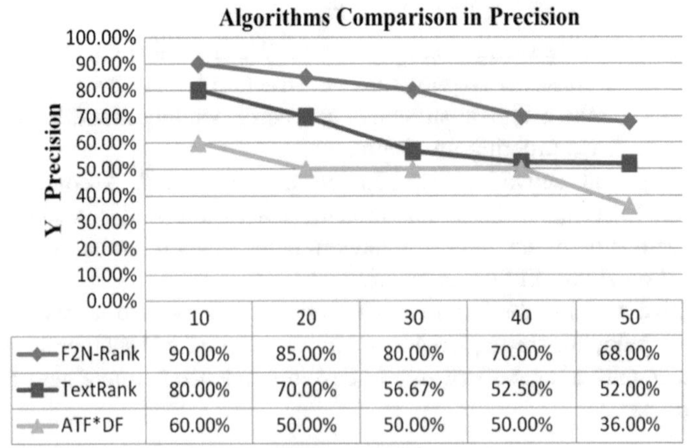

Fig. 3 Algorithm Comparison in Precision

5 Conclusions

Domain Keywords extraction is important for many applications of Natural Language Processing. TextRank is a common graph-based algorithm for keywords extraction. For TextRank, only edge weights are taken into account. We proposed a new text ranking formula that takes into account both edge and node weights of words, named F2N-Rank. The results show our model can obtain coherent sentence orders with higher accuracy in less iterations. The proposed sentence ordering model can be applied in automatic text organization and summarization.

Acknowledgements This work was supported in part by the National Science and Technology Major Project under Grant 2013ZX01033002-003, in part by the National High Technology Research and Development Program of China (863 Program) under Grant 2013AA014603, in part by the National Science Foundation of China under Grant 61300202, in part by the China Postdoctoral Science Foundation under Grant 2014M560085, and in part by the Science Foundation of Shanghai under Grant 13ZR1452900.

References

1. Mihalcea R, Tarau P, "TextRank, Bringing order into texts", Association for Computational Linguistics, 2004.
2. Frank, Eibe, Gordon W. Paynter, Ian H. Witten, Carl Gutwin, and Craig G. Nevill-Manning, "Domain-specific keyphrase extraction", In Proceedings of the 16th International Joint Conference on Artificial Intelligence, pp. 668–673, 1999.
3. Medelyan, Olena, Eibe Frank, and Ian H. Witten, "Human-competitive tagging using automatic keyphrase extraction", In Proceedings of the 2009 Conference on Empirical Methods in Natural Language Processing, pp. 1318–1327, 2009.
4. Tomokiyo, Takashi and Matthew Hurst, "A language model approach to key phrase extraction", In Proceedings of the ACL Workshop on Multiword Expressions, 2003.
5. Turney, Peter, "Learning algorithms for key phrase extraction. Information Retrieval", Vol. 2, pp. 303–336, 2000.
6. L. Flower, J. R. Hayes, A cognitive process theory of writing, College composition and communication (1981) 365–387.
7. A. C. Graesser, M. Singer, T. Trabasso, Constructing inferences during narrative text comprehension., Psychological review 101(3) (1994) 371.
8. X. Luo, J. Zhang, F. Ye, P. Wang, C. Cai, Power series representation model of text knowledge based on human concept learning, Systems, Man, and Cybernetics: Systems, IEEE Transactions on 44(1) (2014) 86–102.
9. C. Hu, Z. Xu, et al. Semantic Link Network based Model for Organizing Multimedia Big Data. IEEE Transactions Emerging Topics in Computing, 2(3): 376–387 (2014).
10. X. Luo, Z. Xu, J. Yu, and X. Chen. Building Association Link Network for Semantic Link on Web Resources. IEEE transactions on automation science and engineering, 8(3):482–494, 2011.
11. Z. Xu, et al. Knowle: a Semantic Link Network based System for Organizing Large Scale Online News Events. Future Generation Comp. Syst. 43–44: 40–50 (2015).
12. Z. Xu et al. Incremental building association link network. Computer systems science and engineering, 26(3):153–162, 2011.
13. Zheng Xu, Xiangfeng Luo, Wenjun Lu. Association Link Network: An Incremental Semantic Data Model on Organizing Web Resources. ICPADS 2009: 793–798.

The Study on Vehicle Detection Based on DPM in Traffic Scenes

Chun Pan, Mingxia Sun and Zhiguo Yan

Abstract After the HoG feature was proposed, a lot of detectors were developed based on the feature. But HoG feature has its defects, as high dimensional data leading to inefficiency, complex scenes leading to poor performances and so on. In this article, we proposed a vehicle detector based on DPM (Deformable Part Model). This detector uses a deformable part model to classify the front and the rear of the vehicles.

Keywords Vehicle detection · DPM · Traffic scenes

1 Introduction

Object detection is one of the most popular researching fields in computer vision, such as vehicle detection, pedestrian detection and so on. Normally, the common detecting solutions are using HoG, Sift or Haar to extract features and using SVM or Adaboost as classifiers. In this article, we propose a solution by using DPM to detect Vehicles. In consideration of the variety of appearance of the vehicles are affected by many factors, as the changes of illuminations or angle of view. The traditional detecting algorithms are hard to overcome the rigid deformations. DMP uses mixture of multiscale deformable part models to describe an object detection system which represents highly variable objects [1], which has better robustness against deformation.

DPM (Deformable Part Model) as one of the most successful detection algorithms, was proposed by Pedro Felzenswalb in 2008 and was awarded the PASCAL VOC "Lifetime Achievement" Prize in 2010. Due to Felzenswalb's paper, the resulting system is both efficient and accurate, achieving state-of-the-art results on PASCAL VOC benchmarks and the INRIA Person dataset in 2007 [2]. The strong low-level features of DPM are based on the HoG (Histograms of

C. Pan · M. Sun (✉) · Z. Yan
The Third Research Institute of the Ministry of Public Security, Shanghai, China
e-mail: mingxiasun@163.com

© Springer Nature Singapore Pte Ltd. 2018

N.Y. Yen and J.C. Hung (eds.), *Frontier Computing*, Lecture Notes in Electrical Engineering 422, DOI 10.1007/978-981-10-3187-8_3

Oriented Gradients) features. So the DPM can be considered as an upgrade of HoG in some ways [3].

As showed in Fig. 1, the upgraded HoG feature in DPM kept the "Cell" concept in HoG feature, but altered the normalization progress. The result shared similarity with the result of HoG feature as the upgraded HoG feature normalized the region which consisted of the target cell and the four surrounding cells. In order to reduce the feature dimension, P. Felzenswalb used PCA (Principal Component Analysis) [4] to analyze the unsigned gradients. As illustrated in Fig. 1, there are 31 dimensional features.

In P. Felzenswalb's work [5], he showed pedestrian detection model as following.

In Fig. 2, figure (A) shows the pedestrian, figure (B) is a root filter, figure (C) shows several part models with high resolution, figure (D) shows the spatial relationships of the part filters.

DPM uses a root filter and several part filters and the corresponding deformable model, the construction of the whole model is based on the pictorial structures. Normally, the part models use higher resolution than the root filter, about two times. Figure 2B, C illustrate the visual structure of the root and part models, it shows the weighted sum of SVM coefficients which oriented in gradient direction, and the brightness is proportional to the value. In order to reduce the complexity of the

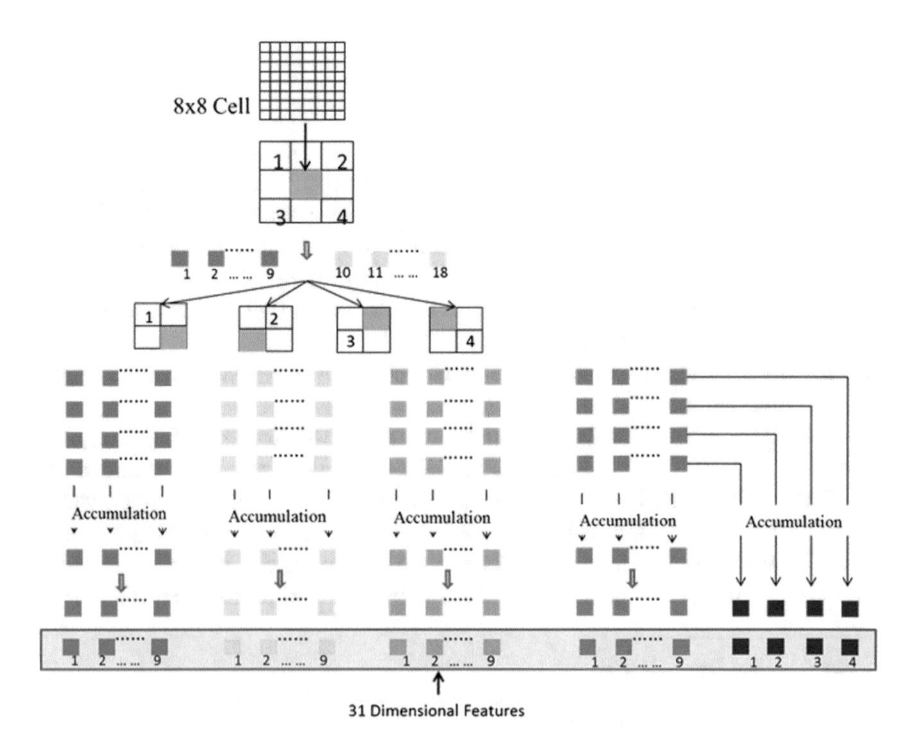

Fig. 1 The upgraded HoG feature in DPM

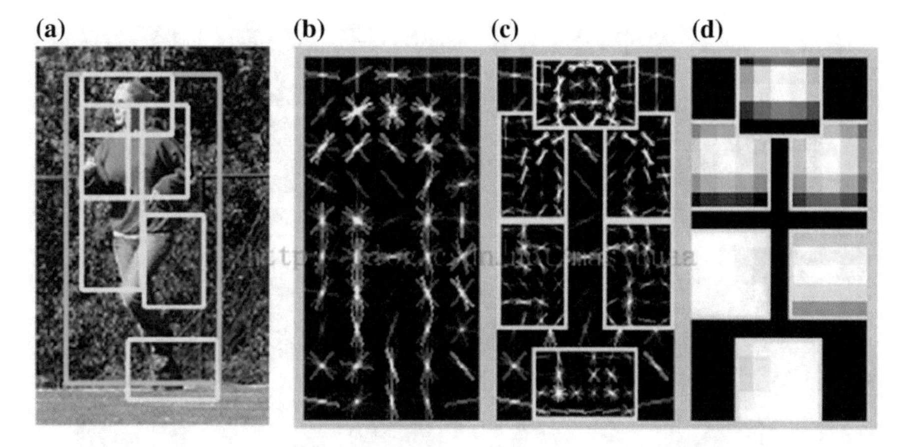

Fig. 2 DPM pedestrian detection model

whole model, the part models are symmetry. Figure 2D shows the deviation cost of the part model. The cost is zero in ideal case; the further the part model deviates, the greater the cost is. Then the target object can be represented by a collection of parts and the relative deformable position of the parts, the parts are connected by certain ways [6]. Each part describes local properties, and the spring-like connections are used to represent the relation between the deformable models [7]. As a single deformable model is not capable enough to describe an object, usually multiple deformable models are according with the request [8–15]. In this article, the variations among different vehicle types are quite significant, so the mixture of deformable models is required.

2 Methodology

In detecting progress, a scale pyramid is constructed and a scan window approach [4] is used to scan different layer of pyramid. Figure 3 shows the detecting process of DPM. In Fig. 3, the score of layer l_0 coordinate (x_0, y_0) can be calculated as follows [1]:

$$score(x_0, y_0, l_0) = R_{0,i_0}(x_0, y_0) + \sum_{i=1}^{n} D_{i,l_0-\lambda}(2(x_0, y_0) + v_i) + b.$$

$R_{0,i_0}(x_0, y_0)$ is the score of root filter, in other words, it expresses the matching degree between model and target. $\sum_{i=1}^{n} D_{i,l_0-\lambda}(2(x_0, y_0) + v_i)$ is the scores of n part filters. b is the root of set which is used to align the components. (x_0, y_0) is the

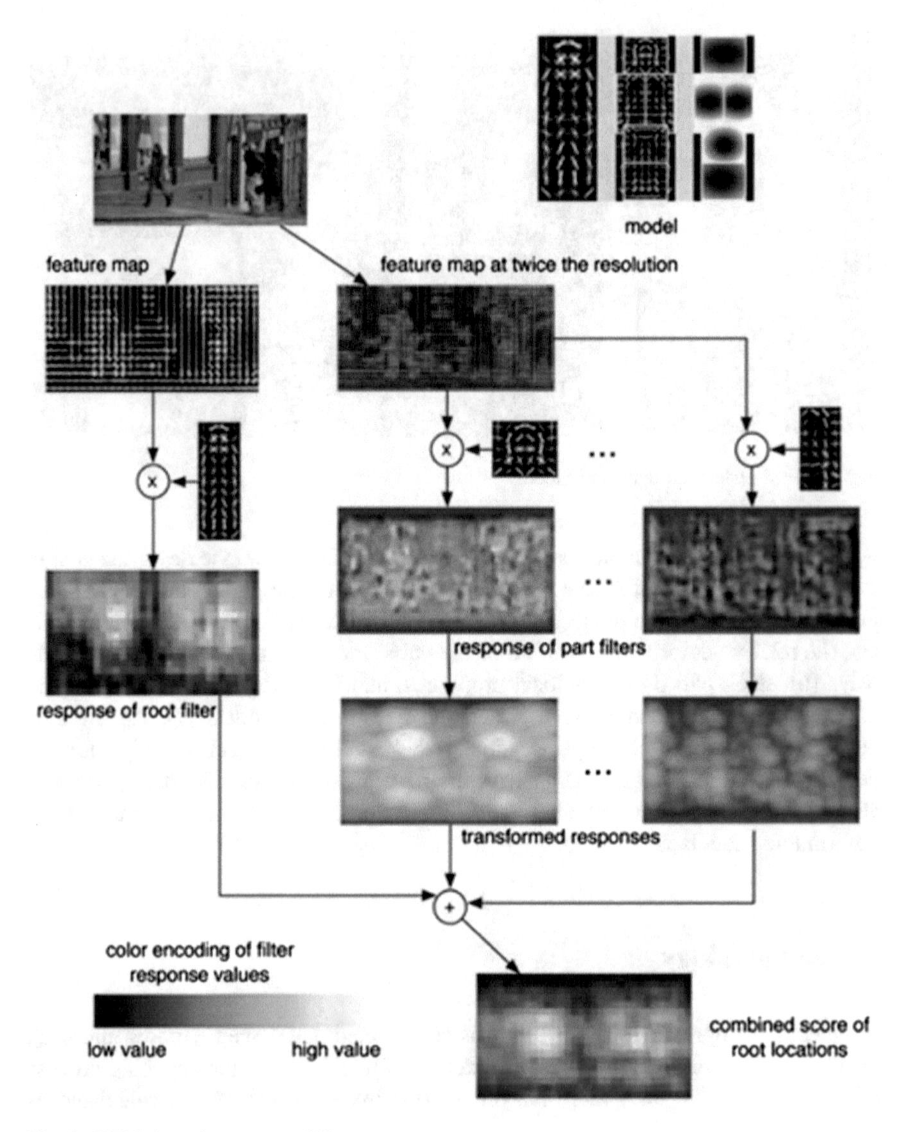

Fig. 3 DPM detection process [1]

coordinate of the root filter's left-top in the root feature map. $(2(x_0, y_0) + v_i)$ is coordinate of the i-th part filter in the root feature map.

The score of part filters can be calculated as follows [1]:

$$D_{i,l}(x, y) = \max_{dx,\, dy}(R_{i,l}(x + dx, y + dy) - d_i \cdot \Phi_d(dx, dy)).$$

$D_{i,l}(x, y)$ is the optimal solution of part filter, namely, it searches the anchor position and within a certain range for a proper location which has combined matching and optimal deformation. (x, y) is the ideal position of the i-th part filter in layer l. (dx, dy) illustrates the relative offset from (x, y). $R_{i,l}(x+dx, y+dy)$ is the matching score in coordinate $(x+dx, y+dy)$. $d_i \cdot \Phi_d(dx, dy)$ expresses the offset loss causing by the offset (dx, dy); $\Phi_d(dx, dy) = (dx, dy, dx^2, dy^2)$, d_i is the coefficient of offset loss, it is to be calculated in the training process. To initialize the model, $d_i = (0, 0, 1, 1)$ is the Euclidean distance between offset location and ideal location, namely the offset loss.

3 Experiment

3.1 Data Preparation

The original image data are captured from traffic surveillance system somewhere in JiangSu province. The training data which are used in DPM illustrated as following:

In training process, positive samples must be labeled with bounding boxes which are illustrated in Fig. 4B. In this experiment, 1700 images of vehicle front and 1900 images of vehicle rear as positive samples which are labeled with bounding boxes and the property files are generated.

3.2 Training Procedure

The training procedure is completed by initializing the structure of a mixture model and learning parameters. The parameters are learned by training LSVM (Latent

Fig. 4 DPM training data: vehicle front

Fig. 5 Training Procedure [1]

Data:
Positive examples $P = \{(I_1, B_1), \ldots, (I_n, B_n)\}$
Negative images $N = \{J_1, \ldots, J_m\}$
Initial model β

Result: New model β

```
1  F_n := ∅
2  for relabel := 1 to num-relabel do
3  |  F_p := ∅
4  |  for i := 1 to n do
5  |      Add detect-best (β,I_i,B_i) to F_p
6  |  end
7  |  for datamine := 1 to num-datamine do
8  |  |  for j := 1 to m do
9  |  |      if |F_n| ≥ memory-limit then break
10 |  |      Add detect-all (β,J_j,−(1+δ)) to F_n
11 |  |  end
12 |  |  β :=gradient-descent (F_p ∪ F_n)
13 |  |  Remove (i,v) with β·v < −(1+δ) from F_n
14 |  end
15 end
```

Procedure `Train`

Support Vector Machine) [16]. The LSVM is trained by gradient descent algorithm and the data-mining approach [17, 18] with a cache of feature vectors (Fig. 5).

4 Results

In this experiment, three models are designed by DPM: vehicle front model is used to recognize the frontal side of a vehicle; vehicle rear model is used to recognize the back side of a vehicle; vehicle mixture model is used to capture either frontal or back side of a vehicle. There are two testing sets in this experiment: 100 vehicle front images and 100 vehicle rear images. The testing results of vehicle front model are illustrated as following:

DPM vehicle front model recognizing testing images of vehicle front			
Total image samples	Correctly recognized samples	Accuracy	Correctly recognized objects
100	93	93%	96

The testing results of vehicle rear model are illustrated as following:

DPM vehicle rear model recognizing testing images of vehicle front			
Total image samples	Correctly recognized samples	Accuracy	Correctly recognized objects
100	96	96%	112

In order to know which model performs better under the same conditions, we used the two models to recognize the same image objects and then outputted the results with the higher confidence degree.

DPM vehicle front and rear models recognizing testing images of vehicle front					
DPM vehicle front model			DPM vehicle rear model		
Total image samples	Correctly recognized samples	Accuracy	Total image samples	Correctly recognized samples	Accuracy
100	39	39%	100	69	69%
DPM vehicle front and rear models recognizing testing images of vehicle rear					
DPM vehicle front model			DPM vehicle rear model		
Total image samples	Correctly recognized samples	Accuracy	Total image samples	Correctly recognized samples	Accuracy
100	47	47%	100	76	76%

As the front and rear side of the same vehicle always share certain similarity, we conjecture the probability of using one model to recognize the two sides of a vehicle. So we used DPM to design a mixture model to capture the vehicles in 2-way lanes. The testing results show as following:

DPM mixture model recognizing testing images of vehicle front			
Total image samples	Correctly recognized samples	Accuracy	Correctly recognized objects
100	70	70%	70

DPM mixture model recognizing testing images of vehicle rear			
Total image samples	Correctly recognized samples	Accuracy	Correctly recognized objects
100	90	90%	203

5 Conclusions

By comparing the results from three DPM vehicle models, the non-mixed models acquired higher accuracy. But they performed not unsatisfactory in mixture test. In order to capture the vehicles in 2-way lane, we proposed the third mixture DPM vehicle model. It is more efficient to capture vehicle vision and shows high versatility.

Acknowledgements This work was supported in part by the National Science and Technology Major Project under Grant 2013ZX01033002-003, in part by the National High Technology Research and Development Program of China (863 Program) under Grant 2013AA014603.

References

1. P. F. Felzenswalb, R. B. Girshick, D. McAllester and D. Ramanan, Object Detection with Discriminatively Trained Part Based Models. IEEE Trans. PAMI, 32(9):1627–1645, 2010.
2. M. Everingham, L. van Gool, C. K. I. Williams, J. Winn, and A. Zisserman, "The PASCAL Visual Object Classes Challenge 2007 (VOC 2007) Results." [Online]. Available: http://www.pascalnetwork.org/challenges/VOC/voc2007/.
3. N. Dalal and B. Triggs, "Histograms of oriented gradients for human detection," in IEEE Conference on Computer Vision and Pattern Recognition, 2005.
4. Y. Ke and R. Sukthankar, "PCA-SIFT: A more distinctive representation for local image descriptors," in IEEE Conference on Computer Vision and Pattern Recognition, 2004.
5. P. Felzenszwalb, D. McAllester, and D. Ramanan. A discriminatively trained, multiscale, deformable part model. In CVPR, 2008.
6. "Pictorial structures for object recognition," International Journal of Computer Vision, vol. 61, no. 1, 2005.
7. M. Fischler and R. Elschlager, "The representation and matching of pictorial structures," IEEE Transactions on Computer, vol. 22, no. 1, 1973.
8. X. Luo, Zheng Xu, J. Yu, and X. Chen. Building Association Link Network for Semantic Link on Web Resources. IEEE transactions on automation science and engineering, 2011, 8 (3):482–494.
9. C. Hu, Zheng Xu, et al. Semantic Link Network based Model for Organizing Multimedia Big Data. IEEE Transactions on Emerging Topics in Computing, 2014, 2(3), 376–387.
10. Zheng Xu et al. Semantic based representing and organizing surveillance big data using video structural description technology. The Journal of Systems and Software, 2015,102, 217–225.
11. Zheng Xu et al. Knowle: a Semantic Link Network based System for Organizing Large Scale Online News Events. Future Generation Computer Systems, 2015, 43–44, 40–50.
12. Zheng Xu et al. Semantic Enhanced Cloud Environment for Surveillance Data Management using Video Structural Description. Computing, 98(1–2):35–54, 2016.
13. C. Hu, Zheng Xu, et al. Video Structured Description Technology for the New Generation Video Surveillance System. Frontiers of Computer Science, 2015, 9(6): 980–989.
14. Zheng Xu et al. Crowd Sensing Based Semantic Annotation of Surveillance Videos, International Journal of Distributed Sensor Networks, Volume 2015 (2015), Article ID 679314, 9 pages.
15. Zheng Xu et al. Crowdsourcing based Description of Urban Emergency Events using Social Media Big Data. IEEE Transactions on Cloud Computing, doi:10.1109/TCC.2016.2517638.
16. S. Andrews, I. Tsochantaridis, and T. Hofmann, "Support vector machines for multiple-instance learning," in Advances in Neural Information Processing Systems, 2003.

17. H. Rowley, S. Baluja, and T. Kanade, "Human face detection in visual scenes," Carnegie Mellon University, Tech. Rep. CMU-CS-95-158R, 1995.
18. K. Sung and T. Poggio, "Example-based learning for view-based human face detection," Massachussets Institute of Technology, Tech. Rep. A.I. Memo No. 1521, 1994.

The Research on Video Security Carving Using Secure Outsourcing Approach

Zheng Xu, Guozi Sun and Yong Ding

Abstract Recovery of fragmented files is an important part of digital forensics. The shared software and hardware resources and information can be provided to the computers and other equipment according to the requirements. A majority of cloud computing services are deployed through outsourcing. Outsourcing computation allows resource-constrained clients to outsource their complex computation workloads to a powerful cloud server which is rich of computation resources. This paper presents a video recovery technique of a fragmented video file using the frame size information in every frame and the index. The proposed method addresses how to extract AVI file fragments from data images and map all the extracted fragments into original order. In this paper, we propose a novel outsourcing algorithm for modular exponentiation based on the new mathematical division under the setting of two non-colluding cloud servers. The base and the power of the outsourced data can be kept private and the efficiency is improved.

Keywords Video carving · Outsourcing computation · Video security

1 Introduction

In the increasingly interconnected society, multimedia social network (MSN) has become a 'mainstream' tool used by online users to connect and share contents with other users 24/7 in real-time. It is, therefore, unsurprising that MSN has become a salient area of inquiry by computer scientists and computer security researchers. For example, researchers need to design intelligent computing and soft computing

Z. Xu (✉)
The Third Research Institute of the Ministry of Public Security, Shanghai, China
e-mail: xuzheng@shu.edu.cn

G. Sun
Nanjing University of Posts and Telecommunications, Nanjing, China

Y. Ding
Guilin University of Electronic Technology, Guilin, Guangxi, China

© Springer Nature Singapore Pte Ltd. 2018

N.Y. Yen and J.C. Hung (eds.), *Frontier Computing*, Lecture Notes in Electrical Engineering 422, DOI 10.1007/978-981-10-3187-8_4

technologies to improve multimedia system functions, efficiency and performance, and improving user's sharing experiences (e.g. using recommendation systems and more effective algorithms). In addition, security of users and data are also an ongoing topic of interest and importance due to the ease in producing and sharing user and multimedia content using MSNs. Digital video equipment, such as digital camera and surveillance cameras have become more and more popular. Large amount of digital videos are produced everyday and record almost every aspect of human life. Due to the limitation of storage space and file system storage strategies, video files may be stored in fragment form and non-sequential order. Therefore recovering video files from storage space is meaningful but challenging work Video always play an important role in about video forensics and video carving is popular these years. Outsourcing computation provides a lot of convenience for people. However, users lose ability to control over the data. The server with powerful computing resources may obtain some sensitive information from the outsourced data. In outsourcing computation, the inputs and the outputs should not be revealed, the results need to be computed correctly, and the correctness of the results can be verified by clients. Thus, the computation outsourced should be blind to the server. Since the user does not compute the outsourced data, when receiving the returned results, user has to do the verifiable computation [1, 2].

Verifiable computation should at least satisfy three basic requirements. Server cannot cheat the user with a random value without computing the outsourced function. Server cannot cheat the user with the computing result of other input values. The verification of client should be efficient [3]. Modular exponentiation is the most expensive computation operation in public key cryptography, which plays an important role in the information era. A large number of modular exponentiations will decrease the execution efficiency of public key cryptographic protocols. It is a good avenue to improve the efficiency by utilizing outsourcing computation. Since a lot of modular exponentiations are computed in polynomials and bilinear pairings. Thus, it is very necessary to study the modular exponentiation outsourcing.

This paper presents a video recovery technique of a fragmented video file using the frame size information in every frame and the index. The proposed method addresses how to extract AVI file fragments from data images and map all the extracted fragments into original order. In this paper, we propose a novel outsourcing algorithm for modular exponentiation based on the new mathematical division under the setting of two non-colluding cloud servers. The base and the power of the outsourced data can be kept private and the efficiency is improved.

2 Related Work

Bifragment File Carving is a carving technique which can recover files which are fragmented into two fragments. This technique use more information than pure signature based carving that it try to find correct combination of data between the header and footer. After finding the file header and footer, a slide size gap of unrelated data is set. The size of the gap is adjusted until the remaining data can be validated as a file. This file carving method are can only cope with files with two fragments and can not work when the gap is large Smart Carving is a general file carving method which is not limited to the number of file fragments. This method identifies fragments and reorders the fragments. Smart Carving consists of three steps: preprocessing, collation and reassembly. The file data is collected on a block base that the total process can overcome the severe fragmentation. The collected file data are classified into different file categories and the associated blocks are merged into a file. Smart Carving is also a file based carving method that it may also fail when part of the file is overwritten.

To enhance the private preserving property [4, 5], the confidential inputs and outputs should not be revealed. In 2005, Hohenberger and Lysyanskaya [6] defined the formal security of outsourcing, and proposed a classical algorithm for outsourcing of modular exponentiations based on precomputation [7] and server-aided computation [8, 9].

In this scheme the inputs and outputs are blind, and the checkability is 12. In 2008, Benjamin and Atallah [10] constructed a verifiable secure outsourcing scheme for linear algebraic calculation by using semantic security based homomorphic encryption. In 2009, Gennaro et al. [11] proposed a non-interactive fully homomorphic encryption scheme for verifiable outsourcing with input privacy. It allows cloud server compute the encrypted data directly without knowing the encryption algorithm and the decryption key. The decryption of the returned results is agreed with the operation results of original data. It gives a theoretically secure outsourcing solution, although the efficiency is low.

3 The Proposed Method

The proposed method extracts frames based on FOURCC and frame size of each substructure. Each frame data is extracted based on the beginning flag and the frame length information. The extraction process regards the frames with continuous physical location as the data of the same file and the adjacent and complete frames constitute one fragment for the next stage. As the data length of AVI header is relatively small, the possibility of fragmentation between the header and the movie data is very low. Therefore the header data will be associated to movie data and the index according to the principle to the nearest. In one fragment the frames range in sequential order and the size of each frame can be acquired in the frame header. In the reordering of extracted fragments, the fragmented videos can be connected and restored based on the frame length information. After the reorder process the playable video or video clips will be carved out. The proposed AVI carving tool is software written in JAVA which realize the function of AVI carving and based on the AVI Carving method above. The overall workflow of the proposed tool is shown below in Fig. 1. The main two functions of the software are extracting video fragments in disk images and reordering the fragments. The other important

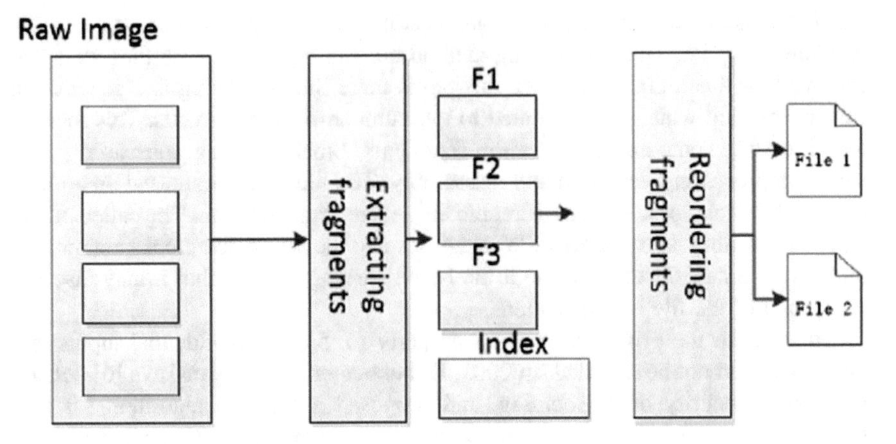

Fig. 1 Overall steps of the proposed technique

function of the software is to show the result of the extracting and reordering function. As a tool which aimed to be used in forensic scene, the software also order some forensic function such as calculating the hash value of the disk and generating reports automatically.

The proposed method extracts frames based on FOURCC and frame size of each substructure. Each frame data is extracted based on the beginning flag and the frame length information. The extraction process regards the frames with continuous physical location as the data of the same file and the adjacent and complete frames constitute one fragment for the next stage. As the data length of AVI header is relatively small, the possibility of fragmentation between the header and the movie data is very low. Therefore the header data will be associated to movie data and the index according to the principle to the nearest. In one fragment the frames range in sequential order and the size of each frame can be acquired in the frame header. In the reordering of extracted fragments, the fragmented videos can be connected and restored based on the frame length information. After the reorder process the playable video or video clips will be carved out.

4 Comparison and Efficiency

We compare the proposed algorithm with the algorithms. Let MM denote modular multiplication, MInv denote modular inversion, Invoke denote the invocation of subroutine. We omit other operations such as modular additions in these algorithms. The comparison of computation workload in the user side is shown in Fig. 2. We know the checkability of our algorithm is greater. Though the checkability is the same as that in the MInv is less. Thus, our algorithm is better. We know that in the same checkability the computation of algorithm is more than that in SExp. The less

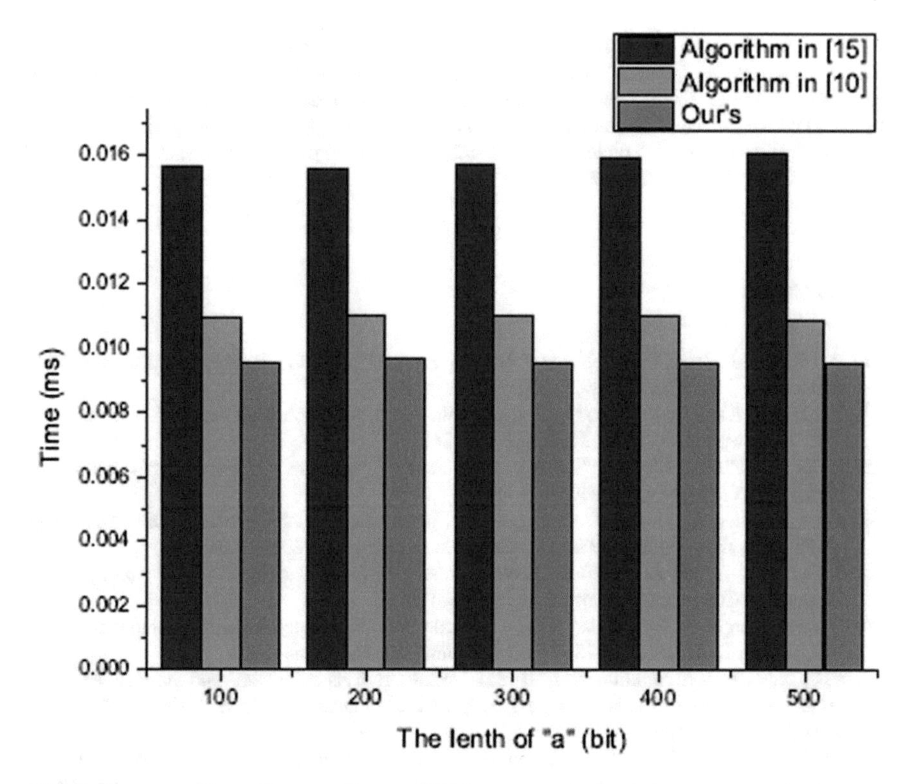

Fig. 2 The comparison of computation workload

MInvs and invocation times of Rand are needed in SExp. Further more, another important advantage of our scheme is that the MInvs and invocation times of Rand keep invariant when the number of outsourcing simultaneous modular exponentiations increasing.

5 Conclusion

This paper propose a video carving method and tool for fragmented and partly overwritten AVI files without using the file system information. The proposed method takes advantage of the feature of AVI files and is applicable to severely damaged video files. Most of the existing video carving techniques rely on file system information and can only cope with few fragments or complete file. The proposed method use signatures and data chunk size information to extract data fragments and reorder the data in a fragment base.

Acknowledgements This work was supported in part by the National Science and Technology Major Project under Grant 2013ZX01033002-003, in part by the National High Technology Research and Development Program of China (863 Program) under Grant 2013AA014603, in part by the National Science Foundation of China under Grant 61300202, in part by the China Post-doctoral Science Foundation under Grant 2014M560085, and in part by the Science Foundation of Shanghai under Grant 13ZR1452900.

References

1. B. Parno, J. Howell, C. Gentry, M. Raykova, Pinocchio: Nearly practical verifiable computation, Springer-Verlag, 2013, pp. 238–252.
2. S. Choi, J. Katz, R. Kumaresan, C. Cid, Multi-client non-interactive verifiable computation, Vol. 7785, Springer-Verlag, 2013, pp. 499–518.
3. J. Ye, H. Zhang, C. Fu, Verifiable delegation of polynomials, International Journal of Network Security 18 (2) (2016) 283–290.
4. J. Li, X. Chen, Q. Huang, D. Wong, Digital provenance: Enabling secure data forensics in cloud computing, Future Generation Computer Systems 37 (2014) 259–266.
5. J. Li, J. Li, X. Chen, Z. Liu, C. Jia, Privacy-preserving data utilization in hybrid clouds, Future Generation Computer Systems 30 (2014) 98–106.
6. S. Hohenberger, A. Lysyanskaya, How to securely outsource cryptographic computations, in: TCC, 2005, LNCS, Vol. 3378, Springer-Verlag, 2005, pp. 264–282.
7. P. Q. Nguyen, I. E. Shparlinski, J. Stern, Distribution of modular sums and the security of the server aided exponentiation, Cryptography and Computational Number Theory 20 (2001) 331–342.
8. M. Girault, D. Lefranc, Server-aided verification: Theory and practice, in: Proc. of 11th International Conference on the Theory and Application of Cryptology and Information Security, Chennai, India, Vol. 3788, Springer-Verlag, 2005, pp. 605–623.
9. W. Wu, Y. Mu, W. Susilo, X. Huang, Server-aided verification signatures: Definitions and new constructions, in: Proc. of Second International Conference, ProvSec 2008, Shanghai, China, Vol. 5324, Springer-Verlag, 2008, pp. 141–155.
10. D. Benjamin, M. Atallah, Private and cheating-free outsourcing of algebraic computations, in: Proc. of the Sixth Annual Conference on Privacy, Security and Trust, PST, Washington, DC, Springer-Verlag, 2008, pp. 483–501.
11. R. Gennaro, C. Gentry, B. Parno, Non-interactive verifiable computing: Outsourcing computation to untrusted workers, in: Advances in Cryptology - CRYPTO 2010, 30th Annual Cryptology Conference, Santa Barbara, CA, USA, August 15–19, 2010. Proceedings, 2010, pp. 465–482.

Community Semantics Recommender Based on Association Link Network

Yang Liu and Xiangfeng Luo

Abstract Community semantics recommender offers the semantic communities for topically homogeneous terms with the user demand, which helps people to acquire personalized information in mass documents and saves users' cognitive energy significantly. However, most of current recommendation models merely focus on users' behavior (e.g., item purchase behavior or user browsing traces) while pay less attention on the knowledge itself, especially the community semantics. In this paper, we present a community semantics recommender based on Association Link Network. We organize and represent the knowledge by Association Link Network (ALN). Given a user demand, we first select topically relevant community based on ALN; then we conduct community semantics activation process on the activated community to produce the topically homogeneous terms. Experimental results demonstrate our model can offer the user topically homogeneous terms and help them to understand the implicit knowledge that hidden in the large scale of documents.

Keywords Community semantics · Recommender system · Association link network

1 Introduction

The knowledge recommendation offers users the knowledge service [1], such as refining the implicit fuzzy demand and extending the domain knowledge, which not only saves users' effort for complex information seeking but also improves users' experience significantly. Community semantics recommender provides the

---

Y. Liu · X. Luo (✉)
School of Computer Engineering and Science, Shanghai University, Shanghai, China
e-mail: luoxf@shu.edu.cn

Y. Liu
e-mail: nmgliuyang@shu.edu.cn

© Springer Nature Singapore Pte Ltd. 2018

N.Y. Yen and J.C. Hung (eds.), *Frontier Computing*, Lecture Notes in Electrical Engineering 422, DOI 10.1007/978-981-10-3187-8_5

semantic communities, which conveys topically homogeneous knowledge to meet users' demands.

Current researches on knowledge recommender mainly focus on recommending documents by users' behavior. The two major recommender models are the collaborative-filtering model [2, 3] and the content-based model [4, 5]. The former explicits relevance feedback from users to update the user profiles [6], which is widely used in the e-commerce sites like Amazon and Taobao. The latter monitors document stream and pushes documents that match a user profile to the corresponding user [7, 8]. Above two kinds of recommender models have two commons: (1) recommend items/documents that are similar to what a user liked in the past; (2) recommend the items/documents which similar users group likes as target user's potential demand. Above two recommender models can hardly provide users their interested knowledge. Another widely used recommender for documents is the general search engines such as Google and Baidu, which recommend the documents that contain the terms in user's queries. The search engines offer different users same document list, if they have the same queries. Current recommenders seldom pay attention on the semantics of knowledge themselves, especially the community semantics, which leads the recommended results having abundant knowledge.

For these reason, this paper proposed an Association Link Network based community semantics recommender, which simulates the recommendation process of human memory based on the study of cognitive science. Specifically, we employ the Association Link Network (ALN) [9], an automatic built semantic link network [10], to organize and represent the knowledge of the documents. Based on ALN, we implement spreading activation [11, 12] model to simulate the knowledge recommendation process of human memory. Given a user demand, we first select topically relevant community based on ALN; then we activated community to produce the topically homogeneous terms. Therefore, our recommender not only can provide the user the community semantics for a better understanding of knowledge scenario, but also offer topically homogeneous terms.

The rest of the paper is organized as follows. The related work is discussed in Sect. 2. Section 3 describes community semantics recommender based on ALN. In Sect. 4, we demonstrate the effectiveness of our approach with comparative empirical results. Finally, we present the conclusions and point out future work in Sect. 5.

2 Related Work

Recommender is changing the way people interact with the Web and being more and more important. The recommender predicts the utility or relevance of the item to a particular user. Considering the different types of information used, the recommender can be divided into content-based recommender and collaborative filtering recommender.

The content-based recommender [13] recommends the documents which are similar to documents in the corresponding user's profile. Content-based recommender creates content-based user profiles using a weighted vector which denote the user feature. User profile can be computed by user's rating score using a variety of techniques such as Bayesian classifiers [14], decision trees [15], and neural networks [16] to estimate the probability that the user like the documents.

Collaborative filtering recommender [17] is based on the opinions of other users. The user ratings of documents are stored and can be used to predict ratings of documents which have not been rated by corresponding users. The user-based collaborative filtering model [18] finds the similar users according to users' rating history and then analyzes similar users' profile to predict the favor of target user. Different metrics have been used to calculate the similarity between users, such as mean squared difference [18], vector similarity [19] and weighted Pearson [20]. Item-based collaborative filtering model [21] finds the similar items by the items history and recommends the similar items to users. The item based model is better than the user based model because the similarity between items tends to be more static than the similarity between users. Taking the advantage of both item-based and user-based collaborative filtering models, a similarity fusion algorithm is proposed that has higher computational-complexity than both two algorithms [22].

Current recommenders pay less attention on the knowledge, especially the community semantics. Content-based recommenders only consider the similarity of items while users' need for association cannot be meeting. The collaborative filtering doesn't consider the characteristic of knowledge itself. The semantic communities are topically homogeneous terms, which help users to obtain personalized knowledge from mass documents and save their cognitive energy. A user friendly knowledge recommender should have following features: (1) Recommender should provide the user the community semantics for a better understanding of knowledge scenario. Human will easily understand the topically relevant knowledge, extend the domain knowledge, and solve specific knowledge questions. (2) It should offer the user topically homogeneous terms, which are relevant terms related to user demand and provide extra choices to users for further interaction. In this paper, we present a community semantics recommender system, where we provide the users relevant semantic communities as well as topically homogeneous terms.

3 Community Semantics Recommendation

In this Section, we will propose the ALN based community semantics recommender system. As Fig. 1 depicted, the framework of our community semantics recommender has two phases: (1) community selection phase, which provides the user the community semantics for a better understanding of knowledge scenario; (2) community semantics activation phase, which offers the user topically homogeneous terms and provide extra choices for the next recommendation interaction. Above two phases are based on the ALN. We employ ALN to organize the

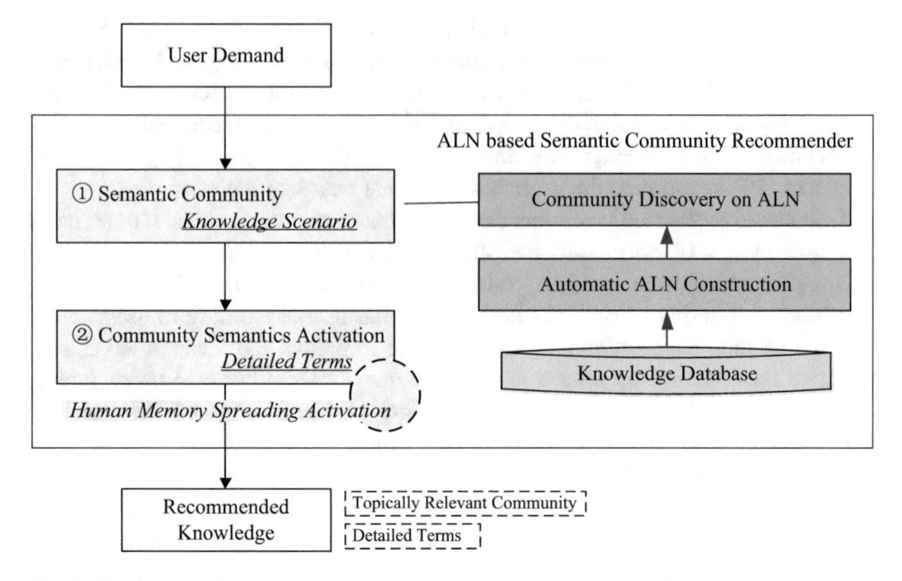

Fig. 1 The framework of ALN based community semantics recommender

semantics of knowledge database by association linked terms. Then, we discover communities on ALN, which can be used in community selection phase. We will detail each phase in the following.

3.1 Automatic ALN Construction

We employ ALN, which is an automatic built semantic link network [9], to organize the association linked nodes. Human knowledge retrieval is naturally directed. We construct ALN using significant terms, and the formula of computing association strength between term a and term b is as follows:

$$w^{a-b} = \frac{Co(a,b)}{\sqrt{DF(a)DF(b)}} \tag{1}$$

where $Co(a,b)$ represents the total co-occurrence times that term a and term b appear in one document. $DF(a)$ is occurrence times of word a in the document set.

The terms correspond to knowledge nodes in the human memory network. Term association has a value in the range [0, 1] to represent semantic related strength which corresponds to the strength between concepts in human memory. Thus, the generated ALN could be a knowledge network to simulating human domain knowledge, and based on it, we offer the user topically homogeneous terms.

ALN is able to organize the associated resources loosely distributed in the knowledge database for effectively supporting the recommender intelligent activities.

3.2 ALN-Based Community Discovery

After constructing ALN using the terms of knowledge database, we discover the community on ALN. The terms in a community are densely linked, and they have strong relationships between each other. We employ community detection algorithm to discover the community on ALN.

Enlightened by the thought of label propagation [23], each node is initialized with a unique label and at every step each node adopts the label that most of its neighbors currently have. In this iterative process, densely connected groups of nodes form a consensus on a unique label to form communities. Consequently, densely connected groups reach a common label quickly. The detailed process is as follows [23]:

(1) *Label Initialization*: we initialize the nodes' labels on ALN, and give each node on the ALN a unique node label.
(2) *Label propagation*: we do label propagation in iteration. For the iteration, each node adopts a label that a maximum number of its neighbours have, with ties broken uniformly randomly. As the labels propagate through the network in this manner, densely connected groups of nodes form a consensus on their labels.
(3) *Community Generation*: at the end of label propagation process, nodes with the same labels are grouped together as the communities discovered on ALN.

The advantage of this algorithm over the other methods is its simplicity and time efficiency. The algorithm uses the network structure to guide its progress and does not optimize any specific chosen measure of community strengths [23]. Furthermore, the community number and their sizes are not known a priori, and can be determined at the end of the algorithm. We conduct community discovery on ALN and provide the topically relevant community to users for a better understanding of knowledge scenario. And human will easily understand the topically relevant knowledge, which extends users' domain knowledge and solves specific knowledge questions.

3.3 Community Selection for Knowledge Scenario

For the user query, we select the topically relevant community to users, which gives the user comprehensive knowledge scenario for their understanding. A community can explain the topics of user demands, which offer the user knowledge context to extend their domain knowledge and solve the specific knowledge questions.

We select the topically relevant community for user demands based on ALN. The community is discovered based on the label propagation process on ALN. We view the terms with the largest degrees as well as association link weights as the most central terms for this community. We activate the most topically relevant community through computing the similarity between the user query and community:

$$I_c = \sum_{a \in (c \cap Q)} \frac{\sum w_c^{a-}}{\sum w_c} \tag{2}$$

where term a is the terms of user query. $\sum w_c^{a-}$ is the total association weight between term a and its adjacent terms on community c. $\sum w_c$ is the sum of all association weights of community c. We select the community with the max I_c value as the knowledge scenario for users.

3.4 Community Semantics Activation for Detailed Terms

After selecting the topically relevant community with user demands, we will offer the user topically homogeneous terms from the community based on the based on the spreading activation of human memory. The spreading activation of human memory tells the way that how human recommend the knowledge in human memory. We provide the relevant terms related and offer extra choices to users for further interaction.

In the human memory spreading activation process, user demand is the source of activation, and we conduct the spreading activation on the ALN-based topically relevant community.

(1) *Energy Initialization*: We give the terms of user queries an activation value represent their activation energy. And other terms of the community are zero.
(2) *Spreading Activation*: The activation energy is spreading in the community to get the topically homogeneous terms. After the spreading activation process, higher value represents user's higher focus.
(3) *Term Generation*: Terms with higher activation energy value will be generated as the topically homogeneous terms for the community semantics.

4 Experiments

In this section, we will make evaluations of the proposed community semantics recommender. We downloaded 1,222 documents from Tencent new channel as the knowledge database. Then, we construct ALN based on this dataset and perform community discovery on the ALN. We discovered 19 communities from the ALN.

Table 1 The Results of Community Semantics Recommendation

Query terms	Recommendation
Arizona, student, shooting	*Topically Relevant Community (Keywords of the Community)*: Shooting, U.S., Obama, Arizona, suspect, death, Tucson, condemn, college, American, head, police, blood, kill, federal, charge
	Recommended Detailed Terms: Obama, suspect, condemn, college, death
Iran, earthquake, Azerbaijan	*Topically Relevant Community (Keywords of the Community)*: Iran, earthquake, death, Azerbaijan, rescue, hospital, China, injure, Armenia, aftershock, shelter, damage, town, Ahar, emergency
	Recommended Detailed Terms: injure, damage, China, Armenia, rescue
Nigeria, flight, crash	*Topically Relevant Community (Keywords of the Community)*: Nigeria, China, death, Dana, passenger, flight, crash, disaster, people, aircraft, fire, damage, Lagos, reason, airframe, time
	Recommended Detailed Terms: China, Dana, passenger, disaster, aircraft
U.S, hurricane, Sandy	*Topically Relevant Community (Keywords of the Community)*: Sandy, storm, U.S, hurricane, flood, Obama, victim, weather, power, people, kill, wind, electric, speed, hospital, Cuba, town
	Recommended Detailed Terms: Victim, Obama, electric, power, power

Table 1 shows parts of recommendation results of our recommender system. Given a user query, our recommender first selects topically relevant community based on ALN, where we show the keywords of the relevant community. The recommended community provides users the knowledge scenario for a better understanding of relevant semantics. For example, the topically relevant community of the query "Arizona, student, shooting" can well explain the context how the Arizona shooting happens. Secondly, we recommend the detailed terms that are related to user demand, and provide extra choices to users for further interaction. For example, the detailed terms of the query "Arizona, student, shooting" describes some detailed information about the "Arizona shooting".

5 Conclusion

In this paper, we present a community semantics recommender, which focuses on the knowledge itself and provides the users knowledge including topically relevant community and related terms. The former provides the comprehensive knowledge scenario for user understanding. The latter offers relevant terms representing detailed community semantics. We employ Association Link Network (ALN) to organize and represent the knowledge. Based on the spreading activation of human memory, we activate topically relevant community and related semantics, and recommend them to users, which can better serve users and will be widely used in the field of e-learning and web service.

Acknowledgements The research work reported in this paper is supported by the National Science Foundation of China (grant no. 61471232), and the Key Innovation Program of Shanghai Municipal Education Commission (grant no. 13ZZ064).

References

1. P. Resnick, and H. R. Varian, "Recommender Systems," Communications of the ACM, vol. 40, no. 3, pp. 56–58, 1997.
2. J. Bobadilla, F. Serradilla and J. Bernal, "A new collaborative filtering metric that improves the behavior of recommender systems," Knowledge-Based Systems, vol. 23, no. 6, pp. 520–528, 2010.
3. Y. Koren, and R. Bell, "Advances in collaborative filtering", Recommender Systems Handbook, Springer, pp. 145–186, 2011.
4. P. Lops, M. de Gemmis and G. Semeraro, "Content-based recommender systems: State of the art and trends," Recommender Systems Handbook, Springer, pp. 3–105, 2011.
5. R. J. Mooney, and L. Roy, "Content-based book recommending using learning for text categorization," Proceedings of the fifth ACM conference on Digital libraries, pp. 195–204, 2000.
6. Y. Zhang, and J. Koren, "Efficient Bayesian and Hierarchical user Modeling for Recommendation Systems," Proceedings of the 30th annual international ACM SIGIR conference on Research and development in information retrieval, pp. 47–54, 2007.
7. J. Bian, A. Dong, X. He, S. Reddy, and Y. Chang, "User Action Interpretation for Online Content Optimization," IEEE Transactions on Knowledge and Data Engineering, vol. 25, no. 9, pp. 2161–2174, 2013.
8. M. J. Pazzani, and D. Billsus, "Content-based recommendation systems," The adaptive web, Springer, Berlin Heidelberg, pp. 325–341, 2007.
9. X. F. Luo, Z. Xu, and J. Yu, "Building association link network for semantic link on web resources," IEEE Transactions on Automation Science and Engineering, vol. 8, no. 3, pp 482–494, 2011.
10. Z. Xu, X. F. Luo, and W. Lu, "Association link network: An incremental semantic data model on organizing web resources," IEEE 15th International Conference on Parallel and Distributed Systems, pp 793–798, 2009.
11. A. M. Collins, and E. F. Loftus, "A spreading-activation theory of semantic processing," Psychological review, vol. 82, no. 6, pp. 407–428, 1975.
12. J. R. Anderson, "A spreading activation theory of memory," Journal of verbal learning and verbal behavior, vol. 22, no. 3, pp. 261–295, 1983.
13. P. W. Foltz and S. T. Dumais, "Personalized information delivery: An analysis of information filtering methods," Communications of the ACM, vol. 35, no. 12, pp 51–60, 2000.
14. C. Chen, L. F. Pau, and P. S. Wang, "Handbook of pattern recognition and computer vision," Imperial College Press, 2010.
15. N. Golbandi, Y. Koren, and R. Lempel, "Adaptive bootstrapping of recommender systems using decision trees," Proceedings of the fourth ACM international conference on Web search and data mining, pp. 595–604, 2011.
16. S. S. Park, K. K. Seo, and D. S. Jang, "Expert system based on artificial neural networks for content-based image retrieval," Expert Systems with Applications, vol. 29, no. 3, pp. 589–597, 2005.

17. U. Shardanand, and P. Maes, "Social information filtering: algorithms for automating 'word of mouth'," Proceedings of the SIGCHI conference on Human factors in computing systems. ACM Press, pp. 210–217, 1995.
18. P. Resnick, N. Iacovou, and M. Suchak, "GroupLens: an open architecture for collaborative filtering of netnews," Proceedings of the 2001 ACM conference on Computer supported cooperative work. ACM, pp. 175–186, 2001.
19. J. S. Breese, D. Heckerman, and C. Kadie, "Empirical analysis of predictive algorithms for collaborative filtering," Proceedings of the Fourteenth conference on Uncertainty in artificial intelligence, Morgan Kaufmann Publishers Inc., pp. 43–52, 1998.
20. J. L. Herlocker, J. A. Konstan, L. G. Terveen, and J. Riedl, "Evaluating collaborative filtering recommender systems," ACM Transactions on Information Systems (TOIS), vol. 22, no. 1, pp 5–53, 2004.
21. B. Sarwar, G. Karypis, J. Konstan, and J. Riedl, "Item-based collaborative filtering recommendation algorithms," Proceedings of the 10th international conference on World Wide Web. ACM, 285–295, 2001.
22. J. Wang, A. P. De Vries, M. J. T. Reinders, "Unifying user-based and item-based collaborative filtering approaches by similarity fusion," Proceedings of the 29th annual international ACM SIGIR conference on Research and development in information retrieval. ACM, pp. 501–508, 2006.
23. U. N. Raghavan, R. Albert and S. Kumara, "Near linear time algorithm to detect community structures in large-scale networks," Physical Review E, Vol. 76, Issue. 3, 2007.

Is Low Coupling an Important Design Principle to KDT Scripts?

Woei-Kae Chen, Chien-Hung Liu, Ping-Hung Chen and Yu Wang

Abstract In keyword-driven testing (KDT), a keyword represents a sequence of actions (events and assertions) and a test case is constructed by using a sequence of keywords. That is, in a KDT script, a test case depends on a number of keywords, and a keyword depends on some other keywords and GUI components. Such dependency is also known as the coupling between test cases, keywords, and components. This paper studies the question of whether low coupling is an important design principle to KDT Scripts. A coupling measure, called unweighted coupling, is proposed. A case study is conducted to assess the maintainability and readability of KDT scripts with different couplings. The results indicated that, when maintaining KDT scripts, a low-coupling script required, on average, less changes than a high-coupling one. On the other hand, a low-coupling script does not necessarily offer a better readability.

Keywords Keyword-driven testing · Test script · Coupling · Maintainability · Readability

1 Introduction

As GUI (Graphical User Interface) is pervasive in all kinds of software applications (Web, mobile, and rich client applications), GUI testing assumes increasing importance nowadays. To automate GUI testing, a tester prepares a test script (or test case), which contains a sequence of actions (events and assertions) that are performed on

W.-K. Chen (✉) · C.-H. Liu · P.-H. Chen · Y. Wang
Department of Computer Science and Information Engineering, National Taipei University of Technology, Taipei, Taiwan
e-mail: wkchen@ntut.edu.tw

C.-H. Liu
e-mail: cliu@ntut.edu.tw

P.-H. Chen
e-mail: cbh@ntut.edu.tw

Y. Wang
e-mail: t99598055@ntut.edu.tw

© Springer Nature Singapore Pte Ltd. 2018

N.Y. Yen and J.C. Hung (eds.), *Frontier Computing*, Lecture Notes in Electrical Engineering 422, DOI 10.1007/978-981-10-3187-8_6

the components (widgets) of the GUI [1, 2]. The test script is then executed to exercise the GUI and perform verifications.

When developing a test script, Keyword-Driven Testing (KDT) approach (e.g., [3–6]) is widely used. In KDT, a keyword performs one or more actions and a test case is constructed by using a sequence of keywords. In general, a keyword can also use (call) some other keywords, creating a hierarchical keyword structure. For example, Fig. 1 shows a simple KDT test script (called simply KDT script or script hereafter) supported by Robot Framework [3]. The keyword `Login` sequentially opens a login page, enters an account name into the `account` widget, enters a password into the `password` widget, and then submit the login form. In this example, `Login` keyword uses another keyword `Open Login page`, which opens a browser with a specified login URL. The advantages of KDT are that test cases are more concise and readable, and keywords are reusable and can be revised to accommodate to the changes in the GUI.

The principle behind KDT is *procedural abstraction*. The role of a keyword/action is analogous to a C language function/statement. Figure 2 depicts a call graph in which test cases use keywords, and keywords in turn use actions to accomplish a test job. Such relationships between test cases, keywords, and actions are also known as the *coupling* between them. Note that, as shown in Fig. 2, when a keyword *K* uses an action *A* performed on component *C*, *K* is in fact coupled to *C*, not *A*. This is because, without the existence of *C*, *A* cannot be used at all (e.g., suppose *K* uses the `click` event of a button, *K* depends on the button, not on the `click` event).

In software design, coupling is a measure of how strongly one software module is connected to, has knowledge of, or relies on other modules [7–10]. It is generally

Keyword	Action	Argument	Argument
Open Login Page	Open Browser	http://host/login.html	
	Title Should Be	Login Page	
Login	Arguments	${account}	${password}
	Open Login Page		
	Input Text	account	${account}
	Input Text	password	${password}
	Submit Form		

Fig. 1 A simple test script

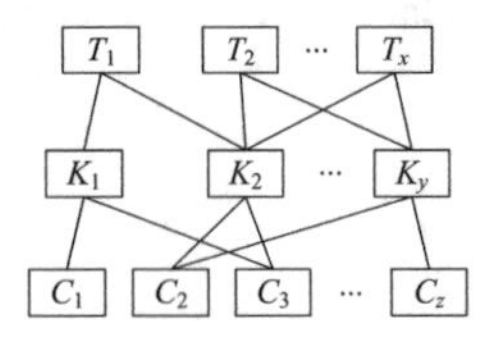

Fig. 2 A general call graph of test cases, keywords, and components. A vertex T_i denotes a test case, a vertex K_i denotes a keyword, and a vertex C_i denotes a component on which an action is performed

known that a software system designed with low coupling supports low change impact and promotes maintainability. However, to our knowledge, so far there are no researches that study the coupling for the special case of KDT scripts. Is low coupling an important design principle to KDT scripts? Since the purpose of a KDT script is to create user interactions, it is not the same as a piece of source code that performs intensive arithmetic/logic computations. Thus, code coupling may not be directly applicable to the case of KDT scripts. In particular, it is not even known whether different KDT scripts may have different degree of couplings.

This paper studies the question of whether low coupling is important to a KDT script when maintainability and readability are of concern. This is important to a tester who needs to constantly develop and maintain KDT scripts. We define a measure, called unweighted coupling, that evaluates the coupling of a KDT script. A case study with three experiments is conducted to address whether coupling is related to maintainability and readability. The results indicated that (1) given exactly the same sequence of actions to perform, different testers created KDT scripts with completely different couplings; (2) when maintaining KDT scripts, a low-coupling script required, on average, less changes (test cases and keywords) than a high-coupling one; and (3) in terms of readability, a low-coupling test script is not necessarily easier to read than a high-coupling one.

The rest of this paper is organized as follows. Section 2 presents a simple coupling measure for KDT scripts. Section 3 reports a case study that addresses whether coupling is related to maintainability and readability. The related work is discussed in Sect. 4. A conclusion is given in Sect. 5.

2 Coupling in KDT Scripts

We define the coupling of a KDT script as the degree to which each module (either a test case, keyword, or component) relies on each one of the other modules. A KDT script is like the source code of a high-level language. However, a typical KDT script contains mainly actions. Its purpose is to drive the GUI and verify its correctness. Therefore, a KDT script does not normally perform complicated arithmetic/logic computations. Consequently, language constructs such as *structure* and *class* are unnecessary and are not supported by many popular KDT enabled tools (e.g., [3, 5]). According to Myers [10], there are five different kinds of code couplings (i.e., data, stamp, control, common, and content couplings). However, since a KDT script is not designed for computations, using sophisticated code coupling metrics can be unrealistic. Therefore, we focus mainly on evaluating the caller-callee relationships between modules.

We use a coupling diagram, like Fig. 2, to illustrate the coupling between modules (for simplicity, we use the term keyword to represent a module hereafter). When a keyword uses another keyword, there is a coupling between them. Therefore, a straightforward measure of coupling is to evaluate the *density* of the coupling diagram—the higher the density, the higher the coupling. More precisely, a coupling diagram is a directed, unweighted graph whose vertices are keywords and there is an edge (K_i, K_j) if and only if K_i uses K_j. Thus, given a coupling diagram $G = (V, E)$ where V is the set of vertices and E the set of edges, we can define *Unweighted Coupling* (UC) as:

$$UC = \frac{|E|}{(|V| - 1)}$$

In this equation, $|E|$ is divided by $|V| - 1$ to obtain the density. Since a normal coupling diagram should have at least $|V| - 1$ edges (i.e., a connected graph without any unused keywords), the minimum value for UC is 1. Figure 3a is an example of 7 vertices (2 test cases and 5 components) and 8 edges (note: as all edge directions are top-down, arrows are not shown in the graph). In this case, $UC = \frac{8}{7-1} = 1.33$. Figure 3b is another example with $UC = 1$.

How does a KDT script have a high coupling? We present three cases that keywords are not properly designed, resulting an increased coupling: (1) using too few keywords, (2) using keywords that are not reusable, and (3) having redundant keywords. The first case appears when many actions are repeatedly executed for several times, yet no keywords are used to simplify the repetition. For example, Fig. 3a shows that both T_1 and T_2 use the same actions from C_2, C_3, and C_4. In this case, by adding a new keyword K_1 (Fig. 3b) to encapsulate these actions, T_1 and T_2 no longer depends directly on C_2, C_3, and C_4. Thus, the overall coupling is reduced from 1.33 to 1.

The second case appears when a keyword cannot easily be reused. Figure 4a shows an example that T_1 uses K_1 to perform actions on C_1, C_2 and C_3. On the surface, T_2 also performs actions on C_1, C_2 and C_3, and thus, it is maybe possible for T_2 to reuse K_1. However, suppose K_1 contains one or more actions that T_2 does not need (e.g., K_1 uses two different actions of C_1, but T_2 needs only one of them), T_2 is forced to use C_1, C_2 and C_3 directly, resulting the diagram shown in Fig. 4a. The fundamental problem in this case is that K_1 is not designed as a common keyword for both T_1 and T_2. Thus, a better design, like Fig. 4b, is to refactor K_1 so that both T_1 and T_2 can use it, and such a refactoring also reduces the overall coupling from 1.4 to 1.2.

The third case appears when a test script grows larger and larger, and the tester is unaware of an existing keyword and creates a new one, or when a keyword is not properly parameterized and cannot be reused. Figure 5a shows an example that K_1

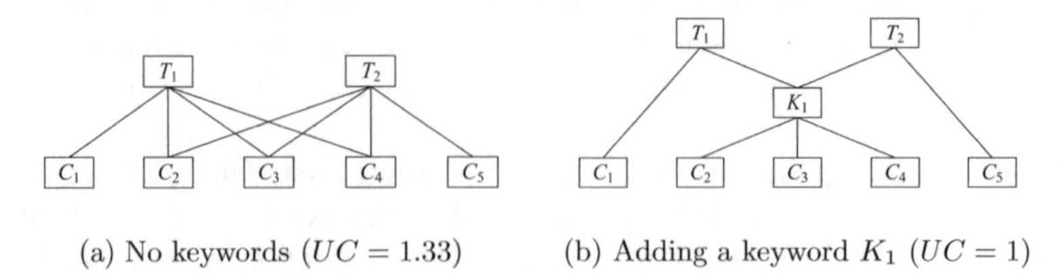

(a) No keywords ($UC = 1.33$) (b) Adding a keyword K_1 ($UC = 1$)

Fig. 3 The coupling diagram of the first case

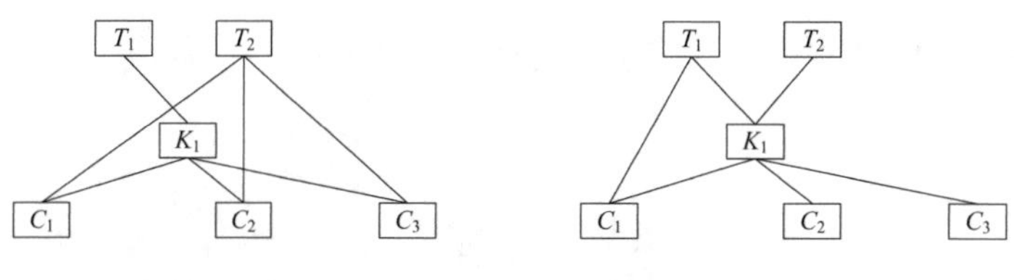

(a) T_2 cannot use K_1 ($UC = 1.4$) (b) K_1 is redesigned ($UC = 1.2$)

Fig. 4 The coupling diagram of the second case

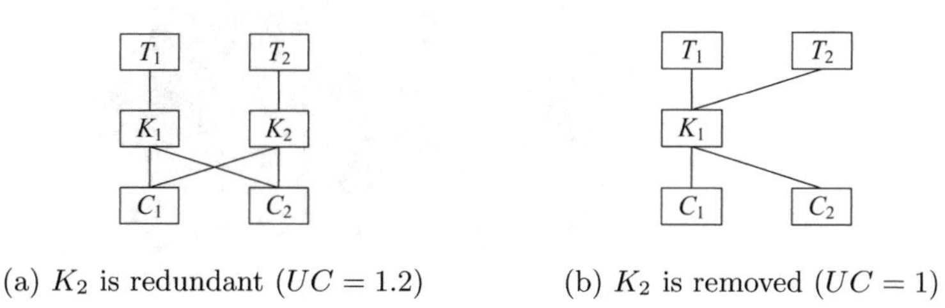

(a) K_2 is redundant ($UC = 1.2$) (b) K_2 is removed ($UC = 1$)

Fig. 5 The coupling diagram of the third case

and K_2 are in fact identical (or identical when the differences between K_1 and K_2 can be eliminated by using parameter substitutions). In this case, T_2 can use K_1 instead of K_2 (Fig. 5b), reducing the overall coupling from 1.2 to 1.

3 Case Study

This section reports a case study that addresses whether low coupling is an important design principle to a KDT script. The research questions are:

RQ1 Given the same test cases (i.e., the same sequence of test actions), do different testers create different keyword designs?
RQ2 Does a different keyword design produce a different degree of coupling?
RQ3 When maintaining a test script, does its coupling affect maintenance cost?
RQ4 When reading a test script, does its coupling affect readability?

We conduct three experiments to answer the above research questions. The experiments use Crossword Sage v.0.3.5 (called simply CS) [11] as the target software under test. CS is a rich client application (Fig. 6) that can be used to create/solve crossword puzzles. We choose CS for our experiments because it has been tested by many GUI testing researches [12–15].

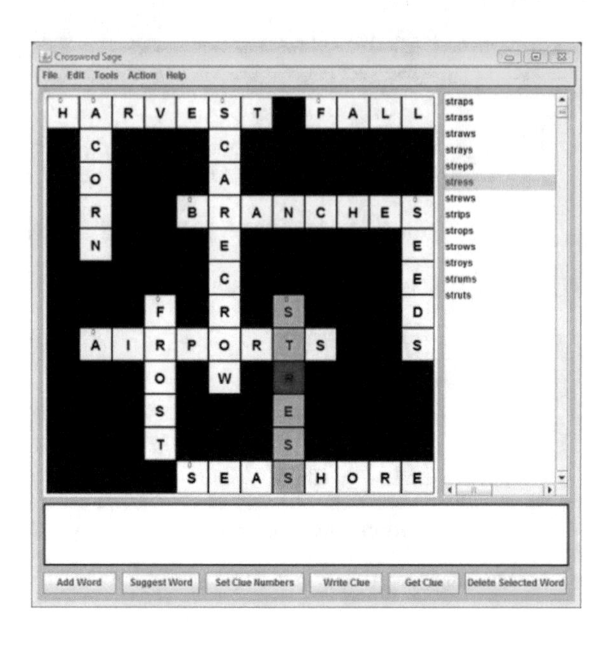

Fig. 6 Crossword Sage (CS)

3.1 *Experiment I*

The first experiment addresses RQ1 and RQ2. We recruit 5 graduate students to participate the experiment. The participants, called P1–P5, act as testers who are requested to develop a KDT script that implements a pre-defined test plan. The test plan contains a total of 9 different test cases, testing the most important user scenarios of CS. Each of the first 7 test cases exercises and tests a single, simple feature. The last 2 test cases, on the other hand, perform integration tests that use several features altogether to create a real crossword puzzle. The test plan specifies the exact sequence of actions that are performed in each test case. A total of 422 actions are required.

Each participant is requested to use Robot Framework [3] to implement a KDT script that can produce exactly the same 422 test actions. When developing a script, the participants are reminded of keeping maintainability and readability in mind. When a participant completed his script, we perform a verification to ensure that the script is implemented correctly. Then, we compute the coupling of the script. The results are shown in Table 1. It can be seen that every participant designed keywords differently. For example, P2 used 15 keywords, P3 used only 14, and P4 used no keywords at all. Thus, the answer to RQ1 is "yes—while the participants were requested to develop exactly the same test cases (and exactly the same sequence of actions), each participant created a completely different keyword design."

Figures 7 and 8 show the coupling diagrams of P2 script and P3 script, respectively (note: for simplicity, we call P2's script simply P2 script). In P3 script (Fig. 8), many test cases directly access components without using any intermediate keywords. P2 script (Fig. 7), on the other hand, had a better keyword hierarchy. Thus,

Table 1 Coupling of the scripts created by participants P1–P5

Participant ID	Keywords used	UC
P1	16	1.94
P2	15	1.87
P3	14	2.21
P4	0	2.44
P5	13	1.95

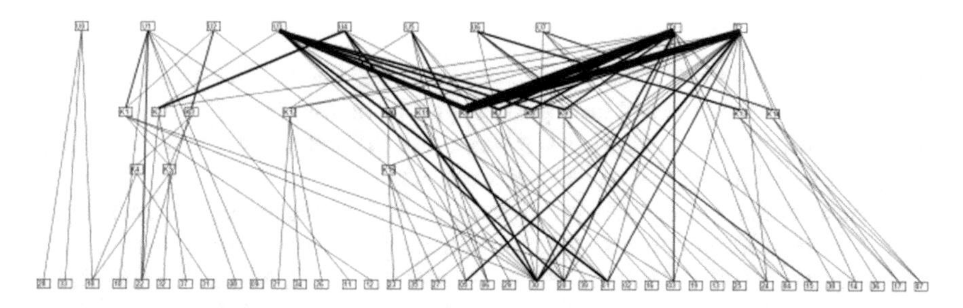

Fig. 7 The coupling diagram of P2 script (note: an edge (K_1, K_2) is drawn with a thicker line when K_1 accesses K_2 for more than one time)

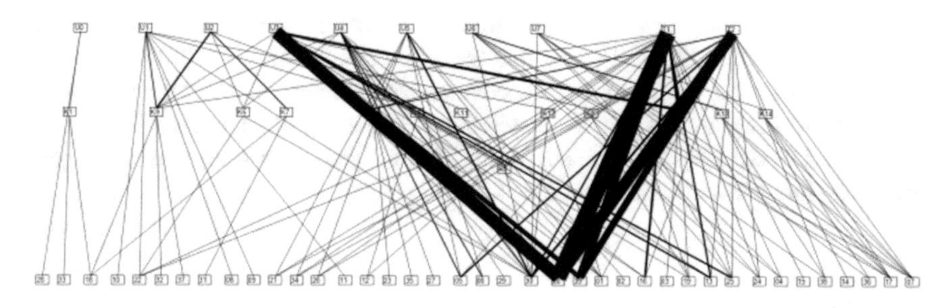

Fig. 8 The coupling diagram of P3 script

Fig. 8 shows an overall higher coupling than Fig. 7. The UC of P2 script was 1.87, lower than that (2.21) of P3 script. To save space, we do not present all five coupling diagrams. Overall speaking, the answer to RQ2 is "yes, a different keyword design did result in a different degree of coupling."

3.2 *Experiment II*

The second experiment addresses RQ3. To evaluate maintainability, we modify the GUI of CS (via modifying its source code). The new GUI results in an improved CS, called CS', which simplifies the following user interactions: (1) remove the "Suggest

Table 2 Test script maintenance cost

Participant ID	Modifications
P1	45
P2	53
P3	93
P4	140
P5	69

Word" button and show the suggested words automatically, (2) remove the "Find Possible Matches" button and show possible matches automatically, (3) remove "Add Word" button and use double click instead, and (4) add a new "Add Word Tips" dialog to remind the user of how to add a new word.

A test script designed for CS no longer works for CS'. We request each participant of the first experiment (P1–P5) to maintain (repair) his own script so that the script can be reused to test CS'. The results are shown in Table 2, where "Modifications" indicates the total number of modifications that was made (including both test case and keyword modifications). The correlation coefficient between UC and modifications was 0.97. The strong correlation indicated that a low-coupling script generally required less modifications, i.e., less maintenance cost. Note that, UC evaluates the overall coupling of a script. In case that there is a small, partial GUI modification, the modification may not necessarily produce a global impact to the entire test script. Therefore, it is reasonable that the correlation between coupling and modification is imperfect. Overall speaking, the answer to RQ3 is "yes, a low-coupling script generally requires a lower maintenance cost."

3.3 Experiment III

The third experiment addresses RQ4 and then revisits RQ3. We recruit another 4 graduate students to participate the experiment. The participants, called P6–P9, are requested to assess the readability of the scripts developed by P1–P5 from experiment I. Each participant (P6–P9) is given all the 5 scripts in random order, and is requested to read each script and then assign a readability grade for each script, based on the grading standard defined in Table 3. Figure 9 shows the average readability grade received by each script. Except for P2 script, which received a higher grade, the rest of the scripts received similar grades, indicating that these scripts had similar readability. The correlation coefficient between UC and readability grade was −0.59. Here, the correlation was not that strong. Our interview with P6–P9 indicated that whether a test script was easy to read depended more on keyword naming

Table 3 Readability grades

Grade	Description
5	The test script is very easy to understand. The actions that are to be executed are very clear.
4	The test script is easy to understand. The actions that are to be executed are clear.
3	Most of the test script can be understood. There are occasionally some actions that are not so clear.
2	The test script is difficult to understand. There are a lot of actions that are not so clear.
1	The test script is very difficult to understand. The actions that are to be executed are not clear at all.

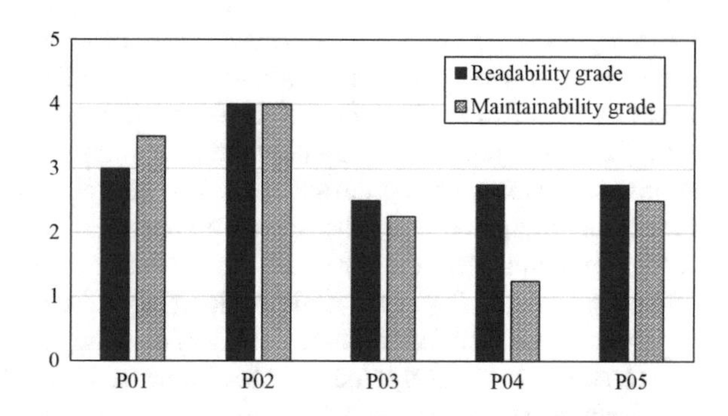

Fig. 9 Average readability and maintainability grade of each script

than on keyword structures. Some participants pointed out that, when a keyword was not properly named, one needed to constantly reread its actions so as to confirm its action sequences; some participants considered that P4 script was not any more difficult to read even though it had no keywords at all. This was mainly because the test scripts contained mostly straightforward actions (without any complicated logics), and therefore, the lack of structure did not pose serious readability problems. Thus, the answer to RQ4 is a vague "yes and no, there is a correlation between coupling and readability, but the correlation is not very strong—keyword naming is also an important factor to readability."

We then revisit RQ3 by requesting each participant (P6–P9) to assess the maintainability of P1–P5 scripts. We inform the participants that the GUI is to be changed from CS to CS' (the same changes as described in Sect. 3.2). Each participant is requested to identify how each of the five scripts can be repaired and reused for CS', and then assign a maintainability grade for each script. The results are shown in Fig. 9. Note that, while P4 script was not particularly bad in readability, it received the worst maintainability grade, indicating that a test script could be easy to read, yet difficult to maintain. The correlation coefficient between UC and maintainability grade was –0.91, a very strong correlation. In other words, from the viewpoints of P6–P9, a script's coupling was highly related to its maintainability. The results reaffirmed the answer to RQ3 reported in the previous section.

4 Related Work

Among various test automation approaches, keyword-driven testing (KDT), of- fered by Fewster and Graham [4] and Kaner et al. [6] among others, has been widely used. KDT is an improvement over data-driven testing. In KDT, both test data and keyword implementations can be taken apart from test scripts and put into external input files. To offer better flexibility, many KDT tools (e.g., [3]) also allow testers to construct high-level keywords. Thus, many testers develop/maintain their own keyword hierarchies by themselves. However, as reported in Sects. 3.2 and 3.3, a KDT script is not necessarily easy to maintain. The current trend in iterative/agile processes [9] promotes evolutionary refinement of plans, requirements, and design. Thus, the test script associated with the design must also be updated frequently and its maintenance becomes important. That brings up the question, when creating keywords, is there a guideline that a tester should follow? The results of this paper suggest that low coupling is an important design principle for the development of KDT scripts.

The measurement of coupling for high-level languages and object-oriented systems has been extensively studied in the literature [7, 8, 10]. However, a KDT script, a procedural abstraction of GUI operations, is not equivalent to a piece of code. Though some KDT tools (e.g., Robotframework [3]) also support keywords that perform loops, conditional expressions, and even arithmetic/logic computations, the bulk of a script is made of sequences of actions. At the extreme, a test script simply stores all the actions linearly. In case that arithmetic/logic computations are necessary, they are normally encapsulated into special keywords and account for only a very small portion of the script. Therefore, the most prominent coupling in a script is the caller-callee relationships between modules (test cases, keywords, and components). Consequently, general code-coupling measures (e.g., data, stamp, control, common, and content couplings [10]) do not capture the essence of a script nicely. To our knowledge, so far there are no previous researches that specifically address the coupling of KDT scripts. This paper fills this gap by offering a simple measure along with an evaluation (Table 4).

Table 4 Maintainability grades

Grade	Description
5	The test script is very easy to maintain. A small GUI change results in only very few keywords (or test cases) need to be changed, and a keyword change does not impact the other keywords (or test cases)
4	The test script is easy to maintain. A small GUI change results in only a small number of keywords (or test cases) need to be changed, and a keyword change does not impact the other keywords (or test cases)
3	The test script is not so easy to maintain. A small GUI change results in many keywords (or test cases) need to be changed, and a keyword change could impact some other keywords (or test cases)
2	The test script is difficult to maintain. A small GUI change results in a large number of keywords (or test cases) need to be changed, and a keyword change could impact many other keywords (or test cases)
1	The test script is very difficult to maintain. A small GUI change results in almost all keywords (or test cases) need to be changed, and a keyword change could impact almost all other keywords (or test cases)

5 Conclusion

This paper studies the question of whether low coupling is an important design principle to KDT scripts. We use the density of a coupling diagram to evaluate the overall coupling of a KDT script. In addition, a case study with three experiments is conducted to assess the relationships between coupling and maintainability/readability. The results indicated that, when maintaining a test script, a low-coupling script required, on average, less maintenance cost. On the other hand, as keyword naming is also an important factor to readability, a low-coupling script does not necessarily offer a better readability than a high-coupling one. Overall, the results suggest that, for a tester who needs to constantly develop/maintain a set of test scripts, the tester should follow the low-coupling principle and use appropriate keyword names.

Acknowledgements This research was partially supported by the Ministry of Science and Technology, R.O.C., under contract number MOST 104-2221-E-027-008, which is gratefully acknowledged.

References

1. Chen, W.K., Shen, Z.W., Tsai, T.H.: Integration of specification-based and cr-based approaches for GUI testing. J. Inf. Sci. Eng. 24(5), 1293–1307 (2008)
2. Memon, A.M., Pollack, M.E., Soffa, M.L.: Hierarchical GUI test case generation using automated planning. IEEE Trans. Softw. Eng. 27(2), 144–155 (2001)
3. Robot framework, http://robotframework.org/, Feb. 1, 2016
4. Fewster, M., Graham, D.: Software Test Automation. Addison-Wesley Professional (1999)

5. HP: Unified functional testing (UFT), http://en.wikipedia.org/wiki/HP_QuickTest_Professional, Feb. 1, 2016
6. Kaner, C., Bach, J., Pettichord, B.: Lessons Learned in Software Testing: A Context-Driven Approach. Wiley (2001)
7. Dhama, H.: Quantitative models of cohesion and coupling in software. Journal of Systems and Software 29(1), 65–74 (1995)
8. Fenton, N., Melton, A.: Deriving structurally based software measures. Journal of Systems and Software 12(3), 177–187 (1990)
9. Larman, C.: Applying UML and Patterns, 3rd Ed. Prentice Hall PTR (2004)
10. Myers, G.J.: Reliable Software Through Composite Design. Petrocelli/Charter (1975)
11. Westgarth, B.: Crossword sage, http://crosswordsage.sourceforge.net, Feb. 1, 2016
12. Chen, W.K., Wang, J.C.: Bad smells and refactoring methods for GUI test scripts. In: Software Engineering, Artificial Intelligence, Networking and Parallel Distributed Computing (SNPD), 2012 13th ACIS International Conference on. pp. 289–294 (Aug 2012)
13. Memon, A.M.: Automatically repairing event sequence-based GUI test suites for regression testing. ACM Trans. on Softw. Eng. and Method. (2008)
14. Xie, Q., Memon, A.M.: Using a pilot study to derive a GUI model for automated testing. ACM Trans. Softw. Eng. Methodol. 18(2) (2008)
15. Yuan, X., Memon, A.M.: Generating event sequence-based test cases using GUI runtime state feedback. IEEE Transactions on Software Engineering 36(1), 81–95 (Jan 2010)

Research and Implementation of FM-DCSK Chaotic Communication System Based on GNU Radio Platform

Bingyan He, Yuanhui Yu and Lizhen Chen

Abstract FM-DCSK is a non-coherent modulation and demodulation technique, the output of which has the inherent wide-frequency performance will cause the estimation interference and the bit error rate decline. This paper designs and implements Frequency-modulated Differential Chaotic Shift Keying (FM-DCSK) Communication system based on GNU Radio platform. In order to get FM-DCSK signal, it needs to combine the FM modulation with the DCSK system. The bit energy of signals remained stable by using FM modulation so that the BER performance in such system can be improved. Since the system is based on software defined radio (SDR) to perform in a real-time wireless transmission, the bitrate, bandwidth and central frequency can be modified at ease. The experimental performance are discussed and compared to the theoretical performance.

Keywords Frequency-modulated differential chaotic shift keying · Chaos · Random-like property · Soft defined radio

1 Introduction

The unique initial value sensitivity, random-like property and unpredictability of chaotic signal can provide the guarantee for the secure transmission of information. The basic idea of chaotic secure communication is that using chaotic signal as a carrier, the transmissing data signal is hidden in the chaotic carrier, the waveform of the modulated chaotic signal is disorderly and irregular, not easy to be stolen. Chaotic signal is seemingly unpredictable, in fact, it is pseudo random signal which can be described using mathematical equations describing the broadband

B. He · Y. Yu · L. Chen (✉)
Computer Engineering College, JiMei University, Xiamen, China
e-mail: lzchen@jmu.edu.cn

B. He
e-mail: iceorspring@sina.com

© Springer Nature Singapore Pte Ltd. 2018

N.Y. Yen and J.C. Hung (eds.), *Frontier Computing*, Lecture Notes in Electrical Engineering 422, DOI 10.1007/978-981-10-3187-8_7

characteristics make it widely used in the fields of multiple access communication and spread spectrum communication, etc.

Chaotic signals are a relatively new field of in communication systems. Potential of this method derived from the advantages offered by chaotic signals, such as robustness in multipath environments and resistance to jamming. Chaotic signals are non-periodic, broadband, and difficult to predict and reconstruct. These are properties which match with requirements for signals used in spread spectrum communication systems [1].

Various types of modulation can be used in direct chaotic communication systems like chaotic on-off keying (COOK). Differential chaotic shift keying (DCSK) and additive chaos modulation (ACM), etc. In all chaotic modulation types, DCSK with orthonormal basis properties offer the best robustness against the multipath and channel imperfections [2].

The performance of chaos-based digital communication systems under additive white Gaussian noise (AWGN) and m-distributed fading channels has been thoroughly studied [3, 4, 5].

SDR (Soft defined radio) is an intelligent wireless communication system, which can automatically sense the wireless environment and the use of spectrum. It's architecture has broken the traditional design pattern based on the special hardware, as much as possible using general hardware as the basic platform, to achieve wireless and communication functions by the software on the general processor, in order to achieve the wireless communication system upgraded and reconfigurable.

In this paper, we implement a FM-DCSK communication system on SDR. We put forward a simple and robust method to implement a Chaotic generator where the period loop length is sufficiently long and exactly known making this generator suitable for real-time transmissions. Considering complex wireless transmission environment, we study The symbol synchronization problem, and a better algorithm is proposed and implemented to make this system operational in real-time transmissions.

2 FM-DCSK Communication Scheme

A block diagram of FM-DCSK communication system is shown in Fig. 1 in this paper, This system uses FM-DCSK as a chaotic carrier to spread the digital signal over a wide frequency band. And modulate the digital signal to achieve the purpose of transmitting information. Firstly, the chaotic signal is generated by Logistic map, then Logistic signal is fed into the FM modulator, and FM-modulated Logistic signal is used as a new carrier. At the receiving end, the self correlation and cross-correlation of the chaotic signal are used to demodulation, and the digital sequence is obtained.

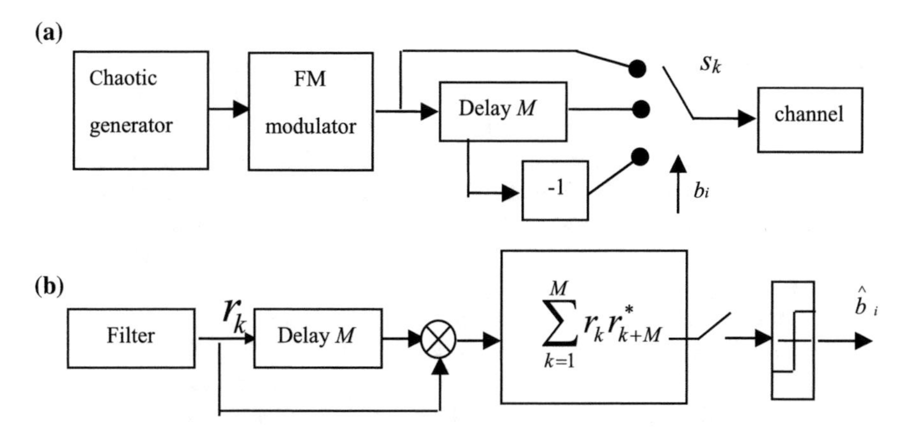

Fig. 1 **a** FM-DCSK transmitter **b** DCSK receiver

2.1 Generation of Chaotic Signal

Various types of chaotic and hyper-chaotic signals which are generated by various types of nonlinear circuits and are suitable for security communication are the hot topics in the field of physics and information science. Chaotic signal in this paper, we use Logistic mapping method to generate chaotic signal. At present, many research achievements have been made, such as Four order variant chaotic circuit, multi-scroll chaotic and super-chaotic circuit [6].

Considering the simplicity and good statistical properties of the logistic [7], the implementation of chaotic signal generator is based on the improved logistic full mapping equation, one-dimensional differential mapping equation is defined as follows:

$$X_{n+1} = 1 - X_n^2 \quad X_n \in (-1, 1) \tag{1}$$

X_0 is the initial value

$$E[X_n] = \lim_{M \to \infty} \frac{1}{M} \sum_{M=0}^{M-1} X_n = \int_{-1}^{1} xf(x)dx = 0 \tag{2}$$

the X_0 has a very small change, After a certain number of iterative computations, it will produce completely different two sequences, making use of this feature, by setting the initial value, can generate a number of unrelated sequences. The DCSK system is not relevant and does not need to generate chaotic signals at the receiving side. In view of this, we use the Linux random number generator as the basic uniform random number generator. Linux random number generator produces random bytes from the entropy pool collected from the host computer activities such as the time interval The user uses the keyboard, the distance and interval the mouse moves and the time interval between the specific interrupts, these host activities for the computer are nondeterministic and unpredictable. The kernel maintains an entropy pool based on these non deterministic device events, and the data in the pool are completely random. When there is a new device event coming, the kernel will estimate the randomness of the newly added data, when we get the data from the entropy pool, the kernel can reduce the estimated value of the entropy. Through the get_random_bytes () interface of the kernel, we can obtain the kernel 32 random integers from the entropy pool. Random integer range in $[-2^{31}, 2^{31}-1]$, these random integers can be converted into $[-1, 1]$ range of floating point number.

The parabola map contains the basic idea of the modern chaos theory, including the basic frame and the model of the nonlinear theory, such as the period to chaos and bifurcation diagram. Then, by using the $F^{-1}(x)$ function to process the values of these uniform distributions, we obtain the random values of the Logistic distribution.

2.2 Modulation and Demodulation of FM-DCSK Signal

In the FM-DCSK [8] modulating system, the system firstly modulate the chaotic signal using FM modulation, and then the FM-modulated chaotic signal is used as the carrier for digital modulation. Since the energy of periodic positive cosine signal is constant in one cycle. So if the bit cycle is several times the cycle of positive cosine signal, you can ensure that the energy of each bit after the modulation of the FM-DCSK is not randomly variable. To realize the above description, as long as the frequency of positive cosine signal used to modulate chaotic signal is much higher than the chaotic signal, it can be easily achieved.

Firstly introducing Binary DCSK modulation, the modulated transmission signal can be expressed as

$$s_k = \begin{cases} c_k & 1 < k \leq M \\ b_i c_{k-M} & M < k \leq 2M \end{cases} \tag{3}$$

c_k is a chaotic reference signal, c_k has M chaotic signal samples, $2\ M$ is the number of chaotic samples sent per bit, M is an integer. A chaotic signal c_k as a reference signal is transmitted in the first half cycle of each bit, the data is transmitted in the second half of the cycle, if the data is 1, still transmitted chaotic signal c_k, if the data is 0, an inverted version of the chaotic signal to be transmitted.

Moreover, at the receiver side, the signal is embedded in an additive white complex Gaussian noise (AWGN) n_k with two side power spectral density equal to $2N_0$. Finally, the received signal is modeled as: $r_k = s_k + n_k$.

At the receiving side, the received quadrature signal r_k delayed by half a bit duration and correlated with undelayed signal. It then passes by a correlator where the reference and corresponding data samples are correlated. At last the sign of the correlator output is computed to estimate the transmitted bit. In a practical implementation, many parameters like synchronization and time sampling correction must be taken into account to correctly achieve demodulation.

The output of the correlator is y_i.

$$\begin{aligned} y_i &= \sum_{k=1}^{M} r_k r_{K+M}^{*} = \sum_{k=1}^{M} [\mathrm{s_k} + \mathrm{n_k}][\mathrm{s_{k+M}} + \mathrm{n}_{k+M}]^{*} \\ &= \sum_{K=1}^{M} [\mathrm{c_k} + \mathrm{n_k}][b_i \mathrm{c_{k+M}} + \mathrm{n}_{k+M}]^{*} \\ &= b_i \sum_{K=1}^{M} \mathrm{c_k} c_k^{*} + \sum_{K=1}^{M} \mathrm{c_k} n_{k+M}^{*} + b_i \sum_{K=1}^{M} \mathrm{c_k^{*}} n_k + \sum_{K=1}^{M} \mathrm{n_k} n_{k+M}^{*} \end{aligned} \quad (4)$$

The first item is a useful signal and the second one is a random noise with zero mean value. It can be seen that the polarity of b_i and y_i is consistent, so the original signal can be demodulated. The decision threshold is 0, y_i is positive, then the data is 1, y_i is negative, the data is 0.

3 FM-DCSK Communication System Based on GNU Radio

GNU Radio [9] is a open and free software radio platform which can run on the common PC. It provides signal processing components to implement SDRs, through combining with the Minimal configuration hardware (mainly USRP), it can define the radio waves transmitting and receiving mode, and construct various wireless communication systems. It uses the two stage design, that data and control channel is separated. C++ is used to describe a variety of signal processing modules, Python is used for the configuration and connection between modules.

Figure 2 shows The whole FM-DCSK communication system model. In the GNU Radio platform, the parameter of the channel_model in the FM-DCSK communication system is as follows:

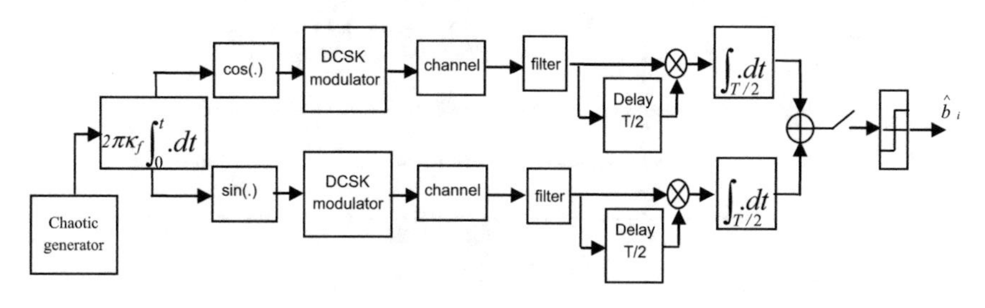

Fig. 2 The FM-DCSK system model

noise_voltage = 0.5; frequency_offset = 1e2/samp_rate; epsilon = 1.0; taps = (1.0 + 1.0j); noise_seed = 0; samp_rate = 100 k.

In the GNU Radio platform, we design and encapsulate several blocks such as logistic Filter, Chaos Generator, Chaos Modulator and Chaos Demodulator, install these blocks to the GNU Radio platform. We can use these blocks to expand our chaotic communication system. The core function of the blocks is implemented by C++. The output of Chaos Generator is 32 bit float number. Chaos Modulator block implement a FM-DCSK modulation, this block has two inputs, one output and a parameter N_samples. One input is Frequency-modulated chaos, another input is data bit, the Frequency-modulated chaos is complex, the data is unsigned char.

Chaos Demodulator implement a demodulation of a chaotic signal, The input of this block is complex signal samples. The output is the value of the best synchronized correlation value. This block also includes a symbol synchronization algorithm in order to correct the phase and frequency errors of the radios. Figure 3 demonstrates the output wave of the Chaotic generator. Figure 4 shows the output wave of the Chaos Modulator.

The main problem of the DCSK receiving module is the symbol synchronization. Considering the USRP2 clocks are not synchronized and not controlled by software method, there are non-negligible drifts between the transmitter and the receiver clocks. This drift has a direct effect on the center frequency and the sampling rate. As the reference and data signals will be equally shifted in frequency, a small portion of noise will be mixed with the useful signal, resulting in a performance degradation. The sampling rate error is a bigger issue. Our synchronization algorithm tries to esynchronize symbols by finding the best auto-correlation value amongst multiple

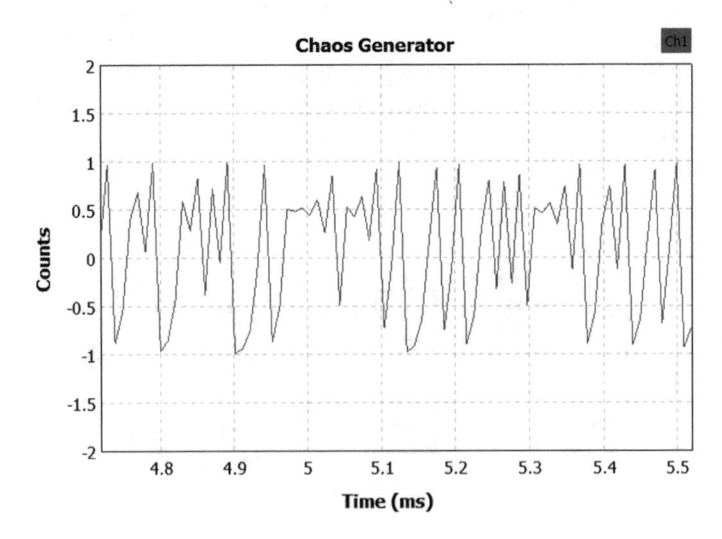

Fig. 3 The output of the Chaotic generator

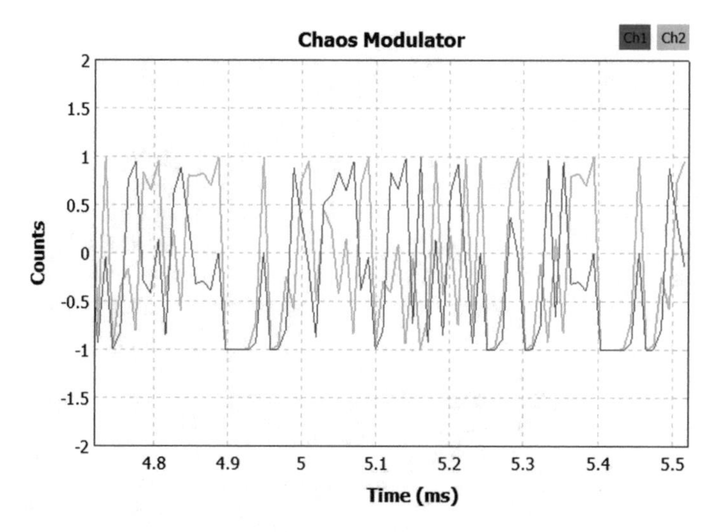

Fig. 4 The output of Chaos Modulator

delayed copies of the received signal. Due to the good auto-correlation properties of the chaotic signal, any symbol misalignment (delay) will result in a very low correlation value between the reference and the data. In order to control the computational complexity and efficiency that this synchronization algorithm required. We set a limiting parameter to the number of delayed copies.

4 Performance Measurement Results of the FM-DCSK System

The analytical BER performance for FM-DCSK system computed under Gaussian approximation [10] is given by:

$$\mathrm{BER} = \frac{1}{2} erfc\left(\sqrt{\frac{E_b}{4N_0}\left(1 + \frac{2}{5M}\frac{E_b}{N_0} + \frac{MN_0}{2E_b}\right)^{-1}}\right) \tag{5}$$

where *erfc* is the complementary error function, $N_0/2$ is the noise variance, E_b is the bit energy computed at the output of chaotic modulator.

Figure 5 shows measured BER and theoretical BER.

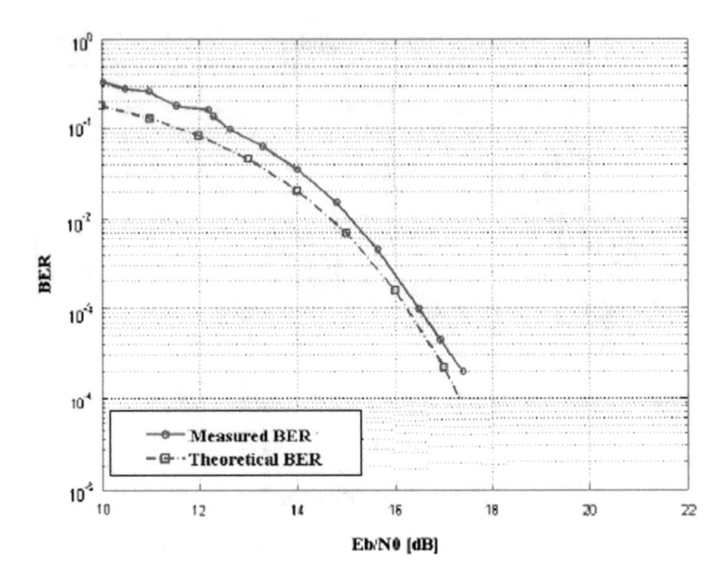

Fig. 5 BER performance of the FM-DCSK system

5 Conclusion

The article describes the implement of a real-time FM-DCSK communication system based on GNU Radio platform, introduces a simple and robust implementation method of chaotic signal generator and symbol synchronization algorithm. System implementation method is suitable for the study and development of the wireless communication system that have a customized requirements about the communication protocol standards and the whole system. It helps to reduce iterative process at a later stage of system development, shorten the development period of the system.

References

1. Hikmat N. Abdullah, Alejandro A. Valenzuela · Performance Evaluation of FM-COOK Chaotic Communication System Journal of Signal and Information Processing 01/2011; 2 (03):175–177. DOI:10.4236/jsip.2011.23023.
2. G. Kaddoum and F. Gagonen, "Error Correction Codes for Secure Chaos-Based Communication System," 25*th Bi-ennial Symposium on Communication*, Kingston, 12–14 May 2010, pp. 193–196.
3. G. Kaddoum, P. Charg´e, and D. Roviras, "A generalized methodology for bit-error-rate prediction in correlation-based communication schemes using chaos," Comm. Letters., vol. 13, no. 8, pp. 567–569, 2009.

4. M. Sushchik, L. S. Tsimring, and A. R. Volkovskii, "Performance analysis of correlation-based communication schemes utilizing chaos,"IEEE Trans. Circuits and Systems, vol. 47, pp. 1684–1691, 2000.
5. Y. Xia, C. K. Tse, and F. C. M. Lau, "Performance of differential chaosshift-keying digital communication systems over a multipath fading channel with delay spread," IEEE Trans. Circuits Syst. II, Express Briefs, vol. 51, pp. 680–684, 2004.
6. Y. Zhang, X. Shen and Y. Ding "Design and Performance Analysis of an FM-QCSK Chaotic Communication System," *2nd International Conference on Wireless Communications, Networking and Mobile Computing,* wuhan, 22–24 september 2006, pp. 1–4.
7. F. C. M. Lau and C. K. Tse, Z Chaos-Based Digital Communication Systems. Springer-Verlag, 2003.
8. X. Li, X. Lin, and D. Guo, "The experimental blind timing acquisition scheme for fm-dcsk communication system," in Anti-Counterfeiting Security and Identification in Communication (ASID), 2010 International Conference on, July 2010, pp. 120–125.
9. "GNU Radio." [Online]. Available: http://gnuradio.org/.
10. M. A. B. Faran, A. Kachouri and M. Samet, "Design of Secure Digital Communication Systems Using DCSK Chaotic Modulation," *DTIS* 2006 *International Confer-ence on Design and Test of Integrated Systems in Nano-scale Technology*, Tunis, 5–7 September 2006, pp. 200–204.

A Survey of Techniques for the Representation of Very Large Access Control Matrices

Garfield Zhiping Wu, Junyi Gu and Jie Dai

Abstract In industry, the efficiency of access control has become the bottleneck of many very large data management systems; however, little work has been done to develop an effective and efficient representation of access control data. We survey a number of relevant techniques, including several sparse matrix compression schemes and bitmap compression schemes. All these techniques can be potentially used to represent very large access control matrices.

Keywords Access control matrix · Data strucuture · Compression

1 Introduction

Efficient access control is critical for a variety of data management applications, such as Enterpise Content Management (ECM) systems, comment management systems [10], and multimedia management systems [13]; however, previous work on access control has been focused on the design of models, and very little work has been done systematically on the implementation side. In industry, the efficiency of access control has become the bottleneck of many very large data management systems. Therefore, the effective and efficient representation of access control data is nowadays worth careful investigation.

All systems' access control data can be represented as an access control matrix. First introduced by Lampson in 1971 [8], an access control matrix is a matrix with each subject represented by a row, and each object represented by a column. A matrix

G.Z. Wu · J. Gu (✉) · J. Dai
The Third Research Institute of the Ministry of Public Security, Shanghai 201204, People's Republic of China
e-mail: junyi.gu@gmail.com

G.Z. Wu
e-mail: zhiping.wu@outlook.com

J. Dai
e-mail: olivierdai@163.com

© Springer Nature Singapore Pte Ltd. 2018

N.Y. Yen and J.C. Hung (eds.), *Frontier Computing*, Lecture Notes in Electrical Engineering 422, DOI 10.1007/978-981-10-3187-8_8

Table 1 An example of an access control matrix

	Object 1	Object 2	Object 3
Subject 1	orwx	rw	
Subject 2		r	r
Subject 3	w	rx	

Table 2 An access control matrix in which the length of data in each cell is pre-defined

	Object1	Object2	Object3
Subject 1	11111111111	00000000000	00000100100
Subject 2	00000000000	00000000000	00010000000

entry M[S, O] is the permissions the subject S has on the object O. Table 1 shows an example of a simple access control matrix.

In some systems (e.g.ECM systems), the number of permission types is pre-defined; thus each cell in the matrix (except the first row and the first column which represent the object IDs and the subject IDs, respectively) contains a fixed number of bits (see Table 2 for an example).

Essentially, what we desire is an approach to representing access control matrices, making them (1) space efficient, and (2) fast for access control operations, including checking permissions, updating permissions, revoking permission, etc. We survey a number of techniques that can be potentially used to represent large access control matrices in this paper.

The remainder of this paper is organized as follows. We review sparse matrix compression techniques in Sect. 2, followed by Sect. 3 in which several bitmap compression schemes are described. We conclude our survey in Sect. 4.

2 Sparse Matrix Compression

An access control matrix is usually sparse. In practice, we require look-up and update of one or more cells to be efficient. Two types of update operations are considered in our context: changing the value in a cell to another value without any pre-knowledge about the value in the cell (whether it is zero or not) or changing a pre-known zero cell to be a non-zero one. The latter type is usually called insertion of a cell. The cost of update is dependent on the cost of look-up and the cost of insertion (a look-up followed by either a change of the value or an insertion). We primarily discuss look-up and insertion in this section. Additionally, a cell with the value of zero and an empty cell are not distinguished in the rest of the survey.

There are many schemes for compressing a sparse matrix. Generally speaking, we can group those schemes into two categories: one category is optimized for space and fast insertion (often slow look-up); the other category is designed for fast matrix operations, such as multiplication (slow insertion). Dictionary of Keys (DOK), List of Lists (LIL), and Coordinate List (COO) fall into the first category, while Compressed Sparse Row (CSR or CRS) and Compressed Sparse Column (CSC or CCS) fall into the second category [5]. Typically, in the community of scientific computing, schemes in the first category are used to construct a sparse matrix, and then the matrix is transformed into the format of a scheme in the second category for further computation [5]. There also exist schemes that aim to balance the efficiency of insertion and other matrix operations. We review several typical schemes below. There are also many schemes designed for special matrices (e.g. banded matrix, diagonal matrix, and symmetrix matrix) [5]; we, however, are not interested in these special schemes.

2.1 List of Lists

List of Lists (LIL) stores one list per row for the non-zero cells, where each entry stores a non-zero cell's column index and value. Four arrays are used to implement the list of lists. The first array, A, contains all non-zero values in the matrix. The second array, C, stores the corresponding column indexes for each element in array A. The third array, NEXT, stores the index of the next element for each element in arrays A and C (-1 if no next element). The last array, R, contains the index of each row's first non-zero element in arrays A and C. For example, a matrix

$$M = \begin{bmatrix} 10 & 20 & 0 & 0 & 0 & 0 \\ 0 & 30 & 0 & 40 & 0 & 0 \\ 0 & 0 & 50 & 60 & 70 & 0 \\ 0 & 0 & 0 & 0 & 0 & 80 \end{bmatrix}$$

may be compressed to be (zero-based index)

$$\begin{array}{ll} A = & [\ 30\ 20\ 10\ 70\ 50\ 60\ 40\ 80\ ] \\ C = & [\ 1\ 1\ 0\ 4\ 2\ 3\ 3\ 5\ ] \\ NEXT = & [\ 6\ -1\ 1\ -1\ 3\ 4\ -1\ -1\ ] \\ R = & [\ 2\ 0\ 5\ 7\ ] \end{array}$$

Logically, the above arrays store four lists as the name of the scheme list of lists indicates. Figure 1 presents how the non-zeros are stored logically.

This scheme is usually used to construct a small matrix. It supports fast insertion, but the lookup of an entry is slow. For insertion at (RowID, ColumnID), we simply have to add the new element at the front of the list of the row RowID. Specifically, we first append the value of the newly inserted entry at the end of array A and the column number ColumnID at the end of array C; second, the value in array R associated with the row RowID (i.e., R(RowID)) is appended to array NEXT; finally, the value of

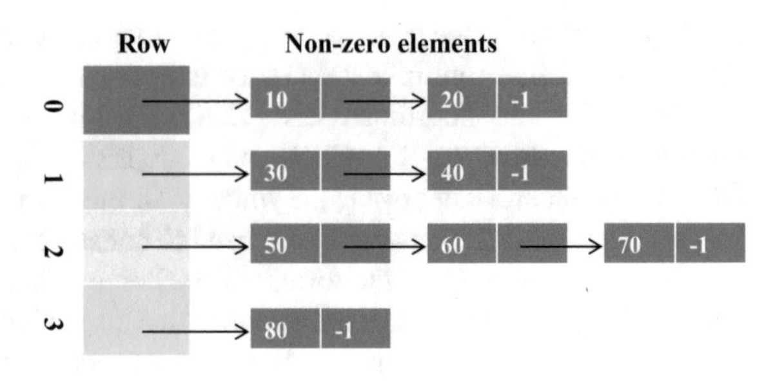

Fig. 1 A logical example of the list of lists (LIL)

Table 3 A matrix compressed by COO

Row	Column	Value
0	0	10
0	1	20
1	3	40
1	1	30
2	3	60
3	5	80
2	4	70
2	2	50

R(RowID) is changed to refer to the last entry in arrays A, C, and NEXT. For lookup of the value at a specific position, we, however, have to linearly search the non-zeros in the corresponding row.

2.2 Coordinate List

The Coordinate List (COO) scheme stores a list of (row, column, value) triples for all non-zero cells. For example, the matrix in Sect. 2.1 may be compressed to the list of triples (zero-based index) in Table 3. Alternatively, we may use three arrays, each of which stores the values of a corresponding row in Table 3 in the same order.

Theoretically, the triples can be in any order. In practice, however, they are usually stored in insertion order since we simply append a triple to the end of the list whenever we are inserting a value into a cell in the matrix (very efficient insertion). Note that storing tuples in insertion order does not help improve the efficiency of look-up, and a linear search is inevitable anyway in order to look up (or then update) the value of a specific cell. This also indicates that a general update of the value of a cell can be slow since we have to go through the list in order to find the corresponding cell first.

Overall, COO is slow for look-up (and general update). It is, however, very efficient for insertion.

2.3 Dictionary of Keys

Dictionary of Keys (DOK) encodes non-zero cells as a dictionary (hash table) mapping <row, column> pairs to values (a <row, column> pair is a key). Obviously, DOK consumes more space than COO does (i.e., extra space for buckets and pointers). Various hash tables may be implemented. This scheme supports fast insertion and look-up (O(1)); however, iterating over non-zero values in sorted order is not well supported since the order of the non-zero cells is random after compression.

2.4 Compressed Sparse Row

Instead of storing both row and column information for each non-zero cell, Compressed Sparse Row (CSR) further compresses the row information. Thus it is more space efficient than COO.

Let NNZ denote the number of non-zero cells in an $m \times n$ matrix M. COO needs a table containing $3 \times$ NNZ cells (or 3 arrays, each of which is of length NNZ) to represent M. Using CSR, 3 arrays are necessary. The first one, A, and the second one, C, are both of length NNZ. The array A holds all non-zero elements of M in strict left-to-right top-to-bottom order; the array C keeps the column index for each element in array A. The last array R is of length $m + 1$, containing the starting pointers to the elements in array A for each row. Therefore, row i contains all elements from A (R(i)) to A (R(i + 1) − 1). For the special case that row i has no non-zero cells, we will have R (i + 1) = R (i). The last element of array R equals to NNZ (zero-based index for array A), which is the ending flag.

Using CSR to compress the matrix in Sect. 2.1, we will get three arrays after compression (zero-based index)

$$\begin{aligned} A &= [\ 10\ 20\ 30\ 40\ 50\ 60\ 70\ 80\ ] \\ C &= [\ 0\ \ 1\ \ 1\ \ 3\ \ 2\ \ 3\ \ 4\ \ 5\ ] \\ R &= [\ 0\ \ 2\ \ 4\ \ 7\ \ 8\ ] \end{aligned}$$

Besides the space-efficiency, look-up is fast using CSR since we can efficiently find elements in a specific row and then use binary search to reach the element in the specific column. The major drawback of this scheme is that insertion is expensive. Suppose a new element needs to be inserted into row i. We have to insert an element in array A and C, respectively (by shifting elements and potentially expanding the arrays). Also, we have to modify the pointers for each row starting from row i + 1. In fact, in scientific computing, a matrix is transformed to CSR format only when it is assumed to be static (no more insertions).

2.5 Compressed Sparse Column

Compressed Sparse Column (CSC) is very similar to CSR with the exception that the column information is compressed instead of row information. Therefore, we need an array A to keep all non-zero elements of a matrix with the strict top-to-bottom left-to-right order, an array R to record the row indexes for each element in A, and an array C to keep pointers to element in array A for each column. For example, the matrix in Sect. 2.4 is compressed into the following arrays

$$
\begin{aligned}
A &= [\ 10\ 20\ 30\ 50\ 40\ 60\ 70\ 80\ ] \\
R &= [\ 0\ \ 0\ \ 1\ \ 2\ \ 1\ \ 2\ \ 2\ \ 3\ ] \\
C &= [\ 0\ \ 1\ \ 3\ \ 4\ \ 6\ \ 7\ \ 8\ ]
\end{aligned}
$$

CSC has the same advantages and drawbacks as CSR does. For an m × n matrix, whether CSC or CSR is more space-efficient depends on m and n (CSR if $m < n$; CSC if $m > n$). Please note that both CSR and CSC may consume more space than COO for matrices containing many rows or columns without any non-zero cells.

2.6 MTL4

While most schemes are optimized for either look-up or insertion, Matrix Template Library 4 (MTL4) provides a scheme that balances look-up and insertion by pre-allocating fixed-sized space for each row or column [6]. It then stores each row's or column's non-zero cells in the pre-allocated space for each row or column (the space has to be big enough for the row or column with the largest number of non-empty entries). For example, we use this scheme to represent the previous matrix in Sect. 2.1. Assuming that we choose a row-based compression and there are at most 4 nonzeros in a row, we will get

$$
\begin{aligned}
A &= [\ 10\ \ 20\ \ \emptyset\ \ \emptyset\ \ 30\ 40\ \emptyset\ \emptyset\ 50\ 60\ 70\ \emptyset\ 80\ \emptyset\ \emptyset\ \emptyset\ ] \\
C &= [\ 0\ \ 1\ \ \emptyset\ \ \emptyset\ \ 1\ \ 3\ \ \emptyset\ \emptyset\ 2\ \ 3\ \ 4\ \ \emptyset\ 5\ \ \emptyset\ \emptyset\ \emptyset\ ] \\
R &= [\ [0,2]\ [4,6]\ [8,11]\ [12,13]\ ]
\end{aligned}
$$

The result is similar to CSR; however, there are two differences. First, we have pre-allocated unused entries (represented by phi). Second, the array, R, is now a 2-dimensional array containing the starting and ending pointers for each row's used entries; for example, the first element [0, 2] of R indicates that Row 0 contains the 0th to 1st (2-1) elements in array A; the 2nd and 3rd elements are also reserved for Row 0.

Obviously, with this scheme look-up is as fast as with CSR. For insertion, we (binary) search the elements in the corresponding row, and then insert the new element into the corresponding space; thus no other rows will be affected. The representation, however, relies on there being a small number of non-zero entries in each row or column. Thus this scheme does not work well if some rows/columns have just a few non-zero entries but some others have a relatively large number of non-zero entries (wasting considerable space).

3 Bitmap Compression

For matrices in which each cell contains a fixed number of bits data, each row or each column can naturally be viewed as a bitmap, and an access control operation can be interpreted as an operation against a bitmap. In this case, a matrix is transformed to a number of bitmaps.

We begin by noting immediately that we are only concerned with lossless bitmap compression, and therefore do not consider lossy image compression (such as JPEG) to be relevant. Research on lossless bitmap compression has been introduced by DB researchers designing column-oriented database systems. At the same time, IR researchers have introduced many techniques for compressing ordered lists of integers representing document IDs, with or without storing positions within the corresponding documents. Because there is an equivalent bitmap for any list of document IDs, we use the term bitmap compression to cover both types of work.

Although compressing a bitmap and compressing an ordered list of integers are logically equivalent problems, DB researchers and IR researchers developed quite different schemes because of their different points of departure. Specifically, the input for compression in column-oriented databases is a bitmap, while the compression schemes for information retrieval take a list of integers as input.

Naturally, the fundamental idea among DB researchers is to compress contiguous 1 s or 0 s into smaller space (e.g. using a byte or word to represent several contiguous 1 s or 0s). In general, we can categorize this work into two groups: byte based schemes, which consider a byte as the smallest unit, and word based schemes, which consider a word as the smallest unit. Byte-aligned Bitmap Code (BBC) [3] and PackBits (PAC) [7] fall into the first group, while Hybrid Run-Length encoding (HRL) [9], Word-aligned Bitmap Code (WBC) [11, 12], Pack Word Code (PWC) [11, 12] and Word-Aligned Hybrid run-length code (WAH) [11, 12] fall into the second group. None of these schemes require decompression in advance for bitwise operations; instead, simple interpretation is sufficient. Researchers have shown that word based schemes are usually faster for both compression/decompression and bitwise operations at the cost of a little extra space, since modern CPUs access data by word [11, 12]. Please refer to the paper by Wu et al. [11] for more detailed comparisons among the schemes mentioned above.

In contrast, the core idea of compressing a list of document IDs is to use less space to represent an integer. The compression procedure is usually broken down into two steps, as follows. The first step is to transform the list of document IDs to a list of differences (d-gaps) so that most elements in the list become smaller integers. The second step is to represent each d-gap using one or more bits, bytes, or a fraction of a word. Variable length sequences of bytes (vbytes) [4] is a standard compression algorithm which contains 7 bits of d-gap and 1 bit indicating whether additional bytes are needed. Simple-9 [1, 2] and its extension Simple-16 [14] encode multiple d-gaps into one word. The four most significant bits of a word are used to indicate the number of d-gaps encoded in the word. Simple-16 has been shown to have faster decompression.

Many other compressions schemes have been proposed with the aim to reduce compression and decompression time. For example, PFOR-Delta [15] encodes d-gaps in batch sizes of some multiple of 32; it is not word-aligned and requires decompression of each batch prior to performing bitwise or lookup operations. Such schemes are not suited to our task.

We review two typical and influential schemes (i.e., WAH and Simple-9) developed by DB researchers and IR researchers, respectively, in detail below.

3.1 Word-Aligned Hybrid

Word-Aligned Hybrid run-length coding (WAH) is a mainstream bitmap compression scheme. WAH encodes long run of contiguous 0 s or 1 s using run-length encoding (called a fill), and represents a mixed-value word in its literal version. Therefore, there are two types of words: fill word and literal word. In WAH, each word represent $(w - 1) \times N$ literal bits (N is a natural number and $N \geq 1$), where w is the length of computer word (e.g., 32 or 64). In any word, the most significant bit (MSB) is used as a flag to distinguish a fill word and a literal word (0 for a literal word; 1 for a fill word). For a literal word, the next w-1 bits is simply a copy of the actual value. For a fill word, the second MSB is called the fill bit which represents the value of the contiguous bits. The rest of a fill word encodes the length of the run (number of w-1 bits); for example, in a 32-bit implementation, 62 contiguous 0 s may be encoded as 10000000000000000000000000000010. Due to its word-aligned requirements, we may not have a full w-1 bits in the last word we want to encode. Thus there is a special tail word to encode the last few bits of the bitmap (a literal word). There is also an additional word to record how many bits are used in the tail word.

Figure 2 shows how a 128-bit bitmap is compressed using a 32-bit WAH [12].

Although WAH requires slightly more space than previous byte aligned schemes, it can better exploit modern CPUs to get better performance of bitwise operations. WAH, however, fails to take into account the efficiency of checking, setting, or clearing given bits in compressed bitmaps.

128 bits	1,20*0,3*1,79*0,25*1			
31-bit groups	1,20*0,3*1,7*0	62*0	10*0,21*1	4*1
groups in hex	40000380	00000000 00000000	001FFFFF	0000000F
WAH (hex)	40000380	80000002	001FFFFF	0000000F 00000004

Fig. 2 An example of WAH encoding [12]

Table 4 Simple-9 encoding options in a 32-bit word

Selector (4 bits)	NO. of coded d-gaps	Length of each code (bits)	NO. of wasted bits
0000	28	1	0
0001	14	2	0
0010	9	3	1
0011	7	4	0
0100	5	5	3
0101	4	7	0
0110	3	9	1
0111	2	14	0
1000	1	28	0

3.2 *Simple-9*

As a scheme designed to compress an ordered list, Simple-9 first transforms the list of positions to a list of d-gaps. It then encodes as many d-gaps (up to 28) into one 32-bit word as possible (a word-aligned scheme). The four most significant bits of a word are used to indicate the number of d-gaps encoded in the word (called a selector). The remaining 28 data bits encode up to 28 d-gaps, each of which occupies exactly the same number of bits. Table 4 shows the 9 possible ways in which a word is partitioned. For some cases, a few bits are wasted.

Interestingly, given a d-gap, instead of encoding its actual value, Simple-9 encodes the actual value minus 1. For example, a list of d-gaps (5, 2, 1) will be encoded as (4, 1, 0). In this case, a bit can be used to encode the value of 1 or 2 (a d-gap is always at least 1).

During the compression, Simple-9 first checks whether the next 28 d-gaps can be encoded into one word; if not, it will check whether the next 14 d-gaps can be encoded into one word; this process will not stop until one of the nine possible ways is found to be appropriate. (Since there are at most 28 bits used to encode a d-gap, any d-gap greater than 2^{28} cannot be encoded.) For example, suppose we have a posting list (4, 11, 12, 13, 16, 21, 22, 29, 30, 42, 65, 66, 76, 94). It will first be converted to a list of d-gaps (4, 7, 1, 1, 3, 5, 1, 7, 1, 12, 23, 1, 10, 18). Then two words will be used to encode these d-gaps. The first word is (0010, 011, 110, 000, 000, 010, 100, 000, 110, 000, ∅), where a ∅ means an unused bit, and the second one is (0100, 01011, 10110, 00000, 01001, 01001, ∅∅∅).

4 Conclusion and Future Work

We surveyed a number of techniques for the representation of large access control matrices (particularly ECM systems' access control matrices), including several mainstream sparse matrix compression schemes and bitmap compression schemes. However, none of them can satisfy our expectation for an effective and efficient representation very well. The primary issue is that permission granting/revocation is typically too expensive using existing techniques. We therefore have to explore new data structures for access control data representation in large data management systems.

In the future, we could potentially take advantage of hash tables to represent access control data. For example, we may build a hash table for each row or each column. Using hashing tables, the time complexity of checking/updating/revoking a certain permission could be reduced to $O(N) + O(1) + O(M)$, where N is the number of subjects/objects (which equals to the number of hash tables built), and M is the number of permission types. As long as we can make N and M relatively small, access control operations are expected to be fast. Additionally, we could take into account some access control data's specific characteristics (e.g. ECM systems' access control data), and design novel data structures for them.

Acknowledgements This work was supported in part by the Canada NSERC Business Intelligence Network and by the University of Waterloo, in part by the National Science and Technology Major Project under Grant 2013ZX01033002-003, in part by the National High Technology Research and Development Program of China (863 Program) under Grant 2013AA014601, in part by the National Science Foundation of China under Grants 61300028, in part by the Project of the Ministry of Public Security under Grant 2014JSYJB009.

References

1. Anh, V.N., Moffat, A.: Index compression using fixed binary codewords. In: Proceedings of the Fifteenth Australian Database Conference. pp. 61–67. Australian Computer Society, Dunedin, New Zealand (2004)
2. Anh, V.N., Moffat, A.: Inverted index compression using word-aligned binary codes. Inf. Retr. 8(1), 151–166 (2005)
3. Antoshenkov, G., Ziauddin, M.: Query processing and optimization in Oracle Rdb. VLDB J. 5(4), 229–237 (1996)
4. Croft, W., Metzler, D., Strohman, T.: Search Engines: Information Retrieval in Practice. Addison Wesley Publishing Company Incorporated (2010), http://books.google.ca/books?id=VVYAPgAACAAJ
5. Golub, G., Van Loan, C.: Matrix Computations. Johns Hopkins Studies in the Mathematical Sciences, Johns Hopkins University Press (2012), http://books.google.ca/books?id=X5YfsuCWpxMC
6. Gottschling, P., Lindbo, D.: Generic compressed sparse matrix insertion: Algorithms and implementations in MTL4 and FEniCS. In: Proceedings of the 8th Workshop on Parallel/High-Performance Object-Oriented Scientific Computing. pp. 2:1–2:8. ACM, Genova, Italy (2009), http://doi.acm.org/10.1145/1595655.1595657
7. Inc., A.C.: Understanding packbits. http://devword.apple.com/technotes/tn/tn1023.html (1996)

8. Lampson, B.W.: Protection. In: Proceedings of the 5th Princeton Conference on Information Sciences and Systems. pp. 437–443. Prinston University, Prinston, NJ, USA (1971)
9. Nelson, M., Gailly, J.L.: The Data Compression Book (2nd ed.). MIS:Press, New York, NY, USA (1996)
10. Wei, X., Luo, X., Q.Li, Zhang, J., Xu, Z.: Online comment-based hotel quality automatic assessment using improved fuzzy comprehensive evaluation and fuzzy cognitive map. IEEE Transactions on Fuzzy Systems. 23(1), 72–84 (2015)
11. Wu, K., Otoo, E.J., Shoshani, A.: A performance comparison of bitmap indexes. In: Proceedings of the 2001 ACM CIKM International Conference on Information and Knowledge Management. pp. 559–561. ACM, Atlanta, Georgia, USA (2001)
12. Wu, K., Otoo, E.J., Shoshani, A., Nordbergi, H.: Notes on design and implementation of compressed bit vectors. Tech. Rep. LBNL/PUB-3161, Lawrence Berkeley National Laboratory, Berkeley, CA (2001)
13. Xu, Z., Mei, L., Liu, Y., Hu, C., Chen, L.: Semantic enhanced cloud environment for surveillance data management using video structural description. Computing 98(1–2), 35–54 (2016), http://dx.doi.org/10.1007/s00607-014-0408-7
14. Zhang, J., Long, X., Suel, T.: Performance of compressed inverted list caching in search engines. In: Proceedings of the 17th International Conference on World Wide Web. pp. 387–396. ACM, Beijing, China (2008)
15. Zukowski, M., Héman, S., Nes, N., Boncz, P.A.: Super-scalar RAM-CPU cache compression. In: Proceedings of the 22nd International Conference on Data Engineering. pp. 59–70. IEEE Computer Society, Atlanta, GA, USA (2006)

A Settling Time Model for Testing Potential Induced Degradation of Solar Cells

Jieh-Ren Chang, Yu-Min Lin and Chi-Hsiang Lo

Abstract In recent years, a barren degradation phenomenon of solar cells in large photovoltaic fields called potential induced degradation (PID) has been intensively investigated and discussed. PID is characterized by the power attenuation under high voltage stress between glass and solar cells through encapsulates. It will cause serious solar module power loss when the terminal voltage is applied on solar modules at outdoor field. For PID testing at solar-cell level, a quick and direct process method is required. In this study, a settling time model is applied for data analysis with an experimental data by a PID tester at solar-cell level to shorten the PID testing time. A settling time model is built up by the trend of measured data of the PID tester process. According the solar cell degradation settling time model, a threshold value is found at the settling time to predict PID or not on solar cell level. The experiment results show that the average test period is only 14.2 h and the hit rate is 94.2% for PID prediction. The proposed method is an efficient approach for reducing PID phenomenon in working field.

Keywords Solar cell · Settling time · Potential induced degradation

1 Introduction

The development of sustainable energy is the common trends around the world. We know the solar power is the most easy to take for using energy, but there are some problems to be solved with photovoltaic (PV) system which is used for generating electricity power. Potential Induced Degradation (PID) is a phenomenon that makes solar cells system lose power seriously under high voltage stress between glass surface and solar cells when PV system is in the working field.

J.-R. Chang (✉) · Y.-M. Lin · C.-H. Lo
Department of Electronics Engineering, National I-Lan University, No.1, Sec. 1, Shennong Rd., Yilan City, Yilan County 260, Taiwan
e-mail: jrchang@niu.edu.tw

© Springer Nature Singapore Pte Ltd. 2018

N.Y. Yen and J.C. Hung (eds.), *Frontier Computing*, Lecture Notes in Electrical Engineering 422, DOI 10.1007/978-981-10-3187-8_9

There were some ways to prevent PID phenomenon with changing the structure or material of encapsulates in recent research [1–5]. S. Pingel and the other authors presented some options on panel and system level for preventing PID and further decreasing overall degradation rates of PV system [1]. PID also can be prevented by using chemically strengthened glass as solar module cover glass that was made in high quality factory [2]. Some optimization techniques had been demonstrated process to improve cell-level PID performance [3]. The Anti-Reflection Coating layer had been optimized for better film quality.

The procedure of manufacture had been improved by using some techniques as mentioned in the previous paragraph but there were some solar cells existed with PID in the manufacture process. So some different test methods which can be used on-site in a PV installation to detect PID in modules were proposed [4]. Options to prevent PID on module and cell levels were found and verified experimentally in the research [5]. Dominik Lausch had proposed an experimental setup for PID testing at solar-cell level in order to avoid PID in the solar module [6].

The testing procedures are time consuming in Dominik's method. It almost costs 24 h for testing each solar cell. In the detection process the test equipment collects the data by the time, and the equipment identifies the solar cell is PID or not. If there is solar cell with PID phenomenon then the resistance would be below 100 Ω in 24 h [6]. In this research, a settling time model is proposed for fast detection of PID phenomena.

In this paper, a PID testing result for the resistance against PID on the solar-cell level is introduced and the sequential resistance data trend is described in Sect. 2. Subsequently, Sect. 3 shows a settling time model for fast detection of PID phenomena. In order to discuss the possibility of the proposed method, an average settling time for shorten the testing process and the optimized threshold for recognizing PID are shown in Sect. 4. Finally, a conclusion is presented in Sect. 5.

2 The Trend of Sequential Shunt Resistance Data of PID Test

In this study, a PID tester is used for measuring the shunt resistance of the solar cell, which had shown in Dominik's research [6]. The solar cell is placed on a temperature-controlled aluminum chuck to control a constant temperature throughout the testing process. The shunt resistance R was measured every sampling time Δt on the solar cell for the duration of the testing procedure.

During the test process, shunt resistance value decreases or keeps stable until the voltage is turned off. This PID shunt resistance degradation of the solar cell by the PID tester had been shown as from PV system at working field experiments [7]. The time dependence of the degradation is described in [6, 7].

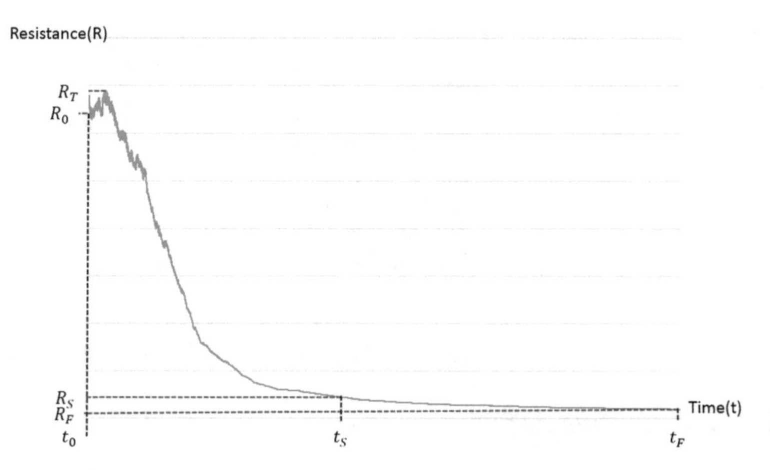

Fig. 1 A typical plot of sequential shunt resistance of a PID solar cell

The testing procedures are time consuming for identifying PID in Dominik's method. It almost costs 24 h for testing each solar cell. Therefore, it is expected to shorten the test time for PID recognition.

In a typical plot of the shunt resistance of PID solar cell, the trend of the curve is a downward by time. In such decay trend, the shunt resistance value declines rapidly in early testing time, but stabilizes in the post. This phenomenon is due to attenuation by the beginning of the high voltage impact on the potential of solar cells that were induced. So the shunt resistance value decreases seriously by the leakage current on some cell regions in the beginning duration. After a while, there is no significant breakdown in other regions, thus shunt resistance is no longer decline, and the resistance curve is nearly horizontal. A typical plot of sequential shunt resistance of a PID solar cell is shown in Fig. 1.

3 Settling Time Analysis

In the process of PID measurement, the shunt resistance changes like the process for charging the capacitor in RC circuit, which the electric charge gradually reaches stable by time. In this section, a settling time model is proposed for prediction of PID by using the time-varying shunt resistance data, which the concept is from [8].

The original method identifies PID if the shunt resistance is less than 100 Ω in the end of test process. For shortening the PID test time, the settling time model is based on the idea that the trend of curve is very stable after settling time. So PID is identified at settling time with a threshold value.

Fig. 2 Settling time training algorithm

```
Input: resistance sequence data base S_trs
Output: Settling Time for PID test
Method:
X_1 = find_all_seqeunces_PID(S_trs)
for each x_1(iΔt) ∈ X_1 {
  R_0 = x_1(0);
  R_F = x_1(FinalΔt);
  R_s = |R_F − R_0| × 5% + R_F;
  Find t_s(R_s) from x_1(t_s) = R_s
add t_s(R_s) to T_s ;
  }
  t_Savg =      T_s ; //Settling Time of PID
```

3.1 Settling Time Algorithm

By the concept of settling time, the corresponding resistance value R_s with settling time t_s is defined as:

$$R_s = |R_F - R_0| \times 5\% + R_F, \tag{1}$$

where initial shunt resistance value R_0 is denoted with the starting time t_0, and the end of detection time is t_F (typically 24 h in [6]) with the final resistance value R_F for the process of the PID test.

In order to find the settling time, some definitions are described as follows. The input of this problem contains a sequence database S, S includes many time-interval sequence data:

$$\mathrm{X} = \{x(i\Delta\mathrm{t}) | i \in \mathrm{N} \cup \{0\}\} = [x(0), x(\Delta\mathrm{t}), x(2\Delta\mathrm{t}), \ldots, x(Final\,\Delta\mathrm{t})]$$

for each solar cell with sampling time Δt time. The database S is grouped into 3 parts, S_{trs} with some PID sequence data for settling time training, S_{trh} with mixing PID and non-PID sequence data for threshold value training, and S_{test} is the rest of data for testing of the accuracy of the proposed method. The pseudo code is shown for searching the settling time in Fig. 2.

4 Searching the Optimized Threshold Value

The threshold value is defined for identifying PID on the solar cell level at the settling time. By searching S_{trh} at t_{Savg}, the corresponding shunt resistance $x_2(t_{Savg})$ is found with the following range:

$$x_2(t_{Savg}) \in (\min(\mathrm{x}(\mathrm{i}\Delta\mathrm{t})) \sim \max(\mathrm{x}(\mathrm{i}\Delta\mathrm{t}))) \tag{2}$$

Fig. 3 Optimized threshold algorithm

Input: resistance sequence database S_{trh} 、 t_{Savg}
Output: Optimized Threshold ΔR is a resistance value
Method:
for each $\Delta R \in (\min(x(i\Delta t)\text{for\_}S_{trh}) \sim \max(x(i\Delta t)\text{for\_}S_{trh})$ {
for each $x_2(i\Delta t) \in S_{trh}$ {
  if $(x_2(t_{Savg}) < \Delta R)$ {
      pridictvalue=PID;
  else
      pridictvalue=NoPID; }
  add pridictvalue to predict;
  }

It is recognized in PID if the shunt resistance value is less than the threshold value ΔR. Otherwise, it is recognized in non-PID. In this study, ΔR is set from the minimum value to maximum value to find an optimum value. The pseudo code is shown for searching the optimum threshold value in Fig. 3.

4.1 Online Testing

After finding the settling time and threshold value, an online testing method is developed as the following flowchart in Fig. 4.

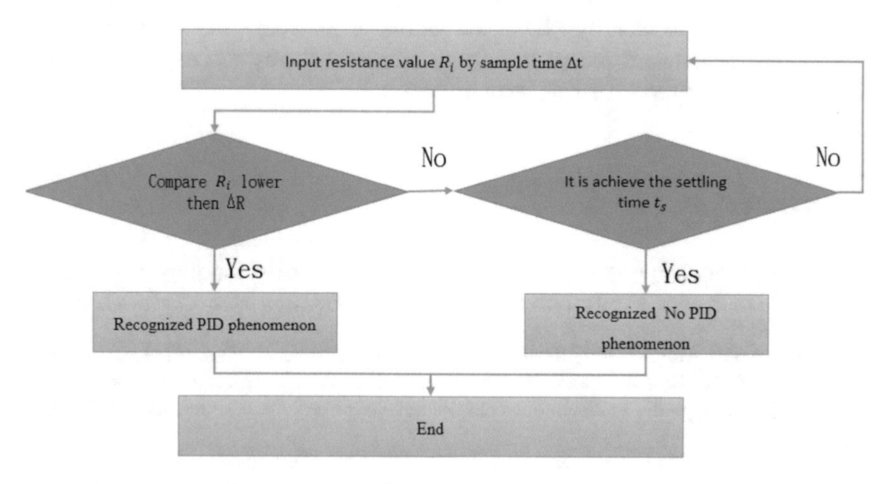

Fig. 4 Online testing flowchart

5 Experiment Results

In this experiment, all data measured from the PID tester which a prototype from [6]. The solar cells tested with the voltage 1000 V and temperature 60 °C which were as occurring under field operation. The test time was 24 h for experiment data. There were 8641 sample data with sampling time 10 s for each solar cell test. There were 48 solar cells with PID for training settling time. After settling time training process, the average settling time for PID test was found at 14.2 h. The optimized threshold value was derived in 511 Ω with 30 solar cells training data at the settling time of 14.2 h.

30 solar cells combined with PID and no PID were randomly selected for testing this proposed method. The results show the average prediction accuracy rate of 94.2% for 10 times test.

6 Conclusions

In this study a settling time model is proposed to predict PID phenomenon on solar cell level. According to the experimental results, the proposed method effectively forecast the PID phenomenon. In this study, the settling time algorithm is not only shorten the solar cell testing time in 14.2 h (the original method was 24 h process), but also predict the PID phenomenon in high hit rate 94.2%. A threshold value 511 Ω is derived to identify PID at the settling time. Some intelligent algorithms such as neural network theory and fuzzy inference theory are expected for prediction of PID on the solar cell level in the future.

References

1. S. Pingel, O. Frank, M. Winkler, S. Daryan, T. Geipel, H. Hoehne and J. Berghold, Potential Induced Degradation of solar cells and panels, Photovoltaic Specialists Conference (PVSC), 2010 35th IEEE.
2. Mika Kambel, Kohjiro Hara, Kazuhiko Mitarai, Satoshi Takeda, Makoto Fukawa, Naohiko Ishimaru, Michio Kondo, Chemically strengthened cover glass for preventing potential induced degradation of crystalline silicon solar cells, Photovoltaic Specialists Conference (PVSC), 2013 IEEE 39th.
3. Ta-Ming Kuan, Chih-Chiang Huang, Li-Guo Wu, Yu-Chih Chan, and Cheng-Yeh Yu, Process Optimization for Potential Induced Degradation Improvement on Cell Level, Photovoltaic Specialists Conference (PVSC), 2013 IEEE 39th.
4. F. Martínez-Moreno, E. Lorenzo, J. Muñoz, R. Parra, T. Espino, On-site tests for the detection of potential induced degradation in modules, 28th European Photovoltaic Solar Energy Conference (PVSEC).
5. M. Schütze, M. Junghänel, M.B. Koentopp, S. Cwikla, S. Friedrich, J.W. Müller, and P. Wawer, Laboratory study of potential induced degradation of silicon photovoltaic modules, Photovoltaic Specialists Conference (PVSC), 2011 37th IEEE.

6. Dominik Lausch, Volker Naumann, Otwin Breitenstein, Jan Bauer, Andreas Graff, J¨oerg Bagdahn, and Christian Hagendorf, Potential-Induced Degradation (PID): Introduction of a Novel Test Approach and Explanation of Increased Depletion Region Recombination, IEEE JOURNAL OF PHOTOVOLTAICS, VOL. 4, NO. 3, MAY 2014.
7. C. Taubitz, M. Sch¨utze, and M. B. K¨ontopp, "Towards a kinetic model of potential-induced shunting," in Proc. 27th Eur. Photovoltaic Sol. Energy *Conf. Exhib.*, 2012, pp. 3172–3176.
8. Demerow, R. "Settling Time of Operational Amplifiers." Analog Dialogue 4.1 1970.

Multi-stage Dictionary Learning for Image Super-Resolution Based on Sparse Representation

Dianbo Li, Wuzhen Shi, Wenfei Wang, Zhizong Wu and Lin Mei

Abstract Sparse representation has been proved successful in solving image super-resolution (SR) problems. It aims to compensate the high-frequency details from a pair of high–low (HL) resolution dictionary which is trained by the corresponding resolution of image patches. This paper presents a novel strategy to generate a super-resolution image via multi-stage HL dictionaries which are trained by a cascade training process. Extensive experiments on image super-resolution validate that the proposed solution can get much better results than some state-of-the-arts ones in terms of PSNR and FSIM.

Keywords Multi-stage dictionary learning · Image super-resolution · Sparse representation

1 Introduction

One single image SR problem has been a concerned issue in image processing for a long time. The goal is to recover the high resolution (HR) image from its low resolution (LR) form. However, it is an ill-posed inverse problem that some prior knowledge is in need to make the solution unique and stable. Lots of articles provide various methods to address this problem, which can be roughly divided into three categories, interpolating based, reconstructing based and learning based. Among them, the third one is more worth being researched than others in trend. For example, example learning based methods [1–4] employ a database consisting of co-occurrence examples from a training set of HR and LR image patches. Since they rely much more on the similarity between the training set and the test set,

D. Li (✉) · W. Wang · Z. Wu · L. Mei
The Third Research Institute of the Ministry of Public Security, Shanghai, China
e-mail: dianxinwu@126.com

W. Shi
School of Computer Science and Technology, Harbin Institute of Technology, Harbin, China

© Springer Nature Singapore Pte Ltd. 2018

N.Y. Yen and J.C. Hung (eds.), *Frontier Computing*, Lecture Notes in Electrical Engineering 422, DOI 10.1007/978-981-10-3187-8_10

they are not very practical in some situations. Another kind of efficient learning-based method [5–8] use the sparse-representation modeling to deal with this problem. Sparse-representation theory assumes that there is a linear relationship between high and low dimension, so that high dimension signal can be restored from their low dimension projection accurately [6]. Besides, [7] found that image patches can be well-represented as a sparse linear combination of elements from an appropriately chosen over-complete dictionary, so they made a compact representation for these patch pairs to capture the co-occurrence prior to improve the speed and the robustness significantly, achieving much better performance. Lately, [5] modified the approach above in various respects including computational complexity and algorithm architecture, which shows to be more efficient and much faster than [7]. Because of the limitation in recovering high-frequency details and the wide gap between the frequency spectrum of the corresponding HR image and that of the initial interpolation, [8] put forwards a dual-dictionary learning method via parse representation for image super-resolution, which consist of two steps to make up the wide gap. First, a main learned high-frequency dictionary was used to reduce the most gap of the frequency spectrum primarily. Then, a residual high-frequency dictionary was trained to recover the lack of residual high-frequency signal. According to [8], it obtained better results than [5] in PSNR.

However, the gap between the frequency spectrum of the corresponding HR image and that of the initial interpolation is so wide that two-layer progressive estimation of high frequency is not enough to recover the whole image high frequency details. In order to alleviate this problem, the multi-stage dictionary learning method is proposed. First, multiple stages of dictionary are trained offline, and each one also contains both high and low resolution parts. After that, high frequency details will be compensated by using these dictionaries via sparse representation stage by stage until the gap is smaller enough. This scheme can be treated as a cascade coarse-to-fine recovering progress, and the final results in the experimental section show that our method is better than expected.

This overall framework is as follows: some methods and research were introduced before in this section. In Sect. 2, the proposed SR scheme are described in detail including dictionary learning in Sect. 2.1 and image restoration in Sect. 2.2. Section 3 shows some experimental results in different views, and Sect. 4 makes some conclusions.

2 Method

When capturing image, it is easy to be affected by some factors such as deformation, blur, noise and down-scaling etc. Assuming that the original capturing image is an HR image, the actual obtained result is a LR image. This process can be described by formulation (1):

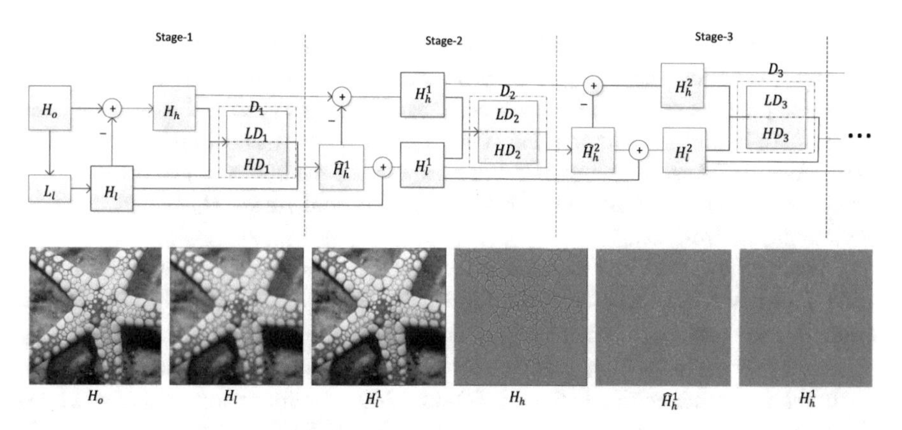

Fig. 1 The frame of dictionary learning stage

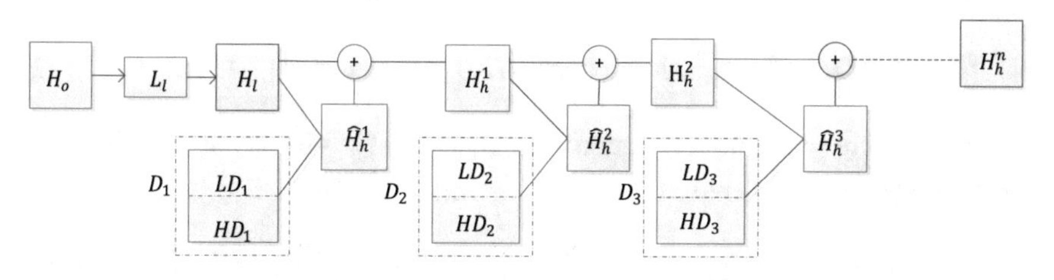

Fig. 2 Frame of image synthesis stage

$$y = GBDx + n \tag{1}$$

where x is the original HR image, y is the observed LR image. G denotes the geometric deformation operator, B denotes a blurring operator, D denotes a down-scaling operator and n is the additive Gaussian noise.

It can be seen that solving x is an ill-posed inverse problem. As a learning-based method, sparse representation method can get the coefficient between LR and HR image via a trained over-complete dictionary, which avoid to solving the equation directly. Both dictionary training and image generation are needed inescapability. We describe the training process as Fig. 1 and image generation progress in Fig. 2.

2.1 *Offline Dictionary Learning*

In this stage, multi-stage dictionaries are trained using sparse representation, i.e. $D_1, D_2, D_3, \ldots$. Each dictionary like D_1 has two parts: low-frequency dictionary (LD_1) and high-frequency dictionary (HD_1), respectively. Our training scheme is similar in spirit to that of [7].

As shown in Fig. 1, H_l and H_h which represent HR low-frequency image and HR high-frequency image is the first pair input to train the first stage dictionary D_1, and some pre-progress to the defined original HR image H_o should have been done to get them before the true training stage. First, we down sampling H_o and get its blur image L_l. Then, applying bi-cubic interpolation method on L_l to construct the image H_l, which is of the same size as H_o. The final image H_h is generated by subtracting H_l from H_o.

Since we have said that each stage dictionary has two coupled sub-dictionaries (LD_1, HD_1), we need to extract the local patches from H_l and H_h to forming the training data $\{pa_l^n, pa_h^n\}$, where pa_h^n is the set of patches extracted from image H_h directly while pa_l^n is built in another way which has been explained in detail in [8].

In order to generate the dictionary LD_1 and HD_1, the following two Eqs. (2), (3) can be used to generate them. Formulation (2) is K-SVD dictionary learning [9] procedure and Formulation (3) is based on the theory of high-dimension image patches can be accurately recovered from their low-dimension projections.

$$\mathrm{LD}, \{q^n\} = \operatorname{argmin} \sum_n \left\| pa_l^n - LD \cdot q^n \right\|_2^2, \text{ s.t.} \|q^n\|_0 \le L, \forall n \tag{2}$$

where $\{q^n\}_n$ are sparse representation vectors, and $\|\cdot\|_0$ is the l_0 norm counting the nonzero entries of a vector.

$$\mathrm{HD} = \operatorname{argmin} \sum_n \left\| pa_h^n - HD \cdot q^n \right\|_2^2 = \operatorname{argmin} \sum_n \|P_h - HD \cdot \mathrm{Q}\|_2^2 \tag{3}$$

where the matrices $P_h = \{pa_h^n\}_n$ and $Q = \{q^n\}_n$, respectively.

So far, the first stage dictionary D_1 has been trained, and we need set a stage number n to train more stage dictionaries. The next stages of dictionary can be built by using the same method of dictionary learning as D_1. As the input training image of the next stage, H_l^1 is generated by adding H_l and $\widehat{H}_h^1$, which contains more details $\left(\widehat{H}_h^1\right)$ than H_l. It is important to note that other stage of dictionary D_i is also consist of two coupled sub-dictionaries: low-frequency residual dictionary (LD_i) and high-frequency residual dictionary (HD_i).

Finally, all the rest stages of dictionary are trained as the same way described above. Theoretically, the back stage of dictionary contains less high-frequency signal than the previous stage and the dimension of the dictionary is higher, and at some point, the dictionary may has little use to compensate the high frequency signal.

2.2 Online Image Generation

After the offline training stage, there are multiple stages of dictionaries were generated. Each stage of dictionary can be used to compensate some high frequency component for the low resolution image. More high frequency details can be got via a cascade compensating strategy in theory. However, too much compensation is not necessary, and even cause a distortion. Generally speaking, the problem of how many stages of dictionary should we use for generating the final HR image is hard to be determined, because we have not the strict evaluation standard to estimate the result. In this paper, we select the PSNR value as an indicator. When the PSNR value decline or stay the same, we stop the next image synthesis stage.

As shown in Fig. 2, $\mathrm{H}_h^{\mathrm{n}}$ is the final synthetic image. $\mathrm{H}_h^1, \mathrm{H}_h^2, \ldots, \mathrm{H}_h^{\mathrm{i}}$ are the intermediate synthetic image after each stage of dictionary representation, which is also used to the next stage input. $\widehat{\mathrm{H}}_h^1, \widehat{\mathrm{H}}_h^2, \ldots, \widehat{\mathrm{H}}_h^{\mathrm{i}}$ are the lost high frequency of each input LR image.

Each stage of image synthesis has the same procedure to restore the loss. For the first stage example, suppose that an input LR image denoted by L_l has been done the same pre-progress in Sect. 2.1 to an HR image. Then, H_l is the first input target of the super resolution. With the use of dictionary D_1 and the method in [5], the first stage high-frequency image is generated $\widehat{\mathrm{H}}_h^1$, which is just contain the lost high frequency signal, and add the input LR image H_l.

First, make sure that H_l is filtered with the same high-pass filters and PCA projection as the training stage, and then is decomposed into overlapped patches $\{\mathrm{pa}_{\mathrm{l}}^n\}_n$. After all, employ the traditional OMP method [8] to generate $\{\mathrm{pa}_{\mathrm{l}}^n\}_n$, and calculate the sparse representation vectors $\{\mathrm{q}^n\}_n$ by allocating L atoms to their representation under LD_1. Next, the HR image patches can be reconstructed by the formulation: $\{\widehat{\mathrm{pa}}_{\mathrm{h}}^n\}_n = \{\mathrm{HD}_1 \cdot \mathrm{q}^n\}_n$. Finally, generate the first high frequency loss $\widehat{\mathrm{H}}_h^1$ by solving the following minimization problem (4):

$$\widehat{\mathrm{H}}_h^1 = \operatorname{argmin} \sum\nolimits_n \left\| R_n \widehat{\mathrm{H}}_h^1 - \widehat{\mathrm{pa}}_{\mathrm{h}}^n \right\|_2^2 \tag{4}$$

More details of the solution can be referred to [8]. Then, the first HR temporary image $\mathrm{H}_{\mathrm{LF}}^1$ containing more details than H_{LF} is built by adding H_l to $\widehat{\mathrm{H}}_h^1$.

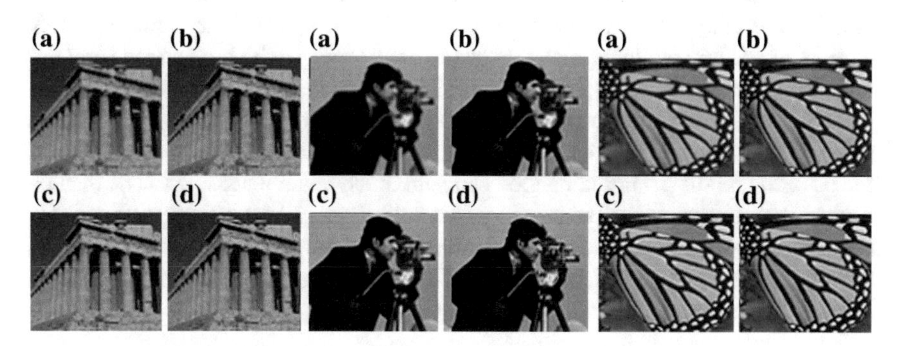

Fig. 3 Some vision comparison by different methods: **a** Bicubic interpolation; **b** J Zhang et al. [8]; **c** our method; **d** original images

In the same way, H_h^2 can be generated by using of H_l^1 and D_2, then, $H_l^3, \ldots, H_l^i$ and so on until reach the certain stopped condition. The last synthesized HR high-frequency image H_l^n contains much more details than the original HR high-frequency H_l. The stopped condition has been explained before in this section. Some synthesized image result is shown in Fig. 3.

3 Experiments

Extensive experiments on image super-resolution by using our method are demonstrated in this section. Bi-cubic interpolation method is a kind of complex interpolation method which is the best method of super resolution based on the interpolation method. It is comparable with sparse representation on the comprehensive performance, as a result, we employ it as a basic correlation method used in this paper. Besides, we take the comparison with the similar method in [8] to illustrate the advantages of our method.

First, we trained 9 stages dictionary as an offline library for the image synthesis step in Sect. 2.2. In order to test our performance with the methods bi-cubic interpolation method and dual-dictionary learning method [8], we take the same parameters as the method [8] including the Gaussian filter size and standard deviation of blurring operator which are set to 5×5 and 1 respectively, down sampling scale factor of decimation operator which is set to 2, and also the size of each level dictionary which is set to 500. Besides, the number of atoms for representing each image patch is fixed to 3, and the size of image patch is 9×9 with overlap of 1 pixel.

Some experimental results are shown in Fig. 4, which separately show the result of PSNR and FSIM with different stages of dictionary to be used in the image synthesis step. Each curve represents a test image, and each point in curve is an evaluation result corresponding to the stage in axis X. From the figures, we can see that in the front several stages, PSNR and FSIM increased significantly, and then remain the same or stay a little shock. In which, PSNR is the most widely used evaluation quality objective measurement and FSIM indicates the similarity of the original image and the interpolated high frequency image which is ranged from 0 to 1. Both of them are the bigger the better in their ranges.

To show the performance of the proposed method intuitively, we draw Table 1 as the compare result between different methods with the evaluation index PSNR. It can be seen that the proposed method can gain much better results of PSNR than the methods mentioned above, which increased 3.45 dB and 0.48 dB, respectively. The last column means how much the proposed method gain over Zhang's method [8], in which it claimed that his approach is better than the state-of-art method [5]. In conclusion, our method is effective in any way.

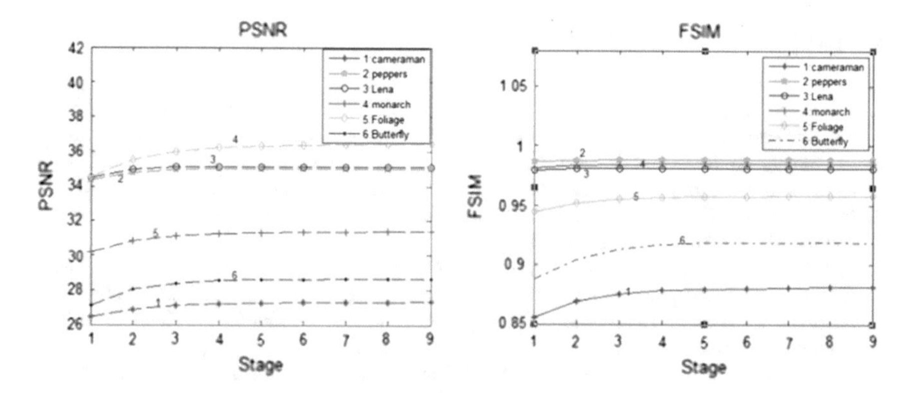

Fig. 4 PSNR and FSIM results on different test images

Table 1 PSNR comparisons with different algorithms (dB)

Images	Bicubic	J. Zhang [8]	Proposed	Gain
Cameraman	24.97	26.88	**27.31**	**0.43**
Foliage	31.65	35.50	**36.45**	**0.95**
Monarch	27.78	30.88	**31.39**	**0.51**
Peppers	32.32	34.78	**35.01**	**0.23**
Lena	32.19	34.96	**35.09**	**0.13**
Butterfly	24.23	28.01	**28.64**	**0.53**
Average	28.86	31.83	**32.31**	**0.48**

4 Conclusions

This paper presents a novel image super-resolution approach via multi-stages dictionaries learning based on sparse representation, which can restore a high-resolution image from a low-resolution one by a series of progressive high-frequency compensation utilizing multi-stages dictionaries. Experimental results show that the proposed method is able to narrow the gap between the frequency spectrum of the corresponding HR image and that of the initial interpolation, hence achieving better results in terms of both PSNR and FSIM. However, our method may spend some time off because of too much compensation in high frequency. Next, we will do some work to improve it.

Acknowledgements This work was supported in part by the Canada NSERC Business Intelligence Network and by the University of Waterloo, in part by the National Science and Technology Major Project under Grant 2013ZX01033002-003, in part by the National High Technology Research and Development Program of China (863 Program) under Grant 2013AA014601, in part by the National Science Foundation of China under Grants 61300028, in part by the Project of the Ministry of Public Security under Grant 2014JSYJB009.

References

1. J. Sun, N. N. Zheng, H. Tao, and H. Shum, "Image hallucination with primal sketch priors," IEEE Conference on Computer Vision and Pattern Recognition, vol. 2, pp. 729–736, 2003.
2. Z. Xiong, X. Sun, and F. Wu, "Image hallucination with feature enhancement," IEEE Conference on Computer Vision and Pattern Classification, vol. 1, pp. 2074–2081, 2009.
3. W. T. Freeman, E. C. Pasztor, and O. T. Carmichael, "Learning low-level vision," International Journal of Computer Vision, vol. 40, no. 1, pp. 25–47, 2000.
4. H. Chang, D.-Y. Yeung, and Y. Xiong, "Super-resolution through neighbor embedding," IEEE Conference on Computer Vision and Pattern Classification, vol. 1, pp. 275–282, 2004.
5. R. Zeyde, M. Elad, and M. Protter, "On Single Image Scale-Up using Sparse-Representations," Curves & Surfaces, Avignon France, June, 24–30, 2010.
6. D. L. Donoho, "Compressed sensing," IEEE Transactions on Information Theory, vol. 52, no. 4, pp. 1289–1306, 2006.
7. J. Yang, J. Wright, T. Huang, and Y. Ma, "Image superresolution via sparse representation," IEEE Trans. on Image Processing, vol. 19, no. 11, pp. 2861–2873, Nov. 2010.
8. J. Zhang, C. Zhao, S.W. Ma, D.B. Zhao."Image Super-Resolution via Dual-Dictionary Learning And Sparse Representation". ISC*AS*, *page 1688–1691. IEEE*, (*2012*).
9. M. Aharon, M. Elad, and A. Bruckstein, "K-SVD: An algorithm for designing over complete dictionaries for sparse representation," IEEE Trans. on Signal Processing, vol. 54, no. 11, pp. 4311–4322, Nov. 2006.

Research and Practice of Genetic Algorithm Theory

Junyi Gu, Zhiping Wu and Xin Wang

Abstract Genetic Algorithm is a class of high collateral, stochastic self-reliance search algorithms which based on mechanism of nature select and nature genetic. The paper introduces the principles of genetic algorithm and its methodology. The algorithm is practiced on the solution to find the maximum value of function in a given interval and the result is satisfied.

Keywords Genetic Algorithm · Nature select · Extrema problem

1 Introduction

Genetic algorithm is a search heuristic that mimics the process of natural selection and routinely used to generate useful solutions to optimization and search problems [1]. The schemata theorem and the implicit parallelism are two basic theoretical principles of genetic algorithms. Genetic algorithm use techniques inspired by natural evolution, such as inheritance, mutation, selection, and crossover. It plays an important role in artificial intelligence field and provides solutions for TSP, Prisoners' Dilemma, etc.

J. Gu · Z. Wu (✉) · X. Wang
The Third Research Institute of Ministry of Public Security,
Shanghai 201204, People's Republic of China
e-mail: zhiping.wu@outlook.com

J. Gu
e-mail: junyi.gu@gmail.com

X. Wang
e-mail: xinwang.xjtu@qq.com

© Springer Nature Singapore Pte Ltd. 2018

N.Y. Yen and J.C. Hung (eds.), *Frontier Computing*, Lecture Notes
in Electrical Engineering 422, DOI 10.1007/978-981-10-3187-8_11

2 Basic Theory of Genetic Algorithm

There is a great variance between Genetic Algorithm and the traditional algorithm and the two main features of Genetic Algorithm are intelligence and parallelism. Genetic Algorithm is a self-organization and self-adaptive search technique with learning ability by simulating a natural evolutionary process and realizing the survival of fitness law. The simple implementation of parallel processing for Genetic Algorithm is performing the genetic calculation of each population on separate computers since basically the evolutionary process of each population is relatively individual.

2.1 Schema Theorem

Holland's schema theorem, also called the fundamental theorem of genetic algorithms, [2] is widely taken to be the foundation for explanations of the power of genetic algorithms. It says that short, low-order schemata with above-average fitness increase exponentially in successive generations. The theorem was proposed by John Holland in the 1970s.

A schema is a template that identifies a subset of strings with similarities at certain string positions. Schemata are a special case of cylinder sets, and hence form a topological space.

Schema Theorem identify that schema with high level of average fitness grow exponentially in competition, and after consideration of other operator, the theorem further suggest that the evolution of schema with high fitness, short defined-length and low-order bring the exponentially growth of the number of solutions.

2.2 Implicit Parallelism

Implicit parallelism is a characteristic of a programming language that allows a compiler or interpreter to automatically exploit the parallelism inherent to the computations expressed by some of the language's constructs. A pure implicitly parallel language does not need special directives, operators or functions to enable parallel execution.

Schema with high fitness, short defined-length and low-order increase exponentially in reproduction process and thus calculation size of each generation is proportional to n3 supposing that the size of population is n.

3 Methodology of Genetic Algorithm

In a genetic algorithm, getting the optimal solution means that a population of candidate solutions evolves toward better solution. The evolution usually starts from a population of randomly generated individuals. It is an iterative process, with the population in each iteration called a generation. In each generation, the fitness of every individual in the population is evaluated; the fitness is usually the value of the objective function in the optimization problem being solved. The more fit individuals are stochastically selected from the current population, and each individual's genome is modified (recombined and possibly randomly mutated) to form a new generation. The new generation of candidate solutions is then used in the next iteration of the algorithm. Commonly, the algorithm terminates when either a maximum number of generations has been produced, or a satisfactory fitness level has been reached for the population.

3.1 Encoding

Since the genetic algorithm cannot operate directly on problem space, we need to establish an mapping from problem space to genetic space, thus to perform the issue object in genetic format which refers to chromosomes.

The common coding methods include binary coding and floating point coding.

3.2 Fitness Function

Genetic Algorithm use fitness function to judge the adaptation degree and population with low degree with got eliminated. The fitness function is normally designed as a relative function to the object function.

3.3 Genetic Operators

Selection

The selection operation is designed to simulate the evolving process of survival of the fitness: individuals with higher fitness are more likely to be selected to pass their genes to the next generation.

The common selection methods include ratio method and arrangement method.

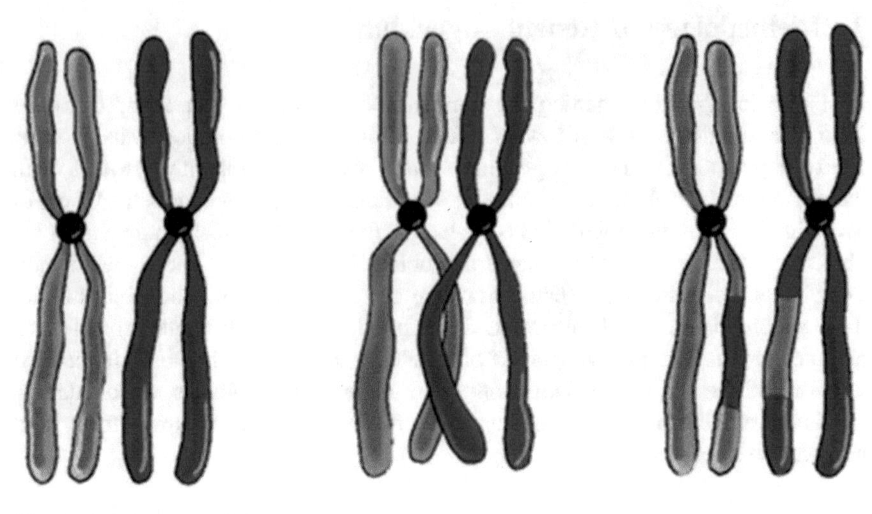

Fig. 1 Crossing-over of chromosomes

Crossover
The crossover operation is main operation in Genetic Algorithm. The paired chromosomes exchange their genes on a random selected position and make a difference between the fathers and children (Fig. 1).

Mutation
The mutation operation refers to generating a new individual by changing genes on random selected positions according to the mutation ratio. It is designed to keep the diversity of population.

Although crossover and mutation are known as the main genetic operators, it is possible to use other operators such as regrouping, colonization-extinction, or migration in genetic algorithms [3].

It is worth tuning parameters such as the mutation probability, crossover probability and population size to find reasonable settings for the problem class being worked on. A very small mutation rate may lead to genetic drift (which is non-ergodic in nature). A recombination rate that is too high may lead to premature convergence of the genetic algorithm. A mutation rate that is too high may lead to loss of good solutions, unless elitist selection is employed.

3.4 Termination Condition

The generational process is repeated until a termination condition has been reached. Common terminating conditions include:

- A solution is found that satisfies minimum criteria
- Fixed number of generations reached
- Allocated budget (computation time/money) reached
- The highest ranking solution's fitness is reaching or has reached a plateau such that successive Iterations no longer produce better results

4 Practices of Genetic Algorithm

To get intuitive understand of Genetic Algorithm, an interesting experiment which called jumping kangaroo was done.

Consider the solution to get the maximum value of function in a given interval, and try to image the below function curve as peaks and valleys. Each possible solution for the problem is like a kangaroo and optimum solution is the kangaroo on the highest peak and the process of finding the optimum solution can be simulated as kangaroos jumping to the highest peak. The first generation of kangaroo is located somewhere among the peaks and valleys and they jump higher or lower and reproduce the next generation. Every several years, the kangaroos in lower places got eliminated to keep the whole number of kangaroos stable. And after a few generations, the survivals will be the highest jump kangaroos (Fig. 2).

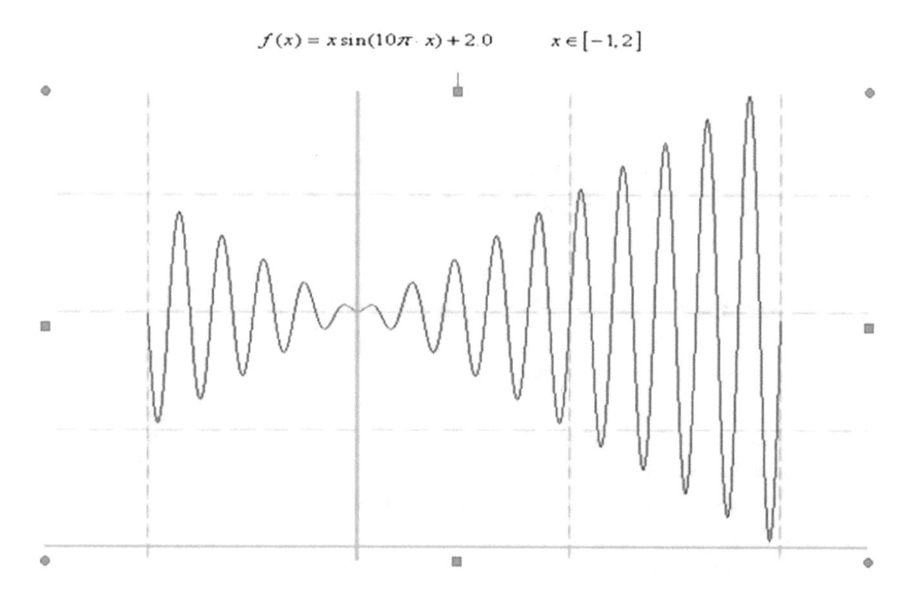

Fig. 2 Function curve

4.1 Encoding and Object Design

The first job is to design the coding of chromosomes for kangaroos. The key feature for these kangaroos is the level of their location, that is to say, coding their x-coordinate (a real number) according to the function curve. Floating point encoding is simpler to represent real numbers.

4.2 Fitness Function and Selection Tragedy

The fitness function is simply the level function of kangaroos since the location of kangaroos is the only consideration for selection and elimination.

And the selection tragedy is roulette wheel selection. Suppose there are three individuals with their adaptive number are 10, 15 and 25 respectively.

$$\mathrm{F} = \sum_{i=1}^{n} f_i$$

So the sum of adaptive number is F = 10 + 15 + 25 = 50.

The selected possibility for each individual is

$$P_1 = \frac{f_1}{F} * 100\%$$

$$P_2 = \frac{f_2}{F} * 100\%$$

$$P_3 = \frac{f_3}{F} * 100\%$$

$$\mathrm{P1} = 10/50 * 100\% = 20\%,$$

$$\mathrm{P2} = 15/50 * 100\% = 30\%,$$

$$\mathrm{P3} = 25/50 * 100\% = 50\%.$$

The realization of Roulette Wheel Selection is as follow.

```
Genome GenAlg:: GetChromoRoulette()
{
  //generate a random number betweeen 0 and the total population
  double Slice = (random()) * totalFitness;
  //This gene will include the selected individuals.
  Genome TheChosenOne;
  double FitnessSoFar = 0;
  for ( int i=0; i<popSize; ++i)
  {
    //Add fitness
```

```
        FitnessSoFar += vecPop[i].fitness;
        //Select this gene if the added score bigger than the random number
        if (FitnessSoFar >= Slice)
        {
          TheChosenOne = vecPop[i];
           break ;
        }
    }
    //return the selected gene
    return TheChosenOne;
}
```

4.3 Parameter Control and Genetic Operations

```
//Code number of each chromosome, here is 1.
int g_numGen = 1;

//Generation of evolution
int g_Generation = 1000;

//Population
int g_popsize = 50;

//Probability of genetic mutation
double g_dMutationRate = 0.8;

//Mutation step size (the maximum distance Kangaroo jump)
double g_dMaxPerturbation = 0.005;
```

The realization of mutation is as follow.

```
void GenAlg::Mutate(vector< double > &chromo)
{
   //Following a predetermined probability of mutation, pro-
cess the gene mutation
   for ( int i=0; i<chromo.size(); ++i)
  {
      //if mutated
      if (random() < mutationRate)
      {
        //increase or decrease a little random number to the weight
```

```
        chromo[i] += ((random()-0.5) * maxPerturbation);
        if (chromo[i] < leftPoint)
        {
          chromo[i] = rightPoint;
        }
        else  if (chromo[i] > rightPoint)
        {
          chromo[i] = leftPoint;
        }
      }
    }
}
```

4.4 Execution and Result

```
void GenEngine:: OnStartGenAlg()
{
    //Generate random numbers
   srand( (unsigned)time( NULL ) );
    //Initialize the genetic algorithm engine
   genAlg.init(g_popsize, g_dMutationRate, g_dCrossoverRate, g_numGen,
g_LeftPoint,g_RightPoint);
    //Empty containers
   m_population.clear();
    //Put random initialized population in to container
   m_population = genAlg.vecPop;
   vector < double > input;
    double  output;
   input.push_back(0);
    for ( int  Generation = 0;Generation <= g_Generation;Generation++)
   {
       //Each chromosome is operated
       for ( int  i=0;i<g_popsize;i++)
      {
         input = m_population[i].vecGenome;
          //Do adaptability evaluation for each individual, the evalua-
tion score is the function value.
         output = ( double )curve.function(input);
         m_population[i].fitness = output;
      }
       //Evolve progeny population from parent population
      genAlg.Epoch(m_population);
```

```
        //if(genAlg.GetBestFitness().fitness>=bestFitness)
        bestSearch=genAlg.GetBestFitness().vecGenome[0];
        bestFitness=genAlg.GetBestFitness().fitness;
        averageFitness=genAlg.GetAverageFitness();
        //cout<<bestSearch<<endl;
        report(Generation+1);
    }
    //return bestSearch;
}
```

The following pictures provide an intuitive show of generation evolution. It is obvious that the final result is not the optimal solution. Since the selection of parameters has influence on the process and a larger initial population and a higher step size may get the optimal solution in 1000 generations which required further experiment and research (Fig. 3).

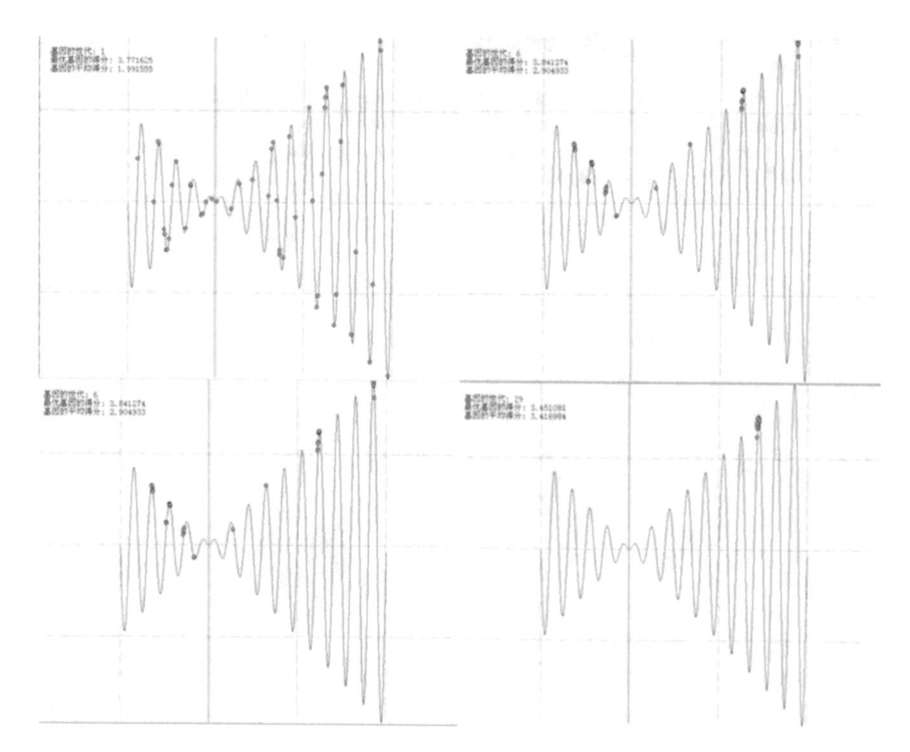

Fig. 3 Process of generation evolution

5 Conclusions

Based on the introduction of the basic principle and theory of Genetic Algorithms, aiming at the machinery and step of genetic algorithms design, the paper elaborates common encoding project, fitness function, selection strategy and the parameter control selection. Genetic Algorithms have good performance to a few applications such as extreme problem, and also have its limitations which required further research and optimization.

Acknowledgements This work was supported in part by the Canada NSERC Business Intelligence Network and by the University of Waterloo, in part by the National Science and Technology Major Project under Grant 2013ZX01033002-003, in part by the National High Technology Research and Development Program of China (863 Program) under Grant 2013AA014601, in part by the National Science Foundation of China under Grants 61300028, in part by the Project of the Ministry of Public Security under Grant 2014JSYJB009.

References

1. Mitchell, Melanie (1996). An Introduction to Genetic Algorithms. Cambridge, MA: MIT Press. ISBN 9780585030944.
2. Bridges, Clayton L.; Goldberg, David E. (1987). An analysis of reproduction and crossover in a binary-coded genetic algorithm. 2nd Int'l Conf. on Genetic Algorithms and their applications.
3. Akbari, Ziarati (2010). "A multilevel evolutionary algorithm for optimizing numerical functions" IJIEC 2 (2011): 419–430 [1].

Micro-blog Friend Recommendation Algorithms Based on Content and Social Relationship

Liangbin Yang, Binyang Li, Xinli Zhou and Yanmei Kang

Abstract First, this paper researches the micro-blog information push, which leads to the concept of user's friends, expounds the reason and meaning of friends recommendation algorithm, and introduces its current research situation, the paper has made the detailed introduction and analysis of existing algorithms and made a comprehensive comparison of the advantages and disadvantages of them. Then we make a recommendation of the micro-blog friend recommendation algorithms, which has two broad categories and three types: the recommendation algorithm based on content, the topology recommendation algorithm based on social relations and the filtering recommendation algorithm. Through the analysis of existing micro-blog friends recommendation algorithm, we represent the process of the algorithm and emphatically elaborated the implementation process, and finally we work out the Reasonable weighting of the three recommendation algorithm, get a sequence of recommended as a result, improved the algorithms, and reached a more comprehensive recommendation method. The improved algorithm could be a more effective way of potentially friends recommended for users.

Keywords Micro-blog · Information push · Social relationship · Friend recommendation · Algorithm

1 Background of the Research

As the personal computer is more and more popular and been spread into common families, the purpose of the Internet has changed from the original military researching tools to a medium which could achieve a convenient contact with others anytime and anywhere. We can use QQ and email to chat with some friend we can't often meet, to exchange work content and achievement with our classmates and

---

L. Yang (✉) · B. Li · X. Zhou · Y. Kang
School of Information Science and Technology, University of International Relations, Beijing 100091, China
e-mail: ylb@uir.cn

© Springer Nature Singapore Pte Ltd. 2018

N.Y. Yen and J.C. Hung (eds.), *Frontier Computing*, Lecture Notes in Electrical Engineering 422, DOI 10.1007/978-981-10-3187-8_12

colleagues, to receive and submit tasks we work on; and we can also input the keywords of what we need to know in the search box and let the searching engine tells; we can even enter an enormous online game platform which have millions of players at the same time, enjoy the pleasure of searching for treasure and upgrading with friends… In a sense, the Internet has greatly changed the way we live, work and entertains.

However, everything is in a constant development. Entering the Web2.0 era, matured technology and human wisdom has given rise to the appearance of the social network. Making changes of the Web1.0 era—when users' just browsing information as a reader, users today could be more than a visitor and disseminator of the information, they are also the maker and publisher of information. Many remarkable sites and applications, such as Facebook, Twitter in the U.S., Flickr, Renren and Kaixin in China, millions kinds of blog and Twitter have miraculously appeared, and at the moment they born, they draw attention of the time sailors, there are tens of millions of registrations, hundreds of millions of views in a few months. Generally, they have a slogan such as to mobile, sharing, interaction; they try their best to provide users with easy, rich and enthusiasm site environment and atmosphere. Users can write down their own mood, upload the photos they take during their vacations, share the funny video they found online, publish something make sense to them or some philosophy essay… whatever they want to tell others, they could publish onto their own social network. At the same time, people shared their information and received information from other users. If someone makes comments on your social network, you can also have a communication with him/her.

Then the question has risen. How could a social networking website which has numerous users to push out the information we published? If we are using a social network, such as Renren, we published a information such as "it's a nice day today, I'm so glad", there is no doubt that this information would be send into the database website. However, would this information appear on the interface of all users? Every user will see our message, but will we see all the information released by users? Let's do a simple calculation (as each social network have a different number of user, we only do a rough hypothesis): the registration of domestic at most social network are mostly about tens of millions now, we assume that users who landing everyday (including new registered users) of a web site is only at a number of five million, only 10% of them (this will be five hundred thousand people) release something new, then we would receive five hundred thousand messages a day, while many of them is talking about meaningless content such as the weather as we described above. Obviously, to push all the information to all people is not realistic.

Ones who used social networking could know that the real information push is pushing the information published by their friends, the ID which publishes something of their interested fields or some wonderful content to us. After log in our own interface, we will only receive messages from those specific accounts. We pay attention to the ID we chose, and the wonderful pieces system deemed are choose by some data such as forwarding number and the number of comments, thumb up evaluation. We add our own friends and it has been confirmed. So, when we want to find someone and add them as our friends, do we have to enter their name in the search box? If every friend is come from this, then the time and energy we spent on it will be too huge to image. Fortunately, today's social networks have a smart friend recommendation algorithm, it can find out the user who you have not added at the time but may have registered in the same social network with you, and these people will be recommend to you as potential friends. The meaning of this paper is to research on how friends' recommendation algorithm working and producing results on micro-blog, and basing on the result, the more comprehensive and reasonable algorithm is put forward.

At present, there are four main method of friend recommendation in micro-blog: firstly, users could be initiative to search for friends they want to add. This is the user's voluntary behavior, which has strong pertinence and high efficiency, but the operation could be tedious, and it needs time and energy; the second way is "people you may interest to". At the beginning the recommended strategy of this area is depend on the number of mutual friend between the two users, there is no order between the recommend ones, maybe someone have a lot of mutual friends would be tail of the queue as a contract. Besides, you can obtain more mutual friends using this kind of method, "change" a lot could also miss the friend you would like to add. Nowadays, most of it is based on the similarity of which area users concern about. The third kind is "what friends concern", which is recommended by the person your friends added but you did not focused on; The fourth is the "community", becoming a member can raise the chance of being recommend, which is a extend of the hot user recommendation before.

Social network is designed to help users establishing and expanding their own social circle online, and it is changing the way people sharing information and communicating today [1]. According to the research, the social network users not only hope to keep in touch with friends who they already known in real life, but also hope to make friends with people who share similar interests and hobbies with them [2]. However, for ordinary users, with the explosive growth of the scale of social networking user, finding familiar friends in real life or friends who have same interests is becoming more and more difficult.

Besides, friend searching methods at mostly social networking site is inhumane. When users searching for a name on social networking sites, the result coming out is usually a random one or a list by first letter of their last name, if there are a lot people share a same name, finding the friend you need will cost a lot of time. Even if some sites considered this problem, they mostly are simply consider whether they are in the same school or if there is a mutual friend and the results generally could not satisfy people at all [3]. Secondly, the list of friends is usually arranged in order of time, rather than according to the relevance [4]. User can't be familiar with every person of the social graph, and who they pay attention to may also be just one part of it. These things have cause many inconvenience to user.

As a form of social network, micro-blog also has similar problems. How to get the information you want, how to expand your circle of friends, these are the core research contents of the current social networking sites and recommendation research [5]. In our daily life, the importance of friends is gradually highlighted;

dividing "friends of friends" into account could largely expand the user communities.

Therefore, basing on the characteristics of micro-blog, friend recommendation mechanism is studied in this paper in order to design a more reasonable personalized recommendation algorithm which can serve the user.

2 The Research Status of Friends Recommendation Algorithm

At present, most of the recommendation algorithms tend to automatically extend some web pages, movies, commodity and label to the user which can meet the demand of their interest [6]. The research of algorithm which referring potential friends to users are relatively few. Lo S. and Lin C. put forward a theory that measure the relationship according to the interaction between friends, the more interaction they make, the better relationship they share. If two users show highly intimacy at the same time to a third user, that means there is a good relationship between the three users, so that they can set up a new friendship [7]. Chin A. tends to recommend friends according to the interaction between any two users [8]. Shen D. and Zheng Y. etc. analyses user's interest model according to the blog which they pay close attention to or the place they have visited, and calculated user interest similarity to find potential friends using the model [9]. Bacon K. put forward to an idea to establish a complete sub-graph according to the relationship between the user, the sub-graph meet the relationship between any two users are friends, and then carried out in accordance with the common user merger between sub-graph, so there must be concluded someone hasn't been in a friend relationship in the picture. Finally we can recommend friends for ones who have not yet establish relationships with others [10]. Wu Z., etc. Come out with a algorithm to recommend friends according to the user's appearance [11]. Yu Haiqun etc. Through doing analysis of the subject of social network user preference, they proposed a algorithm based on the topic of user preferences [12]. Niu QingPeng studied in the blog potential friends' recommendation algorithm [13]. Shi Lingfeng etc. studied the map query algorithm based on the relationship between social network friends [14].

Wen-bing Zhao etc., using metrology method, through the statistical analysis showed that the characteristic of users in micro-blog users are focus on several follows power-law distribution, which is widely attention, while the majority of users' influence is very limited [15]. Gou put forward. SFViz system in visual friends recommend SFViz system was provides a new visual, interactive tools, users not only recommend search results in a single view. SFViz system can support the user interactively explore and seek to have a common interest friend, using social network topology structure and semantic structure of the active data, social tag hierarchy, reflects the interest of users to surf the web. Through multiple perspectives of fused similarity between people in the hierarchy to enable users to seek potential good friend relationship [16]. Xie X. design has realized the online social network system based on biology, which USES the friends of the recommended strategy is actively considering the users of all kinds of information, showing the effectiveness of the design of friends recommend framework [17]. Hannon skillfully using Twitter user information and Twitter users social graph structure, put forward a recommended strategy, an analysis of the different range is more effective to improve the quality of the recommended a friend. The last show out the recommended method is recommended results have better effect [18]. Yan Yu proposed a social network based on hybrid graph friend recommendation algorithm, this algorithm adopts to reset the random walk method based on hybrid graph model, and complete to the users of social networks have a common interest friends recommend. Optimal parameters obtained by many times experiments weight, is the result of the recommendation is most ideal [19]. Java on Twitter, the study found a large number of micro-blog users to know each other before using micro-blog, there are a large number of micro-blog users are after a friend's invitation to try to use micro-blog and add as a friend. So, the current mainstream micro-blog is mainly based on "hot spots" and friends of friends "indirect" recommended mode. Although the two methods provide the user with a huge reference candidate crowd, ignoring the user's own habits and hobbies (a large part of the user wants to find some more and his congenial friends), caused the recommendation information redundancy, accuracy and recall rate is extremely low, more let a lot of users have no choice [20]. Jia-jia Zheng friend recommendation algorithm based on graph theory is proposed. The core idea of the algorithm is used for the user, the user of the tag and searching as well as user's friends have the same concept of semantic space, and then use graph structure to describe the semantic space [21].

Current requirements and trends of the friends recommendation algorithm are deeply analyze each user attention object structure position and role in the whole social networking features, and then from the trend of the overall grasp of the masses of users to use, and dug up each user's behavior, improve the effectiveness of the recommended by the user. Because friend recommendation algorithm has certain targeted mostly, the recommended algorithm described above the recommendation on the object and purpose have bigger difference. So in accordance with the actual situation and features of the micro-blog friend recommended, there is a need to make a comprehensive analysis and research.

3 Micro-blog Friends Recommend Related Technology

Compared with the goods recommendation technology widely used, such as book recommendation system on Amazon, Netflix movie recommendation system of micro-blog friends recommend related research is less. A twitter user's friends is to focus on the user object. In general, micro-blog friends recommendation technology is divided into recommended based on content and based on the recommendations from the social relations two kinds big. In content-based recommendation, Hannon using micro-blog keywords as a characteristic, mining high quality sources of information. Based on the recommendation of social relations, one is based on the recommendations from the social networking topology, Armentano [22] social networking topology is used to find the user interested friends [23]; Two is recommended based on social filtering algorithm, the friends of friends as recommended target.

3.1 Content-Based Recommendation

Friend recommendation algorithm based on content is based on this idea, if two people have similar topics, they may be more willing to get to know each other. That is to say, this algorithm is trying to find and target users with similar interests [24].

In 2010, Hannon Twitter friends recommended method put forward by micro-blog content information as the characteristics of the user, through information retrieval methods are recommended. Algorithm of this paper is divided into two steps: establishing attribute information for each user, with TF-IDF model is recommended.

Build user attribute information User tweets can well reflect the history of user interests, so can according to micro-blog content attribute information to the user. The simplest attribute information source is released the users themselves. UT target users for UT, defined as users recently published micro-blog:

The definition can be further extended, each user attention object set called followers, each user's fans set called followers, are defined as follows: The user can focus on followers, think users interested in followers of micro-blog, so you can use followers micro-blog as the user's profile, are defined as follows: User UT, in turn, the followers of the UT, is likely to be interested in UT's micro-blog, fans of the micro-blog content also has a certain probability and UT interest is consistent, can be based on the attribute information to the user.

Through micro-blog content and social relationships, we have established attribute information for each user, only out of each user's characteristic vector, are defined as follows: The simplest case, can use word frequency as weights. But here the application of scenarios, with TF–IDF model is more appropriate, namely the UT of t_i entry score, and the user is directly proportional to the frequency of the term in the attribute information, and other attribute information in the entry is inversely proportional to the frequency. This leads to appear in the user attribute information frequently appear in other user attributes information don't frequent terms have higher weight. Corresponding are defined as follows: Calculated by TF–IDF model, each user's characteristic vector, when the user is recommended to calculate candidate recommendation object features and user cosine similarity, the similarity of the largest former several objects as recommended.

Using cosine similarity to calculate users a and b two vector similarity of v_a and v_b. Can intuitively think that if users a and b share the same keywords in their daily use, and a few other users to share these keywords, then they have a lot of similarities. Algorithm based on content and link is through the use of social link information in the social network to enhance the accuracy of the matching algorithm based on content.

Algorithm by using the weak constraint and implicit social network users [25], target users are more willing to accept this kind of algorithm. This algorithm and the algorithm of calculation based on the content of similarity method has a lot of similarities. And recommended to the user, however, before a few method to the highest users, if the user effective connection between u and user c, the degree of similarity between users u and c would add 50% of the weight, that is if there was a link between the users u and c, in its recommended recommend that order will be above the recommended item content based similarity. A valid connection is defined as: a number of users in a row, the first users as the target, the last user as recommended users, each between two users a and b must satisfy at least one of the following three conditions:

① a initiative to contact b
② a to b had a comment
③ b initiative to contact a

This definition ensures that the two entities social connections between the user and the minimum that they or their friends are acquaintances or have some interaction. Such as user a c comments to users, and users b and c are friends relationship, we think that the user a and user b between such a valid connection. When recommend using effective connection, also consider the content of the same keyword matching, we can also send a link as a kind of extension, including consider users u and candidate concentrated all the connections between c. In the recommended by users, at least more than 7 cases need to be considered effective connection information.

3.2 The Recommendation Based on Social Topology

Based on the recommendations from the social networking topology In 2011, Armentano et al. proposed a micro-blog friend recommended method based on the social networking topology structure, the method is divided into two steps: search candidate recommended users; According to the characteristics of the different weighting of candidate recommendation ordered by users. (1) Search candidate for users. In order to find the target user U candidate recommended users, the algorithm is based on the assumption that the user U focus fans the attention of the user object is an object candidate recommendation. Simply speaking, users U and F are focus on an object, other objects of attention of the user F users U may also be interested in. The rationality of the hypothesis is that the target user is searching for information on and found he is interested in information source, the other focus on users of the information sources and target users are likely to have a common interest, and other sources have discovered the same topic. Ditto section, are defined as follows, if A focus on the B, then B is a follower, A is B's followers, search candidate user steps are as follows:

(1) Starting from the target user, first, to get its attention to collection of objects, S, (2) for each element of S, its collection of fans, merge each collection (allows repeating element), (3) for each element of L, its focus on a collection of objects, merge each collection, to get the recommended candidates (allows repeating element), (4) remove the user from T U have focused on object, get the final set of candidate recommendation R (allows repeating element), R is the each element in the target user of candidate recommendation object, each element may appear multiple times in the R.

The weighted feature the first characteristic is recommended candidate object media properties, because the user is looking for information, you can assume that there are a lot of fans, such as user but very few users attention. Are defined as follows: The second characteristic is the candidate recommended frequency of object in R list, the list of an object in R x is the number of occurrences of neighbor number of x and the target user. Are defined as follows: The third feature is the target user U and candidate recommendation object x number of common friends, are defined as follows: The combination of these three features: Recommend to target users, from the candidate recommendation object selected top-K are recommended according to the weight.

3.3 The Recommendation Based on Social Filtering Algorithms

Based on the recommendations from the social filtering algorithm Social filtering algorithm (Social filtering) is based on such A premise: if A friend is A friend of B, then it is possible that B friends. The recommended method not only consider the user's interests, also through the analysis of implicit in every friend of information users, to recommend friends to users [26].

First give a definition: in the social network, if user b is a good friend, will be defined. The description of the algorithm are as follows: Assumes that user a and user u, user u recommend friends RC (u) is defined as the candidate set, the user is a user u friends, users c is a good friend at the same time, the user is the recommended candidate for the user u and c concentration of a user, Is defined as a mutual friend set, Through a mutual friend relationship, we added between users u and c. And then through the calculation and collection of mutual friend user can recommend percentage of c, user can recommend percentage and centrally by the candidate to the target user u recommend top-K with the highest scores. The algorithm is more suitable for application in contact with realistic society has great social networking sites. Algorithm is larger deficiency is the accumulation of target users need to have a certain number of friends, for a new friends or a small number of registered users, the algorithm is not applicable.

4 The Improvement of Micro-blog Friend Recommended Algorithm

4.1 *The Basic Thought of Solving the Problem About It*

As the current recommendation algorithms have advantages and disadvantages, in practical applications, the recommended method is generally mixed (Hybrid Recommendation), that is, how to combine different recommended strategies to produce better performance results. The more commonly used method is to use two recommended methods to produce two predictions, and then in some way combine the two results. Although there are many combinations feasible in theory, but not in the specific application of each effect are good. The most important principle for mixing is that the mixing one can avoid or compensate their weaknesses.

On mixing the recommended method, there are the more commonly used ideas:

(1) Weighted: The result for each recommendation algorithm plus weights, then get the final recommendation result.
(2) Mixed [27]: Provide a variety of results of recommendation algorithm to the user reference in a while.
(3) Switching: according to the actual situation of the problem and asked to select the other recommended methods.
(4) Feature combination: combine different characteristics recommendation result, by the use of a recommendation algorithm.
(5) Cascade [28]: Start with a preferred method for generating preliminary recommendation result, re-use the second recommended method recommended earlier on the basis of the results of further recommendations.
(6) Feature augmentation: a recommended method uses additional feature information generated by another recommended method.
(7) Meta-level: a recommended method uses the entire recommended model produced by another recommend method

From the current recommended combination of technology research situation seen on the content and composition studies based on more collaborative filtering. Other recommended combination of fewer major online data privacy reasons. When using a combination of recommended techniques must be noted that a combination of strategies for content-based and collaborative filtering, it must consider the quantity and quality of data in the database.

4.2 The Proposition of the Recommendation Algorithm Which Is Based on Content and Social Relationships

At present, micro-blog friends' recommendation for each user's is in multiple columns to display, each column is determined by the area only a recommendation (some are hot spots recommended members, some are friends' concern), factors to be considered too single, so although net shop is large enough, its recommend efficiency is very low. This article seeks to consolidate different factors, to develop a way to take full account of user preferences and social recommendation algorithm, based on their ideology and that is: each user has their own social network (including the user object of concern and attention to the user fans) and the tweets you post, the two categories of potential users recommend friends have a certain impact. This article is to be recommended and users with the highest similarity and potentially more friends in common with the user already exist.

The similarity between users is different user personal information, interests, and on the geographic location of the similarity. Interests include the user's first concern when the public account login micro-blog; micro-blog users post content and show it in personal information and other items of interest. Location is personal information, hometown name, school name, or the name of the user to fill in the workplace, but also fill the other users on the same locations on the geographical similarity with the user. This part can use the content-based recommendation algorithm come recommended result.

In common between two users, the more there is a great probability that two users described in reality the larger social circle intersecting surface coverage, the higher the possibility the two met. On the other hand, based on social topology is also recommended to use a mutual friend a few sort the results, so this part is used in social relationships recommended.

Finally, the three recommended results are given different weights, and then use the formula to arrive at final recommended top-K friends.

Micro-blog users have friends represent their personal information and network connections, so that each user is unique presence, the proposed recommendation algorithm main reference is the user's personal information, and the user's existing micro-blog content social networks. Through these aspects into consideration the recommendation result obtained.

4.3 Processes of Micro-blog Friends' Recommendation Algorithm Which Is Based on Content and Social Relationships

Micro-blog friends' recommendation algorithm based on content and social relationships implementation process consists of three modules: data collection preprocessing module, recommendation algorithm module implementation and results of the display module. The basic flow chart shows in Fig. 1.

(1) Pre-processing module is divided into three steps of data collection:

First, get the data set. The data set includes a user's personal information, use micro-blog published content and their social networks. Personal information contains both various place names associated with the user, but also contains the user's public concern number. Users can now include social networking friends and fans.

The second step, analyze the data. Get a great probability data set is very large, so the need for these data processing, screening out some repetitive, less significant, for example, data time, fans and so the user select to represent the user's data and structural process.

The third step, read out the processed data for the next module.

(2) Algorithm module is divided into the following sections:

The first part, read the user's personal information and his micro-blog content, using TF-IDF algorithm derived user tags, calculate the similarity between the user and the target user, select one kind of K highest similarity Top-N (N > k) recommended composition of a candidate set 1;

The second part, establish the user's friends relations matrix, identify candidate centralized buddy relationship between users, association rules algorithm to calculate the number of common friends among users, the largest number of elected mutual friend Top-N recommendation candidate set consisting of 2 users;

The third part, use the social filtering algorithms to identify both similar interests have more in common candidate sets 3;

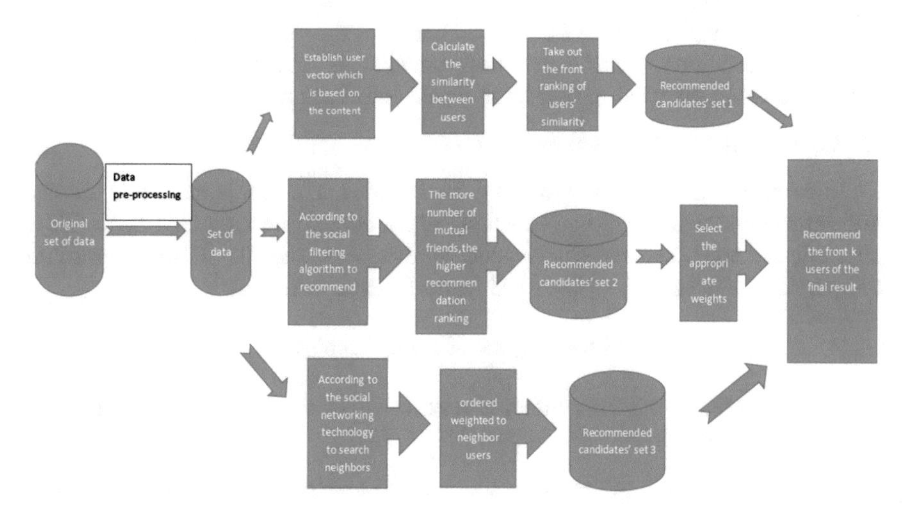

Fig. 1 Based on content and social relationships micro-blog friends recommendation algorithm's flowchart

The fourth part, achieve the content recommendation of micro-blog friends' algorithm which is based on social relationships and share the number of weight values obtained by the three methods feature vector using a combination of weights three candidates set as the target user recommended a comprehensive right friends.

(3) Recommended result display module:

The final Top-K Promoted results are displayed.

4.4 *The Realization of Recommended Algorithm Based on Content and Social Relationships Micro-blog Friends*

Basing on the user's personal characteristic information, the overall objective of the algorithm is to calculate the set of users as similar as to the target user and word feature vector, that is, to produce a recommendation basing on the users' profile in descending order of similarity. Specifically, for the target user u, by his personal feature information and specific similarity function, calculate the K users which is closest to his feature information as the target user u's nearest neighbor set, which is the target user u's Top-K recommended set.

The implementation steps of the algorithm:

(1) Collect users' information

Social networking sites typically require users to describe their own interests and personal information. In micro-blog, the user needs to choose the direction of their selves' interest and other areas of expertise. Residential address filled in the basic information as well as the label is also on behalf of the individual characteristics of the user. In addition, users' micro-blog content and the user's friends are also concerned about the need to collecting users' information.

(2) Create a user feature vectors

This step needs to consider two major categories. The first is the user's interest information and relationship information, which come from the user's micro-blog, interests, and location content associated with the user. Then use TF-IDF algorithm to translate the information into word feature vector written as:

$$\mathrm{Vu} = (w_1, w_2, \ldots, w_i, \ldots, w_m)$$

Wherein m represents the number of user's characteristics, and w_i represents the user's characteristics.

(3) Calculate the similarity between the users to obtain the candidate set 1

After obtaining the user's feature vector, calculate the cosine similarity between the feature vector to obtain the similarity between the user u and the target a,

$$sim(u,a) = \cos(\overline{Vu}, \overline{Va}) = \frac{\overline{Vu} \cdot \overline{Va}}{|\overline{Vu}| \cdot |\overline{Va}|}$$

By similarity calculation, we obtain the Top-N recommendation, which is most similar to the target user u and set as friend recommended candidate set 1.

(4) Find what the user u's friends follow, and compute the number of common friends between the two

As social filtering algorithm shows, first find all the friends which the object of interest user u follows, one of which is located c, and a is a common friend of u and c, and then calculate the number of common friends between the two. As is described in the previous section, a mutual friend set is defined as:

$$MF(u,c) = \{a|F(u,a) \wedge F(a,c)\}$$

By mutual friends, we add contact between user u and user c. Then we can get user c's recommend percentage by calculating the number of mutual friend set MF (u, c).

Here, p is defined as the recommended candidate relationship matrix. In this matrix, if the user i and user j are friends, P_{ij} is 1, 0 otherwise.

$$P = \begin{bmatrix} p_{11} & \cdots & p_{1j} & \cdots & p_{1n} \\ \vdots & \ddots & \vdots & \ddots & \vdots \\ p_{i1} & \cdots & p_{ij} & \cdots & p_{in} \\ \vdots & \ddots & \vdots & \ddots & \vdots \\ p_{n1} & \cdots & p_{nj} & \cdots & p_{nn} \end{bmatrix}$$

Let A equal to matrix P calculated from the correlation matrix, with n users each associate rules confidence. The correlation matrix of A is a matrix of n × n, n is the number of users, a_{ij} is the confidence of association rules. a_{ij} represents both user i's friend, but also the user j's friends proportion of all users in the N.

$$A = \begin{bmatrix} a_{11} & \cdots & a_{1j} & \cdots & a_{1n} \\ \vdots & \ddots & \vdots & \ddots & \vdots \\ a_{i1} & \cdots & a_{ij} & \cdots & a_{in} \\ \vdots & \ddots & \vdots & \ddots & \vdots \\ a_{n1} & \cdots & a_{nj} & \cdots & a_{nn} \end{bmatrix}$$

The preference vector of use u is a matrix of 1 × n. u_{ij} expresses mutual friend relationship between the target user K and J, which is the amount of lateral P matrix. Then recommend for the target user vector s, and it can be got from the correlation matrix A and the target user's preference vector u.

(5) Base on the number of common friends to sort the candidate set of potential friends to give the candidate set 2

According to the number of common friends between the target users and the mutual friend recommended centralized, we can get a mutual friend and recommend candidates for the number of the target user set 2 from more to less friends.

(6) Get the user u's neighbor

Using social topological methods, find the user u's neighbor T, and remove of which there is already a friend of the user to obtain the final user neighbor R.

(7) Weight neighbor users to get the candidate set 3

Weight the R value based on the user's media properties, repetition rate and the number of common friends. Add the calculation results to obtain the candidate set 3 of friends recommend.

(8) Select suitable weights

In micro-blog, each user's behavior is a unique behavior collection which is different from the behavior of other users'. So the three candidate sets generated for each user when considering their weight distribution is necessary to specific conditions. When considering the weight, we first exclude the situation that users u has no friends or social networking in which we can simply recommend the recommended candidate q directly. We consider both the content of the user and the situation of social relations.

Suppose three weights are α (the weight of candidate sets 1), β (the weight of candidate set 2), $1 - \alpha - \beta$ (the weight of candidate sets 3). In the new algorithm, the three values of the weights are based on the user's content feature vector, the friends' friends in social relations and the number of fans concerned about the user to determine the proportion of three. For example, if a user's micro-blog content is rare but it has a huge social networking circle, then α will surely much less than $1 - \alpha$. Let the final recommendation result is Z, then

$$Z = \alpha \times Set1 + \beta \times Set2 + (1 - \alpha - \beta) \times Set3$$

(9) Obtain the final recommendation result

Recommend the recorded Top-K friends to the target user, and complete micro-blog friends' recommendation algorithm.

5 Conclusions

With the rapid development of network, human beings have entered the Internet information age. As the way that people commonly use in the network to contact friends, social network get more and more attention. Friends recommended in the social network as the basic function of the various social networking sites, has become a hot area of research.

This article first introduces the relatively excellent project recommendation algorithm, and then analyzes the status quo and technology of the existing friends in micro-blog recommendation algorithm. Secondly it puts forward a personalized friend recommendation algorithm based on the content and social relationship, and it mainly introduces the realization of the combining user's personal information, micro-blog content and the social relations in the algorithm.

The paper mainly does the following work: the statement of the definition, development history and characteristics of micro-blog, the introduction of the current recommended algorithm, and the analysis of the advantages and disadvantages. It then recommends related technologies for the current micro-blog friends and lay the theoretical foundation for the presenting of the personalized friends recommendation algorithm. It puts forward a personalized friend recommendation algorithm based on the content and social relationship, and extracts the details of the users' personal information, micro-blog content and social relation circles. It can be divided into three parts. First, it calculates the similarity between the users, basing on the recommended content. Second, it gets the users' neighbors according to the users' social networking topology, and it plus the right to obtain candidate sets.

In this paper, there are some shortcomings, in future work, needs to be done: 1, personalized recommendation algorithm proposed in this paper, although the user can solve the problem recommended by friends, but also fully used in the user's personal information, but did not consider dynamic changes to user interests, and therefore need to increase work in the future consideration of user interests dynamics. 2, although the new algorithm by multiple project team members to discuss improvement from, but there are always ill-considered part of, and the practical application and theoretical research will be most performance difference. 3, the pace of development of the Internet much faster than expected, and now the mainstream friends recommendation algorithm may have a new algorithm replaces the future, and therefore in future research should pay attention to research prospective.

Acknowledgements Supported by "the Fundamental Research Funds for the Central Universities" and "National Natural Science Foundation of China", Project No. 3262015T20, 3262016T31, 3262015T70, 3262014T75, 61502115. Project Leader: Liangbin YANG; Binyang LI.

References

1. Wang Binghui. The research of the potential Friends' recommend Algorithm in social network [D]. YUNNAN UNIVERSITY, 2013.
2. Chen J, Geyer W, Dugan C, et al. Make new friends, but keep the old: recommending people on social networking sites [C]. Proceedings of the SIGCHI Conference on Human Factors in Computing Systems. ACM, 2009: 201–210.
3. Tao Jun, Zhang Ning. Classification of collaborative filtering recommendation algorithm based on user interest [J]. COMPUTER APPLYMENT, 2011, 5(11):55–59.
4. Xie Yuan, Feng Lifang. Build your "social graph" [J]. Successful Marketing, 2010, 12 (12):37–38.
5. Massa P, Bhattacharjee B. Using trust in recommender systems: an experimental analysis [M]. Trust Management. Springer Berlin Heidelberg, 2004: 221–235.
6. Lo S, Lin C. Wmr-a graph-based algorithm for friend recommendation [C]. Proceedings of the 2006 IEEE/WIC/ACM International Conference on Web Intelligence. IEEE Computer Society, 2006:121–128.
7. Chin A. Finding cohesive subgroups and relevant members in the Nokia friend view mobile social network [C]. Computational Science and Engineering, 2009. CSE'09. International Conference on. IEEE, 2009, 4: 278–283.
8. Shen D, Sun J T, Yang Q, et al. Latent friend mining from blog data [C]. Data Mining, 2006. ICDM'06. Sixth International Conference on. IEEE, 2006: 552–561.
9. Zheng Y, Chen Y, Xie X, et al. GeoLife2.0: a location-based social networking service [C]. Mobile Data Management: Systems, Services and Middleware, 2009. MDM'09. Tenth International Conference on. IEEE, 2009: 357–358.
10. Bacon K, Dewan P. Towards automatic recommendation of friend lists [C]. Collaborative Computing: Networking, Applications and Worksharing, 2009. CollaborateCom 2009. 5th International Conference on. IEEE, 2009:1–5.
11. Wu Z, Jiang S, Huang Q. Friend recommendation according to appearances on photos [C]. Proceedings of the 17th ACM international conference on Multimedia. ACM, 2009:987–988.
12. Yu Haiqun, Liu Wanjun, Qiu Yunfei. The secondary contacts of social network recommend which is based on the user preference topic [J]. Computer application, 2012, 32(5): 1366–1370.
13. Niu Qingpeng. The study of potential blog friend technology [D]. Shenyang: Northeastern University, 2009.
14. Shi Lingfeng. Query Algorithm Research and Application Based on the relationship of FIG social networking friends [D]. Nanjing: Nanjing University of Science and Technology, 2012.
15. Zhao Wenbing, Zhu Qinghua, Wu Kewen, etc. Micro-blog user characteristics and motivations analysis [J]. Library and Information Technology, 2011, 2.
16. Gou L, You F, Guo J, et al. SFViz: interest-based friends exploration and recommendation in social networks [C]. Proceedings of the 2011 Visual Information Communication-International Symposium. ACM, 2011: 15.
17. Xie X. Potential friend recommendation in online social network [C]. Green Computing and Communications (GreenCom), 2010 IEEE/ACM Int'l Conference on & Int'l Conference on Cyber, Physical and Social Computing (CPSCom). IEEE, 2010: 831–835.
18. Hannon J, Bennett M, Smyth B. Recommending twitter users to follow using content and collaborative filtering approaches [C]. Proceedings of the fourth ACM conference on Recommender systems. ACM, 2010: 199–206.
19. Yu Yan, Qiu Guanghua, Chen Aiping. Recommendation algorithm based on online social network friends mixed graphs [J]. Library and Information Technology, 2011 (11): 54–59.
20. Java A, Song X, Finin T, et al. Why we twitter: understanding micro blogging usage and communities [C]. Proceedings of the 9th WebKDD and 1st SNA-KDD 2007 workshop on Web mining and social network analysis. ACM, 2007: 56–65.

21. Jia-jia Zheng. Social network of friends and Implementation Mechanism recommendation based on FIG. Sort [D]. Zhejiang University, 2011.
22. Armentano, M.G., D.L. Godoy, A.A. Amandi. A topology-based approach for followees recommendation in Twitter, in Workshop chairs.
23. Wu Yanqing. microblog friends' recommendation which is based on heterogeneous data [D]. Zhejiang University, 2013.
24. Yang Honglei. Recommendation algorithm based on content and social filtering Friends [D]. Inner Mongolia University of Science and Technology, 2013.
25. Geyer W, Dugan C, Millen D R, et al. Recommending topics for self-descriptions in online user profiles [C]. Proceedings of the 2008 ACM conference on Recommender systems. ACM, 2008: 59–66.
26. Linden G, Smith B, York J. Amazon.com recommendations: Item-to-item collaborative filtering [J]. Internet Computing, IEEE, 2003, 7(1): 76–80.
27. Peng Tao. Research and achievement of wireless mobile environment image information in recommendation system [D]. Beijing University of Post and Telecommunications, 2010.
28. He Keqin. the research of Recommendation system model and algorithm which is based on label [D]. East China Normal University, 2011.

A Hybrid Methodologies for Intrusion Detection Based Deep Neural Network with Support Vector Machine and Clustering Technique

Tao Ma, Yang Yu, Fen Wang, Qiang Zhang and Xiaoyun Chen

Abstract This paper proposes a novel approach called KDSVM, which utilized the k-mean techniques and advantage of feature learning with deep neural network (DNN) model and strong classifier of support vector machines (SVM) , to detection intrusion networks. KDSVM is composed of two stages. In the first step, the dataset is divided into k subset based on every sample distance by the cluster centers of k-means approach, and in the second step, testing dataset is distanced by the same cluster center and fed into the DNN model with SVM model for intrusion detection. The experimental results show that the KDSVM not only performs better than SVM, BPNN, DBN-SVM (Salama et al., Soft computing in industrial applications, 2011 [21]) and Bayes tree models in terms of detection accuracy and abnormal types of attacks found. It also provides an effective tool for the study and analysis of intrusion detection in the large network.

Keywords Intrusion detection systems · Deep neural network · Hybrid model · K-means clustering · Support vector machine

T. Ma (✉) · Y. Yu · X. Chen (✉)
School of Information Science and Engineering, Lanzhou University,
Lanzhou 730000, China
e-mail: mat13@lzu.edu.cn

X. Chen
e-mail: chen_xiaoyun@yeah.net

T. Ma · F. Wang
School of Mathematical and Computer Science, Ningxia Normal University,
Guyuan, Ningxia 756000, China

Q. Zhang
Statistics & Research Division, China Insurance Regulatory Commission Ningxia Bureau,
Yinchuan, Ningxia 750000, China

© Springer Nature Singapore Pte Ltd. 2018

N.Y. Yen and J.C. Hung (eds.), *Frontier Computing*, Lecture Notes in Electrical Engineering 422, DOI 10.1007/978-981-10-3187-8_13

1 Introduction

Network intrusion detection is a new network security mechanism designed to detect, prevent and repel unauthorized access to a communication or computer network. An intrusion detection system (IDS) plays a crucial role in maintaining a safe and secure network. In recent years, a huge network data is generated due to the application of new network technologies and equipment, which leads to the declining of the defect rates. The intrusion detection process is a difficult and complicated one in terms of detection accuracy, detection speed, the dynamic nature of the networks and the available processing power for processing high volumes of data from distrusted network systems [15]. Recently, many researchers proposed innovative approaches in recent years.

These methods, based on detecting in team of behavior-based and resource type of access, are divided into four categories. The first category is to detect anomalies based on statistical analysis, such as, Bayesian model [3], Decision Tree. Anomaly-based techniques build models of normal network samples and detect the samples that deviate from these models in literature [7]. It can detect new types of attacks via already known normal events. Therefore the anomaly detection approach suffers from a high rate of failure. The second category is anomaly detection approach where most methods require a set of standard normal dataset to train the classifier and check whether new sample fits the model. These principle methods are coined as outlier detection algorithm, such as k-mean, self-organizing maps and unsupervised support vector machines approaches [8]. The third category employing AI techniques to detect attack types by taking advantage of machine learning can prioritize solutions to certain problem, such as, SVM [5], RF [23], genetic algorithm (GA) and artificial neural networks etc. The last category is the hybrid and ensemble detecting methods that integrated advantages of different or same methods in order to incase accuracy of detection. These approaches include bagging, adaboost [19] technology, and the PSO-K-means ensemble approach [16]. The PSO-k-means methods could achieve optimal numbers of clusters and increase the high detection rate which utilized K-means technology to detect attack types in networks. In addition, the SVM-KNN-PSO ensemble method proposed by [1] can obtain the best results, which used advantage of nonlinear processing capability and classification capability based distance for each sample. However, their work is based on binary classification methods, which can distinguish between the two states. Alom et al. [2] combines the deep belief network (DBN) and SVM model, the proposed model utilized DBN to select the feature and SVM to capture the rules from attack process, then the reduction dimension output data by DBN regarded as the input dataset fed SVM into detection intrusion. In above methods, it is supposed that each feature of datasets is independent in all time, but in real world, the feature of intrusion dataset is complex and needed a comprehensive analysis.

Taking above discussions into consideration, this paper proposes the KDSVM model using the k-means algorithm to capture the feature of raw data and divide dataset into different subsets. Then each subset is fed to the improved DNN which top layer instead of SVM model, respectively, and learning different characteristics of the sub dataset. Next, these tested datasets are divided by prior cluster center of training dataset into sub testing datasets. Finally these testing sub datasets are fed to the trained DNN for intrusion detection. Because the DNN can acquire enough information, via prior learning processing and capture more specific rules of attack types in networks based on extracting feature capability for massive and complex data [6, 13]. The DNN model based on theory of deep learning, can solve non-linear problems with complex and large-scale data, and has been successfully applied in the area of weather forecasting and stock prediction [10]. The experimental results based on the knowledge of KDD CUP99 datasets and NLS-KDD datasets [22] show that KDSVM generates better accuracy and more robust than other well-known algorithms, and well supported for parallel computing.

The rest of the paper is organized as follows. The related literature concerning of IDS is reviewed in the Sect. 2. Section 3 presents the proposed approach in detailed and describes it works. Section 4 describes the experimental datasets and illustrates the data preparation, evaluation criteria, results and discussions of experiments. Finally, the conclusions and suggestions for future work are provided in Sect. 5.

2 Literature Review

In this section, the deep learning approach of deep neural network is briefly introduced. As a matter of fact, IDS as classification method is very important for deal with feature of dataset, because the categories learner has acquired knowledge and patterns based on the characteristic of data. Additionally, the level of feature representation is determining the performance of a learner.

2.1 DNN Algorithm

The essence of the deep neural network is to learn more useful feature of machine learning and construct multiple layers in network and vast amounts of training data.

Auto-encoder: An auto-encoder is one type of unsupervised neural networks with three layers [12] and the output target of the auto-encoder is input data. The encoder network transforms the input data from a higher dimensional space to codes in a low dimensional space and the decoder network remodels the inputs from the previous works.

The encoder network is defined as an encoding function denoted by $f_{encoder}$. This function indicates the encoding process:

$$h^m = f_{encoder}(x^m) \quad (1)$$

In which x^m stands for data point from a dataset, h^m is the encoding vector obtained from x^m.

Decoder: The decoder network is defined as a reconstruction function denoted as $f_{decoder}$, this function indicates the decoding process:

$$\hat{x}^m = f_{decoder}(h^m) \quad (2)$$

In which $\hat{x}^m$ is the decoding vector obtained from h^m. There are specific algorithms for several encoding functions and reconstruction functions including:

$$\text{Logsig:} \quad f_{encoder}(x^m) = \frac{1}{1+e^{-x^m}} \quad (3)$$

$$\text{Satline:} \quad f_{encoder}(x^m) = \begin{cases} 0 & if \quad x^m \leq 0 \\ z & if \quad 0 < x^m < 1 \\ 1 & if \quad x^m \leq 0 \end{cases} \quad (4)$$

$$\text{Pureline:} \quad f_{encoder}(x^m) = x^m \quad (5)$$

Pre-training: The process is proceeding in the sequence until the *N*th auto-encoder is trained for initialization the final hidden layer of the DNN. In this way, all the hidden layers of DNN are stored auto-encoder by stacked structure in each training *N* times, and are regarded as pre-trained. This pre-training process is proven to be significantly better than random initialization of the DNN and conducive to achieving generalization in classification [9, 11].

Fine-tuning: Fine-tuning is the process that utilizes the supervised fashion to improve the performance of DNN. The network is retraining and labeled from training data, and the errors by difference between real and predicted values are back propagation with stochastic gradient descent (SGD) method for all multilayer network. The equation of SGD is defined as follows:

$$E = \frac{1}{2}\sum_{i=1}^{n}(y_i - t_i)^2 \quad (6)$$

where, the function E is loss function, y is the real label and t is the output of network. The gradient of weight parameter ω is obtained by derivative the error equation.

$$\frac{\partial y}{\partial \omega_{ij}} = \frac{\partial E}{\partial y_j} \cdot \frac{\partial y_j}{\partial \mu_j} \cdot \frac{\partial \mu_j}{\partial \omega_{ij}} \quad (7)$$

With the gradient of the ω_{ij} the equation of updated SGD is defined as:

$$\omega_{ij}^{new} = \omega_{ij}^{old} - \eta \cdot (y_j - t_j) \cdot y_j(1 - y_j) \cdot h_i \tag{8}$$

In which, the η is the step size and greater zero, h is the hidden layer number in the deep network [4].

This process is tuned and optimized by the weight and threshold based on the real label data in the DNNs, in this way, the deep networks can learn important knowledge for final output and direct the parameter of whole network to detect correct classification [20].

3 Proposed Approach of KDSVM

This section, the proposed approach is used based on clustering methods and deep learning with SVM model to solve above problems. In the first place, the sub training datasets divided the training process into different subsets and calculate center points by each train points. Second, the sub train datasets are trained by *k*th DNNs, the number *k* is the value of clusters, this take DNNs that have learned different characteristic of each cluster centers. Third, the sub testing datasets are divided from the test datasets by k-means algorithm that uses the previous cluster centers in the first step, and these sub testing datasets are applied to detect intrusion attack type by completely trained per DNN which top layer used SVM classifier. Finally, the outputs of every DNN are aggregated for the final results of intrusion detection classifiers.

3.1 The KDSVM Algorithm

The approach in detail is showed the algorithm of KDSVM. The point center and training sets are generated by output of k-means function in line 1, the sub testing sets are obtained by calculating the distance with Huffman function in line 2–6, the *k*th DNNs is trained by training set in line 7–12, the sub testing sets are index and the final results are predicted by the aggregation in line 13–19.

Algorithm 1: KDSVM algorithm

Input: TR-Train Dataset, TE-Test Dataset, K- the number of the cluster, HLN- the number of the hidden layer node, HL- the number of the hidden layer, TSVM-top layer classifier of architecture in each DNN.

Output: classification result- for KDSVM model

```
    /* get the center points and sub-train dataset */
1.    C, subTR ← kmeans ( TR, K)
    /* calculate distance from each data points in TE to center points */
2.  For i = 1 to N
3.     For j = 1 to K
4.        distance (i, j) = huffman (TE_i, C_j)
5.     End
6.  End /* Train the every deep neural network by the each sub-train dataset */
7.  Switch p do
8.    case p do
9.          Train the DNN_p =DNN(subTRp, HLN,HL);
10.         Train the TSVM_p=DNN_p
11.   End
12. End /* get sub-test dataset by the per cluster center Cj*/
13. For m=1 to N
14.    index ← find (min(distance (i))) ;   subTE_index ← TE_i
15. End /* train DNN model for each subTR and test DNN model for subTE */
16. For m=1 to K
17.    prediction_m = model_m(subTE_m)
18. End /* aggregate each prediction result */
19. Return classification result = aggregate (prediction)
```

4 Experiments

The experiments will be examined and compared with other detection engineer models, for instance, SVM, BPNN, DBN-SVM and naive Bayes. The six datasets from the KDDCUP99 and NSL-KDD are used to evaluate the performance of all models. Then, the parameters of the number of clusters and the weights of DNN are discussed and analyzed.

4.1 The Dataset

In this research, six datasets are randomly generated from two datasets, KDD CUP'99 and NSL-KDD, which reduce the amount of data, and called Dataset1 to Dataset6, respectively [18] and show in Table 1.

Table 1 The distribution of training set and testing set are shown in six dataset from the KDD‘99 and NSL-KDD

Data set	Training dataset					Testing dataset				
	Normal	Dos	Prob.	U2R	R2L	Normal	Dos	Prob.	U2R	R2L
Dataset1	9727	39145	4107	52	1126	60593	229853	4166	228	16189
Dataset2	48639	195729	4107	52	1126	60593	229853	4166	228	16189
Dataset3	97278	391458	4107	52	1126	60593	229853	4166	228	16189
Dataset4	13449	9234	2289	11	209	9711	7458	2421	200	2754
Dataset5	33671	22963	11656	52	995	9711	7458	2421	200	2754
Dataset6	13449	9234	2289	11	209	2152	4342	2402	200	2754

The six new datasets are used to evaluate the performance of KDSVM algorithm, and execute to compare the other detection engineering methods, such as SVM, BPNN, DNB-SVM, and Naive Bayes [14].

4.2 Evaluation Methods

In this study, the Accuracy, Recall, and Error Rate (ER) are used to evaluate the performance of the detection models. The formulas of above criteria are calculated as follows [17]:

$$\text{Accuracy} = \frac{TP + TN}{TP + TN + FP + FN} \tag{9}$$

$$\text{Recall} = \frac{TP}{TP + FN} \tag{10}$$

$$\text{Error Rate} = \frac{FP + FN}{FP + TP + TN + FN} \tag{11}$$

In which, True Positives (TP) indicates the number of network attack types distinguishing correct cases, the True Negatives (TN) shows the number of normal network type classifying the correct normal type, the False Negative (FN) is denotes the number of classified attack type detection as normal type, the False Positive (FP) means that the number of classified normal type as attack cases. The step of Accuracy shows the degree of whole correct detection accuracy of dataset and the ER refers to robust of classifier, the Recall indicates the degree of correctly detection attack type in whole attack recodes. In above team, higher accuracy and recall and lower ER is represented good performance.

4.3 Experiments with KDSVM

In this section, cluster number of k is evaluation of KDSVM based on the six dataset, because the area of value of k are different in each dataset and this are serious impact precision of results for KDSVM method. Next, the testing datasets are used to compare the performance of the five models.

Results and Comparisons

In this section, the fusion matrix and the evaluated criterion are calculated with the KDSVM and other four traditional detection engineers in the six datasets respectively. The experiment results of above algorithm in six datasets are shown in Table 2 and Fig. 1.

Table 2 The comparing the results for the intrusion network for six datasets (\%)

Dataset	k	model	Normal	Dos	Probe	U2R	R2L	Acc	Recall	ER
Dataset1	–	SVM	98.21	83	66.01	0.88	3.14	81.52	77.72	18.48
	–	BPNN	96.51	89.49	46.18	9.21	1.93	85.66	83.48	14.34
	–	DBN-SVM	93.65	96.62	59.27	0	0	90.44	91.08	9.56
	–	Bayes	91.51	95.59	61.35	4.39	3.56	89.48	**92.57**	10.52
	2	KDSVM	97.21	96.87	80.32	11.4	6.88	**91.97**	91.68	**8.03**
Dataset2	–	SVM	96.22	97.1	65.84	0	0.05	91.39	90.52	8.61
	–	BPNN	91.44	97.42	62.69	7.02	5.41	90.93	92.88	9.07
	–	DBN-SVM	98.23	96.48	38.26	0	0	90.95	89.51	9.05
	–	Bayes	95.92	95.98	62.55	4.82	4.38	90.69	91.07	9.31
	4	KDSVM	98.42	97.2	70.64	3.51	1.57	**92.03**	**91.35**	**7.97**
Dataset3	–	SVM	95.87	97.23	64.86	0	0.06	91.41	90.59	8.59
	–	BPNN	81.53	96.95	8.81	6.14	7.26	88.03	90.05	11.97
	–	DBN-SVM	99.57	96.57	0	0	0	90.76	89.37	9.24
	–	Bayes	96.38	96.29	59.15	7.02	7.46	91.12	90.95	8.88
	5	KDSVM	97.61	97.23	65.96	4.39	6.59	**92.1**	**92.23**	**7.9**
Dataset4	–	SVM	95.54	70.18	57.37	0	1.63	70.73	53.26	29.27
	–	BPNN	96.35	71.17	65.55	0	0.58	72.16	**57.79**	27.84
	–	DBN-SVM	99.63	63.11	7.23	0	0	64.57	40.45	35.43
	–	Bayes	93.9	72.18	41.02	0	0	68.73	52.78	31.27
	3	KDSVM	96.17	75.84	53.37	3	3.01	**72.64**	57.48	**27.36**
Dataset5	–	SVM	98.57	18.93	49.89	0	0.11	54.1	20.45	45.9
	–	BPNN	91.79	7.63	66.58	1.5	2.43	49.53	27.56	50.47
	–	DBN-SVM	99.69	62.64	48.99	0	0	68.93	46.43	31.07
	–	Bayes	99.06	61.65	35.4	0	0	66.87	44.28	33.13
	3	KDSVM	97.19	74.51	48.37	5	0.62	**71.83**	**55.08**	**28.17**
Dataset6	–	SVM	95.81	41.5	43.67	0	0	41.46	30.6	58.54
	–	BPNN	74.72	4.61	88.67	0	1.53	33.59	30.6	66.41
	–	DBN-SVM	99.72	36.15	6.74	0	0	32.73	18.9	67.27
	–	Bayes	82.16	48.25	28.52	0	0	38.37	30.08	61.63
	5	KDSVM	84.2	50.02	52.66	1.5	0.98	**44.55**	**37.85**	**55.45**

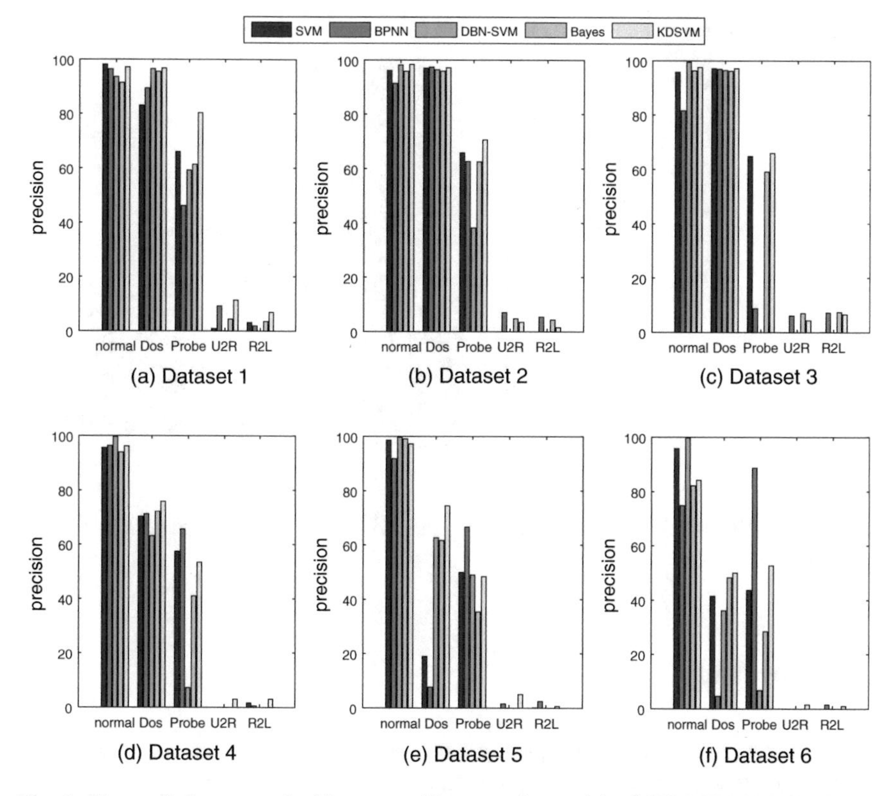

Fig. 1 The prediction accuracies histogram of five types for models of SVM, BPNN, DBN-SVM, Bayes and KDSVM are compared in six datasets in different colors

In which, the columns symbol of ACC in table heads mean the average accuracy for each models. The records are unbalance in six dataset, the types of Normal and Dos have major compositions, the U2R and R2L have sparse distribution, because the last two cases have especially intrusion actions which have obtained advanced user right, it is more covert intrusion for difficultly detection.

From Table 2 and Fig. 1, consideration accuracy, the KDSVM has better accuracy than other four methods, and has the lowest error rates in the datasets.

4.4 Discussion

The overall accuracy is used to generate the histogram and compare distinguished results in six datasets and shown in Fig. 2. This is more detailed to evaluate the classification performance with five types (one normal and four types).

From the above, the results show that KDSVM algorithm is good at detection cases of Normal, Dos and Probe in the six dataset. Therefore, for sparse and difficult

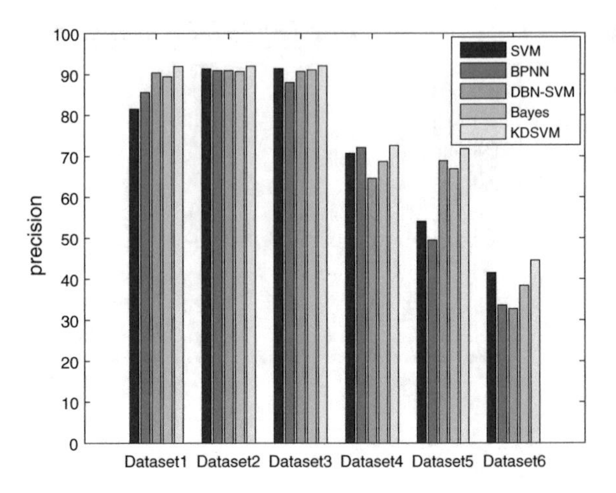

Fig. 2 The histograms of the average precision by the five models are compared with the six datasets

cases of U2R and R2L in six datasets, the KDSVM model also obtains higher accuracy.

5 Conclusion

The attacking events of low frequent are usually difficult to predict and it can cause severe threats to networks. This paper puts forward the innovative approach which takes the advantage of k-means and hybrid deep neural network with top layer used SVM classifier, to detect attack types. In the first stage, the features of the network dataset are clustering and divided into k sub datasets in a bid to find more knowledge and patterns from the similar clusters. Then in the second stage, the highly abstract information is obtained by deep learning networks from the subsets during the clustering process. Finally, the DNNs which used SVM classifier to instead of softmax layer, are used to detect the attack cases with testing subsets. This is an efficient way to improve the accuracy of the detection rates. The results of experiment show that the KDSVM performs better than the SVM, BPNN, DBN-SVM and Bayes with best accuracy over the six datasets. On the other hand, the proposed algorithm is more capable of classifying term of sparse attack cases and effectively improves detection accuracy in real security system. However, limitations of the KDSVM include the DNN parameters of weights and threshold of the every layer, and the SVM parameters that need to be optimized by heuristic algorithms, and it will be study works in the further.

Acknowledgements This work is supported by the National Natural Science Foundation of China (Grant No. 11361046) and the Key Research Fund of Ningxia Normal University (Grant No. NXSFZD1517 NXSFZD1603 and NXSFZD1608), the Natural Science Fund of Ningxia Province (Grant NZ16260) and the Fundamental Research Fund for Senior School of Ningxia Province (Grant No. NGY2015124).

References

1. Aburomman, A.A., Reaz, M.B.I.: A novel svm-knn-pso ensemble method for intrusion detection system. Applied Soft Computing 38, 360–372 (2016)
2. Alom, M.Z., Bontupalli, V., Taha, T.M.: Intrusion detection using deep belief networks. In: 2015 National Aerospace and Electronics Conference (NAECON). pp. 339–344. IEEE (2015)
3. Barbara, D., Wu, N., Jajodia, S.: Detecting novel network intrusions using bayes estimators. In: SDM. pp. 1–17. SIAM (2011)
4. Bengio, Y., Simard, P., Frasconi, P.: Learning long-term dependencies with gradient descent is difficult. Neural Networks, IEEE Transactions on 5(2), 157–166 (1994)
5. Chen, W.H., Hsu, S.H., Shen, H.P.: Application of svm and ann for intrusion detection. Computers & Operations Research 32(10), 2617–2634 (2005)
6. Chilimbi, T., Suzue, Y., Apacible, J., Kalyanaraman, K.: Project adam: Building an efficient and scalable deep learning training system. In: 11th USENIX Symposium on Operating Systems Design and Implementation (OSDI 14). pp. 571–582 (2014)
7. Denning, D.E.: An intrusion-detection model. Software Engineering, IEEE Transactions on SE-13(2), 222–232 (1987)
8. Dokas, P., Ertoz, L., Kumar, V., Lazarevic, A., Srivastava, J., Tan, P.N.: Data mining for network intrusion detection. In: Proc. NSF Workshop on Next Generation Data Mining. pp. 21–30 (2002)
9. Erhan, D., Bengio, Y., Courville, A., Manzagol, P.A., Vincent, P., Bengio, S.: Why does unsupervised pre-training help deep learning? The Journal of Machine Learning Research 11, 625–660 (2010)
10. Grover, A., Kapoor, A., Horvitz, E.: A deep hybrid model for weather forecasting. In: Proceedings of the 21th ACM SIGKDD International Conference on Knowledge Discovery and Data Mining. pp. 379–386. ACM (2015)
11. Hinton, G.E., Osindero, S., Teh, Y.W.: A fast learning algorithm for deep belief nets. Neural computation 18(7), 1527–1554 (2006)
12. Hinton, G.E., Zemel, R.S.: Autoencoders, minimum description length, and helmholtz free energy. Advances in neural information processing systems pp. 3–3 (1994)
13. Huang, P.S., He, X., Gao, J., Deng, L., Acero, A., Heck, L.: Learning deep structured semantic models for web search using click through data. In: Proceedings of the 22nd ACM international Conference on information & knowledge management. pp. 2333–2338. ACM (2013)
14. Japkowicz, N., Shah, M.: Evaluating learning algorithms: a classification perspective. Cambridge University Press (2011)
15. Kabiri, P., Ghorbani, A.A.: Research on intrusion detection and response: A survey. IJ Network Security 1(2), 84–102 (2005)
16. Karami, A., Guerrero-Zapata, M.: A fuzzy anomaly detection system based on hybrid pso-kmeans algorithm in content-centric networks. Neurocomputing 149, 1253–1269 (2015)
17. Kayacik, H.G., Zincir-Heywood, A.N., Heywood, M.I.: A hierarchical som-based intrusion detection system. Engineering Applications of Artificial Intelligence 20(4), 439–451 (2007)
18. Koc, L., Mazzuchi, T.A., Sarkani, S.: A network intrusion detection system based on a hidden naive bayes multiclass classifier. Expert Systems with Applications 39(18), 13492–13500 (2012)
19. Marin, G.: Network security basics. Security & Privacy, IEEE 3(6), 68–72 (2005)
20. Palm, R.B.: Prediction as a candidate for learning deep hierarchical models of data. Technical University of Denmark (2012)
21. Salama, M.A., Eid, H.F., Ramadan, R.A., Darwish, A., Hassanien, A.E.: Hybrid intelligent intrusion detection scheme. In: Soft computing in industrial applications, pp. 293–303. Springer (2011)

22. Tavallaee, M., Bagheri, E., Lu, W., Ghorbani, A.A.: A detailed analysis of the kdd cup 99 data set. In: Proceedings of the Second IEEE Symposium on Computational Intelligence for Security and Defence Applications 2009 (2009)
23. Zhang, J., Zulkernine, M., Haque, A.: Random-forests-based network intrusion detection systems. Systems, Man, and Cybernetics, Part C: Applications and Reviews, IEEE Transactions on 38(5), 649–659 (2008)

PET/CT Imaging Automated Classification, Structure Testing and Inspection of the Human Spinal Cord

Wei-Ming Chen and Chih-Sheng Feng

Abstract For the Positron Emission Tomography/Computed Tomography (PET/CT) technology, the traditional vertebra spinal cord examination tumor shift examination, is decided by doctor itself experience because the experience determination easily to have the fail diagnosis probability, these fail diagnosis make the patient body huge burden, these risks use the Computer-aided diagnosis avoid, vertebra spinal cord and so on specific spot use computer precise computation, the design of calculating method strengthens forecast ability in the tumor position, urges the accuracy promotion, reduces the risk of fail diagnosis, this medical examination service is the indispensable diagnostic assistance. In our system, we can classify characteristics of the vertebra, the spinal cord and the tumor, identify vertebral metastases, and accumulate in the training data, which improve success ratio of the system automatic identification. The experimental results make the doctor's Computer-aided examination and operation use of the three-dimensional model with the corresponding position and the tumor information.

Keywords Positron emission tomography/computed tomography · Magnetic resonance imaging · Vertebral column detection · Backbone detection · Spine detection · Tumor detection · Neoplasm detection

1 Introduction

Along with technical progress and development, the mortality rate of malignant tumor (cancer) stays at the highest level, recent years the World Health Organization announced the human ten big causes of death, the malignant tumor (cancer)

W.-M. Chen · C.-S. Feng (✉)
Department of Computer Science and Information Engineering,
The University of National Ilan, Yilan, Taiwan, ROC
e-mail: wsx5031060310guy@gmail.com

W.-M. Chen
e-mail: wmchen88@gmail.com

© Springer Nature Singapore Pte Ltd. 2018

N.Y. Yen and J.C. Hung (eds.), *Frontier Computing*, Lecture Notes
in Electrical Engineering 422, DOI 10.1007/978-981-10-3187-8_14

placed fifth, the malignant tumor (cancer) still is disease which in the medical system convalesced most with difficulty. Nowadays medical establishment uses Magnetic Resonance Imaging (MRI) to examine the malignant tumor, the extremely few use Positron Emission Tomography/Computed Tomography (PET/CT) to examine the malignant tumor, in the tumor shift vertebra spinal cord part, MRI examination cost higher than PET/CT examination, causes the waste of medical resource, also existed many limit, e.g., the human subject cannot be equipped any metallic decoration in the MRI instrument when processing, this limit leads excessively people cannot get health screening on subjects protection.

Nowadays studies in Lee et al. [1] and Staal et al. [2] proposed the method to solve the rib classification, lies in the automatic spinal canal tracing, but this method is only for chest CT, which is unable to use on the chest outside, André Mastmeyer et al. [3] proposed the mark type of spinal column examination, the mark type of spinal column examination must let doctor to mark, this paper will propose the automatic detection method to be able the rapid effective automatic detection vertebra pathological.

In order to improve above shortcoming, this research uses Positron Emission Tomography/Computed Tomography (PET/CT) to automatic tumor detection. Using PET/CT examination cost lower than Magnetic Resonance Imaging (MRI) examination, with examines the tumor characteristic merit, the more higher analysis rate of the more precise examination result, improves many patients to be obtain the examination safeguard.

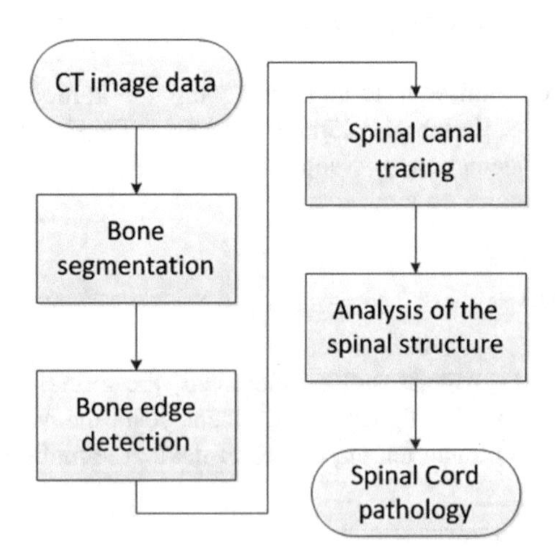

Fig. 1 The experimental flow chart

2 Method

So far, the detection methods have been proposed are all using artificial to detection spinal column. The method proposed in this paper, Fig. 1, can detect the majority of spinal columns automatically by using CT image data, and can narrow down the diagnostic area to get a better efficiency on the system processing speed. Each time our method will train the training data while system make an analysis process, and it will provide the training result to the next phase to improve the analysis accuracy. Moreover, combined with the characteristic the PET tumor detection which can get more effective detection of tumor characteristics and location.

2.1 Bone Segmentation

For the first step, in Fig. 2, the highest intensity area is also a value as HU value + 1000 which indicates the bone range. We can use this characteristic to separate the bone from other parts. Figure 3 shows the result of this separating process which include the head, the body and the buttocks.

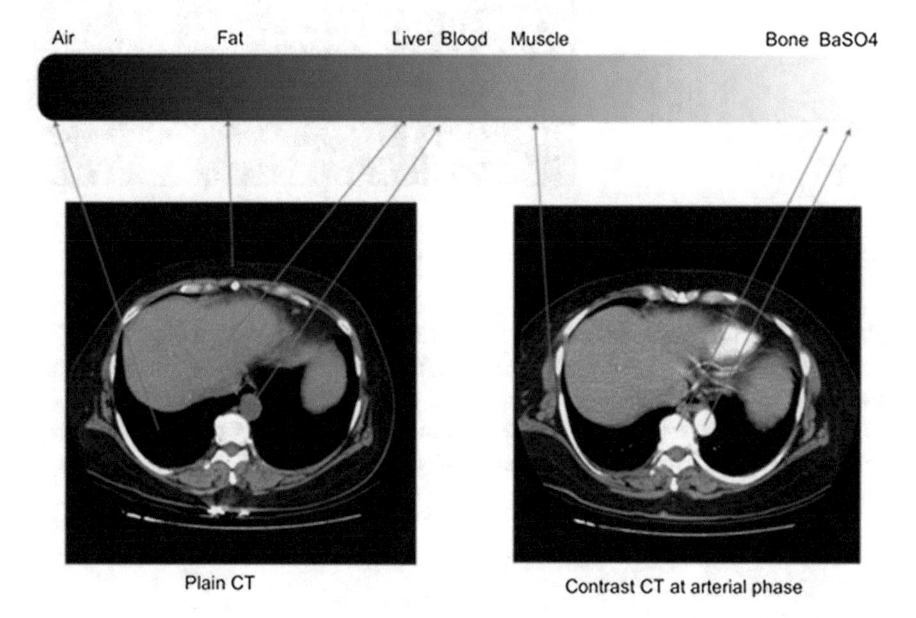

Fig. 2 HU value distributed scope *Source* www.cmu.edu.tw

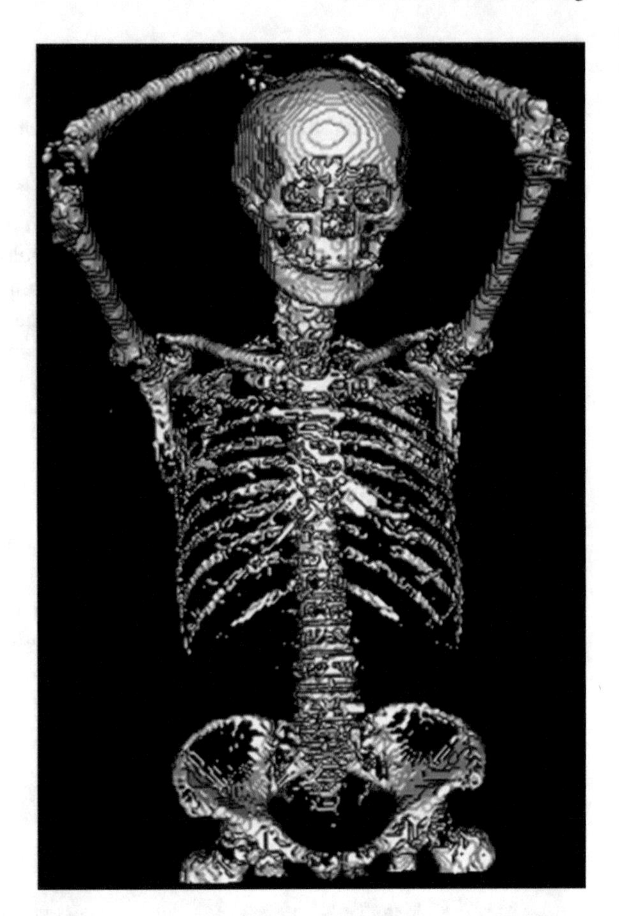

Fig. 3 The result of classify the bone with other part, the bone range including the head, the body, the buttocks

2.2 Bone Edge Detection

This step we use the result of the previous step, which produce many noise. We employ J. Canny [4] et al. edge detection of the bone to solve this problem. We take the result of edge detection to be the initial points. By using Romen Singh, T [5] et al. local adaptive threshold as the threshold value, the disconnect noise around the bones will be removed more easily and clearly. The result is shown as Fig. 4.

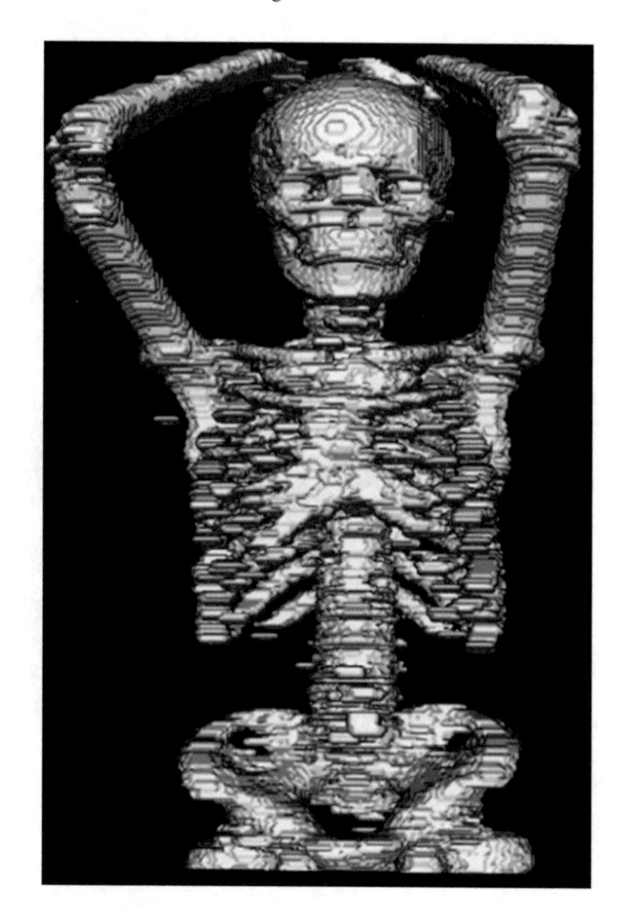

Fig. 4 The edge detection and the inflation result of the bone

2.3 Spinal Canal Tracing

We use the result of previous step to trace the spinal canal, and make statistics each 2D for slice. The top 1% of the statistics result will be marked as vertical position points. Use these points as the initial points to inflate toward the outside, the inflation range will not surpass the result of before step, then use the different slice biggest area occurring together deletion noise. This step method will be as follows:

First, statistics each 2D slice position points (Fig. 5).

Second, uses the mark of vertical position points to be the initial points which inflate outward (Fig. 6).

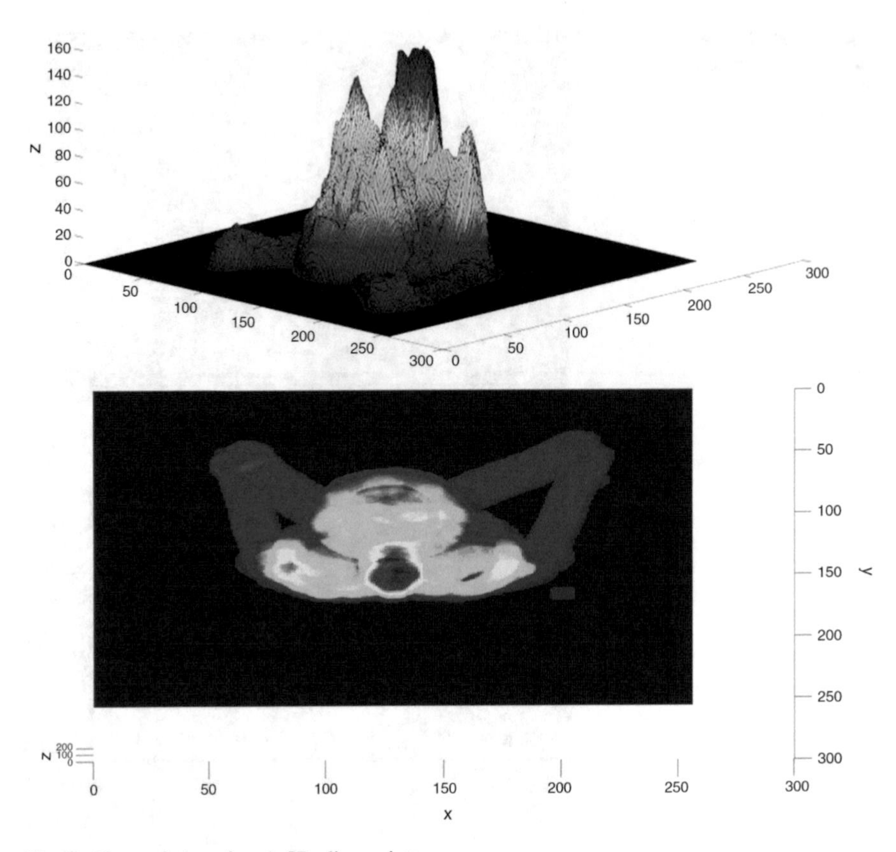

Fig. 5 The statistics of each 2D slice points

Third, find the largest group of different slice and use inflation (Fig. 7).

Fourth, find the average and standard deviation of all the 2D slice statistical middle line (Fig. 8).

After above step completes, examines the spinal column pathological characteristic according to the experimental result, Stokes et al. [6], Kang et al. [7], Steiger et al. [8], Kalender et al. [9], Kalender et al. [10], Steiger et al. [11] proposed the spinal cord pathological characteristic save in the training database, in order to help other analysis.

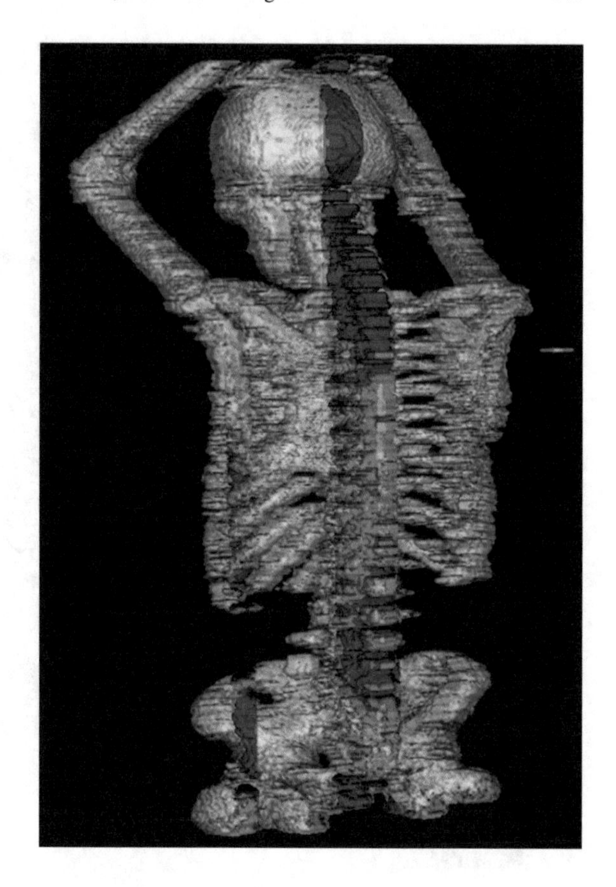

Fig. 6 The result of vertical statistics which inflate outward

2.4 *Analysis of the Spinal Structure*

This paper collects the spinal column disease CT data to analysis (Fig. 9), use on tumor detection and spinal cord pathology.

2.5 *Experiment*

This training data is collected from the China Medical University Hospital, all of the CT scans data use the low radiation dose (120 kV, 29 mAs–35 mAs) with slice thicknesses varying from 3.8 mm. The doctor selects the correct result to compare this paper method detection result.

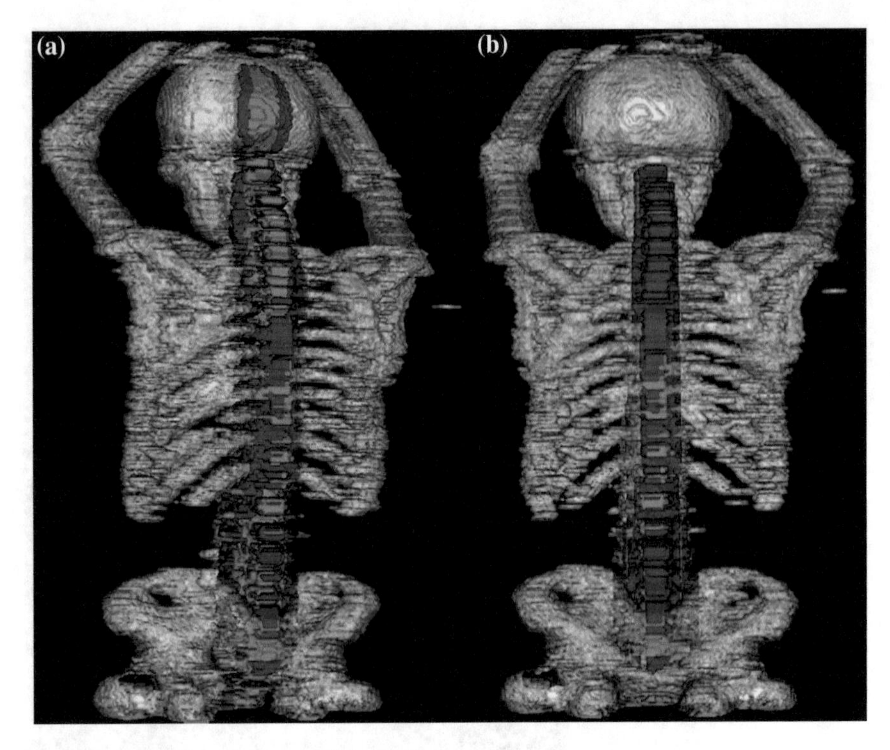

Fig. 7 **a**, **b** The largest group of different slice

3 Results

The experimental results are shown as Figs. 10, 11, 12. The medical pictures in this experimental will be resized automatically into 256 px × 256 px. The doctor selects the target area which have about 20,000 points, and the hit probability probably of our experiment result is above 85%. The incorrect parts located at division vertebra are mainly in cervical vertebra part.

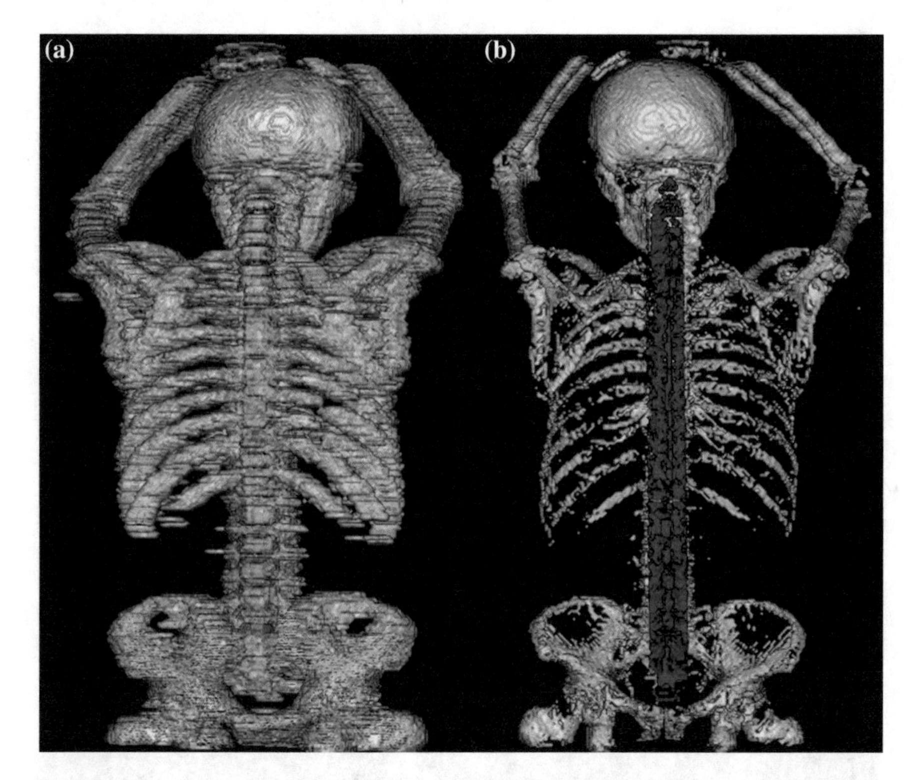

Fig. 8 Statistical average and standard deviation middle line result. **a** Inflation result. **b** Primitive bone result

4 Conclusion

The experimental result can be used on detect spinal cord pathology and intramedullary tumors. This paper proposed the automated spinal cord detection method, patient's posture must place the hand forehead in the CT scan. The preliminary result showed 85% above spinal column are detected. This paper's weaknesses is in cervical vertebra region.

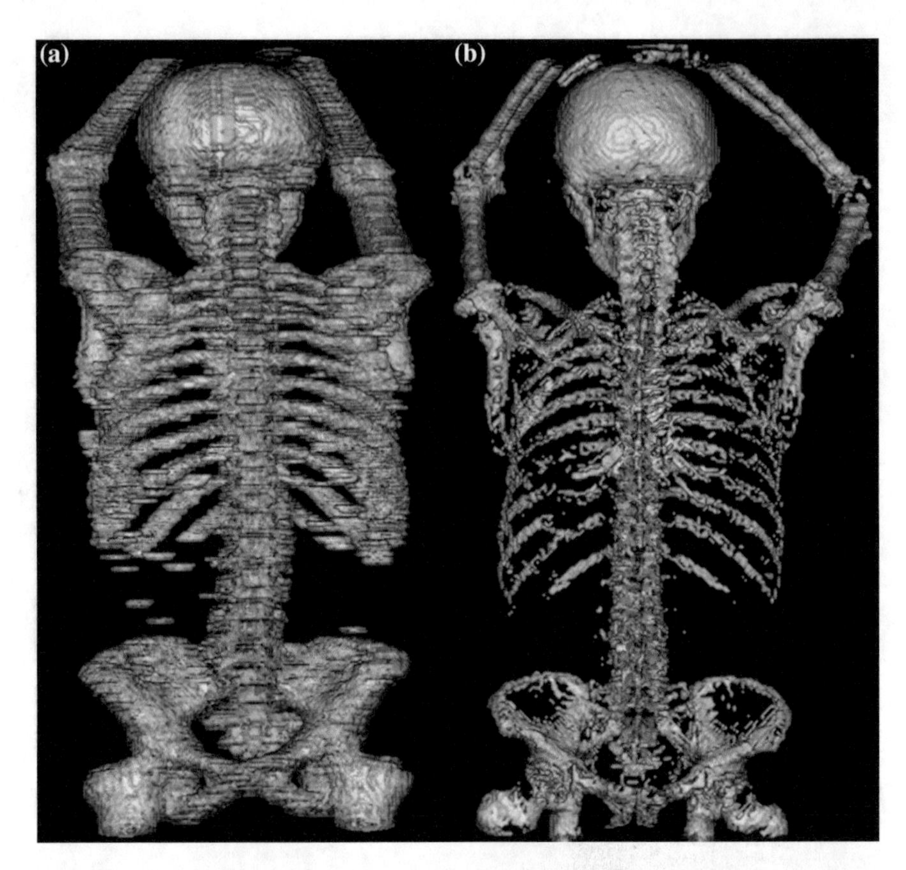

Fig. 9 Spinal cord pathology detection. **a** Scoliosis. **b** Intramedullary tumors

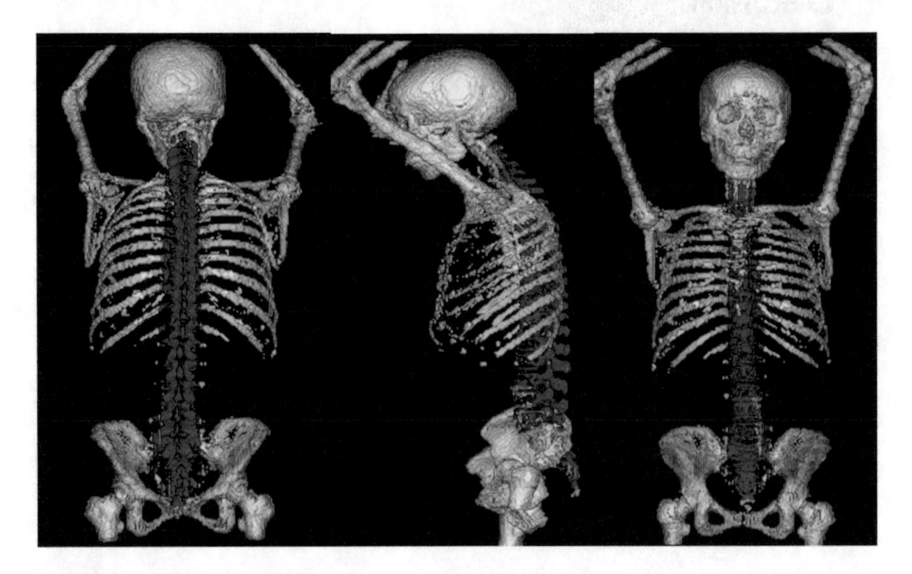

Fig. 10 The experimental result (normal)

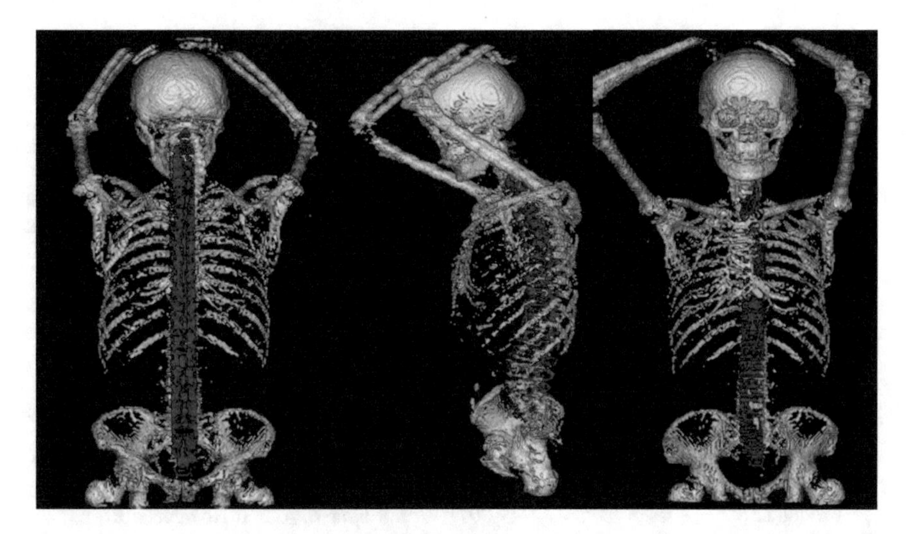

Fig. 11 The experimental result (Intramedullary tumors)

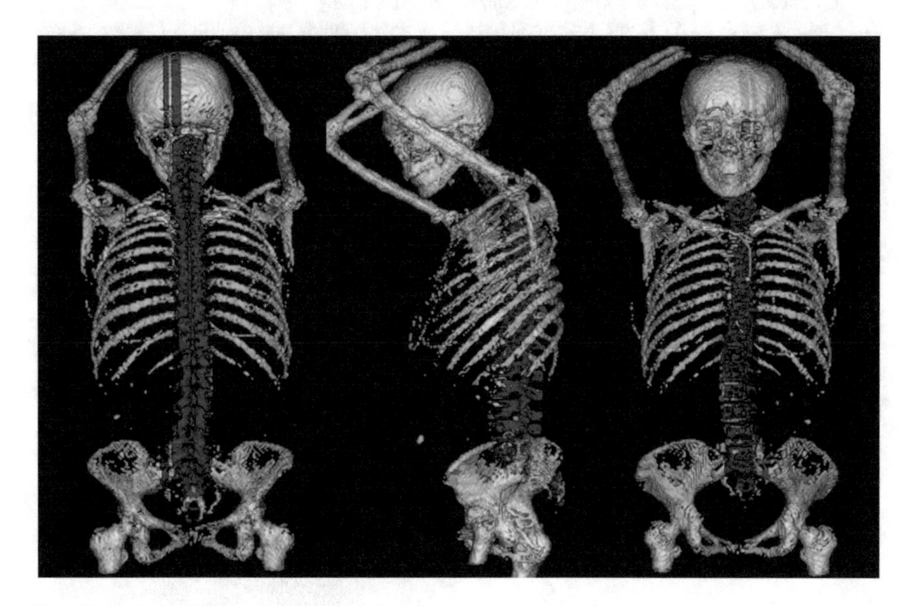

Fig. 12 The experimental result (Scoliosis)

References

1. Jaesung Lee, Anthony P. Reeves.: Segmentation of Individual Ribs from Low-dose Chest CT. In: Proceedings of SPIE—The International Society for Optical Engineering (Impact Factor: 0.2). 03/2010; DOI:10.1117/12.844565.
2. Joes Staal, Bram van Ginneken, Max A. Viergever.: Automatic rib segmentation and labeling in computed tomography scans using a general framework for detection, recognition and segmentation of objects in volumetric data. In: Medical Image Analysis, Volume 11, Issue 1, February 2007, Pages 35–46.
3. André Mastmeyer., Klaus Engelke., Christina Fuchs., Willi A. Kalender.: A hierarchical 3D segmentation method and the definition of vertebral body coordinate systems for QCT of the lumbar spine. In: Medical Image Analysis, Volume 10, Issue 4, August 2006, Pages 560–577, Special Issue on Functional Imaging and Modelling of the Heart (FIMH 2005).
4. J. Canny.: A computational approach to edge detection. In: IEEE Transactions on Pattern Analysis and Machine Intelligence, 8(6):679–698, 1986.
5. Romen Singh, T.; Roy, Sudipta; Imocha Singh, O.; Sinam, Tejmani; Manglem Singh, Kh.: A New Local Adaptive Thresholding Technique in Binarization. In: International Journal of Computer Science Issues (IJCSI); Nov 2011, Vol. 8 Issue 6, p. 271. November 2011.
6. Stokes IA.: Three-dimensional terminology of spinal deformity. A report presented to the Scoliosis Research Society by the Scoliosis Research Society Working Group on 3-D terminology of spinal deformity. In: Spine (Phila Pa 1976). 1994 Jan 15; 19(2):236–48.
7. Kang et al., 2003 Y. Kang, K. Engelke, et al. A new accurate and precise 3-D segmentation method for skeletal structures in volumetric CT data. IEEE Trans. Med. Imag., 22 (5) (2003), pp. 586–598.
8. Steiger et al., 1990 P. Steiger, J.E. Block, et al. Spinal bone mineral density measured with quantitative CT: effect of region of interest, vertebral level, and technique Radiology, 175 (2) (1990), pp. 537–543.
9. W. A. Kalender, D. Felsenberg, H. K. Genant, M. Fischer, J. Dequeker, and J. Reeve, "The European Spine Phantom—a tool for standardization and quality control in spinal bone mineral measurements by DXA and QCT", Eur. J. Radiol., vol. 20, pp. 83–92 1995.
10. W. A. Kalender, E. Klotz, and C. Suess, "Vertebral bone mineral analysis: An integrated approach with CT", Radiology, vol. 164, pp. 419–423 1987.
11. Steiger P., Block JE., Steiger S., Heuck AF., Friedlander A., Ettinger B., Harris ST., Glüer CC., Genant HK.,: Spinal bone mineral density measured with quantitative CT: effect of region of interest, vertebral level, and technique Radiology, 175 (2) (1990), pp. 537–543.

The Improved Gaussian Mixture Model for Real-Time Detection System of Moving Object

Zhiwei Tang, Yunfei Cheng and Longhu Chen

Abstract Detection of moving objects is a kind of segmentation techniques based on regional characteristic such as color, gray, texture, which is the key technology in analyzing and processing of video image. A real-time motion detection method based on improved Gaussian Mixture Model is presented in this paper which is optimized and structure adjusted from Gaussian Mixture Model. Gaussian Mixture Model has been widely used in complex background scene modeling, especially in some occasions with small repetitive motion, such as shaking of the leaves, a rotating fan, bushes, the sea waves, rain, snow, etc.

Keywords Gaussian mixture model · Real-time · Motion detection

1 Introduction

In the ideal background model, the static background and moving targets are not as background, but what kind of specific extraction targets will be decided by the application of the scene. To distinguish between the moving target and background, we should find the deviation of the moving target and the background of some inherent properties. The key problem of the research is how to model the natural attributes of the scene, and how to define the deviation, which is the problem to be solved in the background modeling and moving object extraction.

Z. Tang · Y. Cheng (✉) · L. Chen
The Third Research Institute of Ministry of Public Security, Shanghai, China
e-mail: 99chaoyang@163.com

© Springer Nature Singapore Pte Ltd. 2018

N.Y. Yen and J.C. Hung (eds.), *Frontier Computing*, Lecture Notes in Electrical Engineering 422, DOI 10.1007/978-981-10-3187-8_15

2 The Principle and Implementation of Improved Gaussian Mixture Model

In the video sequences, gray value of each pixel is in line with the gauss distribution. When the target area is difference with background area in gray value, its gray histogram will show as double peak or valley, also one peak corresponds to one target, and another peak corresponds to the background, so the complex image needs to use Gaussian Mixture distribution for its double peak. It can increase the number of gauss model to approximate any complex image, but traditional Gaussian Mixture Model enjoys a large calculation, and the embedded CPU is difficult to achieve real-time detection, so it needs to be simplified.

Assume that the number of Gauss distribution for each pixel is K, and the k Gauss distribution is $\eta_k(\mu_k, \Sigma_k)$, so the k Gauss distribution can be expressed as the formula (1).

$$f_k(\mathbf{x}) = \frac{1}{\sqrt{(2\pi)^2|\Sigma_k|}} \exp\left[-\frac{1}{2}(\mathbf{x}-\boldsymbol{\mu}_k)^T \Sigma_k^{-1}(\mathbf{x}-\boldsymbol{\mu})\right] \tag{1}$$

Where $k = 0, 1, \ldots, K$, x is the pixel value of the current pixel, μ_k is the mean value of Gauss distribution, Σ_k is variance of Gauss distribution, and the weights of K Gauss distribution are $\omega_1, \omega_2, \ldots, \omega_K$ respectively. When reading an image, gauss distribution is set with initial variance Σ_0 and mean value x_t corresponding each pixel value, and assigns initial weight ω_0 [1–9].

When reading a new frame, it needs to follow the steps below to update the gauss distribution.

(1) Computing the Mahalanobis distance

The current frame of each pixel and the k Gauss distribution of the Mahalanobis distance as shown in formula (2).

$$L_k = \sqrt{(x-\mu_k)^T \sum_k^{-1} (x-\mu_k)} \tag{2}$$

For the color image, it has three channels of RGB, so the formula is extended to type (3).

$$L_k = \sqrt{(x-\mu_k)^T \sum_k^{-1} (x-\mu_k)} = (r-\mu_r, g-\mu_g, b-\mu_b)\begin{bmatrix} 1/\sigma_r^2 & 0 & 0 \\ 0 & 1/\sigma_g^2 & 0 \\ 0 & 0 & 1/\sigma_b^2 \end{bmatrix}\begin{bmatrix} r-\mu_r \\ g-\mu_g \\ b-\mu_b \end{bmatrix}$$
$$= \frac{(r-\mu_r)^2}{\sigma_r^2} + \frac{(r-\mu_g)^2}{\sigma_g^2} + \frac{(r-\mu_b)^2}{\sigma_b^2} \tag{3}$$

(2) Background matching

If the Mahalanobis distance calculated is greater than the threshold value of θ_L, the current pixel is determined as the foreground, and judge the effective number of Gauss distribution. If it lacks the number of K, it can add an new Gauss distribution. If the effective Gauss distribution reaches to the number of K, it should remove the smallest-weight Gauss distribution, and add a new Gauss distribution. The new Gauss distribution built up with x_t as mean, and Σ_0 as the initial variance, and ω_0 as initial weight.

The rest of the Gauss distribution weights in current pixel will be updated according to type (4) [10–17].

$$\omega_k = (1 - \alpha)\omega_k \tag{4}$$

Then K weights of Gauss distributions will be normalized, and ranked in descending order.

If the Mahalanobis distance calculated is not greater than the threshold value θ_L, the current Gauss distribution will be marked as k_{min}, and all Gauss distributions in current pixel will be ranked in descending order according to the weight value ω_k. And it will add all weight value of Gauss distribution in current pixel from first to last, until the weight value added is greater than the threshold value of θ_ω. If the Gauss distribution corresponding to k_{min} belongs to the distribution added, the current pixel belongs to background, otherwise the current pixel belongs to foreground.

The weights of existing Gauss distribution will be updated according to formula (5).

$$\omega_k = \begin{cases} (1-\alpha)\omega_k + \alpha, & k = k_{\min} \\ (1-\alpha)\omega_k, & k \neq k_{\min} \end{cases} \tag{5}$$

The mean value and variance of Gauss distribution marked as k_{min} will be updated according to formula (6) and (7).

$$\mu_k = (1-\rho)\mu_k + \rho x \tag{6}$$

$$\Sigma_k = (1-\rho)\Sigma_k + \rho(x-\mu)^{\mathrm{T}}(x-\mu) \tag{7}$$

3 The Optimization System

In this paper, the mixed Gaussian background modeling algorithm was partly improved. When in a pixel background model and a Gaussian distribution with each beat new entrants to the pixel value matching frequency is higher than the threshold, within the next 100–200 frame of video image, the pixel background model of each Gaussian distribution parameters are unchanged, no longer updated. After 100–200 frame, re study the Gauss distribution parameters in the relatively equal conditions, until the Gauss distribution and the new into the pixel value matching frequency is greater than the set threshold, so the cycle repeated. This method can be used to improve the speed of the algorithm, not only in the static region of the monitoring scene, but also in the scene area such as the leaf shaking, the water surface wave, the display flicker, etc. In the repeated movements in the region, the pixel has been recurring for several values, by acquiring new pixel value continuous training Gaussian mixture model, there must be two or three Gaussian functions alternating with newly acquired pixel value matching and the weight of the Gaussian function will larger and less, has maintained a larger, and at the same time several Gaussian functions between the weight difference between the two. When the Gaussian weighting function is greater than a certain threshold, in the back of a period of time that the Gaussian distribution of the mean representing the background, in this period of time no longer to update parameters, after a period of time and then transferred to the Gaussian distribution of the weights is relatively equal status to learn, until the discovery of new Gaussian distribution of the weight sum is greater than the threshold, so the cycle repeated.

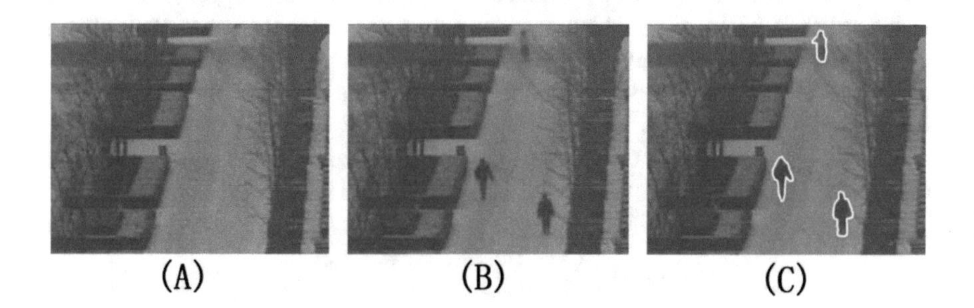

Moving target extraction effect

4 Result

In the surveillance video, if a body position change, or placing new stationary objects in the scene, if not timely update the background of the scene, the shadow will appear. In the mixed Gauss model, usually by adjusting the parameter learning rate, real-time updating background model. Higher learning rate, the background of the establishment of model can quickly adapt to the stationary targets within the scene changes, or changes in light, but may also appear will require moving target learning to become part of the background scene. Therefore, the difficulty is to weigh the various application conditions, and select the most appropriate. May need to use some other information to adjust, such as when the number of foreground pixels, and each frame image of the ratio of the total number of pixels greater than a certain threshold (65%), improve the learning rate, accelerate the speed of updating background model, or reduce the learning rate.

The aiming at the problem of the computation of Gaussian mixture background model, real time poor, through the analysis of the existing Gaussian mixture algorithm the parameter update mechanism defect proposed a fast Gaussian mixture improved algorithm can emerge from the chaos degree depending on the value of each pixel, to take different Gaussian function parameter update mechanism. A large number of experimental results show that the proposed algorithm can improve the processing speed of the hybrid Gauss model without affecting the background modeling and moving object extraction.

Acknowledgements This work was financially supported by National Science and Technology Major Project (No. 2013ZX010033002-003); the Technology Research Program of Ministry of Public Security (No. 2015JSYJC21). And this work has been partially sponsored by the Technology Research Program of Ministry of Public Security of the People's Republic of China (2014QZX005).

References

1. AN Bo-wen, AI Yan. Moving Object Detection And Tracking For Real-Time Video[J]. Computer Simulation. 2012; 29(2): 249–252.
2. Durte Duque, Henrique Santos,ect. Moving Object Detection Unaffected by Cast Shadows. Highlights and Ghosts. In Proc. IEEE Int. Conf. Image Processing. 2005; 413–416.
3. Cui Ying-ying. The Research on vehicle type recognition in intelligent transportation system [D]. University of Electronic Science and Technology of China. 2013 (in Chinese).
4. Zhang Yan. The Method Research and Completion of Identifying the Types of the Military Vehicle[D]. DUT.2007 (in Chinese).
5. Jolly M P D, Lakshmanan S, Jain, A K. Vehicle Segmentation and Classification Using Deformable Templates. IEEE Transaction on Pattern Analysis and Intelligence, 1996, 18(3): 293–308.
6. Huo Wei. The Research on the automatic classification of complex background models [J]. Journal of Qingdao Technological University. 2008, 01:107–110 + 115 (in Chinese).
7. Gorur. P. Speeded up Gaussian Mixture Model algorithm for background subtraction. 2011 8th IEEE International Conference on Advanced Video and Signal-Based Surveillance (AVSS), 2011, 39(51): 386–391.
8. Yong Rui, Huang T.S, Mehrotra S. Exploring video structure beyond the shots, Multimedia Computing and Systems, 2010, pp. 237–240.
9. Albert Ahumada, Maria Chatzigiorgaki. A visual detection model for DCT coefficient quantization [J]. 9th Computing in Aerospace Conference, 2008. 22(8): 809–830.
10. X. Luo, Zheng Xu, J. Yu, and X. Chen. Building Association Link Network for Semantic Link on Web Resources. IEEE transactions on automation science and engineering, 2011, 8 (3), 482–494.
11. C. Hu, Zheng Xu, et al. Semantic Link Network based Model for Organizing Multimedia Big Data. IEEE Transactions on Emerging Topics in Computing, 2014, 2(3), 376–387.
12. Zheng Xu et al. Semantic based representing and organizing surveillance big data using video structural description technology. The Journal of Systems and Software, 2015, 102, 217–225.

13. Zheng Xu et al. Knowle: a Semantic Link Network based System for Organizing Large Scale Online News Events. Future Generation Computer Systems, 2015, 43–44, 40–50.
14. Zheng Xu et al. Semantic Enhanced Cloud Environment for Surveillance Data Management using Video Structural Description. Computing, 98(1–2): 35–54, 2016.
15. C. Hu, Zheng Xu, et al. Video Structured Description Technology for the New Generation Video Surveillance System. Frontiers of Computer Science, 2015, 9(6): 980–989.
16. Zheng Xu et al. Crowd Sensing Based Semantic Annotation of Surveillance Videos, International Journal of Distributed Sensor Networks, Volume 2015 (2015), Article ID 679314, 9 pages.
17. Zheng Xu et al. Crowdsourcing based Description of Urban Emergency Events using Social Media Big Data. IEEE Transactions on Cloud Computing, 10.1109/TCC.2016.2517638.

Music Similarity Evaluation Based on Onsets

Shingchern D. You and Ro-Wei Chao

Abstract This paper describes a music similarity approach based on the time differences between two adjacent onsets. To better detect onsets, temporal and spectral detection methods are employed. Each set of detected features are individually matched by using the rough longest common subsequence (RLCS) algorithm. The final score is a weighted sum of individual scores from each detection method. The simulation results show that, on the average, 85% of the audiences agree that two musical soundtracks are similar if the computed score is greater than 0.3. When compared with an existing approach, it is easier for the proposed approach to set up a threshold to recommend highly similar soundtracks.

Keywords Music similarity · Onsets · Rough longest common subsequence

1 Introduction

With the advances of technology, more and more online soundtracks are available to music lovers. For many instances, a music lover may want to listen to more soundtracks similar to his/her favorite ones. To provide this type of service, techniques for music recommendation can be applied. Currently, there are two approaches to provide the recommendation list. The first one is based on the preference of other users, whereas the second one is based on the temporal and/or spectral similarity of the soundtracks. The first approach is easy to implement. For example, if two soundtracks A and B are frequently downloaded or listened by many users together, then we may assume that A and B are similar. Therefore, if the user requests to recommend soundtracks similar to A, then soundtrack B will be recommended. Though effective, this approach, nevertheless, does not truly recommend "similar" soundtracks to the query soundtrack. Furthermore, this approach

---

S.D. You (✉) · R.-W. Chao
Department of Computer Science and Information Engineering,
National Taipei University of Technology, Taipei, Taiwan
e-mail: you@csie.ntut.edu.tw

© Springer Nature Singapore Pte Ltd. 2018

N.Y. Yen and J.C. Hung (eds.), *Frontier Computing*, Lecture Notes
in Electrical Engineering 422, DOI 10.1007/978-981-10-3187-8_16

almost always recommends most popular soundtracks at the time of query, and ignores any really similar ones with only a few downloads (or browse). Finally, this kind of approach requires Internet connection, which is inconvenient for some situations.

As the second approach assesses the temporal and/or spectral similarity between two soundtracks, the similarity is truly based on the contents of the soundtracks without referring to other users' preferences. For this type of approach, we could either provide a set of musical works to train the similarity evaluation system. Or, we could alternatively use a pre-defended metric to measure the similarity between two soundtracks. In this paper, we only consider approaches without prior training for its ease to use.

According to Wikipedia [1], there are many different criteria to assess whether two pieces of music are similar, such as based on pitched similarity, non-pitched similarity, and semiotic similarity. However, in actual implementation, approaches based on timbral and/or rhythmic pattern similarity are more popular because these approaches match the perceptual intuition of human beings.

E. Pampalk [2] proposed a similarity evaluation system based on MFCC (Mel-frequency cepstrum coefficients) and other features. The overall similarity score is a weighted sum of the feature distances. A release of his program is available in [3]. In this paper, we will compare the simulation results of our approach with Pampalk's approach.

The Austrian Research Institute for Artificial Intelligence (OFAI) has released another music similarity system in its official webpage [4]. This system uses both features from timbre and rhythmic patterns to evaluate the similarity of two soundtracks [5]. The comprehensive version of the program is subject to a license fee, whereas the basic version is open to public [6].

Other than these two systems, there are still other researchers conducting research in this area. One of the well-known competition on music similarity is MIREX (Music Information Retrieval Evaluation eXchange) [7], which attracts many teams to compete each year.

So far, most available music similarity approaches measure the similarity based on spectral and/or rhythmic similarity. The rhythm mentioned here actually means the regularity of temporal repetition of strong energy. Although rhythm is an important factor for similarity measure, it, nevertheless, is insufficient in some situations. In this paper, we use the relative time differences between onsets as features to measure the similarity between two soundtracks to increase the discrimination capability of temporal similarity. The purpose of this paper aims to provide an alternative similarity method other than existing ones. With more variations of similarity evaluation methods, hopefully the user can choose, among the approaches, a particular one to better serve his/her needs in the future.

2 Proposed Approach

According to Wikipedia [8], "*onset refers to the beginning of a musical note or other sound, in which the amplitude rises from zero to an initial peak.*" Currently there are many different approaches to detect onsets, including temporal and spectral approaches [9]. In this paper, both approaches are employed in onset detection.

As shown in Fig. 1, the proposed approach combines four different onset detection methods to find the similarity score. The first method is based on the variation of energy in time domain, denoted as temporal detection in Fig. 1. Onsets detected based on temporal approach provides acceptable accuracy if the audio signal contains strong energy variation. On the other hand, temporal detection is not accurate enough if relatively smooth (or slow) musical waveforms are encountered. For this type of signal, it is better to use spectral-based methods than temporal ones. To this end, we introduce three spectral-based methods in the proposed model, denoted in Fig. 1 as HFC (high-frequency contents) detection, spectral difference detection, and up-count detection. Once the onsets from a particular method are detected, the time difference between two adjacent onsets becomes one feature. The collected features are to be matched with the features in the database by using the

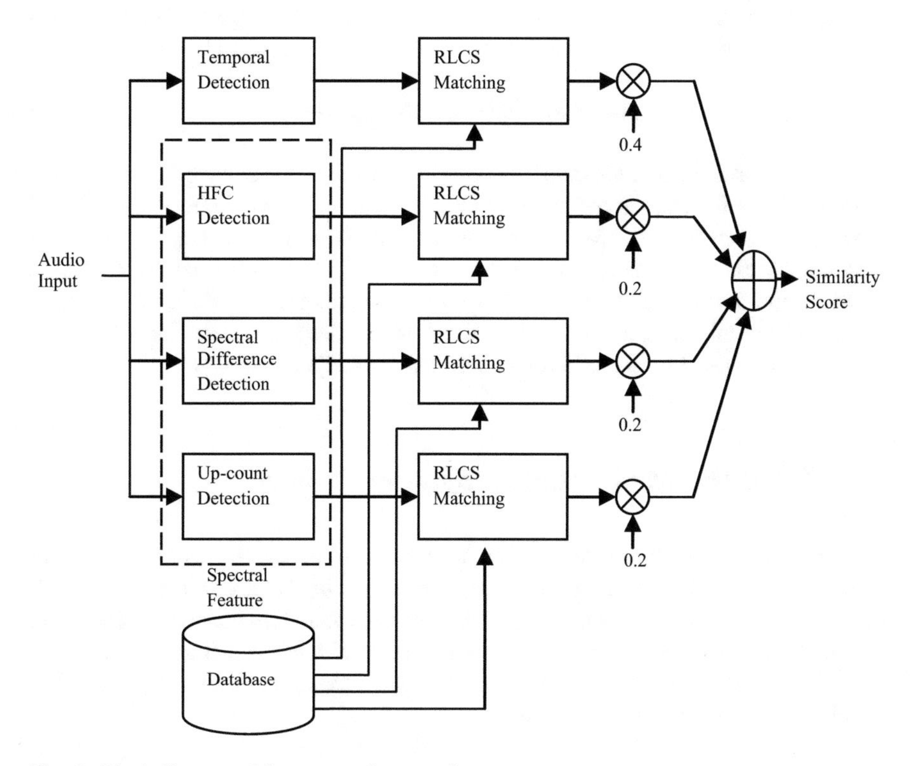

Fig. 1 Block diagram of the proposed approach

RLCS (rough longest common subsequence) algorithm [10]. The final score is a weighted sum of individual scores. The following briefly describes these methods.

2.1 Temporal Onset Detection

The processing flow of the temporal onset detection is given in Fig. 2. The incoming audio samples have a sample rate of 11,250 s/s. When the audio samples pass through a four-band filterbank, four sets of subband samples are obtained. The frequency bands of the filterbank are 0–630 Hz, 630–1720 Hz, 1720–4400 Hz, and higher than 4400 Hz. Let the obtained subband samples be denoted as $x_p(n)$, $1 \leq p \leq 4$. The subband samples are divided into frames of 512 samples. The energy of each frame is computed as follows:

$$E_p(n) = \frac{1}{512} \sum_{m=-256}^{255} |x_p(256n+m)|^2 w(m) \quad (1)$$

where $w(m)$ is the Hamming window. This step is denoted as computing band energy in Fig. 2. As the Hamming window is used, overlapping of 50% samples between successive frames are carried out. The obtained energy E_p is then undergone a first-order difference after taking the logarithm value [11]

$$A_p(n) = \log(E_p(n)) - \log(E_p(n-1)). \quad (2)$$

If $A_p(n_0)$ is a local maximum value, it is an onset candidate. To reduce the number of candidates, we remove any local maximum whose value is less 0.01 of the average amplitude, denoted as min peak threshold in Fig. 2. If $A_p(n_0)$ is a local maximum within 100 ms centered around n_0, then n_0 is a candidate position for an onset. In the decision-making step, if $A_p(n_0)$ is a candidate for all four bands, then $A_p(n_0)$ is determined as an onset [12]. The features used in the similarity comparison are based on the time difference between two adjacent onsets.

2.2 Onset Detection Based on Spectral Domain

This subsection describes the computational steps of the spectral-based onset detection blocks. As shown in Fig. 3, the pre-processing step for all spectral-based methods is to divide the incoming audio samples into frames, with each frame containing 512 samples. Samples in a frame are multiplied by a Hamming window with 50% overlapping. The windowed samples are transformed to spectral domain by FFT (fast Fourier transformation).

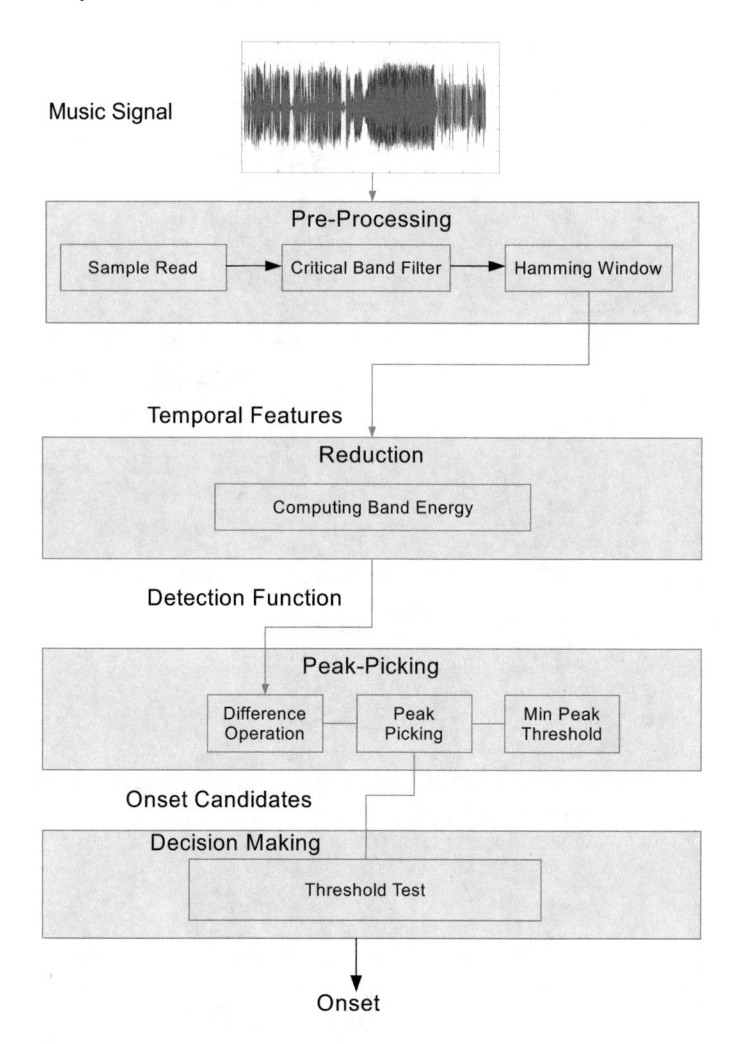

Fig. 2 Block diagram of the temporal onset detection

The HFC (high-frequency component) detection method [13] assumes that the variation of high-frequency energy is strongly correlated with onsets. Specifically, assume that (after FFT) the obtained spectral coefficient for frame n is denoted as $X_n(k)$. Then, the frequency-weighted energy is computed as

$$E_{HFC2}(n) = \sum_{k=1}^{256} k^2 \cdot |X_n(k)| \tag{3}$$

where k is the spectral index. The energy difference is then calculated as

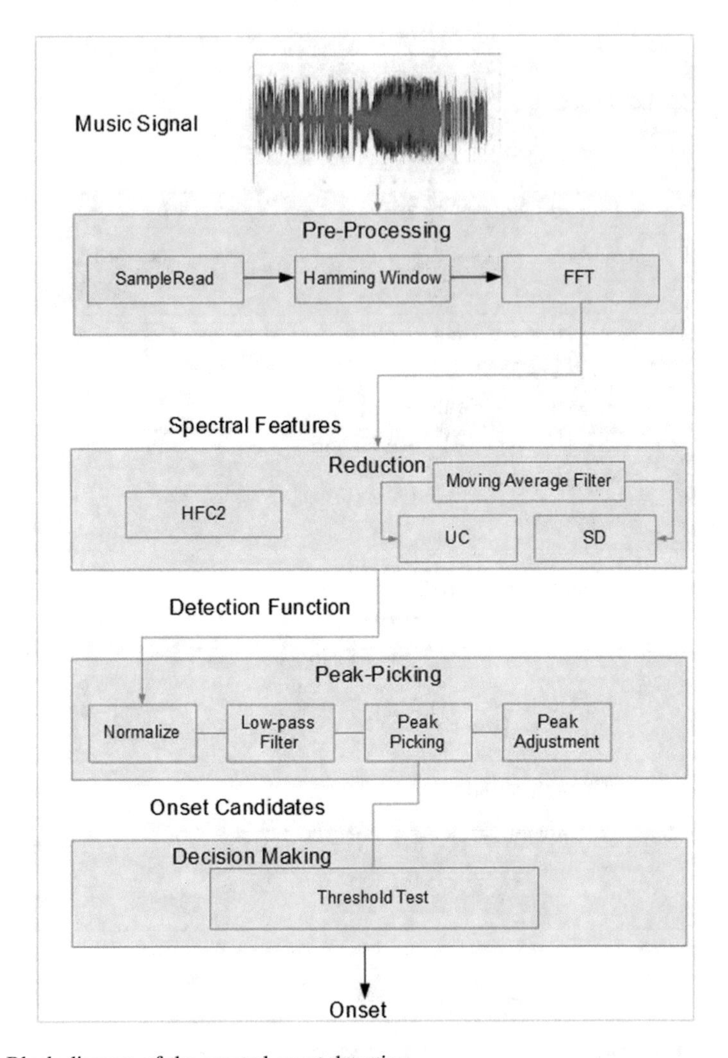

Fig. 3 Block diagram of the spectral onset detection

$$A_{HFC2}(n) = E_{HFC2}(n) - E_{HFC2}(n-1) \tag{4}$$

We will use $A_{HFC2}(n)$ in the decision-making step to determine onset locations.

The spectral difference detection method considers the spectral difference for each spectral index k [14]. To this end, this method computes $A_{SF}(n)$ by

$$A_{SF}(n) = \sum_{k=1}^{256} H\big(|X_n(k)| - \hat{X}_n(k)\big)^2 \tag{5}$$

where

$$\hat{X}_n(k) = \frac{1}{10}\sum_{m=1}^{10} |X_{n-m}(k)| \tag{6}$$

is a simple moving average of the past spectral coefficients to reduce the influence of noise, and

$$H(x) = (x + |x|)/2 \tag{7}$$

returns 0 for any non-positive argument x. Again, $A_{SF}(n)$ is to be used in the decision-making steps.

The up-count detection method is a modified version of the spectral difference method. As the former one is sensitive to noise, a possible modification is to count only the number of spectral lines with increasing energy, and ignores the actual (positive) value. Therefore, we use $A_{UC}(n)$ as the basis to determine the location of an onset:

$$A_{UC}(n) = \sum_{k=1}^{256} G\left(|X_n(k)| - |\hat{X}_n(k)|\right) \tag{8}$$

where

$$G(x) = \begin{cases} 1, & x > 0 \\ 0, & \text{otherwise} \end{cases} \tag{9}$$

Once we obtain $A_X(n)$ (x is either *HFC*2, *SF*, or *UC*), we use a moving average filter to reduce the fluctuation to obtain $\bar{A}_X(n)$. A onset candidate point is a location n_0 with $\bar{A}_X(n_0)$ is a local maximum in the vicinity of 100 ms. An onset is determined as a local maximum with its value exceeding a pre-defined threshold. Finally, the time difference between two adjacent onsets is a feature to be compared by the matching algorithm.

2.3 RLCS Algorithm

In addition to (time difference) features, we also need a matching algorithm to evaluate how similar two sequences of features are. For this purpose, we adopt a string-matching algorithm. Some well-known matching algorithms include dynamic warping, edit distance, and longest common subsequence. In this paper, we use the extension version of longest common subsequence algorithm, called rough longest common subsequence (RLCS) algorithm. Previously, we have used the RLCS algorithm for copy detection of music [15] with satisfactory results, and therefore we again use this algorithm for the proposed approach. For the sake of completeness and clear explanation, we outline the RLCS algorithm below.

Assume that there are two sequences (of strings) given as $A_i = <a_1, \ldots, a_i>, 1 \leq i \leq M$ and $B_j = <b_1, \ldots, b_j>$, $1 \leq j \leq N$ with A_0 及 B_0 as empty sequences. The longest common subsequence can be computed as

$$L_{CS}(A_i, B_j) = \begin{cases} 0, & i \cdot j = 0 \\ L_{CS}(A_{i-1}, B_{j-1}) + \delta, & i \cdot j > 0, a_i \approx b_j \\ \max(L_{CS}(A_i, B_{j-1}), L_{CS}(A_i - 1, B_j)), & i \cdot j > 0, a_i ! \approx b_j \end{cases} \quad (10)$$

where "≈" means $|a_i - b_j| \leq T_d$, "!≈" means $|a_i - b_j| > T_d$, and $\delta = 1 - \frac{|a_i - b_j|}{T_d}$. In the experiment, we use $T_d = 3$. We then compute width across reference (WAR) W_R and width across query (WAQ) W_Q functions as follows:

$$W_R(A_i, B_j) = \begin{cases} 0, & i \cdot j = 0 \\ W_R(A_{i-1}, B_{j-1}) + 1, & i \cdot j > 0, a_i \approx b_j \\ W_R(A_{i-1}, B_j) + 1, & i \cdot j > 0, a_i ! \approx b_j, L_{CS}(A_{i-1}, B_j) \geq L_{CS}(A_i, B_{j-1}), W_R(A_i, B_{j-1}) > 0 \\ 0, & i \cdot j > 0, a_i ! \approx b_j, L_{CS}(A_{i-1}, B_j) \geq L_{CS}(A_i, B_{j-1}), W_R(A_i, B_{j-1}) = 0 \\ W_R(A_i, B_{j+1}), & i \cdot j > 0, a_i ! \approx b_j, L_{CS}(A_{i-1}, B_j) < L_{CS}(A_i, B_{j-1}) \end{cases} \quad (11)$$

and

$$W_Q(A_i, B_j) = \begin{cases} 0, & i \cdot j = 0 \\ W_Q(A_{i-1}, B_{j-1}) + 1, & i \cdot j > 0, a_i \approx b_j \\ W_Q(A_{i-1}, B_j), & i \cdot j > 0, a_i ! \approx b_j, L_{CS}(A_{i-1}, B_j) \geq L_{CS}(A_i, B_{j-1}) \\ W_Q(A_i, B_{j-1}) + 1, & i \cdot j > 0, a_i ! \approx b_j, L_{CS}(A_{i-1}, B_j) < L_{CS}(A_i, B_{j-1}), W_Q(A_i, B_{j-1}) > 0 \\ 0, & i \cdot j > 0, a_i ! \approx b_j, L_{CS}(A_{i-1}, B_j) < L_{CS}(A_i, B_{j-1}), W_Q(A_i, B_{j-1}) = 0 \end{cases} \quad (12)$$

The similarity is given as

$$S_{RLCS}(A, B) = \max_{i,j} score(i,j) \quad (13)$$

where

$$score(i,j) = \begin{cases} \frac{L_{CS}(A_i, B_j)}{N} \cdot \left(\frac{\alpha \cdot L_{CS}(A_i, B_j)}{W_R(A_i, B_j)} + \frac{(1-\alpha) \cdot L_{CS}(A_i, B_j)}{W_Q(A_i, B_j)} \right), L_{CS}(A_i, B_j) \geq \lambda \cdot N \\ 0, \text{otherwise} \end{cases} \quad (14)$$

In the experiment, λ is (1/*N*) and α is 0.5. We know from (14) that the value of $S_{RLCS}(i, j)$ is between 0 and 1 and 1, means perfectly matched.

3 Experiments and Results

To perform the experiments, we collect 38 soundtracks of classic music and 94 soundtracks of pop music from various albums. The duration of each soundtrack is 30 s. The original sample rate for each soundtrack is 44,100 s/s. However, the sample rate is reduced to 11,025 s/s before conducting the experiments.

When the user input a particular soundtrack, the proposed system computes the features for the input soundtrack. The computed features are then compared with the features in the database through the weighted sum S_{WRLCS} of four S_{RLCS} scores.

To understand the correlation between the computed S_{WRLCS} and the perceptual impression of a human listener, we conduct a listening test. In the test, five (5) soundtracks are selected as the input to the system. The computed S_{WRLCS} are divided into four categories: $0 < S_{WRLCS} \leq 0.1$, $0.1 < S_{WRLCS} \leq 0.2$, $0.2 < S_{WRLCS} \leq 0.3$, and $S_{WRLCS} > 0.3$. The soundtrack corresponding to the greatest and smallest scores in each category is selected. Thus, totally eight soundtracks are picked for each testing input. Ten (10) audiences are asked to give opinions regarding whether the testing soundtrack is similar to one of the eight picked soundtracks (individually compared). The experimental results are given in Table 1. It can be observed that if the S_{WRLCS} score is greater than 0.3, on the average, 85% of the audiences feel that both soundtracks are perceptually similar. Therefore, we can use this value as a threshold to recommend soundtracks to a user.

To further investigate the performance of the proposed approach, we compare ours with the approach proposed by Pampalk [3]. We use the same testing soundtracks for listening tests as the input to both systems. The scores for both systems are given in Table 2. For the proposed system, the range of the score is between 0 and 1, and 1 means highest similarity. On the other hand, the scores of

Table 1 Subjective similarity versus S_{WRLCS}

S_{WRLCS} score	Similar percentage (%)	Dissimilar percentage (%)
$S_{WRLCS} > 0.3$	85	15
$0.2 < S_{WRLCS} \leq 0.3$	40	60
$0.1 < S_{WRLCS} \leq 0.2$	0	100
$0 < S_{WRLCS} \leq 0.1$	0	100

Table 2 Similarity scores of both approaches

Query index	Score (proposed)		Score (Pampalk)	
	First	Fifth	First	Fifth
1	0.3552	0.3019	9.2669	9.3085
2	0.3691	0.3111	9.3068	9.3922
3	0.5376	0.2644	9.0575	9.4085
4	0.3691	0.3084	9.3586	9.44
5	0.3102	0.2701	9.2172	9.4061

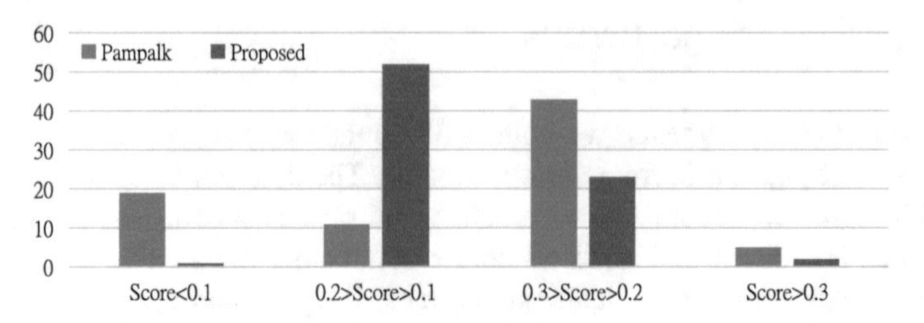

Fig. 4 Cross comparison between the proposed system and the Pampalk system

the Pampalk approach ranges between 9 and 13, with 9 as the highest similarity. It can be seen that the proposed system has larger (normalized to the full range of 1) score differences between the first (best) match and the fifth match, whereas scores obtained by the Pampalk approach have relatively smaller differences (normalized to the full range of 4). Conceptually, a larger difference (wider distribution) means that it is easier to set a threshold to recommend truly similar soundtracks. In this regard, the proposed system is a better choice.

When cross-comparing the number of soundtracks in the four categories mentioned previously, the results becomes apparent. As shown in Fig. 4, the proposed system has many more soundtracks with scores less than 0.2 and fewer soundtracks with scores of greater than 0.3. As a score (in the proposed system) less than 0.2 means that both soundtracks are not similar at all, the proposed system can better discriminate dissimilar soundtracks than the Pampalk approach.

4 Conclusions

This paper describes an approach for music similarity evaluation based on the detected onsets. When combining scores computed from individual onset features with the RLCS algorithm, the proposed approach is able to provide a final, weighted score for two soundtracks. The listening tests confirm that if two soundtracks have a similarity score of 0.3 or higher, these two soundtracks are perceptually similar according to the opinions of the listeners. When compared with an existing system, the proposed approach has a better score distribution to ease the determination of a threshold to recommend highly similar soundtrack titles among the titles in the database. Overall, the proposed approach is a possible choice for users to choose other than existing similarity evaluation methods.

References

1. https://en.wikipedia.org/wiki/Musical_similarity, retrieved Dec. 10, 2015
2. Pampalk, E.: Computational Models of Music Similarity and Their Application in Music Information Retrieval. Doctoral Thesis, Vienna University of Technology, Austria (2006)
3. http://www.pampalk.at/ma/, retrieved Dec. 10, 2015
4. The Austrian Research Institute for Artificial Intelligence, (OFAI), Music Similarity and Recommendation, available at http://www.ofai.at/research/impml/technology/musly.html
5. Pohle, T., et al.: On Rhythm and General Music Similarity. In: 10th International Society for Music Information Retrieval Conference, pp. 525–530 (2009)
6. Audio Music Similarity, available at http://www.musly.org/, retrieved Dec. 10, 2015
7. http://www.music-ir.org/mirex/wiki/MIREX_HOME, retrieved Dec. 10, 2015
8. https://en.wikipedia.org/wiki/Onset_(audio), retrieved Feb. 25, 2016
9. Bello, J. P., et al: A Tutorial on Onset Detection in Music Signals. IEEE Trans. Speech and Audio Processing. 13, 1035–1047 (2005)
10. Lin, H.-J., Wu, H.-H., Wang, C.-W.: Music Matching Based on Rough Longest Common Subsequence. J. Info. Sci. Eng. 27, 95–110 (2011)
11. Klapuri, A.: Sound Onset Detection by Applying Psychoacoustic Knowledge. In: 1999 IEEE International Conference on. Acoustics, Speech, and Signal Processing, pp. 3089–3092. IEEE Press, New York (1999)
12. Ricard, J.: An Implementation of Multi-band Onset Detection. In: 1st Annu. Music Inf. Retrieval Evaluation eXchange (MIREX), pp. 1–4. (2005)
13. Masri, P. and Bateman, A.: Improved Modeling of Attack Transients in Music Analysis-resynthesis. In: 1996 International Computer Music Conference, pp. 100–103. (1996)
14. Duxbury, C., Sandler, M., and Davies, M. A Hybrid Approach to Musical Note Onset Detection. In: 2002 Digital Audio Effects Conf. (DAFX,'02), pp. 33–38. (2002)
15. You, S. D. and Pu, Y.-H.: Using Paired Distances of Signal Peaks in Stereo Channels as Fingerprints for Copy Identification. ACM Trans. Multimedia Comput. Commun. Appl. 12, 1–22 (Aug. 2015)

Eye Tracking as a Tool in Manga-Based Interactive E-book on Reading Comprehension in Japanese Learning

Chun-Chia Wang, Hong-Fa Ho, Guan-An Chen and Hui-Sheng Su

Abstract As indicated by some studies, the problem of prior knowledge often exists when exploring the outcome of reading comprehension in academic language learning. This pilot study aimed to employ eye tracking technology to explore how students with different levels of prior knowledge processed the content of manga-based interactive E-book while learning Japanese language. Students' visual behaviors were tracked and recorded when they read a Japanese conversation with the relationship between graphical manga and interactive textual annotations. According to the pretest scores, 6 university students were categorized into high and low prior knowledge (PK) groups. Using EyeNTNU-120 eye tracker to compare including Total Contact Time (TCT), Number of Fixations (NOF), and Number of Clicks on textual annotations of the two PK groups based on areas of interests (AOIs) was measured. After the eye tracking experiment, students received a posttest of reading comprehension. The results revealed that (1) the high PK students showed longer reading time in graphic AOIs than the low PK students, (2) the low PK students showed longer reading time in text AOIs than the high PK students, (3) the low PK students showed longer reading time in annotation AOIs than the high PK students, (4) the high and low PK students had no significant difference in the whole reading time, (5) the low PK students showed more NOF of texts that the high PK students, (6) the low PK students clicked many of annotations AOIs than the high PK students, and (7) the low PK students had a significant outcome of reading comprehension compared with pretest and posttest scores. This suggests that interactive E-book containing graphical manga attracted students' visual attention and improved students' outcome of reading comprehension. Suggestions are made for future studies and instructional design for interactive E-book learning.

C.-C. Wang (✉)
Department of Information Management, Taipei City University of Science and Technology, Beitou, Taiwan
e-mail: toshihitowang@gmail.com

H.-F. Ho · G.-A. Chen · H.-S. Su
Department of Electrical Engineering, National Taiwan Normal University, Taipei, Taiwan

© Springer Nature Singapore Pte Ltd. 2018

N.Y. Yen and J.C. Hung (eds.), *Frontier Computing*, Lecture Notes in Electrical Engineering 422, DOI 10.1007/978-981-10-3187-8_17

Keywords Eye tracking · Prior knowledge · E-book · Reading comprehension

1 Introduction

According to the report of ministry of education in Taiwan, learning Japanese as a second foreign language (JFL) has gradually increased because not only young people love Japanese popular cultures, but also Taiwan is one of the top consumers of Japanese pop culture, likely due to factors such as geographical proximity and the two country's colonial history. It is worth mentioning, manga (i.e. *Japanese comic*) is form of these popular cultures that derive from Japan, especially among young adults. The hype and enthusiasm of manga is tremendous not just in Japan, but throughout the world [2]. Generally speaking, the adolescents' consumption of manga cannot be regarded as mere entertainment, the notion of manga can provide an emotional intimacy, a visual representation of conversation, and a believable social context for students' own identities as future workers [5, 17, 20]. In the past two decades, manga has begun to receive more scholarly attention from the standpoint of popular culture studies and literacy education [3, 13]. This is due to the fact that the graphic representation and ideologies contained in imported manga may have a more powerful cognitive effect on the group of youths than any formal educational process they undergo. For example, Khurana [9] considered manga as an effective tool for media literacy instruction. Ogawa [14] used educational manga in English language classrooms in a Japanese university to indicate learning and motivational benefits. After a postcourse survey, it revealed that positive overall responses from students with regard to both language and content learning. Adams [1] reported that high school students' reading skills are influenced and heightened due to reading manga.

With the rapid development of technology, college and university students own laptops, and an increasing number of them are purchasing tablets, smartphones, and other handheld devices [19]. Meanwhile, since the state-of-the-art technology development makes students' learning habits change, publishers are offering an increased number of textbooks in digital format, called Electronic-book (E-book), to affect student's adoption, along with the broad content of E-book, such as text, text-speech, music, sound, and animation [10]. Lin [11] investigated that the features of E-book enhance the motivation of students while reading foreign languages. Chou [6] attempted to find out Taiwanese undergraduate students' E-book reading attitudes in both first language (L1—Mandarin) and second language (L2—English) and explored factors that may play a role in explaining students' E-book reading attitude in L2. The results showed that the students demonstrated a slightly more positive e-book reading attitude in L2 than in L1 and indicated that if a reader has a positive reading attitude in an e-book environment when reading in his or her L1, this same attitude can be transferred to an L2 context. Yin et al. [21] analyzed on students' learning behaviors comprise an important thrust in education research and used the E-book system BookLooper to record students' learning behaviors in their daily academic life. The paper found that a number of learning behaviors have a significant relation with students' test scores and showed that the number of pages read correlated with students' scores. Shimada [18] proposed a method to analyze preview behaviors of students using a learning management system (LMS) and an E-book system. The paper collected a large number of operation logs from E-books to analyze the process of learning and reported that students who undertake good preview achieve better scores in quizzes.

In this pilot study, we developed a manga-based interactive E-book with the combination of the effectiveness and advantages of E-book and manga to construct a useful and practical tool for academic learning and promote learning motivation while learning Japanese. On the basis of the Dual-Coding Theory [15] and Cognitive Theory of Multimedia [12], this pilot study utilized an eye tracking as a tool to explore how the effect of the manga-based interactive E-book on students' visual behaviors while reading Japanese language. An eye-tracking technology has been widely applied to studies on human visual behaviors and adopted eye trackers to study the process of reading [16]. There are over ten different types of eye movements, of which the most important ones are fixations, saccades, and smooth pursuit. When the eyes stop to focus it is called a fixation and the movements between these fixations are called saccades. Besides, areas of interests (AOIs) and gaze-related variables are often used when recording visual behavior with eye tracking technology. Researchers often define AOIs according to research questions and main areas of concern before the experiment to observe the relationship between eye movement variables and the main areas of interest of the experiment. The AOI definitions and the visual information of the material are interdependent: the AOIs of various materials and the research questions are defined accordingly. Based on the defined AOIs, the pilot study adopted two common eye movement variables, Number of Fixations (NOF) and Total Contact Time (TCT), used in visual behavior observations on each AOI. The two variables are common models for processing eye movement data [4, 8]. In addition, the third variable, called Number of Clicks, was adopted to count the number of annotations they were clicked to present the meanings of words or phrases in the sentence.

2 Research Questions

This pilot study intends to examine the relevance between different prior knowledge and students' visual attention, the experimental analysis is conducted to observe students' visual patterns. In order to probe in-depth into how students learn Japanese language in the classroom with multimedia materials, we conducted a pilot study that examined students' visual attention in terms of their eye-movement patterns as they were given a conversation presentation with manga-based interactive E-book design in the classroom. Therefore, this study proposes three research questions as follows:

1. How would university students distribute their visual attention to a manga-based interactive E-book with annotation text–picture formats?
2. How would university students make use of annotations to realize what the words and phrases mean in manga-based interactive E-book?
3. How do university students with different prior knowledge backgrounds differ in their outcome of reading comprehension?

3 Methodology

3.1 Participants

A total of 6 university students, with Japanese competence background, were selected from a mid-level Japanese course in Taiwan. According to the pretest scores, the 6 university students were categorized into high and low prior knowledge (PK) groups. All participants passed the eye tracking calibrations. They all consented and were included in an eye tracking experiment with a Japanese conversation used in manga-based interactive E-book reading comprehension.

3.2 Stimuli

The reading stimuli material was prepared a manga-based interactive E-book presentation on the topic of "daily conversation" for the pilot study. The manga-based interactive E-book consisted of 8 slides showing various textual annotation–picture conversation. The content and the design of the interactive E-book presentation were constructed by a content expert who is an Associated Professor of the relevant area and a Japanese education researcher.

3.3 Apparatus

An EyeNTNU-120 eye tracker with a sampling rate of 120 Hz (sampling 120 times per second) was used to track each participant's eye movements while they read the material about the context of scenario. While collecting the movement data, a chin-rest was used in the experiment to reduce the occurrence of invalid or inaccurate data. The error rate of EyeNTNU-120s eye measurement is less than 0.3°, which is sufficient for this experiment. SPSS software were further utilized to store and analyze the eye movement data.

3.4 Data Collection

In order for the participants to be familiar with the software, the researcher gave them a short orientation and overview of the experiment. A paper-and-pencil pretest was used to measure each participant's PK about Japanese competence. All participants received the same pretest and wrote down their answers on paper. The participants were asked to rest their chin on the chin rest while the EyeNTNU-120 eye tracker camera was directed to their right eye. Participants have gone through a nine-point calibration process to ensure data accuracy. After passing an eye tracking calibration, the experiment started by letting the participants view the arranged stimuli with graphical and textual information shown on a computer screen. No time limit was set for the task. Each participant's eye movements were tracked and recorded by EyeNTNU-120 during the whole reading process. After finishing reading the stimuli, participants received a posttest immediately.

4 Results and Discussions

4.1 Results of Total Contact Time and Number of Fixations

Independent sample t tests were employed to examine whether there was any significant difference in the participants' viewing behaviors as follows: (1) total contact time (TCT) and (2) number of fixations (NOF) within the graphic AOIs and with the text AOIs between the higher and lower PK groups. If a significant result was found, an effect size of Cohen's d [7] was then further calculated. The results in Table 1 revealed that the high PK group has more *total contact time* on the graphic AOIs than the low PK group with a large effect size ($t = 2.86$, $p = 0.046$, $d = -2.34$) and the low PK group has more *total contact time* on the text and annotation AOIs than the high PK group with a large effect size ($t = -2.79$, $p = 0.049$, $d = 2.28$) and ($t = -3.72$, $p = 0.021$, $d = 3.03$), respectively. However, with respect to *total contact time* on the whole reading behaviors, no significant difference was found between the two PK groups. It showed that the two PK students paid the same attention to and put the same mental effort into reading the whole texts and graphics in the learning material. Meanwhile, Table 1 showed that the low PK group has more *number of fixations* on the text AOIs than the high PK group with a large effect size ($t = -3.97$, $p = 0.017$, $d = 3.24$). However, with respect to *number of fixations* on the graphic AOIs, no significant difference was found between the two PK groups.

Table 1 Eye tracking measures compared between the high and low groups

Eye tracking measure	PK				t	p	Cohen's d^a
	Higher (n = 3)		Lower (n = 3)				
	Mean	SD	Mean	SD			
TCT (Text)	149.33	47.48	594.67	272.71	−2.79	0.049*	2.28
TCT (Graphic)	2078.33	976.22	390.33	299.41	2.86	0.046*	−2.34
TCT (Annotation)	88.00	61.80	883.33	365.49	−3.72	0.021*	3.03
TCT	2315.67	951.85	1868.33	554.31	0.70	0.521	
NOF (Text)	42.33	31.56	188.33	55.37	−3.97	0.017*	3.24
NOF (Graphic)	45.00	19.28	61.00	42.88	−0.59	0.587	

$^*p < 0.05$, $^{**}p < 0.01$, $^{***}p < 0.001$
$^a|d| > 0.5$ shows a medium effect size, $|d| > 0.8$ shows a large effect size

Table 2 Independent sample *t*-test of number of clicks on annotations

	PK				t	p	Cohen's d^a
	Higher (n = 3)		Lower (n = 3)				
	Mean	SD	Mean	SD			
Number of Click (Annotation)	1.67	1.15	8.00	2.00	−4.75	0.009**	3.88

$^*p < 0.05$, $^{**}p < 0.01$, $^{***}p < 0.001$
$^a|d| > 0.5$ shows a medium effect size, $|d| > 0.8$ shows a large effect size

4.2 Results of Number of Clicks on Annotations

The Table 2 showed that the familiarity with annotations is significantly different between the two PK groups with a large effect size ($t = -4.75$, $p = 0.009$, $d = 3.88$). It means that the low PK group clicked annotations for the need to realize the meaning of words or phrases in Japanese conversation. Due to this pilot study, the *number of click* on annotations was considered as an indicator to promote the outcome of reading comprehension in the posttest.

4.3 Paired-Samples t-Test of Pretest and Posttest Scores

As shown in Table 3, through a paired-samples *t*-test, pretest had a significant difference from posttest for the low PK students. It means that the posttest score is higher than pretest for low PK students after reading manga-based interactive E-book. That is, a manga-based interactive E-book promoted a reading comprehension in this pilot study.

Table 3 Paired-samples *t*-test of pretest and posttest scores

	Pretest		Posttest			95% CI		
	Lower (n = 3)		Lower (n = 3)					
Variable	Mean	SD	Mean	SD	t(2)	p	LL	UL
Score	43.33	5.77	80.00	10.00	−11.00	0.008**	−51.01	−22.32

*$p < 0.05$, **$p < 0.01$, ***$p < 0.001$

5 Conclusions

This pilot study was aimed to employ eye tracking technology to examine how learners with different levels of expertise engaged in reading a Japanese conversation in manga-based interactive E-book that was provided as graphic and textual annotation formats. The findings of this pilot study were as follows. First, the high PK students spent more time observing and investigating the graphic information than the low PK students. Conversely, the low PK students spent more time observing and investigating the textual annotation information than the high PK students. Second, the low PK students showed more NOF of texts that the high PK students and clicked many of textual annotations AOIs than the high PK students. It means that the low PK students spent much time trying to realize the meanings of the annotations what the words and phrases meant. Third, there was a significant difference in score between the pretest and the posttest in the low PK students. That is, the low PK students resulted in better learning efficiency while reading a manga-based interactive E-book in this pilot study. Furthermore, as this was also a pilot study, the sample size might not be enough to find an evident difference in the analysis of eye movement data among students that were with different levels of expertise in Japanese. The future experiment is suggested to conduct a formal study with a bigger population to investigate the effects of manga-based interactive E-book on students' visual attention and learning performance.

Acknowledgements Funding of this research work was supported by Ministry of Science and Technology (MOST), Taiwan, under Grant number: MOST 104-2511-S-149-001. The authors also wish to thank the Aim for the Top University (ATU) project of National Taiwan Normal University (NTNU) on eye-tracking technology.

References

1. Adams, J.: A Critical Study of Comics. Jade 20.2, NSEAD, p.p. 133–143 (2001)
2. Black, R.W.: Access and Affiliation: The Literacy and Composition Practices of English-language Learners in an On-line Fanfiction Community. Journal of Adolescent & Adult Literacy. 49, 2, 118–128 (2005)
3. Bryce, M., Davis, J. & Barber, C.: The Cultural Biographies and Social Lives of Manga: Lessons from the Mangaverse. Retrieved from http://scan.net.au/scan/journal/display.php?journal_id=114

4. Calvo, M.G., Lang, P.J.: Gaze Patterns When Looking at Emotional Pictures: Motivationally Biased Attention. Motivation and Emotion, 28, 3, 221–243 (2004)
5. Cary, S.: Going graphic: Comics at work in the multilingual classroom. Portsmouth, NH: Heinemann (2004)
6. Chou, I.C.: Investigating EFL Students' E-book Reading Attitudes in First and Second Language. US-China Foreign Language, 12, 1, 64–74 (January 2014)
7. Cohen, J.: Statistical Power Analysis for the Behavioral Sciences (2nd ed.). New Jersey: Lawrence Erlbaum Associates (1988)
8. Hewig, J., Trippe, R.H., Hecht, H., Straube, T., Miltner, W.H.R.: Gender differences for specific body regions when looking at men and women. Journal of Nonverbal Behavior, 32, 2, 67–78 (2008)
9. Khurana, S.: So You Want to Be a Superhero? How the Art of Making Comics in an Afterschool Setting Can Develop Young People's Creativity, Literacy and Identity. Afterschool Matters. 4, 1–9 (2005)
10. Korat, O., Shamir, A.: Do Hebrew Electronic Books Differs from Dutch Electronic Books? A Replication of a Dutch Content Analysis. Journal of Computer Assisted Learning. 20, 4, 257–268 (2004)
11. Lin, I.Y.: The Effect of E-books on EFL Learners' Reading Attitude, (National Taiwan Normal University, Taiwan). Retrieved from http://ndltd.ncl.edu.tw/cgi-bin/gs32/gsweb.cgi?o=dnclcde&s=id=%22097NTNU5238009%22.&searchmode=basic
12. Mayer, R.E.: Multimedia Learning (2nd ed.). New York: Cambridge University Press (2001)
13. Moist, K.M., Bartholomew, M.: When Pigs Fly: Anime, Auteurism, and Miyazaki's Porco Rosso, London, Los Angeles, New Delhi & Singapore: SAGE Publications (2007)
14. Ogawa, E.: Educational Manga: An Effective Medium for English Learning. In N. Sonda & A. Krause (Eds.), JALT2012 Conference Proceedings. Tokyo: JALT (2003)
15. Paivio, A.: Mental Representations: A Dual Coding Approach. Oxford. England: Oxford University Press (1986)
16. Rayner, K.: Eye Movements in Reading and Information Processing: 20 Years of Research. Psychological Bulletin. 124, 3, 372–422 (1998)
17. Riley, P.: "Drawing the Threads Together," In D. Little, J. Ridley, & E. Ushioda (Eds.), Learner Autonomy in the Foreign Language Classroom: Teacher, Learner, Curriculum and Assessement. Dublin: Authentik, p.p. 237–252 (2003)
18. Shimada, A., Okubo, F., Yin, C.J., Oi, M., Kojima, K., Yamada, M., Ogata, H.: Analysis of Preview Behavior in E-book System. In Ogata, H. et al. (Eds.): the 23rd International Conference on Computers in Education (ICCE 2015), p.p. 593–600. Hangzhou, China, Nov. 30–Dec. 4 (2015)
19. Smith, S.D., Caruso, J.B.: The ECAR Study of Undergraduate Students and Information Technology, (Research study, vol. 6). Boulder, CO: EDUCAUSE Center for Applied Research. Retrieved from http://www.educause.edu/ers1006
20. Wolfe, P.: Brain Matters: Translating Research into Classroom Practice (2nd ed.) A, VA: Association for Supervision & Curriculum Development (2010)
21. Yin C.J., Okubo, F., Shimada, A., Oi, M., Hirokawa, S., Ogata, H.: Identifying and Analyzing the Learning Behaviors of Students using e-Books. In Ogata, H. et al. (Eds.): the 23rd International Conference on Computers in Education (ICCE 2015), p.p. 118–120. Hangzhou, China, Nov. 30–Dec. 4 (2015)

Based on Wavelet Analysis—Fuzzy Neural Network Real Time Traffic Flow Prediction

Tian Fu and Zhen Wang

Abstract This paper analyses the advantages and disadvantages of Prediction Model of wavelet analysis and fuzzy neural network model in real-time traffic prediction, proposes the traffic flow forecasting model based on wavelet analysis and fuzzy neural networks, uses the frequency components of the model of traffic flow time series for forecasting and finally induces the forecast results. Finally, it verifies the effectiveness of traffic simulation data validation model.

Keywords Wavelet analysis · Fuzzy neural network · Component · ANFIS · Forecast model

1 Overview

With China's rapid economic growth and urbanization, the urban traffic problems have been more serious. The ways of using the Intelligent Transportation Systems (ITS) and effectively solving the traffic congestion has become an important issue of sustainable urban development. So how to finish accurate real-time traffic flow forecasting is the key technology to realize ITS. As the traffic flow with highly non-linear characteristics, domestic and foreign researchers use artificial neural network for nonlinear prediction and achieved certain results [1, 2]. But because the artificial neural network is not unique, it can not meet the demand of the forecast of traffic flow.

Wavelet transform is a time-frequency analysis method. Wavelet analysis can make the stable and random separation of the traffic flow information according to the different frequency components of the data, then the characteristics to be processing and analysis, which can achieve the purpose to improve traffic flow

T. Fu · Z. Wang (✉)
Hainan College of Software Technology, Qionghai, Hainan, China
e-mail: 490345452@qq.com

T. Fu
e-mail: fu-tian@163.com

© Springer Nature Singapore Pte Ltd. 2018

N.Y. Yen and J.C. Hung (eds.), *Frontier Computing*, Lecture Notes in Electrical Engineering 422, DOI 10.1007/978-981-10-3187-8_18

prediction accuracy [3–5]. Adaptive neural fuzzy inference system (adaptive neuro-fuzzy inference system, ANFI), with artificial neural network self-learning function and fuzzy system inference and decision function, has been used in many fields [6–8]. In this paper, the wavelet analysis and adaptive neural network simulation system (ANFIS) are studied, and the combined model is applied to real-time traffic prediction. Because the combination has the characteristics of self adaptation and self adjustment, it can predict the traffic time series well.

2 Model Prediction Theory

First to obtain the traffic flow data with Mallat wavelet transform method for several resolution decomposition of different frequency signal decomposition, then the decomposition sequences by Mallat space reconstruction, dimension calculation, then use ANFIS model is trained, then ANFIS according to the decomposition sequence component prediction sample results. Finally, the prediction samples each component of the synthesis of the final forecasting results are obtained.

Step 1: Firstly, using wavelet transform Mallat algorithm (i.e. a sequence of discrete wavelet transform algorithm) on the traffic flow time series decomposition, the algorithm is as follows:

$$\begin{cases} c_{i+1} = Hc_i \\ d_{i+1} = Gc_i \end{cases}, \ i = 0,\ 1,\ \ldots,\ I \tag{1}$$

In the formula, H is the low frequency component operator and G is the high frequency component operator; c_i and d_i, respectively, as the original signal in the resolution of the approximation signal and the details of the 2^{-i} signal; and the I is the maximum decomposition level.

c_0 is defined as the original signal L, through the algorithm (1) can be decomposed into $x_1, x_2, x_3, \ldots, x_j$ and c_i, each layer of the details of the signal and the approximation signal is the adjacent frequency of the adjacent components of the original signal L.

Step 2: The decomposition sequence of the use of Mallat reconstruction algorithm for space reconstruction, the specific algorithm is as follows:

$$C_i = H^* C_{i+1} + G^* D_{i+1},\ \mathrm{i}{=}I,\ I-1,\ \ldots, 1,\ 0 \tag{2}$$

In the formula: H^* and G^* are the conjugate transpose matrix H and G.

The reconstruction algorithm (2) of the wavelet decomposition of signal reconstruction can increase the signal points, $d_1, d_2, d_3, \ldots, d_I$ and C_I are reconstructed, the reconstructed signal $D_1, D_2, \ldots, D_I$ and C_I have consistent points with the original signal L and

$$X = D_1 + D_2 + \cdots + D_I + C_I \tag{3}$$

Step 3: The component dimension of the component signal is calculated, and the G-P method is used to reconstruct the phase space, and the distance between the Yi and Yj is less than R:

$$C(r) = \frac{1}{N_R^2} \sum_{i=1}^{N_R} \sum_{j=1}^{N_R} H\left(r - \|y_i - y_j\|\right) \tag{4}$$

Based on the Takens theorem, the phase space dimension m is determined, and then the time series can be predicted by using the M theorem.

Step 4: The ANFIS model is established, and then the ANFIS model is trained by using the component signals, and the ANFIS is used to predict the sample data. Finally, the final prediction results are obtained. ANFIS is a kind of structure of fuzzy system and neural network, Its regular form is:

$$y_n = \sum_{j=1}^{m} a_j y_{nj} \Big/ \sum_{j=1}^{m} a_j = \sum_{j=1}^{m} \bar{a}_j y_{nj} \tag{5}$$

In the formula: $n = 1, 2, \ldots k$, according to the T-S model, can be designed as shown in Fig. 1 the ANFIS network structure, the network is a total of six layers.

Each node of the first layer is wholly intact to input variables passed to the next layer, so it is

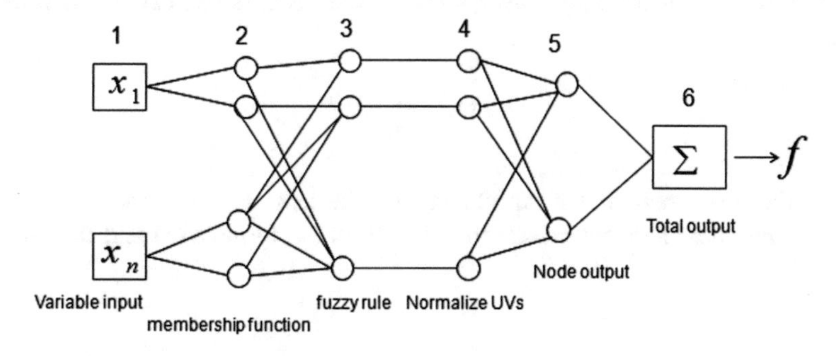

Fig. 1 Structure of ANFIS

$$f_i^{(1)} = x_i \tag{6}$$

In the formula: $i = 1, 2, \ldots n$, represents the sequence of the input nodes in the input layer.

After the fuzzy processing of second layers, the input variables are obtained respectively corresponding to the degree of membership of the fuzzy subset.

$$f_{u_i}^{(2)} = u_i^m(x_i) \tag{7}$$

In the formula: $m = 1, 2, \ldots u_i$, $i = 1, 2, \ldots n$; u_i is fuzzy partition number for x_i.

The third layer is the fuzzy rule base, each node of this layer represents a fuzzy rule, and its function is to calculate the applicability of each rule. The output of this layer is

$$\begin{gathered} f_j^{(3)} = \min\left\{f_{1s_{1j}}^{(2)}, f_{2s_{2j}}^{(2)}, \ldots f_{ns_{mj}}^{(2)}\right\} \\ j = 1, 2, \ldots, m \quad m = \prod_{i=1}^{n} m_i \end{gathered} \tag{8}$$

The number of nodes in the fourth layer is the same as the third layer, which is the normalized calculation. The output of this layer is

$$f_j^{(4)} = f_j^{(3)} / \sum_{j=1}^{m} f_j^{(3)} \tag{9}$$

Each node of the fifth layer is an adaptive node with a node function. The output of this layer is

$$f_k^{(5)} = \sum_{j=1}^{m} y_{kj} f_j^{(4)} \tag{10}$$

In the formula: Parameter $\{w_i\}$ in y_{kj} is the parameter set of the node.

The sixth layer is the output layer, which calculates the sum of all the transmitted signals.

$$f^{(6)} = \sum_{k=1}^{r} f_k^{(5)} \tag{11}$$

For the linear part of the ANFIS output, the least square method is used to identify the linear parameters, which is based on the formula (5) and the formula (10).

$$f_k^{(5)} = \sum_{j=1}^{m} \left[\sum_{i=1}^{n} \left(x_i f_j^{(4)} w_{ji}^k \right) + f_j^{(4)} w_{j0}^k \right] \tag{12}$$

i = 1, 2, …, n; j = 1, 2, …, m; k = 1, 2, …, r. By the formula (12), it is a linear function of the linear set $\left\{ w_{ji}^k, w_{j0}^k \right\}$, which can be identified by the least square method.

3 Urban Traffic Flow Time Series Prediction

This article refers to a road of the city within 2 h of traffic flow monitoring data simulation. First, the adverse events are screened out from the data, such as weather, road maintenance, traffic accidents, and others eliminate unfavorable time factors. Then the Daubechies wavelet analysis is used to deal with the time sequence. The fractal dimension value is 3.6, and the dimension of the phase space is 5.7. In order to test the effect of the forecast model, with the aid of the MATLAB simulation experiments, he traffic flow data by a group of 10 min, call 291 records, first extracted 200 records for training, the remaining 91 records for the validation of the prediction results.

The predicted values of the traffic flow in the same time period are obtained from the trained ANFIS network model (Table 1).

Analysis of training results and measured values (Table 2).

Table 1 Traffic flow forecast value for every 10 min

Time period	9:00–9:10	9:10–9:20	9:20–9:30	9:30–9:40	9:40–9:50	9:50–10:00
Traffic flow	181	164	182	142	170	158

Table 2 The results of traffic flow forecast

Time period	Actual flow	Predicted flow	Prediction error %
1	181	196	−8.2
2	164	150	8.5
3	182	176	3.2
4	142	149	−4.9
5	170	161	5.2
6	158	152	3.7

4 Conclusions

Wavelet analysis has the characteristics of effective analysis in the numerical value. At the same time, ANFIS has strong information storage and learning ability, especially the ability to use knowledge to deal with fuzzy situation. In this paper, wavelet analysis and ANFIS are combined to analyze the traffic time series of a certain intersection. The adaptive fuzzy neural network model (ANFIS) is used to simulate the experiment. The experiment results show that the method is used to predict the traffic flow in a certain period of time.

Acknowledgement This work was supported by Hainan Natural Science Foundation of China (No. 617172).

References

1. Fangce Guo, Polak J.W. Comparison of modelling approaches for short term traffic prediction under normal and abnormal conditions[C]. 13th Intelligent Transportation Systems (ITSC) International IEEE Conference 2010. Funchal, Portugal: Conference Publications. 2010: 1209–1214.
2. ZHU Zhong. YANG Zhao-sheng. A real time traffic flowpredict ion model based on artificial neural network[J]. Journal of Highway and Transport, 1998, 11(4): 89–92.
3. YANG Ij—cai, JIA Lei. Rough neural network and itsapplication on the tFa. ~ c flow forecasting[J]. Journal of Highway and Transportation Research and Development, 2004, 21 (10): 95–98.
4. Theja P.V.V.K. VanajakshiL. Short term prediction of traffic parameters using support vector machines technique[C]. 3rd Emerging Trends in Engineering and Technology (ICETET) International Conference 2010. Goa, India: Conference Publications. 2010: 70–75.
5. SUN Xiao-dian, CHEN Xue-mei. The optimization to the structure of a continuous wavelet neural network[J]. Journal of System Simulation, 2001, 27(8): 156–159.
6. LI Bing, JIANG Wei—sun. Chaos optimization and its applicationlJ J. Control Theory and Applications, 1997, 14(4): 613–615.
7. ISHAK S, AL—DEEK H. Performance evaluation of short—term time—series traffic prediction model[J]. Journal of Transportation Engineering, 2002, 28(6): 490–498.
8. Wusheng Hu, Yuanlin Liu, Li Li. The short-term traffic flow prediction based on neural network[C]. 2nd Future Computer and Communication (ICFCC) International Conference 2010. Wuhan, China: Conference Publications. 2010, 1: 293–296.

Human Activity Recognition with Smart Watch Based on H-SVM

Tao Tang, Lingxiang Zheng, Shaolin Weng, Ao Peng and Huiru Zheng

Abstract Activity recognition allows ubiquitous wearable device like smart watch to simplify the study and experiment. It is very convenient and extensibility that we do study with the accelerometer sensor of a smart watch. In this paper, we use Samsung GEAR smart watch to collect data, then extract features, classify with H-SVM (Hierarchical Support Vector Machine) classifier and identify human activities classification. Experiment results show great effect at low sampling rate, such as 10 and 5 Hz, which will give us the energy saving. In most cases, the accuracies of activity recognition experiment are above 99%.

Keywords Human activity recognition · Smart watch · H-SVM

1 Introduction

In the studies of human activity recognition, there are two main directions. One of them is based on vision sensors, which is not suitable for long-term monitoring in real life because of monitor environmental, equipment price and protection of privacy. The other is based on wearable sensors, which has been widely used because of low cost, small size and low energy consumption.

Mi Zhang did his study by wearing a device around his waist, this device is similar to a pager [1]. Piyush Gupta improved his study on the basis of Mi Zhang's study by wearing three devices around his waist. Thus, the accuracy of human activity recognition is higher [2]. Jennifer R. Kwapisz and his research group used a smart phone to instead of a sensor device to identify different activities in 2011. And his method has a praiseworthy recognition accuracy [3]. There is a higher

T. Tang · L. Zheng (✉) · S. Weng · A. Peng
School of Information Science and Engineering, Xiamen University, Xiamen, China
e-mail: lxzheng@xmu.edu.cn

H. Zheng
School of Computing and Mathematics, University of Ulster, Jordanstown Campus, Shore Road, Newtown Abbey, UK

© Springer Nature Singapore Pte Ltd. 2018

N.Y. Yen and J.C. Hung (eds.), *Frontier Computing*, Lecture Notes in Electrical Engineering 422, DOI 10.1007/978-981-10-3187-8_19

accuracy of SVM than accuracies of other classification algorithms in Davide Anguital's paper [4]. However, the use of smartphones also has its limitations in the study of human activity recognition. It has different results when smartphones are placed in different pockets of clothes. Thus, the smartphones are putted into specified pockets in more and more studies [5]. Now it is so popular to do the study of activity recognition with home-made wrist-mounted devices [6]. Of course, it is very convenient and extensibility that we do study with the accelerometer sensor of a smart watch. In this paper, we use Samsung GEAR smart watch to collect data, then extract features, classify with H-SVM classifier and identify human activities classification.

There is a high accuracy of human activity recognition by using home-made device. But it has no generalizability by using that device. By contrast, it is a lot easier for activity recognition by using smart phone. However, it has different experimental results when smartphones are placed in different pockets of clothes. The experiment conducted by smart watch [7–9], but its identification accuracy is not particularly high. James Amor shows the wonderful walking accuracies at high frequency and low frequency [8]. Our H-SVM algorithm performs better than James Amor's at low frequencies.

The remainder of this paper is structured as follows. Section 2 describes the methodology of H-SVM. Section 3 describes our experiments and results, including data collection, feature extraction and classification performance. Section 4 summarizes our conclusions and discusses areas for future research. Acknowledgement is described in final section.

2 Methodology

The proposed approach is illustrated in Fig. 1. Raw data is collected with high sampling rate (50 Hz) to extract features activity. And the H–SVM classifiers were applied to distinguish human activities [10].

2.1 Sampling Rate

The original frequency of human activities (sitting, standing, walking and running) are 50 Hz.

2.2 Feature Extraction

Four features were extracted to recognize the user behaviors, including the motion acceleration in three axis X, Y and Z, and the RMS (root-mean-square) of the

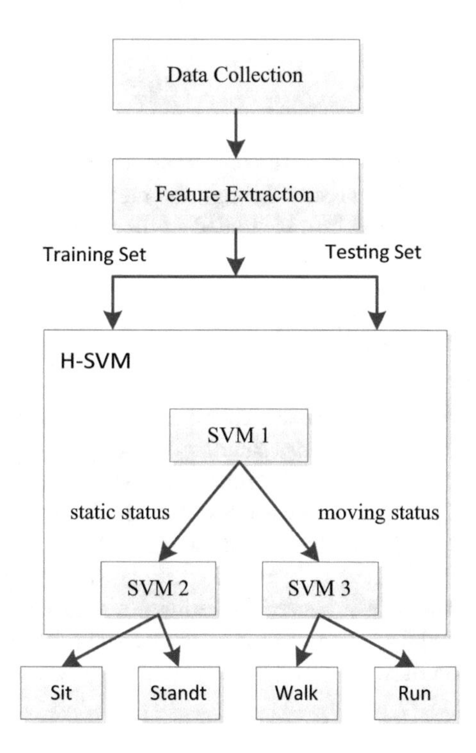

Fig. 1 The system of human activity recognition

changes in acceleration. Three-axis acceleration are the features that reflected the human position. The three-axis acceleration changed during the transition between sitting and standing, so that it can be used for distinguishing sitting and standing. The root-mean-square of the changes in acceleration reflected the amplitude changes of human activities, the acceleration of movement changed significantly while very little during the static status, so it can be used for distinguishing movement and static. It can also be used for distinguishing running and walking because the amplitudes and the variations of the acceleration in running are larger than those in walking.

The root-mean-square value of the dynamic variation of acceleration can be calculated by Eq. (1).

$$acc_t = \sqrt{highX_t^2 + highY_t^2 + highZ_t^2} \tag{1}$$

where $highX_t$, $highY_t$ and $highZ_t$ are the changes of the acceleration in the three axis X, Y and Z at time t. acc_t is the root-mean-square value of the change of acceleration in the three axis at time t.

All the features are extracted from a time window and integrated by using the mean filter described in Eq. (2).

$$\alpha_t = \sum_{i=-b}^{b} (\alpha_{t+i}) / (2b + 1) \tag{2}$$

where $2b + 1$ is the width of the sliding time window, and α is X, Y, Z or acc_t.

After passing through the mean filter, a median filter with the time window of $2b + 1$ (width), is applied to the features before being analyzed by the H-SVM.

2.3 Activity Recognition

An H-SVM classification model was applied in the research to distinguish four activities (sitting, standing, walking, and running) in the daily living for identification and classification. Support vector machine (SVM) is a supervised learning algorithm. The basic SVM model is the probability of a binary classification. The H-SVM includes three basic SVM classifiers: SVM1, SVM2 and SVM3.

The SVM1 is used to distinguish static status and moving status based on acc_t. The SVM2 is used to distinguish standing and sitting activities according to X_t, Y_t and Z_t. The SVM3 is used to distinguish the walking and running activities based on acc_t.

3 Experiments and Results

3.1 Data Collection

We installed a data collection Application on Samsung GEAR Smart Watch. We collect raw data of acceleration sensor by sampling frequency of 50 Hz. Our experiments contains four motions, which are sitting, standing, walking and running. Original sampling frequency is 50 Hz. We divided it into four kinds of sampling frequencies, 50, 25, 10 and 5 Hz in experiments. There are five volunteers participate in our experiments. The five volunteers are all males and age from 24 to 25. The five volunteers are numbered as A, B, C, D and E. Each motion of each person was sampled 4 min (240 s). To avoid the influence of extraneous data, each data set is removed its first 20 s and last 20 s. So each data set is 200 s.

3.2 Feature Extraction

Feature selection methods select the features, which are most discriminative and contribute most to the performance of the classifier, in order to create a subset of the existing features.

Although SVM are powerful neural computing methods, their performance is reduced by too many irrelevant features. Therefore, H-SVM feature selection methods are proposed. We consider an SVM feature selection approach for better system performance.

In this paper, we propose 4 attributes for human activity recognition:

X axis: Filtered data of X axis
Y axis: Filtered data of Y axis
Z axis: Filtered data of Z axis

Root Mean Square (RMS) of Variation: RMS value of the change of accelerations in the three axis.

The filtered data of each axis are different between Fig. 2a, b, so it can identify sitting and standing. The raw data are processed by mean filter and median filter. Figure 2 is the filtered data in 50 Hz.

RMS of variation value is almost the same between Fig. 3a, b, but it has a huge difference between Fig. 3c, d. Thus, this value can be used to distinguish walking and running. Figure 3 is the RMS of variation value of each motion in 50 Hz.

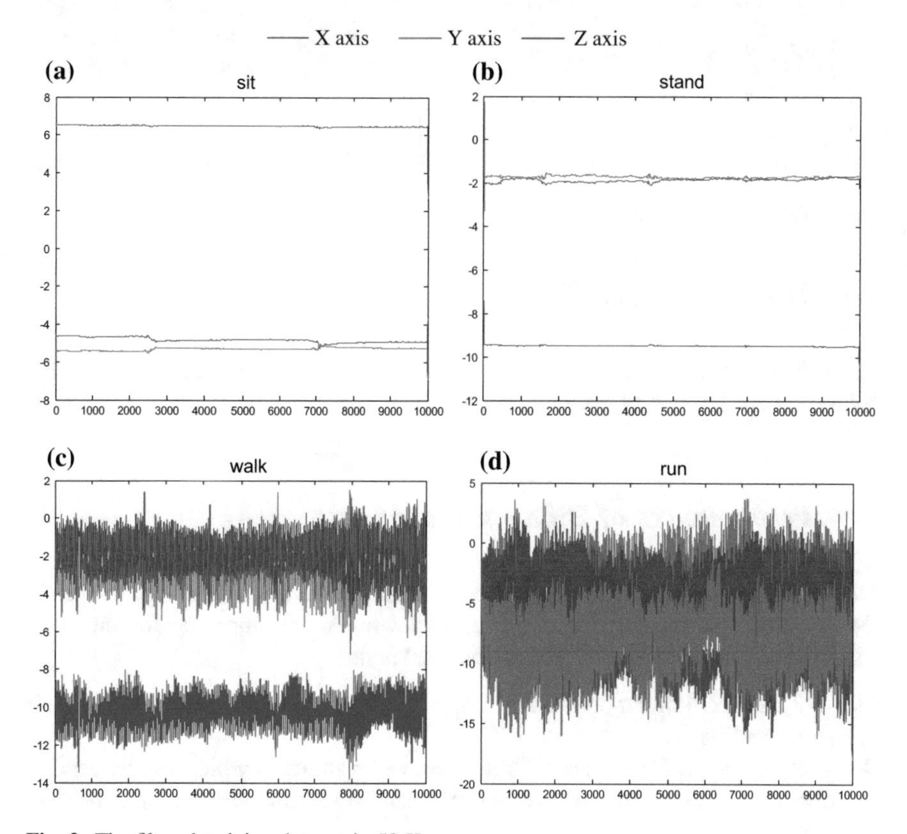

Fig. 2 The filtered training data set in 50 Hz

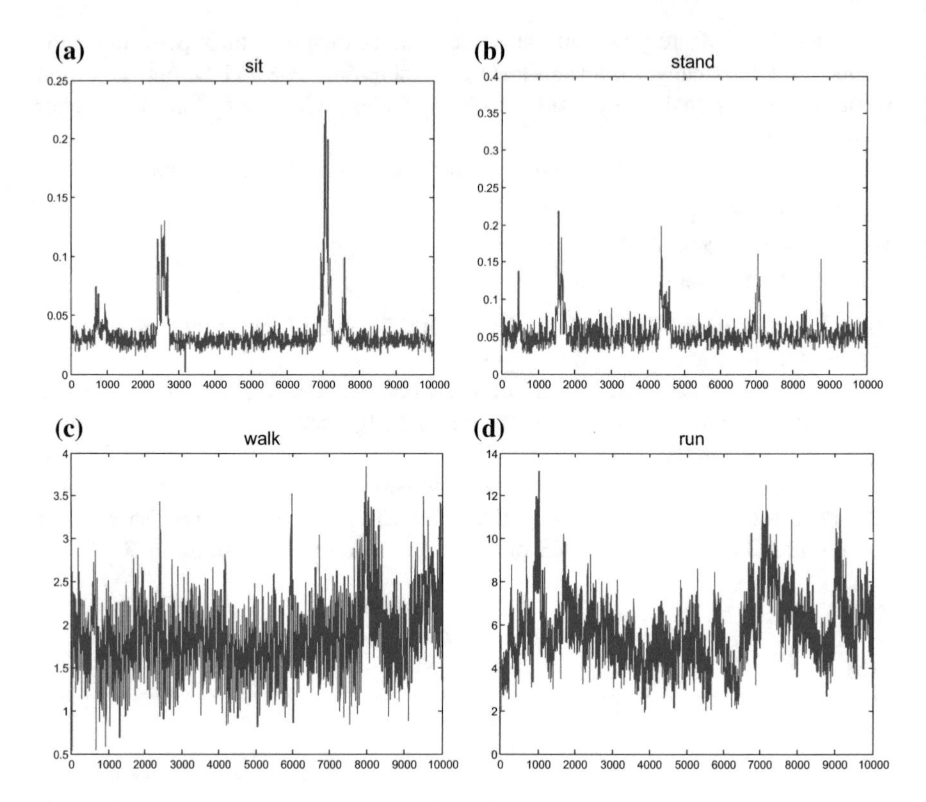

Fig. 3 The RMS of variation value of training data set in 50 Hz

3.3 Training Set and Testing Set

The data set of volunteer A is set as a training data set, while the data of other volunteers are set as a big testing data set.

3.4 Performances of Different Classifiers

The selected or reduced features that create feature sets are used as inputs for the classification and recognition methods. Following are a summary for the most widely used classification and recognition methods.

J48: J48 are decision support tools using a tree-like model of decisions and their outcomes, and costs.
Decision Tables (DT): Decision Tables serve as a structure which can be used to describe a set of decision rules and record decision patterns for making consistent decision.

Table 1 Accuracies of activity recognition

	% of records correctly predicted				
	H-SVM	SVM	J48	NB	DT
Sitting	99.31	24.90	100	2.28	90.60
Standing	98.49	16.25	81.93	0	4.34
Walking	98.99	87.51	83.48	86.04	87.79
Running	99.34	99.54	90.07	99.14	98.48
Overall	99.03	57.05	88.87	46.87	70.30

Naive Bayes (NB): Naive Bayes is a simple probabilistic classifier based on Bayes' theorem.
Support Vector Machine (SVM): SVM is supervised learning methods used for classification.

We put the selected characteristic values into different classifiers. Table 1 shows the accuracies of different classifiers.

As can be seen from Table 1, the accuracies of SVM and NB are very low at the motion of sitting. And the accuracies of SVM, NB and DT are also low at the motion of standing. It can be seen in these attributes, H-SVM and J48 perform wonderful at each motion. On the whole, H-SVM algorithm performs the best between them.

3.5 *Performances of Different Frequencies*

To test the performance of H-SVM algorithm at different frequencies, we extract four different frequency from the raw data as 50, 25, 10 and 5 Hz. Table 2 shows the classification accuracy of each motion at different frequencies.

As can be seen from Table 2, the accuracies of H-SVM performs very well at different frequencies. Even at the low frequency (5 Hz), this classifier can very easy to distinguish different motions, and its overall accuracy is above 99%.

Table 2 Accuracies of activity recognition based on H-SVM

	% of records correctly predicted			
	50 Hz	25 Hz	10 Hz	5 Hz
Sitting	99.31	99.86	99.99	99.87
Standing	98.49	99.02	98.95	98.62
Walking	98.99	99.78	99.91	99.67
Running	99.34	99.90	99.64	99.85
Overall	99.03	99.64	99.62	99.50

4 Conclusions

In this paper, we present an activity recognition approach based on H-SVM. In our experiments, smart watch performs a good classification ability. Smart watch is not lost to other devices in the field of recognition activity. Experiment results show that H-SVM algorithm performs the best between many algorithms. At each motion, H-SVM almost has the highest classification accuracy between those algorithms. Experiment results show great effect at low sampling rate, such as 10 and 5 Hz. In most cases, the accuracies of activity recognition experiment are above 99%. Future work will include more participants, especially elderly users and evaluating the proposed algorithm with data collected at real living environments.

Acknowledgements This work was supported by the Major Science and Technology special project of Fujian Province (No. 2012HZ0003-2).

References

1. Zhang M, Sawchuk A A. Human daily activity recognition with sparse representation using wearable sensors[J]. Biomedical and Health Informatics, IEEE Journal of, 2013, 17(3): 553–560.
2. Gupta P, Dallas T. Feature selection and activity recognition system using a single triaxial accelerometer[J]. Biomedical Engineering, IEEE Transactions on, 2014, 61(6): 1780–1786.
3. Kwapisz J R, Weiss G M, Moore S A. Activity recognition using cell phone accelerometers [J]. ACM SigKDD Explorations Newsletter, 2011, 12(2): 74–82.
4. Anguita D, Ghio A, Oneto L, et al. Human activity recognition on smartphones using a multiclass hardware-friendly support vector machine[M]//Ambient assisted living and home care. Springer Berlin Heidelberg, 2012: 216–223.
5. Anjum A, Ilyas M U. Activity recognition using smartphone sensors[C]//Consumer Communications and Networking Conference (CCNC), 2013 IEEE. IEEE, 2013: 914–919.
6. Dietrich M, van Laerhoven K. A typology of wearable activity recognition and interaction [C]//Proceedings of the 2nd international Workshop on Sensor-based Activity Recognition and Interaction. ACM, 2015: 1.
7. Tan T H, Gochoo M, Chang C S, et al. Fall Detection for Elderly Persons Using Android-Based Platform[J]. Energy, 2013, 2(2): 2.
8. Ahanathapillai V, Amor J D, Goodwin Z, et al. Preliminary study on activity monitoring using an android smart-watch[J]. Healthcare technology letters, 2015, 2(1): 34–39.
9. Jeong Y, Chee Y, Song Y, et al. Smartwatch App as the Chest Compression Depth Feedback Device[C]//World Congress on Medical Physics and Biomedical Engineering, June 7–12, 2015, Toronto, Canada. Springer International Publishing, 2015: 1465–1468.
10. Weng S, Xiang L, Tang W, et al. A low power and high accuracy MEMS sensor based activity recognition algorithm[C]//Bioinformatics and Biomedicine (BIBM), 2014 IEEE International Conference on. IEEE, 2014: 33–38.

Integrating Augmented Reality Technology into Subject Teaching: The Implementation of an Elementary Science Curriculum

Rong-Chi Chang and Ling-Yi Chung

Abstract Augmented Reality (AR) technology has changed abstract science to be presented to the students with concrete image by interoperable way which has provided with new possibilities to reduce students' cognitive load. The nature and composition of materials is an important concept to study Chemistry. However, to the middle school students, it is a great challenge for them to understand both micro and macroscopic world of the whole Chemistry. This study has designed a periodical table based on AR to develop a course of Chemistry changes with interoperation. The study has implemented an experiment targeting on learners to study and research students' learning effects by applying AR technology and how it affects them in learning Chemistry. The result indicates that AR technology can be integrated with teacher's teaching, which has made the students' learning attitudes become positive. By way of interoperation to learn, students' grades have been obviously more improved than those before the experiment. Through the process of the study, the researcher realizes the importance that students study knowledge in science with interactive technology and it has significant help in promoting students' learning achievements as well as their learning motivation.

Keywords Augmented reality · Scientific education · Learning effectiveness · Chemical changes

R.-C. Chang (✉)
Department of Technology Crime Investigation, Taiwan Police College, Taipei, Taiwan, ROC
e-mail: roger@mail.tpa.edu.tw

L.-Y. Chung
Department of Applied English, Chihlee University of Technology, New Taipei City, Taiwan, ROC
e-mail: lychung@chihlee.edu.tw

© Springer Nature Singapore Pte Ltd. 2018

N.Y. Yen and J.C. Hung (eds.), *Frontier Computing*, Lecture Notes in Electrical Engineering 422, DOI 10.1007/978-981-10-3187-8_20

1 Introduction

Chemistry is the scientific study of physical properties, composition, structure and physical changes of matters. The properties and composition are crucial concepts in learning chemistry, acting as the foundation for further education in organic chemistry and material science. However, the abstract concepts in chemistry, such as molecules, atoms and mass, present great challenges for junior high school students in the understanding of the microscopic and macroscopic chemical world. At times, it is difficult for students who lack the necessary mental acuity to construct a mental image of how atoms form matters. Therefore, an enhanced learning method or tool may be needed for students to grasp the scientific knowledge of chemistry.

Augmented reality (AR) is an extension of virtual reality (VR). Compared to traditional VR, AR features a seamless interface that combines the real world and the virtual world. Users can merge real-life scenes with virtual objects to gain a natural and realistic man-machine interactive experience.

With the rapid advances in AR, its applications in disciplinary education have seen significant success stories. AR is best suited in the following two scenarios: (1) when a phenomenon cannot be easily simulated in real life, such as planetary movements in the solar system; and (2) when an experiment has obvious shortcomings or is dangerous in the classroom, such as experiments on the image production using convex lens and lighting of candles.

We believe that AR provides a workable solution to the problem in computer-assisted chemistry learning by solving the microscopic aspect of chemistry courses, e.g. the imperceptible tiny particles that cannot be observed in real life. The purpose of this study is to develop an investigative learning tool powered by AR applications to be used in chemistry courses in junior high schools. The study will explore the effects of AR learning tool on students' cognitive performance, compare the learning effectiveness and explore students' attitudes toward an AR-enhanced learning environment.

The research focused on the "composition of matters" in the chemistry curriculum in junior high school, i.e. exploration of the microscopic structure of our world. Traditional 2D images and texts cannot present the true three-dimensional structure of matters, making the learning process of molecular geometry difficult. Regular chemistry learning software only allows students to observe the structure of the molecules and nothing more. For these reasons, we developed an AR-powered learning tool, where students can use AR markers to manipulate the microscopic particles, and observe molecules or crystals in three-dimensional space from various angles to effectively further their understanding of the composition of matters.

2 Literature Review

Many teachers or researchers have applied computer-assisted learning tools in chemistry teaching. With these tools, researchers design experimental projects and test the effects of students' learning. In recent years, learning tools based on Virtual Reality and Augmented Reality are highly recommended have been applied in Microstructure study of Chemistry.

Dalgarno et al. [1] constructed a virtual laboratory to be provided as distance learning environment for freshmen. Most students believe it is an important preparatory tool and it is suggested that it should be widely applied in the future. The applications of the Virtual Reality are proven to be helpful in learning, however, the interaction of the Virtual Reality is still unnatural and has its restriction.

It is greatly helpful for the students being equipped with space capabilities. It is a daunting task to demand students have visualization in Microstructure of Chemistry. Harle and Towns [2] found in their study that specific students who are lack of visual spatial skills in Chemistry have difficulties in understanding the molecular alternation. Tuckey et al. [3] indicated in their study that many college students still have difficulties in perspective thinking and these difficulties come from their misunderstanding about simple concepts and skills. Merchant et al. [4] studied how 3D Virtual Reality environment effect the simulation leaner's features in Second Life. There inter-features of these application procedures include enlarging, zooming, rotating objects and stylized objects to be interacted with objects in some ways. They found that 3D Virtual environment can promote students to learning Chemistry.

Augmented Reality (AR) is an extension of Virtual Reality (VR). AR provides a seamless joint, which integrates the real world with the virtual world. AR technology is widely applied in the field of engineering, flight training, environmental science, medicine and education. Designers have constructed a series of interactive process so that it allows users to get the most natural and true 3D human-computer interactive experience.

Nunez et al. [5] made a research targeting on college students about spatial relationship of AR teaching system and Chemistry issues. In the experiment, the students can manipulate certain objects such as ZrSiO4 and the crystal structure of markers. However, in the above-mentioned study, only a still image or a structure is rendered. Recent studies indicate that full application of AR technology has made the interaction between the students and the computer more interesting. In addition to subject of science, AR to subject of arts also has a very good effect. In the visual arts curriculum, AR system has a positive impact on high school students' learning motivation [6].

AR help students to explore in the real world [7] and also allows us to experience a chemical reaction which can not be easily carried out in the real world. AR provides concept of real-world virtual element visualization such as operation of air flow or magnetic field, etc. [8]. AR can help students improve their knowledge and skills and is more effective than other techniques [9]. In this way, students can get better investigation skills as well as increase their learning motivation.

Based on these studies, our goal is to eliminate the difficulty of spatial skills when students are learning Chemistry microstructure.

3 Research Methods

3.1 Materials Development and Design

This study aimed to develop an AR-based learning tool for assisted learning in chemistry courses. Chemistry is a large body of knowledge. This study primarily focused on the "composition of matter" in the design of learning content. Prior to the design, chemistry teachers were interviewed to gain a better understanding of the abstract contents students generally have trouble with and lack the initiative to learn. It is hoped the AR tool can arouse students' interests in learning chemistry and encourage them to take initiatives in exploring the knowledge on the composition of matters.

The effectiveness of AR interactive learning tool in chemistry courses was investigated. The AR tool pack consisted of the AR software, six AR markers and a learning sheet. The learning pack included four applications on matter composition: (1) water molecules made up of oxygen and hydrogen atoms, and water from the composition of water molecules; (2) diamond crystal composed of carbon atoms; (3) graphite crystal composed of carbon atoms and; (4) NaCl (table salt) composed of chloride and sodium ions.

AR markers are interactive picture cards printed with different symbols, allowing students to use this AR tool to observe and manipulate the different elements in the composition of matters. After installing the software, students can control and observe the molecules and their structure using different AR markers. With the learning sheet, students can further acquire the concept underlying the composition of matters and gain the relevant knowledge. The software identifies the various AR markers via a camera to render accurate 3D models of the different molecules. The interface is designed to allow human-machine interactions, where the applications trigger different animations to present the changing process of composition of matters, as shown in Fig. 1.

3.2 Research Design

This study adopted a quasi-experimental design. Subjects were divided into the experimental group and the control group. The independent variable of this experiment was the intervention of the learning tool, where the control group

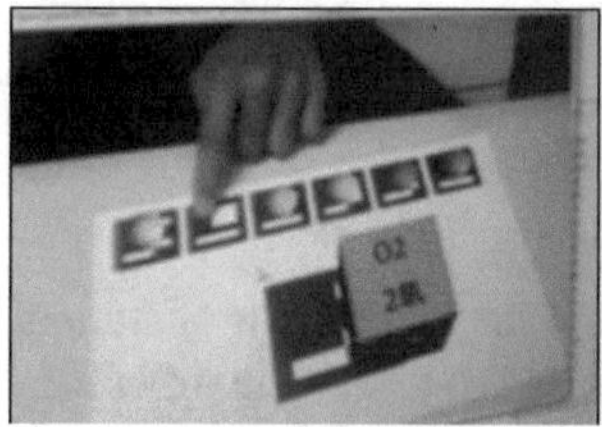

(a) Various forms of elements

(b) Structure of molecules

(c) Changes in composition of matters: diamond crystallization

Fig. 1 Interactive learning process using the AR tool

received conventional textbook learning, while the experimental group received AR-powered assisted learning. The dependent variable was the learning achievement of learners in the acquisition of chemistry knowledge.

The participants comprised a total of 55 seventh-grade students in Taiwan, including 27 in the control group and 28 in the experimental group. Students in the two classes were taught by the same chemistry teacher. Composition of matters was selected as the chemistry learning topic for the experiment, and the learning effectiveness was explored.

The experiment included the learning achievement tests and questionnaires, and was divided into three phases as follows.

1. Pre-test: To assess students' basic knowledge of chemistry and to ensure that no significant differences were observed between the two groups of students prior to the experiment. The teacher designed a quiz containing 10 questions according to the content of their textbooks, and the quiz was administered and lasted 30 min. The participating students tried to answer the questions, and one point was given for each correct question answered, with a full score of 10.
2. Learning exploration activities: Students in the experimental group were given instructions on how to operate the software to make sure they know how to use the AR learning tools. Students in the control group learned the contents via textbook reading and video clips. Students in either group learned the materials on the learning sheet in the absence of teacher instructions.
3. Post-test: According to the learning unit, the test content was designed in a similar fashion to the pre-test, at a total of 10 questions to be completed in 30 min.
4. A questionnaire was conducted after the learning experiment, where students answered questions on a five-point Likert scale regarding learning motivation and attitude, technology acceptance and satisfaction of the learning tools.

4 Analysis and Discussion

4.1 Analysis of Learning Effectiveness

To better understand the similarity in prior knowledge of the two groups of students before the experiment, independent sample t-test was used to assess the differences in their chemistry knowledge. Statistics from the pre-test results showed that the experimental group averaged 6.71 (mean), while the control group averaged 6.76 (mean), and there was no statistically significant difference ($t = 0.421$, $p > 0.1$) between the two groups of students in terms of prior knowledge, meaning the two groups of students had similar level of knowledge on chemistry before the experiment.

To study the learning effectiveness between the experimental and control groups, ANCOVA was performed for the post-test results with the pre-test scores as covariates. For consistency with the assumptions of ANCOVA, tests of homogeneity of variance and homogeneity of within-class regression coefficient were carried out for the two sets of scores. The homogeneity of within-class regression coefficient test showed homogeneity of the regression coefficients ($F = 0.526$, $p = 0.47 > 0.05$), indicating the covariate variable (pre-test scores) and the dependent variable (post-test scores) would not vary due to variations in the independent variable, fulfilling the assumption of homogeneity of within-class regression coefficient for it to be further analyzed by ANCOVA.

Table 1 shows that after excluding the effect of covariate variable on the dependent variable, the two groups of students exhibited significant differences in test scores ($F = 4.259$, $p = 0.039 < 0.05$) in the learning the composition of matters. Mean score of the experimental group was 7.59 ($SD = 2.09$), while the mean score of the control group was 6.18 ($SD = 2.15$). Therefore, it can be concluded that, statistically speaking, AR learning tools were significantly more effective than traditional textbook learning with videos.

4.2 Learning Motivation Analysis

To explore whether AR learning tools were helpful in enhancing the learning motivation for students in the experimental group, a motivation scale was in place

Table 1 ANCOVA result (Post-test) of analysis of learning effectiveness

	Items	Mean	Adjusted average	*SD*	*F*
Control group	27	6.18	6.97	2.15	4.259*
Experimental group	28	7.59	7.61	2.09	

*$P < 0.05$

Table 2 Descriptive statistics for the results of the learning motivation questionnaire

Items	Control group		Experimental group		Cohen's *d*
	Mean	SD	Mean	SD	
Item 1. I like the way the class is being taught today.	3.87	0.93	4.25	0.66	0.49
Item 2. The way the class is taught draws my attention.	3.82	1.00	4.13	0.87	0.31
Item 3. I think the teaching materials are diversified.	3.76	0.75	4.29	0.64	0.69
Item 4. I have more understanding of the structure of molecules.	3.83	0.87	4.25	0.75	0.39
Item 5. I like the strengthening activity that helps me learn about the composition of matters.	3.91	0.91	4.21	0.83	0.41

for the pre-test and the post-test to understand the impact of different learning strategies and activities on students' motivation to learn the basics of chemistry.

The total score of all items on the learning motivation scale (5-point Likert scale) prior to the learning experiment was subjected to independent sample t-test. Mean score for the control group was 4.17 (SD = 1.06) and the mean score for the experimental group was 4.21 (SD = 1.17). The t-test results showed no significant difference between the two groups of students in terms of learning motivation prior to the learning activities ($t = 0.054$, $p = 0.957 > 0.05$).

After the learning activity, the two groups of students took the questionnaire on learning motivation. Their responses to the questionnaire for learning motivation are summarized in Table 2. The results as shown indicate that students in the experimental group rated most items higher than those in the control group, though they were a relatively moderate effect size. Accordingly, compared with digital video, AR technology was particularly useful to stimulate students towards the study of composition of matters.

4.3 Analysis on Technology Acceptance

In order to understand the technology acceptance of the two groups of students in strategically different learning activities, the scale used in this study included 10 questions from aspects of perceived usefulness and perceived ease of use. Using a 5-point Likert scale for scoring, independent sample *t*-test was performed to probe into the scores from the two groups. Table 3 shows the results, where technology acceptance in the experimental group was significantly higher than that of the control group ($p = 0.00 < 0.001$). This suggested that students have a higher acceptance to learn with AR tools compared to general digital learning.

Table 3 t-test result of analysis on technology acceptance

	Items	Mean	*SD*	*t*
Control group	27	4.05	0.42	4.01***
Experimental group	28	4.74	0.61	

***$P < 0.001$

5 Conclusion

Based on the findings on learning effectiveness, students' attitude, and analysis from classroom observation and interview data, the following conclusions were derived. The results from the study found that the integration of AR technology into learning the basics of chemistry was helpful in improving students' learning effectiveness and motivation, and students generally agreed that AR tool was useful in encouraging an active learning attitude and that they enjoyed the exploratory learning experience.

In addition, learners using AR tools were observed to have a significantly higher technology acceptance than their control group counterparts. This study explored the potential and acceptance of AR tools, helping us better understand AR applications in learning. AR learning tools were found to significantly improve students' academic performance in understanding the microscopic structure of chemical matters. We hope to further the use of AR tools, extending their application to other units in the chemistry curriculum, such as the memorization of abstract chemical structures or concepts.

Acknowledgements This work was supported in part by the Ministry of Science and Technology of the Republic of China under contract 102-2511-S-468-005-MY2.

References

1. Dalgarno, B., Bishop, A. G., & Adlong, W.: Effectiveness of a virtual laboratory as a preparatory resource for distance education chemistry students. Computers & Education, Vol. 53, 853–865 (2009)
2. Harle, M., & Towns, M.: A review of spatial ability literature, its connection to chemistry, and implications for instruction. Journal of Chemical Education, Vol. 88, No. 3, 351–360 (2011)
3. Tuckey, H., Selvaratnam, M., & Bradley, J.: Identification and rectification of student difficulties concerning three-dimensional structures, rotation, and reflection. Journal of Chemical Education, Vol. 68, No. 6, 460–464 (1991)
4. Merchant, Z., Goetz, E. T., Keeney-Kennicutt, W., Kwok, O., Cifuentes, L., & Davis, T. J.: The learner characteristics, features of desktop 3D virtual reality environments, and college chemistry instruction: A structural equation modeling. Computers & Education, Vol. 59, 551–568 (2012)
5. Nunez, M., Quiros, R., Nunez, I., Carda, J. B., & Camahort, E.: Collaborative augmented reality for inorganic chemistry education. New Aspects of Engineering Education, 271–277 (2008)
6. Di Serio, Á., Ibáñez, M. B., & Kloos, C. D.: Impact of an augmented reality system on students' motivation for a visual art course. Computers & Education, Vol. 68, 586–596 (2013)

7. Dunleavy, M., Dede, C., Mitchell, R.: Affordances and Limitations of Immersive Participatory Augmented Reality Simulations for Teaching and Learning, Journal of Science Education and Technology, Vol. 18, No. 1, 7–22 (2009)
8. Stull, A. T., Barrett, T., & Hegarty, M.: Usability of concrete and virtual models in chemistry instruction. Computers in Human Behavior. Vol. 29, No.6, 2546–2556 (2013)
9. El Sayed, N. A. M., Zayed, H. H., & Sharawy, M. I.: ARSC: Augmented reality student card an augmented reality solution for the education field. Computers & Education, Vol. 56, 1045–1061 (2011)

Influence of Inclined Angles on the Stability of Inclined Granular Flows Down Rough Bottoms

Guanghui Yang, Sheng Zhang, Ping Lin, Yuan Tian, Jiang-Feng Wan and Lei Yang

Abstract The granular flow down an inclined plate or chute is a potential choice as a high-power spallation target. Here we studied about the stability of the inclined granular flows down rough bottoms through a series of simulations on GPUs. The periodic boundary is used here. Following the previous work, there are some conclusions in this work: (1) the phases of flows with various inclination angles are classified. (2) According to the oscillation modes, the oscillation flow region can be further divided into three sub-regions. (3) The oscillation flow region is a transition region between ordered and disordered flow region, where more details about the self-organization and dilatant are shown.

Keywords Inclined granular flows · DEM · Stability · GPU · Granular target

---

G. Yang · S. Zhang · P. Lin · Y. Tian · J.-F. Wan · L. Yang (✉)
Institute of Modern Physics, Chinese Academy of Science,
509 Nanchang Road, Lanzhou 730000, Gansu, China
e-mail: lyang@impcas.ac.cn

G. Yang
e-mail: yangguanghui@impcas.ac.cn

S. Zhang
e-mail: halifax@gmail.com

P. Lin
e-mail: pinglin@impcas.ac.cn

Y. Tian
e-mail: tianyuan08@impcas.ac.cn

J.-F. Wan
e-mail: jiangfengwan@impcas.ac.cn

G. Yang · J.-F. Wan
University of Chinese Academy of Science,
No. 19A Yuquan Road, Beijing
100049, China

© Springer Nature Singapore Pte Ltd. 2018

N.Y. Yen and J.C. Hung (eds.), *Frontier Computing*, Lecture Notes in Electrical Engineering 422, DOI 10.1007/978-981-10-3187-8_21

1 Introduction

For high power neutron sources, there are various kinds of targets that have been designed, constructed, operated, such as plate targets in IPNS and ISIS, CSNS, rod targets in German SNQ project [1, 2] and SINQ at PSI [3], and liquid metal targets in SNS and JSNS [4, 5]. Following the various spallation targets above, as a new concept, the gravity-driven dense granular-flow target (DGT) in a hopper was proposed lately [6], which combines the advantages of solid and liquid targets. The inclined granular flow is another attractive design because of the wide adjustable range of velocity of grains. The stability of the granular flow is very concerned in the design of granular target system.

Inclined granular flows relate with many phenomena ranging from mudslides and avalanches to dune formations and singing sands. To understand these phenomena more clearly, flows of granular materials down inclined planes or chutes are widely used as laboratory paradigms. The inclined flows are complicated and can be divided into many phase regions according to the microscopic structure and dynamics [7]. To solve the problem theoretically, kinetic theory was developed to describe the inclined collisional flows in the dilute limit [8]. However, numerous verifications have been carried out and there are discrepancies between the theory and experimental or numerical results, especially in dense flows of inelastic particles [9, 10]. In last decades, many experimental tools have been developed to observe the flows and measure the physical quantities directly [11]. For the physical quantities which are hardly to be measured (such as rotation velocity), numerical simulation is able to offer a reliable picture.

A characteristic feature of inclined flows is that the flows will stop unless the inclination angle θ is big enough. The hstop is a decreasing function of the inclination angle θ [12, 13]. In Pouliquen's work, there is a fitted formula: $h_{stop} = Ad\frac{\tan(\delta_2) - \tan(\theta)}{\tan(\theta) - \tan(\delta_1)}$, where δ1 is the minimum angle required for flow, δ2 is the maximum angle at which steady flow is possible, d the particle diameter, and A is a characteristic dimensionless length [14, 15]. Moreover, the Froude number can be expressed in terms of hstop: $F = \frac{u}{\sqrt{gh}} = \beta\frac{h}{h_{stop}} - \gamma$, where β and γ are constants independent of chute inclination and particle size [14]. The formula of hstop also can be predicted theoretically by using shear transformation zone (STZ) theory [16]. From Baran et al.'s simulation it shows an approximate relationship $\lambda_{xx}/d \sim \ln\left(h_{stop}/d\right)$, where λxx is the correlation length in velocity correlation functions [17, 18]. In Silbert et al.'s work, the phase diagram of periodic inclined flows was plotted with different packing height h and θ, the steady flow can be further divided into three sub-phases [19]. Besides the height and inclination angle, phase transition can also be induced by changing the base roughness [20]. Recently, Weinhart et al. presented a deta picture of inclined flows with various base roughness λ (defined as the size ratio of the fixed and the flowing spheres) by using discrete particle method (DPM) simulations [21].

In classical Bagnold's assumption, there is a constitution relation between shear stress and shear rate: $\tau \sim \gamma^2$ [22]. Based on it, the profile of velocity in inclined flows can be deduced:

$$v_x(z) = 2/3B\sqrt{\rho g \sin\theta}\,(\frac{h^{3/2} - (h-z)^{3/2}}{h^{3/2}}) + v_b \quad (1)$$

where B is a parameter which varies with θ [23] and v_b. is the slip velocity at the base. The classical Bagnold scaling was verified by simulation works in periodic inclined flows or vertical chute flows [23, 24]. The Bagnold profiles of velocity were also verified in two-dimensional (2D) or three-dimensional (3D) inclined flows [25–29]. The shear rate in periodic inclined flows was plotted and compared with Bagnold profiles [30]. Reddy and Kumaran investigated Bagnold coefficients in periodic inclined flows with linear spring-dashpot model carefully [31] and the results show the influence of stiffness on the coefficients is negligible. Recent experiment of granular flows down the inclined plane suggested the Bagnold relation may be invalid for those non-spherical grains [32]. In cohesive granular flows, Brewster et al. showed that Bagnold scaling is broke down [33]. The interpretation of experimental results made by Bagnold was doubted by Hunt et al. [34].

Another important quantity in inclined flows is the shear rate, which is always mentioned in the studies of singing sands and Bagnold scaling [23, 35]. In Andreotti et al.'s work, the shear rate is assumed to only depend on the inclination angle in steady flows but not on the height [36, 37]. Whereas from Silbert et al.'s simulations with linear spring-dashpot model, it is found that in an inclined flows, the variations of shear rate is obvious, especially in the chute flows with ordered bases [38]. Moreover, there is a range of inclination angle where there is an oscillation between ordered and disordered states [21, 38, 39]. Tan et al. developed a model to explain this phenomenon [40]. The shear rate decreases with increasing height which is consistent with the experiments by Azanza et al. [11]. Additionally, from previous experiments it shows the shear rate in the surface flows is nearly constant [41, 42].

Here we studied the granular flow down a periodic, inclined planes with hexagonal-ordered base by using Hertz-Mindlin contact model and discrete element method (DEM) simulations. The steady phases with different inclination angles and heights in this system are concerned, with a specific focus on the switches of steady phases (oscillatory flows).

2 Methods

Due to the inhomogeneous and anisotropic properties, it is a big challenge to study granular materials theoretically and computationally [43], and under certain conditions the continuum approximation is not applicable anymore [44, 45]. Thus, the discrete element method (DEM) integrating with molecular dynamics (MD) starts

Table 1 The parameters of the material properties in this work

Quantity	Symbol	Value
Young's modulus (MPa)	E	70
Poisson's ratio	υ	0.25
Sliding friction coefficient	μ	0.5
Density of spheres (kg/m3)	ρ	2500
Coefficient of restitution	E	0.8

to attract increasing interest for simulating the granular dynamics [46, 47]. The simulations in this work were performed on GPUs [48, 49]. Mono-disperse glass spheres were simulated in this work and the interactions between the spheres were given by Hertz-Mindlin contact model [50, 51].

We employ the discrete element method to perform 3D simulations of granular flows down inclined planes. Mostly, a system, consisting of 8,000 mono-disperse spheres (the diameter $d = 1$ mm), is used. The system is periodic in the x (flow) and y (span-wise) directions, and has a hexagonal-ordered base constructed of spheres of the same diameter (i.e., $\lambda = 1$) and material properties. The parameters of material properties are shown in Table 1. The simulation cell dimensions in the x and y direction are 20d and 10d respectively, resulting in an approximately 40d height in the z direction. The time-step in our simulations is $5 \times 10 - 7$ s. There are two ways of box identification to achieve the periodic boundary condition. In Process I, identification of a transfer particle is according to the box number and the box number of the particle in next step is determined by an AND operation [48]. Process II is searching the neighbor boxes to judge if a particle passes through the boundary. The disadvantage of Process I is the space is divided into 2n in the direction of periodic boundary for computing efficient. Here Process II is employed.

Initially, 8,000 spheres are randomly dropped into the horizontal chute and we simulate until a static packing is produced. And then the system is tilted to a high inclination angle (30°) to initiate flow and erase preparation effects, as used in previous work [38]. After 80,000 time-steps, the system is tilted to the desired inclination angle and reaches steady state after a period of time.

3 Results and Discussions

The inclined plane configuration has the advantage to generate steady uniform flows, for suitable parameters, with only two control parameters: the angle of inclination (θ) and the thickness of the flowing granular layer h. Below a thickness threshold (hstop (θ)) the flow stops. In Fig. 1a, we plot the hstop as a function of θ for our system. From simulation results of the system with the hexagonal-ordered base, we observe five major flow regimes for inclination angle in the range $17° \leq \theta \leq 40°$: no flows, steady ordered flows, oscillatory flows, steady disordered flows and gas flows. This phenomenon is similar to the result of [21], where the authors also observed five flow regimes for system with random base and $\lambda = 1/2$. For $\theta \leq 19°$, we observe that granular flow cannot continue running after a period of time, thus name this region no flows. For $\theta \geq 30°$, spheres frequently hit the base and behave like gas (i.e. gas flows). In the region $20° \leq \theta \leq 29°$, we obtain three steady-state flows: steady ordered flows, oscillatory flows (quasi-steady) and steady disordered flows. At small inclinations $20° \leq \theta \leq 23°$, the flow exhibits low-dissipation behavior (steady ordered flows). On the contrary, high-dissipation appears in large inclinations 26° $\theta \leq 29°$ (steady disordered flows). For intermediate range of inclinations 23.5° $\theta \leq 25°$, the kinetic energy of system shows different modes of oscillation between low and high values (oscillation flows). The bulk-average kinetic energy (KE) per particle against time for four flow regimes (with a typical angle respectively and the oscillatory regime excluded), is displayed in Fig. 1b. In this paper, we primarily study the effect of inclination angle on the oscillatory flow regime.

The oscillation between high and low kinetic energy levels for the inclined flows was reported previously in [21, 38]. Here we find, for inclination angles $23.5° \leq \theta \leq 25°$, there are three modes of oscillation in this system: transient low-energy levels and durable high-energy levels (see in Fig. 2a, b); transient high-energy levels and durable low-energy levels (see in Fig. 2d); alternate high-energy and low-energy levels (see in Fig. 2c). For every mode, the high-energy state of this system is similar to the steady ordered flows and the low-energy state of this system is similar to the steady disordered flows. From these figures it is found that the periods of the oscillations are varied over time while the maximum kinetic energy in every oscillation of this system is almost the same, and so is the minimum kinetic energy. The change of height of this system is consistent with the variation of kinetic energy.

In [38], the authors described the transition from ordered to disordered flows is because of the dilatant and there is an 'explode' of the system (see the maximum height in Figs. 3, 4, 5 and 6). Then the transition from disordered to ordered flows is due to the self-organization of this system. Here we plot the development of the

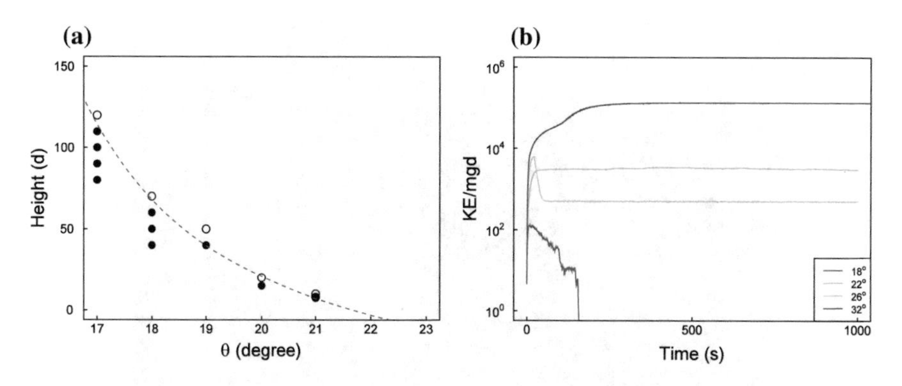

Fig. 1 **a** The h_{stop} as a function of θ. No flow state is shown with an open symbol, steady state with a closed symbol. The demarcation line is fitted to h_{stop} with a fitted formula used in [15] for inclined flows (the *red dash line*). **b** The KE per particle against time for four flow regimes (with a typical angle respectively and the oscillatory regime excluded)

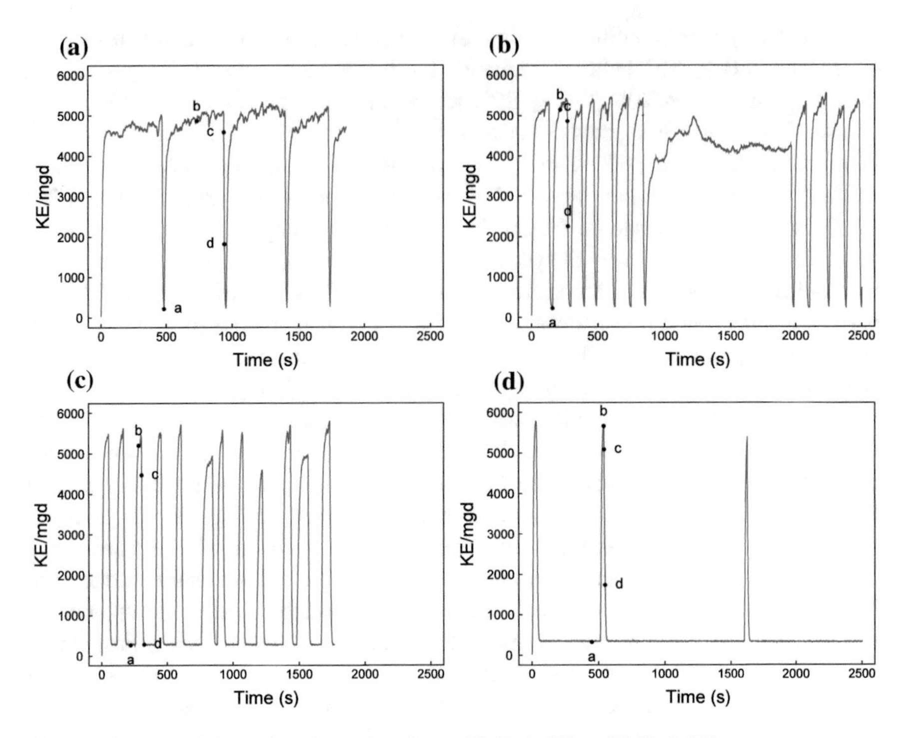

Fig. 2 KE per particle against time when $\theta =$ **a** 23.5°; **b** 24°; **c** 24.5°; **d** 25°

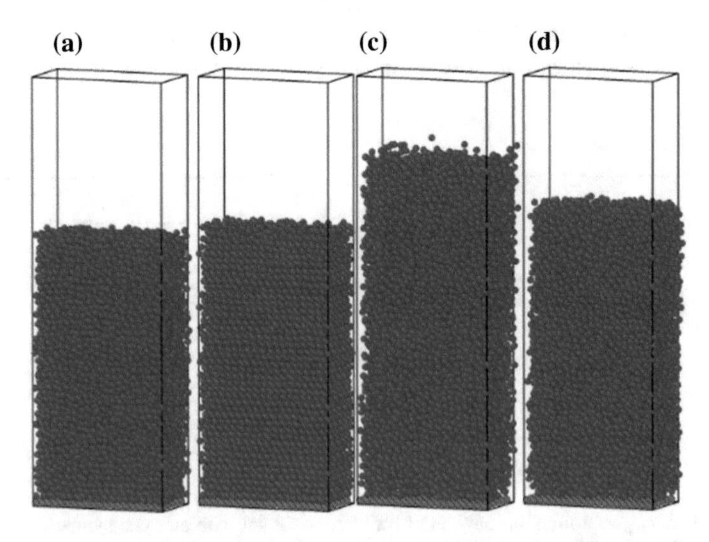

Fig. 3 Configuration snapshots of the system at $\theta = 23.5°$. **a–d** refer to the corresponding time points showed in Fig. 2a

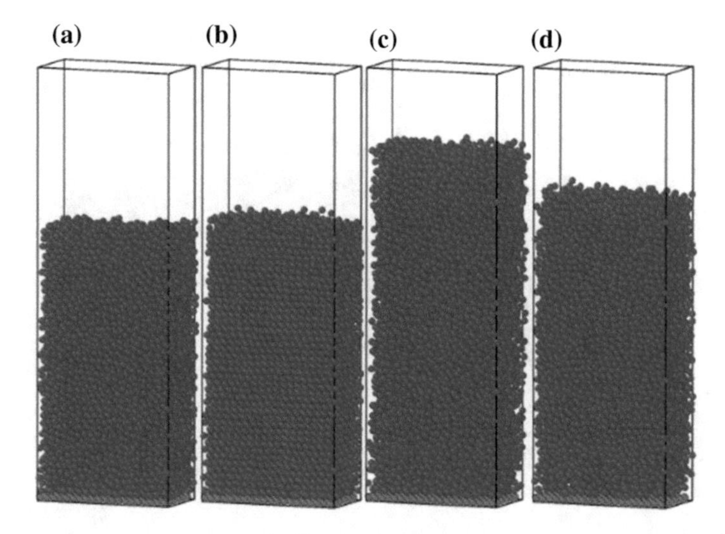

Fig. 4 Configuration snapshots of the system at $\theta = 24°$. **a**–**d** refer to the corresponding time points showed in Fig. 2b

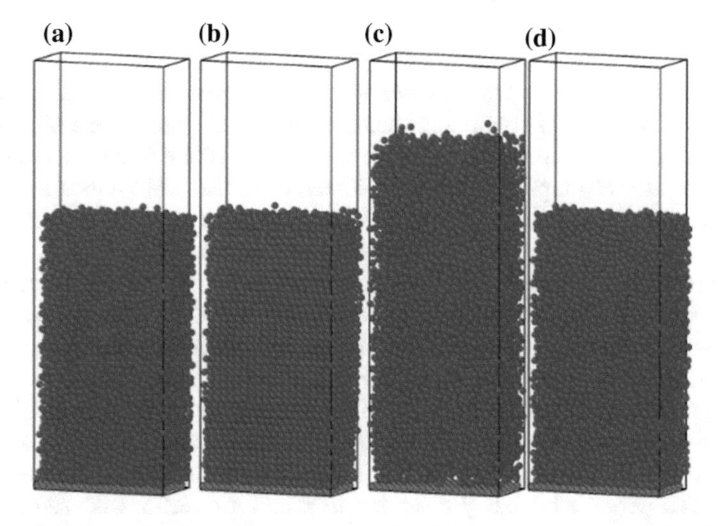

Fig. 5 Configuration snapshots of the system at $\theta = 24.5°$. **a**–**d** refer to the corresponding time points showed in Fig. 2c

profiles of number density and shear rate during the oscillations. The self-organization before the 'explode' of the system is shown in Figs. 2, 3, 4, 5 and 6. The transition from disordered state into ordered state is quick. When the flows are exploding, there is still a big shearing in the system.

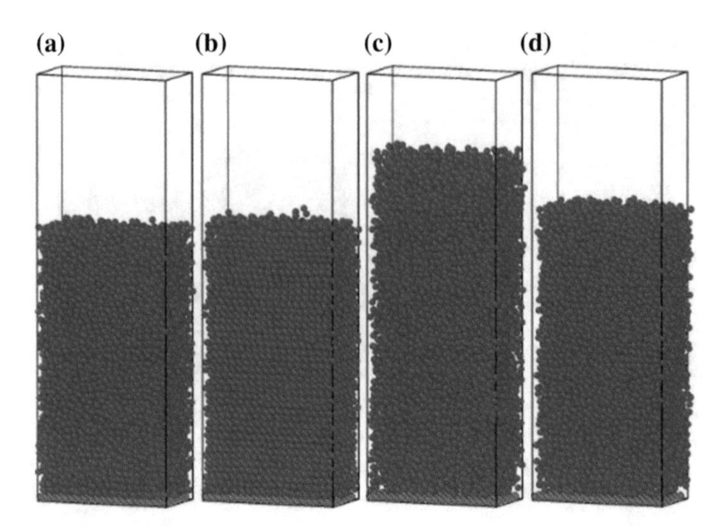

Fig. 6 Configuration snapshots of the system at $\theta = 25°$. **a–d** refer to the corresponding time points showed in Fig. 2d

4 Conclusion

This work is a study about the granular flow down a periodic, inclined plane with rough base by using Hertz-Mindlin contact model and discrete element method (DEM) simulations. The inclination angle is changed and the behaviors of flows are investigated. Based on previous work by Silbert et al. [38] and Weinhart et al. [21], there are some conclusions of this work: (1) the phases of flows with various inclination angles are classified. (2) According to the oscillation modes, the oscillation flow region can be further divided into three sub-regions. (3) The oscillation flow region is a transition region between ordered and disordered flow region, where more details about the self-organization and dilatant are shown.

The simulation results of inclined granular flows in this paper show good agreement with previous results using different simulation methods. In the future there are a lot work need to be done. The previous work used softer sphere to accelerate the simulation. So what will happen when the sphere becomes much harder? The periodic boundary condition in x and y directions is used here, the effect of sidewalls is neglected. For the design and engineering of inclined chute target, the large-scale granular materials (more than 10 000 000 particles) flows down a long, inclined chute has to be simulated further.

Acknowledgements This work is supported by the National Magnetic Confinement Fusion Science Program of China (Grant No. 2014GB104002).

References

1. Bauer, G.S., *Overview on spallation target design concepts and related materials issues.* Journal of Nuclear Materials, 2010. **398**(1–3): p. 19-27.
2. Fu, S., et al., *Status of CSNS Project.* IPAC2013, May, 2013.
3. Wagner, W. *Target development for the SINQ high-power neutron spallation source.* in *HIGH INTENSITY AND HIGH BRIGHTNESS HADRON BEAMS: 20th ICFA Advanced Beam Dynamics Workshop on High Intensity and High Brightness Hadron Beams ICFA-HB2002.* 2002. AIP Publishing.
4. Gabriel, T.A., J.R. Haines, and T.J. McManamy, *Overview of the Spallation Neutron Source (SNS) with emphasis on target systems.* Journal of Nuclear Materials, 2003. **318**: p. 1–13.
5. Center, J.-P., *Technical Design Report of spallation neutron source facility in J-PARC.*
6. Yang, L. and W. Zhan, *New concept for ADS spallation target: Gravity-driven dense granular flow target.* Science China Technological Sciences. **58**(10): p. 1705–1711.
7. Drake, T.G., *STRUCTURAL FEATURES IN GRANULAR FLOWS.* Journal of Geophysical Research-Solid Earth and Planets, 1990. **95**(B6): p. 8681–8696.
8. Savage, S.B. and K. Hutter, *The Motion of a Finite Mass of Granular Material down a Rough Incline.* Journal of Fluid Mechanics, 1989. **199**: p. 177–215.
9. Goldschmidt, M.J.V., R. Beetstra, and J.A.M. Kuipers, *Hydrodynamic modelling of dense gas-fluidised beds: Comparison of the kinetic theory of granular flow with 3D hard-sphere discrete particle simulations.* Chemical Engineering Science, 2002. **57**(11): p. 2059–2075.
10. Kumaran, V., *Dynamics of dense sheared granular flows. Part 1. Structure and diffusion.* Journal of Fluid Mechanics, 2009. **632**: p. 109–144.
11. Azanza, E., F. Chevoir, and P. Moucheront, *Experimental study of collisional granular flows down an inclined plane.* Journal of Fluid Mechanics, 1999. **400**: p. 199–227.
12. Delannay, R., et al., *Towards a theoretical picture of dense granular flows down inclines.* Nature Materials, 2007. **6**(2): p. 99–108.
13. Pouliquen, O., *Scaling laws in granular flows down rough inclined planes.* Physics of Fluids, 1999. **11**(3): p. 542–548.
14. Forterre, Y. and O. Pouliquen, *Long-surface-wave instability in dense granular flows.* Journal of Fluid Mechanics, 2003. **486**: p. 21–50.
15. MiDi, G.D.R., *On dense granular flows.* European Physical Journal E, 2004. **14**(4): p. 341–365.
16. Lemaitre, A., *Origin of a repose angle: Kinetics of rearrangement for granular materials.* Physical Review Letters, 2002. **89**(6).
17. Baran, O., et al., *Velocity correlations in dense gravity-driven granular chute flow.* Physical Review E, 2006. **74**(5).
18. Pouliquen, O., *Velocity correlations in dense granular flows.* Physical Review Letters, 2004. **93**(24).
19. Silbert, L.E., J.W. Landry, and G.S. Grest, *Granular flow down a rough inclined plane: Transition between thin and thick piles.* Physics of Fluids, 2003. **15**(1): p. 1–10.
20. Kumaran, V. and S. Maheshwari, *Transition due to base roughness in a dense granular flow down an inclined plane.* Physics of Fluids, 2012. **24**(5).
21. Weinhart, T., et al., *Closure relations for shallow granular flows from particle simulations.* Granular Matter, 2012. **14**(4): p. 531–552.
22. Bagnold, R.A., *EXPERIMENTS ON A GRAVITY-FREE DISPERSION OF LARGE SOLID SPHERES IN A NEWTONIAN FLUID UNDER SHEAR.* Proceedings of the Royal Society of London Series a-Mathematical and Physical Sciences, 1954. **225**(1160): p. 49–63.
23. Silbert, L.E., et al., *Granular flow down an inclined plane: Bagnold scaling and rheology.* Physical Review E, 2001. **64**(5).

24. Drozd, J.J. and C. Denniston, *Constitutive relations in dense granular flows.* Physical Review E, 2010. **81**(2).
25. Bi, W.T., et al., *Experimental study of two-dimensional, monodisperse, frictional-collisional granular flows down an inclined chute.* Physics of Fluids, 2006. **18**(12).
26. Faug, T., R. Beguin, and B. Chanut, *Mean steady granular force on a wall overflowed by free-surface gravity-driven dense flows.* Physical Review E, 2009. **80**(2).
27. Staron, L., *Friction and the oscillatory motion of granular flows.* Physical Review E, 2012. **86**(4).
28. Bi, W.T., et al., *Two- and three-dimensional confined granular chute flows: experimental and numerical results.* Journal of Physics-Condensed Matter, 2005. **17**(24): p. S2457–S2480.
29. Kumaran, V. and S. Bharathraj, *The effect of base roughness on the development of a dense granular flow down an inclined plane.* Physics of Fluids, 2013. **25**(7).
30. Weinhart, T., et al., *Coarse-grained local and objective continuum description of three-dimensional granular flows down an inclined surface.* Physics of Fluids, 2013. **25**(7).
31. Reddy, K.A. and V. Kumaran, *Applicability of constitutive relations from kinetic theory for dense granular flows.* Physical Review E, 2007. **76**(6).
32. Borzsonyi, T. and R.E. Ecke, *Flow rule of dense granular flows down a rough incline.* Physical Review E, 2007. **76**(3).
33. Brewster, R., et al., *Plug flow and the breakdown of Bagnold scaling in cohesive granular flows.* Physical Review E, 2005. **72**(6).
34. Hunt, M.L., et al., *Revisiting the 1954 suspension experiments of R. A. Bagnold.* Journal of Fluid Mechanics, 2002. **452**: p. 1–24.
35. Andreotti, B., *Sonic sands.* Reports on Progress in Physics, 2012. **75**(2).
36. Andreotti, B., A. Daerr, and S. Douady, *Scaling laws in granular flows down a rough plane.* Physics of Fluids, 2002. **14**(1): p. 415–418.
37. Rajchenbach, J., *Dense, rapid flows of inelastic grains under gravity.* Physical Review Letters, 2003. **90**(14).
38. Silbert, L.E., et al., *Boundary effects and self-organization in dense granular flows.* Physics of Fluids, 2002. **14**(8): p. 2637–2646.
39. Richard, P., S. McNamara, and M. Tankeo, *Relevance of numerical simulations to booming sand.* Physical Review E, 2012. **85**(1).
40. Tan, D., P. Richard, and J.T. Jenkins, *A model for the onset of oscillations near the stopping angle in an inclined granular flow.* The European physical journal. E, Soft matter, 2012. **35** (11): p. 122–122.
41. Khakhar, D.V., et al., *Surface flow of granular materials: model and experiments in heap formation.* Journal of Fluid Mechanics, 2001. **441**: p. 255–264.
42. Jop, P., Y. Forterre, and O. Pouliquen, *Crucial role of sidewalls in granular surface flows: consequences for the rheology.* Journal of Fluid Mechanics, 2005. **541**: p. 167–192.
43. de Gennes, P.G., *Granular matter: a tentative view.* Reviews of Modern Physics, 1999. **71**(2): p. S374-S382.
44. Schaeffer, D.G., *A Mathematical-Model for Localization in Granular Flow.* Proceedings of the Royal Society of London Series a-Mathematical Physical and Engineering Sciences, 1992. **436**(1897): p. 217–250.
45. Ostoja-Starzewski, M., *Random-Fields and Processes in Mechanics of Granular-Materials.* Mechanics of Materials, 1993. **16**(1–2): p. 55-64.
46. Aranson, I.S. and L.S. Tsimring, *Patterns and collective behavior in granular media: Theoretical concepts.* Reviews of Modern Physics, 2006. **78**(2): p. 641–692.
47. Ottino, J.M. and D.V. Khakhar, *Mixing and segregation of granular materials.* Annual Review of Fluid Mechanics, 2000. **32**: p. 55–91.

48. Tian, Y., et al. *A heterogeneous CPU-GPU implementation for discrete elements simulation with multiple GPUs*. in *Awareness Science and Technology and Ubi-Media Computing (iCAST-UMEDIA), 2013 International Joint Conference on*. 2013. IEEE.
49. Qi, J., et al., *GPU-accelerated DEM implementation with CUDA*. International Journal of Computer Science and Engineering, Inderscience, 2015. **11**(3): p. 330–337.
50. Burman, B.C., *A Discrete Numerical-Model for Granular Assemblies*. Geotechnique, 1980. **30**(3): p. 331–334.
51. Johnson, K.L. and K.K.L. Johnson, *Contact mechanics*. 1987: Cambridge university press.

Evaluation of Influences of Frictions in Hopper Flows Through GPU Simulations

Ping Lin, Sheng Zhang, Guanghui Yang, Jiang-Feng Wan, Yuan Tian and Lei Yang

Abstract The applicability of general purpose computing on GPUs for scientific and engineering applications has been growing several folds in recent years. Discrete Element method (DEM) is a way to modelling of particles, in which microscopic understanding of millions of particles is studied through simulation of granular materials such as sand or powders. Taking advantage of the highly data parallel nature of such computations, the benefits of executing DEM simulations have been widely spotlighted, and accelerations of computations through heterogeneous many-core environments are expressive. In this work, we present efficient implementations and investigate the applicability of GPUs to DEM used in particle motion simulation. In our implementations, algorithms such as neighbor list generation and pointer-exchange are performed. For the design of granular flow target, the influence of different frictions in hopper flows are investigated and presented here.

Keywords GPU · DEM · Neighbor list · Friction

P. Lin · S. Zhang · G. Yang · J.-F. Wan · Y. Tian · L. Yang (✉)
Institute of Modern Physics, Chinese Academy of Science, 509 Nanchang Road, Lanzhou 730000, Gansu, China
e-mail: lyang@impcas.ac.cn

P. Lin
e-mail: pinglin@impcas.ac.cn

S. Zhang
e-mail: zhangsheng@impcas.ac.cn

G. Yang
e-mail: yangguanghui@impcas.ac.cn

J.-F. Wan
e-mail: jiangfengwan@impcas.ac.cn

Y. Tian
e-mail: tianyuan08@impcas.ac.cn

G. Yang · J.-F. Wan
University of Chinese Academy of Science, No. 19A Yuquan Road, Beijing 100049, China

© Springer Nature Singapore Pte Ltd. 2018

N.Y. Yen and J.C. Hung (eds.), *Frontier Computing*, Lecture Notes in Electrical Engineering 422, DOI 10.1007/978-981-10-3187-8_22

1 Introduction and Related Studies

Discrete Element Method (DEM) was first proposed by Cundall et al. in 1971 [1] for simulating the behavior of materials consisted of particles. These particles could be rocks, grains, sands or toners. Because of its convenience and validity, DEM has been well developed and is widely used today. Today it has been applied in many fields such as agriculture and food handling, chemical engineering, mineral processing, pharmaceutical, powder metallurgy, and several other fields. GPUs (Graphics Processing Units) have become a powerful alternative for such computationally intensive applications in recent years, and clusters with GPUs as accelerators offer a possibility to tackle with DEM simulations with hundreds of millions particles in reasonable time.

Compared to traditional CPUs, GPUs have a large number of lightweight computational units called streaming processors. It was not popular for people to dispatch calculations into such units till 2002. However during that period, only those scientists with experiences in graphics-oriented libraries such as OpenGL or DirectX could use the so-called GPGPU technology to conduct parallel computations indirectly. Later in 2006, NVIDIA released CUDA (Compute Unified Device Architecture), to provide scientists with a much easier programming paradigm for common computations. In CUDA C, one can easily launch a kernel function on GPU. The kernel function prefixed with modifier __global__ is annotated by adding “≪ <…,…,… ≫>” between the function name and the parameters, which contents “tell” GPU how to organize the computational resources [2]. Since most of these computations are independent and GPUs have large number of light-weighted processors suitable for parallel processing, implementation of DEM on GPUs could be much more efficient than on CPUs. In this paper, we propose a novel algorithm and its efficient implementation that highly exploits parallelism that fully takes advantage of GPU resources.

DEM, which closely relates to MD (molecule dynamics), needs a huge number of calculations. Since CPUs are not good at this, several researchers have driven the implementation of DEM on GPUs to take advantage of a large number of small processors for calculations. Implementations of MD and DEM on GPUs could be much more efficient than its CPU counterpart with high efficiency [3–5]. Liu et al. [6] have accelerated MD simulations using CUDA, while Rapaport proposed several algorithms for MD [7] and later implemented in CUDA. Le Grand et al. [8] discussed a mixed precision model for GPU accelerated MD simulations. Among these implementations, the method in [7] has been widely used. Brown et al. [9] discussed several important issues in porting large molecular dynamics application program to parallel hybrid machines, and then partially used the algorithm in [10]. We investigated DEM on GPUs using the algorithm in [10]. Compared to traditional MD, the DEM cells are much smaller. However, this algorithm did not explore to take advantage of small cells. We derived a new but simpler method to take advantage of DEM with large scale GPUs parallel computation. We found that, within certain conditions such as smaller cell requirements in DEM, the method proposed in this work has large advantages as performance improvement [3].

Recently, Xu et al. [11] and Xiong et al. [12] have implemented DEM on GPUs without including tangential force. Walizer et al. [13] provided a box search algorithm for particles with different sizes. Su et al. [14] developed a robust and accurate algorithm for detecting the interaction between a spherical particle and an arbitrarily complex geometric surface. Ye et al. [15] studied large-scale granular flow scenes on GPUs with application of DEM. Zheng et al. [16] concentrated investigations on the contact detection based on GPUs. Radeke et al. [17] used GPUs to simulate large-scale powder mixer. These previous researches as introduced did not provide enough details of implementation neither comparative benchmark results.

Flowing granular materials usually behave like solid and liquid simultaneously. As high power neutron sources, various kinds of targets have been designed, constructed, operated, such as plate targets in IPNS and ISIS, CSNS, rod targets in German SNQ project [18, 19] and SINQ at PSI [20], liquid metal targets in SNS and JSNS [21, 22]. Following the ideas above, the gravity-driven dense granular-flow target (DGT) was proposed lately [23], which combines advantages of solid and liquid targets and the container is generally like a hopper. The DEM simulations on GPUs are very available for design of the target. The natures of the hopper flows can be investigated by large-scale DEM simulations [4, 5].

In the design of granular target, the roles of friction is a crucial issue, which will affect the flowing behavior of the hopper flow [4, 24, 25]. Moreover, the influernce of wall and bottom on the granular flow should be obtained. In this paper the influence of particle-particle, particle-bottom, particle-wall frictions will be evaluated by large-scale simulations on GPUs.

2 Physical Model

In DEM systems, particle i may interact with its neighboring (or contacting) particles. Many have been used to describe the force when particle i and particle j. Here we use the so-called Hertz Granular Contact Model [26]. In this model if one particle i and another particle j collide at position $(\boldsymbol{r}_i, \boldsymbol{r}_j)$ with velocities $(\boldsymbol{v}_i, \boldsymbol{v}_j)$ and angular velocities $(\boldsymbol{\omega}_i, \boldsymbol{\omega}_j)$, the contact force between them is given by:

$$\boldsymbol{F}_{ij} = \boldsymbol{F}_{n_{ij}} + \boldsymbol{F}_{t_{ij}} \tag{1}$$

$$\boldsymbol{F}_{n_{ij}} = \sqrt{\frac{\delta_n}{d_i}}(k_n \delta_{\boldsymbol{n}_{ij}} - \gamma_n \boldsymbol{v}_{\boldsymbol{n}_{ij}}) \tag{2}$$

$$\boldsymbol{F}_{t_{ij}} = \sqrt{\frac{\delta_n}{d_i}}(-k_t \delta_{t_{ij}} - \gamma_t \boldsymbol{v}_{t_{ij}}) \tag{3}$$

Table 1 Constants

Notation	Meaning
N	Number of particles
$d\tau$	Time step
μ	Coefficient of friction
ε	Coefficient of restitution
k_n	Normal stiffness
k_t	Tangential stiffness
γ_n	Normal viscosity
γ_t	Tangential viscosity
d_i	Diameter of particle i
r_i	Radius of particle i
m_i	Mass of particle i
I_i	Moment of inertia of particle i

Table 2 Variables

Notation	Meaning
$\boldsymbol{r}_i$	Position of particle i
$\boldsymbol{v}_i$	Translation velocity of particle i
$\boldsymbol{\omega}_i$	Angular velocity of particle i
$\boldsymbol{a}_i$	Translation accelerations of particle i
$\boldsymbol{b}_i$	Rotational accelerations of particle i
τ	Current time
τ_0	The time when collision starts

Here the subscripts n and t represent normal and tangential direction. Moreover, k, γ, d_i, v are elastic parameters, viscoelastic parameters, displacements, velocities, respectively.

In Hertz model, all the physical quantities on the right hand of Eqs. (2), (3) can be obtained by two types of parameters. One is constant as the system is running, such as density, Young's modulus, Poisson's ratio, coefficient of restitution, etc., and as a result these constants are inputs according to particle material. All of this kind of parameters are listed in Table 1. The values of these constants can be obtained from the tables of material properties. The other kind varies when spheres change their status, and they are listed in Table 2. Table 3 provides how to calculate other physical quantities on the right hand of Eqs. (2) and (3). Note that $\delta_{t_{ij}}$ is integrated from the moment two particle collided, and after the collision is over it vanishes.

In the other hand, a local Coulomb yield criterion which is $\left\|\boldsymbol{F}_{t_{ij}}\right\| \leq \mu\left\|\boldsymbol{F}_{n_{ij}}\right\|$ must be satisfied where μ is the coefficient of friction. Forces between two contacting particles is obtained by Eqs. (1), (2), (3) and Table 3. In simulations, walls are usually used. The walls used to be considered as spheres with infinite radius and mass. As a result, the forces between walls and particles could also be calculated by the method stated above. It's also necessary to mention that μ, ε and k_n etc. are

Table 3 calculation of relative displacement and velocities in normal and tangential directions

Output(s)	Input(s)	Formula(s)
$\boldsymbol{r}_{ij}, \boldsymbol{n}_{ij}$	$\boldsymbol{r}_i, \boldsymbol{r}_j$	$\boldsymbol{r}_{ij} = \boldsymbol{r}_i - \boldsymbol{r}_j$ $\boldsymbol{n}_{ij} = \boldsymbol{r}_{ij} / \lVert\boldsymbol{r}_{ij}\rVert$
$\boldsymbol{v}_{ij}$	$\boldsymbol{v}_i, \boldsymbol{v}_j$	$\boldsymbol{v}_{ij} = \boldsymbol{v}_i - \boldsymbol{v}_j$
$\boldsymbol{v}_{n_{ij}}$	$\boldsymbol{v}_{ij}, \boldsymbol{n}_{ij}$	$\boldsymbol{v}_{n_{ij}} = (\boldsymbol{v}_{ij}.\boldsymbol{n}_{ij})\boldsymbol{n}_{ij}$
$\boldsymbol{v}_{t_{ij}}$	$\boldsymbol{v}_{ij}, \boldsymbol{v}_{n_{ij}}, r_i, r_j, \boldsymbol{\omega}_i, \boldsymbol{\omega}_j$	$\boldsymbol{v}_{t_{ij}} = \boldsymbol{v}_{ij} - \boldsymbol{v}_{n_{ij}} - (r_i\boldsymbol{\omega}_i + r_j\boldsymbol{\omega}_j) \times \boldsymbol{r}_{ij}$
δ_n	$r_i, r_j, \boldsymbol{r}_{ij}$	$\delta_n = r_i + r_j - \lVert\boldsymbol{r}_{ij}\rVert$
$\boldsymbol{\delta}_{n_{ij}}$	$\delta_n, \boldsymbol{n}_{ij}$	$\boldsymbol{\delta}_{n_{ij}} = \delta_n \boldsymbol{n}_{ij}$
$\boldsymbol{\delta}_{t_{ij}}$	$\boldsymbol{v}_{t_{ij}}, \tau, \tau_0$	$\boldsymbol{\delta}_{t_{ij}} = \int_{\tau_0}^{\tau - \tau_0} \boldsymbol{v}_{ij} d\tau$

parameter between two bodies. Therefore, μ, ε and k_n etc. between particles and those between particle and walls should be distinguished.

Translational and rotational accelerations $\boldsymbol{a}_i$ and $\boldsymbol{b}_i$ are determined by:

$$m_i\boldsymbol{a}_i = \sum_j \boldsymbol{F}_{ij}(+m_i\boldsymbol{g} \text{ when considering gravity}) \quad (4)$$

$$I_i\boldsymbol{b}_i = -r_i \sum_j \boldsymbol{r}_{ij} \times \boldsymbol{F}_{t_{ij}} \quad (5)$$

At last a velocity-Verlet scheme [27] is chosen to integrate the Newtonian equation of particle motion. The specified formulas can be found somewhere else.

3 Implementation and Parameters

In this section, we discuss and present implementation issues on our proposed strategy. In order to develop an efficient parallel application that computes the DEM, one must come up with a good high-level decomposition of the problem. As we showed in Sect. 2, one must have clear understanding of behavior to make most suitable decisions. In order to take advantage of GPUs, computation steps and algorithms used for linked-cell and neighbor lists are employed. Figure 1 is the flow chart of whole process. The processes 1, 2, 4, 6 are simply parallel implemented in GPU, because each particle corresponding to one thread would do. It is necessary to generate neighbor list to accelerate force calculations and other operations. Instead of traditional neighbor list generating algorithm [10, 28], we chose a method which is more suitable for DEM cells. Algorithm 1 demonstrates this implementation to generate the neighbor list. After the neighbor list generated, we use a pointer-exchange algorithm to reduce the time of memory exchange. Tables 1, 2 and 3 give the constants, variables and intermediate variables in the code and the physical parameters needed in the code are listed in Table 4.

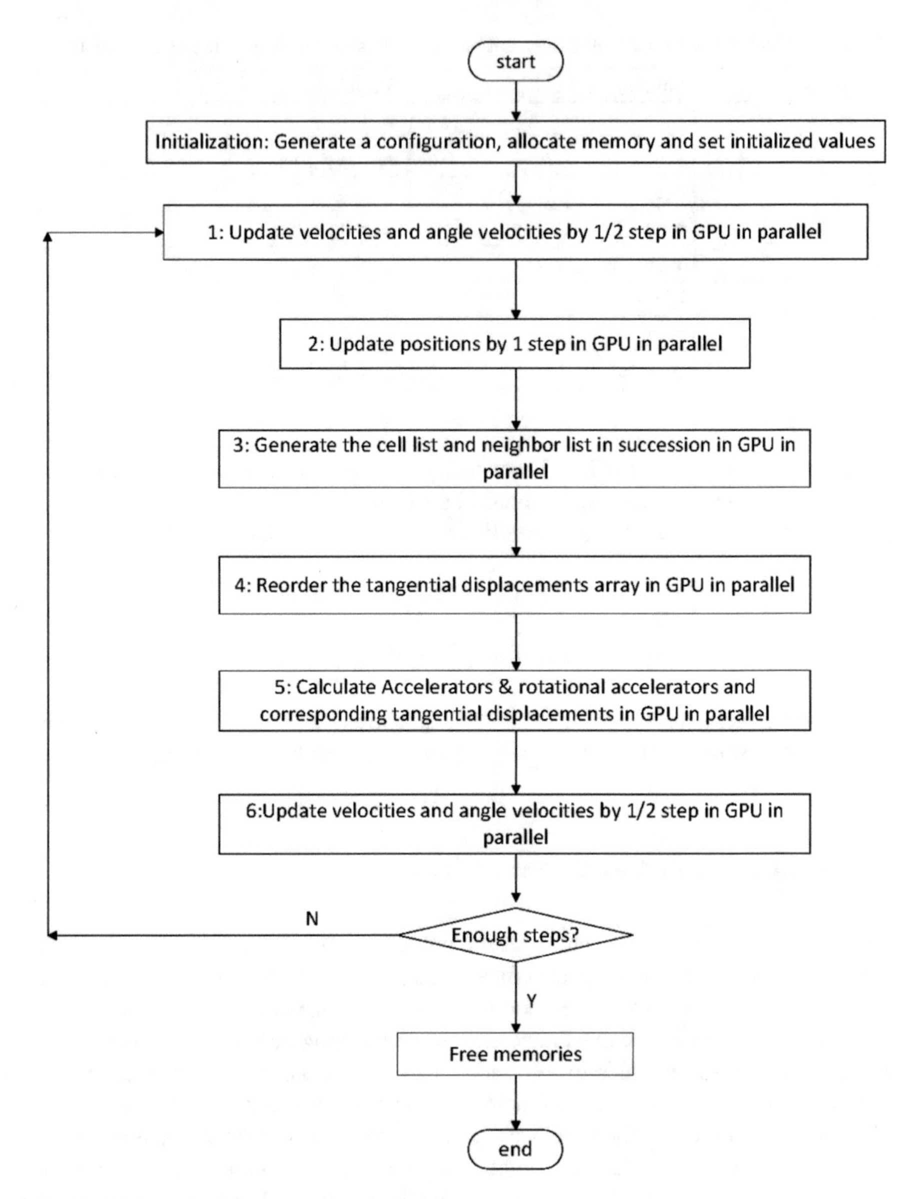

Fig. 1 The complete flow chart for single-GPU

Table 4 Geometrical and material parameters in simulations

Quantity	Notation	Value
Particle diameter	d	1
Particle mass	m	1
Acceleration of gravity	g	1
Normal elastic constant	k_n	2.0×10^5
Tangential elastic constant	k_t	2.0/5.0 k_n
Nomal damping constant	γ_n	50.0
Tangential damping constant	γ_t	30.0
Coefficient of friction	μ	0.0, 0.001, 0.05, 0.5
Coefficient of restitution	ε	0.8
Hopper diameter	D	30
Hopper orifice diameter	D_0	6, 7, 8, 9, 12

Algorithm 1: Generation of neighbor list

```
1: i = blockIdx.x * blockDim.x + threadIdx.x
2: if i < N then
3:  ri = rsorted[i]; iID = pID[i];
4:  define CCoor;
5:  CCoor.x= floor(ri.x - ox)/cellx;
6:  CCoor.y= floor(ri.y - oy)/celly;
7:  CCoor.z= floor(ri.z - oz)/cellz;
8:  cntnbr = 0;
9:  for Cjz = CCoor.z - 1 to CCoor.z + 1 do
10:   for Cjy = CCoor.y - 1 to CCoor.y + 1 do
11:     for Cjx = CCoor.x - 1 to CCoor.x + 1 do
12:       if Cell(Cjx,Cjy,Cjz) is not empty then
13:         startp = CS[Cjz*cellnx*cellny+Cjy*cellnx+Cjx];
14:         endp   = CE[Cjz*cellnx*cellny+Cjy*cellnx+Cjx];
15:         for j = startp to endp do
16:           if i ≠ j then
17:             rj = rsorted[j];
18:             dist = |ri - rj |;
19:             if dist < Ri + Rj then
20:               jID = pID[j];
21:               nbrlst[N * cntnbr + iID] = jID;
22:               cntnbr++;
23:             end if
24:           end if
25:         end for
26:       end if
27:     end for
28:   end for
29:  end for
30:  nbrlstcnt[iID] = cntnbr;
31: end if
```

4 Roles of the Frictions

As mentioned in Sect. 2, the frictions are interactions between 2 bodies. Therefore, in our simulations investigating the influence of frictions in hopper flows, three types of frictions are considered: particle-particle, particle-bottom, particle-wall frictions. The values of these coefficients we used are shown in Table 4, as well as other physical and geometrical parameters. These parameters are dimensionless with d, g and m as the basic dimensionless parameters. The flat-bottomed hopper diameter we used is 30 d and the orifice varies from 6 d to 12 d. Figure 2 gives the results of our simulations. It could be seen that the influence of particle-particle friction on flow rate is the most while particle-wall friction also has a significant influence. There is no obvious influence of particle-bottom friction on flow rate, which should be related with the formation of stagnant region on the bottom [29]. The flow rate can be quantitatively described by the Beverloo's law which has a form $\phi = C\rho_i\sqrt{g}(D_0 - kd)^{2.5}$ for three dimensional hoppers with round orifices [30]. In this equation, C and k are parameters to be fitted, and ρ_i is the equivalent density of granular materials in the hopper. In our simulation results, k is fitted to be 1.4, and C varies from 0.53 to 0.76 for different friction conditions, which is consistent with typical values of k and C [30].

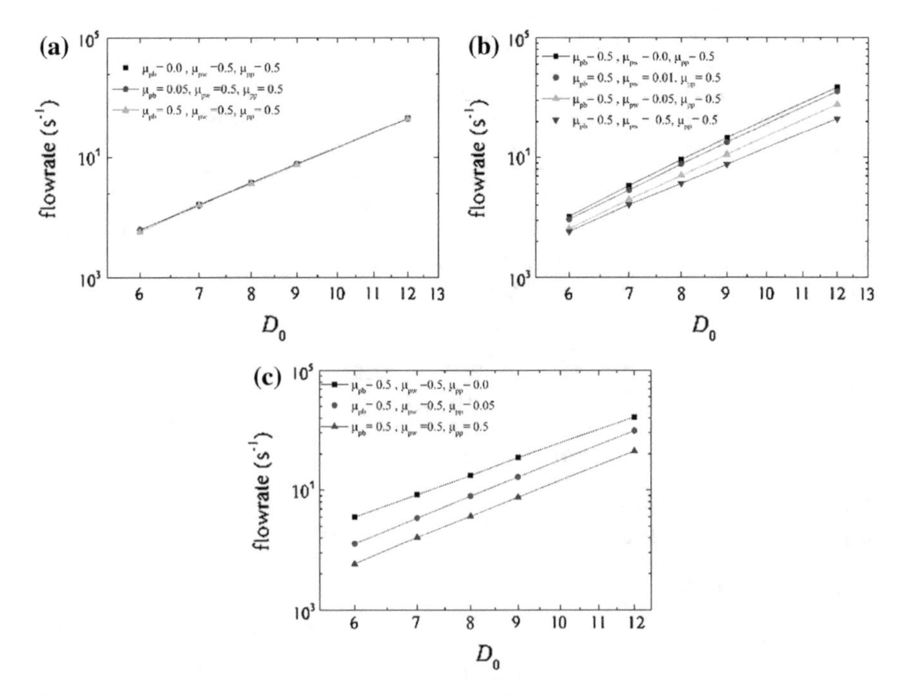

Fig. 2 Flow rate against orifice diameter under condition of **a** varying particle-bottom friction coefficient; **b** varying particle-bottom wall coefficient; **c** varying particle-particle friction coefficients

5 Conclusions and Future Work

DEM combing MD method are implemented on GPUs to simulate granular materials. Because of the features of the simulation system, it is suitable for parallel computation on GPUs and some algorithms such as neighbor list generation and pointer-exchange are used to accelerlate the simulation. Wall are modelled in the GPU code and hopper flows under different friction conditions are simulated. It is found particle-particle and particle-wall friction affect the flow rate obviously. For a full size design, larger systems simulated by multi-GPUs will be performed in the future.

Acknowledgements This work is supported by the National Magnetic Confinement Fusion Science Program of China (Grant No. 2014GB104002) and CAS 125 Informatization Project (No. XXH12503-02-03-2). Thanks to Dr. Ji Qi for valuable suggestions which improved the final manuscript.

References

1. Cundall, P.A. and O.D.L. Strack, *DISCRETE NUMERICAL-MODEL FOR GRANULAR ASSEMBLIES.* Geotechnique, 1979. **29**(1): p. 47–65.
2. NVidia, C., *C programming guide version 4.0.* NVIDIA Corporation, Santa Clara, CA, 2011.
3. Qi, J., et al., *GPU-accelerated DEM implementation with CUDA.* International Journal of Computer Science and Engineering, Inderscience, 2015. **11**(3): p. 330–337.
4. Zhang, S., et al., *Investigating the influence of wall frictions on hopper flows.* Granular Matter, 2014. **16**(6): p. 857–866.
5. Lin, P., et al., *Numerical study of free-fall arches in hopper flows.* Physica a-Statistical Mechanics and Its Applications, 2015. **417**: p. 29–40.
6. Liu, W.G., et al., *Accelerating molecular dynamics simulations using Graphics Processing Units with CUDA.* Computer Physics Communications, 2008. **179**(9): p. 634–641.
7. Rapaport, D.C., *Enhanced molecular dynamics performance with a programmable graphics processor.* Computer Physics Communications, 2011. **182**(4): p. 926–934.
8. Le Grand, S., A.W. Gotz, and R.C. Walker, *SPFP: Speed without compromise-A mixed precision model for GPU accelerated molecular dynamics simulations.* Computer Physics Communications, 2013. **184**(2): p. 374–380.
9. Brown, W.M., et al., *Implementing molecular dynamics on hybrid high performance computers - Particle-particle particle-mesh.* Computer Physics Communications, 2012. **183**(3): p. 449–459.
10. Anderson, J.A., C.D. Lorenz, and A. Travesset, *General purpose molecular dynamics simulations fully implemented on graphics processing units.* Journal of Computational Physics, 2008. **227**(10): p. 5342–5359.
11. Xua, J., et al., *Quasi-real-time simulation of rotating drum using discrete element method with parallel GPU computing.* Particuology, 2011. **9**(4): p. 446–450.
12. Xiong, Q.G., et al., *Large-scale DNS of gas-solid flows on Mole-8.5.* Chemical Engineering Science, 2012. **71**: p. 422–430.
13. Walizer, L.E. and J.F. Peters, *A bounding box search algorithm for DEM simulation.* Computer Physics Communications, 2011. **182**(2): p. 281–288.
14. Su, J.W., Z.L. Gu, and X.Y. Xu, *Discrete element simulation of particle flow in arbitrarily complex geometries.* Chemical Engineering Science, 2011. **66**(23): p. 6069–6088.

15. Ye, J., et al., *Modeling and Rendering of Real-time Large-scale Granular Flow Scene on GPU.* 2011 3rd International Conference on Environmental Science and Information Application Technology Esiat 2011, Vol 10, Pt B, 2011. **10**: p. 1035–1045.
16. Zheng, J.W., X.H. An, and M.S. Huang, *GPU-based parallel algorithm for particle contact detection and its application in self-compacting concrete flow simulations.* Computers & Structures, 2012. **112**: p. 193–204.
17. Radeke, C.A., B.J. Glasser, and J.G. Khinast, *Large-scale powder mixer simulations using massively parallel GPU architectures.* Chemical Engineering Science, 2010. **65**(24): p. 6435–6442.
18. Bauer, G.S., *Overview on spallation target design concepts and related materials issues.* Journal of Nuclear Materials, 2010. **398**(1–3): p. 19-27.
19. Fu, S., et al., *Status of CSNS Project.* IPAC2013, May, 2013.
20. Wagner, W. *Target development for the SINQ high-power neutron spallation source.* in *HIGH INTENSITY AND HIGH BRIGHTNESS HADRON BEAMS: 20th ICFA Advanced Beam Dynamics Workshop on High Intensity and High Brightness Hadron Beams ICFA-HB2002.* 2002. AIP Publishing.
21. Gabriel, T.A., J.R. Haines, and T.J. McManamy, *Overview of the Spallation Neutron Source (SNS) with emphasis on target systems.* Journal of Nuclear Materials, 2003. **318**: p. 1–13.
22. Center, J.-P., *Technical Design Report of spallation neutron source facility in J-PARC.*
23. Yang, L. and W. Zhan, *New concept for ADS spallation target: Gravity-driven dense granular flow target.* Science China Technological Sciences. **58**(10): p. 1705–1711.
24. Ferellec, J.F., et al., *Influence of particle rolling resistance on silo flow in DEM simulations.* Powders and Grains 2001, 2001: p. 409–412.
25. Sandlin, M., *An experimental and numerical study of granular hopper flows.* 2013.
26. Silbert, L.E., et al., *Granular flow down an inclined plane: Bagnold scaling and rheology.* Physical Review E, 2001. **64**(5).
27. Allen, M.P. and D.J. Tildesley, *Computer simulation of liquids*. 1989: Oxford university press.
28. Hrenya, C., *Computational Granular Dynamics - Models and Algorithms, Thorsten Pöschel and Thomas Schwager, Springer, 2005.* Granular Matter. **8**(1): p. 55–55.
29. Nedderman, R.M., *Statics and kinematics of granular materials*. Digitally printed 1st pbk. ed. 2005, Cambridge, UK; New York: Cambridge University Press. xv, 352 p.
30. Nedderman, R.M., et al., *The Flow of Granular-Materials.1. Discharge Rates from Hoppers.* Chemical Engineering Science, 1982. **37**(11): p. 1597–1609.

The Requirement Analysis and Initial Design of a Cloud and Crowd Supported Mathematics Learning Environment for Computer Science Students

Chun-Hsiung Tseng, Jyi-Shane Liu, Yung-Hui Chen, Lin Hui, Yan-Ru Jiang and Jia-Rou Lin

Abstract Math learning has never been easy and math learning in computer science is not an exception. However, the important of math can never be underestimated. In computer science, it was found that students learning performance in math is strongly connected with the development of the following capabilities: problem solving, programming, computer hardware and architecture design, computer science theory understanding, and software engineering and system analysis. The goal of this research is to develop a method based on cloud technologies and crowd intelligence to enhance students learning performance of math in computer science.

Keywords Crowdsourcing · CSCL · Elearning

1 Introduction

Mathematics plays an important role in many learning and research fields, and computer science is not an exception. Several mathematics topics are considered required for students who choose computer science as their major. For example, most students have to pass the training of basic statistics and calculus. Additionally, as shown

C.-H. Tseng · Y.-R. Jiang · J.-R. Lin
Department of Information Management, Nanhua University, Chiayi, Taiwan
e-mail: lendle_tseng@seed.net.tw

J.-S. Liu
Department of Computer Science, National Chengchi University, Taipei, Taiwan
e-mail: jsliu@nccu.edu.tw

Y.-H. Chen (✉)
Department of Computer Information and Network Engineering,
Lunghwa University of Science and Technology, Taoyuan, Taiwan
e-mail: cyh@mail.lhu.edu.tw

L. Hui
Department of Innovative Information and Technology, Tamkang University,
New Taipei City, Taiwan
e-mail: 121678@mail.tku.edu.tw

© Springer Nature Singapore Pte Ltd. 2018

N.Y. Yen and J.C. Hung (eds.), *Frontier Computing*, Lecture Notes
in Electrical Engineering 422, DOI 10.1007/978-981-10-3187-8_23

in several surveys, how university students performed in mathematics classes also affects their working performance. For those who want to be good programmers after being graduated, a solid mathematics background is usually needed for writing error-proof programs. For those who want to participate in research jobs in computer science, the importance of good mathematics background is even higher. Some hot research topics, such as big data, put high demand on mathematics capabilities. However, teaching and learning mathematics are never easy tasks. The situation motivates this research.

According to the survey made by Bravaco et al. [1], there are several reasons about why students majoring in computer science do not perform well in mathematics [1]:

1. students have wide range of mathematical abilities, so course design is difficult
2. some students do not see the importance of the linkage between mathematics and their major
3. it is difficult to get students to take their courses in the best order.

To deal with the challenges shown above, the researchers believe that several information technologies can help. To ease the design difficulties of course materials due to students wide range of mathematical abilities, e-learning technologies can be adopted. A possible solution is to augment the traditional computer supported collaborative learning (CSCL [2]) methods with crowd technologies. In such a way, we can incorporate the advantages of computer supported collaborative personalized learning methods into CSCL by organizing collaborative groups dynamically according to students abilities. Then, the more flexible CSCL groups become stronger supports for students with different mathematics backgrounds. Furthermore, to help students realize the importance of linkages between mathematics and their major, we have to give them more practical materials. For instance, to have students understand the importance and use cases of linear regression, including real data prediction cases in class materials will be helpful.

2 Related Works

For students majoring in computer science, mathematics is a very important course. The survey made by Konvalina et al. showed that mathematical reasoning ability and mathematical background has very important effect for the potential success in computer science [3]. The work of Henderson et al. showed a similar result and stated that mathematics is an important tool for problem-solving and conceptual understanding in computing [4]. The work of Beaubouef summarized several fields in computer science in which mathematics is essential [5]. However, learning mathematics has never been an easy task. Fleming wrote an article on about.com, which said The thing that makes mathematics difficult for many students is that it takes patience and persistence. For many students, mathematics is not something that comes intuitively or automatically - it takes plenty of effort. It is a subject that sometimes requires students to devote lots and lots of time and energy. Shermans article summarized several factors about why students struggle in mathematics [6]. These factors included: instruction, curricular materials, the gap between learner and subject, locus of control, memory ability, attention span, and mathematics language understanding. Although former studies draw to different conclusions about why learning mathematics is difficult, the importance and difficulties of mathematics learning was stated clearly. Focusing on mathematics classes in computer science, the work of Bravaco et al. listed the difficulties encountered in teaching [1]. Many pioneering researchers have devoted their works to make learning and teaching mathematics easier. Some researchers focused on game-based learning. For example, Zanchi et al. described a Next Generation Preschool Mathematics project in which researchers and media developers joined their works to develop mathematics curriculum supplement that supports young childrens learning of subitizing and equipartitioning [7]. Kes study indicated that gaming goal structures, beyond the games themselves, yield significant effects on participants mathematics learning attitudes [8]. Kes another work argued that using computer-based educational game as a motivational tool for cooperative learning is more convincing than using it as a cognitive or metacognitive one [9]. In addition to game-related methods, researchers developed various ways of benefiting from computer technologies to aid learning mathematics. Niess highlighted that Mathematics teachers are challenged to think about scaffolding students learning about spreadsheets while they are also learning mathematics [10]. Stahl found that mathematics can be accomplished collaboratively, even by small groups of novice mathematics students helping each other, building sequentially on each others moves and exploring together, even across session, and proposed a concept named as virtual mathematics teams [11]. Although not specifically targeted at mathematics learning, Lambropoulos, N. and Romero considered the personalised information retrieval in a CSCL task through the use of a Group-Awareness widget and achieved excellent results [12]. The work of Edrees proposed e-Learning 2.0, which integrated web 2.0 technologies and tools into educational and institutional practice [13]. In this research, in addition to CSCL-based technologies, the researchers would like to benefit from the intelligence of the crowd. The concept is similar with crowd sourcing, and its importance was pointed out by Greengard in his research work [14].

3 Cloud and Crowd Supported Mathematics Learning

In this research, the researchers propose a cloud and crowd supported mathematics learning method which focuses on mathematics classes in computer science. As stated in previous section, since computer science students are usually familiar and feel at ease with information technologies, the researchers will design an e-learning system to facilitate the adoption of the proposed method. The system will utilize crowd intelligence to augment the traditional CSCL method to help students with various mathematics abilities benefit from group learning. The system will also utilize crowd intelligence for the construction of scaffoldings of topic flows and course

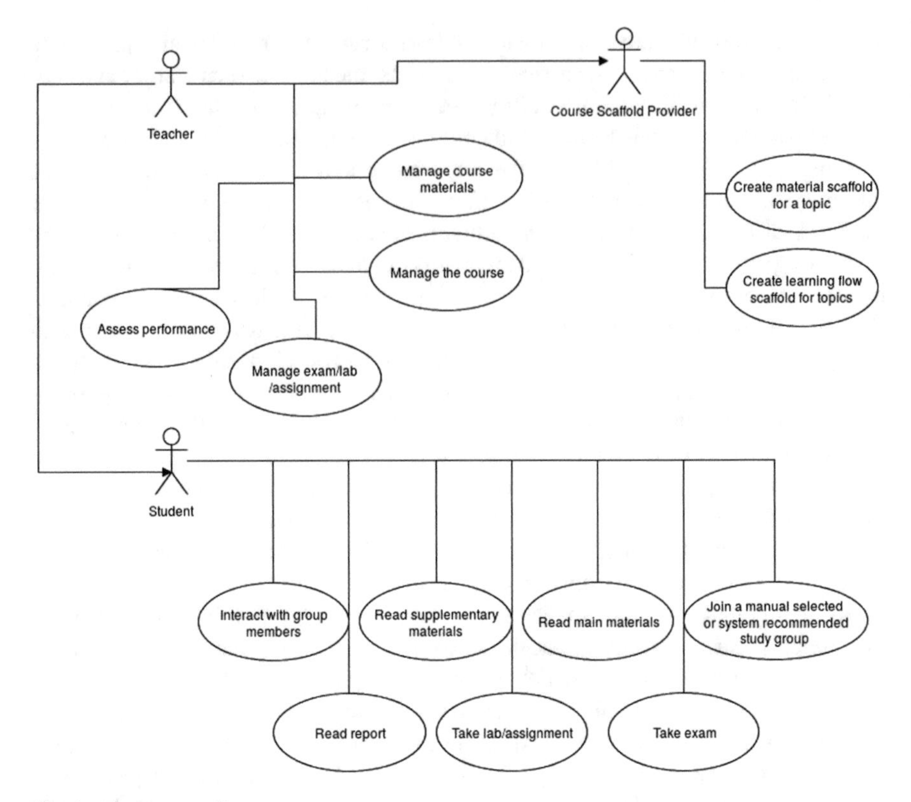

Fig. 1 The usecase diagram

contents. Furthermore, the system will utilize information extraction technologies to obtain real world supplementary materials from the cloud.

There are three types of users involved in the proposed system: teachers, students, and course scaffold providers. Teachers are the main mediators of a course and are responsible for the preparation of course materials, the management of courses, and assessments. On the other hand, students are main players in a course. In most cases, students follow the flow designed by teachers. To benefit from group learning, the system includes various group interaction utilities. Furthermore, in addition to main materials, students can read supplementary materials that either contributed by teachers or automatically collected by the system. Course scaffold providers are responsible for designing scaffolds for course materials or flows of topics. The designed scaffolds can be used by teachers to aid the design of the course. Note that teachers can also play the role as students or course scaffold providers. Figure 1 shows the complete use case diagram.

There are 13 use cases included in the design:

1. Create material scaffold for a topic: a material scaffold defines what should be include in a class, e.g. exams, main materials, and the criteria of supplementary materials; additionally, a material scaffold can also include preferred assessment method for this class
2. Create learning flow scaffold for topics: a flow scaffold defines the flow between several topics for a class
3. Read main materials: students can read the main materials of a class
4. Read supplementary materials: students can read the supplementary materials of a class; supplementary materials can be provided by teachers or automatically collected by the system
5. Take exam: students can take exams provided by the teacher; note that if collaboration is allowed and needed, students should execute the interact with group members use case
6. Take lab/assignment: students can take labs or assignments provided by the teacher; note that if collaboration is allowed and needed, students should execute the interact with group members use case
7. Join a manual selected or system recommended study group: classes adopting CSCL benefit from interaction among group members; however, how well a learning group is formed will definitely affects the learning performance; by including a learning group recommendation module, the proposed system can automatically recommend suitable groups for students
8. Interact with group members: after joining a learning group, students can interact with group members; applications such as discussion rooms and collaboration environments will be provided
9. Read report: students can read their assessment reports of the learning performance for their participated classes
10. Manage course materials: teachers can manage both main and supplementary materials for a course; note that for automatically extracting supplementary from the Web, teachers have to specify proper information sources and extraction rules; when managing course materials, teachers can use existing material scaffolds as templates
11. Manage courses: teachers create, update, modify, and delete courses with this functionality; when managing courses, teachers can use existing flow scaffolds as basis to design the learning flow among topics; besides, logs of the courses are also available
12. Manage exams/labs/assignments: exams, labs, and assignments are important for students to practice concepts learned from classes and for teachers to evaluate the learning performance of students; in this use case, teachers will create, update, modify, and delete exams, labs, and assignments; also, teachers can correct exams, labs, and assignments completed by students; note that in some cases, collaboration may be allowed and required to complete exams, labs, and assignments; in such cases, students should execute the interact with group members use case

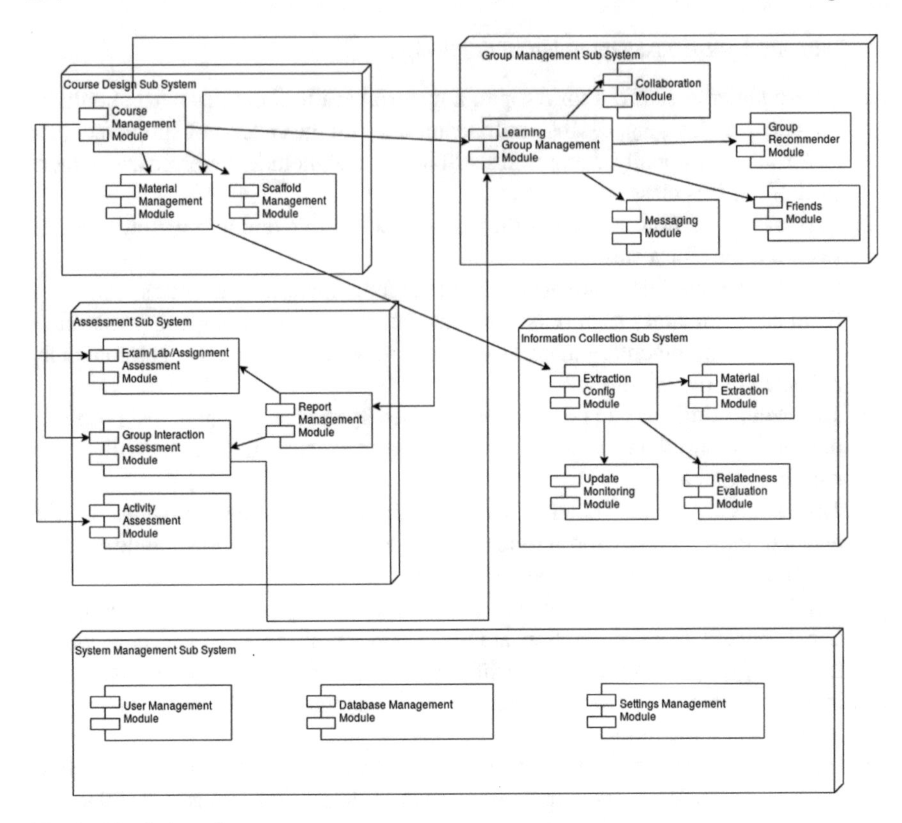

Fig. 2 The design of system components

13. Assess performance: in this use case, teachers will assess students learning performance; four types of assessment are available: exam/lab/assignment assessment, manual assessment, group interaction assessment, and activity assessment; exam/lab/assignment assessments are based on students performance on exam/lab/assignment; manual assessments allow teachers to assess students performance according to their empirical impression; group interaction assessments are based on students involvement in group activities; activity assessments come from analyzing students overall activities such as how many times students read course materials, etc.

To implement these use cases, Fig. 2 illustrates the design and relationships of system components. Five sub systems are included: the System Management Sub System, the Course Design Sub System, the Group Management Sub System, the Assessment Sub System, and the Information Collection Sub System. These sub systems are described below:

1. System Management Sub System: handle the underlying functionalities of the whole system

2. Course Design Sub System: for teachers to design courses
3. Group Management Sub System: for teachers to design and manage learning groups
4. Information Collection Sub System: for collecting information from the cloud
5. Assessment Sub System: for assessing students learning performance; each individual assessment sub modules assess a certain type of performance and teachers can specify the weight.

4 Conclusions and Future Work

In this manuscript, we propose the initial design of a cloud and crowd based mathematics learning environment targeting the students majoring in computer science. To teach computer science students mathematics is challenging since they have diverse mathematics backgrounds. In this research, we listed 13 use cases along with three system components. In the future, we have the following goals:

1. complete the listed sub systems
2. integrate the sub systems with an existing e-learning system
3. incorporate affective learning concepts into the learning groups.

Acknowledgements The manuscript is prepared with all the co-authers and my lab members in Department of Information Management in Nanhua University The author would like to acknowledge the support of the research project "104-2511-S-343-003", which is founded by MOST.

References

1. Bravaco, R., Kim, B., Gousie, M.B., and Wilkens, L.: Math connections in computer science. J. Comput. Sci. Coll. 24, 6 (June 2009), 57–61.
2. Koschmann, T. D. CSCL, theory and practice of an emerging paradigm. Routledge.
3. Konvalina, J., Wileman, S.A., and Stephens, L.J.. 1983. Math proficiency: a key to success for computer science students. Commun. ACM 26, 5 (May 1983), 377–382.
4. Henderson, P. B., Baldwin, D., Dasigi, V., Dupras, M., Fritz, J., Ginat, D., and Walker, H. (December 2001). Striving for mathematical thinking. In Working group reports from ITiCSE on Innovation and technology in computer science education, 114–124.
5. Beaubouef, T. 2002. Why computer science students need math. SIGCSE Bull. 34, 4 (December 2002), 57–59.
6. Sherman, H.J., Richardson, L.I., and Yard, G.J. Why students struggle with mathematics. Pearson Allyn Bacon Prentice Hall, 2014.
7. Zanchi, C., Presser, A. L., and Vahey, P. (2013, June). Next generation preschool math demo: tablet games for preschool classrooms. In Proceedings of the 12th International Conference on Interaction Design and Children, 527–530
8. Ke, F. (2006, June). Classroom goal structures for educational math game application. In Proceedings of the 7th international conference on Learning sciences, 314–320.
9. Ke, F. (2007, July). Using computer-based math games as an anchor for cooperative learning. In Proceedings of the 8th iternational conference on Computer supported collaborative learning, 357–359.

10. Niess, M. L. (2005). Scaffolding Math Learning with Spreadsheets. Learning Connections–Mathematics. Learning & Leading with Technology, 32(5), 24.
11. Stahl, G. (2010, June). Computer mediation of collaborative mathematical exploration. In Proceedings of the 9th International Conference of the Learning Sciences-Volume 2, 30–33
12. Lambropoulos, N. and Romero, M. (2011). Personalised information retrieval through the use of a collaboration awareness tool, a chat and a forum tool in a computer supported collaborative learning task. International Journal of Knowledge and Web Intelligence, 2(2-3), 219–229.
13. Edrees, M. E. (2013, May). eLearning 2.0: Learning management systems readiness. In e-Learning "Best Practices in Management, Design and Development of e-Courses: Standards of Excellence and Creativity", 2013 Fourth International Conference, 90–96.
14. Greengard, S. (2011). Following the crowd. Communications of the ACM, 54(2), 20–22.

The Design of Music Conducting System Using Kinect and Dynamic Time Warping

Chuan-Feng Chiu, Chia-Ling Tsai and Yu-Chin Hsu

Abstract This paper propose a design of music conducting system by analyzing the human hand gesture and translate to music playing command. The proposed system uses Kinect device as the human computer interface to recognize the hand gesture. In order to recognize the hand gesture we propose the virtual grid to track the path of the gesture and use Dynamic Time Warping to recognize the gesture in this paper. And the proposed mechanism allows music conductor to control the music feeling by hand gesture command and to be as a tool for training music conducting gesture.

Keywords Music conducting system · Kinect · Dynamic time warping · Virtual grid

1 Introduction

Human Computer Interaction(HCI) is a popular area in recent years. HCI technology leads users to have a newly and nature way to make interaction with electronic device including computers and so on. Microsoft Kinect is the one of popular devices to realize human computer interaction(HCI) and has been applied to areas including entertainment, education etc. In this paper we take the advantage of Kinect technology and propose music conducting system based on Kinect technology. Music conductor is the major player in an orchestra and responsible for controlling the playing and revealing the emotion of the song via the hand gesture, facial expression or human action of the body. However a music conductor must practice the song playing with all music players and need to make all music players together, so that the music conductor can present his/her own music style to all music players. Making all music players together is difficult because of many music

C.-F. Chiu (✉) · C.-L. Tsai · Y.-C. Hsu
Department of Information Management, Minghsin University of Science and Technology, Hsinchu, Taiwan
e-mail: cfchiu@must.edu.tw

© Springer Nature Singapore Pte Ltd. 2018

N.Y. Yen and J.C. Hung (eds.), *Frontier Computing*, Lecture Notes in Electrical Engineering 422, DOI 10.1007/978-981-10-3187-8_24

players are involved in the orchestra and be in different locations. On the other hand, young music conductors have less experience for conducting music song, they need more and more experience. So a tool for practicing music conducting gesture is useful for most of young music conductors.

Therefore, in this paper we propose a serial design for practicing and simulating for music conducting that includes the following features:

- Designing the beat recognition to lead music conductor to have more flexibility to display their emotion in the song
- Designing the alternative gesture to lead music conductor to have more music surrounding feeling control.
- Providing a design of the tool for music conductor to practice or simulate the playing situation that music conductors expect.

In the following we describe the related technology and research regarding the music conducting system in Sect. 2. In Sect. 3 we reveal our proposed method and design of the music conducting system. Finally a brief conclusion and future works is described in Sect. 4.

2 Related Works

There exist interactive music conducting system based on radio baton, Kinect etc. In this section, we reveal the literatures regarding related technology and past researches.

Mathews [1] proposed the first music conducting system in 1991. The system uses the radio baton to conduct music playing. Based on the tacking of the moving baton, music is played according the command from the moving batons. Digital Baton [2] and Conductor's Jacket [3] also propose music conducting system based music batons, and propose more parameters or sensors to capture more information and translate to have more music commands. Eric Lee et al. [4] proposed the music conduct system for children only and do not acquire priori experience of music conducting. Toh et al. [5–7] also propose the music conducting system using Kinect device to capture the human hand gesture to translate the music command the beat control. Lim and Yeo [8] propose the music conducting system using smartphone as the controller and [9, 10] use Wii controller to change the music tempo.

Kinect [11] is device developed by Microsoft Inc. The Kinect device consists RGB camera to capture RGB image and video, IR camera to get the depth data to understand the distance information between object and camera, and audio module for capture the voice. Based on the development modules, tracking or recognizing the object have been became possible. Many applications have been studied in the past. Chang et al. [12, 13] propose the medical application to assists people to do rehabilitation using Kinect. [14, 15] use Kinect to be a natural user interface to manipulate the system. Izadi et al. [1] use Kinect to capture the object scenes and reconstruct it as a 3D model. [11] studies more applications using Kinect and give a comprehensive review.

3 Proposed Design

Before describing the detail of the design of our proposed method, we illustrate the basic concept of a music conducting processing. A sheet of music consists several measures that include beats and notes. In the front of the sheet of music, there exists two numbers as shows in Fig. 1 called as time signature. The top number denotes how many beats in each measure and the bottom number denotes the basic unit of a beat. Basically the time signature and beat pattern are same in each measure. A music conductor would use hand gesture to present the beat pattern and control the speed of the playing. On the other hand, the music conductor would add some hints by hand gesture alone with the beat pattern to music players to convey the music conductor's expectation of the surrounding feeling of the playing to all audiences. According to the above description we proposed a Kinect-based music conducting system that conforms the requirements of the music conducting process. In the following sub-sections we will reveal the design of the system.

3.1 *System Overview*

In this section we describe the overview of the proposed design of music conducting system. In our proposed system, we have a Kinect device to capture the user behavior. After gathering the user behavior, the music conducting system would analyze the user hand gesture to identify the conducting action. The detail of the behavior identification mechanism would be described in next section. The proposed system has two major functions including training module and styling module which providing the tool for music conductors to be familiar with music conducting gestures and cooperating with a virtual orchestra before playing music

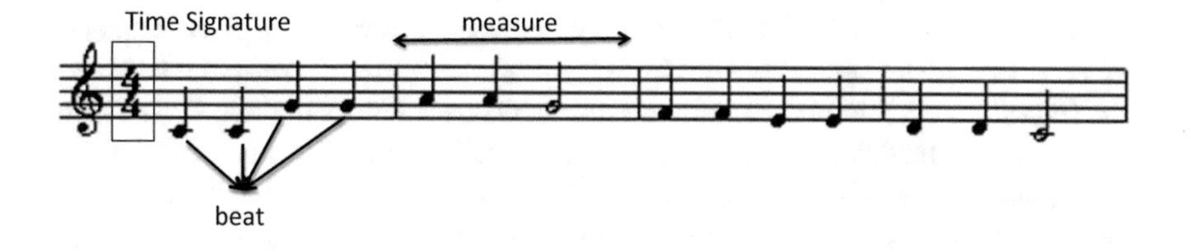

Fig. 1 The basic concept of music sheet from the English lullaby "Twinkle, Twinkle, little Star" [20, 21]

with orchestra to enhance performance. The proposed system also concludes virtual orchestra that is used for simulating possible music instruments to lead music conductors to understand the resulting music performance. The music conductor can select necessary music instruments via our system user interface. Besides virtual orchestra, we also provide the pre-recorded music sheet in the system. Using the pre-installed music sheet, music conductor can edit the music and corresponding music instruments easily. In the following we will describe the detail design regarding the two major functions that are described above.

3.2 Music Conducting Training Module

In order to training young music conductor to be familiar with the gesture of music conducting, we need to provide a tool for this purpose. The major functionality of the tool is to recognize the hand gesture of young music conductor and identify the beat. The beat hand gesture of music conducting is a hand moving sequence. Different gesture denotes different beat. First we use Kinect device to capture human hand gesture by analyzing the human skeleton especially for human's hand. The human skeleton captured by Kinect is showed in Fig. 2. And then we investigate the music conducting gesture regarding the basic theory of music and find the sequence will start from a position and end with the same position. The different moving sequence is resides between the starting and ending position. Therefore, we propose the Virtual Beat Grid that contains 5×4 cells to identify the hand moving sequence. When a music conductor start to use the training system, the proposed system would calibrate with the user's position and put the Virtual Beat Grid over the human body. The central column of the Virtual Beat Grid will be aligned with the SPINE-HEAD line of the human skeleton capturing from Kinect. Figure 3 shows the proposed Virtual Beat Grid and corresponding position human skeleton. The conducting gesture sequence will be tracking through the cells in the Virtual Beat Grid and our proposed system would record the sequence to recognize the action. In order to recognize different actions with the Virtual Beat Grid, we design moving sequence pattern according to each possible music conducting beat gesture. Figure 4a–i shows the basic beat sequence pattern alone with the Virtual Beat Grid.

We use Dynamic Time Warping (DTW) [16–19] to find the user's real-time hand moving sequence with different beat sequence to identify the action. DTW is a popular technology that is used for finding the optimal sequence with the given sequence pattern. For given template $T = (t_1, t_2, \ldots, t_m)$ and user's real-time sequence $C = (c_1, c_2, \ldots, c_n)$, to find a minimum sum of distance between T and C. The minimum sum is calculated alone with the warping path $W = (w_1, w_2, \ldots, w_l)$ that is a series points with smaller distance between T and C. The warping point of the warping path $w_k = (t_i, c_j)$ is a pair indices including template indices and user's real-time sequence indices. The template and user's real-time sequence formed a $m \times n$ cost table. In the cots table, y-axis denotes the given template T and x-axis denotes the

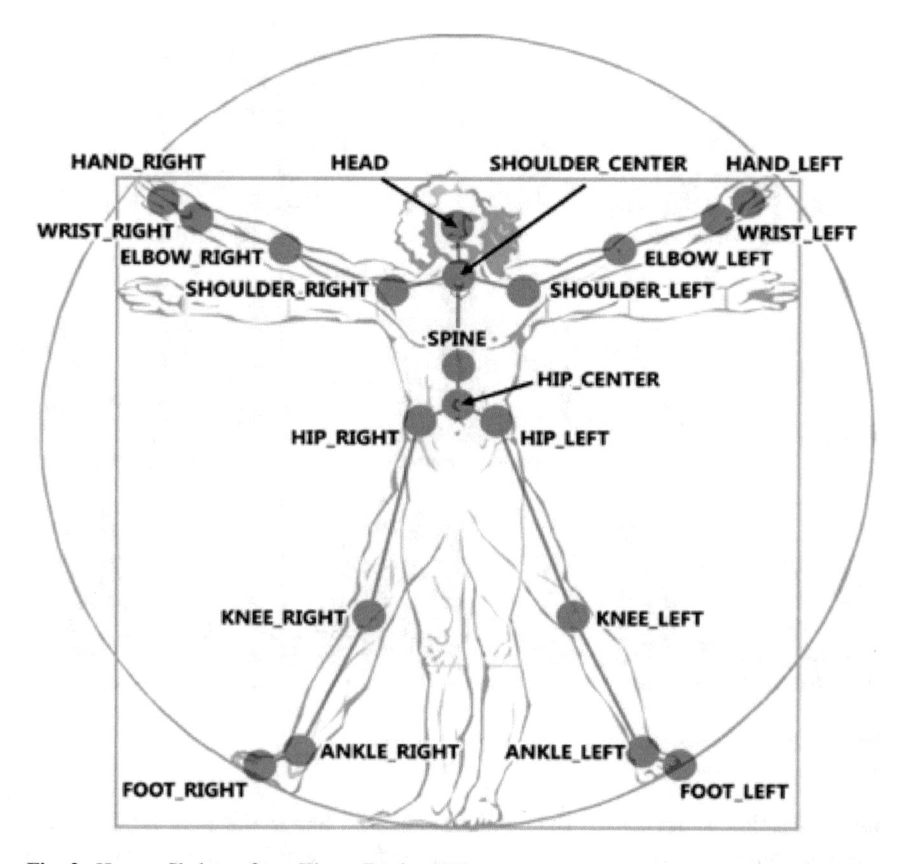

Fig. 2 Human Skeleton from Kinect Device [11]

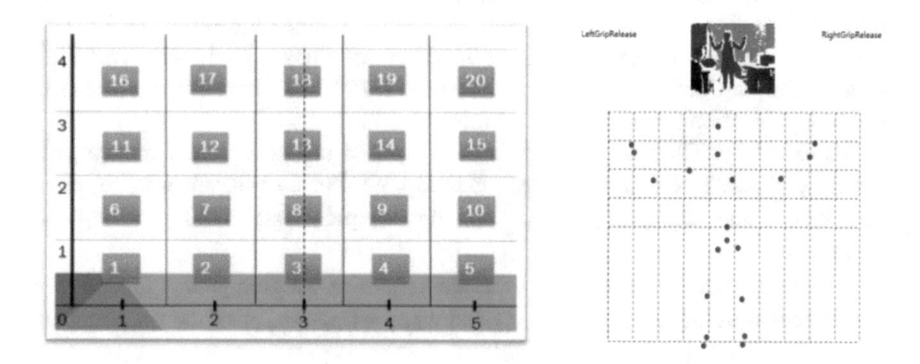

Fig. 3 The proposed virtual beat grid and corresponding user interface

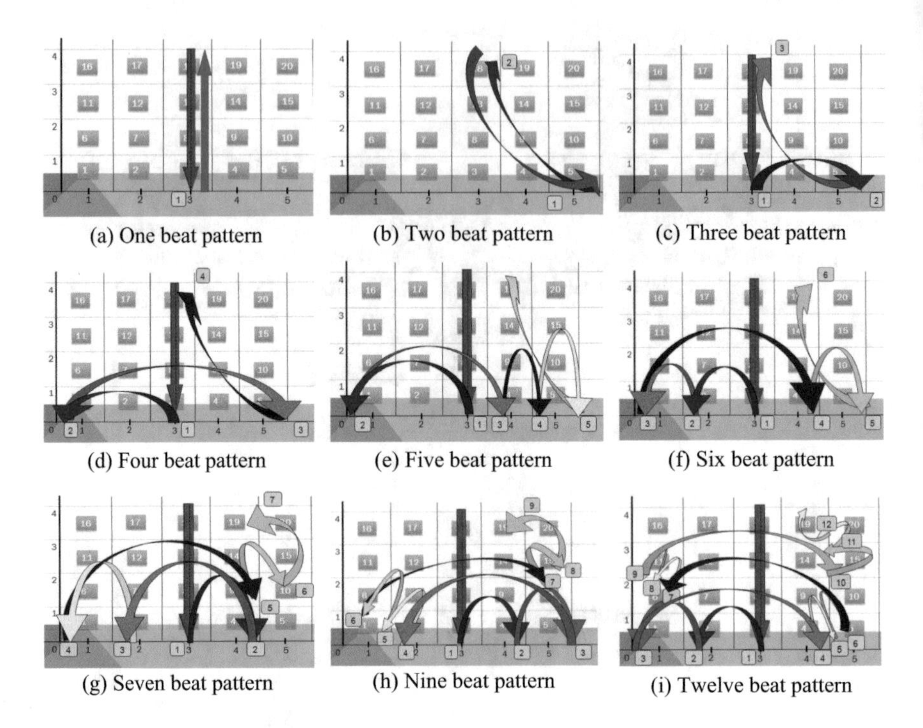
(a) One beat pattern (b) Two beat pattern (c) Three beat pattern
(d) Four beat pattern (e) Five beat pattern (f) Six beat pattern
(g) Seven beat pattern (h) Nine beat pattern (i) Twelve beat pattern

Fig. 4 The beat pattern based on music conducting theory

user's real-time sequence *C*. The continue sequence line in the $m \times n$ table is the warping path *W*. Each cell in the table denotes the distance of the sample point between *T* and *C*. The overall distance between *T* and *C* is the summation of each distance of the corresponding point alone the warping path *W*. Therefore, to find the matching template is to find the minimum distance between real-time sequence and templates. According the classical Dynamic Time Warping technology, it needs to avoid malformed warping path generation. So classical DTW had given the constrains to warping generation and showed in the following:

Monotonicity: The warping path indices have an increasing order for template indices and user's real-time sequence indices both. For an adjacent warping point $w_k = (t_i, c_j)$ and $w_{k+1} = (t_{i+1}, c_{j+1})$ of the warping path *W*, $t_{i+1} >= t_i$ and $c_{j+1} >= c_j$.

Continuity: The increment step of success point of warping path is limited. For an adjacent warping point $w_k = (t_i, c_j)$ and $w_{k+1} = (t_{i+1}, c_{j+1})$ of the warping path *W*, $t_{i+1} - t_i <= 1$ and $c_{j+1} - c_j <= 1$.

Boundaries: Start and end point of warping path is the start and end point of template T and user's real-time sequence C. For a warping path $W = (w_1, w_2, \ldots, w_l)$, the start point $w_1 = (t_1, c_1)$ and the end point $w_k = (t_m, c_n)$.

Based on the Dynamic Time Warping technology, we design the action recognition mechanism as describing in the following.

- A proposed Virtual Beat Grid $G_{x,y}$, x is the x-axis indices of the Virtual Beat Grid and *1 <=x <=5*, *y* is the *y*-axis indices of the Virtual Beat Grid and *1 <=y <=4*.
- For a beat Template $T = \{t_i\,(gx, gy) \in G_{x,y} | 1 \le i \le m,\ 1 \le gx \le 5,\ 1 \le gy \le 4\}$
- For a user's sequence $C = \{c_j\,(gx, gy) \in G_{x,y} | 1 \le j \le n,\ 1 \le gx \le 5,\ 1 \le gy \le 4\}$

First we calculate the $m \times n$ cost table using the local cost function $D_{i,j}$ showed in Eq. (1) which is denoted the distance between sample point T_i of beat pattern and C_j of user's sequence.

$$D_{i,j} = \left(\begin{array}{c} |t_i(gy) - c_j(gy)|, \mathit{if}\ t_i(gx) = c_j(gx) \\ |t_i(gx) - c_j(gx)|, \mathit{if}\ t_i(gy) = c_j(gy) \\ \sqrt{(t_i(gx) - c_j(gx))^2 + (t_i(gy) - c_j(gy))^2},\ \mathit{otherwise} \end{array} \right) \quad (1)$$

And then we can find the warping path with respect to the cost table to find the minimum cost between beat template and user's real-time sequence and the corresponding beat action.

3.3 *Music Styling Module*

Besides the training function for young music conductor, we also propose the music styling function for senior music conductor. Senior music conductors have been familiar with the beat conducting, but they intent to provide the customized feeling to the music playing. Therefore, we also propose the styling module for senior music conductors to give a hint to music players to present of the music feeling to audiences. In the styling module, the action duration is different comparing with beat action. We design alternative mechanism to recognize the styling action to present music feeling. We also use the similar concept of Virtual Beat Grid and call it as Virtual Style Grid. The proposed Virtual Style Grid contains 10 × 4 cells showed in Fig. 5 and the Virtual Style Grid is aligned with central SPINE-HEAD line of the human skeleton also.

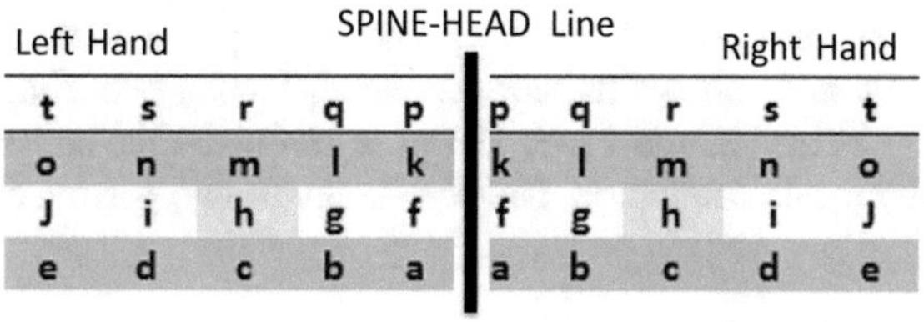

Fig. 5 The proposed Virtual Style Grid

Left Hand | SPINE-HEAD Line | Right Hand

t	s	r	q	p	p	q	r	s	t
o	n	m	l	k	k	l	m	n	o
J	i	h	g	f	f	g	h	i	J
e	d	c	b	a	a	b	c	d	e

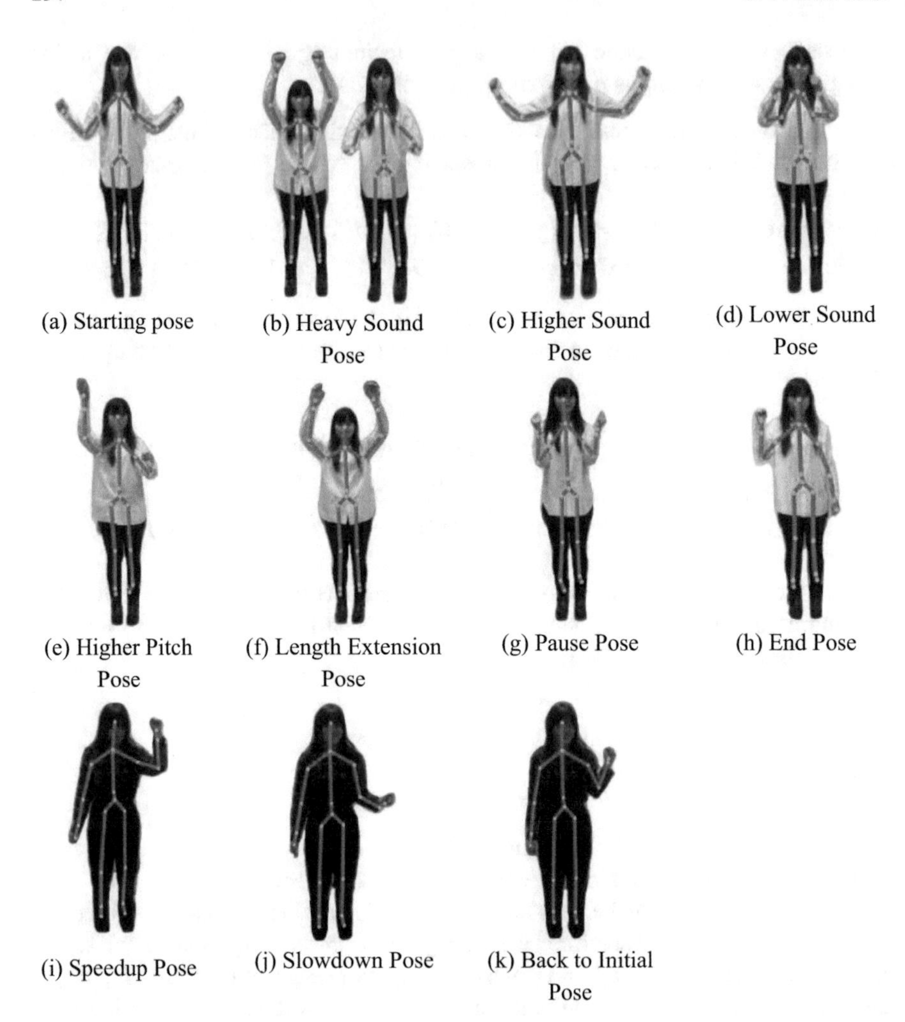

Fig. 6 The proposed music styling gesture pattern based on Human Skeleton

According to the proposed Virtual Style Grid, we provide eleven music feeling style to customize by senior music conductor and design the recognition rule for each styling action which are described in the following.

Starting Pose: The pose is the preparation starting point for other action. When the user's right hand and left hand are reside on the '**h**' block in Virtual Style Grid, the music start to playing and be a preparation pose for some styling action. Figure 6a shows the corresponding human behavior.

Heavy Sound Pose: The pose can make the stress of music sound immediately by music conductor. The pose includes two gesture command and need to move as downward direction. One is for trigger and the other is for backing to original. When the user's right hand and left hand move from top of the proposed Virtual Style Grid to the '**k**', '**f**' or '**a**' area in Virtual Style Grid after the Starting Pose, the heavy sound would be triggered. After heavy action triggered, music conductor need to back to original music phase. When user's right hand and left hand move to the '**f**' or '**a**' area in Virtual Style Grid after the trigger pose of Heavy Sound. Figure 6b shows the corresponding human behavior.

Higher Sound Pose: When user's right and left hand are resided on the '**n**' block in Virtual Style Grid. The volume of the music would be increased smoothly. Figure 6c shows the corresponding human behavior.

Lower Sound Pose: When user's right and left hand are resided on the central SPINE line and within the vertical range of 80 in Virtual Style Grid. The volume of the music would be decreased smoothly. Figure 6d shows the corresponding human behavior.

Higher Pitch Pose: The gesture of this pose would be moved as upward direction after Starting Pose. The action is used for changing the frequency of the music and making the different feeling by audience's ears. When user's left hand is move to '**s**', '**r**', '**n**', '**m**', '**i**', '**h**', '**d**' and '**c**' in Virtual Style Grid after the Starting Pose, the higher pitch action would be triggered. Figure 6e shows the corresponding human behavior.

Length Extension Pose: Music conductors would use this pose to extent the duration of music beat for different feeling of the music. The gesture of this pose would be moved as upward direction after starting pose. When user's left hand and right hand are move to '**s**', '**r**', '**n**', '**m**', '**i**', '**h**', '**d**' and '**c**' in Virtual Style Grid after the Starting Pose both, the length extension action would be triggered. Figure 6f shows the corresponding human behavior.

Pause Pose: When user's left hand make a first in the '**l**' block in Virtual Style Grid, the pause action would be triggered for stop the playing for a while until music conductors have Starting Pose. Figure 6g shows the corresponding human behavior.

End Pose: The action is used for ending the playing of the music. When user's left hand and right hand make a first in the '**l**' block in Virtual Style Grid at the same time, the pause action would be triggered. Figure 6h shows the corresponding human behavior.

Speedup Pose: When user's right hand make a first, the position is resided on the '**m**' block of Virtual Style Grid and the speed of music measure would be increased in one unit. Figure 6i shows the corresponding human behavior.

Slowdown Pose: When the position is resided on '**c**' block of Virtual Style Grid, the speed of music measure would be decreased in one unit. Figure 6j shows the corresponding human behavior.

Back to Initial Pose: When the position is resided on '**h**' block of Virtual Style Grid, the speed of music measure would be back to the initial of the music. Figure 6k shows the corresponding human behavior.

Therefore, based on the design of styling gesture mechanism, music conductor could simulate the surrounding feeling of music playing without spatial and temporal constrain.

4 Conclusion

In this paper we propose the virtual grid to track the human hand gesture and use the Kinect device as the human computer interface. We analyze the music conducting gesture based on the theory of music conducting and design the basic music conducting patterns. We track the human hand gesture by recording the path as the sequence based on virtual grid. And we use Dynamic Time Warping to find the music conducting command by comparing the captured hand gesture and the analyzed music conducting pattern. Based on the proposed the design, we also implement the system that allows music conductor to simulate the music feeling without making all music players together and be a training tool for practicing the music conducting gesture. We still work on improving the system. The current implementation does not emphasis different music instruments. Therefore the emphasis module for music conductor by human face or human skeleton direction needs to be proposed to make more natural feeling of music. On the other hand, the music tempo would be changed by music conductor, to make the audio playing more smooth is valuable for the future music conducting system.

References

1. M. V. Mathews: Current Directions in Computer Music Research, chapter The Conductor Program and Mechanical Baton. MIT Press, Cambridge (1991)
2. T. Marrin: Possibilities for the digital baton as a general-purpose gestural interface. CHI, pages 311–312. ACM (1997)
3. T. Marrin Nakra: Inside the Conductor's Jacket: Analysis, interpretation and musical synthesis of expressive gesture. PhD thesis, Massachusetts Institute of Technology (2000)
4. Eric Lee, Teresa Marrin Nakra, and Jan Borchers: You're the conductor: a realistic interactive conducting system for children. In Proceedings of the 2004 conference on New interfaces for musical expression (NIME '04), Singapore, 68–73 (2004)
5. Leng-Wee Toh, W. Chao and Yi-Shin Chen: An interactive conducting system using Kinect. Multimedia and Expo (ICME), 2013 IEEE International Conference on, San Jose, CA, pp. 1–6 (2013)
6. Rosa-Pujazon, A., Barbancho, I., Tardon, L. J., & Barbancho, A. M.: Conducting a virtual ensemble with a kinect device. In Proceedings of the Sound and Music Computing Conference(SMC 2013), Logos Verlag Berlin, Stockholm, Sweden, p. 284–291 (2013)
7. Chen, S., Maeda, Y. & Takahashi, Y.: Chaotic Music Generation System Using Music Conductor Gesture. Journal of Advanced Computational Intelligence and Intelligent Informatics (JACIII), 17, 194–200 (2013)

8. Yang Kyu Lim, Woon Seung Yeo: Smartphone-based Music Conducting. In Proceedings of the International Conference on New Interfaces for Musical Expression(NIME'14), June 30 – July 03 (2014)
9. D. Bradshaw and K. Ng.: Analyzing a conductors gestures with the wiimote. In Proceedings of EVA London 2008: the International Conference of Electronic Visualisation and the Arts (2008)
10. T. Nakra, Y. Ivanov, P. Smaragdis, and C. Ault: The ubs virtual maestro: An interactive conducting sys- tem. NIME2009, pp. 250–255 (2009)
11. Kinect, https://dev.windows.com/en-us/kinect/
12. Y.-J. Chang, S.-F. Chen, and J.-D. Huang: A Kinect-based system for physical rehabilitation: A pilot study for young adults with motor disabilities. Research in developmental disabilities, vol. 32, no. 6, pp. 2566–2570 (2011)
13. B. Lange, C.-Y. Chang, E. Suma, B. Newman, A. S. Rizzo, and M. Bolas: Development and evaluation of low cost game-based balance rehabilitation tool using the Microsoft Kinect sensor. In Proceeding of IEEE International Conf. Eng. Med. Biol. Soc., pp. 1831–1834 (2011)
14. M. N. K. Boulos, B. J. Blanchard, C. Walker, J. Montero, A. Tripathy, and R. Gutierrez-Osuna: Web GIS in practice X: a Microsoft Kinect natural user interface for Google Earth navigation. International journal of health geographics, vol. 10, no. 1, pp. 45 (2011)
15. H. Richards-Rissetto, F. Remondino, G. Agugiaro, J. von Schwerin, J. Robertsson, and G. Girardi: Kinect and 3D GIS in Archaeology. In Proceedings of the 18th International Conference on Virtual Systems and Multimedia, pp. 331–337.
16. B. Hartmann and N. Link: Gesture recognition with inertial sensors and optimized DTW prototypes. In Proceedings of IEEE International Conference on Systems Man and Cybernetics (SMC), Istanbul, pp. 2102–2109 (2011)
17. E. Keogh and M. Pazzani: Derivative Dynamic Time Warping. In First SIAM International Conference on Data Mining. pp. 1–11, Chicago, Illionis (2001)
18. H. Sakoe and S. Chiba: Dynamic programming algorithm optimization for spoken word recognition. Acoustics, Speech and Signal Processing, IEEE Transactions on, vol. 26, no. 1, pp. 43–49 (1978)
19. Meinard Müller: Information Retrieval for Music and Motion, Springer Berlin Heidelberg (2007)
20. Wiki, https://en.wikipedia.org/wiki/Twinkle,Twinkle,Little_Star
21. Conducting Course Book and Audio Examples, https://www.lds.org/music/conducting-music/conducting-course-book-and-audio-examples?lang=eng
22. M. G. S. Beleboni: A brief overview of Microsoft Kinect and its applications. In Interactive Multimedia Conference, University of Southampton, UK (2014)

A Model for Predicting Vehicle Parking in Fog Networks

Meng-Yen Hsieh, Yongxuan Lai, Hua Yi Lin and Kuan-Ching Li

Abstract This paper proposes a prediction model with driver's personal parking preferences for vehicle parking in parking lots. A parking preference adopted in the model involves not only parking fee, but also time required for parking, space waiting, and destination to the space allocated. This model advances and optimizes the usage of parking lots, also satisfying individual parking requirements. Unlike other studies acting solely with vacant parking spaces, this one also applies to parking lots with full state to the assignment of parking space to each parking request, while each of the lots has the probability of releasing occupied spaces soon. The request a vehicle parking space corresponds to the parking fee and time consumption of related parking operations. A fog network is defined to realize the mode assisting vehicles in search of an appropriate lot. Result analysis indicates that the proposed mode is reliable and efficient in search of a parking space.

Keywords Fog · VANET · Parking · Prediction

M.-Y. Hsieh (✉) · K.-C. Li
Department of Computer Science and Information Engineering,
Providence University, Taichung, Taiwan
e-mail: mengyen@pu.edu.tw

K.-C. Li
e-mail: kuancli@pu.edu.tw

Y. Lai · K.-C. Li
Software School, Xiamen University, Xiamen, China
e-mail: laiyx@xmu.edu.cn

H.Y. Lin
Department of Information Management, China University of Technology,
Taipei, Taiwan
e-mail: calvan.lin@msa.hinet.net

© Springer Nature Singapore Pte Ltd. 2018

N.Y. Yen and J.C. Hung (eds.), *Frontier Computing*, Lecture Notes
in Electrical Engineering 422, DOI 10.1007/978-981-10-3187-8_25

1 Introduction

Fog computing, as extension of concepts that defined Cloud computing, it enables data to be analyzed and managed locally before having it transferred to the Cloud. In this text, it refers to a virtual layer between vehicles and a cloud network to generate the features assistant to vehicular transmission such as low latency, location awareness, geographical distribution, a large number of nodes, predominant role of wireless access, and others. If fog computing is associated to vehicular networks, a fog network can be divided into two modes, infrastructure-based and ad-hoc. In infrastructure-based mode, fog nodes could be fixed or long-term unmoved devices, possibly along the roadside or located in business buildings. The fog nodes not only supply/accept useful data to/from moving vehicles, but also assist neighbor vehicles in message delivery. In the latter mode, one vehicle could be a fog node supporting ad hoc data transmission and computing to communicate between vehicles directly without any medium. There are a large number of vehicular applications that may utilize fog computing, such as traffic light scheduling, congestion mitigation, pre-caution sharing, parking facility management, traffic information sharing, etc. Most of them are appropriate to infrastructure-based fog networks, while others in the ad hoc mode are suitable to urban environment with slow-moving vehicles [1]. This paper focuses on a parking problem that vehicles attempt to find a parking spot in crowded, busy parking places. The infrastructure mode with fog servers attached to parking places or buildings is more appropriate to this research than others.

Various applications in fog networks always require significant components such as authentication and authorization, offloading management, location service, system monitor, and resource management and scheduling. The vehicular transmission operations have a number of types, vehicle to vehicle, Vehicle to Access Point, and Access Point to Access Point [2], while access points could be equipped in roadside units or traffic devices. Fog nodes can deliver not only instant and rich traffic-related and geo-distributed information to vehicles, but also be a medium for intercommunication between two vehicles and location-awareness business service. Extra devices such as sensors are required for all traffic-detected assistance, while fog nodes disguise themselves as traffic light devices to improve vehicular traffic communication.

When living in crowded areas, planning unoccupied parking space wisely will improve effectively traffic flow and reduce the cost of vehicle parking. Most of parking lots are attended beside city buildings such as apartment, market mall, bank, office building, and several others, or be neighbors with commercial streets and areas. A number of these parking places are organized as a local fog network to share parking-related information, while each of fog nodes can manage one or more parking lots. Those parking lots with remaining vacant spaces are involved for vehicle parking. For instance, a many-to-one matching game in [3] is adopted to provide lists of vacant parking lots to vehicles in an associated city area based on the parking cost and price for instant parking requirements from vehicles. However, the problem of missing some parking lots with implicit vacant spots could appear, while vehicles park in these lots at the present time, but will leave from occupied spaces shortly, in short period of time.

This paper involves with implicit parking spots for vehicle parking, while the fog network is able to calculate the probabilities of parking state changes between occupied and vacant. For example, a vehicle can determine whether going for parking in a lot with no vacant spots and waiting for a spot that would be changed from occupied to vacant state based on the vehicle-leaving probability of the lot.

After simple instructions about the vehicle parking with features and problems as depicted in this section, a number of fog applications related to vehicle parking are presented in Sect. 2. Section 3 describes the proposed system model with communication and computation designs, while vehicles require parking spaces in the fog network, influencing by predictable parking probabilities. In Sect. 4, the analysis results for the model are given, while the two lots are illustrated. Finally, conclusion remarks are delivered in Sect. 5.

2 Related Work

Fog computing technology supports large range of applications, and we may list as example, smart vehicles, smart grid, smart home, health data management, among several others. Basically, fog computing often assists end-users to retrieve information in time, such as traffic-related data, intercommunication data between neighbor vehicles, or data from servers built in near roadside and buildings. The network types for fog computing can be defined as a local cloud, Cloudlet, mobile edge computing, and mobile cloud computing [4]. Although a fog network is designed with particular features, latency, efficiency, and generality, a number of challenges in designing a fog computing platform, such as choice of virtualization technology, latency, network management, security and privacy.

Research issues related to vehicular parking [1, 3, 5] keep on increasing. Localize parked vehicles [1, 5] can play as fog nodes to support computing and communication and to service moving others. However, security and privacy issues must be required to protect message delivery from selfish fog nodes. In addition, parked vehicles as infrastructures must face the potential crisis of unstable connection and insufficient endurance ability.

In [3], authors proposed the parking problem in the view of IoT (Internet of Things) by a Many-to-One matching mechanism. Fog computing is an infrastructure-based network where fog nodes are installed at parking areas, such as banks and shopping halls. Vehicles connected to the fog network to gain the information of available parking lots by neighbor RSUs (road side unit) along the roadside. That means localized RSUs are the media to aid the communication between vehicles and fog servers. In addition, the status of parking spots, vacant or reserved, is provided from fog servers to vehicles. Fog nodes in a local area are organized as a fog network. Parking space assignment for the vehicles searching park places is as a matching game, where each vehicle will be assigned to one parking slot, at most. The matching mechanism is performed with only explicit parking slots available to vehicles at present. In VFC architecture [1], vehicles are treated as an infrastructure-based fog network, while fog nodes are as slow-moving or parked vehicles, supporting mobility and geographical distribution. Different distributed vehicles forms particular infrastructures with different communication and computing among vehicles. Two application scenarios of VANETs and Jam-Cloud are given, while moving vehicles are as infrastructures; moreover, other scenarios of relay packets and parking lot are described, while parked vehicles are as infrastructures.

Malandrino et al. [5] exploited parked vehicles to extend the transmission service of RSUs for content downloading during inter-vehicle communication. In this system model, the location and movements of vehicles and RSUs are well known. They utilized a time-expanded graph to represent the dynamic network with temporal and spatial changing based on parked vehicles and fixed RSUs. The performance of the vehicle that receives beacon transmissions from other vehicles is offered, and the Veins simulator is used to simulate the network.

Vehicles in a parking lot are organized as a vehicular data center (VDC) [6] to assist other vehicles in taking low cost to gain content service from a local VDC instead of the remote data center. The two-tier data center architecture with three VDC management policies are proposed to leverage the storage resources on localized parked vehicles.

3 System Model

3.1 Assumption and Architecture

This research is proposed based on the following assumptions: (1) a fog network with a number of fog nodes manages a number of parking places to reply to vehicles' parking requests, while one of the fog nodes can 'hear' the updates of the parking lots; (2) regional fog networks have been connected to share parking status of each of parking spots by cloud computation and communication; (3) Each RSU is equipped with wireless connection, whilst each fog node could be equipped with wireless connection to service the vehicles having reached the entries of the managed parking lots; (4) Each vehicle can easily gain parking related information by RSU or fog node after sending the parking request. The system architecture with possible parking scenarios is depicted in Fig. 1, where each vehicle can ask for parking by sending out a request. Each RSU is only deployed with the responsibility of delivering parking requests and responses between fog network and vehicle.

The possible states of one parking spot are reserved, vacant, and occupied [7]. The reserved spaces are not involved in the fog network. Based on the assumptions given, the states of all spots are shared in the fog network, and their updates are

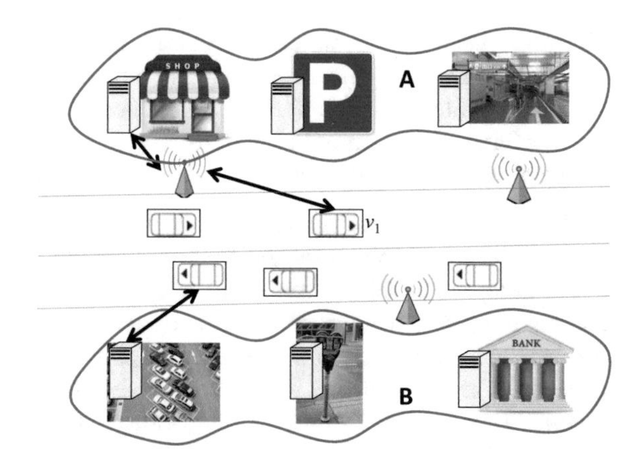

Fig. 1 System architecture in a fog network

immediately synchronized in all nodes of the fog network. The fog network learns about the two parking probabilities of each one of the managed parking lots, which means that each lot has the probability that a vacant parking space is changed into occupied, and the probability of that occupied park space is changed into vacant during an assigned period, denoting P_{v2o} and P_{o2v}, individually. These probabilities can easily be calculated by parking statistics.

3.2 Communication and Computation Design

A fog network must be trusted by vehicles or RSUs to offer correct and convenient parking service. Vehicles are equipped with wireless devices to connect RSUs or fog nodes. In addition, RSUs easily connect nearby fog nodes to reply to vehicle parking requests. The fog network is organized with localized fog nodes attending some deterministic parking places or lots. A vehicle with the basic profile can send a parking request to request for a parking spot. Basic profile of a vehicle is composed by current time, the current GPS and the driving destination that can be retrieved easily and quickly by car navigation system. Specifically, the fog network managing parking slots from a number of parking lots has the following features, while the two ways are offered to users for inquiring whether there is any space for parking:

(1) In the automatic way, the network receives a parking request of vehicles from themselves directly or through RSUs indirectly. The request with a basic profile and individual preference as the weights of the driving, walking, and waiting costs that the driver would like to give. According to each of individual requirement with user preference, the fog node nearest to the vehicle can offer a list of possible parking spaces by increasing order of the parking cost.

(2) In the manual way of searching parking lots, a real-time service for parking query is required. After connecting to a RSU or one of the fog nodes, only the basic profile consisting of the current time, the vehicular GPS and destination is delivered. Based on the profile, the fog network returns all park lots that still have vacant parking spaces to a vehicle in limited geographic range.

A parking request from a vehicle includes not only the basic driving profile, but also the three preferred costs and the two durations that driver desired for parking and waiting. Each vehicle can give the distinct weights and durations for distinct parking request. When drivers always consider the price and time consumption for parking their vehicles, the algorithm of a parking decision adopts these costs as parameters to Algorithm 1, given as basic decision process to associate a vehicle sending a park request with possible parking lots. These parameters for the algorithm are described as follows:

(1) A vehicle, v_i, sends a request, for searching a parking spot to one RSU or fog node. The message for requiring a parking spot is divided into the few fields. The request is formatted as $Req(T_{stamp}, Prof_i, \{\alpha, \beta, \gamma\}_{\text{weight}}, ti, tp)$. The timestamp, T_{stamp}, denotes the time at that the request is sent. The vehicular profile, $Prof_i$, is composed of the vehicular GPS and the travel destination where drivers will achieve, represented as $Prof_i = \langle GPS_i, Dest_i \rangle$. The next field contains the three weights for computing the cost of driver's mind feeling, when $\alpha, \beta \in \{0, 1\}$, and $\gamma \in \text{R}: 0 < \gamma < 1$. The two durations, ti and tp, are given for parking and waiting that the driver desired.
(2) Price cost for parking: a short-term parking payment adopted is to charge user parking price at a unit time such hour or day. For example, parking can be charged at hour rate. The price cost of one lot for a parking request from a vehicle, v_i, with a predictable spending time, ti, is calculated by the parking price of the lot divided by the maximum price of all the lots, denoted as follows.

$$PriceCost_{lj} = \frac{Price(l_j, ti)}{MAX(\{Price(l_i, ti)\})}, when\, Price(l_j, ti) = \Delta p \times \frac{ti}{\Delta t}, \tag{1}$$

where Δp represents a unit parking fee per unit time, denoted as Δt, and ti is the spending time that the driver needs to take to park in a specific lot. However, there could be various equations of calculating parking price in different countries. Suppose that a set of n parking lots, denoting $L = \{l_1, l_2, \ldots, l_n\}$, has a set of parking fee for each vehicular request, denoting $\{Price(l_1, ti)\}$ in the fog network.

(3) Time cost for parking: the other cost is a period time that a vehicle, v_i, has to take for parking a vehicle to Lot l_j, defined as follows:

$$TimeCost(v_i, l_j) = T_d + T_w + \gamma \times T_q, \quad when\, T_q = C_q \times tp, \tag{2}$$

where T_d is the time cost that users must take for driving from a current position to l_j, T_w is the time cost for walking from parking position to the destination, and T_q is the virtual cost for waiting for a possible vacant park spot of l_j. Cost T_d and T_w are calculated actually as the travel time duration, calculated in a specific travel mode using the Google Maps Distance Matrix API, a.k.a. GMAP [8]; moreover the travel mode is supported as one of all modes, driving, walking, bicycling, and transit. Cost T_d and T_w are acquired for driving and walking modes, individually. However, T_q is computing based on a set of vehicular requests, denoting $\{Req\}$, sent to l_j in a time period, *tp*, and the number of its remainder vacant spots, denoting $\{Spot\}$, while Set $\{Req\}$ is composed of the request of v_i and the others during the time period. Duration *tp* is the max length of time that the vehicle can wait for. However, C_q is available in three calculation modes, defined as follows:

$$C_q^1 = \frac{(P_{v2o}^{tp})^{|\{Spot\}| - |\{Req\}|}}{|\{Spot\}|}, \tag{3}$$

where the number of $\{Req\}$ is smaller than the number of $\{Spot\}$, and the number of $\{Spot\}$ is greater than or equal to one, represented as $|\{Spot\}| >= |\{Req\}| > 1$

$$C_q^2 = \frac{(|\{Req\}| + |\{Spot\}|)}{|\{Spot\}|} \times (P_{v2o}^{tp})^{|\{Spot\}|}, \tag{4}$$

where only the request of v_i, and there is only one vacant spot right now, represented as $|\{Req\}| >= |\{Spot\}| >= 1$

$$C_q^3 = \frac{|\{Req\}|}{P_{o2v}^{tp}(1 - P_{v2o}^{tp})}, \tag{5}$$

where the number of $\{Req\}$ is larger than the number of $\{Spot\}$, and there is no vacant spot, represented as $|\{Req\}| > |\{Spot\}| = 0$. This cost is greater than one, and the others are less than one.

Generally, the cost calculation has a feature of that C_q^3 must be greater than C_q^2, and C_q^2 must be greater than C_q^1. The weight, γ, denotes user preference on waiting for a park space, since γ is a value in the interval [0, 1].

3.3 Spot Association

Algorithm 1 is proposed to offer a recommended list of parking lots to a vehicle that has submitted a parking request. The parking costs, calculated from the involved park lots neighboring to the travel destination are referred to user preference on either price or time cost for parking. For each cost, only two states, "like" and "dislike" denote the user's rank with "1" and "0". Based on Eqs. (1) and (2), the sorting mechanism is mainly adopted to recommend a list of parking lots.

Algorithm 1. The Spot Association of Vehicle Parking in the fog networks
Input: $Req(T_{stamp}, Prof_i, \{\alpha, \beta, \gamma\}, ti, tp)$ is a request for Vehicle v_i;
  $\{L\}_c$ is the set of candidate parking lots;
Output: a final list of parking spot, $\{L\}_f$
Initialization: $\{L\}_f$ is null;
1: Send the *Req* to regional fog nodes neighboring to $Dest_i$.
2: For each of $\{L\}_c$, l_i:
3: Calculate *PriceCost* and *TimeCost* by Eq. (1) and Eq. (2), individually;
4: End for
5: If α=1 and β=0: //v_i prefers *price*
6: $\{L\}_p \leftarrow$ OrderBy*PriceCost* ($\{L\}_c$, Ascending);
7: For each subset, $\{l_i\} \in \{L\}_p$, with the same *PriceCost*:
8: $\{l_i\} \leftarrow$ OrderBy*TimeCost* ($\{l_i\}$, Ascending);
9: $\{L\}_f \leftarrow \{L\}_f \cup \{l_i\}$
10: End For
11: Else If β=1 and α=0: //v_i prefers *TimeCost*
12: $\{L\}_T \leftarrow$ OrderBy*TimeCost* ($\{L\}_c$, Ascending);
13: For each subset, $\{l_i\} \in \{L\}_T$, with the same *TimeCost*:
14: $\{l_i\} \leftarrow$ OrderBy*PriceCost* ($\{l_i\}$, Ascending);
15: $\{L\}_f \leftarrow \{L\}_f \cup \{l_i\}$
16: End For
17: Else
18: For each of $\{L\}_c$, l_i:
19: Calculate the cost by $PTC_{i,j} = PriceCost_{i,j} + TimeCost_{i,j}$
20: End for
21: $\{L\}_f \leftarrow$ OrderBy*PTC* ($\{L\}_c$, Ascending);
22: End if
23: Send $\{L\}_f$ to Vehicle, v_i.

4 Model Analysis

For analysis, we estimate the model how to effectively provide the park lots to vehicles, sending parking requests. Suppose that the basic information of the two lots is given in Table 1. When the probabilities of vehicle-coming and vehicle-leaving in a lot are independent, the P_{o2v} and P_{v2o} are given for a specific short-term duration, *tp*. A parking request sent by v_1 to the fog network includes the predictable values, $t_i = 3$ (hours), $t_p = 10$ (mins), $\alpha = 1$, $\beta = 1$, and $\gamma = 1$. Since the fog network can calculate the first two time costs by GMAP, corresponding to the distances from the position of v_1 to a lot, and from the lot to the destination, the time costs (T_d, T_w) for l_1 and l_2, are assumed to be (7, 5) and (10, 4) min. The two *Price* for l_1 and l_2, calculated by Eq. (1) are $20 \times 180/30 = 120$, $50 \times 180/60 = 150$, then the *PriceCosts* values are 0.8 and 1, when the only two lots are all, receiving the request. T_q for a lot could be one of the three possible values, while the lot gains the different number of vacant spaces and parking requests during the period, *tp*, after the time, T_{stamp}. Suppose that the three possible amounts in the sets, ($|\{Spot\}|$, $|\{Req\}|$) are (3, 2), (2, 3), and (0, 5), applied to the lots at the same time. For the three sets, Cost T_q of l_1 are 0.67, 1, and 78.12, and Cost T_q of l_2 are 2, 9, and 625. Table 2 shows the calculation results of *PTC*, when the weights are set to 1.

The cost of PTC_2 of the set (0, 5) is highest, since the l_2 has not only no vacant spaces, but also the five parking requests during the period. In addition, the probabilities of vehicle-coming and vehicle-leaving in l_2 are higher and lower than others.

A statistical graph is shown in Fig. 2, when we focus on the *TimeCost* by only Eq. (5), with no vacant spot and an increasing of parking requests. According to the various pair of P_{o2v} and P_{v2o}, more requests drives up more time cost, as also larger P_{o2v} too.

Table 1 Basic parking info. of the two parking lots

Lot	UnitFee ($)/unitTime (mins)	$P_{v2o}(tp)$	$P_{o2v}(tp)$
l_1	20/30	0.2	0.8
l_2	50/60	0.6	0.2

Table 2 Example of cost *PTC* for l_1 and l_2, when $\alpha = 1$, $\beta = 1$, $\gamma = 1$, $ti = 3$, and $tp = 10$

($\|\{Spot\}\|$, $\|\{Req\}\|$)	$PriceCost_1$	$PriceCost_2$	$TimeCost_1$	$TimeCost_2$	PTC_1	PTC_2
(3,2)	0.8	1	0.67	2	1.47	3
(2,3)	0.8	1	1	9	1.8	10
(0,5)	0.8	1	78.12	625	78.92	626

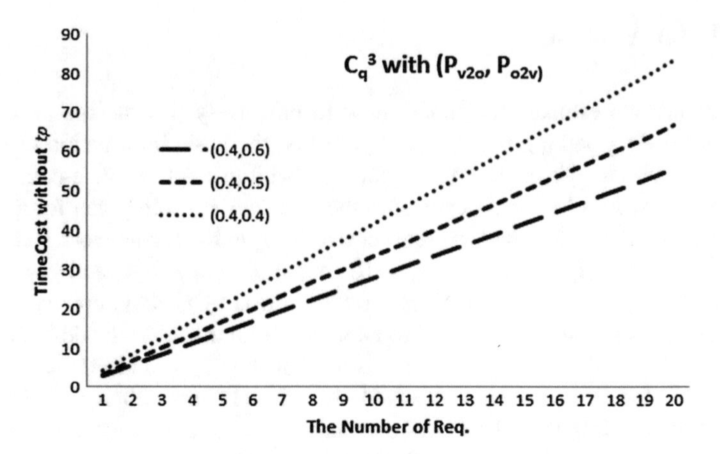

Fig. 2 Illustration of *TimeCost* by Eq. (5), with various P_{o2v} and P_{v2o}

5 Conclusions and Future Work

Searching for a vacant parking space is difficult in a crowded city, since most of parking lots could be already-crowded. This paper proposes a prediction model for vehicle parking to determine a parking space for vehicles based on predictable parking time and waiting time that drivers can endure for a parking. With this model, fog computing is taken into infrastructure-based vehicle networks, so that the fog network, managed park lots, have to record the probabilities of vehicle-coming and vehicle-leaving of each of these lots. The costs related to parking fee and time consumption can be predictable for each parking request of vehicles that contact the network ahead to ask for a parking space. The formula for the model are estimated with a spot association algorithm, efficiently offering confirmed and practical parking lots to vehicles.

As future work, given that the deployment of suitable wireless sensor networks [9] to parking lots is a trend, vacant parking spaces may be detected automatically to supply real-time data to the fog network. A parking recommendation service can be established to suggest user's parking to the most appropriate vacant space, after learning the sensor records and user's driving behavior. Taking into account on developing mobile applications [10, 11] is aimed in providing convenient GUIs of parking service. In addition, encryption and authentication issues [12, 13] will be required for delivering securely correct parking information to users.

Acknowledgements This investigation was supported by Ministry of Science and Technology, Taiwan, under grant no. MOST 104-2221-E-126-007, Providence University research grant, and the National Key Technology Support Program of China (2015BAH16F00/F01/F02).

References

1. Xueshi Hou, Yong Li, Min Chen, Di Wu, Depeng Jin, and Sheng Chen. "VehicuMlar Fog Computing: A Viewpoint of Vehicles as the Infrastructures", IEEE Transactions on Vehicular Technology, 2016.
2. Flavio Bonomi, Rodolfo Milito, Jiang Zhu, Sateesh Addepalli "Fog computing and its Role in the Internet of Things", MCC '12 Proceedings of the first edition of the MCC workshop on Mobile cloud computing pp. 13–16, 2012.
3. Oanh Tran Thi Kim, et al., "A shared parking model in vehicular network using fog and cloud environment". In 17th Asia-Pacific Network Operations and Management Symposium, APNOMS, August 19–21, pp. 321–326, IEEE, 2015.
4. Shanhe Yi, Zijiang Hao, Zhengrui Qin, and Qun Li, "Fog Computing: Platform and Applications", Third IEEE Workshop on Hot Topics in Web Systems and Technologies, 2015.
5. F. Malandrino, C. Casetti, C.-F. Chiasserini, C. Sommer and F. Dressler, "The Role of Parked Cars in Content Downloading for Vehicular Networks," IEEE Transactions on Vehicular Technology, vol. 63 (9), pp. 4606–4617, 2014.
6. L. Gu, D. Zeng, S. Guo, and B. Ye, "Leverage Parking Cars in a Two-tier Data Center," in Proceedings of the 2013 IEEE Wireless Communications and Networking Conference (WCNC '13), Apr. 2013, pp. 4752–4757.
7. G. GGYU, G. ADAI, and K. RPS, "A Smart Vehicle Parking Management Solution", Proceedings of 8th International Research Conference, KDU, Published November 2015, pp. 106–110.
8. Google Maps Distance Matrix API, 2016, [online], Available: https://developers.google.com/maps.
9. Meng-Yen Hsieh, Jen-Wen Ding. Dynamic scheduling with energy-efficient transmissions in hierarchical wireless sensor networks. Telecommunication Systems, 2015, Vol. 60, No. 1, pp. 95 –105.
10. Meng-Yen Hsieh, Tien-Chi Huang, Jason C Hung, Kuan-Ching Li (2015/1). Analysis of Gesture Combos for Social Activity on Smartphone. Lecture Notes in Electrical Engineering, Vol. 329.
11. Hsieh, M.Y., Yeh, C.H., Tsai, Y.T., Li, K.C.: Toward a Mobile Application for Social Sharing Context. Lecture Notes in Electrical Engineering, vol. 274, pp. 93–98, (2014).
12. H.Y. Lin, M.Y. Hsieh and K.C. Li, "Flexible Group Key Management and Secure Data Transmission in Mobile Network Communications using Elliptic Curve Diffie-Hellman", International Journal of Computational Science and Engineering, Vol. 11, No. 1, 2016.
13. H.Y. Lin, M.Y. Hsieh and K.C. Li "Secured Map Reduce Computing Based on Virtual Machine Using Threshold Secret Sharing and Group Signature Mechanisms in Cloud Computing", Telecommunication Systems, Vol. 50, No. 146, 2015.

An Assisted Forklift Pallet Detection with Adaptive Structure Feature Algorithm for Automated Storage and Retrieval Systems

Jia-Liang Syu, Hsin-Ting Li, Jen-Shiun Chiang, Chih-Hsien Hsia, Po-Han Wu and Chi-Fang Hsieh

Abstract This paper is about automatically guided vehicle (AGV) system in the automated-storage-and-retrieval-system (ASRS). In ASRS, it usually uses AGV system to transport materials, because it not only efficient but can cost down logistic cost. However, the major problem of the application about AGV is how to find the position of the pallets due to the difficulties to locating the pallet position on a complicated factory environment. In this work, Haar like-based Adaboost scheme with adaptive structure feature of pallets algorithm to detect pallets is presented, and by combining direction weighted overlapping (DWO) ratio, it can avoid those non-optimal candidates in object tracking. The experimental result shows this method can remove most of the non-stationary background and can increase the average pallet detection rate by 95%.

Keywords Industry 4.0 · Automated storage and retrieval systems · Forklift · Adaboost · Pallet detection

---

J.-L. Syu · J.-S. Chiang
Department of Electrical and Computer Engineering, Tamkang University,
New Taipei City, Taiwan
e-mail: mike22490924@gmail.com

J.-S. Chiang
e-mail: chiang@mail.tku.edu.tw

H.-T. Li · P.-H. Wu · C.-F. Hsieh
Engineering Center, Boltun Corporation, Tainan City, Taiwan
e-mail: jay_19890811@hotmail.com

P.-H. Wu
e-mail: phwu.freeideatech@gmail.com

C.-F. Hsieh
e-mail: hardman761110@gmail.com

C.-H. Hsia (✉)
Department of Electrical Engineering, Chinese Culture University, Taipei City, Taiwan
e-mail: chhsia625@gmail.com

© Springer Nature Singapore Pte Ltd. 2018

N.Y. Yen and J.C. Hung (eds.), *Frontier Computing*, Lecture Notes
in Electrical Engineering 422, DOI 10.1007/978-981-10-3187-8_26

1 Introduction

Nowadays in the development of industry, most of the factories are gradually moving forward to industry 4.0. Among them, ASRS play a very important role. Currently, the transportation of materials on the automated storage system usually used AGV (automatically guided vehicle) because it can rapidly transport and dramatically decrease the logistics costs. However, the biggest problem of the application about AGV is how to get the position of pallet. Hence accurate pallet recognition is necessary. There are many different methods about pallet detection have been proposed. Basically, there are two categories: image-based method and sensor-based method. The image-based method usually employs cameras as an input device and separates the pallet and background by some pallet natural features [1–3]. The sensor-based method usually employs laser sensors or RFID tag to recognition pallets, but such methods will cost a lot of money. So, we proposed a brand new image-based method which only uses one camera as an input in this paper. This method applies Haar like-based Adaboost with cascade classifier scheme to detect pallet, and base on the appearance feature and structure feature of pallets to increase the accuracy.

2 Proposal Method

This pallet detection includes two parts: detection and tracking parts is shown in Fig. 1. At detection part, it combines Haar-like feature and Adaoost with cascade classifier schema, using pallet appearance feature and adaptive structure feature to optimize all result. At tracking part, we will track the detected objects in a period time, and the latest detection refers to the previous frame to compare the direction weighted overlapping ratio to select the best candidate.

2.1 Setting a ROI

In our experiment, we set the single-lens camera on the back upper side of forks on the position of the red circle shown as Fig. 2. So, it can capture the complete information and avoid getting the burden information as shown in Fig. 3. But, it is unavoidable that there still have a few parts of fork in the scene. Here, we set up a region of interest (ROI) as a restriction (the red rectangle region) shown in Fig. 3 for further processing.

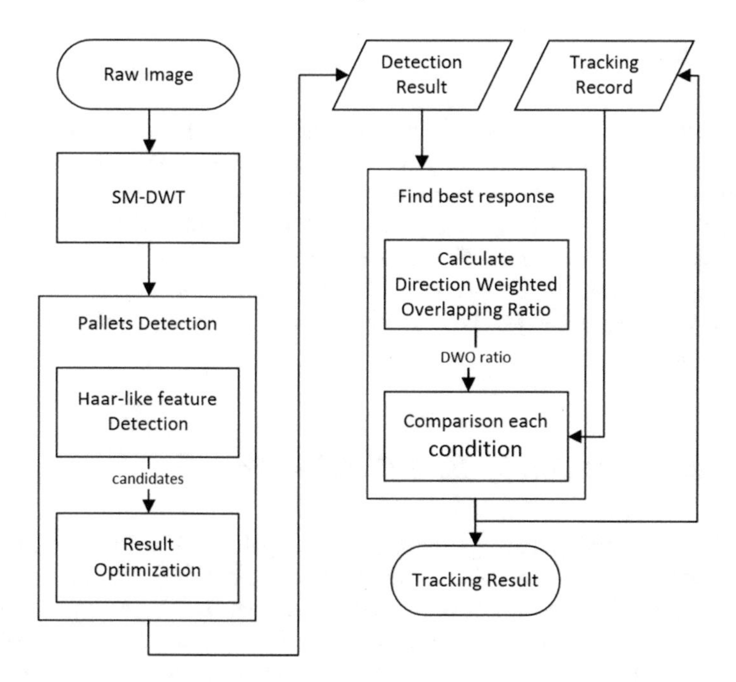

Fig. 1 System flowchart

Fig. 2 Camera setting

Fig. 3 Input image

2.2 Down-Sampling

Object detection need to scan all images at different scale, so the detection task usually spends a lot of computing time. So, in this step, we will reduce image resolution to speed up post-processing time. Hsia et al. proposed a brand new method to calculate DWT, which is called SMDWT [4]. SM-DWT improves some critical flaws of traditional DWT and is more convenient in two-dimensional multi-stage operation. It also has some advantages such as shorter critical path, fast compute and independent sub-band processing, etc. Hence we employ SMDWT to down resolution.

2.3 Pallet Detection by Haar-like Feature

In previous proposed method, people separate the pallet and background by certain pallet natural features and color distribution [1–3], but they did not use feature descriptors to detect pallet. We think that appearance features of pallets are very simple which consists of some rectangles, and the light distribution of pallet has certain rules under the normal light source. Therefore, Haar-like feature might be suitable for describing pallet appearance. In this step, we apply the scheme proposed by Viola et al. [5] to detect the pallet.

2.4 Optimization Result

Because, the natural feature of a pallet is very simple, so there are many objects may have the same feature value. In order to decrease the miss classification, we apply the variance as a feature and employ the adaptive structure feature of a pallet to optimize those results generated by the Haar-like detection as shown in Fig. 4. Figure 5 is the pattern of pallet fixed goods in the real factory. We can divide it into two parts to analyze, pallet and goods. For the part of pallet, the appearance of pallet has dramatic shading. Because of the distinctive feature, it makes the gray value within the pallet ROI become discrete distributions. For the part of goods, its surface is very smooth, so brightness changes are not significant, and its color is also very simple. Hence, we calculate the gray value variance of pallet and goods, and Hue (from HSV) value variance of goods. It uses these three values as a threshold to help our system define which one is pallet.

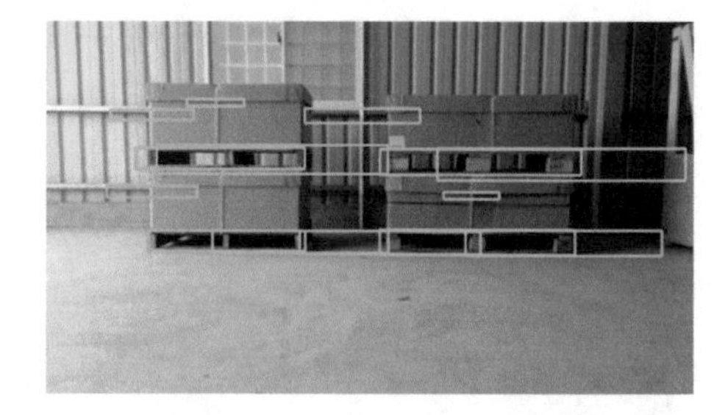

Fig. 4 Result of Haar-like feature detection

Fig. 5 Sample of Pallet

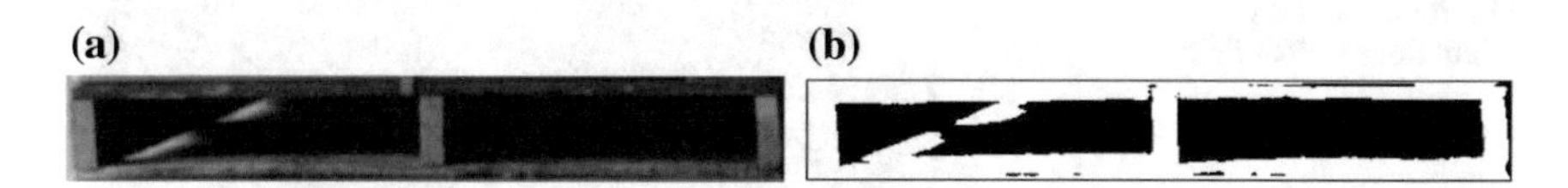

Fig. 6 **a** Standard pallet pattern in *gray scale*. **b** Woods part of pallet (Binarization result)

2.5 Adaptive Structure Feature

When using some image feature descriptors, the disadvantage is that it only considers the statistics as the feature number, it does not have the information about feature position. Hence, we propose an adaptive structure feature algorithm for pallets to compensate for lack that only used general descriptors. Figure 6a is a standard pallet pattern. The wood part of each pallet is about 45–55% of total window, which receives more light. So, we then accumulate the percentage of the gray scale value from the largest value to zero and stop at the accumulation sum reaching 50%. By this way, we can obtain an adaptive binarization threshold. This threshold can accurately separate the woods region in the pallet candidates as shown in Fig. 6b. And, each pallet has three logs, in left, right, and middle places, respectively, and it is a key structural feature. Hence, we calculate the curve of horizontal brightness distribution shown in Fig. 6b by using block-based vertical projection method. The curve of candidates belongs to pallet will have three peak at left, right and middle, as shown in Fig. 7. Using the characteristics of this curve, we can separate those non-pallet objects.

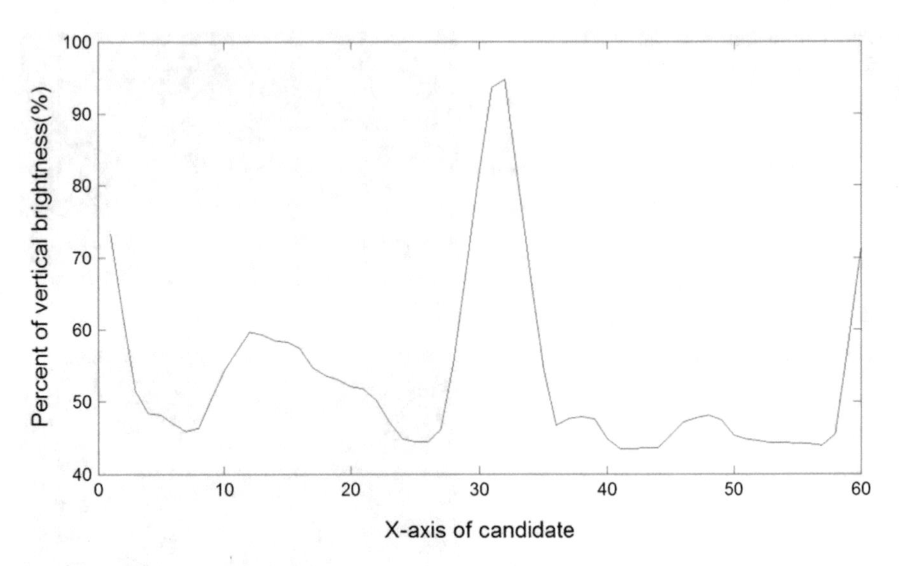

Fig. 7 The curve of brightness distribution

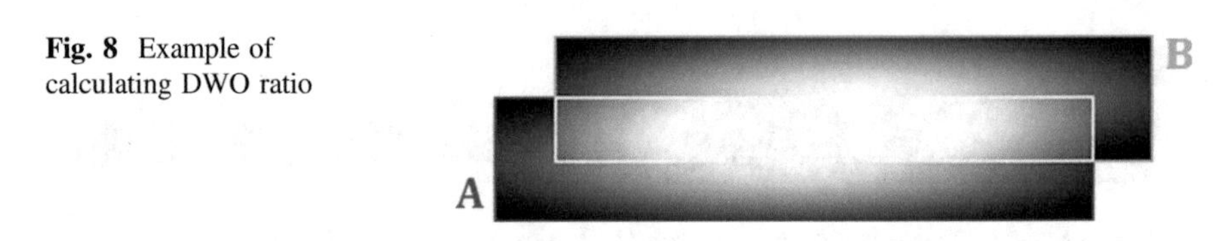

Fig. 8 Example of calculating DWO ratio

2.6 Tracking

In this step, we will record the information of all detection results of every frame, track every object detected and follow the characteristics of continuity of time and space to associate each tracking record with each detection result.

2.7 Direction Weighted Overlapping Ratio (DWO)

A tracking system usually uses the overlapping ratio to assist in selecting the most appropriate candidate as the objective. However, the position where the object appears should be assumed to be random as Gaussian probability distribution. So, this paper proposed direction weighted overlapping (DWO) ratio, which combines the traditional overlapping ratio and Gaussian probability distribution to make it more rationalize when comparing each tracking record and each candidate. It also can enhance the overlapping ratio of some special overlapping condition by adjusting the vertical or horizontal weight to make the candidate more suitable for the real moving object and further to find the optimal solution. Figure 8 is an example of overlapping, and Eq. (1) is the calculation of DWO ratio.

Table 1 Comparison of traditional overlapping ratio and DWO ratio

No.	Overlapping condition	Traditional overlapping ratio (OR) (%)	DWO ratio (%)
1.		100	100
2.		85.239	99.648
3.		66.777	96.786
4.		67.213	86.911
5.		72.114	94.659

$$DWO\,Ratio = \frac{\sum G(A_{A\cap B})\sum G(B_{A\cap B})}{\sum G(A)\sum G(B)} \tag{1}$$

where $G(x)$ is the rectangular Gaussian probability value, and the calculation references to [6].

Table 1 is a comparison and analysis between traditional overlapping ratio and DWO ratio. In this example, we decrease horizontal weight to calculate DWO ratio that means the horizontal probability will concentrate toward the center of rectangle. So we can see that case No. 3 has a similar overlapping area to case No. 4, but the DWO ratio of case No. 3 is greater than that of case No. 4. That is to say, our method has the characteristics of direction when selects the optimal result.

2.8 Comparison Each Condition

In the previous section, we have calculated the DWO individually between each tracking record and each detection result, next we will analyze each overlapping condition. Basically, up to the DWO ratio, there will be three situations as shown in Fig. 9.

And, there are about three characteristics in factory object tracking: (1) In video sequence, objects have the characteristics of continuity of time and space, so each object at time t will appear nearby the same position at time (t−1). (2) Tracking object will not overlap each other. (3) Each response at t only overlaps with one tracking record at time (t−1) in the same time. If there a response at time t overlaps with a tracking record at time (t−1), then the response would not overlap with any other tracking record at time (t−1). Follow these three characteristics and DWO ratio, we can make each tracking record correspond to one detection result.

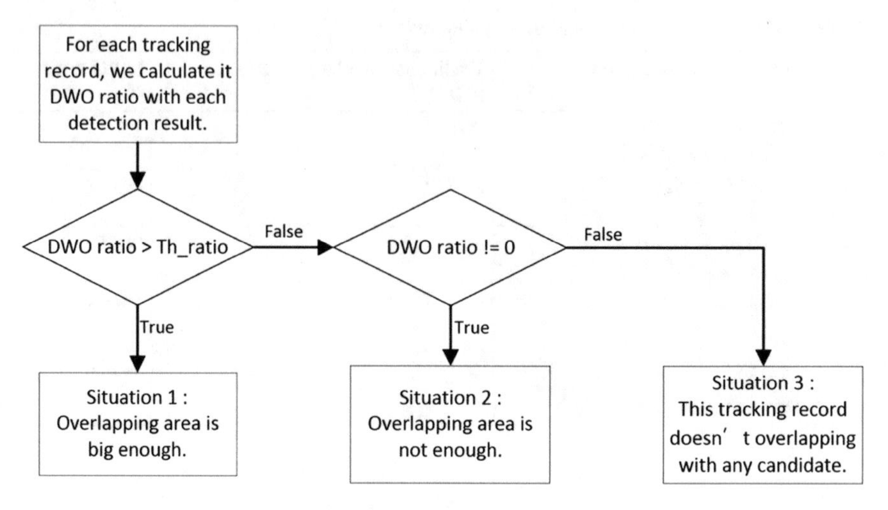

Fig. 9 The flowchart for finding the best candidate

3 Experimental Result

In this work, the specification of the computer for simulation is Intel Core i7-3770 CPU with 3.4 GHz working frequency and 8G RAM. We implemented this research by using Microsoft Visual Studio 2008 combining with OpenCV 2.4.3 in Windows platform and the input device is Logitech HD webcam C310 with 720p. Experimental environment is in the factory warehouse provided by Real-TouchApp Corp. Ltd. which we cooperate with. In the training data, we have about 3000 pallet positive patterns which are created under the real factory environment and about 9000 pallet negative patterns from the database and real factory environment [7].

In our experiment, we tested the pattern videos with different vehicle speed and different light source. Table 2 is our experimental result, where Video 1–4, 4–8 and 9–10 each is independent scene respectively, and each of them is shot with different speed and light source. Video resolution 1–8 are 1280 * 720, the processing time of each frame is about 100 ms. Video resolution 9–10 are 1920 * 1080, the processing time of each frame is about 200 ms. The change of brightness is based on the natural light or incandescent light in factory environment. Although, video 5–8 have high performance in bright and dark environment, but we can see that the higher brightness basically has a higher detection rate. In the ASRD, the faster cargo transport, the more virtual cost it will save. Therefore we join the speed factor. From Table 2 we can see that our algorithm has high detection rate, and the average error rate is only 1.79%.

Table 2 Experimental result

	Frame number	Detection rate (%)	Luminance	Vehicle speed(cm/s)
Video 1	158	96.835	Dark	50.2
Video 2	182	100.000	Bright	50.2
Video 3	118	98.728	Dark	87.0
Video 4	119	100.000	Bright	87.0
Video 5	109	100.000	Dark	87.0
Video 6	126	100.000	Bright	87.0
Video 7	143	100.000	Dark	50.2
Video 8	115	100.000	Bright	50.2
Video 9	172	94.775	Bright	–
Video 10	169	96.578	Bright	–

4 Conclusion

In this paper, we propose a new pallet detection based on adaptive structure feature and DWO to aid forklifts do accurately pallet recognition. Unlike other past method used expensive sensor, we only use camera erected at the forklift. By using Haar-like feature with Adaboost combining cascade classifier schema and adaptive structure feature of pallet, we have high performance at pallet detection in industrial environment and removing those miss classification candidates. In more detail, the proposed adaptive structure feature can accurately obtain the pallet structure and recognize pallet well, and the proposed DWO ratio is more suitable for moving object detection. Overall, we propose pallet feature based on structure and combine detection and tracking making pallet detection more accurate.

References

1. Cui, G.Z., Lu, L.S., He, Z.D., Yao, L.N., Yang, C.X., Huang, B.Y., Hu, Z.H.: A robust autonomous mobile forklift pallet recognition. Informatics in Control, Automation and Robotics (CAR), 2010 2nd International Asia Conference on 3, 286–290 (2010).
2. Seelinger, M., Yoder, J.D.: Automatic pallet engagment by a vision guided forklift. Proceedings of the 2005 IEEE International Conference on Robotics and Automation, 4068–4073 (2005).
3. Chen, G., Peng, R., Wang, Z., Zhao, W.: Pallet recognition and localization method for vision guided forklift. Wireless Communications, Networking and Mobile Computing (WiCOM), 2012 8th International Conference on, 1–4 (2012).
4. Hsia, C.H., Guo, J.M., Chiang, J.S.: Improved low-complexity algorithm for 2-D integer lifting-based discrete wavelet transform using symmetric mask-based scheme. IEEE Transactions on Circuits and Systems for Video Technology 19, 1202–1208 (2009).
5. Viola, P., Jones, M.: Rapid object detection using a boosted cascade of simple features. Computer Vision and Pattern Recognition, 2001. CVPR 2001. Proceedings of the 2001 IEEE Computer Society Conference on 1, I-511-I-518 vol. 1 (2001).

6. Horng, Y.R., Tseng, Y.C., Chang, T.S.: Stereoscopic images generation with directional gaussian filter. Proceedings of 2010 IEEE International Symposium on Circuits and Systems, 2650–2653 (2010).
7. http://tutorial-haartraining.googlecode.com/svn/trunk/data/negatives/.

Distributed Resource Allocation Approach on Federated Clouds

Yi-Hsuan Lee, Kuo-Chan Huang, Meng-Ru Shieh and Kuan-Chou Lai

Abstract This paper addresses the distributed resource allocation problem in the federated cloud environment, in which the deployment and management of multiple clouds aim to meet the clients' requirement. In such an environment, users could optimize service delivery by selecting the most suitable provider, in terms of cost, efficiency, flexibility, and availability of services, to deploy applications. However, different providers have different resource allocation strategies. This paper proposes a distributed resource allocation approach to solve resource competition in the federated cloud environment. Experimental results show that the cloud provider could obtain more profits by outsourcing resources in the federated cloud with enough resources.

Keywords Cloud computing · Federated cloud · Outsourcing · Resource allocation · Communication overhead · Marginal cost

1 Introduction

Cloud computing is emerging as the next generation computing paradigm, which adopts virtualization to enable rapid on-demand resource provisioning for organizing their shared physical resources with minimal management efforts. Therefore, cloud computing empowers users to scale up and down their resources on demands as workload variation. As more and more clients outsourcing their applications to the cloud platform, resource allocation becomes the most important issue in cloud computing.

Because one cloud provider could not provide unlimited services with limited physical resources, resource outsourcing becomes one important issue. There has been also a trend toward the federation of multiple clouds in order to deploy

Y.-H. Lee · K.-C. Huang · M.-R. Shieh · K.-C. Lai (✉)
Department of Computer Science, National Taichung University of Education, Taichung, ROC
e-mail: kclai@mail.ntcu.edu.tw

© Springer Nature Singapore Pte Ltd. 2018

N.Y. Yen and J.C. Hung (eds.), *Frontier Computing*, Lecture Notes in Electrical Engineering 422, DOI 10.1007/978-981-10-3187-8_27

applications and distribute workloads across vendors. In the federated cloud, clients could optimize service delivery by selecting the most appropriate provider, in terms of cost, efficiency, flexibility, and availability of services, to deploy applications and distribute workloads across vendors. Participants in the cloud federation could transact (buy and sell) resources on demand across multiple providers/vendors. Therefore, how to deploy an application to find the best resource allocation becomes the most important issue.

In general, different cloud providers own different pricing and allocation strategies by selfish policies. The federation of clouds has to consider the resource competition problem cause by the selfish policy in each cloud. When the resource is insufficient, the cloud provider rents resources from other clouds. If there are lots of outsourcing requests, competition for some limited shared resources will occur. Such a problem could be modeled as a game with multiple players.

This study proposes the *Distributed Resource Allocation* (DRA) approach to deploy applications over federated clouds to improve system performance in terms of reliability, availability and cost. Our objectives are to find the minimal job execution time with the minimal cost. Simulated results demonstrate that the proposed approach could get better profits of the federated cloud. This paper is organized as follows. Section 2 discusses related works. Section 3 presents the proposed strategy. Section 4 presents experiment results. Final section gives the conclusions and future works.

2 Related Works

Cloud computing [1] adopts the pay-as-you-go model by using the virtualization technology which generates the abstract layer above the cloud OS framework. Therefore, users can use cloud services in any place. As cloud computing continues to gain in popularity, the federation of multiple volunteer clouds becomes one of important issues to meet clients' requests. A federated cloud [2] is the deployment and management of multiple public and private clouds to meet clients' needs.

In cloud computing, resource management is a significant issue. There are various studies on resource management in cloud computing. One previous work [3] adopted horizontal elasticity scaling and vertical elasticity scaling to enhance system performance. Another study [4] also focused on resource re-packing in vertical/horizontal scaling fashions. Hussain et al. [5] introduced a lot of resource allocation approaches for clusters and grids in high performance distributed computing system. Wei et al. [6] proposed the concept of partial mapping which uses the network bandwidth between virtual nodes. To avoid the selfish policy in each provider, Palmieri et al. [7] proposed a distributed scheduling approach. This distributed scheduling approach aims to reduce the resource-conflict by using the game theory. Their approach considers the marginal cost.

In general, the game theory is a good approach to solve the resource competition problem, which could be categorized to the cooperative and non-cooperative approaches. The non-cooperative game theory could be further categorized into static complete information, dynamic complete information, static incomplete information, and dynamic incomplete information ones. Ye and Chen [8] studied the non-cooperative approach for the virtual machine placement problem. The authors focus on the optimal allocation strategy according to the Nash equilibrium. Hassan et al. [9] proposed an approach for the distributed resource allocation problem in cooperative and non-cooperative approaches. Wooldridge [10] also introduced a simple game theory example to define the game theory models. Mao et al. [11] presented a cloud service deployment approach. Their approach deploys the cloud resources by the price and service requirement according to the game theory. But, their approach does not consider the network latency between computing resources. To avoid the selfish policy in each provider, Palmieri et al. [7] proposed a distributed scheduling approach. This distributed scheduling approach aims to reduce the resource-conflict by using the game theory. Their approach considers the marginal cost.

3 Distributed Resource Allocation (DRA) Approach

This section introduces some of definitions and the proposed approach. Figure 1 shows the architecture of federated clouds. In this architecture, let C be a set of cloud providers, i.e., $C = \{C_1, C_2, \ldots, C_n\}$. Each cloud provider $C_i \in C$ could provide a set of virtual machines $V_i = \{V_{i1}, V_{i2}, \ldots V_{im}\}$. Each $V_{ij} \in V_i$ has a cost, $Cost_i$, which is the same in the same cloud. Each provider could decide whether accept requests or not. Let J be a set of jobs, i.e., $J = \{J_1, J_2, \ldots, J_k\}$. Each $J_i \in J$ has the profit, P_i and the number of requested virtual machines (i.e., resources), R_i. Each

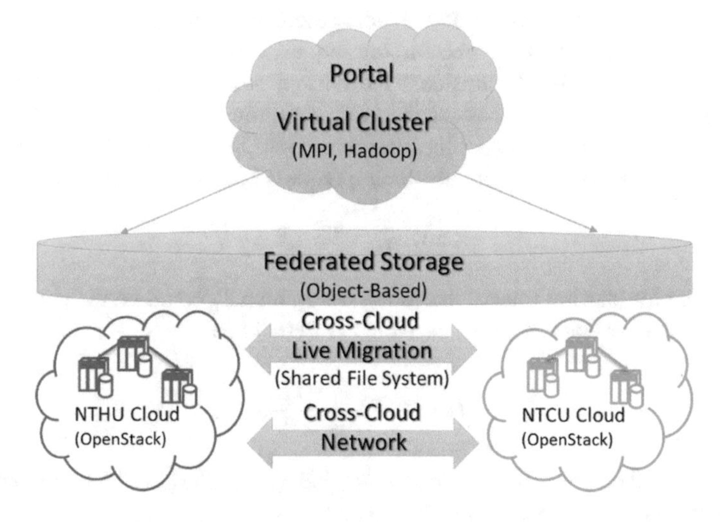

Fig. 1 Architecture of federated clouds

job consists of tasks grouped by the same data pattern, i.e., $J_i = \{G_{i1}, G_{i2},\ldots, G_{il}\}$. Each $G_{ij} \in J_i$ has the resource requirement, R_{ij} (the number of VMs).

The proposed approach decides which resource could be rented when the outsourcing occurs. When one cloud provider has enough resources, it should provide its resources to local service requests first. When the local resource is not enough, the provider will try to find the resource with the minimal renting cost in other providers, and send the outsourcing request to this resource. If the requested resource won't accept the outsourcing request, the provider will send the outsourcing request to next resource with the second smallest renting cost till the outsourcing request is accepted or aborted when there is no enough resource in the federated cloud.

Let R_{oi} be the number of outsourcing resources for the job i (i.e., the *i*th outsourcing request), then the bound of the renting cost for resource outsourcing is defined as

$$\text{Outsourcing\_bound}_i = P_i * \frac{R_{oi}}{R_i}. \quad (1)$$

When one service sends a request to the cloud provider without enough resource, this cloud provider will search matching resources with a lower price in the federated clouds. If there are too many providers which try to get the same resource, increasing the resource price as time goes on is a simple way to force the outsourcing requests to other cloud providers. Let CLevel_c be the competition degree of cloud provider c.

$$\text{CLevel}_c = \left\lceil \left(\sum_{i=1}^{n} R_{oi} \right) / V_c \right\rceil, \quad (2)$$

where V_c indicates the number of available resources in cloud C. If CLevel_c is less than one, it indicates that the outsourcing request could be satisfied.

The proposed approach considers the current available resource, current outsourcing requirement, and the possible resource requirement in the future to calculate the marginal cost for resource allocation with the minimal cost. So, the remaining resource after accepting the *i*th outsourcing request in cloud C is defined as

$$\text{Remain}_{c,i} = V_c - R_{oi} \quad (3)$$

Let T_i be the starting time of executing job J_i, then T_n could be the possible time period ready to execute next job n, which is defined as

$$T_n = \frac{\sum_{i=2}^{n} T_i - T_{i-1}}{n-1} \quad (4)$$

Let T_{ei} be the time period of executing job J_i, then T_{en} could be the possible execution time of next job n, which is defined as

$$T_{en} = \left(\sum_{i=1}^{n-1} T_{ei}\right)/(n-1) \quad (5)$$

The proposed approach takes the marginal cost to be the decision criterion. The marginal cost [12] is defined to reflect the opportunity of renting remainder resources. In general, the marginal cost is the change in the total cost that arises when one unit of resources is obtained. In this study, $\mathrm{Mcost}_{c,i}$ is defined as the marginal cost for outsourcing request i to cloud provider c; and, Mcost_c is the average marginal cost for all outsourcing requests to cloud provider c.

Let the competition level indicate the degree of resource-conflict. When the competition level is greater than one, the cloud provider will increase the price of resources. Otherwise, when there is no resource competition, the price of resources is obtained from the cost-driven pricing approach [13], therefore, the price is defined as follows.

$$\mathrm{Price}_c = \begin{cases} \mathrm{Price}_c^{-1} + \mathrm{CLevel}_c{*}\mathrm{Mcost}_c, & \text{if } \mathrm{CLevel}_c > 1 \\ \mathrm{Cost}_c{*}\mathrm{rate}_c, & \text{otherwise} \end{cases} \quad (6)$$

where rate_c is the rate of applying cost-driven pricing for cloud C.

Let $P_{c,i}$ be the leased-out probability of the remaining available resource after cloud C accepts the *i*th outsourcing request. When V_c is equal to R_{oi}, the outsourcing request R_{oi} consumes all available resource in cloud C, and no extra overhead has to be taken. However, when V_c is not equal to R_{oi}, the marginal cost presents the wasted cost when the remaining resource couldn't be leased out. Therefore, $\mathrm{Mcost}_{c,i}$ is defined as follows.

$$Mcost_{c,i} = \begin{cases} 0, & \text{if } V_c = R_i \\ (1 - P_{c,i}){*}Remain_{c,i}{*}Price_c{*}T_n, & \text{if } V_c \neq R_i \text{ and } P_{c,i} > 50\% \\ Remain_{c,i}{*}Price_c{*}T_{en}, & \text{if } V_c \neq R_i \text{ and } P_{c,i} \leq 50\% \end{cases} \quad (7)$$

where V_c indicates the number of available resources. When the outsourcing request is accepted, the remaining resource is supposed to be unusable this time.

$$\mathrm{Mcost}_c = \sum_{i=1}^{n} (Mcost_{c,i}{*}Cost_c) / \sum_{i=1}^{n} R_i \quad (8)$$

In order to improve the performance of the proposed approach, this study also adopts the 0/1 knapsack algorithm to pack the outsourcing requests into sets. The pseudocode is listed as the following Algorithm 1. In the modified 0/1 knapsack algorithm, each outsourcing request represents one item; the number of outsourcing resources, R_{oi}, represents the weight of each item; the number of available resources, V_c, represents the capacity of knapsack; and the cost of each resource, Cost_i, represents the cost of each item.

Algorithm 1. Modified 0/1 knapsack algorithm

Input: remaining resources (V_c), and outsourcing requests (e.g., R_{o1}, R_{o2}, R_{o3}, R_{o4}, R_{o5})
Output: allocation pattern (e.g.,{R_{o1}, R_{o3}, R_{o4}, R_{o5}},{R_{o2}, R_{o3}}))
{ Minimal_cost=maximal value
  Knapsack (R_{o1})
  Return allocation pattern.}

Knapsack (R_{oi})
{ New a fork for R_{oi}
  Select a non-repetitive R_{oj} to allocation pattern
  If accumulated outsourcing requests <= V_c
    {Calculate the marginal cost, $Mcost_{c,i}$.
     If Minimal_cost > $Mcost_{c,i}$ then
        Update allocation pattern and Minimal_cost.
        Knapsack(next R_{oi});}
  else
     fork end }

The next pseudocode, Algorithm 2, decides the resource allocation pattern for each outsourcing request. When the current remaining resource could support the outsourcing request, the provider accepts the outsourcing request; otherwise, the proposed approach applies the modified 0/1 knapsack algorithm to get the allocation pattern with the minimal marginal cost. If there are more than two allocation patterns with the same minimal marginal cost, the proposed approach selects the allocation pattern with higher resource requirement first. After determining the allocation pattern, the proposed approach will allocate resources according this pattern, and then update the resource status.

Algorithm 2. Distributed Resource Allocation algorithm

Input: remaining resources (V_c), and outsourcing requests (e.g., R_{o1}, R_{o2}, R_{o3}, R_{o4}, R_{o5})
Output: resource allocation for outsourcing request. (e.g., {R_{o2}, R_{o3}})
{ If V_c could satisfy all outsourcing requests then
    accept all outsourcing requests.
  else
    get possible allocation patterns by the modified 0/1 knapsack algorithm
    If there are more than two allocation patterns with minimal marginal cost
      find the allocation pattern with the maximal resource requirement
      when there are more than two patterns with the maximal resource requirement
        select any one allocation pattern randomly
    else
      find the allocation pattern with minimal marginal cost
  update resource status and response requests. }

4 Experimental Results

In order to demonstrate the performance of the proposed approach for the larger scale environment, this study applies the simulation approach to show the performance comparison with the Maui [14] algorithm. The simulation environment is shown in Table 1. The number of job requests is between 40 and 100.

In this experiment, this study assumes there are five clouds in the federated cloud environment. Each cloud has its resource capability and the rate of applying cost-driven pricing. In every second, each cloud will issue two job requests with different resource requirement. Each simulation will be held in 10 min. The average experiment results are simulated by ten times.

Figure 2 shows the comparison of profits obtained by DRA and Maui approaches in the federated clouds. The experimental results demonstrate that the proposed DRA approach can obtain better profits from outsourcing jobs. When the number of jobs is 80, the proposed approach obtains the maximum extra profit by outsourcing jobs.

Table 1 Simulation environment

Clouds	Cloud 1	Cloud 2	Cloud 3	Cloud 4	Cloud 5
Resource capability	30	30	30	50	60
Rates (Eq. 6)	1	1	1	2.3	2.7
# of tasks per job	1 ~ 12				
Job profit	# of task * random rate				

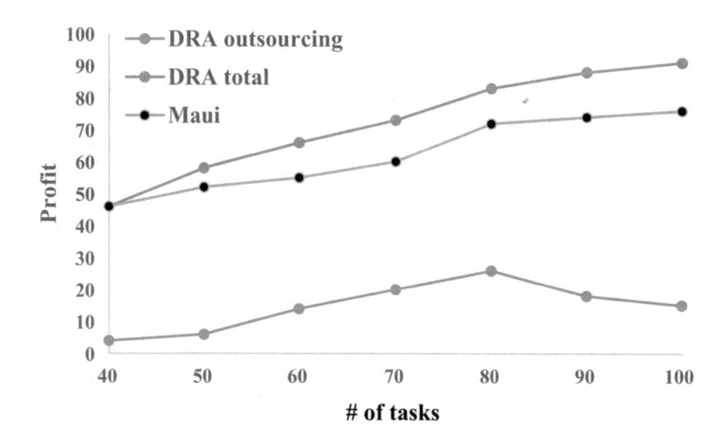

Fig. 2 Comparison of profits by DRA and Maui

5 Conclusions

This paper proposes a distributed resource allocation approach for solving resource competition in the federated cloud. The proposed approach considers the marginal cost and competition degree when the resource-conflict occurs. The 0/1 knapsack algorithm is also applied to improve further performance. Experimental results show the superiority in the profit obtained by DRA. In the future, this work will consider various resource types in a larger scale simulation environment.

Acknowledgements This study was sponsored by the Ministry of Science and Technology, Taiwan, R.O.C., under contract numbers: MOST 103-2218-E-007-021 and MOST 103-2221-E-142-001-MY3.

References

1. Mell, P., and Grance, T., The NIST Definition of Cloud Computing. NIST Special Publication 800–145, 2011.
2. Moreno-Vozmediano, R., Montero, R.S., Llorente, I. M., "IaaS Cloud Architecture: From Virtualized Datacenters to Federated Cloud Infrastructures," IEEE Computer, Vol. 45, No. 12, 65– 72, 2012.
3. Calheiros, R. N., Toosi, A. N., Vecchiola, C., Buyya, R., "A Coordinator for Scaling Elastic Applications across Multiple Clouds," Future Generation Computer Systems, Vol. 28, No. 8, 2012, pp. 1350–1362.
4. Sedaghat, M., Hernandez-Rodriguez, F., Elmroth, E., "A virtual machine re-packing approach to the horizontal vs. vertical elasticity trade-off for cloud autoscaling," Proceedings of the 2013 ACM Cloud and Autonomic Computing, Article No. 6, 2013.
5. Hussain, H., et al., "A survey on resource allocation in high performance distributed computing systems," Parallel Computing, Vol. 39, No. 11, pp. 709–736, 2013.
6. Wei, X., Li, H., Yang, K., Zou, L., "Topology-aware Partial Virtual Cluster Mapping Algorithm on Shared Distributed Infrastructures," IEEE Transactions on Parallel and Distributed Systems, Vol. 25, No. 10, pp. 2721–2730, 2014.
7. Palmieri, F., Buonanno, L., Venticinque, S., Aversa, R. and Martino, B.D., "A distributed scheduling framework based on selfish autonomous agents for federated cloud environments," Future Generation Computer Systems, Vol. 29, No. 6, pp. 1461–1472, 2013.
8. Ye, D. and Chen, J., "Non-cooperative games on multidimensional resource allocation," Future Generation Computer Systems, Vol. 29, No. 6, 2013, pp. 1345–1352.
9. Hassan, M., Song, B. and Huh, E.N., "Game-based distributed resource allocation in horizontal dynamic cloud federation platform," in Algorithms and Architectures for Parallel Processing, Y. Xiang, A. Cuzzocrea, M. Hobbs, and W. Zhou, Eds., Vol. 7016 of Lecture Notes in Computer Science, pp. 194–205, Springer, 2011.
10. Wooldridge, M., "Does Game Theory Work?" IEEE Intelligent Systems, pp. 76–80, November, 2012.
11. Mao, Z., Yang, J., Shang, Y., Liu, C. and Chen, J., "A game theory of cloud service deployment," 2013 IEEE World Congress on Services (SERVICES), pp. 436–443, June 2013.

12. Sullivan, A., Sheffrin S.M., Economics: Principles in action. Upper Saddle River, New Jersey 07458: Pearson Prentice Hall. 2003.
13. Trent, R.J. and Monczka, R.M., "Cost-Driven Pricing: An Innovative Approach for Managing Supply Chain Costs," Supply Chain Forum: an International Journal, Vol. 4, No. 1, pp. 2–10, 2003.
14. Maui, http://docs.adaptivecomputing.com/maui/.

Utilizing Software Defined Wireless Network to Improve Ship Area Network Performance

Shih-Hao Chang and Kuang-Chi Lee

Abstract In this paper, we combined Self-Organizing Time Division Multiple Access (SOTDMA) and software defined wireless network (SDWN) to apply in the next generation ship area network (SAN). We introduced network controller policy and resource pool to reduce network delay and achieve high performance. We sketch out changes and extensions to controller platforms, and wireless AP stations to enable software defined wireless networks. The complete design of a SDWN architecture using SOTDMA is an exciting avenue for future ship area network. Our evaluation result shows that for one wireless AP station can obtain the expected 3 Mbps throughput.

Keywords SAN · SOTDMA · SDWN · QoS · AIS

1 Introduction

In today's global cargo transportation, more than 75% of the total amount of international trade is the use of maritime transport [1]. Since ship in the sea will suffer natural climate and harsh environment impact, it will be an essential option to provide wireless sensors to monitoring the currently situation of the shipboard itself. A wireless transmission between ship instruments and sensor devices become an essential option with respect to energy efficiency, system deployment cost, and recovery and management convenience. Besides, using traditional ship area network cannot enable intelligent and autonomous maintenance services [2].

S.-H. Chang (✉)
Department of Computer Science and Information Engineering, Tamkang University, New Taipei City, Taiwan
e-mail: shhchang@mail.tku.edu.tw

K.-C. Lee
Department of Electrical and Computer Engineering, Tamkang University, New Taipei City, Taiwan
e-mail: kelvin@mail.tku.edu.tw

© Springer Nature Singapore Pte Ltd. 2018

N.Y. Yen and J.C. Hung (eds.), *Frontier Computing*, Lecture Notes in Electrical Engineering 422, DOI 10.1007/978-981-10-3187-8_28

Many researchers recently have been focusing on developing an integrated ship area network (SAN) applying state of the art IEEE 802.15.x based wireless communication technologies such as wireless personal network (WPAN) as well as WiMedia based SAN network as in [3, 4]. These research works relay operation in MAC layer network protocol as well as energy efficient transmission scheme.

One of recent research works in MAC layer network protocol in [5], the author utilize Time Division Multiple Access (TDMA) based Wireless Personal Area Network (WPAN) application model. In [5], the author goal is chosen because it can guarantee energy efficient operations with high date rate transmission in a wireless network environment. However, one problem in the TDMA communication system is that if any user obviously violates the relevant standards or rules, such as acts against the spirit underlying the standards and rules will result in an inability to ensure efficient and safe navigation of ships [5]. Moreover, increase of the channel load in TDMA could lead to a situation of no availability of slots. In contrast, Self-Organizing Time Division Multiple Access (SOTDMA) guarantees an upper bound on the channel access delay defined by the Selection Interval length (SI). The SOTDMA algorithm can intervene channel by reducing in a controlled way to effective operational range of the system [6].

On the other hand, autonomous network control is required in SAN but its not feasible until network virtualization is implemented. Since the network maintenance work is hardly implemented by ship crew, hence, autonomous network control could bring the benefits like reduced maintenance cost and improved reliability in SAWN. Many researchers recently also focus on developing Software Defined Wireless Networking (SDWN) [7, 8]. SDWN is a newborn wireless networks architecture that can be used for conveniently operating, controlling, and managing wireless networks. These features enable SAWN to deliver control data through the wireless network without human interaction. Moreover, in the SDWN architecture, the radio access networks are enhanced with programmability, supporting multiple functionality levels to enable scalability and variability. The contributions of this paper are: proposed wireless ship area network architecture. In Sect. 2, background and related works are briefly described. In Sect. 3, we first described the design criteria and present our software defined wireless network architecture and related mechanisms in Sect. 4 and our simulation results will be present in Sect. 5. Finally, we conclude our works.

2 Background and Related Work

In this section, we described the ongoing researches and key technologies, and challenges issues. First, ship Area Network (SAN) is one of the international standard is mainly based on wired networks such as dedicated connections, instrument networks and shipboard control networks (Ethernet based connection). This network delivers main operations such as sensing and control shipboard systems and management of crucial information for safety and navigation. In last decade, SAN has integrated with navigation system on board using National Marine Electronics Association (NMEA) 2000 standard [1]. It is primarily designed to provide two-way communication between the ship's navigational equipment such as radar, GPS receiver, automatic identification system (AIS) [2, 3], etc. NMEA 2000 is based on controller area network (CAN), which is standardized by ISO and also became the international standard under IEC. However, the maximum throughput of current NMEA 2000 bus connection is limited to 125 kbps; it cannot satisfy the increasing need for large amount of data transmission among the bulk of instruments on the ship [4].

Since the PHY and MAC layers technologies in SDWN are much more complicated than in the wired scenarios, it is indispensable for SDWN to achieve cross layer software defining. In the presence of unconstrained IEEE 802.11 stations, the majority of the above-referred real-time communication approaches need to be able to provide QoS guarantees whenever unconstrained stations try to access the shared communication medium [5]. As TDMA solution has been proposed as an access scheme for multi-hop radio networks where real-time service guarantees are important [6]. The idea is to allow several radio terminals to use the same time slot when possible. A time slot can be shared when the radio units are geographically separated such that small interference is obtained. However, TDMA can lead to the so-called hidden terminal problem. A way around this is to first transmit a short request-to-send (RTS) and then only send the message if a clear-to-send (CTS) is received.

Software-defined networking (SDN) decouples the data and control planes, removes the control plane from network hardware and implements it in software instead, which enables programmatic access and, as a result, makes network administration much more flexible. In past 6 years, SDWN [7, 8], an innovative paradigm, and become one of the latest and hottest topics in network fields. Before year 2011, the SDWN standards and OpenFlow protocol had been designed by Stanford University. After year 2011, various research groups are working towards designing SDN and OpenFlow specification, configuration, management protocols and so on. It is important to note that OpenFlow has become the core protocol of SDN. OpenFlow was proposed by Stanford University to enable the flow control in vSwitch. The SDWN layers can be divided into control plane and data plane.

3 Design Criteria

In this section, we identify several design criteria in SAWN, and explain why integrate SOTDMA into SDWN is a perfect solution. Now, to represent this as a radio network each of wireless AP stations as shown as in Fig. 1, giving us a 9-node network, a node in the middle picture (circles). Direct communication is usually possible if the units are close to each other, but objects like body of ship can prevent direct communication even when units are close, see for example nodes 6 and 8 in Fig. 1. Furthermore, we can see that communication between node 1 and 9

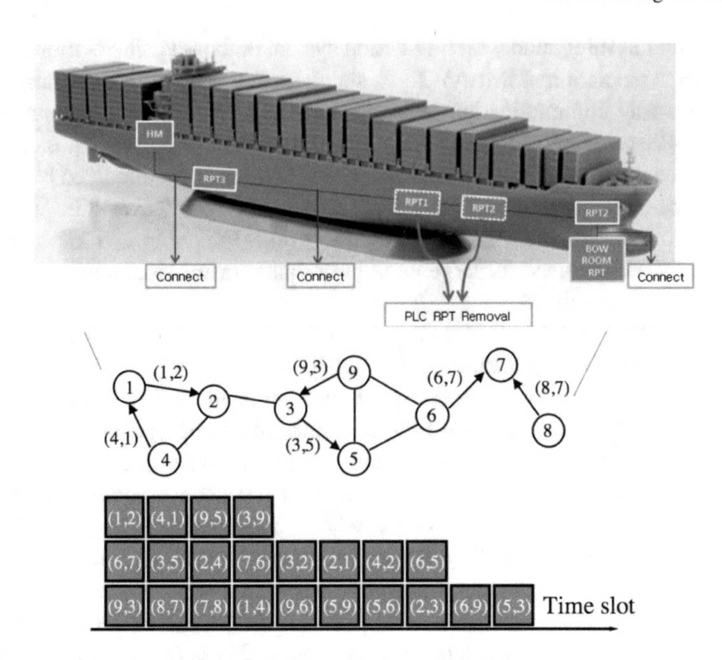

Fig. 1 Monitoring platform for maritime activities

must be relayed by node 2 and 3. Hence, a communication schedule could assign link (1, 2), (9, 3), and (6, 7) to transmit simultaneously, since they are sufficiently far from each other. The relaying of data traffic can enable communication between units further away. However, it also introduced the problem of transmission and end-to-end delay. These two problems are generally referred to performance issue in multi-hop wireless network. To solve these two problems in SAWN system, we claimed that both network (Layer 2) and routing (Layer 3) protocols will be required to guarantee data transmission. In consideration of network layer performance issue, SAWN focused on MAC coding/decoding techniques. In routing protocol, we consider end-to-end path(s) to achieve real-time guarantee. Thus, we utilized SOTDMA MAC protocol and coordinate with SDWN controller to provide a centralized routing which reduce the transmission delay. Here, we listed three design criteria as below:

Low Latency Allowance: The number of instruments and devices of a general container vessel is more than 460 items and these devices are normally connected to an isolated wired network [8]. These devices are connected to the SAN and transmit traffics through the network. Most data traffic within vessel carry control and status information, these data traffic must be lossless with low latency allowance.

High Data Throughput: As mentioned before, data throughput of the conventional NMEA 2000 is not sufficient to satisfy the increasing amount of data caused by a bunch of various machines, devices on the ship. Therefore, a high data rate supported system is necessary for providing a degree of freedom in deploying wireless APs with many instruments and devices on the ship at the lowest level.

Dynamic Channel Assignment: The network traffic generated from number of instruments and devices from the vessel sources such as throttle control, transmission control, engine computer and compass is also considered. Different types of network traffic must be able to handle simultaneously. For example, node assignment is used for broadcast traffic, and link assignment is used for unicast traffic. Hence, it would be preferable if one assignment strategy can be used for all traffic types.

As the SOTDMA provide the transmission rights for each time slot, it can be very efficient if a centralized nod has time information about the network. Due to each wireless AP station will be notified its own transmission schedule (slot), based upon data link traffic history and knowledge of future actions by other wireless AP stations. This can be done in a centralized manner, i.e. all information is collected into a central node, which calculates a new schedule and then propagated throughout the network. The SOTDMA schedule must be updated whenever something changes in the network and synchronized with each wireless AP stations. Hence, the design goal to achieve real-time SOTDMA schedules, e.g. minimizing delay or being able to update the schedules in a centralized fashion.

Another important parameter, which will affect the usefulness of a radio network, is the delay of information from source to destination. Since, we have global time information and each local AP's status information, we can discuss the delay for unicast traffic and how this differs from broadcast traffic. As seen from a network level, this can be translated to the network delay. Network delay is the expected time, in time slots, from the arrival of a packet at the buffer of the entry node to the arrival of the packet at the exit node, averaged over all origin-destination pairs. This is the first parameter we will investigate. When we want to broadcast message to each AP, each arriving packet has all other nodes as destination. We can now use several different definitions of the packet delay since a packet will arrive at its destinations at different times. One example would be the maximum delay, i.e. the delay of a packet would be the delay until the arrival at the last node. However, most nodes will experience a much lower delay. So, we instead will use the average delay of a packet. The end-to-end delay of a packet that originated at node *vi* can be written as:

$$D_i = \frac{1}{N-1} \sum_{j:\,(i,j) \in N^2} D_{ij} \tag{1}$$

where *Dij* can be described as the path delay between AP node *vi* and node *vj*. This is the path in the tree with root *vi* given by the routing algorithm for broadcast routing described in the previous section. The network transmission delay, *D*, can thus be described with the following expression which is similar to the expression for unicast traffic.

$$D = E\left[\frac{1}{N}\sum_{i \in N} D_i = \frac{1}{N(N-1)}\sum_{(i,j) \in N^2} D_{ij}\right] \quad (2)$$

4 Software Defined Wireless Network Architecture

The proposed SDWN architecture as shown in Fig. 2 offers fine grain, real-time control without sacrificing scalability. Working our way down from the controller platforms to the base stations, we propose four main extensions to enable SDWN. First, controller applications should be able to express policy in terms of subscriber attributes, rather than IP addresses or physical locations, as captured in a subscriber information base. Second, to improve control-plane scalability, each switch should run a local control agent that performs simple actions (such as polling traffic counters and comparing against a threshold), at the behest of the controller. Third, switches should support more flexible data-plane functionality, such as deep packet inspection and header compression. Fourth, base stations should support remote control of virtualized wireless resources to enable flexible cell management. Each device is a wireless transceiver connected to each navigational instrument or devices.

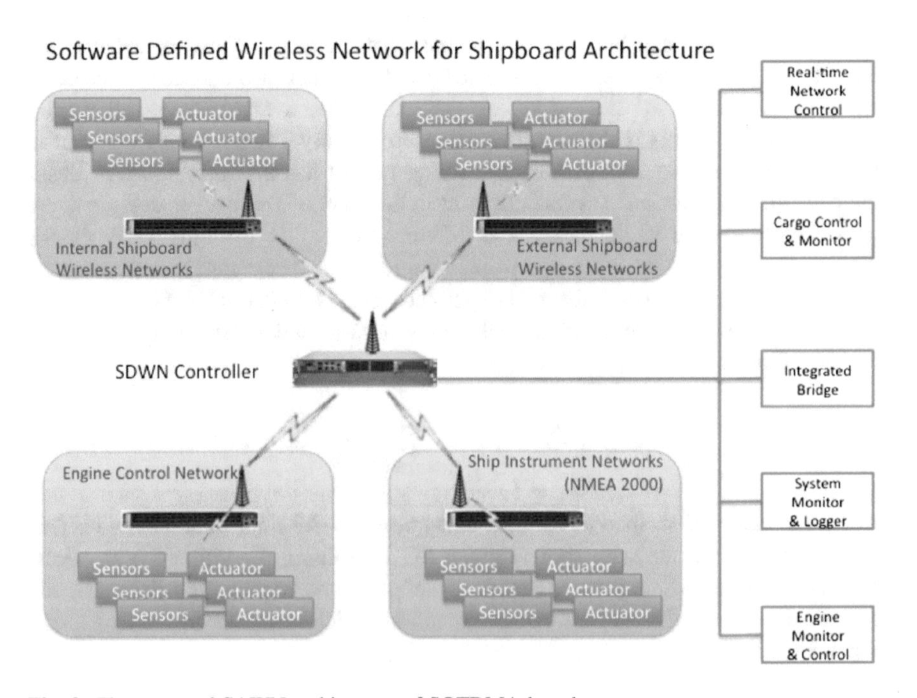

Fig. 2 The proposed SAWN architecture of SOTDMA based system

4.1 Network Controller Policies

A controller application running on the controller can spread access-control rules over multiple switches, and manage the scheduling of traffic by QoS classes across multiple hops in the network. SDWN enables centralized control of base stations. A SDWN controller will have a global view of the current power and subcarrier allocation profile of base stations. In addition, a SDN controller running on a commodity server would have much more computing resources than most base stations. As a result, a SDN controller can make a more efficient allocation of radio resources to handle new requests. The SDWN controller consists of a Network Operating System (NOS) running a collection of application modules, such as CPU, Memory, Bandwidth resource management, mobility management, and routing. The handling of individual packets often depends on multiple modules. For example, the flow of traffic through the network depends on the subscriber's location (determined by the mobility manager) and the paths between pairs of network elements (determined by infrastructure routing), and traffic monitoring and packet scheduling depend on the policy and charging rule function. As such, the NOS should support composition to combine the results of multiple modules into a single set of packet-handling rules in each switch.

The NOS can translate policies expressed in terms of subscriber attributes into switch rules that match on packet headers. Similarly, the NOS can translate network measurements (such as traffic counters) to the appropriate (sets of) subscribers, to allow the application modules to focus on subscribers and their attributes. The controller can also dynamically divide the network into "slices" that handle all traffic matching some predicate on the subscriber attributes. This allows the network provider to isolate roaming traffic, isolate dumb traffic using legacy protocols. To enable scalable slicing of semantic space, the controller can instruct ingress switches to mark incoming packets (e.g., using an MPLS label or VLAN tag) sent to or from subscribers with particular attributes (Fig. 3).

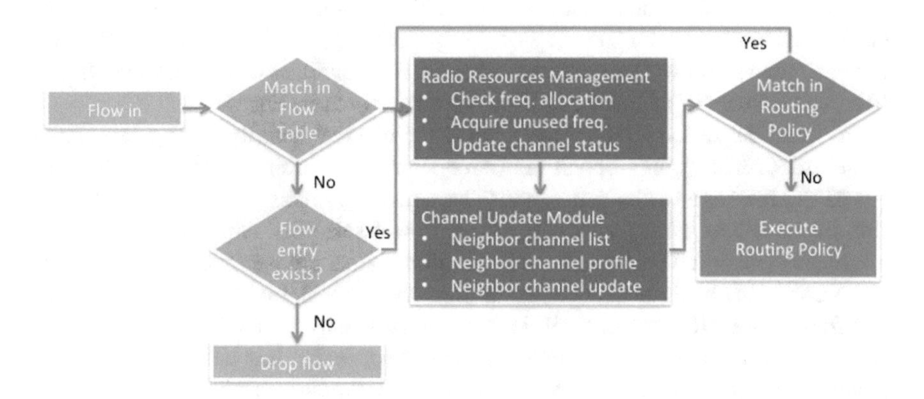

Fig. 3 The proposed SDWN algorithm to control packet flow

4.2 Resource Pool

As we consider the resource requirement in wireless AP stations which can be denoted by *R = (CR, M, B, C, T)*, which includes information such as communication region *CR*, the amount of *CPU* resource *C*, the amount of memory resource *M*, required bandwidth B, and the maximum latency *T*. Each communication between two devices on ship has a specific maximum latency T that should not be exceeded threshold *d*. The *SDWN* Controller could enhance its capabilities by allocating resources to remote Wireless AP or renting resources from the local resource pool. The resources allocation is based on resource requests.

$$X_K^l = a\frac{M_k}{Max\, M_k^l} + b\frac{B_k}{Max\, B_k^l} + c\frac{C_k}{Max\, C_k^l} \quad (3)$$

where Ck is CPU resource, Mk is the memory resource, Bk is the bandwidth resource and X_K^l is the weighted sum of resources. a, b, and c are fixed coefficients of three kinds of resources (memory resource, bandwidth resource, CPU resource).

4.3 Emergency Data Transmission

According to our literature survey, using of traditional TDMA has some disadvantages mainly because the "time delay". Therefore, we applied SOTDMA algorithm in our communication algorithm for sending emergency messages. SOTDMA is the most complex TDMA access scheme defined for AIS and also provides the backbone for autonomous operation of the network offshore. SOTDMA can provide for dynamic and autonomous management of capacity in busy areas. It allows the slots occupied by the stations most distant to a particular mobile station to be re-used for its own transmissions. This effectively reduces the size of an AIS 'cell' and ensures that position reports from the nearest vessels (which are most relevant to safety of navigation) are not affected. The SOTDMA frame structure follows four different states as (I) Initialization, (II) Network Entry, (III) First Frame (IV) Continuous Operation as shown in Fig. 4.

In the initialization state, a node will listen channel activity among one frame (called super-frame) then obtain the frame structure. Within this time, it also will record the message of the active nodes transmitting, obtain position of each time slot. In the network entry state (II), the node will select its own slot based on the information acquired in previous state. If all the slots are occupied, the node will make use of the position knowledge to transmit in the slot of the farthest node. In first frame state (III), the node will join the network actively and begin transmit in a slot which determined by the network entry state; In the last phase, namely continuous operation (IV), the node falls into continuous operation, transmitting periodically message in a specific slot which has been assigned before. Once the

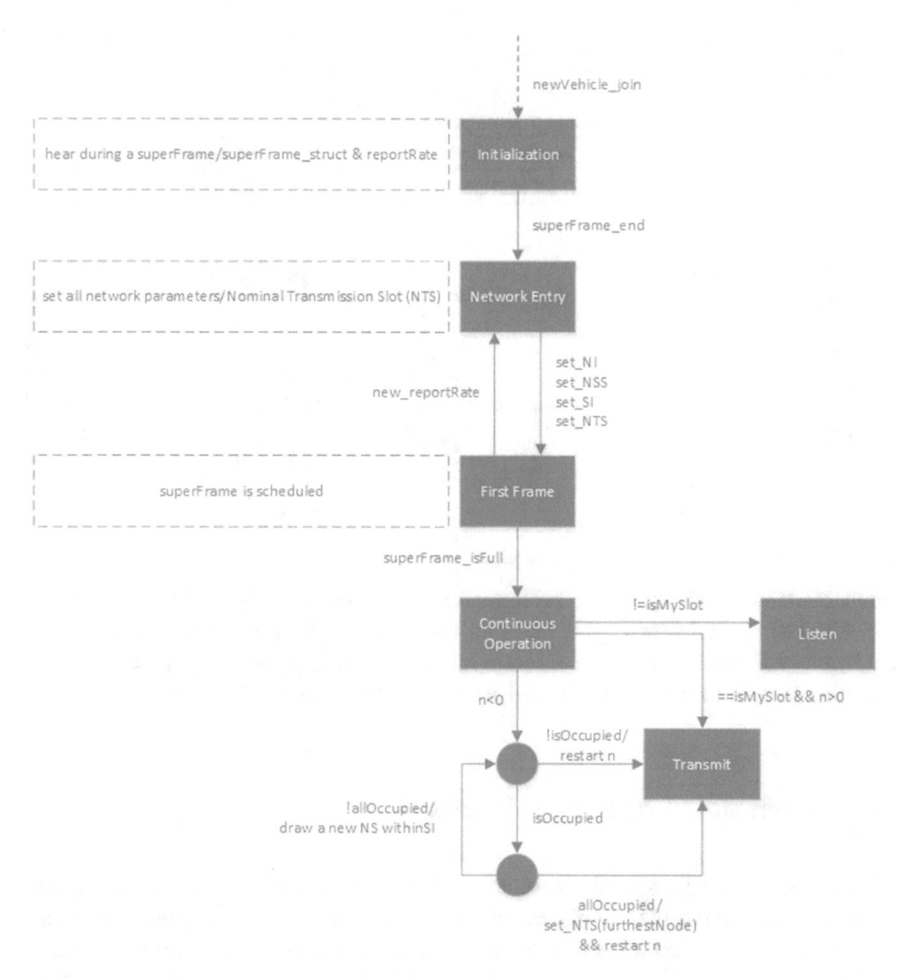

Fig. 4 The SOTDMA algorithm implementation procedure

topology change in the network, the slot allocation should change dynamically. Thus, in first frame state (III), the node will draw a random integer (e.g. n in the Fig. 4) for each time slot assignment. Then, it will determine for how many consecutive frames this particular slot will be used.

5 Evaluation

In this section, we present results different scenarios. The simulation parameters used are shown in Table 1. In a SAN where each super-frame has got a transfer rate of 5 Mbps, each car transmits 500 byte long messages every 100 ms and there can

Table 1 Parameter setting for simulation

Parameter	Value
Packet size	500 byte
Super-frame period	1 s
Transfer rate	5 Mbps
Heartbeat rate	10 Hz

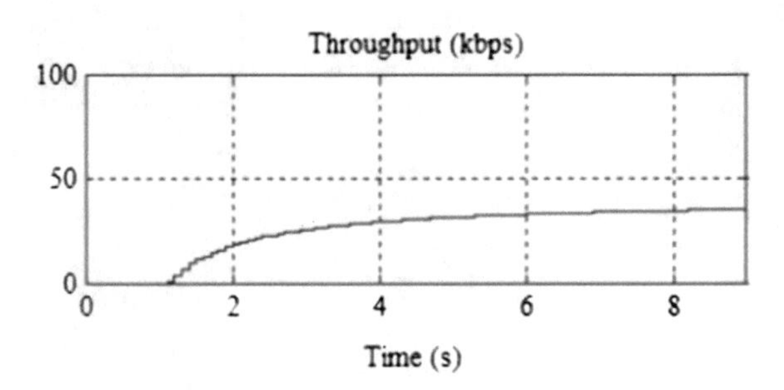

Fig. 5 The throughput for 1 wireless AP

cohabitate up to 75 equipment within 1 s super-frame. The Fig. 5 depicts the traffic in kbps of a single aggregate sends to the channel. Now we calculate from this wireless AP station the total throughput for the 75 equipment aggregates transmitting in a full traffic scenario and we obtain the expected 3 Mbps throughput.

6 Conclusion and Future Works

In this article, we presented combination of SOTDMA and SDWN can make ship area network much high performance issue and easier to manage. We introduced network controller policy and resource pool to reduce network delay and achieve high performance. We sketch out changes and extensions to controller platforms, and wireless AP stations to enable software defined wireless networks. The complete design of a SDWN architecture using SOTDMA is an exciting avenue for future ship area network. Our evaluation result shows that for one wireless AP station can obtain the expected 3 Mbps throughput.

References

1. S. Krile, D. Kezić and F. Dimc, "NMEA Communication Standard for Shipboard Data Architecture", In Proc. of International Journal of Maritime Science & Technology, vol. 60, no. 3, 2013.
2. "Maritime Information T Standards (MiTS)", available at http://www.mits-forum.org/architecture.html.

3. S. Lee, Y. Lee and S. Lee, "Multi-helper Relay Based WUSB/DRD/WLP Protocol in WiMedia Distributed MAC Systems", International Journal of Multimedia and Ubiquitous Engineering, vol. 9, no. 3, 2014.
4. D. Jeon and Y. Lee, "Performance Evaluation of a WiMedia based Wireless Bridge using Relay Cooperative Transmission", Advanced Science and Technology Letters, (Mobile and Wireless 2014), vol. 60, 2014.
5. D. Jeon and Y. Lee, "Performance Evaluation of Ship Area Network with TDMA-based HR-WPAN for a M2M Application", Proceeding of The 8th International Conference on Future Generation Communication and Networking, Dec. 20–23, 2014, Hainan, China.
6. ETSI. Intelligent Transport Systems (ITS); STDMA recommended Parameters and Settings for Cooperative ITS; Access Layer Part. Technical report, European Telecommunications Standards Institute, 2012.
7. M. Jarschel, S. Oechsner, D. Schlosser, R. Pries, S. Goll, and P. Tran-Gia. "Modeling and performance evaluation of an openflow architecture", In Proc. in 23rd International Teletraffic Congress (ITC), Sept. 2011.
8. P. Dely, A. Kassler, and N. Bayer. "Openflow for wireless mesh networks.", In Proc. in 20th International Conference on Computer Communications and Networks (ICCCN), Aug. 2011.

A Quality of Experience Aware Adaptive Streaming Service for SDN Supported 5G Mobile Networks

Chin-Feng Lai, Wei-Ting Chen, Chian-Hao Chen, Chia-Yun Kuo and Ying-Hsun Lai

Abstract In recent years, as the number of mobile users increases sharply, the mobile communication industry rapidly develops, and mobile communication network equipments are added on a large scale; consequently, the transmission problem of mobile communication increasingly occurs, which creates bottlenecks in the development of mobile communications. How to effectively break through the yokes of energy efficiency and spectral efficiency is key to the development of the future 5G mobile communication network. On the other hand, with the emerging technology Software-Defined Networking in the Internet, this new type of network architecture changes the mode of traditional network control, thus, the mobile data transmission of Heterogeneous Multi-Networks is severely challenged. It is important to design a service-oriented Heterogeneous Multi-Networks collaborative architecture in the case of Heterogeneous Multi-Networks, and to create a cooperative transmission strategy according to user requirement, network resource state, and the differences in server capacity and time variations. This paper proposes a bandwidth evaluation mechanism based on the dynamic streaming protocol, where the network bandwidth is timely adjusted to provide users with uninterrupted viewing, and a streaming mechanism of video play with better Quality of Experience is provided in the same network bandwidth.

Keywords QoE-Aware · Adaptive streaming · Software defined network · 5G

C.-F. Lai (✉) · W.-T. Chen · C.-H. Chen · C.-Y. Kuo
Department of Computer Science and Information Engineering,
National Chung Cheng University, Chiayi County, Taiwan
e-mail: cinfon@ieee.org

Y.-H. Lai
Smart Network System Institute, Institute for Information Industry,
Taipei, Taiwan

© Springer Nature Singapore Pte Ltd. 2018
N.Y. Yen and J.C. Hung (eds.), *Frontier Computing*, Lecture Notes
in Electrical Engineering 422, DOI 10.1007/978-981-10-3187-8_29

1 Introduction

In past decades, wireless communication was rapidly developed in both technological research and development and marketing functions; however, the users and requirements for flow of traffic continuously increased, and while the LTE and LTE Advanced (the fourth generation cellular mobile communication system) systems were beginning to be extensively deployed, the establishment of the next generation of the 5G standard has been initiated with great speed. Aiming at the development of a new generation of 5G network techniques, many studies indicated that the future 5G technology shall simultaneously attain three major objectives: (1) wireless transmission capacity expansion; (2) low power requirements; (3) connecting all items [1–3]. Looking forward to the fifth generation, or 5G mobile networks, the requirements for high speed connection will be more urgent. With the development of mobile communication network technology, Software-Defined Networking is regarded as a revolutionary technology overturning the traditional Internet. As SDN separates the Control Plane from the Data Plane, the traditional network equipment can concentrate only on the data transmission of the Data Plane, while the Centralized Controller is in charge of the control and management works of the Control Plane, according to the programmable concept, in order that the layout of new network services is free from the original network equipment; while the optimization engineering of the Switching/Routing mechanism, Traffic Engineering, Network Virtualization, and Network Function Virtualization, can be implemented to obtain agility and flexibility [4, 5]. SDN also establishes the Service-Level-Agreement (SLA) for different usage requirements, meaning that users can obtain the service quality when accessing services.

The two emerging network techniques largely expand the network connection territory, meaning that we can access and share all "things" anywhere, and at any time. In addition to simple speech and data services, the vision of "connect anything all the time" is implemented, meaning the demand for the data services of mobile and fixed networks increasingly grow. However, as the data transmission capacity increases, the quantity of more complicated equipment requiring more data increases, and users have increasingly strict requirements for network services. The "seamless connection" becomes a basic requirement, i.e. smoothly transmitting applications among various mobile devices, for example, transmitting network content without interruption among the tablet PC, mobile phones, and home amusement centers. To meet the aforesaid requirements, users shall access and control content via the Heterogeneous Multi-Networks of the internet and mobile communication network as shown as Fig. 1. Previous studies regarding Cloud multimedia networks often placed emphasis on the dynamic resource scheduling of cloud data centers and high efficiency network transmission technology, including the transfer of virtual machines in local Cloud centers or other Cloud centers, the dynamic resource allocation of virtual networks, and guaranteed Quality of Service of dataflow under dynamic network requirements [6–8]. However, in the aforesaid emerging Heterogeneous Multi-Networks architecture, transmission quality cannot

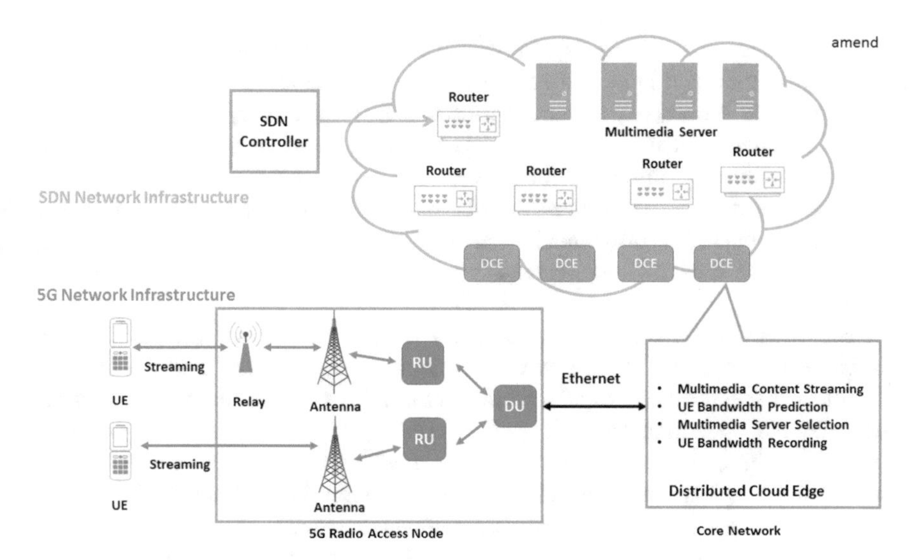

Fig. 1 Multimedia services over heterogeneous multi-networks

be satisfied only by the special network topology of a cloud data center or new Ethernet protocol. Therefore, this paper proposes a bandwidth evaluation mechanism based on the dynamic streaming protocol, where the Distributed Cloud Edge of a Cloud center in SDN records the online bandwidth and response time of 5G mobile communication network users in order to calculate the available multimedia streaming quality for the user [9–11]. The OpenFlow instruction of an SDN controller controls the cloud data center to transmit the streaming data instantly by SDN-enabled switching, and timely adjusts the network bandwidth without interrupting the user's viewing, thus, multimedia streaming service, which is able to satisfy the user's quality of experience, is provided for the 5G mobile communication network user in the same network bandwidth.

2 Quality of Experience of Adaptive Streaming Service

Quality of Experience (QoE) is a sort of evaluation method for surveying user satisfaction with the overall system [12]. Quality of Service (QoS) is often heard and used, and is correlated with QoE; however, there are slight differences, the QoS evaluates two measurable indices, including hardware and software, in order to enhance and improve the overall QoS. The QoE displays the user's subjective and objective satisfaction. The evaluation mechanism of QoE can be applied to commercial activities or services related to users, while QoE is usually used as an index in information technology and consumer electronics.

QoE collects the consumers' experience and appraisal mostly by market survey or questionnaire, and is judged according to several critical indices, such as reliability, privacy, safety, expenditure, performance, and efficiency. These evaluations have key environmental factors, such as hardware facilities and application scenarios, which may influence the overall score. For example, there are many application scenarios that can be discussed merely for QoE testing of video, such as video conference, live programming, or VoD streaming. The application scenarios correspond to different requirements, and discussions regarding QoE are more extensive.

For Adaptive Streaming Service, many experts have proposed guidelines to improve user experience, and even the organizations of 3GPP and MPEG have established specifications stipulating the factor indices of QoE in the specifications of Adaptive Streaming technology. The HAS specifications of 3GPP provide a trigger for the QoE measurement mechanism, which evaluates the Client communication protocol and multimedia format transmission for feedback to the network server, and is a free choice during Client implementation.

Some indices of QoE are defined in the HAS specification of TS 26.247, as established by the 3GPP organization. Seven major indices, and their contents, are listed, as follows:

- Representation Switch Events

This index is used to report the switch event of Representation, where this switch event signal is transmitted to the player in order to switch to the adaptive mechanism adjusted Representation. The Client shall measure and report the new Representation identification code, the time of response to request switching content, and the time from receiving the new Representation media segments to playing the first new media segment.

- Average Throughput

The Client records the byte number of content, activity time, and measured interval length, as received by HTTP Response during measurement. The activity time is defined as; there is at least one HTTP GET Request not yet completed after the measurement begins to the time interval without any request, and the state during the inactive time shall be recorded and reported, such as suspended play or downloading of media segments.

- Initial Playout Delay

This index literally refers to the delay time from when the user presses Play to when the picture or sound appears. However, this is misleading, as there is a time delay of network access, such as looking for MPD and downloading media segments after Play is pressed, as well as the operating time of analyzing the MPD list and multimedia format. Therefore, this index actually measures the time from a request to play the first multimedia content segment (non-initial segment) to obtaining the content from the Buffer.

- Buffer Level

Basically, this index is applicable to measuring and reporting the remaining play time, i.e. detecting the quantity of media segments in the Buffer and the play length. The total time is the remaining multimedia length. As these multimedia have been temporarily stored in the Buffer by network transmission, if the network is disconnected in the future, these buffered media segments can still be played smoothly. The developer chooses the segments suitable for temporary storage, and this point is one of factors to be examined in this paper.

3 A Quality of Experience Aware Adaptive Streaming Service for SDN Supported 5G Mobile Networks

Most adaptive streaming adjustment mechanisms select the optimal or corresponding bit rate media segment for streaming play according to the variance in bandwidth; however, this adjustment mechanism does not improve the user's comfort in viewing. Therefore, a dynamic adaptive streaming mechanism, as based on QoE, is proposed in this paper by researching user satisfaction with two video factors, which aim to create a dynamic adjustment mechanism according to network conditions. The user's discomfort in viewing is reduced by adjustment, as possible, in order that the user perceives little change in video quality when adaptive streaming is implemented according to bandwidth.

3.1 UE Bandwidth Prediction of Distributed Cloud Edge

On the Distributed Cloud Edge side, in order to effectively evaluate the condition of the connection between a 5G mobile communication network user and the Core Network, this study stored several previous packets in the cyclic queue, where the number of packets in the cyclic queue is based on the number of packets to be transferred per media segment of the current streaming media bit rate. When new packet information joins the cyclic queue, the total transmission quantity and overall transmission time are updated, and whether the cyclic queue is full is judged. If it is full, the earliest packet of information is deleted and the total transmission quantity and overall transmission time are updated. If it is not full, the present total transmission quantity is divided by transmission time as the bandwidth information. Thus, the Server can maintain the latest transmitted packet information to master the connection with the Client.

When the server obtains the bandwidth connection with a Client, it analyzes the bandwidth condition for further transcoding. However, as the network condition may be unstable at any time, this paper defines three modes to cope with different network conditions, and uses a mode shift state machine for bandwidth evaluation and error compensation. The three modes are Normal Mode, Active Mode, and Conservative Mode. The strategies adopted by the modes are described as follows.

Normal Mode: In this mode, the server-side uses the bandwidth downloading current media segment as the bit rate for transcoding, the computing mode is described as below, n_p is the number of packets, B_i is the bandwidth value measured by using tcpdump to record packets in the annular queue, B_{next} is the target bit rate of the next media segment, and B_{avg} is the average bit rate of this segment download.

$$B_{next} = B_{avg} = \sum_{i=1}^{n_p} B_i/n_p \tag{1}$$

- Active Mode:

In this mode, the server compares the last bandwidth condition evaluated with the bandwidth condition currently evaluated to determine the trend of bandwidth variation in this period. If the trend is positive, namely, the bandwidth increases with time, the slope of the change trend is multiplied by the remaining time (segment time span minus download time), plus current average bandwidth, as the target bit rate of next segment. The computing mode is described as below, where a is the slope of change trend, B_{end} is the bit rate of the end of this segment download, and $T_{remained}$ is the remaining time to next download:

$$B_{next} = B_{end} + a * T_{remained}, \quad if \quad a > 0 \tag{2}$$

$$B_{next} = B_{avg}, o.w. \tag{3}$$

- Conservative Mode:

In this mode, the server compares the last bandwidth condition evaluated with the bandwidth condition currently evaluated to determine the trend of bandwidth variation in this period. If the trend is negative, namely, the bandwidth decreases with time, the slope of the change trend is multiplied by the remaining time (segment time span minus download time), plus current average bandwidth, as the target bit rate of next segment. The computing mode is described as below:

$$B_{next} = B_{end} + a * T_{remained}, \quad if \quad a < 0 \tag{4}$$

$$B_{next} = B_{avg}, o.w. \tag{5}$$

3.2 Quality of Experience Awareness Adaptive Streaming

In terms of user experience in viewing videos, the video quality is most directly related to the user's impressions. The peak signal-to-noise ratio (PSNR) is the most extensively known index. Many studies of user experience use PSNR and Mean Opinion Score (MOS) for analysis. The conversion between PSNR and MOS is as shown in Table 1.

However, the reference factors of PSNR are complex, not intuitive, and are subjective for the user. Therefore, this paper analyzes and studies two factors that most directly influence image quality in image coding, and uses the result to create the most suitable picture quality for the user. The two factors are the resolution of the video and the Quantization Parameter of coding. The open source ffmpeg is used for image coding. As this paper does not discuss the differences in the type of codec or the image profile, the H264 codec and High Profile are used in implementation, packaged in the MPEG-4 format, and played at a refresh rate of 30 frames per second. The videos for the experiment are from the Xiph Community, and the video information is as shown in Table 2.

The experimental content is to watch two videos continuously, where the former is the control group, and the latter is the experimental group. Whether the latter is better or worse than the former is judged according to the user's subjective determination of picture quality. The experimental design is graded −1–10 points, where −1 means the latter is worse than the former, 0 points represent no difference, and 1–10 represents the difference in the user's subjective determination.

Resolution Factor:
The length and width of resolution in coding will directly influence the bit rate of the encoded video, and the relationship is expressed as Eq. 6. The bit rate is in multiple relationships to product of length and width, and because the bit rate is related to the selection of network streaming, the streaming mechanism in this paper will refer to this part to choose the maximum C/P value.

$$\frac{Bitrate_1}{Bitrate_2} = \frac{W_1}{W_2} \times \frac{H_1}{H_2} \tag{6}$$

Table 1 Comparison table of PSNR and MOS

PSNR (dB)	MOS	Quality	Impairment
>37	5	Excellent	Imperceptible
31–36.9	4	Good	Perceptible but not annoying
25–30.9	3	Fair	Slightly annoying
20–24.9	2	Poor	Annoying
<19.9	1	Bad	Very annoying

Table 2 Original video information

Original videos	Names	Bitrates (kbps)	Content types	Resolution (pixel * pixel)	Length (frames)
1	Ducks_take_off	746,496	Slight movement	1920 × 1080	500
2	Park_joy	746,496	Rapid movement	1920 × 1080	500
3	Sunflower	746,496	Slight movement	1920 × 1080	500
4	Tractor	746,496	Rapid movement	1920 × 1080	690

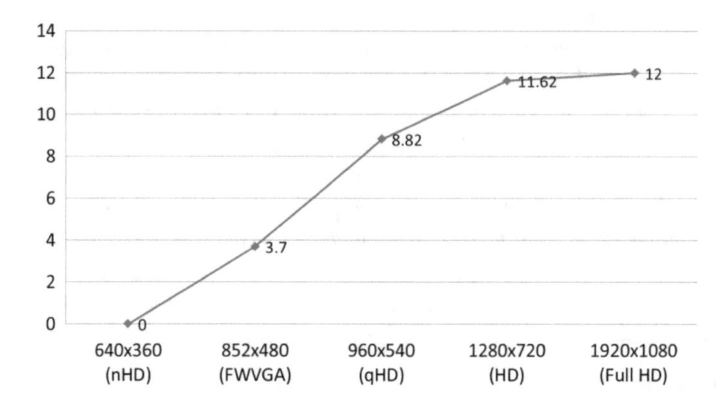

Fig. 2 QoE resolution factor scores cumulation

There are 50 data collected from experimentation and the different scores are averaged, and as quality is gradually upgraded to obtain the difference, the intervals are accumulated to form the line chart shown in Fig. 2. It is observed that it is smooth at resolution 1280 × 720, meaning most subjects are unlikely to perceive the difference between 720 and 1080p, they feel the two picture qualities are close to each other, and 720p is even better than 1080p.

Quantization Parameter Factor:

When the Quantization Parameter (QP) value of H.264 video coding increases by 6 units, the bit rate decreases by 50%, and the equation is Eqs. (2)–(4). The Quantization Parameter range of the ffmpeg coder is 0–51, the default is 23, the smaller the number is, the better the quality, and the Quantization Parameter between 18 and 28 is the range that human eyes cannot recognize the distortion rate. The Quantization Parameter is 11–47 in the experimental design, and the videos of different Quantization Parameters are coded in the unit of 23 ± 3 in order to survey user satisfaction. There are 13 videos with different Quantization Parameters.

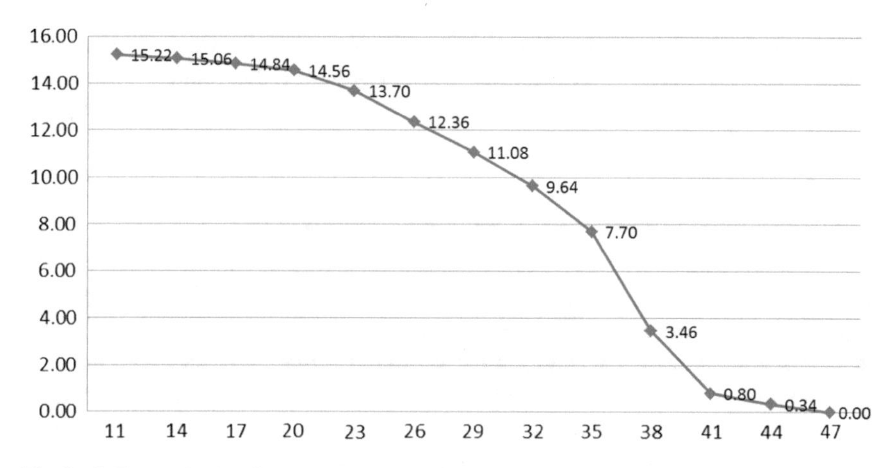

Fig. 3 QoE quantization factor scores cumulation

$$log\frac{Bitrate_1}{Bitrate_2} = -\left(\frac{Q_1 - Q_2}{6}\right) \times log\,2 \tag{7}$$

This part of the experimental process is the same as the content of the resolution factor, where there are 50 test data collected, and the scores are accumulated, as shown in Fig. 3. Two blocks of scores have slight changes, which are 11–20 and 41–47, respectively. According to this experiment, when the Quantization Parameter is set as the two intervals, the quality is too good or too bad, thus, the subjects are unlikely to perceive the difference. Therefore, it is relatively smooth after Quantization Parameter 35 in the remaining intervals, and this score is the threshold of subsequent system implementation.

As shown in the above figure, users have an interval of scores for the resolution factor or Quantization Parameter factor, which is smooth, meaning that users cannot easily recognize the differences in videos at these scores, thus, it is believed that these coding parameters have very slight influence on video quality. However, these adjustment parameters significantly influence the bit rate and file size of the encoded videos, which is related to network bandwidth for streaming.

4 Experimental Results

The adjustment mechanism shall implement adjustment according to the user's feedback content regarding two quality factors of the videos; therefore, this paper researches and analyzes the QoE factor of videos, designs an experiment, and asks the subjects to feed back their experiences and satisfaction of watching videos, and the corrected score and threshold meeting universal conditions are concluded from these data. The subsequent system implementation evaluates the bandwidth and chooses the optimum media segments according to the data obtained from this experiment.

4.1 Network Scenario of Stable Bandwidth

For this system, there are two cases of stable network bandwidth. One is that the bandwidth is higher than the network condition for a Basic Level video. In this case, this system keeps streaming in the Buffering Mode by prestoring a video of good quality. The other one is that the bandwidth is lower than the condition for Basic Level streaming. In this case, the system enters into the Streaming Mode to evaluate current network condition and adjust the streaming content. The high bandwidth is 2048 kbps and the low bandwidth is 512 kbps for the test, and the scenario is as shown in Fig. 4.

4.2 Drop-Rise Network Scenario

The drop-rise scenario is as shown in Fig. 5. The network bandwidth is switched at intervals of about 10 s. In this test, three bandwidth combinations are switched in order to observe how this system chooses streaming under these conditions. Scenario 5 is the network scenario designed for discussing two mode switching.

4.3 Rise-Drop Network Scenario

The last type of network condition is rise-drop. This part only discusses the adjustment mode for a bandwidth above the Basic Level, including scaling up from

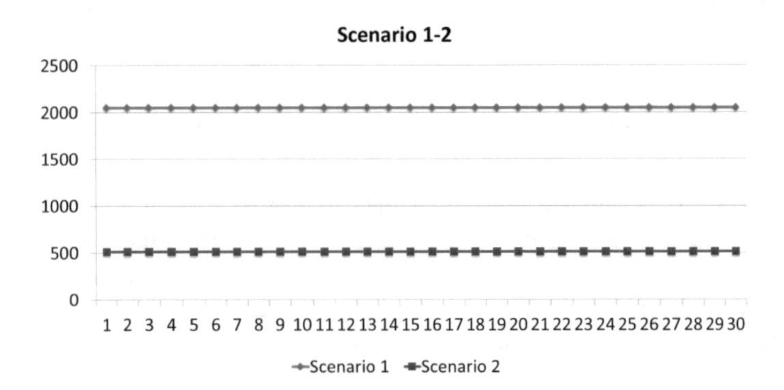

Fig. 4 Scenario 1–2 bandwidth schematic

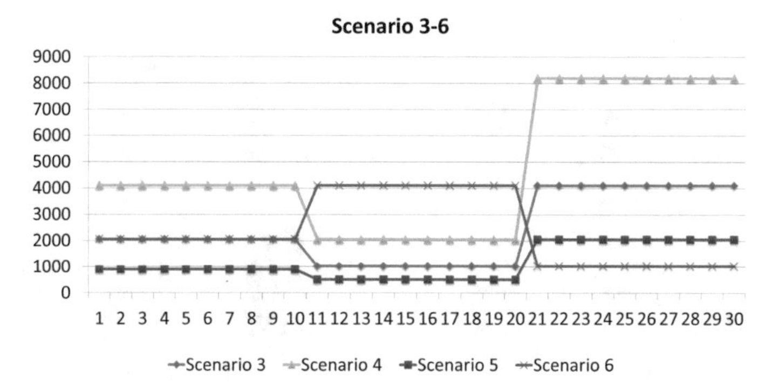

Fig. 5 Scenario 3–6 bandwidth schematic

2048 to 4096 kbps, and then scaling down the bandwidth by four times to 1024 kbps.

In terms of a decision mechanism, this paper discusses the received media segment information, as well as the changes in the decisions of the aforesaid network scenarios. Finally, viewing satisfaction with the streamed video, according to the proposed mechanism in this paper, as well as the video streamed by the general streaming mechanism, are discussed.

In the analysis of viewer experience, the streaming conditions of various network scenarios are recorded in detail and the content of streaming is restored in series, where the videos generated in these scenarios are played by hand-held devices and graded by 20 subjects. The MOS grading method is used to investigate two experiences, which are picture quality and fluency of videos. The fluency is graded according to two factors, including the time waiting for a download in the streaming process and video skip, i.e. picture loss.

Figure 6 shows the scores of streamed videos in six network scenarios. In the case of stable bandwidth, i.e. Scenario 1 and Scenario 2, there are slight differences between the two streaming modes, especially the performance in fluency. As the two mechanisms hardly adjust the picture quality in stable bandwidth, there is difference only in the initialization stage. As HLS initially chooses the video of lowest quality, there will be obvious quality skip in comparison to the architecture proposed in this paper.

In terms of quality, regardless of mode, it is directly related to the bandwidth size. However, as HLS initially downloads low quality video, the overall quality score is reduced. The less the required bandwidth, the more likely the gap is to be observed, and too many low quality videos are downloaded when accumulating the first 100 network bandwidth data for bandwidth forecast due to low requirements, thus, the user sees the worst quality longer. In comparison, this paper presets that the streaming basic level video can instantly adjust the bandwidth in the initialization stage in order to obtain better picture quality.

In the drastically changeful bandwidth performance, e.g. Scenarios 3–6, fluency and quality are greatly enhanced. As Android HLS is insensitive to the bandwidth

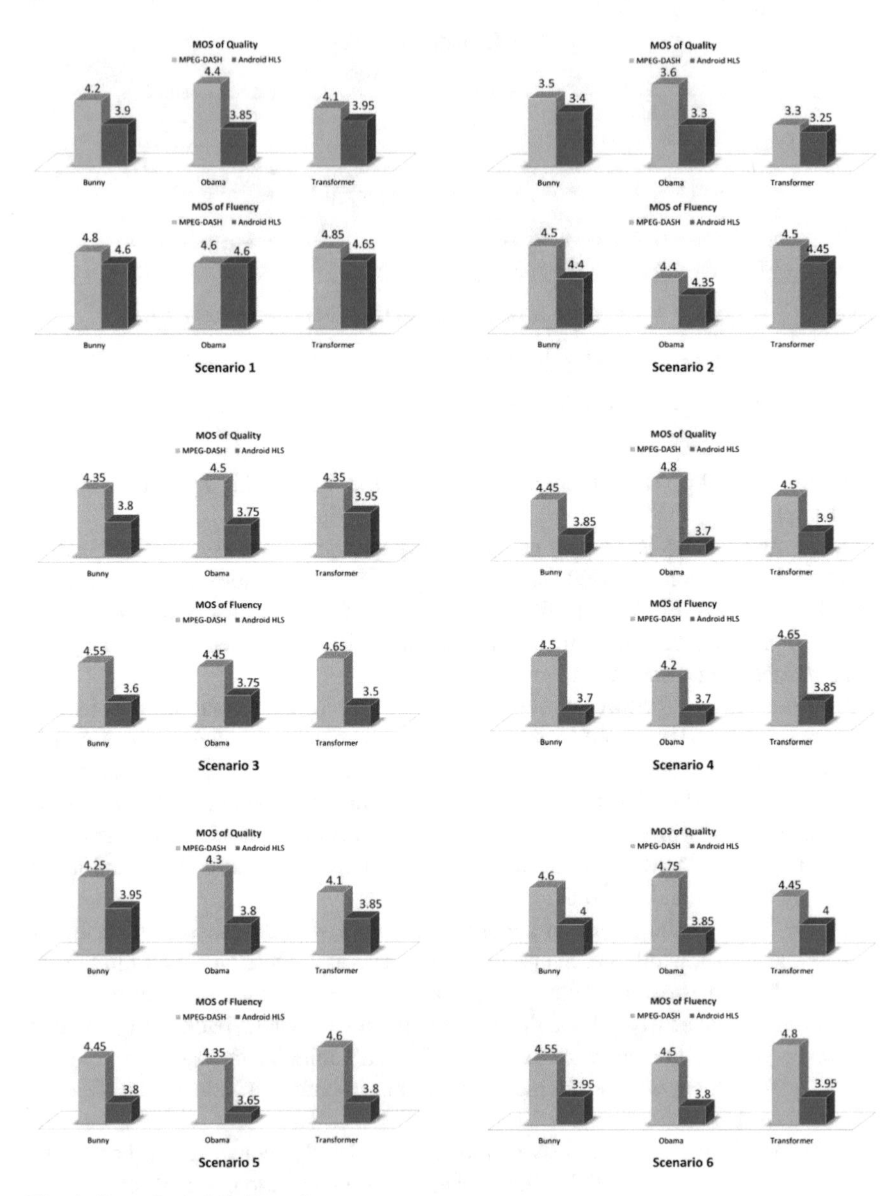

Fig. 6 Scenario 1–6 QoE results

forecast mechanism, a drastically changeful network environment often causes waiting or picture loss, which reduces fluency satisfaction. In comparison to this system, the designed buffering mode can preload videos in the environment with better bandwidth, meaning that users spend less time waiting, and can smoothly view the video. As the quality is mostly maintained above the basic level, the overall picture quality is better than the quality received by HLS.

Finally, based on the statistical results of three kinds of videos, the enhancement of quality is opposite to the enhancement of fluency, meaning the quality enhancement of a video with a low bandwidth requirement is apparent; however, the performance in fluency is worse than a drastically changeful video requiring high bandwidth, as it may cause an apparent quality skip of a static video, and there is difference in the scale-up strategy of this mechanism in the same network environment. Generally speaking, there are stable and dynamic network environments. In a dynamic network environment, fluency can be improved by 20%, and quality can be improved by 15%.

5 Conclusions and Future Work

This study designed a dynamically adjusted adaptive streaming mechanism by surveying satisfaction with video QoE, which can instantly make adjustments according to the changes in the buffered media segments, and stream a video with the best QoE. The proposed adaptive streaming mechanism and decision model are implemented and corrected, and multiple network scenarios and Adaptive Streaming mechanisms are designed to compare satisfaction.

References

1. W.-H. Chin, Z. Fan and R. Haines, "Emerging Technologies and Research Challenges for 5G Wireless Networks," IEEE Wireless Communications, Vol. 21, No. 2, April 2014, pp. 106–112.
2. X. Wang, M. Chen, T. Taleb, A. Ksentini and V. Leung, "Cache in The Air: Exploiting Content Caching and Delivery Techniques for 5G Systems," IEEE Communications Magazine, Vol. 52, No. 2, February 2014, pp. 131–139.
3. T.S. Rappaport et al., "Millimeter Wave Mobile Communications for 5G Cellular: It Will Work!," IEEE Access, Vol. 1, No. 1, May 2013, pp. 335–49.
4. ONF White Paper, "Software-Defined Networking: The New Norm for Network", April 13, 2012.
5. Open Networking Foundation, "OpenFlow switch specification, version 1.4.0", Oct 14, 2013.
6. C.-F. Lai, H. Wang, H.-C. Chao and G. Nan, "A Network and Device Aware QoS Approach for Cloud Mobile Streaming," IEEE Transactions on Multimedia, Vol. 15, No. 4, May 2013, pp. 747–757.
7. S.Y. Wu, and C.E. He, QoS-Aware Dynamic Adaptation for Cooperative Media Streaming in Mobile Environments, IEEE Trans. Parallel and Distributed Systems, vol. 22, no. 3, Mar. 2011, pp. 439–450.
8. C. Xu, T. Liu, J. Guan, H. Zhang and G.-M. Muntean, "CMT-QA: Quality-Aware Adaptive Concurrent Multipath Data Transfer in Heterogeneous Wireless Networks," IEEE Transactions on Mobile Computing, Vol. 12, No. 11, Nov. 2013, pp. 2193–2205.
9. M. Ismail, A. Abdrabou and W. Zhuang, Cooperative Decentralized Resource Allocation in Heterogeneous Wireless Access Medium, IEEE Transactions on Wireless Communications, Vol. 12, No. 2, Feb. 2013, pp. 714–724.

10. H. Kim and N. Feamster, "Improving Network Management with Software Defined Networking," IEEE Communications Magazine, Vol. 51, No. 2, February 2013, pp. 114–119.
11. C.-F. Lai, H.-C. Chao, Y.-X. Lai and J. Wan, "Cloud-Assisted Real-Time Transrating for adaptive streaming," IEEE Wireless Communications, Vol. 20, No. 3, June 2013, pp. 62–70.
12. M. Volk, J. Sterle, U. Sedlar, and A. Kos, "An Approach to Modeling and Control of QoE in Next Generation Networks," IEEE Communications Magazine, Vol. 48, No. 8, August 2010, pp. 126–135.

Exploring the Learning Effectiveness Under Different Stress Scenarios

Hsiu-Sen Chiang

Abstract In recent years, increased research attention has focused on understanding challenges to effective classroom learning, and many studies have explored the effect of stress on learning effectiveness. Appropriate stress can inspire learners to achieve their true learning potential, but this raises the issue of how to best assess learners' tolerance for stress. Teachers can use stress quantification methods to measure the personal stress loading for individual students. By adjusting teaching methods to fit varying levels of tolerance to learning stress, teachers can not only enhance learning effectiveness, but also reduce the possibility of physical and psychological damage to learners. This study uses the experimental design research method to evaluate learning effectiveness under different treatment variables: gender (male and female), logical reasoning ability (high and low) and stress level (high and low). The results found that stress level can affect learning effectiveness ($p < 0.05$), and a positive linear relationship exists between stress level and learning effectiveness. The results also find that the impact that of stress on learning effectiveness is more prominent among boys. Moreover, logical reasoning ability had no significant influence on learning outcomes under different stress levels.

1 Introduction

In recent years, students' learning generally is easily distracted, and many studies have indicated that stress is an important factor which affects the learning effectiveness [1, 2]. The relationship between stress and productivity takes the form of an inverted U-shaped curve [3]. Appropriate stress can stimulate the work productivity and learning effectiveness. Conversely, stress levels set too high or too low can have a negative psychological or physiological impact, thus impairing learning effectiveness [4–6]. Excessive amounts of stress can have serious

H.-S. Chiang (✉)
Department of Information Management, National Taichung University of Science and Technology, Sec 3, Sanmin Rd., Taichung, Taiwan
e-mail: hschiang@nutc.edu.tw

© Springer Nature Singapore Pte Ltd. 2018

N.Y. Yen and J.C. Hung (eds.), *Frontier Computing*, Lecture Notes in Electrical Engineering 422, DOI 10.1007/978-981-10-3187-8_30

consequences, including aggressive or self-destructive behavior. Therefore, how to determine the appropriate stress load for individual learners is a key issue.

This study assesses the impact of stress on learning effectiveness through a range of scenarios for web programming instruction. Three independent variables (gender, logical reasoning ability and stress level) are applied to assess the influence on the dependent variable (learning effectiveness) under treatment level. The results can potentially provide suggestions for improving learning effectiveness in the classroom.

2 Literature Review

2.1 The Effect of Stress on Learning

In students, psychological stress can result from intrinsic and extrinsic factors, and excessive stress can leave students prone to anxiety, frustration, and tension, which can have a negative effect on learning effectiveness, as well as physical and psychological discomfort.

There is a significant but weak negative relationship between student stress levels and academic achievement [2]. Gmelch [3] graphed the relationship between stress and productivity as an inverted U-shaped curve, in which both very low and high levels of stress can result in poor learning effectiveness, while an appropriate amount of stress could raise learning effectiveness through improve learning motivation. However, excessive stress may cause physical and psychological damage. Prior studies have suggested that low intrinsic academic satisfaction can cause feelings of hopelessness, emptiness and the low self-worth among adolescents [7]. Long-term excessive stress can create feelings of being overwhelmed and unable to meet challenges.

Stress can have direct and indirect effects on physical and mental health [1], but appropriate levels of stress can help students realize their full potential to overcome obstacles. Past studies have indicated that choosing the appropriate stress modality is helpful to improve learning effectiveness [4, 5].

2.2 Learning Effectiveness

Assessment is a critical element for ensuring knowledge acquisition in nearly all disciplines. Instruction in computer programming in particular is procedural, logic-oriented and very practical. In other words, it can be said that programming assessment should be more oriented towards practice than theory. Previous attempts to assess learning outcomes in computer programming have included written exams and implementation of the resulting program [8–10]. However, reliance on written examination test along no longer meets new demands of programming language education [11].

This study designed a web based programming lesson to allow learners to implement a sample program and allow classmates to assess each other's work. Learner work was also assessed by experienced programming teachers.

3 Methodology

This study assesses the impact of stress conditions and logical reasoning ability on learning effectiveness using a measurement scale and a review by programming experts.

3.1 Logical Reasoning Ability Test

Logical reasoning ability can affect learning effectiveness for programming languages. This study uses a 10 question online logic test to assess the subjects' aptitude for logical reasoning, (https://www.123test.com/logical-reasoning-test), in which learners are asked to make inferences from a set of images. To prevent excessive bias in terms of logical reasoning capability, two groups were created with an equal distribution of high and low-scoring subjects.

3.2 Programming Capability Measurement

The subjects were asked to learn HTML, CSS and JQuery through making a web-based image carousel, as shown in Fig. 1. Table 1 summarizes the evaluation criteria for learning outcomes. Assessment by expert programmers accounts for 25% of the total score.

Fig. 1 Web-based image carousel

Table 1 Rating scale for programming capability

Items	Description	Score
Adaptive image carousel	Automatic calculation of number of images	10
Automatic image carousel	Adjustable carousel speed	10
Return to first image	From the last image, the user can return to the first image	15
Left and right buttons to switch images	User can use left and right arrows to browse images	15
Hover effect	Implement hover effect based on left and right mouse buttons	10
Mouse movement pauses slideshow	Slideshow stops when user moves the mouse over the image	15
Programming quality	Assessment by programming experts	25

4 Experimental Design

The experimental design was used to evaluate the impact of treatment variables (gender, logical reasoning ability and stress level) on learning effectiveness.

4.1 Subjects

Prior to the start of the experiment, all respondents provided informed consent, completed a personal data questionnaire, and had their programming capability assessed through expert assessment.

From a pool of 14 undergraduates majoring in information management, six subjects were selected based on their logical reasoning ability and assessment by programming experts. The six subjects were randomly divided into experimental and control groups, with each group having one female and two male participants.

4.2 Experimental Process

Two groups of students attended a programming course under different instructional stress conditions, with treatment variables modified to assess the impact on learning effectiveness. Figure 2 shows the experimental design and flow.

As shown in Fig. 3, both groups were instructed in the necessary programming skills in 90 min daily sessions for three consecutive days. The low stress group was conducted without tests or homework, while the high stress group was given a 30 min test at the end of each day on the last day; all learners were required to build a web-based image carousel as a means to evaluate learning effectiveness.

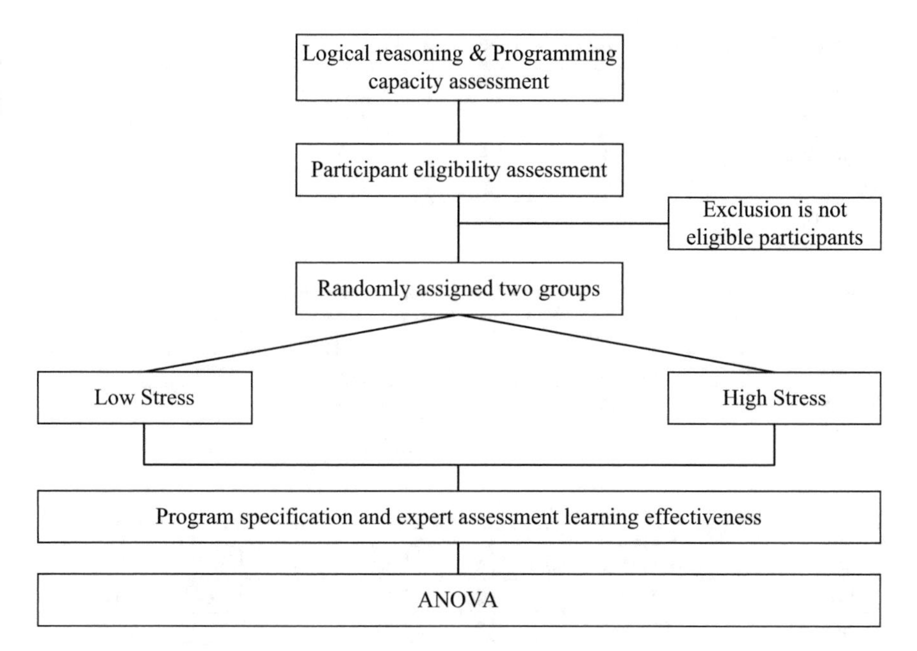

Fig. 2 Experimental flow chart

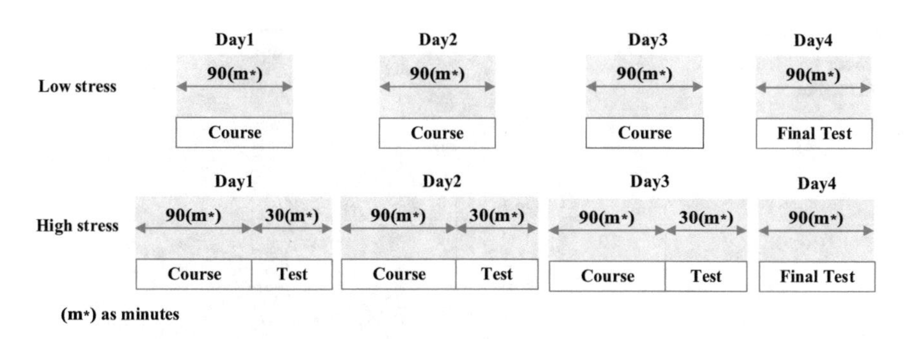

Fig. 3 Experimental scenario

5 Experimental Results

To avoid bias, the chi-square test of homogeneity was used to ensure no significant differences existed between the subjects ($P > 0.05$). To assess learning effectiveness for the various treatment variables (gender, high/low logical reasoning ability and high/low stress level), ANOVA was implemented with gender and stress level as independent variables and learning effectiveness as the dependent variable. Tables 2, 3, and 4 show a significant difference on stress level ($F = 19.074$, $P < 0.05$), but no significant difference for gender ($F = 0.241$, $P > 0.05$) or logical

Table 2 ANOVA results for the impact of stress level on learning effectiveness

Source	Sum of squares	df	Mean squares	F	P
Between groups	1287.735	1	1287.735	19.074	0.012*
Within groups	270.053	4	67.513		
Total	1557.788	5			

*P < 0.05

Table 3 ANOVA results for the impact of gender on learning effectiveness

Source	Sum of squares	df	Mean squares	F	P
Between groups	88.563	1	88.563	0.241	0.649
Within groups	1469.225	4	367.306		
Total	1557.788	5			

Table 4 ANOVA results for the impact of logical reasoning on learning effectiveness

Source	Sum of squares	df	Mean squares	F	P
Between groups	1256.208	2	628.104	6.248	0.085
Within groups	301.580	3	100.527		
Total	1557.788	5			

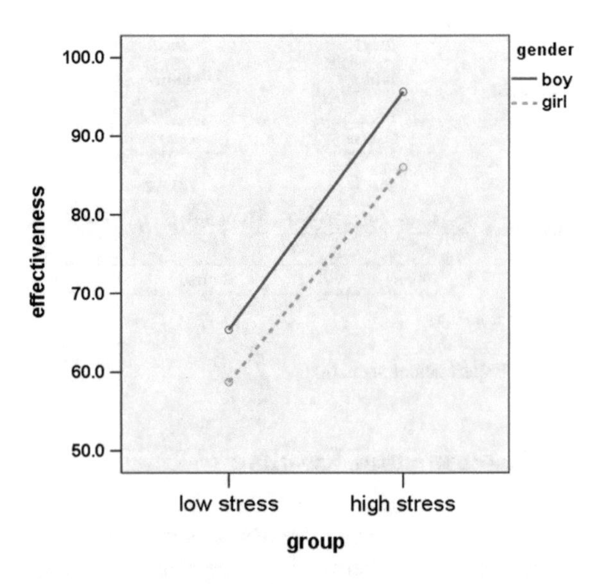

Fig. 4 Relationship between gender and learning effectiveness

reasoning ability (F = 6.248, P > 0.05). A positive correlation was thus found between stress level and learning effectiveness. This positive relationship in boys is more obvious than in girls, as shown in Fig. 4.

6 Conclusion

The correlation analysis results are consistent with previous findings suggesting that learning effectiveness is positively and significantly correlated with stress level [2, 3], suggesting that appropriate stress levels can enhance learning effectiveness. This finding provides some supports for positive psychology theories [7] and has some implications for classroom practice in terms of helping teachers determine appropriate stress levels. While the impact of gender on the impact of stress on learning outcomes was not statistically significant, boys were still more susceptible than girls to overstress. Moreover, the logical reasoning ability had not significant influences on learning outcomes under stress teaching situation. The members of the high stress group will face examinations and major assignments following the course, causing them to focus their attention and make an extra effort to improve their performance. Members of the low stress group were relatively more relaxed and paid less attention to the course content.

Acknowledgements The authors would like to thank the Ministry of Science and Technology of Taiwan for Grants MOST 104-2410-H-025-023 which supported part of this research.

References

1. Lazarus, R. S., and Folkman, S.: "Stress, Appraisal, and coping," pp. 725, (1984).
2. Elias, H., Ping, W. S., & Abdullah, M. C.: "Stress and academic achievement among undergraduate students in Universiti Putra Malaysia," Procedia-Social and Behavioral Sciences, vol. 29, (2011) 646–655.
3. Gmelch, W. H.: "Beyond stress to effect management," New York: John, (1982).
4. Kaiser, J. S., & Polczynski, J. J.: "Educational stress: Sources reactions, preventions," Peabody Journal of Education, vol. 59, no. 2, (1982) 127–136.
5. Ovcharchyn, C. A., Johnson, H. H., & Petzel, T. P.: "Type A behavior, academic aspirations, and academic success1," Journal of Personality, vol. 49, no. 3, (1981) 248–256.
6. Chiang, H. S.: "ECG-based Mental Stress Assessment Using Fuzzy Computing and Associative Petri Net," Journal of Medical and Biological Engineering, Vol. 35, Iss. 6 (2015), pp. 833–844.
7. Dukes, R. L., & Lorch, B. D.: "The effects of school, family, self-concept, and deviant behaviour on adolescent suicide ideation," Journal of Adolescence, vol. 12, no. 3, (1989) 239–251.
8. Lau, W. W., & Yuen, A. H.: "Modelling programming performance: Beyond the influence of learner characteristics," Computers & Education, vol. 57, no. 1, (2011) 1202–1213.
9. Kordaki, M.: "A drawing and multi-representational computer environment for beginners' learning of programming using C: Design and pilot formative evaluation," Computers & Education, vol. 54, no. 1, (2010) 69–87.
10. Rubio, M. A., Romero-Zaliz, R., Mañoso, C., & Angel, P.: "Closing the gender gap in an introductory programming course," Computers & Education, vol. 82, (2012) 409–420.
11. Wang, Y., Li, H., Feng, Y., Jiang, Y., & Liu, Y.: "Assessment of programming language learning based on peer code review model: Implementation and experience report," Computers & Education, vol. 59, no. 2, (2012) 412–422.

The Analysis for Online Game Addiction

Jason C. Hung, Min-Hui Ding, Wen-Hsing Kao and Pi-Chung Wang

Abstract Due to the convenience and popularity, Internet has changed our cognition since always. It is closely linked people's whole daily life. Internet Addiction has become a worth noting issue. Game Addiction has become a big social problem. This paper aims to investigate the interpersonal communication difficulties and abnormal life schedule, even behavior deviation which are caused by Game Addiction of some people. This paper is based on Internet Addiction theory of Dr. Kimberly S. Young. Using his Internet Addiction Test (IAT) in the theory and adapt it becomes "Internet Game Addiction Survey Questionnaire". We analyze the addiction status caused by some recent popular games for finding addiction extent of each game's time interval, therefore, we can calculate addiction index. The questionnaire is divided in five categories with total 20 questions. Thus we can analyze the answer scores, average and percentage gaming time by this questionnaires. Then get the line chart of addiction extent in each game's time interval. Through the chart, we can figure out how long a person gets game addiction. Further we can get the game addiction index and grade game addiction extent in different levels, then tell whether the player is easy to get game addiction or not by this survey. We expect to effectively prevent any user become game addiction by this research.

Keywords Internet addiction · Game addiction · Internet addiction evaluation form · Addiction index

---

J.C. Hung · W.-H. Kao · P.-C. Wang
Department of Information Technology, Overseas Chinese University, Taichung, Taiwan
e-mail: jhung@ocu.edu.tw

W.-H. Kao
e-mail: christine8018458@gmail.com

P.-C. Wang
e-mail: pcwang@nchu.edu.tw

M.-H. Ding (✉) · W.-H. Kao · P.-C. Wang
Department of Computer Science and Engineering, National Chung Hsing University, Taichung, Taiwan
e-mail: star@ocu.edu.tw

© Springer Nature Singapore Pte Ltd. 2018

N.Y. Yen and J.C. Hung (eds.), *Frontier Computing*, Lecture Notes
in Electrical Engineering 422, DOI 10.1007/978-981-10-3187-8_31

1 Background

In 21st century, “Internet” is the most diverse and most growing technology in life. Internet which applies multiple application, constantly improved and innovation impacts all the human’s aspects of life. Otaku (Otaku in meaning as it is used to refer to someone who stays at home all the time and doesn’t have a life) and Smartphone Addicts people’s behavior issues was discussed by news media, newspapers and magazines. Some people who died suddenly in playing game, teens addict internet in cyber cafe, use smartphone while walking to make the danger and even fantasize himself is the leading character in the online game to stabbed passengers on the metro, that reported in the newspaper society pages. According to Taiwan Network Information Center (TWNIC), survey released [1]: May, 2014 Taiwan broadband internet use survey results showed that: There are 1,764 million people online, internet ratio of 75.43%, 12 years people over the age of the Internet has 1,623 million people, use broadband Internet are 1,612 million people, use wireless Internet are 1,277 million people, use mobile Internet are 1,003 million people, as shown in Fig. 1. The number of people that can use the internet increased year by year. Therefore, we need to set Addicting defined to this type of game classification. among them which game content to the user has a strong influence, According to the results of that sample there are 80% game aggression and violence is a means to win, It is to produce the basic elements of the game Attacks, hostility, attitudes and behaviors may become the basis for violence [2]. Online game for the

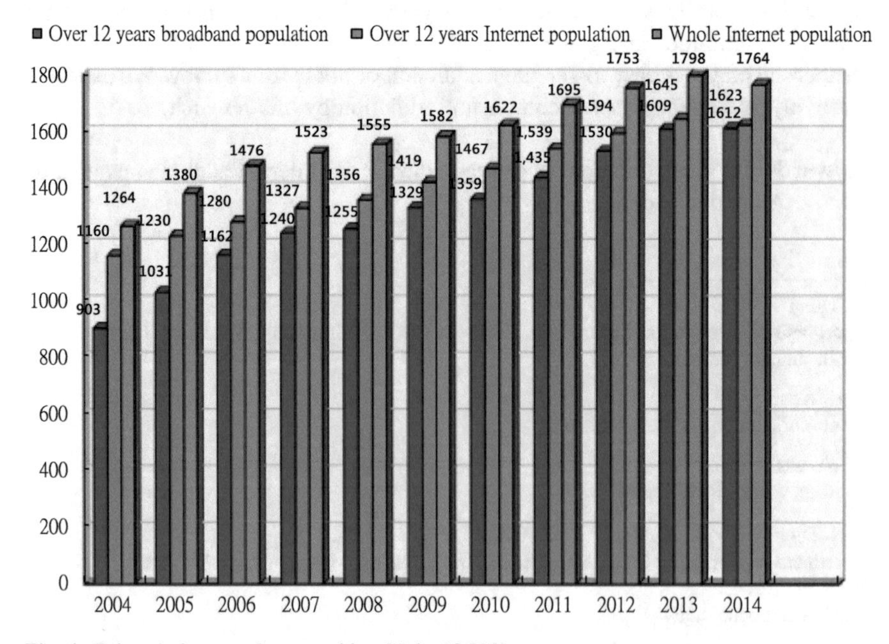

Fig. 1 Taiwan’s internet demographics (Unit: 10,000)

people is recreation and pass the time even though it can't be proven the relation to the above all the issues with game addiction. It is a worth issue to investigate the game addiction that can transform human's personality and life. Game Addiction is an after-effects generated by technology developed. There are many reasons for addiction. Conventional solutions are forced abstinence. Its success rate is not high and the follow-up will extend more problems. If user have a system that is able to inform the user in advance to play the game content and addiction degree. Remind, inform the user whether the addictive, presumably will come more easily than many users addicted to quit.

2 Motivation and Specific Objective

"Game addiction" is a well-known term nevertheless less people to squarely indicate how big its harm and is a modern plagues with astonishing growth to become a heath killer of all ages. It must be seriously discover "game addiction" influence, so draft these purpose for investigating in preventively.

A. Explore "game addiction" forming reason and influence on body, mind, health and work effectiveness.
B. Questionnaire survey to explore "game addiction" in the current situation, and analyze the characteristics of different games, according to their degree of analysis addictive draw of the table game time interval addiction, to find causes and definition of addiction range.
C. Find a way to calculate the "game addiction" and set the degree of addiction.

This model can count game addiction, user's and parents more easily understand the game content and addiction. Game Addiction degree, sequentially classified as mild, moderate, severe, In addition which users can play this game is addictive, or warn users the game play more than how long it must to become addicted.

3 Method

3.1 Schema

The process of the research is based on internet addiction theory of Dr. Kimberly S. Young in psychology and infrastructure, adapted for game addiction questionnaire research flow chart as shown in Fig. 2.

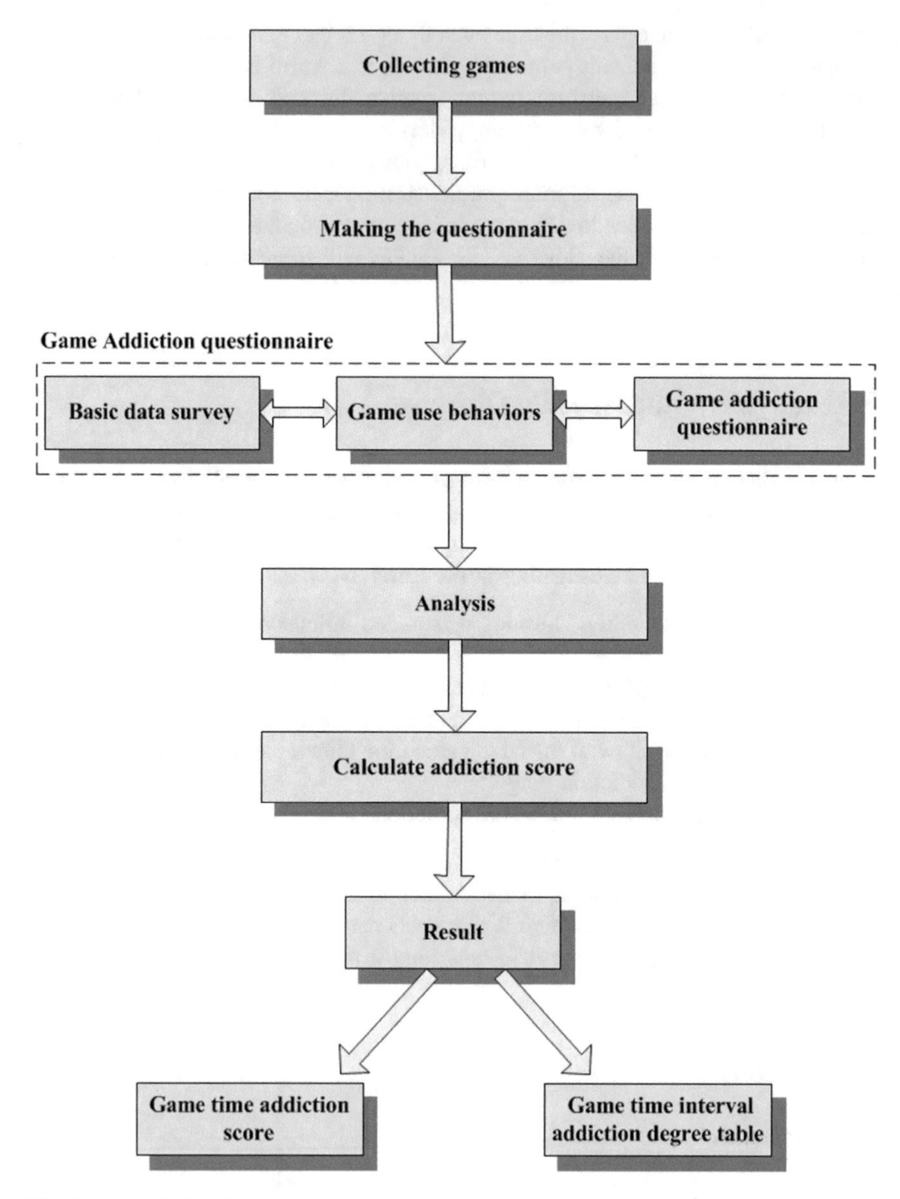

Fig. 2 Research flowchart

3.2 *Steps and Analysis*

This research proposal used questionnaires; the content of game addiction questionnaire contains "personal data", "personal game use behaviors", and "game addiction". The process of questionnaire experienced collection of popular games, the preparation of questionnaires adapted subject and reliability analysis using SPSS, choose to 20 questions, take these five points to calculate the score, through the above to draw the chart, then get the time interval of game addiction table. The procedures step by step instructions [3–6]:

(A) Collecting games

The questionnaire uses to top ten of the games for and analyzing the characteristics after collecting, fill in the other fields of the game, and the list will also analysis.

(B) Making the questionnaire

According to Internet Addiction Scale of Dr. Kimberly S. Young is adapted into a "game addiction" questionnaire.

1. The basic data survey include: name, school, department, grade, gender, birthdate, accommodation.
2. Behaviors of game using: average time to play games online in each day, last game played most often (multiple choices), based on the questions you most often preference, most attract you games, do you most often play games on one day to play the game of how much time?
3. Game addiction questionnaire shown in Table 1 that is according to five points to answer 20 questions. In order to assess the degree of addiction, the user needs to answer the following questions measuring stick.

(C) Analysis

There are 33 questions in questionnaire without filtered, and after reliability analysis, exclusion of the subject of reliability is less than 0.7, the remaining 20 questions is used to five point calculate the answer score. With reference to the Internet Addiction Test (IAT) of Kimberly S. Young to provide addiction range [7], the range values are as follows:

a. Normal range: 0–30 Score.
b. Mild: 31–49 Score.
c. Moderate: 50–79 Score.
d. Severe: 80–100 Score.

According to addiction factor type classified as: salience, mood changes, tolerance, and conflict, time limits, which have salient factor: Q4, Q11, Q12, Q14, mood changes: Q3, Q9, Q10, Q18, Q19, Q20, tolerance: Q2, Q7, Q15, Q16, Q17, conflict: Q5, Q6, Q8, time limit: Q1, Q13, as shown in Fig. 3.

(D) Calculate addiction score

The gaming time interval addiction degree table was drawn through the information collected and analyzed in each game of the degree addiction, Calculation Method addiction scores as follows:

First, the time interval from addiction fig observed every hour of game addiction.

Table 1 Game addiction questionnaire

No.	Question
1	How often would you play game exceed to originally anticipated time?
2	How often do you put aside the completed or executed things and use time to play game?
3	How often do you play games get the excitement even interpersonal intimate interaction?
4	How often do you make new friends in the online game?
5	How often do you spend too much time to playing online game and being around people who complain or blame?
6	Do you spend too much time on the game which (have begun to cause, academic setback?
7	How often do you have to do something else before opening the game?
8	How often do you have to open start the game before you do anything else?
9	How often do you play the game by recall pleasant thing to stop thinking troubled things?
10	How often do you expect to be able to start yet game to play?
11	How often do you fear less the game, life becomes boring, empty?
12	How often do you play the game sacrificing sleep at night?
13	When you playing the game, how often you tell yourself "just a few minutes"?
14	Have you ever ordered at bedtime will finish off the game before falling asleep?
15	Have you ever found yourself playing the game, in fact, don't really feel interesting?
16	Do you feel like you have to spend more and more time in online game?
17	As long as there is free times will want to play the game?
18	When you finally have access to the game, feel happier and joyful; cheerful; delighted?
19	When I tried to cut down or stop using the internet, I would feel down, depressed or cranky
20	Suddenly you want to terminate the game, make you feel very bad

Referring to Dr. Kimberly S. Young of Internet addiction scorecard (Internet Addiction Test, IAT) set range, 30, 50, 80 points, respectively, mild, moderate, severe, and the ups and downs from the line graph changes in performance that users of emotional behavior. Then the amount of time from the use of observation, when the user to play the game longer representative of the more addicted. Game addiction score formula is: Average hourly addiction degree × The number of peple (h)/Total people number (h) × Every hour time-weighted number of addiction, Time is the total time of the day to play the game number. Define the scope of addiction: The maximum score Addiction—Addiction score min/100, find score range adjustment variables. When the score reaches the interval as the adjustment variable, add 1 point of action, then addiction score—benchmark point value (min), obtain score range value. Finally, the score range value/score range adjustment variables, you get addicted to define the range of score.

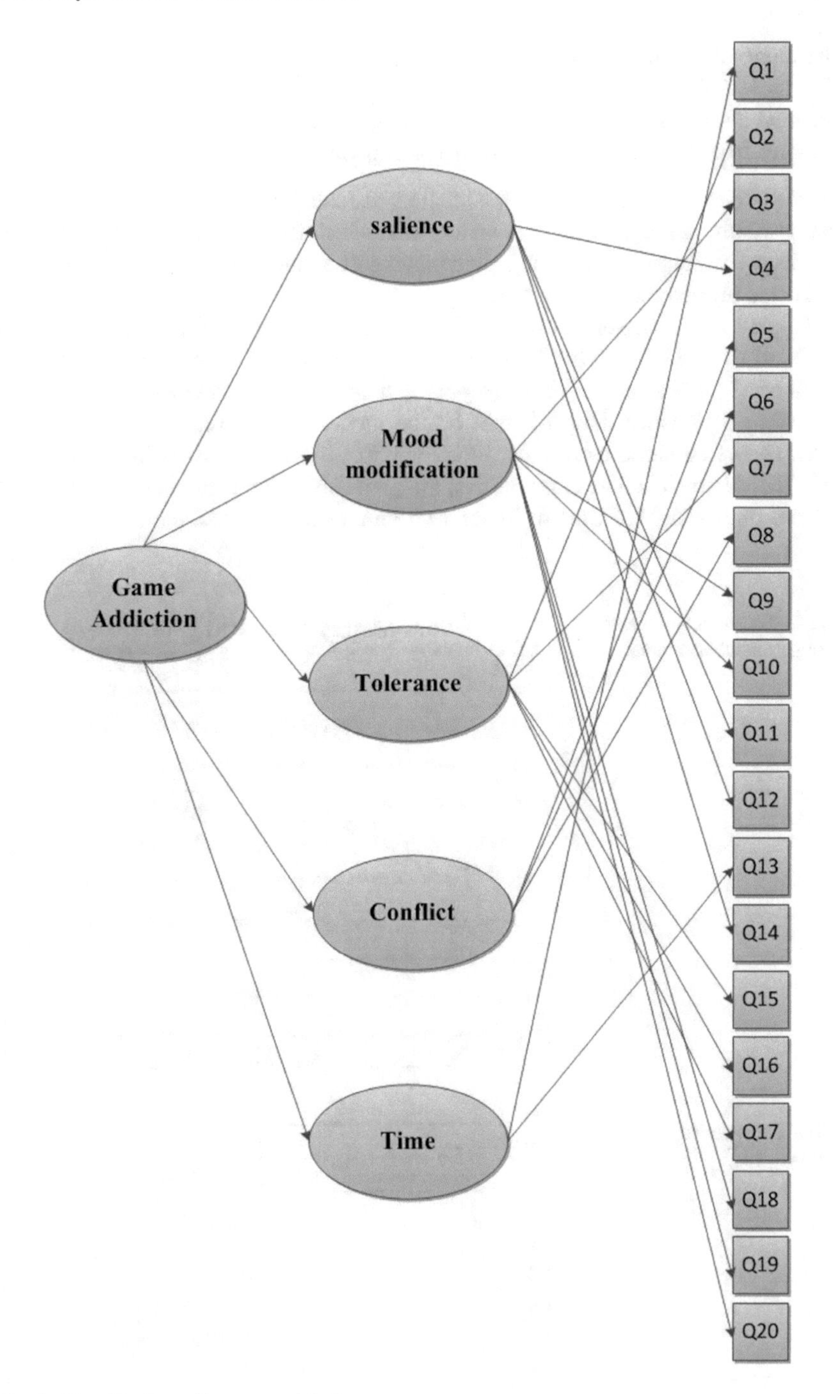

Fig. 3 Classification title game addiction

The range of values as follows:
Slightly less than 33.3% Mild Addicting Games
Between 33.3 and 66.6% moderate Addicting Games
Ranging from 66.6 to 100% severe Addicting Games

Hour time-weighted value of addiction, calculation is divided into light and heavy weight, the light weight is mainly calculated using the mobile phone games, The heavy weighting is for long time playing computer games.

Tower of Saviors, as shown in Fig. 4.

To play 1 h in Tower of saviours the average degree of addiction degree is 38.4, 2 h is 44, 3 h is 51, 4 h is 55, 5 h is 56.5, 6 h is 58, 7 h is 59.5, 8 h is 62.

Figure 3 indicates the while playing 1 h in Tower of saviours, will be mild addiction, play more than 3 h will be the moderate addiction, and its addictive index showed a slope increasing, more emotional ups and downs in performance, and other factors such as anger others, the game addiction score is to get the result set into mild weighted time of about 49.9 min, as shown in Table 2.

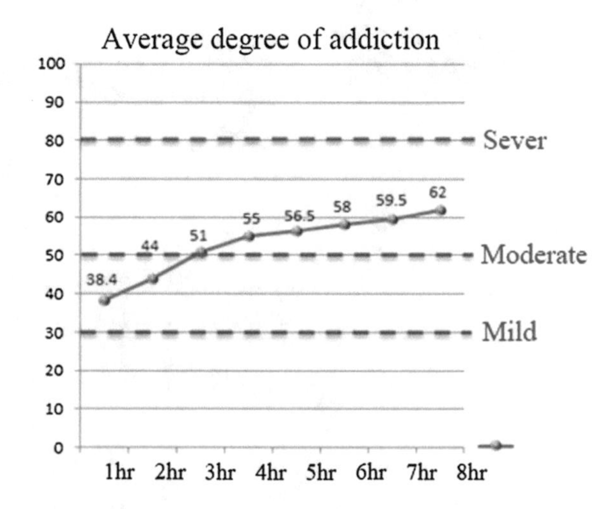

Fig. 4 Tower of saviors play average addiction graphs

Table 2 Tower of saviors time-weighted

Time-weighted	Score
1	19.81935484
1.1	14.0516129
1.2	1.974193548
1.3	0
1.4	5.103225806
1.5	5.612903226
1.6	0
Total score	49.96129032

Then define addiction Score Range, first find scores range adjustment variables: 49.9 − 39.2 = 0.107, score interval values: 49.9 − 39.2 = 10.7, therefore addiction is defined as the range of scores 10.7/0.107 = 100, Game addiction to one-third ratio range mode, Respectively less than 33.3% of mild addiction, ranging from 33.3 to 66.6% moderate addiction, ranging from 66.6 to 100% severe addiction, as shown in Tables 3 and 4.

Finally, the Tower of Saviors severe addicting games, be careful game sequela, such as spending too much time and effort in the game, resulting in daily routine has changed, deterioration of eyesight, affect assignments and so on.

(E) Addiction model present

Figure 5 shows the search interface, the current analysis of games Tower of Saviors, LINE Running Gingerbread Man, national baseball 2014, Candy Crush, Poko Pang, League, Heaven, Puzzle & Dragons, the king of AVA Battlefield, LINE Get Rich etc.

Figure 6 shows the results of analysis, accord with the user searching for a game, the content contains features, play, game time interval chart addiction, addiction scores, degree of addiction, addiction through time interval graph can hourly addiction know, the most important inform users of this addictive game play will be how long, and how high degree of addiction.

Table 3 Calculation games

Game name	Game addiction score	Score interval adjustment variables	Score interval value	Range (0 100)
Tower of saviors	49.9	0.107	10.7	100
League of legends	46.6	0.107	7.4	69.1588785
LINE cookie run	39.2	0.107	0	0
LINE get rich	39.3	0.107	0.1	0.934579439
Candy crush	42	0.107	2.8	26.1682243
A.V.A: alliance of valiant arms	43.8	0.107	4.6	42.99065421

Table 4 Addiction range definition

Addiction range (%)	Addiction
Slightly less than 33.3	Mild addicting games
Between 33.3 and 66.6	Moderate addicting games
Ranging from 66.6 to 100	Severe addicting games

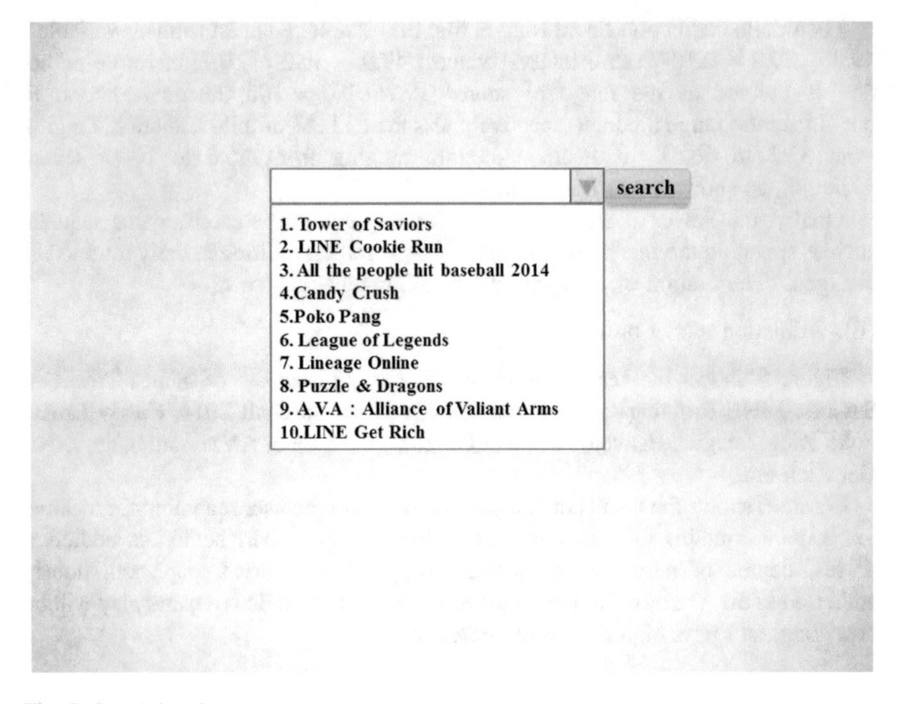

Fig. 5 Search interface

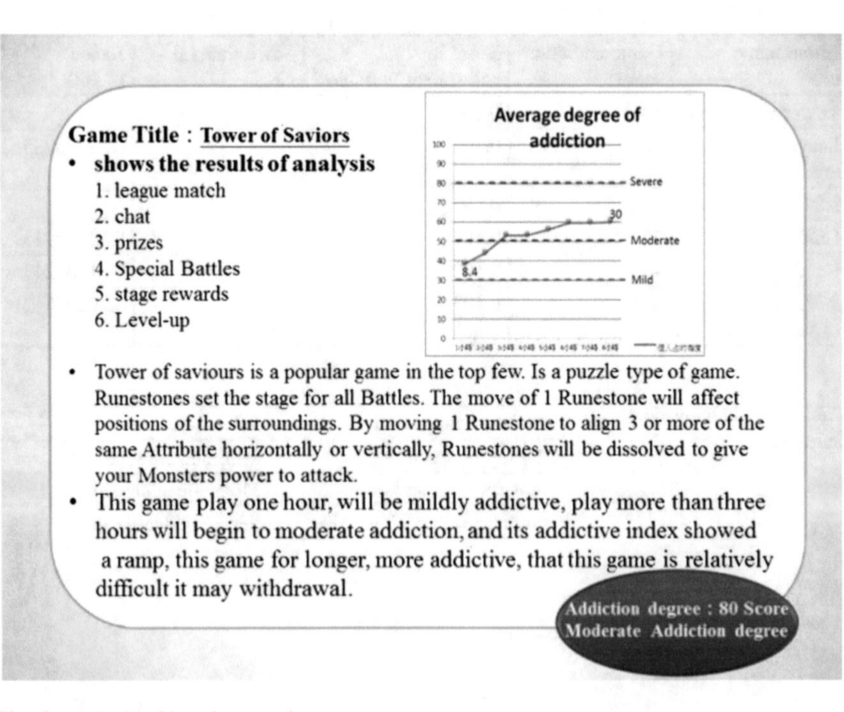

Fig. 6 Analysis of interface results

4 Conclusion

This model is divided into two kinds of "game addiction grade" and "personal game addiction grade", the difference is that the game belongs to the public of addiction scores, a personal addiction classification is in accordance with the user's data do their own analysis of addiction.

Game's rating should not only ordinary level, protection level, parental guidance, restricted, but also play this game by the degree of addiction do classification.

In this paper, expect to establish a model and practical application in the game to prevent a starting point, the contribution of their paper as the following:

(A) Model in which users can play this game is addictive, or warn users the game play more than how long it must to become addicted.
(B) Let parents know their children play the game content through inquiry system.
(C) The game ratings, according to their degree of addiction do classification (mild, moderate, severe addiction).

5 Future Research

Future research will focus on the content of individual game addiction, according to the type of subject classification: time constraints, salience, tolerance, emotion changes, and conflict, analyze the impact of this game due to the user and to find addiction factor, expectations you can achieve the following objectives:

(A) The model is just the idea, hoping to make this idea to reality, and developed into the web page or APP, APP directly to the record of the use of mobile phone users to play games of the time, frequency, will be able to more accurately calculate the degree of addiction for users use at any time.
(B) The model of addiction grading only do the follow-up hoping to be the type of personal information and subject classification, thrust reversers find addiction factor, that is able to find a way of its treatment.
(C) Future Addiction formula will give different weights or gravity based game-play, consider adding more factor in strengthening their accuracy, and community, the daily incentive mechanism.

References

1. The establishment of the Taiwan Network Information Center (TWNIC) Available: www.twnic.net.tw/
2. N. Yee, "Motivations for play in online games," CyberPsychology & behavior, vol. 9, pp. 772–775, 2006.

3. Jeroen S. Lemmens a , Patti M. Valkenburg a & Jochen Peter, "Development and Validation of a Game Addiction Scale for Adolescents,"Media Psychology, 12:77–95, 2009
4. Yao-Lung Tsai,Chin-Shih Tsai and Jia-Min Chuo, "Impact of Facebook Game's Involvement and Addiction on Physical and Mental Health and Learning Attitude-A Case Study from University Students in Madou," Journal of Sport and Recreation Research, pp. 16–38, 2012
5. Hsiao-Mei Wang and Min-Tung Kuo,"The Comparison of Imputation Methods withInternet Addiction Scale," Journal of Quantitative Management, vol. 10, no. 2, pp. 35–46, 2013
6. Sue-Huei Chen, Li-Jen Weng, Yi-Jen Su, Ho-Mao Wu and Pin-Feng Yang, "Development of a Chinese Internet Addiction Scale and Its Psychometric Study," Chinese Journal of Psychology, pp. 279–294, 2003
7. K. S. Young and C. N. de Abreu, Internet addiction: A handbook and guide to evaluation and treatment: John Wiley & Sons, 2010.
8. Rune Aune Mentzoni, Psy.D., Geir Scott Brunborg, M.Sc., Helge Molde, Ph.D., Helga Myrseth, Psy.D., Knut Joachim Ma° r Skouverøe, B.Sc., Jørn Hetland, Ph.D., and Sta° le Pallesen, Ph.D., "Problematic Video Game Use: Estimated Prevalence and Associations with Mental and Physical Health," CYBERPSYCHOLOGY, BEHAVIOR, AND SOCIAL NETWORKING Volume 14, Number 10, 2011
9. TING-JUI CHOU, Ph.D., 1 and CHIH-CHEN TING, M.B.A. 2, "The Role of Flow Experience in Cyber-Game Addiction," CYBERPSYCHOLOGY & BEHAVIOR Volume 6, Number 6, 2003
10. S.M. GRÜSSER, Ph.D. 1 R. THALEMANN, Ph.D. (C), 1 and M.D. GRIFFITHS, Ph.D. 2, "Excessive Computer Game Playing: Evidence for Addiction and Aggression?," CYBERPSYCHOLOGY & BEHAVIOR Volume 10, Number 2, 2007

Noise Estimation for Speech Enhancement Using Minimum-Spectral-Average and Vowel-Presence Detection Approach

Ching-Ta Lu, Yung-Yue Chen, Jun-Hong Shen, Ling-Ling Wang and Chung-Lin Lei

Abstract The accuracy of noise estimation is important for the performance of a speech enhancement system. This study proposes using variable segment length for noise tracking and variable thresholds for the determination of speech-presence probability. Initially, the fundamental frequency is estimated to determine whether a frame is a vowel. In the case of a vowel frame, the segment length increases; meanwhile the threshold for speech-presence is decreased. So the noise magnitude is adequately underestimated. The speech distortion is accordingly reduced in enhanced speech. Conversely, the segment length is rapidly decreased during noise-dominant regions. This enables the noise estimate to be updated quickly and the noise variation to be well tracked, yielding background noise being efficiently removed by the process of speech enhancement. Experimental results show that the proposed method can efficiently track the variation of background noise, enabling the performance of speech enhancement to be improved.

Keywords Noise estimation · Variable segment length · Speech enhancement · Harmonic adaptation · Minimum-Spectral-Average

---

C.-T. Lu (✉) · J.-H. Shen · L.-L. Wang · C.-L. Lei
Department of Information Communication, Asia University, Taichung, Taiwan, ROC
e-mail: Lucas1@ms26.hinet.net

J.-H. Shen
e-mail: shenjh@asia.edu.tw

L.-L. Wang
e-mail: ling@asia.edu.tw

C.-L. Lei
e-mail: BligsLcl@gmail.com

Y.-Y. Chen
Department of Systems and Naval Mechatronics Engineering,
National Cheng Kung University, Tainan, Taiwan, ROC
e-mail: yungyuchen@mail.ncku.edu.tw

© Springer Nature Singapore Pte Ltd. 2018
N.Y. Yen and J.C. Hung (eds.), *Frontier Computing*, Lecture Notes
in Electrical Engineering 422, DOI 10.1007/978-981-10-3187-8_32

1 Introduction

The usage of speech communication devices, such as mobile phones and hearing aids, often suffers from deterioration of speech quality and intelligibility in the presence of background noise. It is important to have speech enhancement methods that offer satisfactory noise reduction. The accuracy of noise estimation significantly affects the performance of speech enhancement. In particular, the mobile phone is very sensitive to adverse environments involving non-stationary noise and low input SNR ratio. An incorrect noise estimate will give a direct impact on the quality of the enhanced signal.

Many studies have been proposed to estimate noise [1–10]. Kianyfar and Abutalebi [1] proposed a noise estimator which employed speech-presence probability to update noise variance. Krawczyk-Becker et al. [2] proposed incorporating spectro-temporal correlations to improve the performance for noise tracking. A minima-controlled-recursive-averaging (MCRA) approach was computationally efficient and was therefore one of the successful noise estimation approaches for speech enhancement [3, 4]. The noise component was estimated by the speech-presence probability and speech absence probability within the subbands of a frame. To model the temporal variation of noises, the noise component was estimated by averaging the past spectral power values with a smoothing parameter that was adjusted by the speech-presence probability within the subbands. However, the beginning of non-stationary noise was independent to previous frames and as a result such an approach was insensitive to the beginning of non-stationary noise. Many new methods have been proposed to improve the performance of the MCRA methods [5–8]. Fan et al. [5] proposed a method to shorten time delay for the detection of abrupt changing noise. Noise update criterion was also additionally controlled to reduce speech leakage for the MCRA algorithm. Kum and Chang [6] proposed employing the second-order conditional maximum a posteriori criterion to improve the performance of the MCRA method. Wu et al. [7] proposed modifying the mechanism of the time variant recursive averaging of the MCRA algorithm by utilizing both noise and speech segments. In addition, speech enhancement residue was employed as an approximation to true noise and used in the recursive averaging procedure to update noise spectra in the case of speech-presence.

Based on the above findings, how to improve the performance for noise estimation is important. Most of the noise estimation methods do not consider speech properties in noise estimation. In this study, we employ the harmonic properties of a vowel to determine the segment length for tracking minimum statistics in the MCRA algorithm. In the case of a vowel frame and its neighbors, the increment of segment length and the decrement of threshold for speech-presence are performed. This enables the MCRA algorithm to pick up lower magnitude as a noise level. The noise estimate tends to be underestimated. This yields speech distortion being reduced in enhanced speech. The quality of enhanced speech is then improved. Conversely, the segment length is decreased during noise-dominant frames. This enables the MCRA algorithm to quickly update the level of noise estimate; while the noise variation being well tracked. A great quantity of background noise can be efficiently removed by the process of speech enhancement. The quantity of musical residual noise is reduced. Accordingly, enhanced speech by using the proposed noise estimator sounds more comfortable than that using the MCRA algorithm. The performance of the MCRA algorithm is therefore improved.

The rest of this paper is organized as follows. Section 2 reviews the MCRA algorithm. Section 3 describes the proposed modifications in the MCRA method. Section 4 demonstrates the experimental results. Conclusions are finally drawn in Sect. 5.

2 Review of MCRA Noise Estimator

A noisy speech signal $y(m, n)$ can be modeled as the sum of clean speech $s(m, n)$ and additive noise $d(m,n)$ in the frame m of the time domain, given as

$$y(m,n) = s(m,n) + d(m,n) \tag{1}$$

where n is the sample index in a frame.

The observed signal $y(m, n)$ is analyzed and transformed to the frequency domain using the short-time Fourier transform (STFT), given as

$$Y(m,k) = \sum_{n=0}^{N-1} y(n+mM) \cdot h(n) \cdot e^{-j(2\pi/N)nk} \tag{2}$$

where k is the frequency bin index, h is an analysis window of size N, and M is the frame update step in time.

$H_0(m,k)$ and $H_1(m,k)$ are two hypotheses which indicate speech absence and presence [3], respectively. They can be expressed by

$$\begin{aligned} &H_0(m,k)\colon Y(m,k) = D(m,k) \\ &H_1(m,k)\colon Y(m,k) = S(m,k) + D(m,k) \end{aligned} \tag{3}$$

where $S(m,k)$ and $D(m,k)$ represent the spectrum of clean speech and additive noise, respectively.

Let $\lambda_d(m,k) = |D(m,k)|^2$ denote the variance of the noise. The noise estimates for speech absence and presence can be obtained, given as

$$\begin{aligned} &H_0'(m,k)\colon \hat{\lambda}_d(m,k) = \alpha_d\, \hat{\lambda}_d(m-1,k) + (1-\alpha_d) \cdot \left|Y(m,k)^2\right| \\ &H_1'(m,k)\colon \hat{\lambda}_d(m,k) = \hat{\lambda}_d(m-1,k) \end{aligned} \tag{4}$$

where $\alpha_d (0 < \alpha_d < 1)$ is a smoothing parameter. H_0' and H_1' designate hypothetical speech absence and presence, respectively.

Let $p^{'}(m,k) \triangleq p(H_1^{'}(m,k)|\, Y(m,k))$ denote the conditional signal-presence probability. The noise estimate given in Eq. (4) can be obtained, given as

$$\begin{aligned}\lambda_d(m,k) &= \lambda_d(m-1,k) \cdot p^{'}(m,k) \\ &\quad + [\alpha_d \cdot \lambda_d(m-1,k) + (1-\alpha_d) \cdot |Y(m,k)|^2] \cdot [1 - p^{'}(m,k)]\end{aligned} \tag{5}$$

The speech-presence probability $p^{'}(m,k)$ in Eq. (5) can be obtained by

$$\hat{p}^{'}(m,k) = \alpha_p \cdot \hat{p}^{'}(m-1,k) + \left(1-\alpha_p\right) \cdot I(m,k) \tag{6}$$

where $\alpha_p(\alpha_p = 0.2)$ is a smoothing parameter for speech-presence probability. $I(m,k)$ denotes an indicator function for speech-presence, given as

$$I(m,k) = \begin{cases} 1, & \text{if } \gamma(m,k) > \delta_\gamma \\ 0, & \text{otherwise} \end{cases} \tag{7}$$

where δ_γ represents a threshold for speech-presence. $\gamma(m,k)$ represents the ratio between the local energy of the noisy speech $P_{Local}(m,k)$ and its derived minimum $P_{\min}(m,k)$, given as

$$\gamma(m,k) = P_{Local}(m,k)/P_{\min}(m,k) \tag{8}$$

where

$$P_{Local}(m,k) = \sum_{i=-\omega_1}^{\omega_1} b(i) \cdot |Y(m,k-i)|^2 \tag{9}$$

The smoothed version of the local energy $P_{Local}^{S}(m,k)$ can be computed by a first order recursive averaging, given as

$$P_{Local}^{S}(m,k) = \alpha_s P_{Local}^{S}(m-1,k) + (1-\alpha_s) P_{Local}(m,k) \tag{10}$$

The minimum $P_{\min}(m,k)$ and a temporary variable $P_{tmp}(m,k)$ are initialized by $P_{\min}(0,k) = P(0,k)$ and $P_{tmp}(0,k) = P(0,k)$. Then, a sample-wise comparison of the local energy and the minimum value of the previous frame yield the minimum value for the current frame, given as

$$P_{\min}(m,k) = \min\left\{P_{\min}(m-1,k), P_{Local}^{S}(m,k)\right\} \tag{11}$$

$$P_{tmp}(m,k) = \min\left\{P_{tmp}(m-1,k), P_{Local}^{S}(m,k)\right\} \tag{12}$$

Whenever L ($L = 64$) frames have been read, i.e., m is divisible by L, the temporary variable $P_{tmp}(m,k)$ is employed and initialized by

$$P_{tmp}(m,k) = P^S_{Local}(m,k) \tag{13}$$

$$P_{\min}(m,k) = \min\{P_{tmp}(m-1,k), P^S_{Local}(m,k)\} \tag{14}$$

The minimum $P_{\min}(m,k)$ is employed to determine the value of speech indicator given in Eqs. (7) and (8) for the MCRA method [3].

3 Proposed Modification for MCRA Algorithm

In order to improve the performance of the MCRA algorithm, we employ harmonic properties of a vowel to determine the segment length L and the threshold for speech-presence of each subband. In the case of vowel regions, the segment length is increased. This enables the modified MCRA algorithm to select a smaller minimum value as a noise reference than that of the original MCRA algorithm. Meanwhile, the threshold of speech-presence is adjusted to be smaller in a vowel and its neighbor frames, enabling weak vowels and consonant components to be classified as speech. So the weak vowels and consonants can be well preserved by the process of speech enhancement. The quality of enhanced speech is accordingly improved.

Harmonic properties are employed to determine the segment length L which controls the period for the update of noise estimate. Initially, the number of harmonic spectra is utilized to determine whether a frame is a vowel. If the frame is detected as a vowel, the segment length L is increased. This increases the period for the search of spectral minimum as given in Eq. (14), yielding noise spectrum being underestimated. Conversely, the segment length L is decreased when a noise-dominant frame is detected. The segment length can be expressed by

$$L(l) = \begin{cases} L(l) + L_1, & \text{if } F^v(m) = 1 \\ L(l) + L_2, & \text{if } \sum\limits_{t=-\varepsilon}^{\varepsilon} F^v(m+t) > 0 \\ \beta * L(l), & \text{otherwise} \end{cases} \tag{15}$$

where l is the segment index. L_1 and L_2 represent the increment of segment length for a vowel and the corresponding neighbor regions. They are empirically chosen to be 63 and 12, respectively. ε controls the number of neighbor frames to be included for the regions of onset, offset, and consonants. It is set to be 3. β controls the decrement ratio of segment length for noise regions. It is empirically chosen to be 0.9. $F^v(m)$ is a vowel flag, given as

$$F^v(m) = \begin{cases} 1, & \text{if } m\text{th frame} \in \text{vowel} \\ 0, & \text{otherwise} \end{cases} \tag{16}$$

In Eq. (15), the segment length is significantly increased with L_1 frames when a frame is detected as a vowel. Conversely, the segment length is significantly decreased with ratio 0.9 of current segment length, i.e., $0.9L$, when a frame is detected as speech absence. A consonant may appear in the precedence of a vowel for Mandarin Chinese spoken language. We slightly increase the segment length with L_2 frames, yielding noise magnitude being underestimated. This enables a consonant to be preserved by the process of speech enhancement. In the regions of onset and offset during a vowel, the segment length is also slightly increased with length L_2.

In Eq. (7), the threshold of speech-presence δ_γ is a constant in the MCRA algorithm [3]. If the value of δ_γ is too high, a great quantity of weak speech spectra, such as weak vowels and consonants, would be classified as noise. The value of speech indicator function $I(m,k)$ is falsely set to zero. Although the quantity of residual noise can be reduced, speech distortion is increased in enhanced speech. The quality of enhanced speech is deteriorated. Conversely, if the value of δ_γ is too small, a great quantity of noise spectra would be classified as speech. The value of speech indicator function $I(m,k)$ is falsely set to unity. Although the speech distortion can be reduced, the quantity of musical residual noise increases. So the enhanced speech sounds annoying and uncomfortable.

Here we employ harmonic properties of a vowel to adapt the threshold of speech-presence δ_γ, given as

$$\delta_r(m) = \begin{cases} \delta_V, & \text{if } F^v(m) = 1 \\ \delta_{Neighbor}, & \text{if } \sum_{t=-\varepsilon}^{\varepsilon} F^v(m+t) > 0 \\ \delta_N, & \text{otherwise} \end{cases} \quad (17)$$

where δ_V, $\delta_{Neighbor}$, and δ_N represent the thresholds of speech-presence for a vowel, the neighbor frames of a vowel, and noise-dominant regions, respectively. The values of the threshold are empirically chosen to be 1.5, 1, and 5, respectively.

In Eq. (17), the values δ_V, $\delta_{Neighbor}$, and δ_N are determined in white noise corruption with various input SNRs at which the selected values can obtain the highest score of average segmental SNR improvement. In the cases of a vowel and the corresponding neighbor frames, the thresholds are small. This prevents weak vowels and consonants being classified as noise, and then being removed by the process of speech enhancement. Accordingly, the quality of enhanced speech can be improved by using the harmonic properties of a vowel to adapt the threshold of speech-presence.

4 Experimental Results

In the experiments, speech signals were Mandarin Chinese spoken by five female and five male speakers. These speech signals were corrupted by various kinds of additive noise, such as white, F16-cockpit, factory, helicopter-cockpit, and car noise signals which were extracted from the Noisex-92 database. Three input average segmental SNR levels, including 0 dB, 5 dB and 10 dB, were used to evaluate the performance of a noise estimator. In order to evaluate the performance of the proposed method, the original MCRA noise estimator [3], the forward-backward MCRA (FB_MCRA) noise estimator [8] were conducted for performance comparison. A three-step-decision gain factor [11] is employed to perform speech enhancement for various noise estimators. The following parameters are used in the experiments: (1) sampling frequency is 8 kHz; (2) the frame size is 256 with 50% overlap; (3) Hanning window is utilized; (4) total number of critical bands is 18, the central frequency and the corresponding bandwidth of each critical band can be found in [12].

The amount of noise reduction, residual noise and speech distortion can be measured by the average segmental SNR improvement (Avg_SegSNR_Imp). The Avg_SegSNR_Imp is computed by subtracting the Avg_SegSNR of noisy speech from that of enhanced speech. Table 1 presents the performance comparisons in terms of the Avg_SegSNR_Imp for various noise estimation methods. It can be found that the proposed method outperforms the MCRA and the FB_MCRA algorithms in most conditions. The major reason is due to a quantity of consonants and weak vowels being preserved by the underestimation of background noise. These results are achieved by increasing the segment length to track the minimum

Table 1 Comparison of SegSNR improvement for the enhanced speech in various noises

Noise type	SNR	Average SegSNR improvement		
	(dB)	MCRA	FB_MCRA	Proposed
	0	6.95	7.15	7.83
White	5	4.44	4.70	5.64
	10	1.57	2.08	3.44
	0	5.81	5.73	5.98
F16	5	3.78	3.83	4.53
	10	1.44	1.76	2.83
	0	5.41	5.35	5.62
Factory	5	3.43	3.47	4.17
	10	1.14	1.46	2.53
	0	6.22	6.29	6.34
Helicopter	5	4.13	4.32	4.99
	10	1.76	2.25	3.28
	0	7.87	10.08	9.86
Car	5	5.70	8.19	9.08
	10	3.10	5.97	7.05

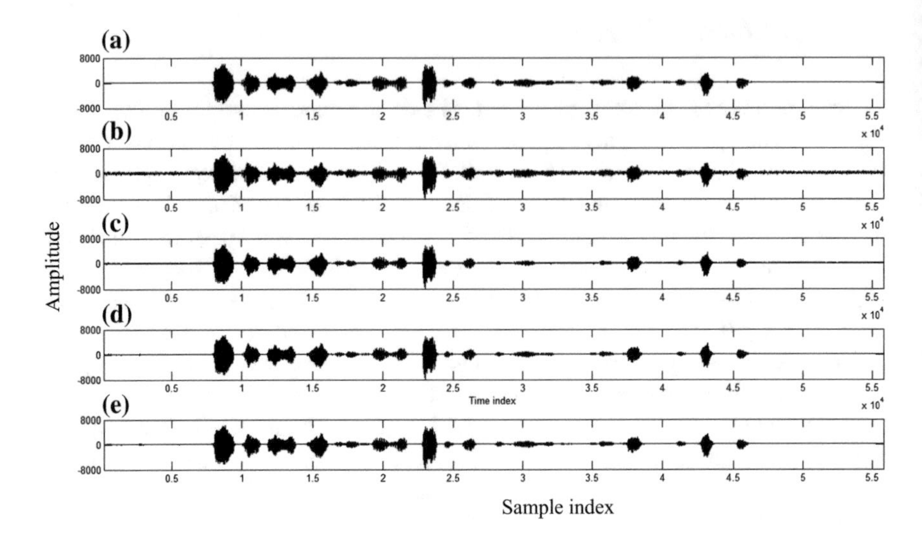

Fig. 1 Example of speech signal spoken in Mandarin Chinese by a male speaker. (From *top* to *bottom*) **a** clean speech, **b** noisy speech corrupted by helicopter noise with an average SegSNR = 5 dB, **c** enhanced speech using MCRA noise estimator, **d** enhanced speech using forward-backward MCRA noise estimator, **e** enhanced speech using proposed noise estimator

spectral magnitude of noisy speech. In addition, the segment length is reduced during speech-pause regions. This enables the spectrum magnitude of noisy speech to be updated quickly, yielding noisy spectra being efficiently removed by the process of speech enhancement. Accordingly, the proposed method can obtain higher scores of average segmental-SNR improvement than the other methods.

Figure 1 demonstrates an example of waveform plots for performance comparisons. A speech signal uttered by a male speaker was corrupted by helicopter-cockpit noise with Avg_SegSNR = 5 dB. In Figs. 1c–e, a clipped signal is absent at the output waveforms of the enhanced speech. This is attributed to all noise estimators do not over-estimate the level of noise power spectra for each subband, yielding enhanced speech not suffering serious speech distortion. Comparing Figs. 1c–e, background noise can be efficiently removed by using the three noise estimators. The proposed method can preserve a greater quantity of speech components than the other two methods during speech-presence regions, including weak vowels, the onset and offset of a vowel, and consonants. This is due to the adaptation of harmonic properties for the determination of segment length and the thresholds for speech-presence.

Figure 2 shows the spectrograms of a speech signal. A speech signal is corrupted by helicopter-cockpit noise signals with Avg_SegSNR = 10 dB (Fig. 2b). Comparing Figs. 2c–e, the level of background noise can be well estimated by the three noise estimators, enabling corruption noise to be efficiently removed by the process of speech enhancement. Employing the proposed method is better able to reserve

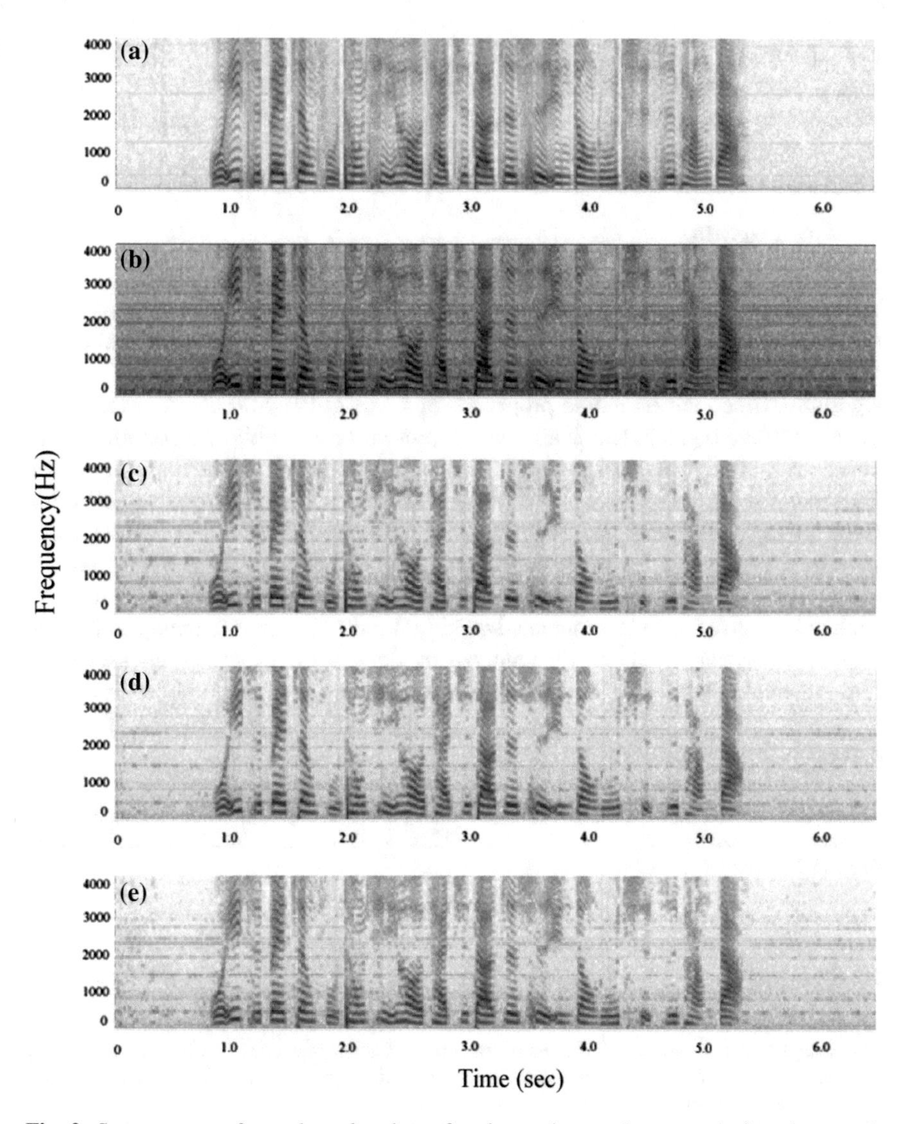

Fig. 2 Spectrograms of speech spoken by a female speaker, **a** clean speech, **b** noisy speech corrupted by helicopter-cockpit noise with average SegSNR = 10 dB, **c** enhanced speech using MCRA noise estimator, **d** enhanced speech using forward-backward MCRA noise estimator, **e** enhanced speech using proposed noise estimator

weak vowels and speech components in enhanced speech during speech-presence regions. So the harmonic structure of a vowel by using the proposed method is better than the other two methods. The quality of enhanced speech is improved. The major reason is attributed to increase the value of speech-presence probability for weak vowels and consonants, yielding the level of noise spectra being underestimated. The quantity of noisy speech been suppressed by the process of speech enhancement is decreased. Speech distortion is then reduced, enabling enhanced speech to sound more comfortable than the other two methods.

5 Conclusions

This paper proposed using variable segment length for the update of noise magnitude and variable thresholds for the determination of speech-presence probability to improve the performance of the minima-controlled-recursive-averaging (MCRA) algorithm. Since the harmonic properties of a vowel are considered in the determination of the segment length and speech-presence probability, the performance of noise estimation can be improved. The segment length is increased and the threshold for speech-presence is decreased in speech-dominant regions, enabling noise to be underestimated. So the speech distortion is decreased in enhanced speech. Conversely, the segment length is decreased and the threshold for speech-presence probability is kept at a high level during noise-dominant regions, enabling noise estimate to be updated quickly. So the background noise can be well estimated and can be efficiently removed by the process of speech enhancement. Experimental results show that the proposed method can efficiently improve the performance of the MCRA algorithm. Consequently, the performance of speech enhancement is improved.

Acknowledgements This research was supported by the Ministry of Science and Technology, Taiwan, under contract numbers MOST 104-2221-E-468-007, and MOST 104-2628-E-006-012-MY3.

References

1. Kianyfar, A., Abutalebi, H. R.: Improved Speech Enhancement Method Based on Auditory Filterbank and Fast Noise Estimation. In: International Symposium on Telecommunications, pp. 441–445 (2014)
2. Krawczyk-Becker, M., Fischer, D., Gerkmann, T.: Utilizing Spectro-Temporal Correlations for an Improved Speech Presence Probability Based Noise Power Estimation. In: IEEE International Conference on Acoustics, Speech, and Signal Processing, pp. 365–369 (2015)
3. Cohen, I. Berdugo, B.: Noise Estimation By Minima Controlled Recursive Averaging for Robust Speech Enhancement. IEEE Signal Process. Lett., vol. 9, no. 1, pp. 12–15 (2002)
4. Cohen, I.: Noise Spectrum Estimation in Adverse Environments: Improved Minima Controlled Recursive Averaging. IEEE Trans. Speech Audio Process., vol. 11, no. 5, pp. 466–475 (2003)
5. Fan, N., Rosca, J., Balan, R.: Speech Noise Estimation Using Enhanced Minima Controlled Recursive Averaging. In: IEEE International Conference on Acoustics, Speech, and Signal Processing, vol. 4, pp. 581–584 (2007)

6. Kum, J. M., Chang, J. H.: Speech Enhancement Based on Minima Controlled Recursive Averaging Incorporating Second-Order Conditional Map Criterion. IEEE Signal Process. Lett., vol. 16, no. 7, pp. 624–627 (2009)
7. Wu, D., Zhu, W. P., Swamy, M. N. S.: Noise Spectrum Estimation with Improved Minimum Controlled Recursive Averaging Based on Speech Enhancement Residue. In: IEEE International Midwest Symposium on Circuits and Systems, pp. 948–951 (2012)
8. Chen, Y. J., Wu, J. L.: Forward-Backward Minima Controlled Recursive Averaging to Speech Enhancement. In IEEE International Symposium on Computational Intelligence for Multimedia, Signal Vision Processing, pp. 49–52 (2013)
9. Yong, P. C., Nordoholm, S., Dam, H. H.: Noise Estimation with Low Complexity for Speech Enhancement. In IEEE Workshop on Applications of Signal Processing to Audio and Acoustics, pp. 109–112(2011)
10. Mai, V. K., Pastor, D., Aissa-EI-Bey, A., Le-Bidan, R.: Robust Estimation of Non-stationary Noise Power Spectrum for Speech Enhancement. IEEE/ACM Trans. Audio, Speech, Lang. Process., vol. 23, no. 4, pp. 670–682 (2015)
11. Lu, C. –T., Shen, J. –H., Tseng, K. –F..: Speech Enhancement Using Three-Step- Decision Gain Factor with Optimal Smoothing. Int. J. Electr. Eng., vol. 18, no. 5, pp. 209–221 (2011)
12. Virag, N.: Single Channel Speech Enhancement Based on Masking Properties of the Human Auditory System. IEEE Trans Speech Audio Process., vol. 7, no. 2, pp. 126–137 (1999)

Implementation and Design of a Middleware Platform Based on WCF

Hua Yi Lin, Jiann-Gwo Doong, Meng-Yen Hsieh and Kuan-Ching Li

Abstract With the advances in computing and communication technologies, message-oriented middleware becomes an important and universal tool in enterprises nowadays. WCF (Windows Communication Foundation) exploits a loosely coupled way that client-side and server-side can accelerate the development and operation. Although, WCF contains most of lower-layer communication components. However, under specific applications, developers unavoidably develop extra connection functions and client-to-serve managing software in practice. Because of the practical need, we prepare to construct a WCF-based middleware operational prototype for future research reference. In this study, we build up a practical prototype of middleware to integrate with heterologous systems using WCF for accumulating our experience on cross-platform systems. The implementation includes developing specific APIs (Application Program Interface) of middleware for the system requirement, achieving middleware with broadcast functions to complete information-exchange between two specific functions within WCF. Our enhanced WCF framework supports a lots of flexibility in APIs. Developers do not need to modify any of the underlying WCF components, and unrestrictedly customize the API attributes and API logic.

Keywords Middleware · WCF · API · Heterologous system

H.Y. Lin (✉) · J.-G. Doong
Department of Information Management, China University of Technology, Taipei, Taiwan
e-mail: calvan.lin@cute.edu.tw

J.-G. Doong
e-mail: sdoong@cute.edu.tw

M.-Y. Hsieh · K.-C. Li
Department of Computer Science and Information Engineering, Providence University, Taichung, Taiwan
e-mail: mengyen@pu.edu.tw

K.-C. Li
e-mail: kuancli@pu.edu.tw

© Springer Nature Singapore Pte Ltd. 2018

N.Y. Yen and J.C. Hung (eds.), *Frontier Computing*, Lecture Notes in Electrical Engineering 422, DOI 10.1007/978-981-10-3187-8_33

1 Introduction

Middleware means providing a set of interfaces to connect between application software and system software. In this way, the various components in a software can communicate with each other, especially in application software for centralized system. Nowadays, website services and architectures popularly take advantages with such integration. Apache Tomcat, database, IBM WebSphere, and BEA web-logic are example of such middleware [1]. Recently, the mobile computing becomes even more popular, where the main function of the mobile middleware is not only to build a communication bridge between various mobile platforms and heterogeneous enterprise platforms, but to create a standard protocol by a common set of data format for the variety of mobile platforms [2]. With the rapid progress of Internet, the developing trend of the Message-Oriented Middleware became the most common method for information transfer in a conglomerate of resources.

From the MSDN (Microsoft Developer Network), we realized that WCF is a framework to construct service-oriented applications [3]. Through it, asynchronous way to send messages from an endpoint to other is technically possible. The service endpoint can be made of IIS (Internet Information Service) or hosting one application. Clients can also have an endpoint to present the requests. In addition, the message exchanged can be a character, XML (Extensible Markup Language) or a complicated binary stream. WCF has several features, listed as (1) Service-oriented: WCF allows programmers to create object-oriented applications. Thus, sending and receiving data rely on SOA (Service-Oriented Architecture) throughout the network. Besides, the advantage of WCF is providing a loose coupling methodology to avoid writing a dedicated code. As long as they have contracts, they can be any client's heterogeneous platforms to connect different services, (2) WCF offers multiple message modes, such as (a) Request/response mode, (b) Unidirectional mode, and (c) Full duplex mode. MSDN also mentions that the service-oriented communication is capable to improve the software development. Regardless of SOAP or other methods, the service-oriented model becomes a typical model for interaction between applications, and MSDN shows such the usage of WCF [4]. We also realize that MSDN exploits WCF functions to simplify operations and develop server-end or client-end applications. When client-end connects with server-end, then client-end is able to send a service request to server-end, and obtains the response from server-end. Additionally, one important aspect is that WCF defines three functions to achieve loose coupling methodology namely (a) Address, where the programmers can use the URL format to define server or client communication address, (b) Binding, the programmers can use Internet protocols as WCF communication methods, and (c) Contract, it defines functions and a parameter format for functions.

Subsequently, this study concentrates on building up a loose coupling methodology via "Address", "Binding" and "Contract" functions within WCF, and supplies several required methods to strengthen insufficient procedures to achieve a more complete middleware. Eventually, this study exploits the designed functions to perform broadcast and transaction examines between server-end and client-end, and verify the connection between the server-end and client-end.

The rest of this paper is structured as follows. Section 2 introduces the detailed function design and specification. Section 3 describes our system development functions. Section 4 presents the system testing. Conclusions and future work are finally drawn in Sect. 5.

2 System Architecture and Implementation

2.1 Function Design

Other than original middleware functions, we need to design and consider new requirements. Therefore, this system implementation will be an experiment as well as occur with a lot of trials or errors, a series of parameters, or program modifications that repeat testing constantly. In order to enhance the effectiveness, the study requires an immediate or a simple method to adjust instruction parameters. From the above opinions, we suggest the following designs.

(1) Server-end instruction: Some new instructions are given, including EXIT for closing a system), SET for parameter setting, and READ for reading system instructions from files and executing them.
(2) Server-end transaction: One of the middleware functions is proposed to provide developers with a friendly interface for transaction function.
(3) Broadcast mechanism: An interface is given that client-ends or server-ends can receive messages each other.
(4) Operational records of server-end: The system procedures are recorded for the further system test.
(5) Connection component: It is a public component which client-end connects with server-end. Client-ends send transaction requests, broadcast requests, and receiving broadcast requests using this component. Thus, WCF simplifies the related setting of endpoints.
(6) Client-end instruction: Several instructions are given, including EXIT for closing a system), SET for parameter setting, READ for reading system instructions from files and executing them, BROADCAST for system broadcast.
(7) Client transaction request: A client-end broadcasts a transaction request to a server-end, and displays the results.
(8) Client-end broadcast request: Client-ends broadcast requests to server-ends. Client-ends can receive and display the broadcast messages form server-ends.
(9) Broadcast monitoring and automatic execution: When client-ends or server-ends receive a dedicated message, they automatically perform transactions, and broadcast the results. Thus, the system can simulate the auto-response functions.

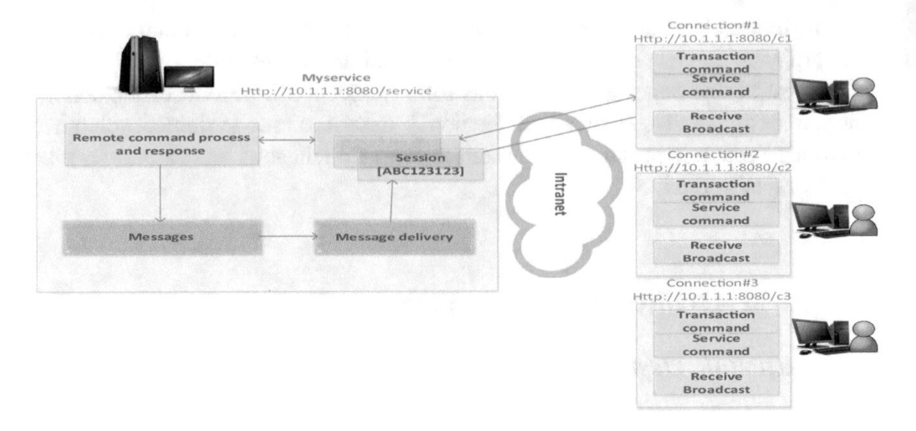

Fig. 1 The system architecture diagram

2.2 System Implementation

A Visual Studio 2012 tool for C# is adopted to develop the system. Server-ends and client-ends are running in console mode for displaying the executing results by texts. This study will test the system broadcast and transaction [5]. In addition, the internal server mechanism must be established some functions to provide related services. The detailed function description is as follows, and our system architecture is shown in Fig. 1.

(1) Session management: When a client-end connects with a server-end, it will automatically create a session, and establish a list of connectors until the session has closed.
(2) Instruction process: If any session receives the instruction from client-ends, server-ends automatically parse and execute the command.
(3) Message queue: If server-ends receive broadcast requests, they put the received messages into the message queue and wait for sending them.
(4) Message delivery: Server-end takes the message from the message queue in a period, and send the message to client-ends according to the type of commands.

2.3 Instruction Design

In this section, this study adds several instructions to supply the insufficient parts of WCF, and adopts the added instruction to complete the proposed middleware platform. Furthermore, we give a brief description of the detailed instruction format, instruction name, parameter, and the usage of new added instructions. Table 1 describes the instruction format, and gives the usage of parameter. Table 2 presents

Table 1 Items in each instruction format

<table>
<tr><th>Item</th><th>Description</th></tr>
<tr><td>Command</td><td>An instruction, not case sensitive</td></tr>
<tr><td>@@</td><td>The delimiter between commands and parameters</td></tr>
<tr><td>Parm 1~N</td><td>A parameter name with a symbol "||" to separate each parameter</td></tr>
<tr><td>@=</td><td>The delimiter of each parameter name and parameter value</td></tr>
<tr><td>||</td><td>The symbol to separate each parameter</td></tr>
</table>

Table 2 The instructions for the server-end

Instruction	Parameter
SET	The parameter is to set up the environment, for example, SET @@ LOCAL_HOST_02_URL @= http://127.0.0.1:8080/mes_02
EXIT	System shutdown
HOSTREAD	The parameter is to take a file as input, then parses and executes the listed instructions in this file, for example, HOSTREAD @@ FILE @= C:\wcfHost\hostCommandList.txt
HELLO	A test instruction, when receiving this instruction, then replying 'Hi, how are you?'
CLS	Clean the console screen
CONN	Connect to other servers, ex: Conn @@ url @= net.tcp://127.0.0.1:8082/iishost02
CONNPARTNER	Using the parameter partner's URL to connect a server, Ex: SET @@ PARTNER_01 @= net.tcp://127.0.0.1:8082/IISHOST02

Table 3 The instructions for the transaction type attribute of server-end

Item	Description
_ModuleName	A transaction of DLL (Dynamic-link library)
_TxnName	A transaction function and its corresponding class
_TxnID	The identity of one transaction, named GUID (Globally Unique IDentifier)。
_ReplyID	The respondent's GUID and server-end's GUID
_ReplyComments	A notation of a transaction
_ReplyTime	The reply time of a finished transaction
_TxnResultId	A message code of a transaction result
_ToServer_Data1	The summand (minuend) of the addition (subtraction) transaction
_ToServer_Data2	The addend (subtrahend) of the addition (subtraction) transaction
_ToClient_Result	The result of an addition (subtraction) transaction
_ClientID	The ID of a transaction requester

the instruction for the server-end, including the environmental parameter setting, a method of connecting a server-end via URL, and clearance of a connection between the server-end and the client-end. Table 3 describes the instruction for server-end

Table 4 The items for broadcasting broadcasting

Item	Description
_ModuleName	A message model name
_TxnName	A message transaction name
_TxnID	The transaction GUID
_SendID	The transaction sender's GUID
_SendTime	The sending time of a transaction
_SendComments	A notation of a sending transaction for recording the content of broadcast
_ServerID	When forwarding this message from a server-end, it records the server's ID in this column

Table 5 The functions of the connecting component

Functions	Description
HostCommand	Client-ends can execute commands on server-ends
ServiceCommand	Client-ends send transaction requests to server-ends
Broadcast	Client-ends send broadcast requests to server-ends

Table 6 The instructions for a client-end

Instruction	Parameter
SET	The server-end instruction name
EXIT	Shutdown system
READ	Read and parse the list of instruction in a file Ex: READ @@ FILE @= C:\Client\APP01_TO_HOST01.txt
CONN	Manually link to another server, ex: Conn @@ url @= net.tcp://127.0.0.1:8082/iishost02
CLS	Clean a console screen
TXNADDITION	Send addition transaction requests to server-ends Ex: txnAddition @@ data1@=1 || data2@=2
TXNSUBTRATION	Send subtraction transaction requests to server-ends Ex: txnSubtration @@ data1@=1 || data2@=2
BROADCAST	Send broadcasts to server-ends Ex: broadcast @@ msg @= hello, this is a broadcast test
/	Execute server-end instructions Ex: execute/ HELO, then client-end receives Hi, how are you? Or execute/CLS to clean the server's console screen

transactions. Here we add two type of transactions that are the attribute of addition and subtraction transactions. Table 4 presents the required class for broadcasting XML contents. Table 5 describes connection components, and the usage of a client-end executing a server-end's command and sending service requests to a server-end. Table 6 depicts the instructions of the client-end. The detailed description is as follows.

(1) Instruction design for a server-end: Table 1 describes the details of an instruction format, for example, one instruction can be represented as "Command@@ parm1@=value1 || parmN@=valueN".
(2) Server-end transaction: This study is designed with one addition transaction, "txnAddition", and one subtraction transaction, "txnSubtraction". Two transactions have the same basic messages, (only the corresponding class is different from client-ends). During one transaction, the client-end delivers XML-formatted messages to the server-end [6, 7]. After execution, the server-end sends the result back to the client-end using XML formats. Finally, the server-end broadcasts the transacation messages and results to client-ends. The transaction type attribute of "txnAddition" and "txnSubtraction" transactions is shown in Table 3.
(3) Broadcast design: The myTxn class is used to describe and comprise a XML message, as shown in Table 4.
(4) System operational record: A common component records data in a text file. The component writes messages into a log file using. Net functions, and displays them in a console.
(5) Connecting component design: The following functions are required, as shown in Table 5.
(6) Client-end instruction design: An instruction set is shown in Table 6.
(7) Client-end transaction design: The client-end sends a transaction request to the server-end using a connection object. This transaction parameter in Table 3, refers to the server-end transaction, and the TXNADDITION and TXNSUBTRATION instructions in client-end instructions are designed in Table 6.
(8) Client-end broadcast request design: The client-end uses a shared connection object to send broadcast messages from the server-end to broadcast. The broadcast parameters refer to broadcast and client-end instructions, and are individually designed in Tables 4 and 6.
(9) Server-end broadcast monitor and automatic execution function: The client-end exploits hard-code methods to develop a program. When a client-end's ID is APP03_TO_HOST02 and the received message is txnAddition. The Data1 and Data2 messages are automatically used to generate a txnSubtration transaction, and sends txnSubtration to the server-end to execute transactions.

3 System Development

In order to build up this middleware platform, this study construct the system framework as show in Fig. 2. The system is divided into three parts. (1) Client-end: users is able to broadcast and perform transactions with server-end via system connection interface; (2) Server-end: it possess the capability to deal with transactions and delivery broadcasting messages. At the same time, the serve is able to reach Internet and access database, and (3) System connection interface: all

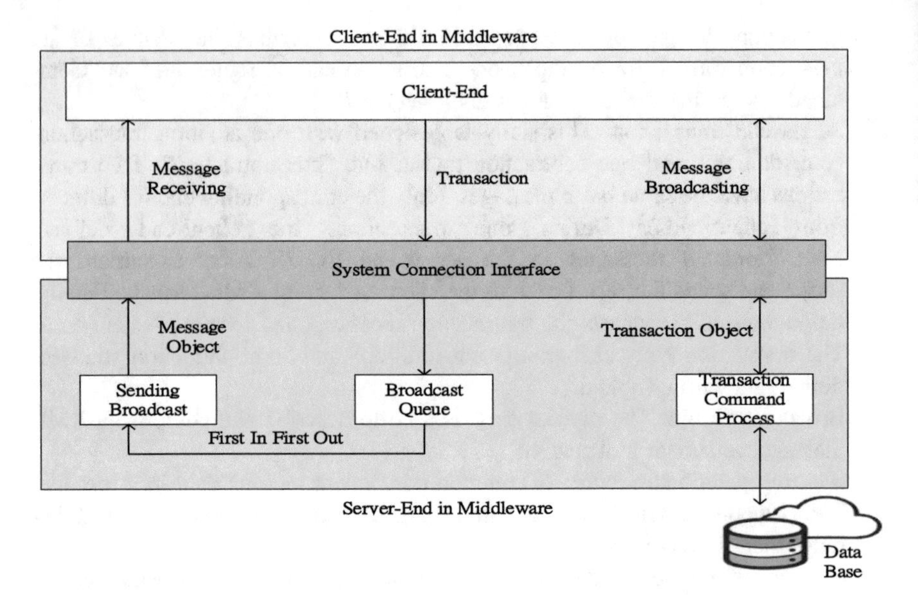

Fig. 2 The middleware framework

Fig. 3 The endpoint service design

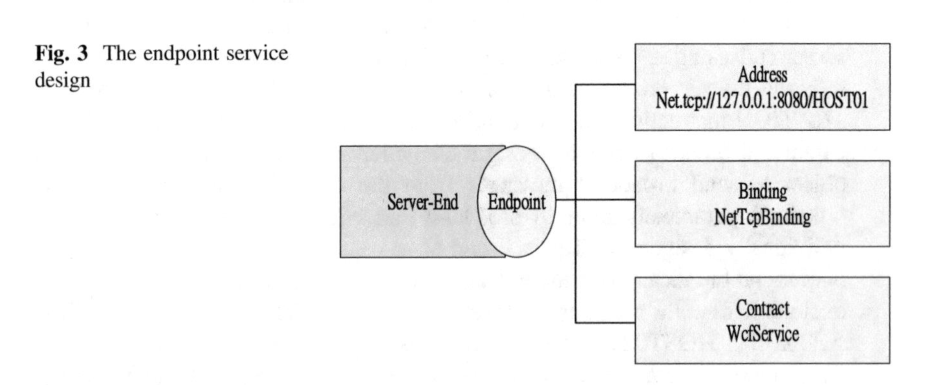

client-ends use this common interface to perform transactions and broadcasts with server-end. Additionally, WCF already provided the communication functions of session layer and presentation layer. Thus, this study merely needs to define the endpoint of server-end and client-end. In another word, we define the Address of endpoint, Binding and Contract, and then can complete connection between client-end and server-end. The endpoint view is as shown in Fig. 3.

For the reason of unified service interface and avoidance of usually alterative interface, this study declares an array byte[] to operate Contract. Since VS (virtual studio) gives serialization functions, the client-end serializes the object as a message and delivers this message to server-end via Internet. When server-end receives the delivered message, then returns it to the original object. Thus, object is able to be delivered in network environments through serialization and nonserialization

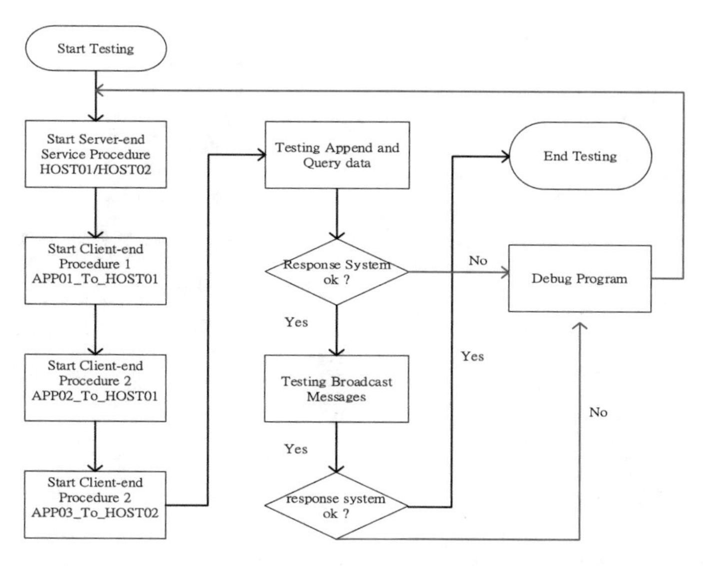

Fig. 4 The system testing flow between sever-end and client-end

functions within VS. Furthermore, according to the requirement of middleware software and WCF, we propose a new infrastructure of WCF.

According to the proposed new WCF functions, we must provide developers \with the message interface (wcfMessage), service interface (wcfService), call-back interface (wcfServiceCallback) and transaction interface (wcfTransaction) [8, 9]. Therefore, we create several projects to implement the server-end program (wcfService), connection interface program (wcfConnection) between server-end and client-end, and defines transaction and broadcast types, and client program (testClient) [10]. Moreover, this study develops related projects and classifies each function and interface into a dedicated project for further management. The project name HostConsol stores the application for server-end and defines WCF services. The project name ClientConsol stores the application for client-end and displays information on a console.

This study assumes windows 7 as the operation system. Before testing, we start to require server and client. The testing flow is as shown in Fig. 4, and the corresponding connection is as shown in Fig. 5. In addition, this study needs to prepare the initial parameter file for the server-end and client-end. At the beginning, this study gives the server name HOST01, URL address and cooperative partner HOST02. The detailed parameter setting for server-end is as shown in Fig. 6. Subsequently, this study sets up the parameter for client-end, and give the client ID, URL address and the target server-end URL, the detailed setting is as shown in Fig. 7.

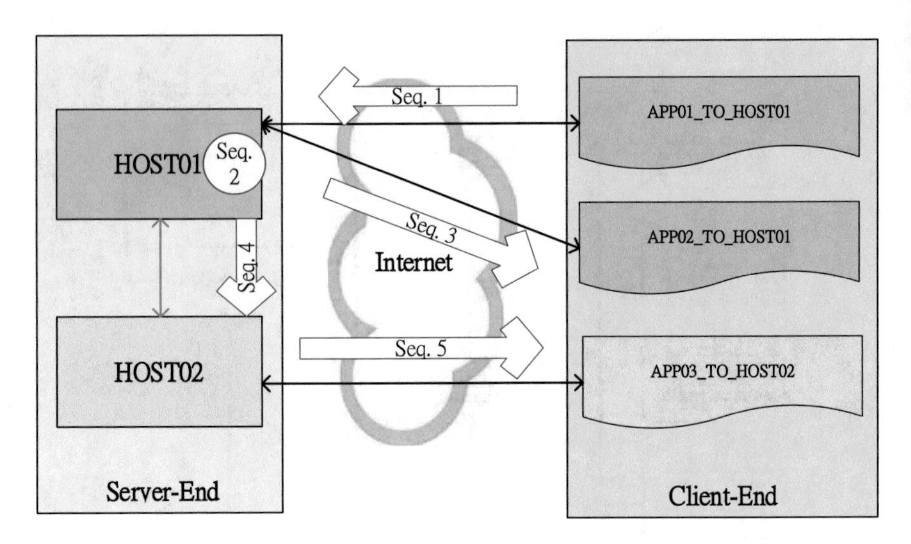

Fig. 5 Initiate communication between server-end and client-end

Fig. 6 Examples of server-end's parameter

```
SET @@LOCAL_NAME @= HOST01
SET @@LOCAL_URL @= http://127.0.0.1:8080/HOST01/HOST01
SET @@EndPointUrl_01 @= net.tcp://127.0.0.1:8081/HOST01
SET @@PARTNER_01 @= net.tcp://127.0.0.1:8082/HOST02
```

Fig. 7 Examples of client-end's parameters

```
SET @@CLIENT_ID @= APP01_TO_HOST01
SET @@CLIENT_URL @= http://127.0.0.1:8080/APP01_TO_HOST01
SET @@HOST_URL @= net.tcp://127.0.0.1:8081/HOST01
CONN
```

4 System Testing

This study assumes windows 7 as the operation system. Before testing, we start to require server and client. The testing flow is as shown in Fig. 4, and the corresponding connection is as shown in Fig. 5. In addition, this study needs to prepare the initial parameter file for the server-end and client-end. At the beginning, this study gives the server name HOST01, URL address and cooperative partner HOST02. The detailed parameter setting for server-end is as shown in Fig. 6. Subsequently, this study sets up the parameter for client-end, and give the client ID, URL address and the target server-end URL, the detailed setting is as shown in Fig. 7.

4.1 Broadcast Examine

We exploit APP01_TO_HOST01 to send the broadcast command, and observe the results. The sequence of the operational process and results is depicted in corresponding figures, as shown in Fig. 5.

4.2 Transaction Examine

In this session, we exploit APP01_TO_HOST01 to send the addition transaction and observe the results. The sequence of the message procedure, as shown in Fig. 5, and returned results are as shown in Figs. 8 and 9.

5 Conclusions and Future Work

This study exploits WCF to build up a prototype of the middleware platform. First of all, this study investigates the role of what a middleware can perform between application components and operation systems. Subsequently, we use rapid

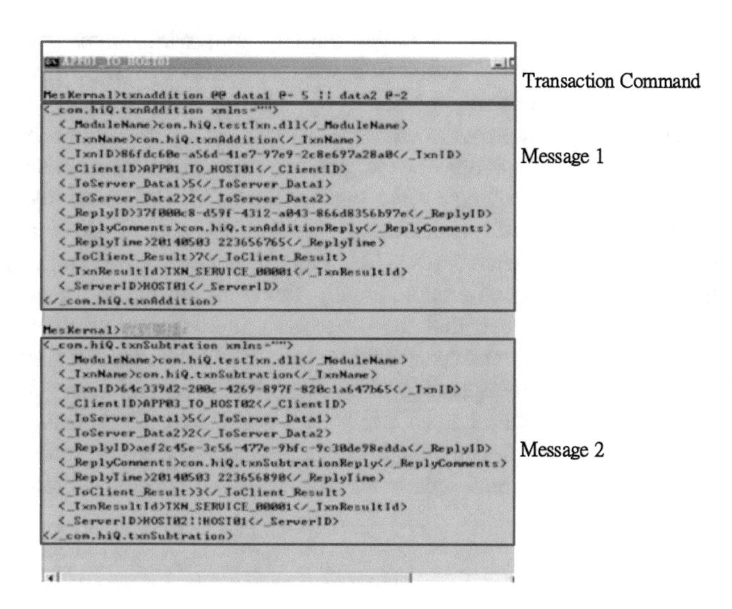

Fig. 8 Transaction requests of the APP01_TO_HOST01

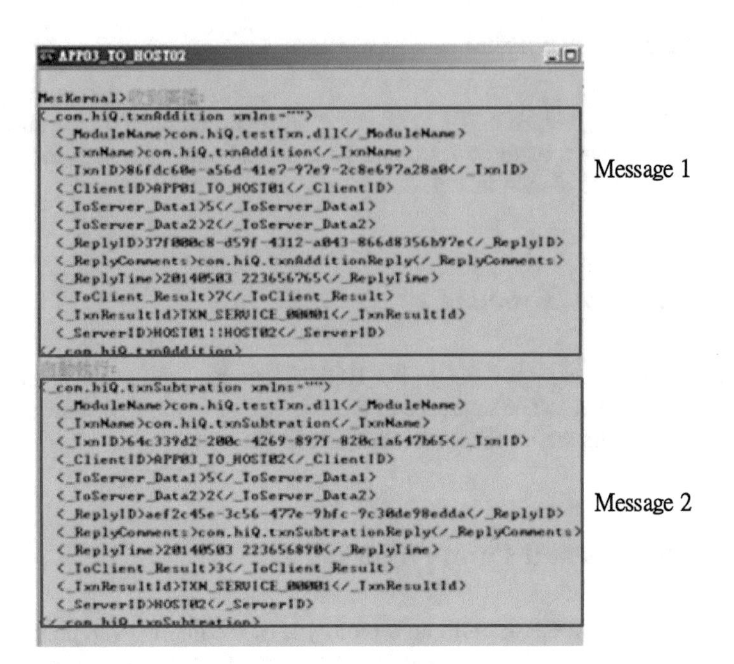

Fig. 9 APP03_TO_HOST02 automatically executes

prototyping method to develop a middleware to deal with some troublesome problems during broadcast and transaction processes. Although WCF gives lower layer communication functions, however we still need to develop additional client-to-server application program interface (API) to satisfy specific needs. In this study, we implement a middleware to achieve the information exchange of broadcasting between two specific functions within WCF. Usages of WCF gives programmers lots of flexibility to develop specific API, and we do not need to consider the underlying element of WCF. Thus developers concentrate on programming the special purpose function. Finally, this study establishes broadcast and transaction examines to prove that the proposed middleware platform is indeed practicable [11–13]. Thus, our experimental results are helpful for integrating with heterologous systems. Although, this study implements several basic functions, including broadcast and transactions. But it isn't considered the performance and security issues on above procedures. In the future work, we will study how to improve the performance and secure the transactions between server-end and client-end.

References

1. Y. Purwati1 and F. S. Utomo, "Windows Communication Foundation for Banyumas Tourism and Culinary Tourism and Culinary Information System", TELKOMNIKA: Telcommunication, computing, Electronics and Control, Vol. 12, pp. 1031–1038, No. 4, Dec. 2014.
2. E. Anupriya and N. Ch. S. N. I Yengar, "Dynamic composition of Peer-to-Peer Web Service using WCF", 3rd IEEE International Conference on Computer Science and Information Technology, pp 570–574, July, 2010.
3. Y. Liu and S. Truong and S. Chen and L. Zhu, "Composing Adaptive Web Services on COTS Middleware", IEEE International Conference on Web Services, pp. 377–384 Sept., 2008.
4. Y. Liu, I. Gorton, and V. K. Lee, "The architecture of event correlation service in adaptive middleware based applications", Journal of Software and System, 2008.
5. Bjarke N. Rahbek, Christian M. Vraa, and Stefan R. Wagner, "Performance Comparison of Selected Object Oriented Middleware Technologies for the. NET Platform.
6. Yuebin Bai, Haixing Ji, Qingmian Han, Jun Huang, Zhiyuan Zhang, "Towards a service-oriented Middleware Enabling Context Awareness for Smart Environment", Jorunal of Ad Hoc and Ubiquitous Computing, Jorunal of Ad Hoc and Ubiquitous Computing, 2009-Vol. 4, No. 1 pp. 24–35, 2009.
7. Y. Tanaka, T. Enokido and M. Takizawa,"Transactional agents on distributed object systems", International Journal of High Performance Computing and Networking, pp. 148–159, Vol. 6 No. 2, 2009.
8. J. F Myoupo and V. K. Tchendji, "Parallel dynamic programming for solving the optimal search binary tree problem on CGM", International Journal of High Performance Computing and Networking, pp. 269–280, Vol. 7 No. 4, 2014.
9. D. Chen, X. Zhu, W. Dai and R. Zhang, "Socially aware mobile application integrations in heterogeneous environments", International Journal of High Performance Computing and Networking, pp. 61–70, Vol. 8 No. 1, 2015.
10. C. Li and G. Hains, "SGL: towards a bridging model for heterogeneous hierarchical platforms", International Journal of High Performance Computing and Networking, pp. 139–151, Vol. 7, No. 2, 2012.
11. Hua-Yi Lin, Meng-Yen Hsieh and Kuan-Ching Li, "Flexible Group Key Management and Secure Data Transmission in Mobile Network Communications using Elliptic Curve Diffie-Hellman", International Journal of Computational Science and Engineering, Vol. 11, No. 1, 2016.
12. Hua-Yi Lin, Meng-Yen Hsieh and Kuan-Ching Li, "Secured Map Reduce Computing Based on Virtual Machine Using Threshold Secret Sharing and Group Signature Mechanisms in Cloud Computing", Telecommunication Systems, Vol. 50, No. 146, 2015.
13. Hsieh, Meng-Yen; Ding, Jen-Wen, "Dynamic scheduling with energy-efficient transmissions in hierarchical wireless sensor networks", Telecommunication Systems, First online: 18 December 2014, DOI:10.1007/s11235-014-9924-1, Vol. 60, No. 1, p. 95, Sep., 2015.

A Comparative Study on Machine Classification Model in Lung Cancer Cases Analysis

Jing Li, Zhisheng Zhao, Yang Liu, Zhiwei Cheng and Xiaozheng Wang

Abstract Due to the differences of machine classification models in the application of medical data, this paper selected different classification methods to study lung cancer data collected from HIS system with experimental analysis, applying the R language on decision tree algorithm, Bagging algorithm, Adaboost algorithm, SVM, KNN and neural network algorithm for lung cancer data analysis, in order to explore the advantages and disadvantages of each machine classification algorithm. The results confirmed that in lung cancer data research, Adaboost algorithm and neural network algorithm have relatively high accuracy, with a good diagnostic performance.

Keywords Machine classification model · Cross validation · Adaboost algorithm · Neural network

1 Introduction

Today, data mining technology in medical field is widely used, achieving gratifying results in the diagnosis and treatment of disease, genetic research, image analysis, and drug development and other aspects like those. With the promotion of EMR, HIS system, medical data is getting reliable and stable, which is a very favorable condition for maintenance of mining results and improvement of tap patterns quality. It is highly valuable of using various data mining techniques on HIS system data set to understand law of mutual relations and development in the various diseases, and summarizing the various treatment regimens for disease diagnosis, treatment and medical research.

Classification means classification according to the type, class or nature; its most important feature is the category known in advance. The application of different

---

J. Li · Z. Zhao (✉) · Y. Liu · Z. Cheng · X. Wang
School of Information Science and Engineering, Hebei North University, Zhangjiakou, China
e-mail: zhaozhisheng_cn@sina.com

© Springer Nature Singapore Pte Ltd. 2018

N.Y. Yen and J.C. Hung (eds.), *Frontier Computing*, Lecture Notes in Electrical Engineering 422, DOI 10.1007/978-981-10-3187-8_34

classification algorithms may vary in results, so different machine classification models of medical classification of different diseases has different application scenarios. For different diseases, choice of what type of model relies on comparative study selectivity and improvement the efficiency of each algorithm in different medical situations.

This paper selected different classification methods to study lung cancer data collected from HIS system, and applied the R language on decision tree algorithm, Bagging algorithm, Adaboost algorithm, SVM, KNN and neural network algorithm classification algorithms in lung cancer data analysis, in order to explore the advantages and disadvantages of each machine classification algorithm. Adaboost algorithm and neural network algorithm have relative high accuracy, suitable for this type of cancer disease analysis.

2 Comparison of Machine Model Methods

2.1 Adaboost Algorithm

Adaboost is an iteration and classification algorithm, continuous improvement classifier by resampling and weighting. Each iteration amends misclassification in previous classification; usually increase the weight of misclassification observations during sample putting back of those observations, which is equivalent to reducing the weight of correct classification observations, thus forming a new classifier into the next iteration. Each iteration gives an error rate in generated classifier, the end result is generated by weighted error rate votes from each classifiers in various stages [1].

Algorithm steps:

Given $(x1, y_1), \ldots, (xm, ym)$ where $xi \in X, yi \in Y = \{-1, +1\}$
Initialize $D1(i) = 1/m$.
For t = 1,…, T;
Train weak learner using distribution Dt.
Get weak hypothesis: X $\rightarrow \{-1, +1\}$ with error

$$e_t = \Pr_i \sim_{Dt}[ht(xi) \neq y_i] \quad \text{Choose} \quad \alpha_t = \frac{1}{2}\ln(\frac{1-et}{et}).$$

Update:

$$D_{t+1}(x) = \frac{D_t(i)}{Z_t} * \begin{cases} e^{-at} if h_t(x_i) = yi \\ e^{at} if h_t(x_i) \neq yi \end{cases}$$
$$= \frac{D_t(i)\exp(-\alpha_t y_i h_t(x_i))}{Zt}$$

where Z_t is a normalization factor (chosen so that D_{t+1} will be a distribution).

Output the final hypothesis:

$$H_{(x)} = sign(\sum_{t=1}^{T} \alpha_t h_t(x))$$

Now boost algorithm has been greatly developed, occurring many other boost algorithms, such as: Logitboost algorithm, real Adaboost algorithm. This paper studies mainly the implementation process of Adaboost algorithm.

2.2 Artificial Neural Networks

Artificial Neural Network (Artificial Neural Networks) is a natural mimic neural network [2]; it can effectively solve the complex problem consisting of variables related regression and classification, we can also use it to establish a classification model of clinical data sets. Algorithm steps:

(1) A set of connection. Respectively from each output X_i, the weight in each connecting line is w_{ki}. The first subscript refers to the current neurons, the second refers to synapsis inputs pointed by weights.
(2) Adder. Added the input signal X_i corresponding to the synaptic weight w_{ki} multiplied to accumulate. The artificial creates a linear adder.
(3) Activation function f_0. Limited neuron output value y_k amplitude.
(4) An external bias b_k. Deviation may increase or decrease the net input of activation function, depending on the bias is negative or positive (Fig. 1).

2.3 Decision Tree Algorithm

Decision tree is a typical classification algorithms of approximate and discrete function values, In essence, it is the process of using induction algorithm to generate readable rules and decision trees, through a series of rules to classify data. Decision Tree algorithm mainly has classical CART, C4.5 algorithm. This paper use CART algorithm.

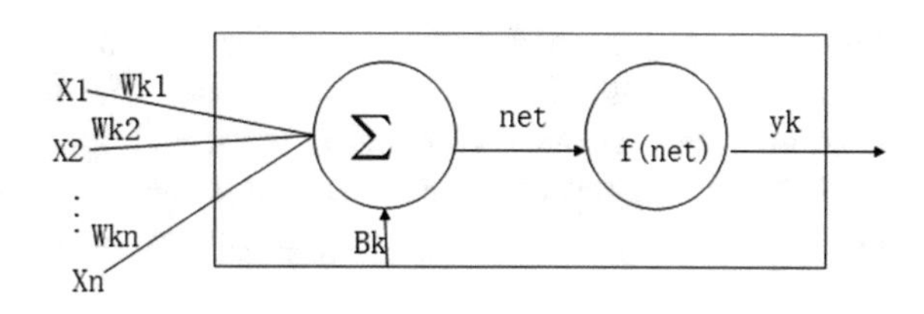

Fig. 1 Neural network algorithm steps

CART (Classification And Regression Tree) algorithm [3] uses a binary recursive partitioning technology, dividing the current sample set into two sub-sample sets, such that each non-leaf node has produced two branches. Therefore, CART algorithm generates simple binary decision tree structure. When the decision tree structuring, many branches reflect anomalies in the data because of the training data noise or acnodes, making decision tree to classify the category unknown data. Classification accuracy is low, CART algorithm often use afterwards pruning.

2.4 Bagging Algorithm

Bagging algorithm [4] is one of the methods used to improve the accuracy of the learning algorithm, it constructed a prediction function series, and then combined them into a predictive function in a certain way, which will improve the weak learning algorithm to an arbitrary precision strong learning algorithm. Bagging utilizes Bootstrap (bootstrap) sampling replacement. It do training sample many times (such as k times) sampling replacement, each time extracts the same observation with the sample ones, so have k different samples. Then, generate a decision tree for each sample. Thus, each tree generate a prediction for each new observation, and classification resulted from these trees majority generates bagging classification.

Bagging algorithm is described in detail:

(1) for t = 1,..., T Do;
(2) Randomly extracted M input from the training set:(x_1, y_1),(x_2, y_2), A,(x_m, y_m);

① got training model h_1;
② replaced into the training sample;

(3) Output: $H(x) = sign(\sum ht(x))$ Breiman noted that stability is the key factor for whether Bagging could improve prediction accuracy rate or not: Bagging can improve the prediction accuracy for unstable learning algorithm, while for stable learning algorithm Bagging cannot obviously improve, even reduced accuracy.

2.5 SVM and KNN

Support vector machine (Support Vector Machine) is a classification algorithm, which is a second-class classification model. The basic model is defined as the maximum interval linear classifier in the feature space, which means that learning methods for supporting vector machine is the maximum interval, and ultimately could be transformed into solving a convex quadratic programming problem.

Table 1 Comparison on advantage and disadvantage of machine classification algorithms

Classification Algorithm	Advantages	Disadvantages
Decision Tree	Able to handle the data type and the conventional type, attributes can analyze and evaluate model by the static test, understandable and interpretable. In a short time can make a well performed and feasible result for large data sources	Ignore correlation between attributes; have difficulty in dealing with missing data; the result in favor of those features have more values
SVM	Outstanding in solving high-dimensional problem and nonlinear problems, avoiding neural network structure selection and local minimum problems. Improving the generalization performance. SVM learning problems can be represented as a convex optimization problem, could utilize known algorithms effectively to find the global minimum of the objective function	Sensitive to missing data; lack of a universal solution for nonlinear problem; Kernel function must be chosen for carefully handle
Adaboost	Could classify different algorithms as weak classifiers, and take advantage of the weak classifier cascade well. With high precision; comparing to the bagging algorithm and Random Forest algorithm, AdaBoost take full account of the weight of each classifier	AdaBoost iterations set difficultly; data imbalance classification causes accuracy decreased; training is time-consuming
Neutral network	Mapping any complex nonlinear relationship, learning rules is simple. Has a strong nonlinear fitting ability, robustness, memory capacity, nonlinear mapping ability and strong self-learning ability	Weak in explanation reasoning process and reasoning ability based on the difference. Cannot work when data is not sufficient. Aspect of numerical problematic values lead to loss of information
KNN	No need to estimate parameters, take training. Simple algorithm, easy to understand and implement; suitable for rare events classification, KNN is better than SVM in multi-classification problems. For overlap or class field for samples have more points to be set, KNN method is more suitable than other methods	Large amount of calculation; Weak output interpretability, cannot give rules like decision tree; category score is not normalized. Smaller sample size class field more prone to misclassification
Bagging	Bagging training set is random, each training set is independent, Each prediction function (weak hypothesis) has no weight, Bagging of prediction functions can be generated in parallel, extremely time-consuming learning methods such as neural networks, Bagging can save a lot of time through a parallel training	Algorithm instability, small changes in the data set can make a significant change in the classification results

Support vector machine exhibited many unique advantages in solving small sample, nonlinear and high dimensional pattern recognition, and can be extended to other machine learning problems such as function fitting. Support vector machine method is based on the VC dimension theory and structural risk minimization principle from statistical learning theory. According t o a complexity of the model of limited sample, SVM finds the best compromise between the, namely learning accuracy and learning ability from specific training samples, in order to obtain the best generalization ability [5].

K nearest neighbor method is a simple and effective method of non-parametric classification algorithm. The basic algorithm is to do category test by voting sample k units nearest sample point distance to determine sample points category. It occupies a significant situation in the classification machine learning algorithms. It is a mature approach theoretically, one of simplest methods based on learning the most basic instance and best text classification algorithm.

Below we list advantages and disadvantages of various machine classification algorithm, as shown in Table 1.

3 Comparative Study and Assessment on Machine Classification Models

3.1 Data Sources

Data set used is coming from artificial lung cancer datasets systematically collected by HIS. Each dataset connection record has 10 properties, 3 discrete attributes and 7 continuous attributes. There have 168 data sets in total, randomly selected 112 as the training data set, and the rest as a test data set. Part of data sample showed below.

3.2 Model Evaluation Method

Error situation in the new data set cannot be reflected by errors on the training set data in model. In order to better estimate model error rate on the new data set, this paper uses a more sophisticated method called cross-validation (cross validation), which is strictly using the training data set to evaluate the accuracy of the model on a new data set. Cross-validation provides us an estimate of the accuracy of the method. It allows us to select the most suitable model scenarios. The most commonly used cross-validation method is holdout method and K-fold cross-validation method. Holdout method is the same training data for both fitting data and accuracy of the assessment, which will lead to excessive optimism. The easiest workaround is to separately preparing the training and testing of two sub-data sets, a training

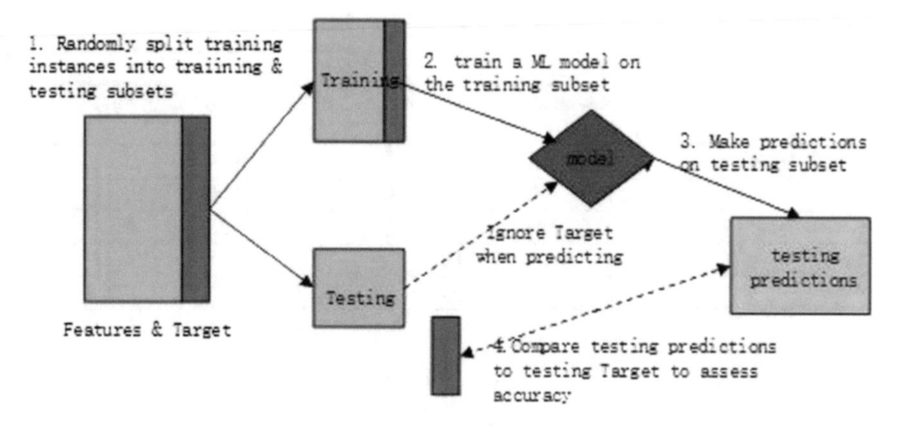

Fig. 2 Flowchart of cross validation

Table 2 Part of data sample

Inst	Time	Status	Age	Sex	ph. ecog	ph. karno	pat. karno	meal.cal	wt. loss
3	455	2	68	1	0	90	90	1225	15
5	210	2	57	1	1	90	60	1150	11
12	1022	1	74	1	1	50	80	513	0
7	310	2	68	2	2	70	60	384	10
11	361	2	71	2	2	60	80	538	1
6	81	2	49	2	0	100	70	1175	−8
…	…	…	…	…	…	…	…	…	…

subset is for fitting the model, another subset of the test used to assess the accuracy of the model. Randomly selected portion of the training data is used only for the training process, usually reserved 30% for test data. Another cross-validation method is K-fold cross-validation. K-fold cross-validation is also dependent on many independent subsets in a number of training data. The main difference is that K-fold cross-validation outset randomly divides the data into the K disjoint subsets. Each time, a set of data used for test, the rest for training model. Experiments uses the Holdout method [6], structure proceeding is showed below (Fig. 2 and Table 2).

4 Experiment and Result Analysis

4.1 Cross Validation of Various Model Algorithms

Now supposed only two classes for classify targets, one is positive and one is negative cases, confusion matrix as shown in Table 3, among them, TP (True

Table 4 Column connection table in Adaboost algorithm

	Predicted classification							
	Training set classification result				Testing set classification result			
Actual classification	0	1	2	3	0	1	2	3
0	15	3	0	0	22	1	0	0
1	10	35	2	0	0	38	0	0
2	0	3	16	0	0	1	20	0
3	0	0	0	0	0	0	0	0

Table 3 Confusion matrix

	Predicted classification			
Actual classification		Yes	No	Total
	Yes	TP	FN	P (actually Yes)
	No	FP	TN	N (actually No)
	Total	P (divided into Yes)	N (divided into No)	P + N

positives) is positive cases and is the number of instances divided into by classifier. FP (False positives) is negative cases and is the instance number divided into positive cases by classifier; FN (False negatives) is positive cases and is the instance number divided into negative cases by classifier; TN (True negatives) is negative cases and is the instance number divided into negative cases by classifier.

- **Adaboost Algorithm**

R languages applied code showed:

```
weka.adab=boosting(ph.ecog~.,data=a[samp,],mfinal=15,control=rpart.control
(maxdepth=5))

weka.pred<-predict.boosting(weka.adab,newdata=a[-samp,])

weka.pred[−1

weka.predt<-predict.boosting(weka.adab,newdata=a[samp,])

weka.predt[−1]
```

As shown in Table 4, this paper uses R language in test set to analyze the factors through cross classification, and the table counting column established on combination of each factor level. In the test, most of the data set appears in the diagonal, but the data belong to the group of 0 class 1 groups are divided into 1 categories; data belongs to the group of 2 class 1 groups are divided into 1 categories.

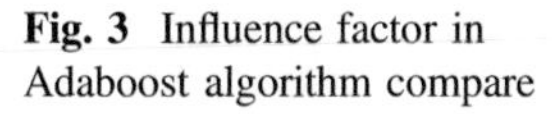

Fig. 3 Influence factor in Adaboost algorithm compare

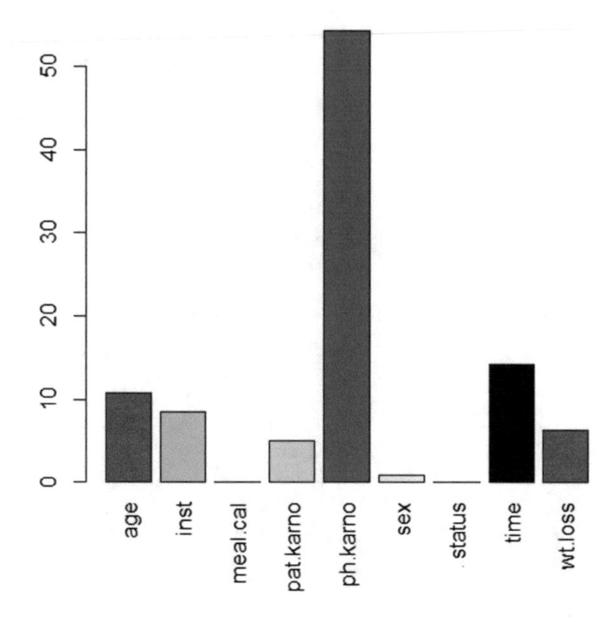

Figure 3 is the compare for influence factors in the AdaBoost algorithm, is similar to bagging algorithm. While ph.karno still occupy a high proportion, age, time, and wt.loss also occupied a certain proportion. But ph.karno occupies a decisive advantage. The accuracy of it plays an important role in AdaBoost algorithm.

- **Decision Tree Algorithm**

Applied R language into code showed in Fig. 4.

Figure 4 shows decision tree draws 83 sets of data. With the condition that the ph.karno is not less than 75, there are 61 sets of data, their ph.ecog survival status value was 0.67 (smaller values indicate better living conditions); With the condition that the ph.karno is not less than 95, there are 15 groups of data accounted for 18%, their ph.ecoh survival status vale was 0.067; With the condition that ph.karno is not less than 85 and Inst is not less than 12, there are 8 groups accounted for 10%, the survival status value average was 0.38; There are 15 sets of data with Inst less than 12, the survival state of the average value of 0.8. Other survival status value was 1.1; With condition that ph.karno is less than 75 and the weight reduction of 7.5, there are 7 groups, survival status value was 1.6; With the condition that decreased in weight is more than 7.5, there are 15 sets of data survival status average value of 2.1. Overall speaking, survival condition is good.

This paper uses R language in test set to analyze the factors through cross classification, and the table counting column established on combination of each factor level. The results were shown in Table 5. If all categories are correct, the

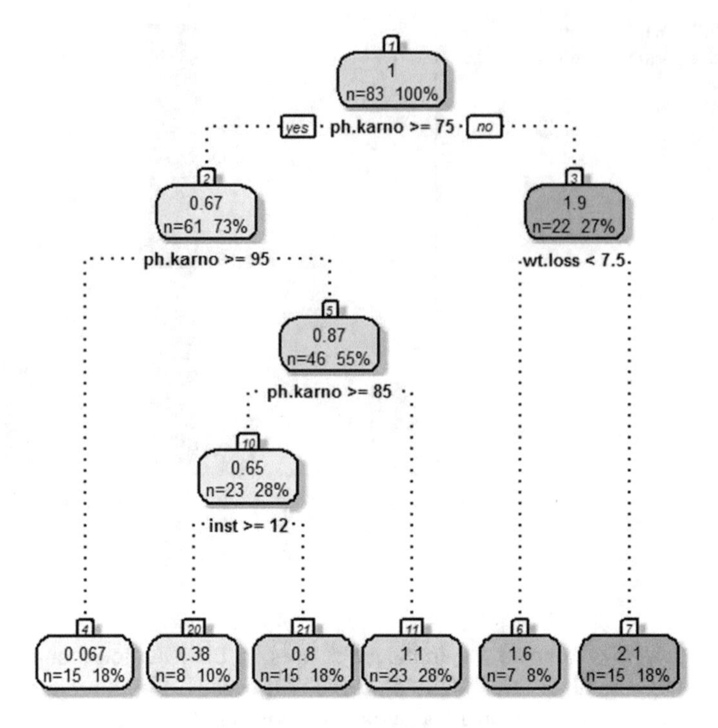

Fig. 4 Applying R language into decision tree

Table 5 Column connection table in decision tree algorithm

	Predicted classification							
	Training set classification result				Training set classification result			
Actual classification	0	1	2	3	0	1	2	3
0	11	2	0	0	14	1	0	0
1	14	34	1	0	8	36	2	0
2	0	5	17	0	0	3	18	1
3	0	0	0	0	0	0	0	0

number on the diagonal is all bigger than 0, and the number other than the diagonal is 0. In the test set, the vast most number appear on the diagonal, but for data set belongs to the class 0, there are 3 group counts into class 1; for data set belongs to the class 1, there are 4 groups counts into class 0 and 3 are counts into class 3; For data set belongs to the class 2, there are 2 groups counts into 1; For data set belongs class 3, there are 1 group was counts into 2.

Table 6 Column connection table in Bagging algorithm

	Predicted classification							
	Training set classification result				Training set classification result			
Actual classification	0	1	2	3	0	1	2	3
0	16	9	0	0	19	3	0	0
1	7	29	5	0	4	33	3	0
2	0	1	17	0	0	2	18	0
3	0	0	0	0	0	0	1	0

- **Bagging Classification**

This paper uses R language in test set to analyze the factors through cross classification, and the table counting column established on combination of each factor level. The results were shown in Table 6. In test set, most of the data set on the diagonal, but data set belongs to 0 class 1 counts into category 1; For data set belong to class 1, there are 8 group counts into category 0 and 2 groups counts into category 2; For data set belong to 2 class, there are 3 groups counts into category 1 and 1 group counts into category 3.

Figure 5 is the influence factors of bagging algorithm that illustrates the influence from ph. Karno algorithm for the bagging is far greater than the influence of other properties.

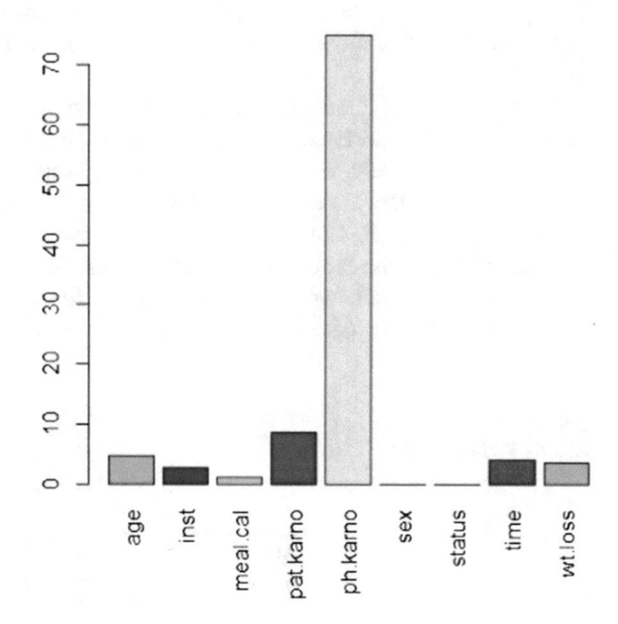

Fig. 5 Influence factor in Bagging algorithm compare

Table 7 Column connection table in neural network algorithm

	Predicted classification							
	Training set classification result				Training set classification result			
Actual classification	0	1	2	3	0	1	2	3
0	12	12	1	0	22	0	0	0
1	8	28	4	1	1	39	0	0
2	1	6	11	0	0	0	20	0
3	0	0	0	0	0	0	0	1

Table 8 Column connection table in KNN algorithm

	Predicted classification							
	Training set classification result				Training set classification result			
Actual classification	0	1	2	3	0	1	2	3
0	11	10	3	0	10	13	2	0
1	1	40	0	1	2	37	2	0
2		8	11	0	0	7	11	0
3	0	0	0	0	0	0	0	0

- **The Neural Network Algorithm, the KNN Algorithm and SVM Algorithm**

This paper using R language applied code implementation of neural network algorithm, the KNN algorithm, SVM as shown in Tables 7, 8 and 9, the code is omitted.

This paper uses R language in test set to analyze the factors through cross classification, and the table counting column established on combination of each factor level. The results were shown in Table 7. In the test set, most of the data set on the diagonal, data set belongs to class 1 have 1 group counts into category 0.

This paper uses R language in test set to analyze the factors through cross classification, and the table counting column established on combination of each factor level. The results were shown in Table 8. In test set, most of the data set on the diagonal, but for data set belongs to class 0, there is 13 group counts into

Table 9 Column connection table in SVM algorithm

	Predicted classification							
	Training set classification result				Training set classification result			
Actual classification	0	1	2	3	0	1	2	3
0	18	1	0	0	9	1	0	0
1	4	38	1	0	15	35	1	0
2	0	1	19	1	1	5	17	0
3	0	0	0	0	0	0	0	0

category 1 while 2 groups into 2; For data set belong to class 1, there are 2 groups counts into 0 and 2 group counts into 2; For data set belong to class 2, there is 7 group counts into 1.

This paper uses R language in test set to analyze the factors through cross classification, and the table counting column established on combination of each factor level. The results were shown in Table 9. In test set, most of the data set on the diagonal, but for data set belongs to class 0, there is 1 group counts into category 1; For data set belong to class 1, there are 15 groups counts into 0 and 1 group counts into 3; For data set belong to class 2, there is 1 group counts into 0 while 5 into 1.

4.2 Comparison of Each Algorithm's Predicted Rates

Through the confusion matrix to calculate following evaluation index, got six kinds of algorithms of error rate, error rate and correct rate are opposite concepts, describing the error percentage by classifier. For example, right and wrong are mutually exclusive events, and lower error rates, classification better.

Table 10 shows the error rate of the 6 algorithms, Adaboost and neural network error rate is 2.4%, decision tree and bagging error rate is below 20%, SVM and KNN's is about 30%. For a more intuitive performance, results as shown in Fig. 6:

Table 10 Error rate of six algorithm

Algorithm	Error rate
Adaboost	0.024
ANN	0.024
Decision tree	0.158
Bagging	0.18
SVM	0.273
KNN	0.309

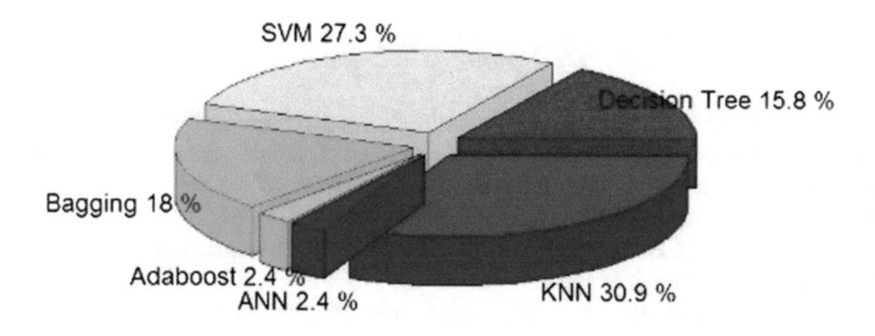

Fig. 6 Error rate of six algorithm

According to the forecast results of the experimental data found: Adaboost algorithm and the neural network accuracy as high as 97.5%, above, and the decision tree and Bagging algorithm accuracy above 80%, and K is the most adjacent and SVM, the worst performance accuracy is about 70%.

For classification model, we introduced several model performance index used in the above process step 3. These techniques include simple calculation of accuracy rate, and the confusion matrix.

5 Conclusion and Future Work

This paper illustrates research and comparison on machine classification models in lung cancer data analysis, and finds out Adaboost and neural network prediction rate is above 95%, decision tree and Bagging algorithm prediction rate is above 80%, KNN and SVM has the worst performance with prediction rate at about 70%. Ph. Ecog occupies a decisive influence factor, to some extent, according to ph. Ecog score, could do disease diagnosis for patients, and the result has some reference value on clinical diagnosis. On the other hand, the process of selecting the classification model should be based on the real data set, trying to use different classification methods, while using cross validation method to test the model, and to choose the classification model with high accuracy, in order to realize the highest value of the data. Different machine classification method has different application scenarios in different diseases. Choosing different types according to different disease model under different situation, can improve the efficiency of medical data application, so as to improve the disease diagnosis accuracy.

Acknowledgements 1. Major Scientific Research Project in Higher School in Hebei Province (Grant No. ZD201310.-85), 2. Funding Project of Science and Technology Research and Development in Hebei North University (Grant No. ZD201301), 3. Major Projects in Hebei Food and Drug Administration (Grant No. ZD2015017), 4. Major Funding Project in Hebei Health Department (Grant No. ZL20140127), 5. Youth Funding Science and Technology Projects in Hebei Higher School (Grant No. QN2016192), with Hebei Province Population Health Information Engineering Technology Research Center.

References

1. Freund Y, SehaPireRE, ADecision-Theoretie Generalization of Online Leaming and an Application to Boosting [J]. Journal of ComPuter and System Seiences, 2010, 55(l):119–139
2. Dong Jun, HU Shang-xu. The progress and prospects of neural network research [J]. Information and Control, 20057, 26(5):360–368.
3. Breiman L, Friedman J H, Olshen R A et al. Classification and regression trees [C]. California: Wadsworth, 1999
4. Peter B, Yu Bin. Analysing Bagging[J]. Annals of Statistics, 2002, 30(4):927–961.

5. Zhang Xue-gong. Statistical learning theory and support vector machines [J]. Acta Automatica Sinica, 2013, 26(1):32–41.
6. Robert E. Schapier, Yoram Singer. Improved Boosting Algorithms Using Confidence-rated Predictions, Machine Learning, 1999, 37(3):297–336.
7. Guoping Zhang, Cihua Liu, Xuesi Ma. Bayes sequential estimation of Lognormal Population distribution parameters [J]. Statistics and Decision.2006(11). p 7–p8
8. Wangyong Lv, Yaoguo Wu, Hong Ma. Lognormal distribution parameter estimation based on the EM algorithm [J]. Statistics and Decision.2007(06), p 21–p23
9. Lijun Wang. An approximate method for three-parameter distribution parameter estimation log_normal [J]. Statistics and principle. Vol.18, 2.1999(01). p 40–p43 、

Evaluation of Smart City Developmental Level Based on Principal Component Analysis and GA-BP Neural Network

Chao Shi, Mengqiao Han and Yanbing Ju

Abstract Smart city assessment issue is an important component of smart city construction. On one hand, it can help the government to guide and direct the activities of Smart-city Construction, on the other hand, it can reflect and give feedbacks to the audience. In this paper, according to the existing evaluation system of Smart City at home and abroad and the division standard of the latest cities in China, we create a more complete and comprehensive evaluation system. At first, we use the Principal Component Analysis (PCA) to reduce index that is according to design the evaluation index of smart city developmental level. Then, these index after reducing let input BP neural network optimized by Genetic Algorithm to train and simulate, find the error of between the actual output value and expected value reach the expected goal. At last, we use directly BP neural network and compare the errors and find using GA-BP neural network prefer. Thus further proves the scientificity and rationality of the evaluation method.

Keywords Smart city · Developmental level · PCA theory · GA-BP neural network

1 Introduction

Smart city is another flyby of city life quality after the industrialization, electrification and digitization of city development. Smart city is a new idea and new mode to contribute to the urban plan, establishment, management and service intelligence through using cloud computing, big data, Internet of Things, mobile Internet, spatial

C. Shi
School of Management and Economics, Beijing Institute of Technology, Beijing 100081, China

M. Han · Y. Ju (✉)
School of Computer Science and Technology, Beijing Institute of Technology, Beijing 100081, China
e-mail: juyb@bit.edu.cn

© Springer Nature Singapore Pte Ltd. 2018

N.Y. Yen and J.C. Hung (eds.), *Frontier Computing*, Lecture Notes in Electrical Engineering 422, DOI 10.1007/978-981-10-3187-8_35

geographic information integration and other new generation of information technology. Many countries and cities have launched their own smart city projects to resolve urbanization issues and challenges. The USA was one of the first countries to launch a smart city project with a high compliment of smarter planet notions from President Barack Obama [1]. The Digital Agenda initiative of the European Commission promotes smart cities in Europe, and a corresponding Smart Cities and Communities initiative, focusing on energy efficient cities, has also been launched and has achieved a great deal to date [2]. With more than 200 pilot smart cities, China has invested heavily more than 2 trillion RMB in smart cities (both research and implementation) in 2015 alone, to sustain its largely urban population [3]. After these cities became the first batch of pilot cities for national smart city establishment, like Beijing, Shanghai, Guangzhou, Shenzhen, Wuhan, Kunming, Chengdu, Wuxi, Ningbo and Foshan, the smart city establishment has become a kind of urban label like mushrooms after rain [4]. Although the smart city establishment is in full swing, the establishment modes copy each other, the location of quite a few cities is not clear and the infrastructures for establishment are incomplete. Besides, driven by the government performance examination, cities blindly copy and compete with each other. At the same time, people fail to clearly recognize the city elements of the smart city establishment and the existing development potentials of cities, so there are too many challenges for the smart city establishment and too many new urban problems.

The smart city is faced with two challenges mainly when it gets started in China soon. On the one hand, people need to know the real requirements of the smart city quickly and accurately and determine the establishment objective of the smart city. Currently, another question to be explored is to set up a set of assessment index system which is scientific and reasonable and can meet the market requirements as well as the spirit of the time. On the other hand, it becomes anther problem how to assess the quality and level of urbanization scientifically based on the established assessment index system. George Cristian Lazaroiu [5] has come up with the establishment concept model of the smart city establishment plan and assessed the implementation effects of the smart city establishment plan by setting up model assessment model. The comprehensive analysis of the existing research results shows that although the scholars have come up with the smart city assessment index system, there are still few assessment about the smart city and no systematic assessment has been conducted for the smart city development potential.

Therefore, according to the domestic smart city development situation, the paper chooses the smart city assessment index system as the research object. Based on the Chinese newest city division standard, it proposes to use the expert voting stimulating data to build a set of smart city development potential assessment system. Principal Component Analysis is used to process the dimension reduction of the second-class index in the assessment index system. Then GA-BP Neural Network assessment method shall be built to evaluate and predict the development potential of the smart city. The PCA theory is used to simplify the original index system based on the principal components which are formed by computing the secondary indicator contribution rate. At the same time, the newly formed indexes (principal component) shall be independent and be input into BP Neural Network as the variables. The genetic algorithm is adopted to optimize the weight value and threshold value of BP Neural Network. The learning and training needs to be conducted through BP Neural Network. The output results need to be compared with the urban division standard. The error shall be controlled within a reasonable range. We can better understand the foundation of the smart city establishment and master the development potential and establishment capacity of the smart city, provide reference for overcoming the blind establishment of the smart city and the unitary establishment pattern by assessing the development potential of smart city. Finally, it shall give the necessary decision reference and practical guidance for setting up the reasonable and scientific development plan of the smart city.

In this paper, we give a new evaluation of smart city developmental level based on principal component analysis and GA-BP Neural Network. Section 2 presents building smart city developmental level evaluation system. Section 3 expounds in detail the basic theory of PCA and GA-BP Neural Network. Section 4 founds the model based on the new evaluation system of smart city and the skills (PCA for attribute reduction and GA-BP Neural Network for forecasting and evaluating). Finally, Sect. 5 concludes the paper.

2 Structure of Evaluation System for Smart City

It is a key part to evaluate the smart city that we establish the assessment index system of the smart city. It shall directly affect the assessment quality whether the index system is comprehensive and the hierarchical structure is clear or reasonable. In 2009, the regional science center of Vienna University of Technology firstly came up with the six dimensions to show the city intelligence: economic growth, mobility and convenience, comfortable environment, people intelligence, life safety, fair treatment. The establishment of the smart city assessment index system shall fully consider the urban information network infrastructures' development level, comprehensive competitiveness, policies and laws, green and low carbon, humanity culture and science technology and many other elements.

In this paper, we study and analyze the definitions of smart city from the following four perspectives:

Technical infrastructure. Application domain. System integration. Data processing [6]. In order to guide general smart city planning, especially for particular systems or applications, it is necessary to let the designers know clearly in which areas their projects are situated, and what aspects are to be considered and covered. In this paper, we continue to use the four perspectives cited in our smart city definition as a basis for analyzing application domains.

- Government: Improving the internal and external efficiency of the government; enabling citizens and other relevant organizations to access official documents and policies; ensuring that public services work efficiently; monitoring and

Table 1 Classification of smart city application domains

Domain	Sub-domain	Domain	Sub-domain
Government (A) (more efficient)	E-government(A_1) government(A_2) Public service(A_3) Public safety(A_4) City monitoring(A_5) Emergency(A_6) Response(A_7) Transparent (A_8)	Business(C) (more prosperous)	Enterprise (C_1) Management(C_2) Logistics(C_3) Supply chain(C_4) Transaction(C_5) Advertisement(C_6) Innovation(C_7) Entrepreneurship(C_8) Agriculture(C_9)
Citizen(B) (happier)	Public transport(B_1) Smart traffic(B_2) Tourism(B_3) Entertainment(B_4) Healthcare(B_5) Education(B_6) Consumption(B_7) Social cohesion(B_8)	Environment(D) (more sustainable)	Smart grid(D_1) Renewable energy(D_2) Water management(D_3) Waste management(D_4) Pollution control(D_5) Building(D_6) Housing(D_7) Community(D_8) Public space(D_9)

managing public safety; responding quickly and effectively in emergency situations.

- Citizen: Traveling and moving more efficiently; accessing contextualized, precise, real-time information in daily life; high-quality essential public services such as education, healthcare and sport; enriching spare time activities, communicating and sharing more with others.
- Business: Traveling and moving more efficiently; accessing contextualized, precise, real-time information in daily life; high-quality essential public services such as education, healthcare and sport; enriching spare time activities, communicating and sharing more with others.
- Environment: Delivering more sustainable, economic and secure energy and water supplies by taking into account citizens' behavior; using more green or renewable energy; recycling and treating waste efficiently and safety; reducing and preventing pollution in the city; offering mobility, telecommunication, information and all other facilities in different city spaces.

Our classification of smart city application domains in shown in Table 1.

3 PCA and GA-BP Neural Network Model

3.1 Principal Component Analysis

Principal component analysis (PCA) is one of the most popular unsupervised dimensionality reduction methods which tries to find a subspace where the average reconstruction error of the training data is minimized. It is useful in representation of input data in a low dimensional space and it has been successfully applied to face recognition, visual tracking, clustering, and so on. Now it is mostly used as a tool in exploratory data analysis and for making predictive models. PCA involves the calculation of the eigenvalue decomposition of a data covariance matrix or singular value decomposition of a data matrix, usually after mean centering the data for each attribute [7]. The first principal component accounts for as much of the variability in the data as possible, and each succeeding component accounts for as much of the remaining variability as possible. The goal of PCA is to reduce the dimensionality of the data while retaining as much as possible of the variation present in the original data set.

Supposing X is a data matrix with n samples and m variables.

$$X = \begin{bmatrix} x_{11} & x_{12} & \cdots & x_{1m} \\ x_{21} & x_{22} & \cdots & x_{2m} \\ \vdots & \vdots & \ddots & \vdots \\ x_{n1} & x_{n2} & \cdots & x_{nm} \end{bmatrix} \triangleq \begin{bmatrix} x_{(1)}^T \\ x_{(2)}^T \\ \vdots \\ x_{(n)}^T \end{bmatrix} \triangleq [x_1, x_2, \ldots, x_m] \in R^{n\times m}$$

- Original data will be standardized (Z-Score Standardization)

Class and quantity in order to eliminate the impact of different dimension, first of all original data on the standardization of treatment (standardized value of the posttreatment x_{ij}^*

$$x_{ij}^* = \frac{x_{ij} - \overline{x_j}}{S_j}$$

where: $\overline{x_j}$ and S_j, respectively, are the mean and standard deviation of the jth target sample, and $\overline{x_j} = \frac{1}{n}\sum_{i=1}^{n} x_{ij}$, $S_j = \left[\frac{1}{n-1}\sum_{i=1}^{n}(x_{ij} - \overline{x_j})^2\right]^{\frac{1}{2}}$

- Calculation of correlation between the matrix

Based on the standardized data matrix $X^* = (x_{ij}^*)$, calculated the correlation coefficient matrix $R = (r_{ij})_{m\times m}$. Where, r_{ij} are the correlation coefficient between the x_i and x_j target factor.

$$r_{ij} = \frac{1}{n-1}\sum_{k=1}^{n} x_{ki}^* x_{kj}^* = \frac{\sum_{k=1}^{n}(x_{ki} - \overline{x_i})(x_{kj} - \overline{x_j})}{\sqrt{\sum_{k=1}^{n}(x_{ki} - \overline{x_i})^2(x_{kj} - \overline{x_j})^2}}$$

- Solving eigenvalue of the correlation matrix and eigenvectors

Calculating the characteristic equation $|R - \lambda I| = 0$, obtained all of the eigenvalue $\lambda_1 \geq \lambda_2 \geq \cdots \geq \lambda_n$, and the corresponding Tikhonov unit eigenvector $t_j = \left(t_{1j}, t_{2j}, \ldots, t_{mj}\right)$

$$Y_j = \sum_{k=1}^{m} t_{kj} \bullet x_k^*$$

- To determine the number of principle components

Selecting r principal components in the m principal components that have been identified to finally realize the evaluation analysis. In general, the contribution rate of variance m principal components that have been identified to finally realize the evaluation analysis. In general, the contribution rate of variance $e_j = \lambda_j / \sum_{k=1}^{m} \lambda_k$ could explain that principal component Y_j reflects the amount of information size. R is determined by the principle that accumulated contribution value $G(r) = \sum_{k=1}^{r} e_k$ is large enough (typically more than 85%).K is kth measured values of the ith and jth factor, $k = 1, 2, \ldots, r$.

3.2 GA-BP Neural Network

Artificial neural network is a new information processing system which basis on a preliminary understanding of human brain structure, activity system. Now it is widely used in pattern recognition, data prediction, system identification, image processing, speech understanding, and function fitting, and other fields [8]. Using BP algorithm of multilayer network model is called BP neural network. BP artificial neural network is the most widely used neural network, the most studied networks. The BP network is also called an error back-propagation network, which is composed of input, hidden and output layers.

In this paper, the input layer is shown in $X = (\mathbf{x}_1, \mathbf{x}_2, \ldots, \mathbf{x}_m)^T$, the output layer is shown in $Y = (\mathrm{y}_1, \mathrm{y}_2, \ldots, \mathrm{y}_n)^T$, the hidden layer is shown in $O = (\mathrm{o}_1, \mathrm{o}_2, \ldots, \mathrm{o}_l)^T$. The initialized weight $W_{lm}^{(1)}$ represents the weight of connection between the input layers and the hidden layer, and $W_{nl}^{(2)}$ represents the weight of connection between the hidden layer and the output layer. The threshold $\theta^{(1)} = \left(\theta_1^{(1)}, \theta_2^{(1)}, \ldots, \theta_l^{(1)}\right)$ indicates the threshold value in the hidden layers during the jth learning process, while $\theta^{(2)} = \left(\theta_1^{(2)}, \theta_2^{(2)}, \ldots, \theta_n^{(2)}\right)$ means the threshold value in the output layer at the same learning process.

net_j states for the Status of network O_j, and $net_j = \sum_{i=1}^{m} w_{ji}^{(1)} x_i - \theta_j^{(1)}, j = 1, 2, \ldots, l$

$$O_j = f(net_j) = f(\sum_{i=1}^{m} w_{ji}^{(1)} x_i - \theta_j^{(1)})$$

Formula as follows computes its output z_k in output layer as a function f of the sum, viz:

$$net_k = \sum_{j=1}^{l} w_{kj}^{(2)} O_j - \theta_k^{(2)}, k = 1, 2, \ldots, n; z_k = f(net_k) = f(\sum_{j=1}^{l} w_{kj}^{(2)} O_j - \theta_k^{(2)})$$

For the ith learning sample, its square error is $E_i = \frac{1}{2} \sum_{k=1}^{n} (y_k - z_k)^2$.

The average error of the system (m represents numbers of learning sample): $E = \sum_{i=1}^{m} E_i$.

If error cannot satisfy the request, we should adjust the weight $W_{lm}^{(1)}$, $W_{nl}^{(2)}$ and the threshold $\theta^{(1)}$, $\theta^{(2)}$ iteratively until system average error meets the request.

The weight adjusts:

$$w_{ji}^{(1)}(t+1) = w_{ji}^{(1)}(t) + \Delta w_{ji}^{(1)} = w_{ji}^{(1)}(t) + \left(-\eta^{(1)} \cdot \frac{\partial E_i}{\partial w_{ji}^{(1)}}\right)$$

Among which:

$$\frac{\partial E}{\partial w_{ji}^{(1)}} = \frac{\partial E}{\partial z_k} \bullet \frac{\partial z_k}{\partial O_j} \bullet \frac{\partial O_j}{\partial w_{ji}^{(1)}} = -\sum_{k=1}^{n} (y_k - z_k) \cdot f'(net_k) \cdot \sum_{j=1}^{l} w_{kj}^{(2)} \cdot f'(net_j) \cdot \sum_{i=1}^{m} x_i$$

Put adjusted weight $w_{ji}^{(1)}(t+1)$ and $w_{kj}^{(2)}(t+1)$ and adjusted threshold $\theta_j^{(1)}(t+1)$ and $\theta_j^{(2)}(t+1)$ in the network (Fig. 1).

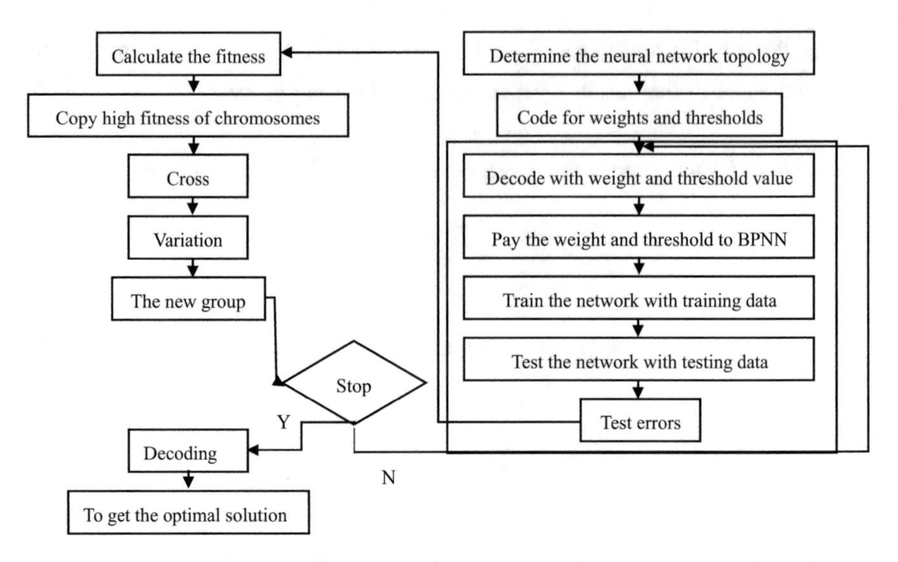

Fig. 1 GA-BP neural network algorithm process

4 Evaluation System Based on PCA-GABP Model in Smart City

4.1 Sampling and Data Collecting

This paper aims for researching on the development level of Chinese smart cities. Based on the division the Chinese cities in 2015, five cities from each division were selected randomly as training and testing samples and the results are listed as follows: first-tier cities-Beijing, Shanghai, Guangzhou, Shenzhen and Tianjin; second-tier developed cities-Hangzhou, Nanjing, Jinan, Chongqing and Qingdao; second-tier medium-developed cities-Chengdu, Wuhan, Shenyang, Xian and Changchun; second-tier less developed cities-Hefei, Nanchang, Nanning, Kunming and Wenzhou; third-tier developed cities-Urumqi, Guiyang, Haikou, Lanzhou and Yinchuan; third-tier medium-developed cities-Shaoxing, Jining, Yancheng, Handan and Linyi; third-tier less developed cities-Liuzhou, Baoji, Zhuhai, Mianyang and Heze; fourth-tier cities-Zhuzhou, Zaozhuang, Xuchang, Tongliao and Xianyang. For the first and second tier cities, with high level of economic development, the urban infrastructure is relatively ample, while when it comes to third and fourth tier cities, the development level and infrastructure is left far behind. However, in order to maintain the diversity and integrity of samples, we still selected samples from all the four tiers. The main data sources are 2014 statistical bulletin of national economic and social development, as well as 2014 statistical monitoring results of scientific and technological progress and statistical bulletin of science and technology, while other data is from the Internet. The symbols $SC_1, SC_2, SC_3, \ldots, SC_{40}$ were used to denote the cities selected.

4.2 Index Reduction

In order to eliminate the dimension in raw data, Min-max normalization theory [9] was applied to convert the data. Dimensionless processing was conducted on the original data. Since each first grade index can reflect one aspect of smart city, principal component analysis was conducted on the secondary indexes derived from each first grade index. Thus, we got the principal components of each first grade index and the cumulative contribution rate of 98%. After analyzing the dataset, for the first grade index Government, the top three characteristic roots are 39.21, 1.21 and 0.50 and the cumulative contribution rate is 98.36%, which is greater than 98%. According to the principle of principal component extraction, the top three principal components were selected as the secondary indexes of Government. And for the first grade index Citizen, the top four characteristic roots are 44.84, 1.05, 0.54 and 0.44 and the cumulative contribution rate is 98.23%, which is also greater than 98%. Thus, the top four principal components were chosen as the secondary indexes of Citizen. For the first grade index Business, the top five characteristic roots are 41.88, 0.82, 0.72, 0.52 and 0.42 and the cumulative contribution rate is 98.90%, much greater than 98%. Thus, the top five principal components were chosen as the secondary indexes of Business. For the first grade index Environment, the top six characteristic roots are 46.01, 0.64, 0.55, 0.48, 0.43 and 0.37 and the cumulative contribution rate is 98.69%, also greater than 98%. Thus, the top six principal components were chosen as the secondary indexes of Environment. Therefore, after implementing the index reduction on the secondary indexes of the four first grade indexes, 18 principal components were obtained, which formed the new evaluation index. The city division, acted as the data label, was used for standardizing the new evaluation index system and the results were shown in the Table 2.

4.3 GA-BP Neural Network Algorithm

Three-layer BP neural network was utilized in this paper. Input vectors have 18 dimensions while output vector has only 1 dimension. As we know that the

Table 2 Part of data dealt with PCA and max-min normalization

Index	A_1	A_2	A_3	B_1	…	D_2	D_3	D_4	D_5
SC_1	0.894	0.958	0.905	0.924	…	0.898	0.551	0.957	0.615
SC_2	1.000	0.995	0.916	0.898	…	0.683	0.769	0.921	0.991
SC_3	0.776	0.882	0.698	0.813	…	0.840	0.862	0.829	0.774
SC_4	0.862	0.729	0.784	0.659	…	0.776	0.708	0.721	0.774
…	…	…	…	…	…	…	…	…	…
SC_{39}	0.109	0.330	0.207	0.216	…	0.243	0.179	0.214	0.222
SC_{40}	0.000	0.055	0.213	0.221	…	0.218	0.312	0.127	0.143

numbers of nodes on hidden layer and input layer have the following relationship: $n_2 = 2 \times n_1 + 1$, Thus, the number of node on hidden layer is 37 and the chosen structure of the neural network is 18-37-1, which means that input layer has 15 nodes, hidden layer has 31 nodes and output layer has 3 nodes. As the total number of weights is 666($18 \times 37 + 37 = 666$), and the total number of thresholds is 38 ($37 + 1 = 38$), we got 704($666 + 38 = 704$) parameters for Genetic algorithm optimization. 32 samples from Table 1 were taken as training data and the rest 8 samples as testing data. The norm of test error can be used for measuring the network generalization ability. When calculating the individual fitness value by error norm, the lower the error norm is, the higher the individual fitness value and the better the data. This paper chose the s-shaped tangent function transig() as the transfer function for hidden layer and s-shaped logarithmic function logsig () as the transfer function for output layer. The parameters for training function trainlm() are net.trainParam.epochs = 1000, net.trainParam.goal = 0.01 and LP.lr = 0.1. Genetic algorithm optimization BP neural network utilizes genetic algorithm to optimize the initial weights and thresholds of BP neural network, which can improve the prediction performance of BP neural network. The major components of genetic algorithm optimization BP neural network are population initialization, fitness function, selection operator, crossover operator and mutation operator. The number of weights and thresholds were shown in the (Table 3).

Assume that both weights and thresholds were 10-digit binary numbers and thus the individual binary code length is 7410. Among them, the first 6660 digits are output layer and hidden layer linking weights coding; 6661–7030 are hidden layer threshold coding; 7031–7400 are hidden layer and output layer linking weights coding; 7401–7410 are output layer threshold coding. Fitness function utilized the sorted fitness assignment function:FitnV = ranking (obj), of which obj is the output of object function. Then, stochastic universal selection (SUS) was adopted after selecting operators. The simplest single-point crossover operator was employed and the parameter setting was listed in the (Table 4).

After running 50 generations of genetic algorithm, the outputs are matrix of weights and thresholds (not listed here since the amount of data is too big). The minimum error is 0.111 and the evolution curve is shown in the Fig. 2.

Table 3 Weights and thresholds for BP neural network

Output layer and hidden layer link weights	Hidden layer threshold	Hidden layer and output layer link weights	Output layer threshold
666	37	37	1

Table 4 Parameter settings for genetic algorithm

Population size	Maximum number of generations	Binary digits of variables	Crossover probability	Mutation probability	Generation gap
40	50	10	0.7	0.01	0.95

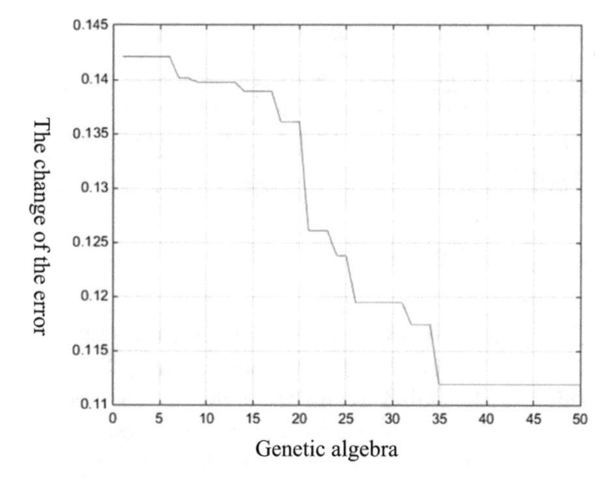

Fig. 2 Evolution curve for genetic algorithm

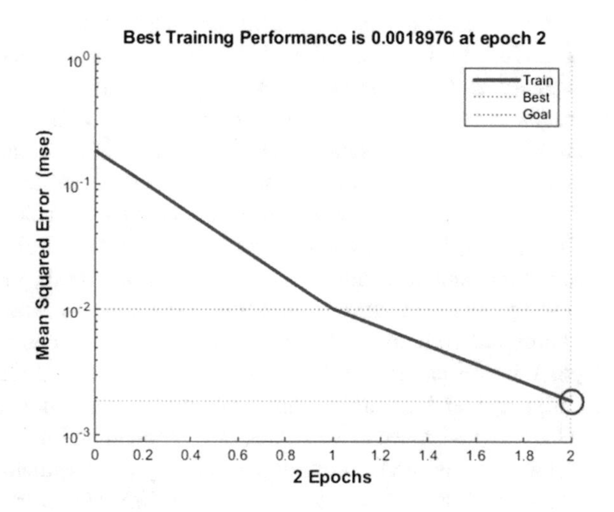

Fig. 3 Best training performance

As such, we got two training error curves, one from random weights and thresholds and the other from optimized weights and thresholds. Then, prediction values, prediction errors and training errors were listed out. From the comparison we can see that, by optimizing initial weights and thresholds, the test error decreased from 0.325 to 0.111 and the training error decreased from 0.448 to 0.238, which suggests that the performance for both training and testing was improved greatly. The actual output was very close to what is expected, and the error between them has met the set objective (Figure 3 and Table 5).

Table 5 Compare the results with BP neural network and GA-BP neural network

Y	1.000	0.857	0.857	0.714	0.571	0.429	0.429	0.286	0.143	0
BP	0.971	0.953	0.889	0.712	0.834	0.375	0.305	0.376	0.189	0.052
GA-BP	0.935	0.851	0.848	0.694	0.616	0.391	0.426	0.287	0.149	0.065
Error for BP:0.325 error for GA-BP:0.112										

5 Conclusion

This paper proposed a smart city development evaluation method based on principal component analysis and the BP neural network optimized by genetic algorithm. First, initial evaluation index system was established by first grade and secondary indexes and then index reduction was conducted for each of the first grade indexes. The obtained 18 principal components acted as input variables to train the model in BP neural network. Thus, we got a PCA-BP neural network smart city development capacity evaluation system, which was then optimized by genetic algorithm on the weights and thresholds to strengthen the algorithm's adaptive capacity and to improve its training performance. Finally, after comparing the results through simulation test, we found that both training error rate and testing error rate were acceptable. As such, this PCA-BP neural network method performs well for evaluating the development capacity of smart city.

According to the evaluation results, cities, such as Beijing, Shanghai and Guangzhou, have relatively complete information infrastructure and high level of social management and public service. Furthermore, their science and technology development and innovation ability rank the top among all the cities in China. Therefore, with the high quality of city construction, these cities have greater potential for smart city development. However, for inland cities such as Lanzhou, Guiyang and Kunming, their smart city construction pace is much slower with relatively low speed of economic development and incomplete urban infrastructure. In this case, in order to improve smart city development ability, it is recommended that cities invoke the potential of innovation and enhance smart city construction quality.

References

1. Bronstein Z. Industry and the smart city. Dissent, 2009, 56: 27–34
2. Digital Agenda Scoreboard 2015: Most targets reached, time has come to lift digital borders. Website of Digital Agenda for Europe. http://ec.europa.eu/digital-agenda/en
3. Liu P, Peng Z. China's smart city pilots: a progress report. Computer, 2014, 47: 72–8
4. Xu Q R, Wu Z Y. The Vision, Architecture and Research Models of Smart City. Journal of Industrial Engineering and Engineering Management, 2012(4): 1–7
5. GEORGE CRISTIAN LAZAROIU A, MARIACRISTINA ROSCIA. Definition methodology for the smart cities model [J]. Energy, 2012(47):326–332

6. Yin C T, Wang J Y. A literature survey on smart cities. Information Science (SCIENCE CHINA), 2015, 100102(18), doi:10.1007/s11432-015-5397-4.
7. Shi H W. Application of Principal Component Analysis to General Contracting Risk Assessment, 2009: 53–56.
8. Yong F R, Guang C X. Study on deformation prediction of landslide based on genetic algorithm and improved BP neural network. 2010, kybernetes, Vol.39 Iss 8 pp. 1245–1254.
9. Arpan Kumar Kar. A hybrid group decision support system for supplier selection using analytic hierarchy process, fuzzy set theory and neural network. Journal of Computational Science 2015 (6):23–33

A Security Reactive Routing Strategy for Ad Hoc Network

Ye Yongfei, Liu Minghe, Sun Xinghua and Zhang Xiao

Abstract Routing protocol algorithm is the basis for establishing a communication link between nodes, and its performance affects the survival of Ad Hoc networks directly. If the malicious node control network communication link, it will launch a variety of attacks, in order to achieve the purpose of stealing network data or destroy the network environment. Considering the weak security of traditional reactive routing protocol in Ad Hoc network, a secure strategy is proposed to resist the attack behavior. The secure reactive routing strategy consists of two parts, the routing request and the routing response. The route establishment based on the successful authenticated for relevant nodes, and then the data is encrypted and transmitted. The algorithm is combined with the identity authentication, hash function, public key encryption and other security strategy, reduce the control probability for the routing information by malicious nodes effectively, and ensure the data communication between the source node and the destination node can be on a safe and reliable route.

Keywords Secure reactive routing strategy · Ad Hoc networks · Identity authentication · Hash chain · Distributed CA

Y. Yongfei · S. Xinghua · Z. Xiao (✉)
School of Information Science and Engineering, Hebei North University, Zhangjiakou, Hebei Province, China
e-mail: 780117251@qq.com

Y. Yongfei
e-mail: yeyongfei005@126.com

L. Minghe
School of Economics and Management, Hebei North University, Zhangjiakou, Hebei Province, China

© Springer Nature Singapore Pte Ltd. 2018

N.Y. Yen and J.C. Hung (eds.), *Frontier Computing*, Lecture Notes in Electrical Engineering 422, DOI 10.1007/978-981-10-3187-8_36

1 Introduction[1]

In Ad Hoc network, each network node plays two roles: the host and the router. When there are routing messages and data packets need to be transmitted, they should cooperate with each other in the network. The frequent moving of mobile terminal equipment forms a dynamic network topology. Because of the network based on poor reliability of wireless communication channels, so the network nodes are easily captured, and the network suffer from serious security threats. Routing between nodes is the basis of data security and reliable transmission and its security is very important to the availability of the network. Therefore, it is the basic guarantee for the secure transmission of data packets in the network to establish the effective routing for the legal nodes.

There are more than ten routing protocol standards for Ad Hoc network have been designed, typical routing protocols are DSDV, AODV and DSR, etc. The common characteristic of these routing protocols is to establish the communication link between nodes to adapt to the dynamic and fast topology changes of Ad Hoc network and improve the speed of routing information forwarding. The route establishment between all nodes is based on a hypothetical: all nodes have a strong trust relationship and can collaborate on data forwarding work perfectly. These routing protocols have provided the opportunity for the invaders and suffer from many kinds of attacks because of they rarely take into account the security problem, such as flooding attack, Sybil attacks, wormhole attack. So far, the improvement research work for Ad Hoc network routing protocols has made great progress [1–3] at home and abroad, and there are some research results on secure routing, such as analysis the existing security issues in Ad Hoc network routing and the quality of service (QoS) and media access control (MAC) protocol is also analyzed. Some researchers have proposed using IPSec (IP Security) to solve the routing security problem in Ad Hoc network, because of the mechanism needs to maintain multiple databases and a large amount of calculation, but the energy and storage are limited in Ad Hoc network, so it has a poor feasibility.

Ad Hoc network is a wireless mobile network with a high degree of autonomy; so traditional network security technologies cannot adapt to the application without any modify. The previous studies have certain aspects of security risks, and cannot provide a safe algorithm to establish routing between nodes. So we need to design security mechanisms to ensure the normal operation of the network.

This paper proposes a secure reactive routing protocol to establish the route between all nodes of the network, aims to resist the related attacks.

---

[1](1) Population Health Information in Hebei Province Engineering Technology Research Center; (2) Medical Information in Hebei Universities Application Technology Research and Development Center.

2 Correlation Algorithm

In the secure reactive routing protocol algorithm, using the thought of Shamir threshold scheme [4] to realize the distributed management of Certificate Authority CA (certificate authority), in order to authenticate the identity of a new node and issue the certification for it [5]. By using asymmetric encryption algorithms to encrypt the routing information to establish the secure routing between nodes and the symmetric encryption algorithm to encrypt the data packages [6].

2.1 Distributed Management of CA (Certificate Authority, CA)

In the public key cryptosystem, the network uses a single authentication management center CA (Authoring Certification) to realize the identity authentication of a new joining node and issue the certificate. Because Ad Hoc network has a poor security, if using the centralized CA to authenticate nodes, once the CA is captured, the entire network system will be into chaos and collapse in the end. The scheme adopts the mechanism of distributed CA to realize the effective authentication of node's identity.

(1) **Shamir Threshold Scheme (t, n)**

In 1979, Shamir first proposed threshold secret sharing scheme. The main idea is: a secret S is divided into n known to share of the secret, and to distribute to n participants. When you want to restore the secret S, you need at least any t secret share portfolio in collaboration to complete the task, and any combination of less than t secret share cannot get any information about S. The numerical t called the threshold in Shamir (t, n) threshold scheme is the key to restore the secret S, and there is a certain requirement for its value.

In 1999, Hass and Zhou for the first time apply the idea of secret sharing be put forward by Shamir to the certified private key management of certification management center CA (Authoring Certification). The algorithm of using threshold mechanism to manage the authenticate private key is: $SK(SK \geq 0)$ is the system private key for node's identity authentication in the network, n is the number of nodes, t is the appointed threshold value, the secret key SK is divided into n secret shares allocated to each node in the network, there any equal to or more than t shares of a secret combination can be restored the system private key SK, while less than t secret shares cooperation cannot be reconstructed. The algorithm implementation mainly through the following two processes to complete:

(2) **Secret Partition System Private Key**

$$y_i \equiv g(x_i) mod\, p', \, i = 1, 2, \, \ldots, n \tag{1}$$

Select a prime $p' > max(SK, n)$, a secret prime a_j $(0 \le a_j \le p' - 1, 0 \le j \le t - 1)$ randomly, and generate a (n − 1) order polynomial g(x). By using formula (1) to calculate the share y_i of the private key *SK* and send (x_i, y_i) to node N_i, at the same time, announced the prime p', and destroy the coefficient a_j.

(3) **Restore System Private key SK**

According to the Lagrange interpolation function, select any *t* different point values (x_i, y_i) and plug into the formula (2) to reconstruct the polynomial, and then use the formula (3) to calculate the private key *SK*:

$$g(x) = \sum_{i=1}^{t} (y_i \prod_{\substack{j=1 \\ j \neq i}}^{t} \frac{x - x_i}{x_i - x_j})(\bmod p') \tag{2}$$

$$SK = g(0) = \sum_{i=1}^{t} (y_i \prod_{\substack{j=1 \\ j \neq i}}^{t} \frac{-x_i}{x_i - x_j})(\bmod p') \tag{3}$$

The calculations show that only at least *t* secret shares holder cooperation to recover the system private key *SK*. When the system private key is used to sign the identity authentication of the node, the system public key is used to verify whether the secret partition is correct or not.

In the network system CA and each node have their own a public/private key pair, used to ensure the security of data information exchange. Each node has its own certificate, mainly by the node identity mark IDi, node public key PKi, request authentication time stamp Ti, for example the certification of node Ni is expressed as: $[ID_i|PK_iT_i]$. After the certificate is generated, the node needs to issue a certificate of identity to the system CA in the network. When node Ni applies for certification services to the network, as long as the threshold t nodes were used to have their own valid secret share of SK to apply for a node certificate valid signatures. Then the algorithm will certificate synthesis, so as to realize the distributed CA signature certification on node Ni certificate.

2.2 *A One-Way Hash Function*

One-way hash function made the input of arbitrary length message on *M* into fixed length output string *h*: $h = H(M)$, and is calculated as a function of the irreversible which is unable to from the output on h get input message on M. The popular hash functions are MD5 (Message Digest Algorithm 5), SHA (Secure Hash Algorithm) and MAC (Message Authentication Code).

The source node that wants to establish route with other node in the network generates the initial value x, and use of one-way hash function MD5 on it to generate the hash value of H_1: $H_1 = H(x)$. Then the next neighbor nodes on the route H_1 using one-way hash function on H_1 to generate the hash value of H_2: $H_2 = H(H_1)$, followed by the same operation until the destination node. Which constitutes a one-way hash chain: $H_1, H_2, H_3, \ldots, H_n$, the number of H_1 functions to perform the hash is the hop number from the source node to reach the destination node in the network.

2.3 Diffie-Hellman Arithmetic

This key exchange algorithm is mainly used to securely exchange symmetric key between two nodes, which is used to ensure the secure transmission of data between nodes in the later stage. If the symmetric key is exchanged between the nodes S and D, the exchange process is as follows:

- Select the global public prime q and an integer original root a of q.
- Node S selects a random number SK_S ($SK_S < q$) as his private key; the corresponding public key PK_S is calculated by: $PKS = r^{SK_S} modq$. The node S stores the private key SKs, and announces his public key PK_S to the network.
- Node D selects a random number SK_D ($SK_D < q$) as his private key; the corresponding public key PK_D is calculated. The node D stores the private key SK_D, and announces his public key PK_D to the network.
- Node S calculates by the following formula: $Key_{SD} = PK_D^{SK_S} modq$, the node D calculates by the following formula: $Key_{SD} = PK_D^{SK_S} modq$. It can be proved that these two values are the same, that is, the symmetric key Key_{SD} is used to ensure the data transmission between nodes S and D safely.

Based on the secret storage of SK_S and SK_D and hard to solve discrete logarithm, it is difficult for an attacker to solve the symmetric key from the public key PKs and PK_D. Because the symmetric key Key_{SD} only belong to node S and node D, so after receiving the encrypted data packets, you can also determine the source of the message, so as to achieve the authentication of the message.

When the identity of the nodes is authenticate by the distributed CA in the network, in order to carry out safe and reliable communication needs to establish the link between nodes. The routing request and routing discovery two processes realize the secure reactive routing algorithm. The following example of establishing the route between the source node S and destination node D to display the work flow of this algorithm. Table 1 is the symbol description used in this algorithm.

Table 1 The used symbol and description

Symbol	Description
S	Source node
D	Destination node
PK_S/SK_S	The public/private key pare of source node S
PK_D/SK_D	The public/private key pare of destination node D
ID_i	The identity of node i
IP_S	The IP address of source node S
t_s	Time-stamping
H()	Hash function
Key_{AB}	The share symmetric key between node A and B
bc_id	Broadcast IP address
de_add	The IP address of destination node
tp_id	The temporary id for sending RREQ package
hop_cnt	Hop count
cip_text	Cipher text with the public key of the destination node
R	Random number
hash_chain	Generated one-way hash chain: $H_1, H_2, H_3, \ldots, H_n$

3 Routing Request Mechanism

Before send message to the destination node *D*, the source node *S* should search the path to there in his own local routing table according to the routing protocol algorithm. If there is no related routing information, the source node *S* will broadcast routing request message of RREQ to establish the route with the destination node hop by hop. Assuming that the path between the source node *S* and destination node *D* is (S, A, B, C, D), the routing request process is performed by the algorithm as follows:

- The source node *S* generates a random number *R*, and then it is connected with the identity of its own identity ID_S and destination node ID_D to calculate the initial value of the one-way hash chain H_1 according to the formula (4):

$$H_1 = H(ID_S|ID_D|R) \tag{4}$$

- The format of a routing request message RREQ is shown as (5). The source node *S* generates his own routing request message RREQ to reach the destination node D as (6).

$$RREQ = \{bc\_id,\ de\_add,\ tp\_id,\ hop\_cnt,\ hash\_chain,\ cip\_text\} \tag{5}$$

$$RREQ = \{H(ID_S|ID_D|R|t_s),\ IP_D,\ id_S,\ R,\ H_1,\ PK_D(ID_S|ID_D|R|t_s)\} \tag{6}$$

Here the {bc_id, de_add} is used to identify a routing request message RREQ uniquely. The random number R, which is generated by the source node, can be used as the hop count, which can hide the true hop count to against targeted attacks.

- The source node *S* attached the request message RREQ with his identity certification signed by the distributed CA to generate a request message $SRREQ_1$ shown as (7), and then broadcast it to all one hop neighbor nodes.

$$SRREQ_1 = RREQ|SK_S(Id_S|PK_S|T_S) \tag{7}$$

Each network node needs to save a pre-processing routing table besides his original formal routing table; pre-processing routing table is mainly used for the storage of incomplete routing information before the complete path from the source node to the destination node be established, and generate a random identity ID_i to index the received routing information.

- When the neighbor node A receives the routing request message SRREQ from the source node *S*, it will verify the authentication certificate by using the public key of the system, which also verifies the validity of the node *S*. If the identity of node *S* is legal, node *A* will give node *S* a route request response message RSRREQ and producing a random id_a to index partial routing information that is stored in the pre processing routing table. The reply message RSRREQ is produced by encrypting the time-stamping t_s with the private key SK_A from node A, here the time-stamping t_s is received from the routing request message SRREQ, as shown in the formula (8). If the identity of the node *S* is not authenticated, the routing request message SRREQ will be discarded.

$$RSRREQ = SK_A[t_S] \tag{8}$$

- When the source node *S* receives the reply message RSRREQ from the node *A*, he will uses the public key PK_A to decrypt the message, and determine the identity of the node *A* by comparing the time stamp t_s. After verification, the source node *S* sends a complete routing request message and data package that is encrypted by Key_{SD} named EnRREQ to the node A, as shown in (9).

$$EnRREQ = RREQ|PK_{CA}[ID_S|K_{CS}|T_S]|PK_S[ID_S|ID_D|Y_S]|Key_{SD}[data] \tag{9}$$

- When receives the message EnRREQ, node *A* updates the content of message RREQ, the value of hop count R will be added 1, and then calculate the H_2: $H_2 = H(H_1)$, and replaced H_1 to H_2. The updated routing request message RREQ, as shown in the formula (10).

$$RREQ = \{H(IF_S|ID_D|R|t_s), IP_D, id_s, R+1, H_2, PK_D(ID_S|ID_D|R|t_s)\} \quad (10)$$

- It is assumed that the node A wants to forward the data packets to the node B, it needs to carry out an authentication operation, the process of identity authentication same as the process of the node S and node A.
- After the completion of the identity authentication, the node A will send the message EnRREQ to its one hop neighbor node B.
- After receiving the routing request message EnRREQ, Node B also need to do similar update operation for message RREQ to the node A. Through mutual authentication with the next hop node C successful, then the message EnRREQ sent to the node C. Followed by execution, until the message EnRREQ is eventually sent to the destination node D.

Because all the intermediate nodes on the path only store partial routing information, it is impossible to produce routing response packet RREP (Route Reply).

4 Routing Response Mechanism

When the destination node D receives the message ERREQ, it only responds to the first request node. The routing response process algorithm works as follows:

- The destination D decrypts the message and obtains the information about the source node *A*, the random number *R* generated by the source node *S* and the time-stamping t_s. According to the value of *R* and *hop_cnt*, the destination node D can calculate the really hop count (hop_cnt-R) form itself to the source node *A*. The destination node D calculate the hash chain by using (11): H_1, H_2, H_3, …, H_N and compare with the received hash chain message. If the two are different, the message must encountered attack in the transmission process, so the node D should refuse to respond; if the two hash chain values are same, the destination node D will perform the next step.

$$H_N = H(H_{N-1}) \quad (11)$$

Note: N = 2, 3, …, hop_cnt-R, $H_1 = H(ID_S|ID_D|R)$.

- Through a variety of authentication, the destination node D decrypt the received message with its own public key PK_D and obtain the data package (data) from the source node A. And then through the following steps to complete the process of establish the path from the source node *S* to the destination node *D*.
- Assuming the node C is the first node to send message EnRREQ to destination node D, then the destination node D will send a response message $SRREP_1$ to node C. The response message $SRREP_1$ is composed of routing response message RREP and encrypted certificate (IdD|PKD|TD) issued by the

distributed CA with its own private key SK_D, as shown in (12), which is then sent to the adjacent node C.

$$RREP = \{idC, IPS, R, hop\_cnt = 0\}$$

$$\mathrm{SRREP}_1 = \mathrm{RREP}|\mathrm{SK_D}(\mathrm{Id_D}|\mathrm{PK_D}|\mathrm{T_D}) \tag{12}$$

Id_C is the temporary identity of node C; IP_S is the IP address of the source node S. R is a random number that is sent from the source node to hide the true number of hops in the routing request message. Hop_cnt is used to record the number of hops from the destination node to the source node, the initial value is 0, and the value is increased by 1 as each new node is sent to a new node.

- When the neighbor node C receives the route reply message SRREP from the destination node D, the system public key PK_{CA}, which verifies the encrypted authentication certificate, verifies the validity of the node D. If the identity of the node D authenticate successfully, node C will send a route response message RSRREP to the node D with his own authentication certificate. If the identity authenticate of the node D is failed, the routing reply message SRREP will be discarded. The reply message RSRREP as shown in the formula (13).

$$RSRREP = SK_C[t_D]|SK_{CA}[ID_C|PK_C|T_C] \tag{13}$$

- Destination node D receives the reply message RSRREP from the node C, with the public key PK_C for its decryption, by comparing the time-stamping t_s to determine the identity of the node C. If the authentication success, the destination node D sends a complete routing request message EnRREP to the node C, as shown in the formula (14). If authentication fails, the node D needs to accept the routing request message EnRREQ from other neighbor nodes.

$$EnRREP = RREP|SK_{CA}[ID_D|PK_D|TD]|PKS[ID_S|ID_D|Y_D] \tag{14}$$

- When receive the routing reply packet EnRREP, node C will update the content of RREP, the value of the hop number will be increased by 1, the temporary identity id_C replaced by the real identity ID_C. The updated routing request message RREP, as shown in the formula (15).

$$RREP = \{H(ID_S|ID_D|R|t_s), IP_D, ID_C, 1, PK_S(ID_S|ID_D|R|t_s)\} \tag{15}$$

When receives route reply message EnRREP, node C also need to update the message RREP similar to what the destination node D has done. With the successful mutual authentication with the neighbor node B, then the routing request packet EnRREP is sent to the node B. Obey the same way, until the routing request message EnRREP sent to the source node S eventually.

The intermediate nodes on the path from the destination node D to the source node S receive route request message EnRREP, integrate information into the complete routing information in the routing table, and record the complete path. The intermediate nodes will calculate the hash value according to the received information and compare with their original values stored in the routing information, then to determine whether there was a tampering attack. Source node S also need to verify the received route reply message, after successful verification, a data transmission rout from the source node S to the destination node D is established and all the revolved nodes record the valid rout information.

5 Conclusion

Routing protocol is the key to build a secure Ad Hoc network environment; it will facilitate the attacker without the safe control scheme. If malicious nodes control the communication path, data transmission is in danger. In this paper, the security reactive routing mechanism is designed for the weak security of Ad Hoc network, which is composed of two parts, the routing request and the routing response. The identity authentication, hash value comparison, public key encryption strategy all strategies resist the control by malicious nodes, and ensure a safe and reliable rout between the source and destination nodes to be established.

References

1. Y. Hu, A. Perrig, and D. Johnson. Ariadne. A Secure On-demand Routing Protocol for Ad Hoc Networks [J]. ACM MOBICOM, 2005, (1): 21–38.
2. Venkatramaman Lakshmi, Arrival Dharma P. Strategies for enhancing routing security in protocols for mobile ad hoc networks [J]. Journal of Parallel and Distributed Computing, 2003, 63(2): 214–227.
3. Zapata M Z, Asokan N. Securing Ad-Hoc Routing Protocols [A]. In: Proc. of the 2002 ACM Workshop on Wireless Security (WiSe2002) [C]. Singapore: IEEE Computer Society Press, 2002. 1–10.
4. Shamir A. How to Share a Secret [J]. Communications of the ACM, 1979, 24(11): 612–613.
5. Zhou L, Haas Z J. Securing Ad Hoc Networks [J]. IEEE Network, 1999, (13): 24–30
6. W Diffie, M Hellman. New directions in cryptography [J]. IEEE Transaction on Information Theory, 1976, 22(6): 644–654.

A Study of Fuzzy Clustering Ensemble Algorithm Focusing on Medical Data Analysis

Zhisheng Zhao, Yang Liu, Jing Li, Jiawei Wang and Xiaozheng Wang

Abstract Unitary clustering algorithm, not well adapted for fuzzy medical data sets, may result in low clustering accuracy and other problems. This paper investigates and compares the effects of various clustering methods to achieve improvements. First, unitary clustering algorithms such as k-means, FANNY, FCM, and etc. are achieved, then FCM algorithm was improved into CFCM algorithm, which increases the accuracy to a certain extent. Second, on this basis, in order to better adapt to the diversity of characteristics of fuzzy medical data, weighted co-association matrix is adopted to achieve integration, and consistency function is designed to present a fuzzy clustering ensemble algorithm. Finally, experiments shows that the Fuzzy Clustering Ensemble Algorithm can solve the problem of low accuracy in unitary clustering algorithm with higher stability, accuracy and robustness.

Keywords Medical data · Fuzzy clustering ensemble algorithm · Fuzzy clustering · Clustering ensemble

1 Introduction

As known, discovering different types of diseases and classifying medical samples accurately are very important for successfully diagnosing and treating disease. For specific medical data, how to choose the appropriate clustering algorithm has always been the focus of the study. However, most of the studies are focusing on a unitary clustering algorithm for clustering medical data, which lacks robustness, stability and accuracy. However medical data has its unique characteristics, such as high-dimensional, vague and diverse. Therefore, it is important to carry out a comparative study on clustering algorithms for specific medical data.

Z. Zhao (✉) · Y. Liu · J. Li · J. Wang · X. Wang
School of Information Science and Engineering, Hebei North University, Zhangjiakou, China
e-mail: zhaozhisheng_cn@sina.com

© Springer Nature Singapore Pte Ltd. 2018

N.Y. Yen and J.C. Hung (eds.), *Frontier Computing*, Lecture Notes in Electrical Engineering 422, DOI 10.1007/978-981-10-3187-8_37

When clustering these data objects, any single type of attribute information is not enough to fully convey the data object. These different types of attribute information can complement each other and describe the entire data objects, thus it's necessary to consider the integration problem among different types of feature during the clustering process. Traditional method weights different types of feature into an attribute vector to form a cluster when calculating the similarity of two data objects, resulting into local optimal solution under normal circumstances instead of the global one, which requires the development of efficient algorithm for data with high computational complexity. And many data samples do not have strict attributes, presenting betweenness in nature and generic terms. Under this circumstance, it is more suitable for soft partition to solve the fuzziness of data. The uncertainty of classification in fuzzy clustering illustrates the betweenness of the sample attributes, objectively reflects the diversity of clustering problems encountered in reality. In this case, targeted at specific data, it is inevitable for the study on fuzzy clustering ensemble algorithm.

This paper studies and compares the effect of various algorithms on different data collection focusing on fuzzy medical data sets. First, unitary clustering algorithms such as k-means, FANNY, FCM, and etc. are achieved, then FCM algorithm was improved into CFCM algorithm for comparison. In order to better adapt to the diversity of characteristics of fuzzy medical data, weighted co-association matrix is adopted to achieve integration, and consistency function is designed to present a fuzzy clustering ensemble algorithm. Experiments shows that the Fuzzy Clustering Ensemble Algorithm can solve the problem of low accuracy in unitary clustering algorithm with higher stability, accuracy and robustness.

2 Data Sets and Data Sources

The medical data set used in this paper comes from studying website UCI, including 5 data sets, Breast-cancer-Wisconsin, dermatology, hepatitis, WDBC, WPBC and other medical data sets (Table 1).

For Breast-cancer-Wisconsin, except ID field, other all fields are clustering data. And for Dermatology data set, except age field, other all are clustering. Hepatitis has 5 fields are clustering, and same clustering fields occurred in WDBC and WPBC.

Table 1 Medical data set in studying website UCI

Data	Data sources	Classes	Samples	Attributes
Breast-cancer-Wisconsin	UCI	2	700	10
Dermatology	UCI	6	358	34
Hepatitis	UCI	2	155	19
WDBC	UCI	2	569	32
WPBC	UCI	2	198	34

3 Unicity Clustering Analysis

3.1 K-means Algorithm

K-means divided n points (may be a single observation or a sample instance) into k clusters, such that every point belongs to the nearest mean, which is the cluster of corresponding cluster center, making it as a standard clustering. In conclusion, it is a problem for data space to be divided into Voronoi cells. The selection for k initial cluster center has a greater influence on the results of the clustering. Algorithm gives each remaining objects in data set during each iteration nearest cluster, according to the distance to each cluster. The landing square error is minimized for the algorithm K division. Results perform better when the cluster is dense, and the difference between classes is obviously. For large data sets, this algorithm is relatively scalable and efficient. K-means algorithm is suitable for spherical data sets, but performs poor for any shape data clustering results.

Input: $k, data[n]$;

(1) Select the k initial centers, such as; $c[0] = data[0], \cdots, c[k-1] = data[k-1]$
(2) For the $data[0], \cdots, data[n]$ respectively compare with $c[0] \cdots c[k-1]$, and assume the difference is minimum, marked as i;
(3) For all points marked, recalculate $c[i] = \{$ all sum of$\}$/from the $data[j]$ marked i, and the number of i marked;
(4) Repeat (2) (3) until all changes in the value of $c[i]$ is less than a given threshold.

3.2 FCM Fuzzy Clustering Algorithm

Because this paper selects most medical data are clustering data, so this paper applies Fuzzy C-means clustering algorithm for experiment. Fuzzy clustering analysis is one of main technologies for machine study without supervision, which bases on characteristics, closeness level, and similarity, clustering objective objects through establishing clustering similarity relationships. For now, the Fuzzy C-means algorithm is the most popular among fuzzy clustering algorithms. FCM applies the minimum weighted average standard deviation, divides data into different sub categories according to different attribute level, which make subordinating degree stands for the closeness level from each data to each clustering center. The purpose for the FCM is to find the subordinating degree value and clustering center optimizing the targeted function. Now, most fuzzy clustering algorithm are based on FCM. FCM algorithm is sensitive for isolated point in data, so it can not be applied for situation that have more isolated points.

Assumed that sample set $X = \{x_1, x_2, \ldots x_n\}$, divides into fuzzy collections c, and works out the clustering center $J_c(j = 1, 2, \ldots c)$ in each collection, making target function minimum. Definition showed below:

$$J_c = \sum_{j=1}^{c} \sum_{i=1}^{n} \mu_{ij}^{\alpha} \|x_i - c_j\|^2, \; 1 \le \alpha \le \infty$$

while satisfied:

$$\sum_{j=1}^{c} \mu_{ij} = 1, \forall i = 1, 2, \ldots, n$$

$\mu_{ij} \in [0, 1]$ stands for: the subordinating degree for the number i data point closing to the clustering center j; c_j is the fuzzy clustering center j, initial selects in training sample set randomly; α is the fuzzy degree, with the amplifying for α, the clustering fuzzy level is amplifying. There, $\alpha = 2$. Fuzzy clustering is achieved through optimization targeted function iteration J_c, which is a process for optimizing. And, the subordinating fuzzy degree μ_{ij} and clustering center c_j severally obtained using the below formulas:

$$\mu_{ij} = \frac{1}{\sum_{j=1}^{c} \left(\frac{\|x_i - c_i\|^2}{\|x_i - c_j\|^2} \right)^{2(\alpha - 1)}}$$

$$c_j = \frac{\sum_{i=1}^{n} \mu_{ij}^{\alpha} x_i}{\sum_{i=1}^{n} \mu_{ij}^{\alpha}}$$

This process gets convergence at partial minimum point J_c. From an random clustering center, through searching targeted function minimum point, adjusts continuously clustering center and each subordinating fuzzy degree in sample, to confirm sample categories.

3.3 Modified FCM Algorithm and CFCM Algorithm Design and Application

Because the relatively high level of depending degree of FCM algorithm to initial clustering center, isolated points do not obviously show up during the solving process for the targeted function, and easier to become the partial extremely points during the clustering process, causing that optimizing result cannot be achieved and other shortages. This paper improves the FCM algorithm based on the FCM, presents a new algorithm named CFCM (Chance Fuzzy C-means).

Detailed speaking, improve the subordinating degree and weight methods to mend the FCM algorithm, we make all samples have the total subordinating degree is n, which is:

$$\sum_{j=1}^{c}\sum_{i=1}^{n}\mu_{ij}=n,\forall i=1,2,3\ldots,n$$

Under this new circumstance, subordinating degree function is

$$\mu_{ij}=\frac{n\left(\frac{1}{\left\|x_i-c_j\right\|^2}\right)^{\frac{1}{\alpha-1}}}{\sum_{i=1}^{c}\sum_{j=1}^{n}\left(\frac{1}{\left\|x_i-c_j\right\|}\right)^{\frac{1}{\alpha-1}}},i=1,2,3,\ \ldots,n,j=1,2,3,\ \ldots,c$$

In the data base included data objects n, each data objects works differently from others for information obtaining. In other words, sample points is working differently to data set classification, some sample points means more significant for classification. FCM cannot reflect the different signification level for sample points such in noise sample set. In order to distinguish the difference, assigned value ω_i for each sample points, which means that the higher weighted level stands for the higher level for similarity. For intensive data points, they have a close distance to center and the weighted level is close too. Noise and isolated points has lower weighted level, so to eliminate the influence of them. CFCM improves the accuracy, so the new targeted function is defined below:

$$J_c=\sum_{j=1}^{c}\sum_{i=1}^{n}\omega_i\mu_{ij}^{\alpha}\left\|x_i-c_j\right\|^2,1\leq\alpha\leq\infty$$

The clustering center formula is:

$$c_j=\frac{\sum_{i=1}^{n}\omega_i\mu_{ij}^{\alpha}x_i}{\sum_{i=1}^{n}\omega_i\mu_{ij}^{\alpha}}$$

3.4 Fuzzy Clustering FANNY Algorithm

FANNY method is a new fuzzy clustering algorithm based on the FCM, presented by Kaufman and other scientists. It target is to minimize the targeted function, the targeted function is showed below:

$$J_c = \min \sum_{j=1}^{n} \frac{\sum_{i=1}^{n} \sum_{j=1}^{n} u_{ij}^2 \|x_i - x_j\|^2}{2 \sum_{j=1}^{n} u_{ij}^2}$$

while satisfied:

$$\sum_{j=1}^{n} u_{ij} = 1,\ i = 1, 2, 3, \ldots, n$$

From the calculation could get the subordinating degree and clustering centers in sample, comparing to general FCM algorithm, FANNY algorithm has lower sensitive degree for wrong and abnormal data, and preforms better in non spherical cluster.

3.5 Comparison of Each Algorithm

In order to compare the effects of K-means, FCM, FANNY, CFCM and other fuzzy clustering algorithm on data clustering. Selecting Breast-cancer-Wisconsin data set to apply using R language, as showed in Figs. 1, 2, 3, and 4.

Through the compare result of K-means and FANNY algorithms, finding that FANNY preforms better in clustering this data set than K-means.

Through the compare on scatter diagram matrix showed above, we could get: K-means performs poor in clustering on Breast-cancer-Wisconsin data set, while the FCM and CFCM algorithms performs better than K-means, still could not get a good result on clustering. FANNY is the best one among these algorithms, which illustrated the obvious boundary from the scatter diagram matrix, and the clustering figure using R language to apply FANNY showed in Fig. 5.

From the Fig. 5, the abnormal values is existed in data, and FANNY cannot well deal with this abnormal situation, although it has a relatively high classification effect, FANNY divided the abnormal one into one category. Scatter diagram matrix cannot well illustrates the clustering results for each algorithm, means that it is hard to get more information in scatter diagram matrix. So calculating the different clustering accuracy as Fig. 6.

From Fig. 6, the clustering accuracy could be showed well in different algorithms, K-means algorithm performs poor in fuzzy medical data, and the traditional clustering algorithm is not suitable to be applied on the fuzzy data, and for the frequently used algorithm FCM, although performs better than K-means, still only have 60% accuracy, and the CFCM algorithm could not improve the clustering accuracy. Compare to these statics, the FANNY algorithm has an accuracy of 98%, which illustrates the application effect is better than other algorithms.

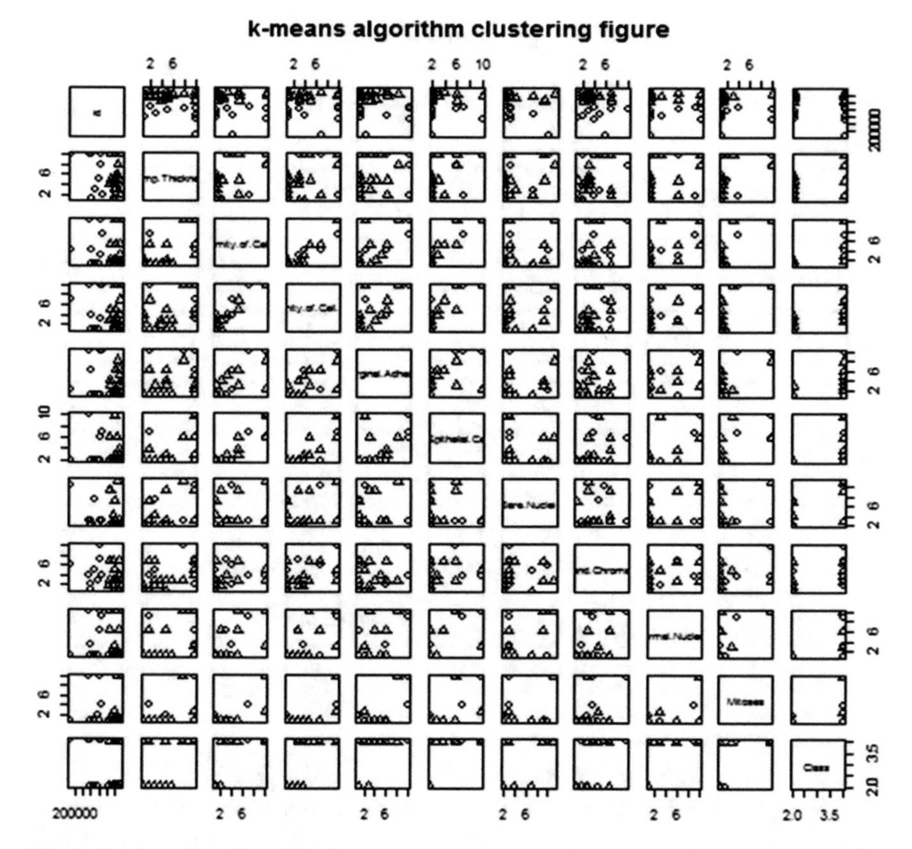

Fig. 1 K-means algorithm clustering

FCM algorithm could not produce the best possible results in Breast-cancer-Wisconsin data set, to prove the FCM power in fuzzy clustering, this paper selects the other data set to apply respectively different clustering algorithms, and draw the accuracy of clustering effects of different algorithms on difference data set:

From the 5 data set, apply the different fuzzy clustering algorithms. Traditional K-means algorithm preforms poor on clustering accuracy from fuzzy data; FANNY performs moderately on classification in partial data, but the error rate is relatively high on more fields data. The FCM performs well on most data, the clustering effect is not well but compare with K-means and FANNY, it is better. For CFCM, the clustering effect on data is best, which means FCM has some shortages, and through our improvement, reduced the error rate, improved the recognition and judgement on abnormal point, and improves the clustering accuracy (Fig. 7).

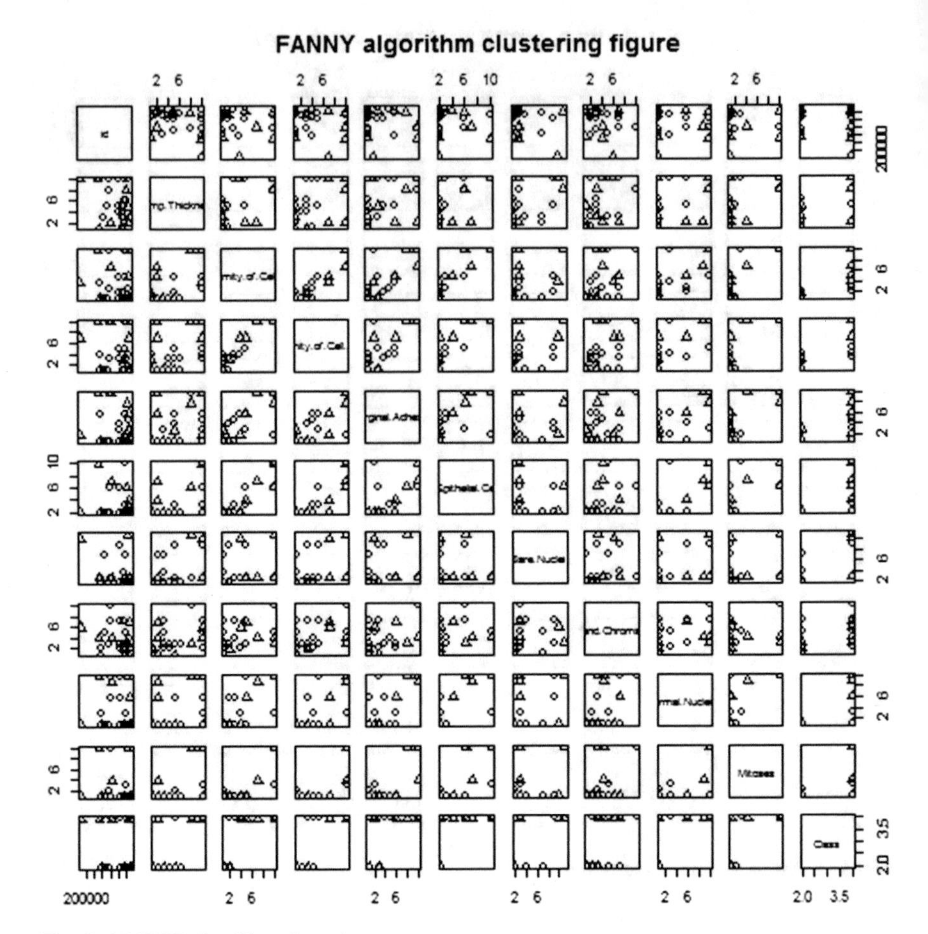

Fig. 2 FANNY algorithm clustering

4 The Achievement of Fuzzy Clustering Ensemble Algorithm

From the above result, as far as achieving the best clustering result of CFCM in unicity clustering algorithms, while when cluster these data objects, the unicity type of attribute information is not sufficient enough to describe data objects. Also, these different types of attribute information cannot be mixed to describe the whole fuzzy medical data objects together. So, when considering the mix problem of different types characteristics, this paper design and complete the Fuzzy Clustering Ensemble Algorithm. The application effect of unicity clustering analysis on different data sets is depending on the data itself, and distribution, dealt situation, and lack situation of data and other factors are all influent on the unicity clustering analysis. It is such a difficulty to select clustering analysis method for data itself,

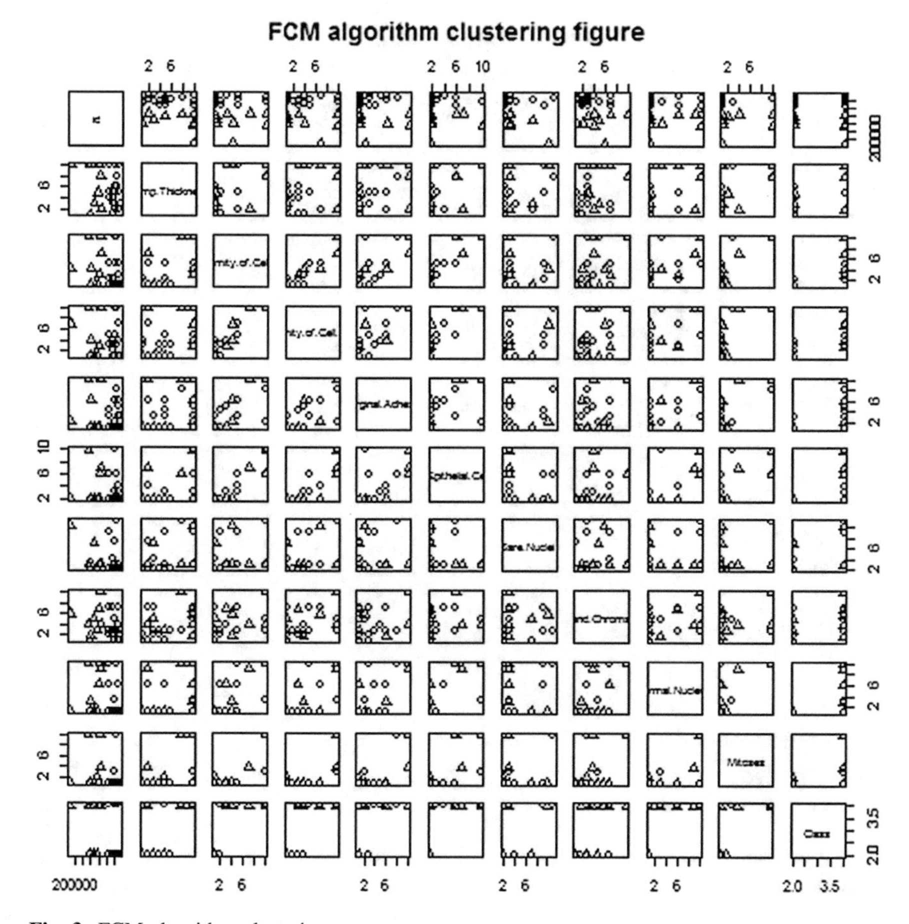

Fig. 3 FCM algorithm clustering

based on the fact that different unicity clustering methods applied in different situation, could result in different clustering outcomes. clustering ensemble could fix this problem to some extent, and it has advantages as below:

1. Clustering Ensemble Algorithm could adapt to the data set, most of data set could be dealt; reflecting the synthesization characteristics, excellent average function, and improvement of robustness in clustering;
2. Clustering Ensemble Algorithm is affected less from missing value, abnormal value, and noise, the accuracy for the clustering is high, robustness is strong;
3. Clustering Ensemble Algorithm could ensemble all clustering algorithms to deal with differently distributed and structured data set, and is suitable for different situations.

This paper uses the clustering ensemble method to experiment on the data set based on total co-relation matrix, the methods is showed in Fig. 8. First, do several

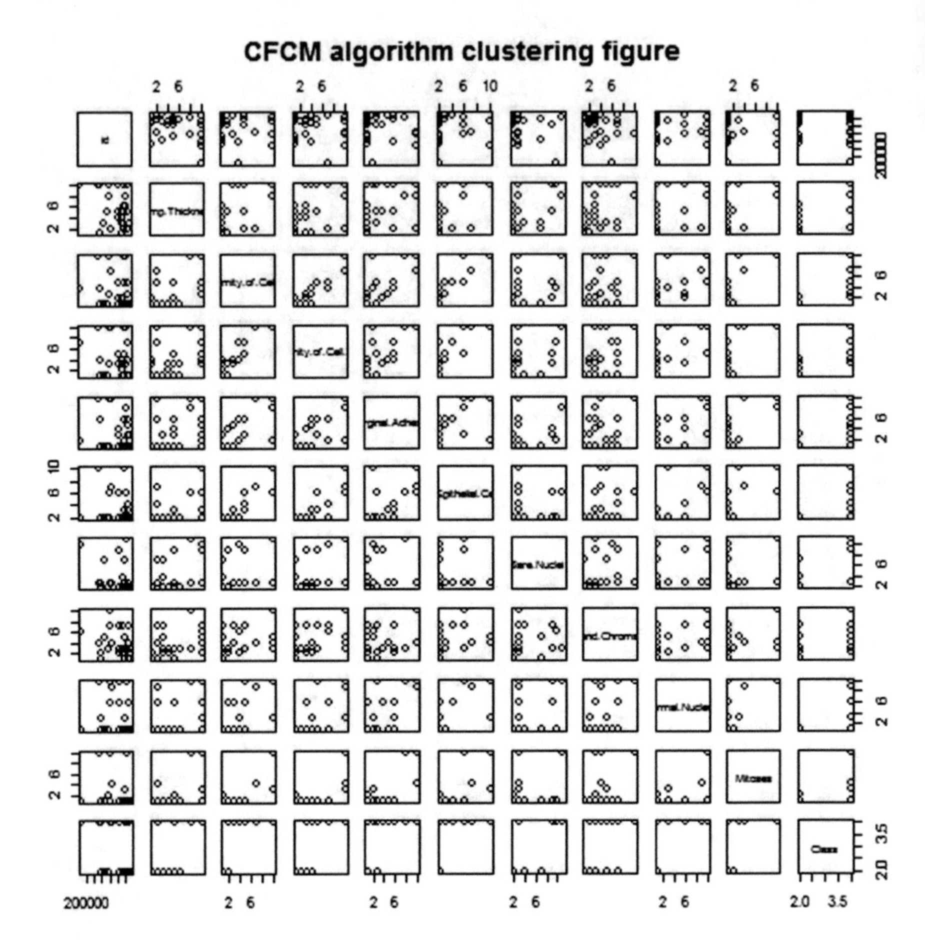

Fig. 4 CFCM algorithm clustering

times of the unicity clustering analysis on initial data set, to generate many clustering subset, then use the consistency function to mix, finally to produce new clustering result with clustering analysis algorithm.

Here is the process:

1. Clustering subset

This paper use CFCM to cluster many times to generate clustering subsets, using different initial k value for N number, then respectively cluster N number to get different cluster subsets.

$$N = \{\pi_i, \pi_i, \pi_i, \ldots, \pi_i\}, \pi_i = \{C_1, C_2, C_3, \ldots, C_i\}$$

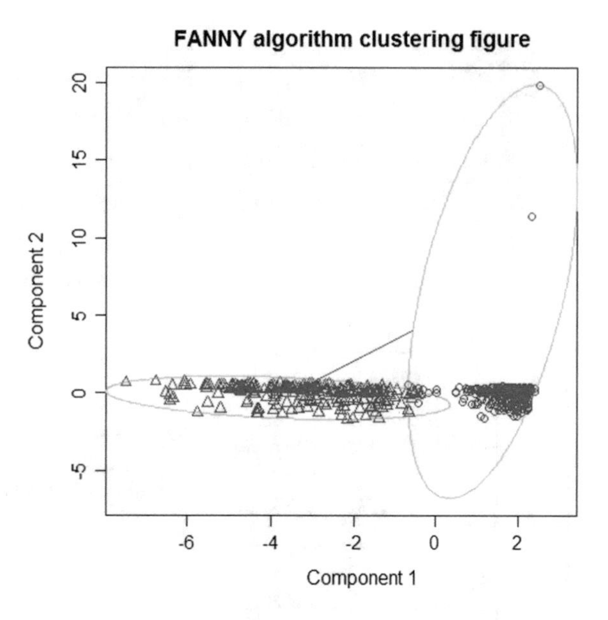

Fig. 5 FANNY algorithm clustering figure

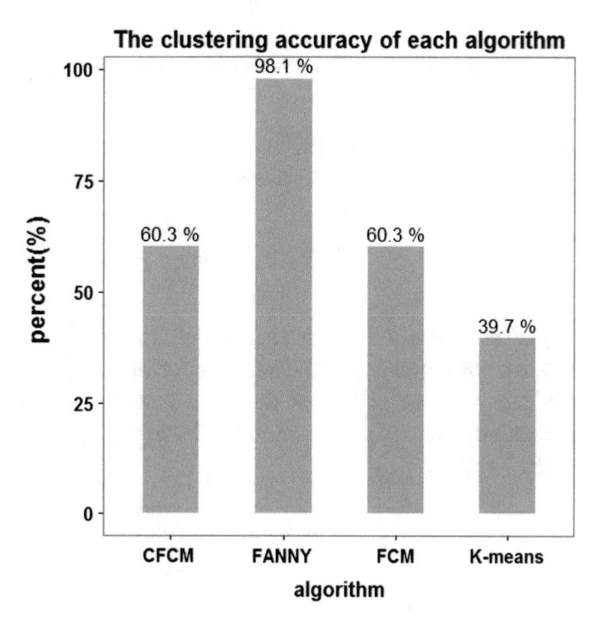

Fig. 6 Compare of the clustering accuracy of each algorithm

2. Consistency function design

Based on the total co-relation matrix (Total co-relation matrix, TCM) to design clustering ensemble algorithm, then recognize the subsets as a data set, then calculate the similarity between each objects in this data set, and to utilize these

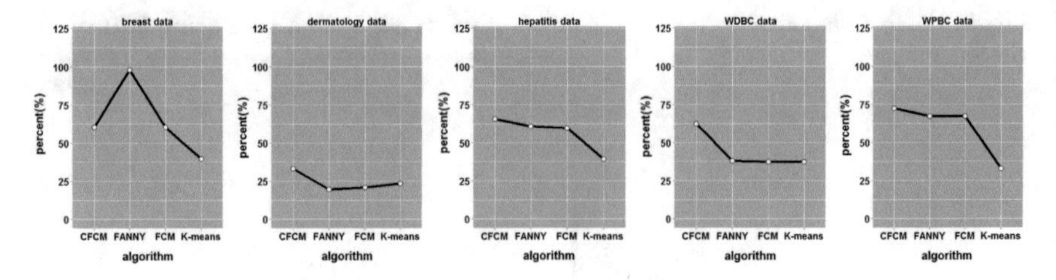

Fig. 7 Accuracy of each clustering algorithm applied on data

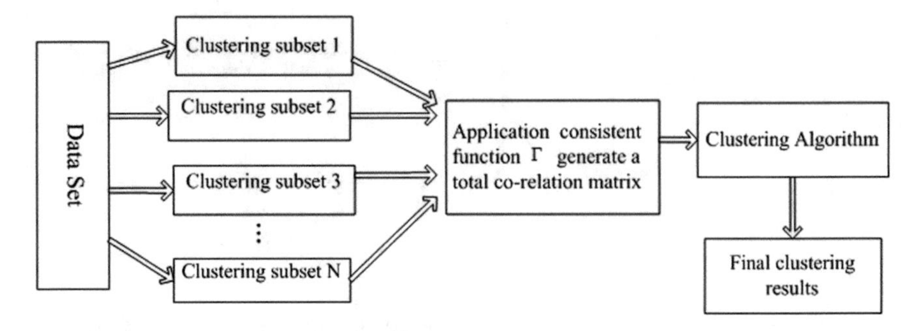

Fig. 8 Clustering ensemble methods based on total co-relation matrix

similarities to establish total co-relation matrix. All N number average weight in relation matrix to generate one ensemble relation matrix S, defined as below:

$$S = \frac{1}{N} \sum_{i=1}^{N} \omega_i S^{(i)}$$

In this formula, ω_i is average weight in i number cluster subsets. This ensemble relation matrix is to get the number for the couple data concentrated points to be divided into one collection during the period forming N number cluster. Which means any factor S_{ij} in S could be described as:

In formula:

$$M_{uv} = \frac{1}{N} \sum_{i=1}^{N} \delta(C_i(u), C_i(v))$$

$$\delta(a,b) \begin{Bmatrix} 1, & if \quad a = b \\ 0, & if \quad a \neq b \end{Bmatrix}$$

u, v are 2 data point, $C_i(v)$ and $C_i(u)$ stands for the category of point u, v in cluster subset C_i, to get cluster ensemble consistency function $\Gamma(x_i) = M_{uv}$. This

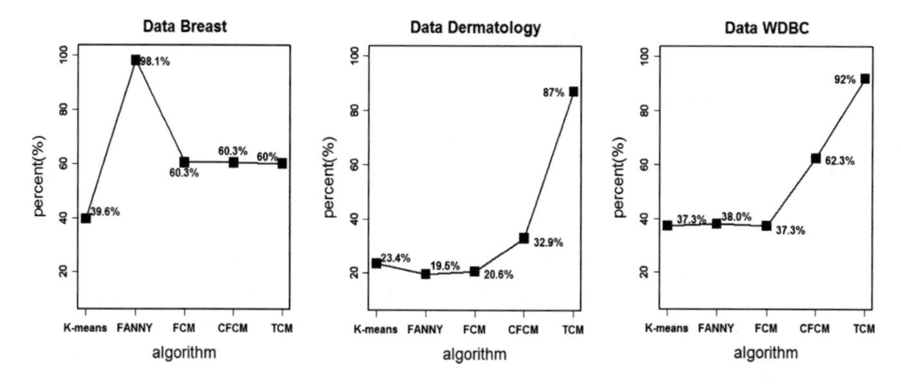

Fig. 9 Compare of accuracy of cluster ensemble TCM Algorithm and unicity cluster algorithm

paper through consistency function Γ to generate co-similarity matrix, then utilize CFCM algorithm to cluster this matrix, to get the final cluster ensemble result.

3. The test compare of algorithm

This paper compare the accuracy of Fuzzy Cluster Ensemble Algorithm and the unicity cluster methods specified on data sets of Dermatology, breast and WDBC, showed in Fig. 9.

Through compare the cluster result from unicity cluster algorithms, cluster algorithm improve the cluster result on medical data set. Doing the unicity cluster analysis in Dermatology data set, the accuracy of result is relatively low, and CFCM could improve the cluster accuracy to 32%, and after using the Fuzzy Cluster Ensemble Algorithm, the accuracy result is improved to 87%, the cluster effect is obvious. Meanwhile, data set has the poor cluster result in unicity cluster result, gets the better result in Fuzzy Cluster Ensemble Algorithm, which shows that it could solve the adaption problem of unicity algorithm in cluster medical data, improve the quality of cluster, and fix the poor result of unicity cluster problem.

5 Conclusion

Targeted at the problem of poor adaption ability of unitary clustering algorithm for various and fuzzy medical data, this paper uses R language to discuss and improve the effect of various cluster methods. First, unitary clustering algorithms such as k-means, FANNY, FCM, and etc. are achieved, then FCM algorithm was improved into CFCM algorithm, which increases the accuracy to a certain extent. In order to better adapt to the diversity of characteristics of fuzzy medical data, weighted co-association matrix is adopted to achieve integration, and consistency function is designed to present a fuzzy clustering ensemble algorithm. Finally, experiments shows that the Fuzzy Clustering Ensemble Algorithm can solve the problem of low accuracy in unitary clustering algorithm with higher stability, accuracy and robustness.

Discovering different types of diseases and classifying medical samples accurately are very important for successfully diagnosing and treating disease. For complicated and diverse medical data, how to choose the appropriate clustering algorithm suitable in specific situation with better robustness, stability and accuracy has always been the focus of the research.

The future work will focus on confirming the key parameters, setting principles of selecting different cluster algorithms, and developing clustering ensemble expansible algorithm. And at the same time, method of reducing dimensions for high dimensional medical data, and clustering ensemble and increment algorithm will also be researched deeply.

Acknowledgements 1. Major Scientific Research Project in Higher School in Hebei Province (Grant No. ZD201310-85), 2. Funding Project of Science & Technology Research and Development in Hebei North University (Grant No. ZD201301), 3. Major Projects in Hebei Food and Drug Administration (Grant No. ZD2015017), 4. Major Funding Project in Hebei Health Department (Grant No. ZL20140127), 5. Youth Funding Science and Technology Projects in Hebei Higher School (Grant No. QN2016192), with Hebei Province Population Health Information Engineering Technology Research Center.

References

1. Kaufman; Rousseeuw Finding groups in data: An introduction to cluster analysis 2008
2. Liubingyi, Clustering algorithm application research Nanjing Science and Technology University, 2012(03)
3. Zengshan, fuzzy clustering algorithm research, Huazhong University of Science and Technology 2012(08)
4. Shiyun, Chenzhong, Sunbing, Based on the mean clustering analysis and multilayer core set condensation algorithm combining network intrusion detection Computer Research application 2016(02) p 518–530
5. Liulimin, Fanxiaoping, Liaozhifang. Selective clustering Ensemble Research Computer Project and application 2012(48(10) p1–p15)
6. Chenhantao, Clustering ensemble Research South ST University 2015(05)
7. Wuxiaoxuan, Nizhiwei, Niliping, Zhangchen, Selective clustering ensemble algorithm based on information and Fractal dimension, Model recognition and AI 2014(9) p 847–p855
8. Luohuijie, clustering ensemble technology research, Zhejiang University 2007(10)
9. Linagrongde, Liubo, clustering ensemble algorithm experiment methods, wireless technology 2015(4) p 127–130
10. Fengchenfei, Yangyan, Wanghongjun, Xuyingge, Wangta One fuzzy ensemble method semi-supervision based on data relation computer Science 2015(6) p 41–45
11. Lilingli, One FCM clustering ensemble method based on attribute decomposition, Computer application and software, 2013(30) p 65–67
12. Lihuimin, Yanjianzhuo, Fangliying, Wangyu, One longitudinal data fuzzy clustering method based on Eros distance, Beijing Industrial University 2013(8) p 1161–1165
13. Lvhui, Research and Experiment on clustering methods based on big data and high dimensional data, Yunnan University 2015(05)

Robot Path Planning Based on Dijkstra's Algorithm and Genetic Algorithm

Yang Liu, Junhua Liang, Jing Li and Zhisheng Zhao

Abstract In this paper, a robot path planning method based on Dijkstra's Algorithm and Genetic Algorithms is proposed. This method works for any appointed start point and end point in an arbitrary rectangular workspace with arbitrary rectangular obstacles whose edges are perpendicular or parallel to the walls of the workspace. Firstly, use Dijkstra's Algorithm to find a path among all the midpoints of obstacles' vertices and the walls of the workspace. Then use Genetic Algorithm to optimize the path gotten from Dijkstra's Algorithm and finally generate an optimal path for the robot from the start point to the end point.

Keywords Robot path planning · Dijkstra's algorithm · Genetic algorithm

1 Introduction

A basic path planning problem is to produce a continuous motion that connects a start configuration S and a goal configuration G, while avoiding collision with known obstacles. The robot and obstacle geometry is described in a 2D or 3D workspace.

Now we have a single point (zero-sized) robot in a 2-dimensional plane (the workspace), we need to find a path for the robot from the start point to the end point avoiding all the arbitrary rectangular obstacles whose edges are perpendicular or parallel to the walls of the workspace.

Y. Liu · J. Liang · J. Li · Z. Zhao (✉)
School of Information Science and Engineering, Hebei North University, Zhangjiakou, China
e-mail: zhaozhisheng_cn@sina.com

© Springer Nature Singapore Pte Ltd. 2018

N.Y. Yen and J.C. Hung (eds.), *Frontier Computing*, Lecture Notes in Electrical Engineering 422, DOI 10.1007/978-981-10-3187-8_38

2 Background

2.1 Dijkstra's Algorithm

An algorithm for finding a graph geodesic, i.e., the shortest path between two graph vertices in a graph. It functions by constructing a shortest-path tree from the initial vertex to every other vertex in the graph. The basic steps of Dijkstra's Algorithm are:

(a) Assign to every node a tentative distance value: set it to zero for our initial node and to infinity for all other nodes.
(b) Mark all nodes unvisited. Set the initial node as current. Create a set of the unvisited nodes of all the nodes except the initial node.
(c) For the current node, consider all of its unvisited neighbors and calculate their distances. If this distance is less than the previously recorded distance, then overwrite that.
(d) After considering all of the neighbors of the current node, mark the current node as visited. A visited node will never be checked again.
(e) If the unvisited set is empty, then stop. The algorithm has finished.
(f) Set the unvisited node marked with the smallest tentative distance as the next "current node" and go back to step (c).
Pseudocode for Dijkstra's Algorithm:
Initialize the cost of each node to ∞ and the cost of the source to 0
While there are unvisited nodes left in the graph
Select an unvisited node b with the lowest cost and mark b as visited
For each node a adjacent b
If b's cost+cost of (b,a)<a's old cost
a's cost=b's cost+cost of (b,a)
a's prev path node=b

End For

End

2.2 Genetic Algorithm

In a genetic algorithm, a population of chromosomes evolves toward better solutions. The evolution usually starts from a population of randomly generated individuals. In each generation multiple individuals are randomly selected from the current population based on their fitness, and modified through crossover and mutation to form a new population. Commonly, the algorithm terminates when either a maximum number of generations has been produced, or a satisfactory fitness level has been reached for the population.

Pseusocode for Genetic Algorithm:
Initialize random population
While not terminated

Evaluate each individual's fitness
Select pairs to mate from the best-ranked individuals
Generate new population using selected pairs

- **Crossover**
- **Mutation**

End

3 Procedure

The overall procedure of this method is shown below (Fig. 1).

3.1 Input Data

Before path planning, we need to input data about the robot's workspace, start and end points, and the obstacles in the workspace.

Firstly, we need to input the size of the workspace, i.e. the maximum coordinates of the 2-D plane. Then, input the coordinates of the start point and end point. After that, input coordinates of the four vertices of a rectangular obstacle. We can set as many obstacles as we want and the obstacles can overlap each other as long as they are rectangular and their edges are perpendicular or parallel to the walls of the workspace.

After inputting all the data we need, the first task is to judge whether the data is correct. If the start point or end point is within or on the edge of any of the obstacles, or the start point, end point or any of the obstacles exceeds the workspace area, then the program will return as "Data Wrong"; otherwise, we can implement the next step.

3.2 Generate Matrix of Nodes

For each obstacle, there are four vertices and eight projection points of the vertices on the walls of the workspace. All of those points of all the obstacles and the four vertices of the workspace make up the matrix of fundamental points. There is a line section between each two points of the matrix. If the section doesn't cross or touch any of the obstacles, then it is an available section and the midpoint of this section is also available for later use.

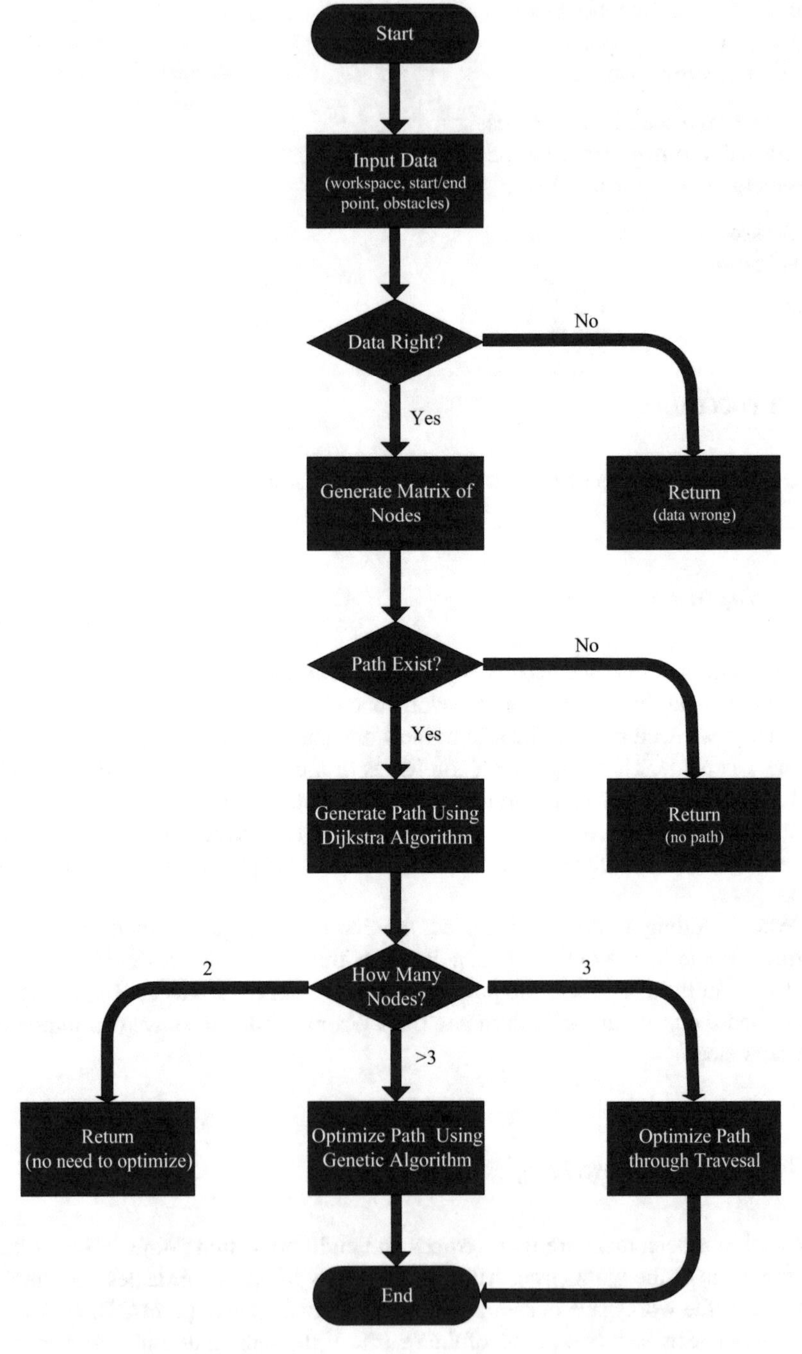

Fig. 1 Flowchart of the method

After detecting the whole matrix of fundamental points, we get a bunch of available midpoints. Those midpoints, the start point and the end point make up the matrix of nodes. We also need to record the two endpoints of each available section corresponding to the midpoint. Then, remove the iterant nodes in the matrix and rearrange the orders of the remaining nodes based on their distance to the start point, with the start point at the first and the end point at the last.

Suppose now there are N nodes in the matrix, create an N × N matrix D. For each node ni (i = 1, 2, …, N), connect ni with each nj (j = i + 1, i + 2, …, N). If the line section between ni and nj doesn't cross or touch any of the obstacles, mark D(i, j) = D(j, i) = 1; otherwise, mark D(i, j) = D(j, i) = 0.

3.3 Generate Path Using Dijkstra's Algorithm

Inplement Dijkstra's Algorithm according to matrix of nodes and matrix D and we can get a shortest path from the start point to the end point through nodes. The positions where the path making a turn are recorded. If there is no available path, the program will return as "No Path".

3.4 Optimize Path Using Genetic Algorithm

As all the nodes we use are the midpoints of those fundamental points, the path we get from Dijkstra's Algorithm may be restricted. We hope to use Genetic Algorithm to get a shorter path, i.e. optimize the original path.

Suppose the original path contains n nodes (included the start point and the end point). For each node (midpoint) P_i, (i = 1, 2, … n),

$$P_i = P_{i1} + t_i \times (P_{i2} - P_{i1})$$

where $t \in [0, 1]$, P_{i1} and P_{i2} are the fundamental points of P_i, i.e. the two end points of the node P_i. For the original path, $t_i = 0.5$ for all nodes. The task of this step is to optimize t_i (i = 2, …, n−1)

If n = 2, i.e. the original path connected the start point and the end point straightly, then we don't need to use Genetic Algorithm to optimize the path as it is already the shortest.

If n = 3, i.e. the original path turns only once at the point p, then for t = (0, 0.01, 0.02, … 0.98, 0.99, 1.00), we calculate the distance for each t and choose the shortest one among the paths that don't cross or touch any of the obstacles as the final result.

If n > 3, then we have more than one t_i to optimize. Now we can use Genetic Algorithm. The fitness function is the distance of the new path. The chromosome contains i genes and each one of them is a real number ranged from 0 to 1 represented one 't'. For each chromosome, we need to judge whether the whole path crosses or touches any of the obstacles. If it does, this chromosome is not available. The final result is the minimum of all the results we get from the whole evolution. There may be some situations where the original path gotten from Dijkstra's Algorithm is the best result, i.e. Genetic Algorithm will not give us a better solution, so we set all genes as 0.5 for the first chromosome of the first generation (initialization) and the final result is the minimum of the whole evolution. By doing this, we can avoid the situation where the path gotten after using Genetic Algorithm is longer than the original one.

4 Experimental Results

Since all the useful data in the program are arbitrary, here we just illustrate some featured situations. In this part, L_1 represents the distance of the path gotten from Dijkstra's Algorithm and L_2 represents the distance of the path after optimization using Genetic Algorithm.

4.1 *Situation 1*

input:

maximum coordinates of the workspace (100,100)
start point (0 0)
end point (100,100)
1 obstacle: x_1~[40,60], y_1~[40,60]
resutls:
L_1=148.1915
L_2=147.3932

(See Fig. 2).

4.2 *Situation 2*

input:

maximum coordinates of the workspace (100,100)
start point (0,0)
end point (100,0)
1 obstacle: x_1~[20,80], y_1~[0,80]
results:
L_1=243.7073

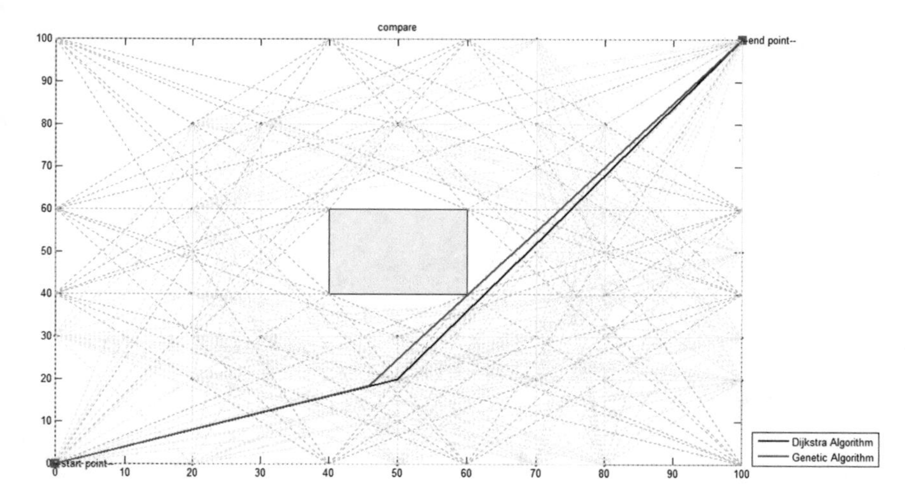

Fig. 2 Experimental result 1

L_2=230.5509

(See Fig. 3).

4.3 Situation 3

input:

maximum coordinates of the workspace (100,100)

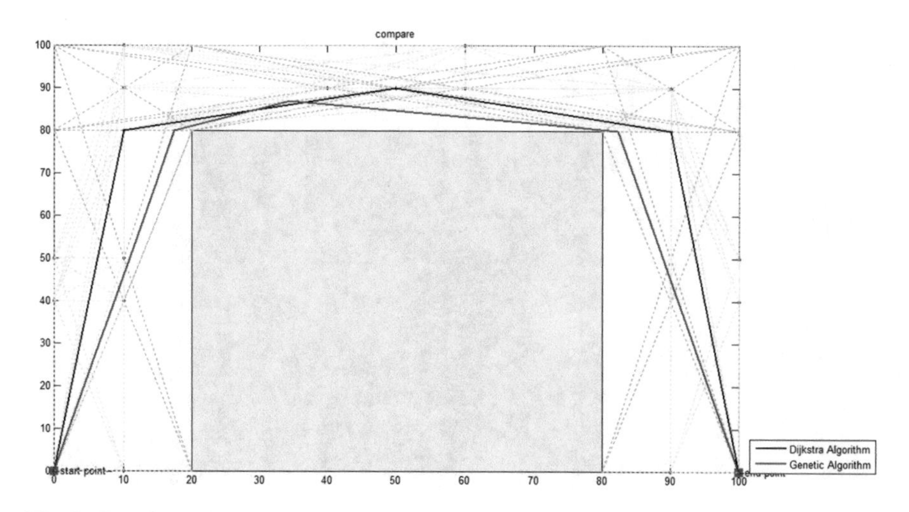

Fig. 3 Experimental result 2

start point (0,0)
end point (100,100)
2 obstacles: $x_1 \sim [20,40]$, $y_1 \sim [0,50]$
$x_2 \sim [60,80]$, $y_2 \sim [50,100]$
results:
$L_1 = 230.0429$
$L_2 = 201.4115$

(See Fig. 4).

4.4 Situation 4

input:

maximum coordinates of the workspace (100,100)
start point (0,0)
end point (100,100)
2 obstacles: $x_1 \sim [0,40]$, $y_1 \sim [40,60]$
$x_2 \sim [60,100]$, $y_2 \sim [40,60]$
results:
$L_1 = 144.2221$
$L_2 = 142.5992$

(See Fig. 5).

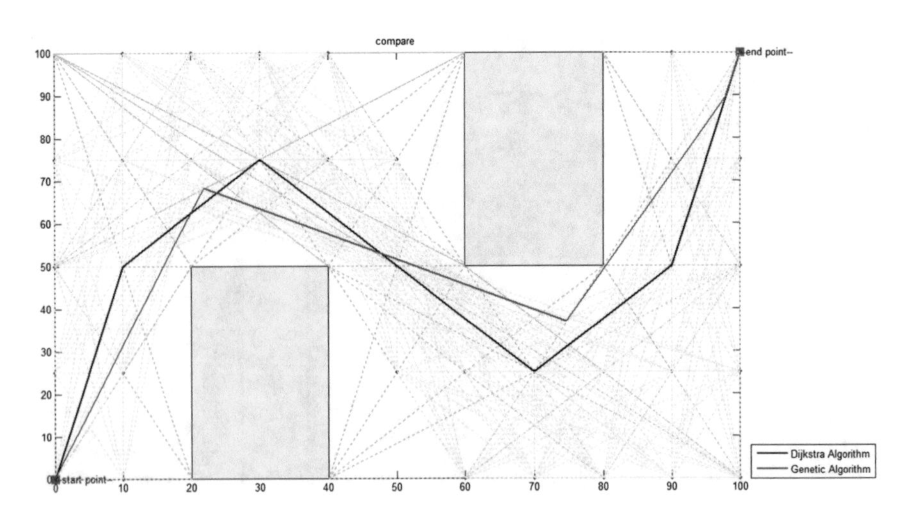

Fig. 4 Experimental result 3

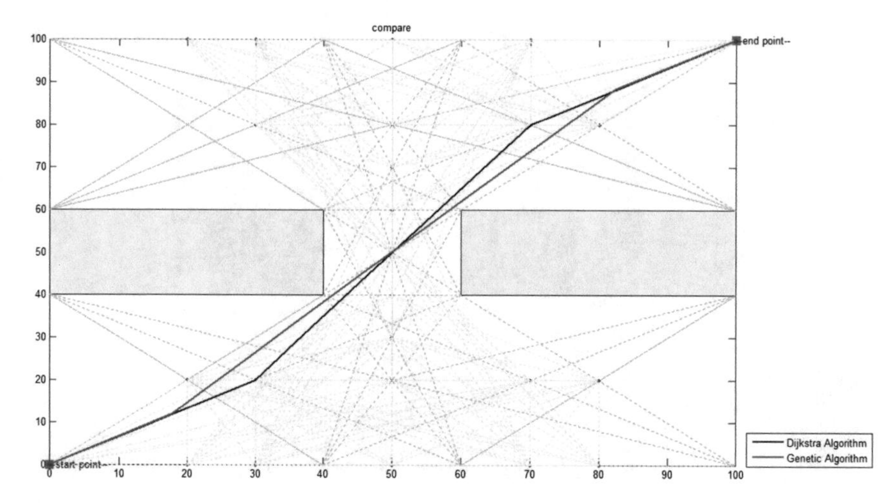

Fig. 5 Experimental result 4

4.5 Situation 5

input:

maximum coordinates of the workspace (50,100)
start point (5,90)
end point (45,10)
4 obstacles: x_1~[0,15], y_1~[0,40]
x_2~[35,50], y_2~[60,100]
x_3~[10,20], y_3~[60,80]
x_4~[20,30], y_4~[20,40]
results:
L_1=94.6422
L_2=94.0922

(See Fig. 6).

4.6 Situation 6

input:

maximum coordinates of the workspace (100,50)
start point (10,10)
end point (90,40)
4 obstacles: x_1~[15,40], y_1~[0,20]

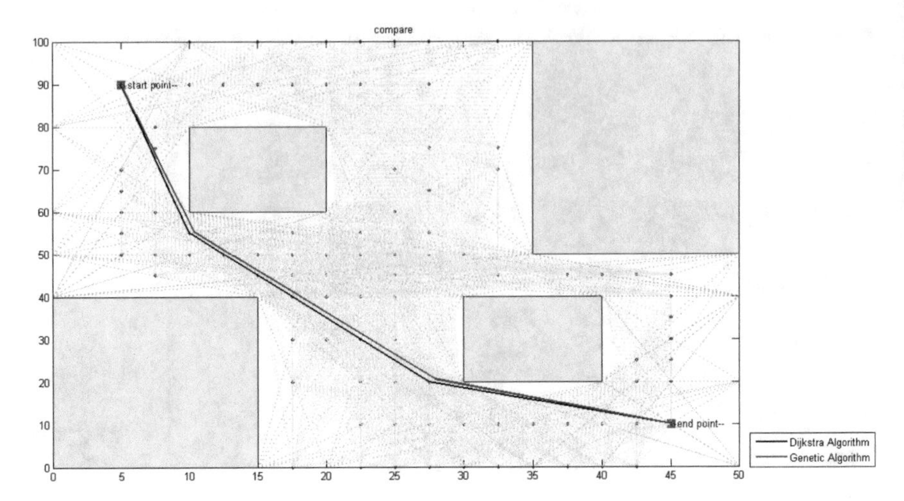

Fig. 6 Experimental result 5

x_2~[20,45], y_2~[35,50]
x_3~[50,80], y_3~[10,35]
x_4~[70,85], y_4~[40,50]
results:
L_1=92.0379
L_2=91.9641

(See Fig. 7).

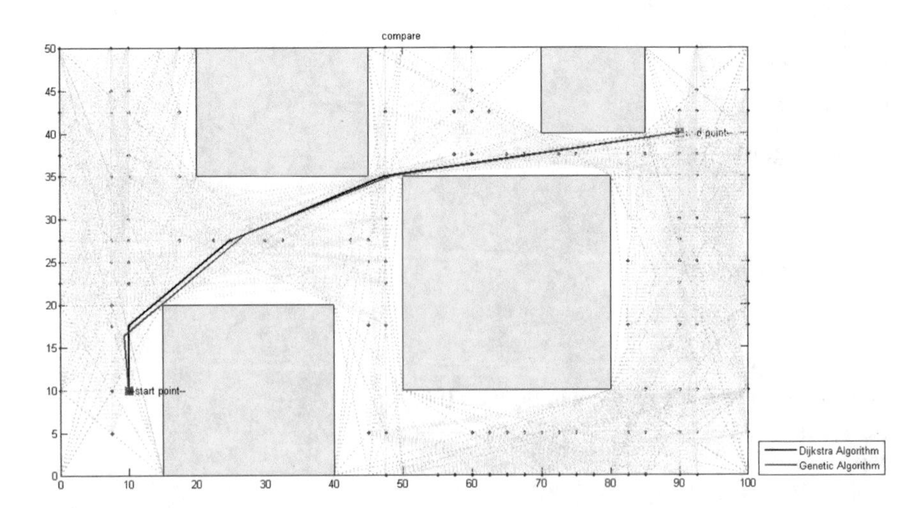

Fig. 7 Experimental result 6

4.7 Situation 7

input:

maximum coordinates of the workspace (100,100)
start point (0,0)
end point (100,100)
9 obstacles: x_1~[10,30], y_1~[10,30]
x_2~[10,30], y_2~[40,60]
x_3~[10,30], y_3~[70,90]
x_4~[40,60], y_4~[10,30]

x5~[40,60], y5~[40,60]

x_6~[40,60], y_6~[70,90]
x_7~[70,90], y_7~[10,30]

x8~[70,90], y8~[40,60]

x_9~[70,90], y_9~[70,90]
results:
L_1=173.2780
L_2=173.2780

(See Fig. 8).

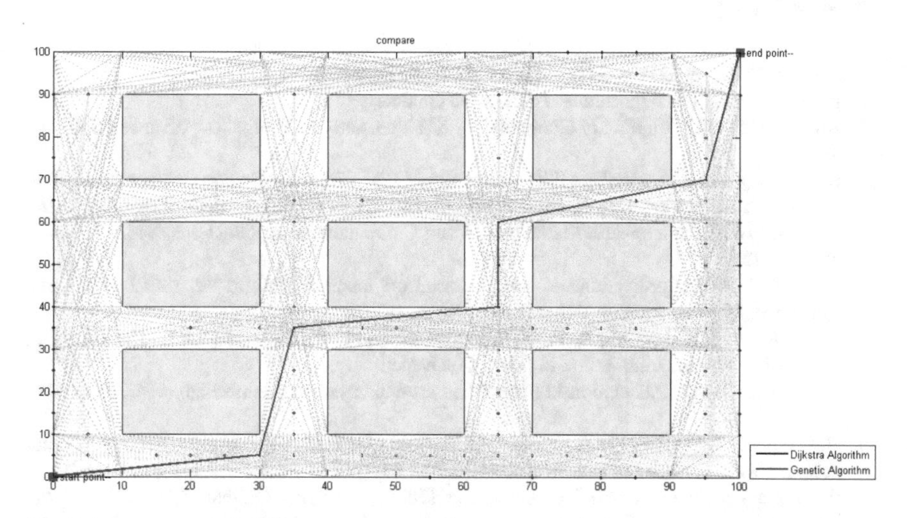

Fig. 8 Experimental result 7

5 Conclusion and Future Work

Dijkstra's Algorithm is an effective and efficient algorithm to for finding the shortest path between two graph vertices in a graph. However, by using Genetic Algorithm to optimize the result we get from Dijkstra's Algorithm, sometimes it works well, and sometimes the path doesn't shorten at all. The main reason is that we optimize each node's position between the two end points of the node, which is a big restriction for moving the node. Overall, this robot path planning method is effective and can find a good path for the robot.

This method only works for rectangular obstacles whose edges are perpendicular or parallel to the walls of the workspace. To improve the method, the obstacles should be more optional. And for the Genetic Algorithm step, if the positions of the nodes that we need to optimize won't be restricted to a line section, the overall result will be better.

Acknowledgements This work was supported by Major Scientific Research Project in Higher School in Hebei Province (Grant No. ZD20131085), Funding Project of Science & Technology Research and Development in Hebei North University (Grant No. ZD201301), Major Projects in Hebei Food and Drug Administration (Grant No. ZD2015017), Major Funding Project in Hebei Health Department (Grant No. ZL20140127), Youth Funding Science and Technology Projects in Hebei Higher School (Grant No. QN2016192), with Hebei Province Population Health Information Engineering Technology Research Center.

Bibliography

1. Shortest Path Algorithms, MIT's Introduction to Algorithms
2. Basic Shortest path Algorithms, Andrew V. Goldberg
3. An Overview of Evolutionary Computation, Xin Yao, Chinese Journal of Advanced Software Research
4. R. Anthony and E. D. Reilly (1993), Encyclopedia of Computer Science, Chapman & Hall. U. Aybars, K. Serdar, C. Ali, Muhammed C. and Ali A (2009), Genetic Algorithm based solution of TSP on a sphere, Mathematical and Computational Applications, Vol. 14, No. 3, pp. 219–228.
5. B. Korte (1988), Applications of combinatorial optimization, talk at the 13th International Mathematical Programming Symposium, Tokyo.
6. E. L. Lawler, J. K. Lenstra, A. H. G. RinnooyKan, and D. B. Shmoys (1985), The Traveling Salesman Problem, John Wiley & Sons, Chichester.
7. H. Holland (1992), Adaptation in natural and artificial systems, Cambridge, MA, USA: MIT Press.
8. H. Nazif and L.S. Lee (2010), Optimized CrosSover Genetic Algorithm for Vehicle Routing Problem with Time Windows, American Journal of Applied Sciences 7 (1): pg. 95–101.
9. R. Braune, S. Wagner and M. Affenzeller (2005), Applying Genetic Algorithms to the Optimization of Production Planning in a real world Manufacturing Environment, Institute of Systems Theory and Simulation Johannes Kepler University.
10. Z. Ismail, W. Rohaizad and W. Ibrahim (2008), Travelling Salesman problem for solving Petrol Distribution using Simulated Annealing, American Journal of Applied Sciences 5(11): 1543–1546.

The Research of Impact of Resident Consumption Level and Unemployment Rate on Population Natural Increase Rate

Zhisheng Zhao, Xueqiong Wei and XiaoLiang Bai

Abstract On the bases of empirical analysis to the statistical data of born rate, unemployment rate and consumption level collected from 31 provinces or cities, and by the method of R language, this thesis subjects to analyze the relationship of each administration district's above elements with its population increase rate, and also the regional difference and moving trend of the relationship or interaction among different region in the last decade. And this is helpful to local government in their macro policy decision and development strategy consideration.

Keywords Resident consumption lever · Unemployment rate · Population natural increase rate · R language · Regional difference

1 Introduction

Being as the largest developing country of world, population has been still one of our main social problems. Since 2013, new trend of serious unbalance of population composition has emerged and impact negatively to social development of whole society as well as the economy.

Therefore it becomes very necessary to analysis the data of relevant society issues, this research will provide strong support to making decisions to improve population policies. Analysis all the factors which have impact on the population natural increase rate will help the government introduces differentiation of macroeconomic regulation and control policy. At this stage there are many articles focus on thesis factors that have impact on the population natural increase, but most of them are based on local areas, the research generalize the principles that based on

Z. Zhao (✉) · X. Bai
School of Information Science and Engineering, Hebei North University, Zhangjiakou, China
e-mail: zhaozhisheng_cn@sina.com

X. Wei
EDHEC Business School, Nice, France

© Springer Nature Singapore Pte Ltd. 2018

N.Y. Yen and J.C. Hung (eds.), *Frontier Computing*, Lecture Notes in Electrical Engineering 422, DOI 10.1007/978-981-10-3187-8_39

the national level and did not focus on the differences among different areas population natural increase rate. From the point of this article the area differences is one factor causes the different increasing rate.

Based on the analysis of the statistical data of born rate, unemployment rate and consumption level collected from 31 provinces or cities real samples from year 2005–2014, this article analysis the impact between the population increase rate and resident consumption lever, unemployment rate. The result will support the government to introduce differentiation of macroeconomic regulation polices.

2 Research Methods and Sources of Data

2.1 Research Method

Given the fact the data collect from 31 provinces or cities cross 10 yeas time zone, and mainly focus on 3 indicator, these data is known as many dimensions ration data.

Therefore, the writer uses hierarchical clustering method to group the data then run the relevance analysis by regression model.

1. Hierarchical clustering

Hierarchical clustering describes the process which divides the data into sub group. Each sub group is a cluster, thus all the data in the same cluster has common nature. As a data mining function tool, hierarchical clustering can be used independently to observe the spread of data, study their pattern of each cluster and focus on the clusters that has value to do a higher level study. Besides that hierarchical clustering method can be used as pre-processing step, these calculations can be run to find unique of different cluster. Also it can be used the detecting the outliers. In most ceases detecting the outlier is more helpful than common data study. Hierarchical clustering has been widely used in many areas such as business intelligence, image pattern recognition, biology and security.

2. linear regression model

Linear regression is using the equation of linear regression last-squares penalty method to set up a regression model of one or several variables. Regression analysis it to find the dependency of objective factors quantity. Based on the impact of independent variables and dependent variables, it can be further divided to linear and nonlinear regression.

Table 1 Beijing three indicators comprehensive data

	Population growth	Consumption levels	Unemployment rate
2005	1.09	1.47	2.1
2006	1.29	1.65	2.0
……	……	……	……
2013	4.41	3.33	1.2
2014	4.83	3.61	1.3

2.2 *Sources of Data*

All the data from the official website of national bureau of statistics, which includes the population natural increase rate, resident consumption level, unemployment rate, as shown in the table blow. Each group of data has 10 attributes and all of them are discrete attributes. Three indicators have 930 data banners, the natural increase rate is the birth rate deduct death rate. Pre processing: based on the different regions the data can be divided into 31 sub group of data as seen in the Table 1.

3 Hierarchy Clustering

Using hierarchy clustering method run the data from 31 different provinces and cities based on R language. We used euclidean method to measure the distance of two samples. Using ward method to measure the distance of two group. The result shown in Table 1.

As we can see, the increasing rate can be grouped into 3 groups:

First group including: Liaonig, Jilin, Heilongjian, Shanghai, Sichuan, Tianjin, Jiangsu, Chongqing, Neimongu, Shanxi, Beijing, Hubei.

Second group including: Shandong, Zhejiang, Shanxi, Henan,Yunnan, Guangdong, Guizhou, Hunan, Gansu, Fujian, Hebei, Anhui.

The third group: Xizang, Xinjiang, Jiangxi, Guangxi, Ningxia, Hainan, Qinghai.

In order to better understand the difference of 3 levels, we sort the data in descending order by R square function, as seen in Fig. 2.

Based on Fig. 2, the regions has the high increasing rate is mostly in the third group. The third group has higher increasing rate, all over 8% increasing, Xizang and Xinjiang have more than 10% increasing every year.

From the bottom of the Fig. 2 the first group has lower increasing rate for the past decade. The second group is in the middle level of population increasing rate. To compare the max and min value from Figs. 1 and 2, the difference between three groups is significant, the highest provinces is Xinjiang which has over 10% increase and Liaoning province even incurred negative growth. Therefore we have to base on different areas to analysis the impact of consumption level and employment rate.

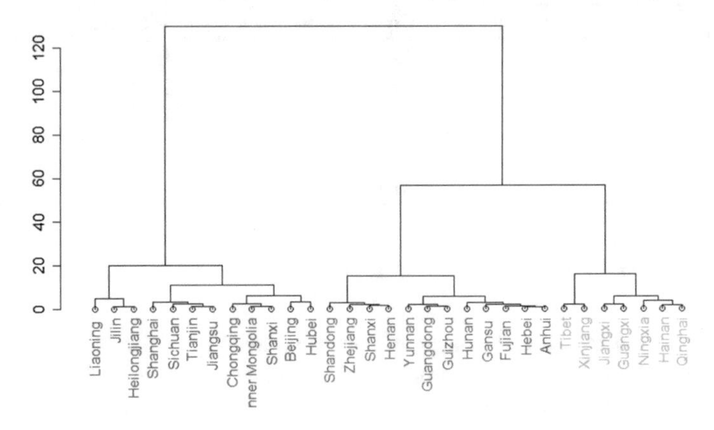

Fig. 1 31 regions hierarchical clustering

```
> cbind(a1,a2,a3,a4,a5,a6,a7,a8,a9,a0)
```

	area	2005年	area	2006年	area	2007年	area	2008年	area	2009年
31	Xinjiang	11.38	Tibet	11.70	Xinjiang	11.78	Xinjiang	11.17	Xinjiang	10.56
30	Ningxia	10.98	Xinjiang	10.76	Tibet	11.30	Tibet	10.30	Tibet	10.24
26	Tibet	10.79	Ningxia	10.69	Ningxia	9.76	Ningxia	9.69	Ningxia	9.68
29	Qinghai	9.49	Qinghai	8.97	Hainan	8.91	Hainan	8.99	Hainan	8.96
21	Hainan	8.93	Hainan	8.86	Qinghai	8.80	Guangxi	8.70	Guangxi	8.53
20	Guangxi	8.16	Guangxi	8.34	Guangxi	8.20	Qinghai	8.35	Qinghai	8.32
25	Yunnan	7.97	Jiangxi	7.79	Jiangxi	7.87	Jiangxi	7.91	Jiangxi	7.89
14	Jiangxi	7.83	Guangdong	7.29	Guangdong	7.30	Guangdong	7.25	Guangdong	7.26
24	Guizhou	7.38	Guizhou	7.26	Yunnan	6.86	Guizhou	6.72	Guizhou	6.96
19	Guangdong	7.02	Yunnan	6.90	Guizhou	6.68	Hebei	6.55	Gansu	6.61
12	Anhui	6.20	Anhui	6.30	Hebei	6.55	Gansu	6.54	Hebei	6.50
3	Hebei	6.09	Fujian	6.25	Gansu	6.49	Anhui	6.45	Anhui	6.47

Fig. 2 Descending order data

4 Linear Regression Analysis

After dividing the data into 3 different groups, we set up linear regression model for each group, using cor()function to inspect the Pearson correlation index of independent variables(consumption level, unemployment rate), dependent variables (natural population increase rate), the result is shown in Tables 2, 3, 4.

4.1 The First Group Linear Regression Analysis

For the first group areas, the Pearson correlation index is in Fig. 3, Jilin Heilongjian Liaoning Neimenggu have strong negative correlations; Beijing Hubei have strong positive correlations. The correlations between Tianjin Shanghai Jiangsu Chongqing Sichuan Shanxi is not significant.

The Pearson correlation index between natural population increasing rate and unemployment is in Fig. 4. Like Beijing Hubei have strong negative correlations; Neimenggu Jilin have strong positive correlation; areas like Tianjin Liaoning Heilongjiang Shanghai Jiangsu Sichuan Shanxi do not have significant correlations

Table 2 Correlation coefficient table of provinces in first cluster

Area	The correlation coefficient of natural population growth rate and consumption levels	The correlation coefficient of natural population growth rate and unemployment rate
Beijing	0.8743829	−0.8780299
Tianjin	0.5490919	−0.3993607
Inner Mongolia	−0.8432606	0.8688636
Liaoning	−0.8078138	0.5993429
Jilin	−0.9610217	0.8419758
Heilongjiang	−0.9272006	0.02568655
Shanghai	0.5806994	−0.5917763
Jiangsu	0.3049923	−0.433648
Hubei	0.9093148	−0.7591772
Chongqing	0.2565875	−0.2901613
Sichuan	0.4510569	−0.2580176
Shanxi	−0.5989794	0.4896338

Table 3 Correlation coefficient table of provinces in second cluster

Area	The correlation coefficient of natural population growth rate and consumption levels	The correlation coefficient of natural population growth rate and unemployment rate
Hebei	0.4137252	−0.1801075
Shanxi	−0.6977895	−0.6480324
Zhejiang	−0.3183632	0.391441
Anhui	0.8416055	−0.864882
Fujian	0.7000388	−0.7617373
Shandong	0.3398286	−0.103708
Henan	0.5718229	−0.6767492
Hunan	0.9048155	−0.7265949
Guangdong	−0.8048701	0.8010081
Guizhou	−0.8623833	0.7609273
Yunnan	−0.6559915	0.3730274
Gansu	−0.4379153	0.4103145

By drawing scatter diagram between two independent variables and dependent variable, and add the linear trend line, we can tell the two independent variables (unemployment rate and consumption level) and dependent variable (natural population increasing rate) moves along one stright line. Which can be prove there is obvious linear relation between these variables.

In order to further analysis the impact of two variables on population increasing rate, we build a multiple linear regressing model and used OLS estimation model to solve the parameter. During the verification of the model, we fund an affirmative

Table 4 Correlation coefficient table of provinces in third cluster

Area	The correlation coefficient of natural population growth rate and consumption levels	The correlation coefficient of natural population growth rate and unemployment rate
Jiangxi	−0.9229607	0.5624926
Guangxi	−0.5673763	0.5228675
Hainan	−0.7280163	0.4418149
Tibet	−0.4996785	0.3903512
Qinghai	−0.6840923	0.5663429
Ningxia	0.3398286	−0.103708
Xinjiang	−0.1173422	0.3447459

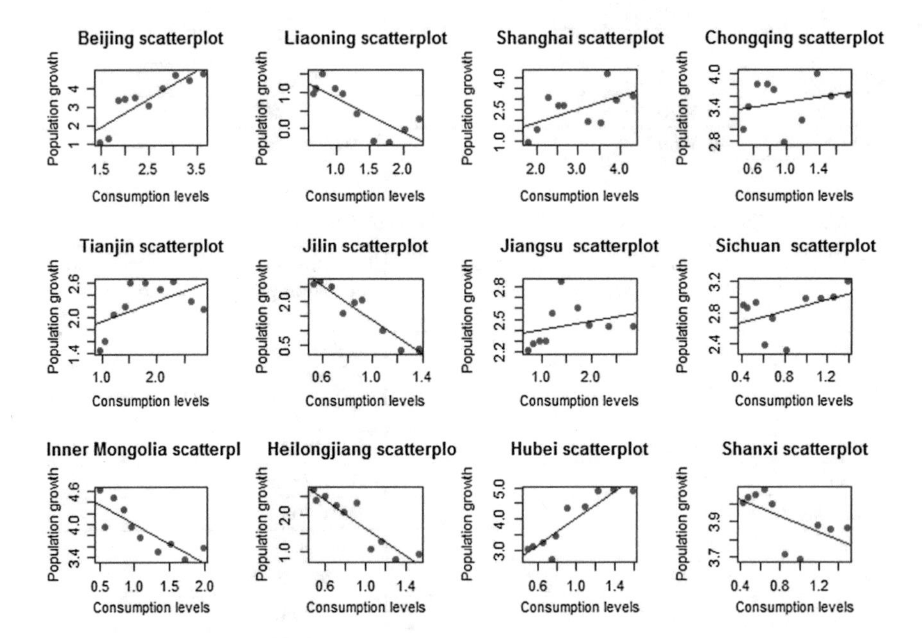

Fig. 3 Scatterplot of growth rate and consumption levels of provinces in first cluster

coefficient is bigger but the regression coefficient is not significant, which means multiple linear regressing model has multiple-colinearity, which also means the variables has linear relation. This violates the classic assumptions. Hence, we set up unary linear regression model for each variable.

$$Y_1 = \beta_1 + \beta_2 x + u \tag{1}$$

$$Y_2 = \beta_1 + \beta_2 x + u \tag{2}$$

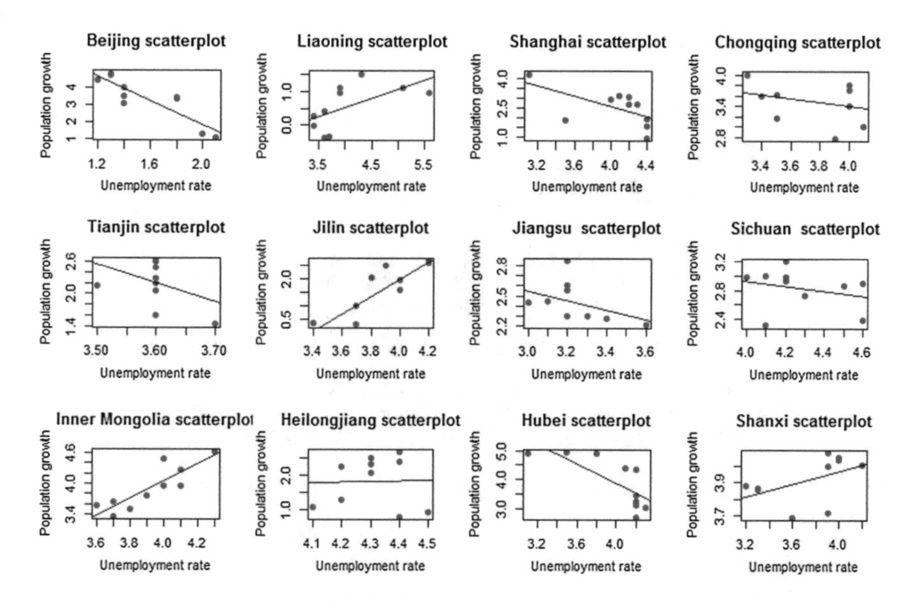

Fig. 4 Scatterplot of growth rate and unemployment rate of provinces in first cluster

Using lm() function to get a conclusion mixed linear regression model, using summary() functions parameter list and the test result. For example, in Beijing sample the unary linear regression model is

$$Y_1 = -0.4272 + 1.5559x + u$$

$$Y_2 = 8.9077 - 3.5227x + u$$

Run goodness of fit test in model 1 got the coefficient of determination is 0.7645, adjusted coefficient of determination R^2 is 0.7351, which means the explanatory ability of the regression model is 76.45%; to do the over all significance test the F test statistic is 25.98, the two degrees of freedom is 1 and 8, P value is 0 therefore the model is effective. To do significance test of individual variable, t test result is in the result of summary function parameter table the P value of t test is 0.597 not significant. The P value of t test of variable X is 0 which means the consumption level has significant impact on natural population increasing rate; the consumption level increased by RMB10,000.00 the increasing rate will increased by 1.5559

Figures 5 and 6 is the parameter table and test result, coefficients is the estimated, Multiple R-squared is R^2, Adjusted R-squared is $\overline{R^2}$, at the bottom is the F test result.

Run goodness of fit test in model 2 got the coefficient of determination is 0.7709, adjusted coefficient of determination is 0.7423, which means the explanatory ability of the regression model is 77.09%; to do the overall significance test the F test statistic is 26.92, the two degrees of freedom is 1 and 8, the P value is 0.00083

```
> summary(a1)

Call:
lm(formula = a[, 1] ~ a[, 2])

Residuals:
    Min      1Q  Median      3Q     Max
-0.8501 -0.3843 -0.1111  0.4875  0.9332

Coefficients:
            Estimate Std. Error t value Pr(>|t|)
(Intercept)  -0.4272     0.7757  -0.551 0.596888
a[, 2]        1.5559     0.3053   5.097 0.000934 ***
---
Signif. codes:  0 '***' 0.001 '**' 0.01 '*' 0.05 '.' 0.1 ' ' 1

Residual standard error: 0.6677 on 8 degrees of freedom
Multiple R-squared:  0.7645,    Adjusted R-squared:  0.7351
F-statistic: 25.98 on 1 and 8 DF,  p-value: 0.0009336
```

Fig. 5 Regression model 1 parameters table and the test results of Beijing

```
> summary(a1)

Call:
lm(formula = a[, 1] ~ a[, 3])

Residuals:
    Min      1Q  Median      3Q     Max
-0.9059 -0.4619 -0.1131  0.4794  0.8532

Coefficients:
            Estimate Std. Error t value Pr(>|t|)
(Intercept)   8.9077     1.0860   8.202 3.65e-05 ***
a[, 3]       -3.5227     0.6789  -5.189 0.000834 ***
---
Signif. codes:  0 '***' 0.001 '**' 0.01 '*' 0.05 '.' 0.1 ' ' 1

Residual standard error: 0.6586 on 8 degrees of freedom
Multiple R-squared:  0.7709,    Adjusted R-squared:  0.7423
F-statistic: 26.92 on 1 and 8 DF,  p-value: 0.0008336
```

Fig. 6 Regression model 2 parameters table and the test rest of Beijing

therefore the model is effective. To do significance test of individual variable, t test result is in the result of summary function parameter table the P value of t test is 0.597 not significant. The P value of t test of variable X is less than 0.05 which means the consumption level has significant impact on natural population increasing rate; the consumption level increased by 1 the increasing rate will decreased by 3.5227.

In the class I districts, there is better degree of fitting in some provinces or cities for regression model 1 like Beijing, Inner Mongolia, Liaoning, Jilin, Heilongjiang, Hubei, etc. The results of Test t and test F are remarkable. This indicates consumption level is a significant factor on natural population growth rate in these areas. While other areas are at a low level of degree of fitting, which is not remarkable for significance of variable and equation test. This shows there is no linear relationship between consumption level and the natural population growth rate, and independent variable of consumption level is not remarkable factor to the natural population growth rate.

In the class I districts, there is better degree of fitting for regression model 2 like Beijing, Jilin, Hubei, Inner Mongolia, etc. The results of test t and test F are remarkable, which indicates that generally there is a linear relationship for explanatory variables and explained variables in these areas, and independent variable of unemployment rate has significant impact on the population natural growth rate. There is at a low level of degree of fitting for regression model 2 like Tianjin, Liaoning, Heilongjiang, Shanghai, Jiangsu, Chongqing, Sichuan, Shanxi, etc. The results of test t and test F are not remarkable, showing that there is not a linear relationship between unemployment rate and the population natural growth rate, and unemployment rate of dependent variable has no significant impact on the population natural growth rate. Inner Mongolia has a good degree of fitting for the regression model 2. The result of test F is remarkable, and only parameter t is not verified. It demonstrates a linear relationship between the population natural growth rate of the region and unemployment rate is not remarkable. Therefore, excluding parameter t, the new result from the regression model 2 calculation is $Y_2 = 1.00016x + u$, coefficient of determination is 0.9961, p-value of test t and test F is 0, which indicates unemployment rate of Inner Mongolia has a significant impact on the population natural growth rate.

In the class I districts, there is no satisfactory degree of fitting for Tianjin, Shaanxi, Jiangsu, Sichuan, Chongqing and Shanghai. The level of correlation coefficient is low, then there is no a linear relationship between variables. Further analysis is shown in Fig. 7.

As it is seen, consumption level is increasing year by year in these six districts. The consumption level of Shanghai, Tianjin, Jiangsu, etc. is growing faster than that of Shaanxi, Sichuan, Chongqing and others in western areas. As far as unemployment rate, Tianjin is at the level of around 3.6, however, other districts are in the fluctuation. During the past decade, population growth rate fluctuates largely in Shanghai region, during which the ratio of population growth is the highest, but unemployment rate is the lowest in 2012. While there is the lowest unemployment rate in 2005, but the population growth rate is at the lowest level. Overall, the population growth rate of fluctuation range is wider than that of unemployment rate in Shanghai, Sichuan, Chongqing during last decade. While unemployment rate of Shaanxi fluctuates larger than the range of the population growth rate.

4.2 The Type II Linear Regression Analysis

In the class II districts, the Pearson correlation coefficient of the population natural growth rate and consumption level is shown below. There is substantial correlation between the two variables in Shanxi, Guangdong, Guizhou, Yunnan, etc. There is an positive correlation between two variables in Anhui, Fujian, Hunan, etc. It shows

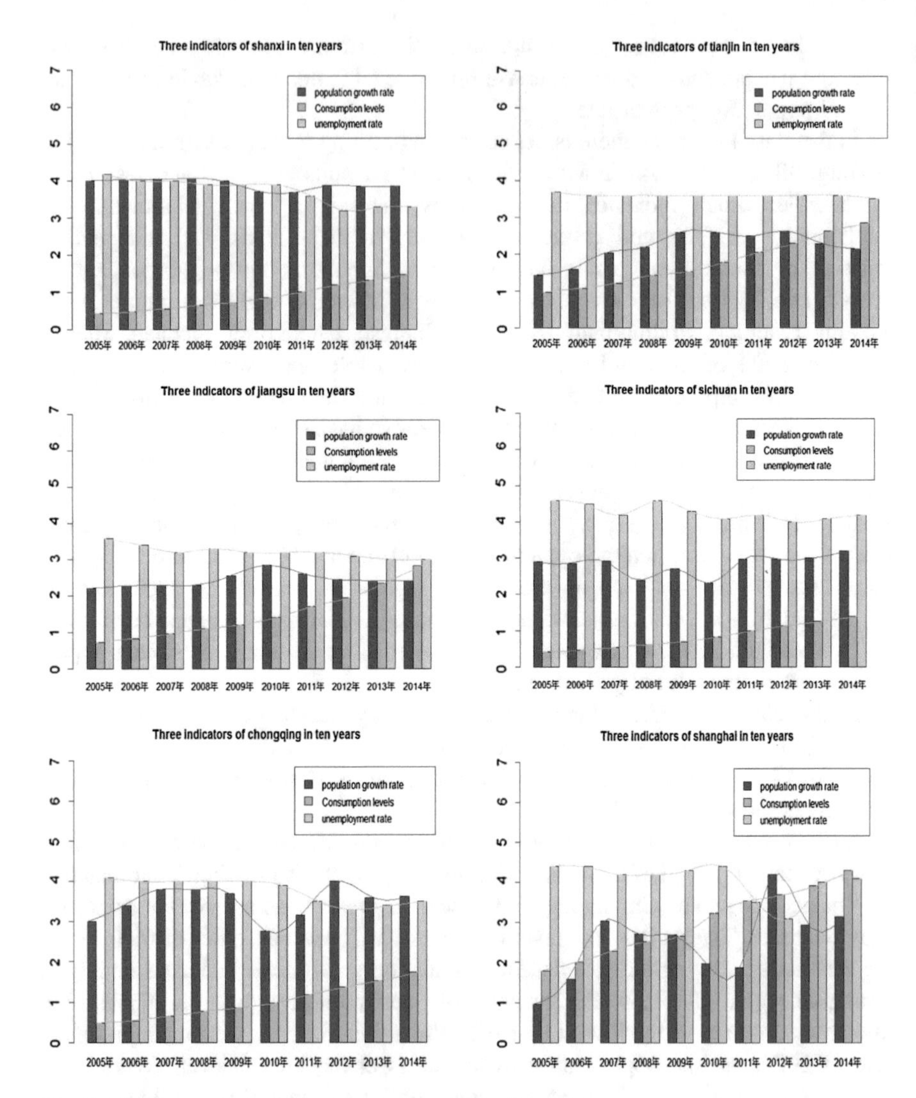

Fig. 7 Nonlinear relationship trends graphs of first cluster

weak correlation between the population natural growth rate and unemployment rate in Hebei, Zhejiang, Shandong, Henan, Gansu (Table 5).

As shown, in the class II districts, there is substantial negative correlation between two variables in Shanxi, Anhui, Fujian, Henan, Hunan, etc. There is positive correlation for those two variables in Guangdong, Guizhou. Correlation is weak between two variables in Hebei, Zhejiang, Shandong, Yunnan, Gansu, etc.

Draw two independent and dependent variables and scatter plot linear trend line as shown in Figs. 8 and 9.

Table 5 Model parameters table of provinces in first cluster

Provinces	Intercept	Coefficient	Multiple R-quared	p-value of T/F test	Intercept	Coefficient	Multiple R-quared	p-value of T/F test
Beijing	−0.4272	1.5559	0.7645	Intercept > 0.05	8.9077	−3.5227	0.7709	All p-value < 0.05
Tianjin	1.5850	0.3470	0.3015	Intercept < 0.05	14.982	−3.550	0.1595	All p-value > 0.05
Inner Mongolia	4.7167	−0.7169	0.7111	All p-value < 0.05	−2.6899	1.6844	0.7549	Intercept > 0.05
Liaoning	1.8011	−0.9489	0.6526	All p-value < 0.05	−1.6236	0.5389	0.3592	p值均大于0.05
Jilin	4.2393	−2.8868	0.9236	All p-value < 0.05	−10.791	3.1954	0.7089	All p-value < 0.05
Heilongjiang	3.5568	−1.9224	0.8597	All p-value < 0.05	1.1474	0.1558	0.0006598	All p-value > 0.05
Shanghai	0.6382	0.6262	0.3372	All p-value > 0.05	7.6695	−1.2701	0.3502	Intercept < 0.05
Jiangsu	2.31682	0.08329	0.09302	Intercept < 0.05	3.9149	−0.4574	0.1881	Intercept < 0.05
Hubei	1.8885	2.1431	0.8269	All p-value < 0.05	10.573	−1.676	0.5763	All p-value < 0.05
Chongqing	3.2519	0.2332	0.06584	Intercept < 0.05	4.8945	−0.3736	0.08419	Intercept < 0.05
Sichuan	2.5246	0.3637	0.2035	Intercept < 0.05	4.2551	−0.3341	0.06657	Intercept < 0.05
Shanxi	4.1137	−0.22375	0.3588	Intercept < 0.05	3.2046	0.1918	0.2397	Intercept < 0.05

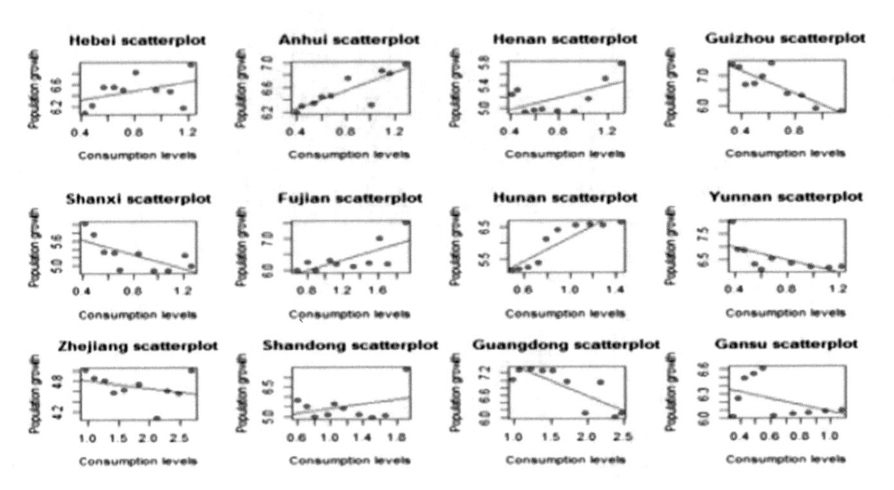

Fig. 8 Scatterplot of growth rate and consumption levels of provinces in second cluster

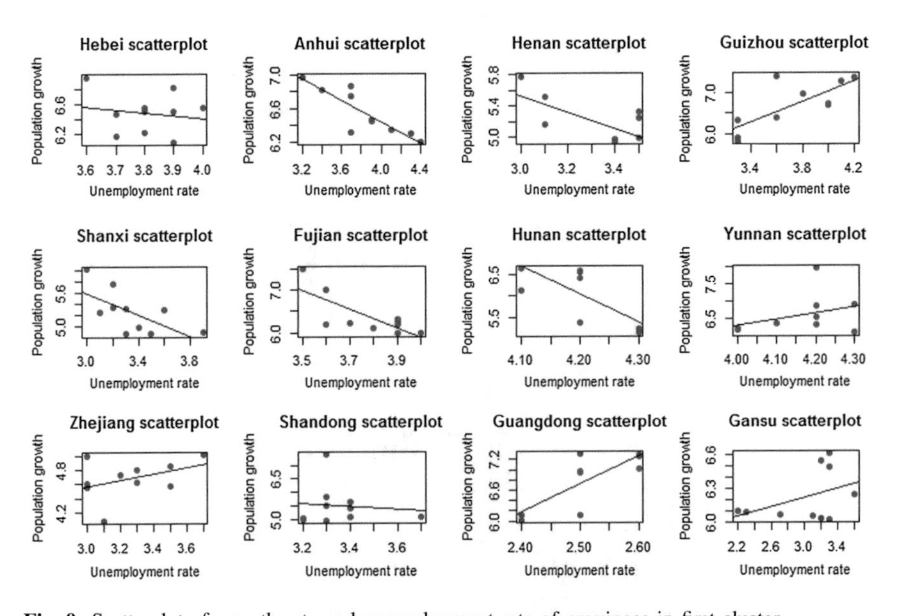

Fig. 9 Scatterplot of growth rate and unemployment rate of provinces in first cluster

The following are parameters of linear regression model for the class 2 districts.

In the class II districts, there is better degree of fitting for regression model 1 like Shanxi, Anhui, Fujian, Hunan, Guangdong, Guizhou, Yunnan, etc. The both p-values of Test t and test F are under 0.05. This indicates consumption level is a significant factor on natural population growth rate in these areas. While other districts are at a lower level of degree of fitting, which is not remarkable for significance of variables and equation test. This shows there is no linear relationship between consumption level and the population natural growth rate, and independent variable of consumption level is not remarkable factor to the natural population growth rate (Table 6).

In the class II districts, there is better degree of fitting for regression model 2 like Shanxi, Anhui, Fujian, Henan, Hunan, Guangdong, Guizhou, etc. The results of test t and test F are remarkable, and there is a linear relationship for explanatory variables and explained variables. It indicates unemployment rate has significant impact on the population natural growth rate. There is at a low level of degree of fitting for regression model 2 like Hebei, Zhejiang, Shandong, Yunnan, Gansu, etc. The results of test t and test F are not remarkable, which demonstrates that there is not a linear relationship between unemployment rate and the population natural growth rate, and unemployment rate of dependent variable has no significant impact on the population natural growth rate. Independent variable of unemployment rate is not a remarkable factor to dependent variable of the population natural growth rate.

In the class II districts, there is no satisfactory degree of fitting on two models for Hebei, Zhejiang, Shandong, Gansu. The level of correlation coefficient is low, and there is no linear relationship between variables. Further analysis is shown in Fig. 10.

As it is shown, for Hebei region, consumption level steps up year by year, while unemployment rate is stable at the level of 3.8 approximately during the past decade, the population growth rate increases a little. For Zhejiang region, the consumption level is also up and it is on the top of whole country. Unemployment rate is declined every year, there is limited increase on the population growth rate. For Shandong region, consumption level steps up for ten years, while unemployment rate change a little at around 3.8, there is a lot increase on the population growth rate. For Gansu region, consumption level increase for ten years, but it is lower than eastern areas. Unemployment rate is dropped and a small change is on the population growth rate. This shows it is fruitful policy on development of west regions.

4.3 *The Type III Linear Regression Analysis*

The type III is analyzed same as above, conclusion illustrated below:

There is strong negative correlation between the two variables in Jiangxi, Guangxi, Hainan, Qinghai, etc. It is very weak correlation between those two variables in Tibet, Ningxia, Xinjiang, etc.

There is substantial positive correlation between two variables in Jiangxi, Guangxi, Qinghai, etc. It is weaker correlation between those two variables in Hainan, Tibet, Ningxia, Xinjiang, etc.

Table 6 Model parameters table of provinces in second cluster

Provinces	Intercept	Coefficient	Multiple R-squared	p-value of T/F test	Intercept	Coefficient	Multiple R-squared	p-value of T/F test
Hebei	6.1638	0.3938	0.1712	Intercept < 0.05	8.0237	−0.4047	0.03244	Intercept < 0.05
Shanxi	5.9910	−0.8986	0.4869	All p-value < 0.05	8.4452	−0.9520	0.4199	All p-value < 0.05
Zhejiang	4.9462	−0.1458	0.1014	Intercept < 0.05	3.2619	0.4371	0.1532	Intercept < 0.05
Anhui	5.9717	0.7198	0.7083	All p-value < 0.05	8.9809	−0.6350	0.748	All p-value < 0.05
Fujian	5.3357	0.8242	0.4901	All p-value < 0.05	14.7176	−2.2070	0.5802	All p-value < 0.05
Shandong	4.8142	0.5739	0.1155	Intercept < 0.05	7.2626	−0.5297	0.01076	All p-value < 0.05
Henan	4.7727	0.5201	0.327	Intercept < 0.05	8.6075	−1.0290	0.458	All p-value < 0.05
Hunan	4.3387	1.8165	0.8187	All p-value < 0.05	32.863	−6.386	0.5279	All p-value < 0.05
Guangdong	8.1799	−0.8026	0.6478	All p-value < 0.05	−6.890	5.443	0.6416	Intercept > 0.05
Guizhou	7.9032	−1.8906	0.7437	All p-value < 0.05	1.9587	1.2692	0.579	Intercept > 0.05
Yunnan	7.4688	−1.2521	0.4303	All p-value < 0.05	−0.9256	1.8040	0.1391	All p-value > 0.05
Gansu	6.4845	−0.4070	0.1918	Intercept < 0.05	5.5971	0.2069	0.1684	Intercept < 0.05

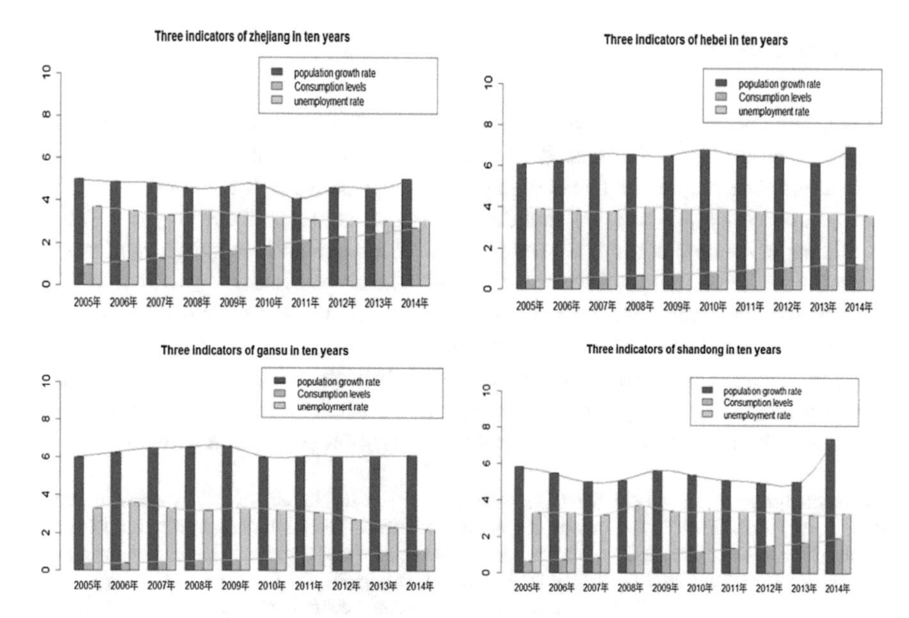

Fig. 10 Nonlinear relationship trends graphs of second cluster

The following are parameters of linear regression model for the class III districts.

In the class III districts, there is better degree of fitting for regression model 1 like Jiangxi, Hainan, Qinghai, etc. The consumption level is a significant factor on population natural growth rate in these areas. While other four districts are at a lower level of degree of fitting, which is not remarkable for significance of variables and equation test. This shows there is no linear relationship between consumption level and the population natural growth rate, and independent variable of consumption level is not remarkable factor to the natural population growth rate.

In the class III districts, there is low degree of fitting for regression model 2 for seven districts. The results of test t and test F are not remarkable, and there is not a linear relationship between unemployment rate and population natural growth rate. Unemployment rate of dependent variable has no important impact on the population natural growth rate.

In the class III districts, there is no satisfactory degree of fitting on two models for Tibet, Ningxia, Xinjiang. There is no a linear relationship between variables.

As it is shown, for Guangxi and Ningxia region, there is small increase on consumption level for a decade, and unemployment rate declines year by year. While the population natural growth rate fluctuates. For Xinjiang and Tibet region, there is very slow growth on the consumption level, and it is ranked bottom of whole country. However, the population growth rate is more than 10%, unemployment rate decreases.

5 Conclusion

This paper researches on the natural population growth rate based on districts and analyses three classes districts. It is clear to prove that there are correlation coefficient and one parameter linear regression model between the natural population growth and consumption level, and the unemployment rate. According to the result in this paper, people could learn the regional difference and changing trend among consumption level, unemployment rate and natural population growth rate. This paper provides support for government to make policies and development strategy and to improve population policy and structure.

What needs to be further researched are the districts, which show no correlation between the natural population growth rate and consumption level, unemployment rate, and take account of other impacts on the natural population growth.

Acknowledgements 1. Major Scientific Research Project in Higher School in Hebei Province (Grant No. ZD20131085), 2. Funding Project of Science & Technology Research and Development in Hebei North University (Grant No. ZD201301), 3. Major Projects in Hebei Food and Drug Administration (Grant No. ZD2015017), 4. Major Funding Project in Hebei Health Department (Grant No. ZL20140127), 5. Youth Funding Science and Technology Projects in Hebei Higher School (Grant No. QN2016192), with Hebei Province Population Health Information Engineering Technology Research Center.

Bibliography

1. Robert I. Kabacoff (2013) Gao Tao, Xiao Nan, Chen Gang(translation) R language.
2. Fang Kuangnan, Zhu Jianping, Jiang Yefei (2015), R data analysis method and case research.
3. JiaweiHai, MichelineKamber, Jian Pei (2012), Data exploration: Notion & Technique (third version in Chinese).
4. Wang Binhui (2010), Multiple statistic analysis and building R language.
5. Peng Yifeng, Categorized research on housing price, resident income and payment ability for housing - based on dimension data research.

How to Help Students Learn Class Combination in the Course of Java Language

Yuan Yali, Yin Feng Qin and Zhang Xiao

Abstract The combination of the class is a common method in software reuse. The teacher carried out a detailed analysis of class combination related difficulties such as object memory model representation, parameter transmission of method, the UML diagram of classes, effectively made the class combination less difficult and abstract, improved the classroom teaching effect, and facilitated students to master the core course of The Java Language Programming faster and better.

Keywords Object memory distribution · Parameter transmission of method · The UML diagram

Classification No:TP393 Reference Identification No:A
Software reuse is significant in reducing costs of development and maintenance as well as improving efficiency and quality in the course of software development. Object technique can provide technical support for software reuse. Java language is one of the popular object procedure design languages, while class combination is a widely used approach of software reuse in Java language.

Abstraction and encapsulation characterize object language. Abstraction forms class, and fills member variables and methods into a class. If a member of one class turns to be an object in another one, then, this is a class combination. A class combination comprises several classes with each class fulfilling one particular function, and classes coordinate and turn into a pipeline. In teaching, we find out that the beginners are likely to make mistakes when writing a program concerning class combination, and they cannot successfully read a progress when learning pipeline program. So it is necessary to analyze certain points, and highlight com-

---

Y. Yali · Y.F. Qin · Z. Xiao (✉)
School of Information Science and Engineering, Hebei Northern University, Zhangjiakou 075000, Hebei, China
e-mail: 780117251@qq.com

Y. Yali
e-mail: yylqmks@163.com

© Springer Nature Singapore Pte Ltd. 2018

N.Y. Yen and J.C. Hung (eds.), *Frontier Computing*, Lecture Notes in Electrical Engineering 422, DOI 10.1007/978-981-10-3187-8_40

bination relations by drawing model and UML whereby pointing out causes of mistakes [1].

The following is a discussion concerning class combination.

1 Memory Allocation for the Object

It is necessary to distinguish the object statement from the building of memory model for the understanding of such complex program as class combination. Take Example 1 as an object memory model.

1.1 Memory Model When Stating an Object

The memory model is shown as in Fig. 1. When statement object is c1 which contains no data, it is then an empty object which is not usable, if it is used, NullPointerException will work abnormally. Therefore objects must be substantiated [2].

1.2 Memory Model When an Object Is Created

Creating an object is to render it an entity, as shown in Example 1, when c1 = new Circle (5.0), the system is the distributed space for variables radius and area, then the statement in the construction method is to be carried out. If the member variable, when declared, does not specify an initial value, and the construction method does not initialize the member variable, then the default value of the transformation variable is 0, that of the floating point number (浮点数) is 0.0, the default value c in relation to the reference is null. At the same time, the object of c1, obtains a reference, namely, the memory name, by referring to access members of the member variables and methods.

The memory model transforms from Fig. 1, when stating an object, into substantiated memory model 2, after the object is distributed variables. The arrow shows that an object can operate its member variables and methods (Fig. 2).

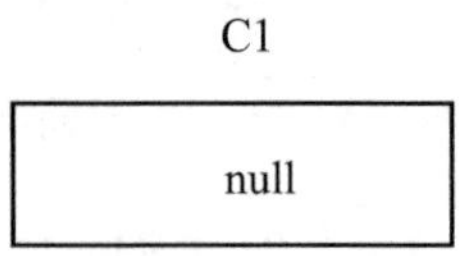

Fig. 1 An object without entity

C1

null

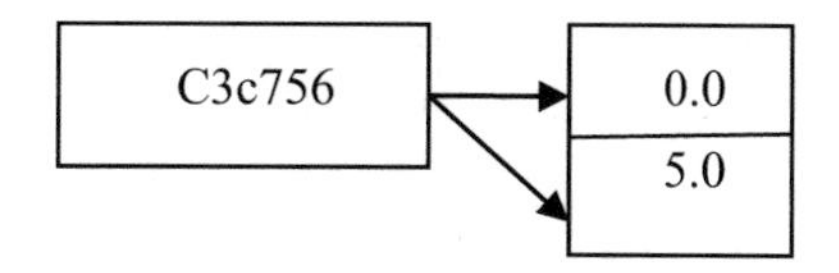

Fig. 2 An allocated object

```
Example1: Circle  c1; //statement of an object
c1=new Circle(5.0);//allocate an entity for an object
public class Circle {
   double radius,area;
   Circle(double r) {
      radius = r;}
   }
```

1.3 The Memory Model of the Class Combination

One class may take objects of another class as its member. If object a is combined with object b, then object a can delegate object b to maneuver object b's methods and member variables. The indirect operation is one of software re-usage. Given that beginners could not understand the reuse, in the following, Case 2 of code fragments is provided by drawing class combination memory model to facilitate students to understand better. This case defines three classes, i.e., Circular, Circle, Example2, Circular is combined with Circle, the volume of a cone is the area of a Circle multiplied by its height, then divided by 3. The memory model demonstrated by Case 2 is shown in Fig. 3.

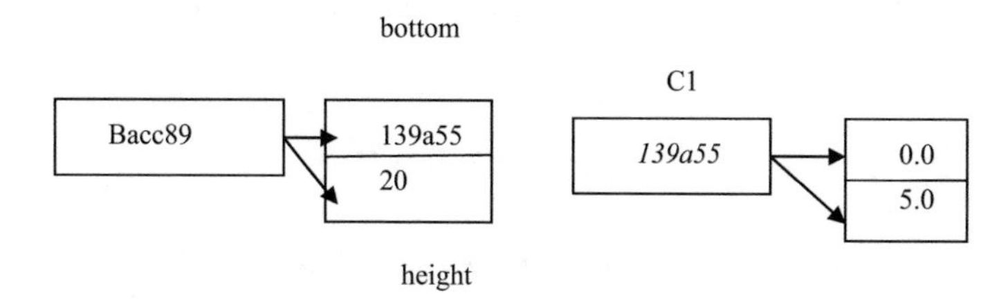

Fig. 3 Memory object model after executing code 2

Example 2 [3]:

```
public class Circle{

   double radius,area;
   Circle(double r) {
      radius = r;}
double getArea(){
       area=3.14*radius*radius;
       return area;    }
}
public class Circular {
   Circle bottom;
   double height;
   Circular(Circle c,double h) {
      bottom = c;// 【code1】
      height = h;}
double getVolme() {
      return bottom.getArea()*height/3.0;    }
}
public class Example2 {
   public static void main(String args[]) {
      Circle c1 = new Circle(5);
     Circular circular = new Circular(c1,20);
//【code 2:】: establishing objects of circular
}
}
```

In Case 2, Circular Circular = new Circular (c1, 20) is a method call, the statement marked by calling code 1, when calling code 1 statement, concerns parameter passing of a method, which constitutes another difficulty for students. The following is an introduction of the parameter passing of a method.

2 The Parameter Passing of a Method

Class is the encapsulation of data and method, while data describes the static characteristics of the class, and the method its dynamic characteristics. An object realizes its function by calling the method to behave. One of the most important parts of method is the parameter which is a local variable. When an object calls a method, the parameter is allocated memory space, and it also requires the caller to pass value to the parameter. Put simply, parameter variables must have a specific value when a method is called.

2.1 The Passing of Basic Data Parameters

In Java, the passing of basic data parameter is accompanied by "the passing of value", that is to say, the value of the parameter variable of a method is a copy of the specified value in calling. For example, when passing an int value towards int x parameter of a method, what the parameter x gets is a copy of the transferred value. Therefore, a method will not change the parameter variables when it changes the parameter value, that is, when the real parameter passes data to the form parameter, the change of the form parameter will not affect the real one.

2.2 The Passing of a Reference Parameter

Reference data in Java include number groups, objects, interfaces and so on. When a parameter is of a reference type, what gets passed is a "reference" stored in the variables, rather than the real one referred by variables. Two reference variables of the same type will use the same entity if they have the same reference. Therefore, if entity referred by parameter variables is changed, the entity of the original variable will be changed as well. To put it simply, when the real parameter passes data to a form parameter, the real parameter change referred by the form parameter will affect the entity referred by the real parameter.

In case 2, the code 1 bottom = c is the passing of instance reference of Circle class to the bottom by c1, the Circle class instances and the bottom have the same reference 139 a55, equivalent to the weight of the Circular bottom towards Circle, which is key to a combination and a prerequisite for Circular to entrust Circle to use its properties and call its methods in combination, otherwise, the chain pointer

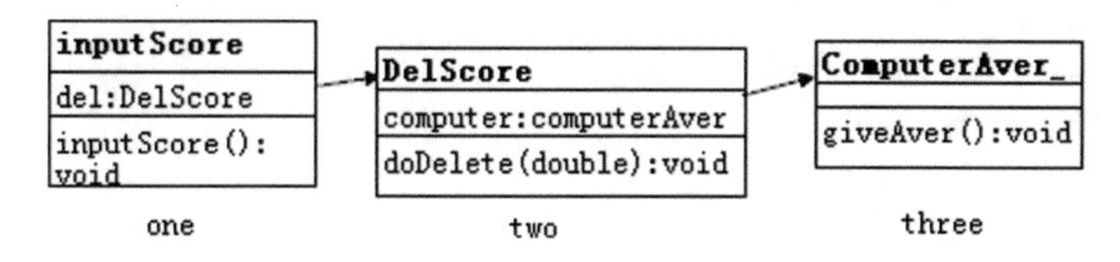

Fig. 4 UML

between the bottom and the c1 in Fig. 3 will disconnect, then these two classes are not truly connected. Remove the code bottom = c, Java. Lang. NullPointerException will occur, indicating that this is an abnormal pointer. This is because that the radius at the bottom of the cone is that of the circle, and if the statement bottom = c fails, bottom turns into a class type, and if no value to keep the agreed empty reference, cone bottom = null, then if the null reference is still used, mistakes are sure to happen. So here is another point meriting our attention.

3 UML Diagram

The so-called pipeline program means that, if an object contains the reference of object b, and object b contains a reference of object C, then objects a, b, C can form a pipeline, namely, a class with the combination of objects a, b and c gets established. The pipeline works only after users key in the data to it, who will in turn order the objects to process the data: first, object a processes the data, and passes automatically the result to object b. Then object b processes the result, and passes automatically the result to object c. In the singing contest, for example, it just needs to key in the judges' scores to the programmed pipeline, the final scores of the players can be obtained. Along the pipeline, the first object needs to feed in the judges' scores for the players, the second object removes a maximum point and a minimum one, the last object calculates the average.

Case 3 is an example of calculating points with the help of a pipeline. Objects of InputScore feed in the points, and class InputScore takes in objects of DelScore. Objects of DelScore delete a maximum point and a minimum one, and class DelScore combines objects of ComputerAver. Objects of ComputerAver work out the average in the end. Class line combines instances of InputScore, DelScoren and ComputerAver. The program runs like Fig. 4.

Example 3 [3]:

```
1:InputScore.java
importjava.util.Scanner;
public class InputScore {
DelScoredel ;
InputScore(DelScore del) {
this.del = del;//B
   }
```

```
public void inputScore() {
System.out.println("key in the number of judges ");
 Scanner read=new Scanner(System.in);
int count = read.nextInt();
System.out.println("key in the scores of every judge ");
double []a = new double[count];
for(int i=0;i<count;i++) {
a[i]=read.nextDouble();
     }
del.doDelete(a);
   }
}
```

2、DelScore.java

```
public class DelScore {
ComputerAvercomputer ;
DelScore(ComputerAver computer) {
this.computer = computer;//A
   }
public void doDelete(double [] a) {
java.util.Arrays.sort(a);  // number group a from big to small
System.out.print("delete a maximum point:"+a[a.length-1]+", ");
System.out.print("delete a minimum point:"+a[0]+"。");
double b[] =new double[a.length-2];
     for(int i=1;i<a.length-1;i++) { // delete a maximum point and a
minimum one
 (b[i-1] = a[i];
     }
computer.giveAver(b);
   }
}
```

3.ComputerAver.java

```
public class ComputerAver {
public void giveAver(double [] b) {
double sum=0;
for(int i =0;i<b.length;i++) {
  sum = sum+ b[i];
      }
  double aver=sum/b.length;
  System.out.println("the final score "+aver);()
    }
  }
```

```
4、Line.java
public class Line {
InputScore one;
DelScore two;
ComputerAver three;
Line(){
three=new ComputerAver();
two=new DelScore(three);
one=new InputScore(two);
    }
public void givePersonScore(){
one.inputScore();
    }
}
5、SingGame.java
public class SingGame {
public static void main(String args[]){
     Line line=new  Line();
line.givePersonScore();
  }
}
```

In this case study, given that most of the students feel difficult, three kinds of UML diagrams are drawn out. As shown in Fig. 4, Three is the object of ComputerAver, Two is the object of DelScore in carrying out the statement two = new DelScore (three). The relationship between two and three is established, namely, computer, the member variable of DelScore points to Three, to put it in another way, computer has the same reference as Three. One is the object of InputScore, when executing the statement one = new InputScore (two). Thus, One is connected with two, namely, del, the member variable of InputScore points to two, to put it simply, del has the same reference as two. The above relations is completed after Line Line = new Line in the main method.

As this combination process shows, the three class objects are established by way of the gradual outward combination of object three, object two, object one, with the outer layer being called the upper and the inner layer the lower. One is the root, if this structure is taken as a tree in data structure course.

The next statement to be executed is line. GivePersonScore (). It starts from the root one, inputScore () is to enter the points. In the next statement to be executed, i.e., del. DoDelete (a), del has the same reference as two, so doDelete method can be utilized to remove a minimum point and a maximum one. In the next statement, Computer giveAver (b), computer and Three are with the same reference, so giveAver (b) method can be used to work out the average.

4 Conclusion

In this paper, Java class combination is studied from memory allocation, parameter transmission, and object passing. The relationship between classes is clarified with the help of UML diagrams, reasons and solutions for a null reference specified.

References

1. Chen Baiqiang Guo Tao, Ruan Hui, Yan Jun. A static analysis of passing and null pointer errors in the Java program [J]. Computer Application, 2009 (29) (5): 1377–1379.
2. Yang Yi. A Study of Java Memory Allocation Mode [J]. Science and Technology Information, 2012 (35).
3. Geng Xiangyi Zhang Yueping. Experiment Guidance and Exercises Analysis for Java 2 Practical Course Book (3rd edition) [M]. Beijing: Tsinghua University Press, 2012.

An Exploratory Study for Evaluating Mathematical Anxiety in Calculus E-Assessment Platform by Using Physiological Signal Analysis

Chih-Hung Wu, Wei-Ting Lin and Shu-Chuan Shih

Abstract Few studies focused on integrating physiological signal analysis, and types of questions in e-learning assessment for understanding detail learning behavior, learning diagnosis, and level of anxiety. Anxiety is a feeling of uncertain, helpless, and fear about upcoming future. A learner's level of anxiety influences his/her learning performance, working memory process, and concentration. Past studies proved that mathematical anxiety interactively influences learners' self-efficacy and learning performance; besides, learners would raise their computer anxiety while learning via e-learning and testing system. Therefore, this research focuses on evaluating and comparing learners' mathematical anxiety level. Furthermore, we will also measure learners' Electroencephalography (EEG, brainwave signal) and Electrocardiography (ECG, heart beat rate) signal to understand their psychological status. This study will recruit University students to join our experiment. The result will help researchers realize the interactive effects among mathematical anxiety level, types of test questions (constructive and multiple choice question), and learning performance in improving the effects of Calculus e-learning tool.

Keywords Mathematical anxiety · Electroencephalography (EEG) · Electrocardiography (ECG) · Calculus E-assessment platform

C.-H. Wu (✉) · S.-C. Shih
National Taichung University of Education, Taichung, Taiwan, ROC
e-mail: chwu@mail.ntcu.edu.tw

S.-C. Shih
e-mail: ssc@mail.ntcu.edu.tw

W.-T. Lin
National Chiao Tung University, Hsinchu, Taiwan, ROC
e-mail: deacon.lovland@gmail.com

© Springer Nature Singapore Pte Ltd. 2018

N.Y. Yen and J.C. Hung (eds.), *Frontier Computing*, Lecture Notes in Electrical Engineering 422, DOI 10.1007/978-981-10-3187-8_41

1 Introduction

Mathematical anxiety commonly happens to college students in their Calculus, algebra, and other math courses [1]. Calculus is a major basic subject in many fields, such as engineering, science, and business management. students' higher mathematical anxiety will cause lower learning performance, [2–4], lower motivation of mathematical thinking, negative learning attitude [5], higher avoidance of math [6], and higher test anxiety [1]. In Mji and Mwambakana [4] research, they investigated college freshmen and pre-college students' mathematical anxiety status while learning basic algebra and Calculus via questionnaire survey, and discovered that students got lower grade if they have higher math anxiety. Thus, understanding students' mathematical anxiety status would assist instructors help students deal with their anxiety and improve learning performance. Previous research usually adopted mathematical anxiety in questionnaire scales. However, there is a lack of objective mathematical anxiety evaluation tool such as physiological analysis for evaluating college students while in Calculus math exam in e-learning planform. As a result, the major purpose of this paper is to explore the possibility of adopting physiological signals for evaluating students' mathematical anxiety in Calculus. We believed this objective evaluation tool will be a useful tool to college instructors to understand students' anxiety in e-learning planform.

2 The Physiological Input Signals for Emotion Recognition

Physiological signal analysis which can provide real-time physiological reaction information has been widely used in emotion [7–10], memory [8], clinic and health [11]. Affective computing techniques integrate physiological signal analysis and artificial intelligence algorithm that can explore the relationship between human's complex emotion status (e.g. anxiety) and physiological signals such as Electroencephalography (EEG) [12, 13], Electrocardiogram (ECG), Heart Rate (HR) [14–20], Skin Conductance Response (SCR) [19–22].

3 Research Design

This study will adopt two physiological signal detection devices to evaluate the anxiety in Calculus E-assessment planform. NeuroSky is comprised of a complex combination of artifact rejection and data classification methods. According the

Table 1 Physiological signals and variables in past studies

Research	Factors
EEG	
Price and Budzynski [12]	Anxiety
Lee, Wu [23]	Math learning performance
Luo, Wang [24]	Flow
Gruzelier, Foks [25]	Attention and music performance
De Pascalis, Varriale [26]	attention and learning ability
Staufenbiel , Brouwer [27]	Cognitive ability and memory
Nan, Rodrigues [28]	Memory
Sun [29]	Learning motivation and performance, attention
Wu, Tzeng [7]	Emotion and attention
Rashid, Taib [9]	Learning style, pressure, and intelligence
Wang and Hsieh [30]	Attention, working memory
Rashid, Taib [31]	Learning style and personality traits
Wu, Tzeng [14]	Game-based learning performance, emotion, cognitive load
Muldner and Burleson [32]	Creativity
ECG & HR	
Wu, Tzeng [7]	Emotion, attention
Friedman and Thayer [15]	Anxiety, distractor literature review
Wu, Tzeng [14]	Game-based learning performance, emotion, cognitive load
Dishman, Nakamura [16]	Anxiety and pressure
Crowley, McKinley [18]	Anxiety, pressure, and breath
Croft, Gonsalvez [19]	Public speaking anxiety
Carrillo, Moya-Albiol [20]	Anxiety and mood status
SCR	
Muldner and Burleson [32]	Creativity
Hofmann and Kim [21]	Anxiety in speech-anxious males
Croft, Gonsalvez [19]	Public speaking anxiety
Naveteur, Buisine [22]	Anxiety
Tang [33]	Learning performance, cognitive style, cognitive load, and fatigue
Carrillo, Moya-Albiol [20]	Anxiety and mood status

NeuroSky proprietary Attention and Meditation eSense algorithms, NeuroSky can report the wearer's 10 brainwave state (attention score, meditation score, delta1, theta, beta1, beta2, alpha1, alpha2, gamma1, gamma2) [34]. This study used the emWave Desktop system that uses an ear sensor to determine HRV, based on human pulse signals, in order to identify human emotion. The emWave system has a patented emotion recognition algorithm for identifying emotional states. With this

Table 2 Physiological signals in this study

Variable	Range	Operational definition	Meaning	Formula	References
Brainwave Mediation	0–100	Average value of mediation	The lower mediation, the higher anxiety	The algorithm of Neurosky's mediation	[35]
Normalized Brainwave Mediation	0–1	Average value/Average value in 5 min before experiment	The lower mediation, the higher anxiety	$= \left(\frac{\sum_{t=301}^{T_i} (Meditation_{it})/T_i}{\sum_{t=1}^{300} (Mediation_{it})/300} \right)$ *i: ith participant; t:* second $T_{i:}$: *ith* learner's finish time in second	
Heart beat rate (HR)	0–200	The number of heart beat in 1 min	The higher heart beat, the higher anxiety	The number of heart beat in 1 min by emwave device	[36, 37]
Normalized heart beat rate		Average value/Average value in 5 min before experiment	The higher heart beat, the higher anxiety	$= \left(\frac{\sum_{t=301}^{T_i} (HR_{it})/T_i}{\sum_{t=1}^{300} (HR_{it})/300} \right)$ *i: ith participant; t:* second $T_{i:}$: *ith* learner's finish time in second	

system, we can record the heart beating, and also can output the HRV data for following analysis [34] (Table 2).

4 System Development and Assessment

The Calculus E-assessment platform will include two types of questions—multiple choice question and construction question as shown in Figs. 1 and 2. The system will be tested on volunteer participants in university students. Participants will recruit university students, aged from 19 to 26. They will randomly be assigned to one of two types of questions in e-assessment platform. All of them were taken the Calculus course in this semester. Therefore, they already possessed some prior knowledge for solving Calculus problem. The physiological signals data will be collected in participant in Calculus E-assessment platform in real-time.

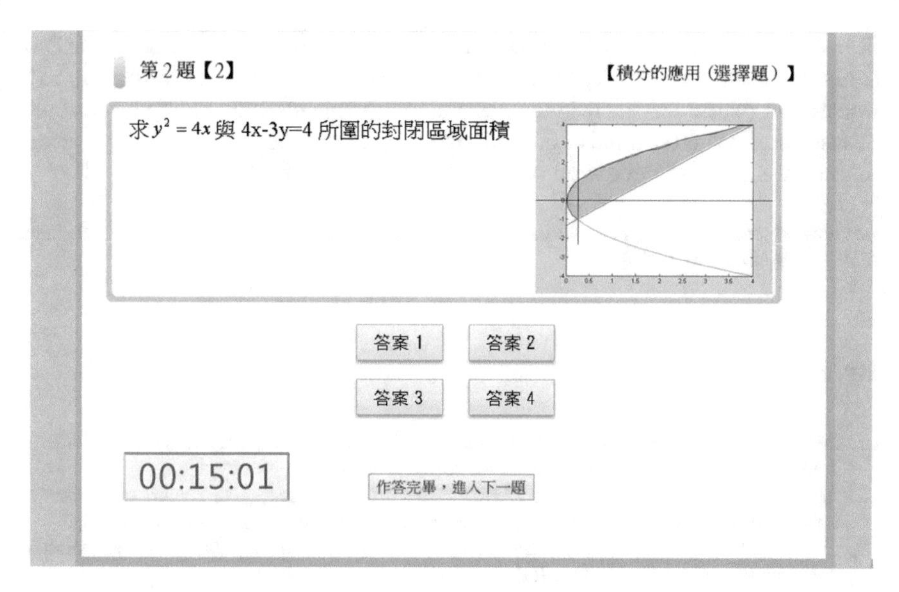

Fig. 1 An example of multiple choice question

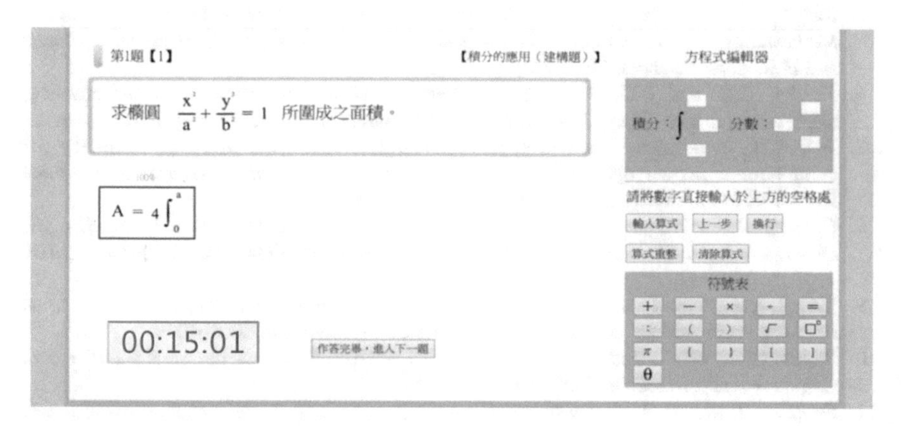

Fig. 2 An example of constructive question

5 Conclusion

The purpose of this study is to explore the method of using physiological signal analysis for evaluating learners' mathematical anxiety level in Calculus E-assessment platform. This study will adopt two types of physiological signals–Electroencephalography (EEG, brainwave signal) and Electrocardiography (ECG, heart beat rate) signal to understand their psychological status regard to their mathematical anxiety level. This study designs two types of questions—multiple choice and constructive question in our Calculus E-assessment platform to compare the differences between students' mathematical level in two types of questions. This study will recruit University students to join our experiment. The result will help researchers realize the interactive effects among mathematical anxiety level, types of test questions (constructive and multiple choice question), and learning performance in improving the design of Calculus E-assessment platform.

Acknowledgements Authors thank the Ministry of Science and Technology for support (grants MOST 104-2511-S-142-010).

References

1. Betz, N.E., *Prevalence, distribution, and correlates of math anxiety in college students*. Journal of Counseling Psychology, 1978. **25**(5): p. 441–448.
2. Beilock, S.L., et al., *Female teachers' math anxiety affects girls' math achievement*. Proceedings of the National Academy of Sciences, 2010. **107**(5): p. 1860–1863.
3. Hembree, R., *The Nature, Effects, and Relief of Mathematics Anxiety*. Journal for Research in Mathematics Education, 1990. **21**(1): p. 33–46.
4. Mji, A. and J. Mwambakana, *Is mathematics anxiety a factor?: First-year university students provide answers*. Africa Education Review, 2008. **5**(1): p. 20–29.
5. Kargar, M., R.A. Tarmizi, and S. Bayat, *Relationship between Mathematical Thinking, Mathematics Anxiety and Mathematics Attitudes among University Students*. Procedia - Social and Behavioral Sciences, 2010. **8**(0): p. 537–542.
6. Hendel, D.D. and S.O. Davis, *Effectiveness of an intervention strategy for reducing mathematics anxiety*. Journal of Counseling Psychology, 1978. **25**(5): p. 429–434.
7. Wu, C.H., et al., *Integration of affective computing techniques and soft computing for developing a human affective recognition system for U–learning systems*. International Journal of Mobile Learning and Organisation, 2014. **8**(1): p. 50–66.
8. Gruzelier, J.H., *EEG-neurofeedback for optimising performance. I: A review of cognitive and affective outcome in healthy participants*. Neuroscience & Biobehavioral Reviews, 2014. **44**(0): p. 124–141.
9. Rashid, N.A., et al., *Learners' Learning Style Classification related to IQ and Stress based on EEG*. Procedia—Social and Behavioral Sciences, 2011. **29**(0): p. 1061–1070.
10. Raymond, J., et al., *The effects of alpha/theta neurofeedback on personality and mood*. Cognitive Brain Research, 2005. **23**(2–3): p. 287–292.
11. Simpson, D.M., et al., *Demystifying Biomedical Signals: A student centred approach to learning signal processing*. Medical Engineering & Physics, 2005. **27**(7): p. 583–589.
12. Price, J. and T. Budzynski, *Chapter 17—Anxiety, EEG patterns, and neurofeedback, in Introduction to Quantitative EEG and Neurofeedback (Second Edition)*, T.H. Budzynski, et al., Editors. 2009, Academic Press: San Diego. p. 453–472.
13. Coan, J.A. and J.J.B. Allen, *Frontal EEG asymmetry as a moderator and mediator of emotion*. Biological Psychology, 2004. **67**(1–2): p. 7–50.
14. Wu, C.-H., Y.-L. Tzeng, and Y. Huang, *Understanding the relationship between physiological signals and digital game-based learning outcome*. Journal of Computers in Education, 2014. **1**(1): p. 81–97.
15. Friedman, B.H. and J.F. Thayer, *Autonomic balance revisited: Panic anxiety and heart rate variability*. Journal of Psychosomatic Research, 1998. **44**(1): p. 133–151.
16. Dishman, R.K., et al., *Heart rate variability, trait anxiety, and perceived stress among physically fit men and women*. International Journal of Psychophysiology, 2000. **37**(2): p. 121–133.

17. Lane, R.D., et al., *Neural correlates of heart rate variability during emotion.* NeuroImage, 2009. **44**(1): p. 213–222.
18. Crowley, O.V., et al., *The interactive effect of change in perceived stress and trait anxiety on vagal recovery from cognitive challenge.* International Journal of Psychophysiology, 2011. **82**(3): p. 225–232.
19. Croft, R.J., et al., *Differential relations between heart rate and skin conductance, and public speaking anxiety.* Journal of Behavior Therapy and Experimental Psychiatry, 2004. **35**(3): p. 259–271.
20. Carrillo, E., et al., *Gender differences in cardiovascular and electrodermal responses to public speaking task: the role of anxiety and mood states.* International Journal of Psychophysiology, 2001. **42**(3): p. 253–264.
21. Hofmann, S.G. and H.-J. Kim, *Anxiety goes under the skin: Behavioral inhibition, anxiety, and autonomic arousal in speech-anxious males.* Personality and Individual Differences, 2006. **40**(7): p. 1441–1451.
22. Naveteur, J., S. Buisine, and J.H. Gruzelier, *The influence of anxiety on electrodermal responses to distractors.* International Journal of Psychophysiology, 2005. **56**(3): p. 261–269.
23. Lee, T.-W., et al., *A smarter brain is associated with stronger neural interaction in healthy young females: A resting EEG coherence study.* Intelligence, 2012. **40**(1): p. 38–48.
24. Luo, X., F. Wang, and Z. Luo, *Investigation and analysis of mathematics anxiety in middle school students.* Journal of Mathematics Education, 2009. **2**(2): p. 12–19.
25. Gruzelier, J.H., et al., *Beneficial outcome from EEG-neurofeedback on creative music performance, attention and well-being in school children.* Biological Psychology, 2014. **95**(0): p. 86–95.
26. De Pascalis, V., V. Varriale, and M. Rotonda, *EEG oscillatory activity associated to monetary gain and loss signals in a learning task: Effects of attentional impulsivity and learning ability.* International Journal of Psychophysiology, 2012. **85**(1): p. 68-78.
27. Staufenbiel, S.M., et al., *Effect of beta and gamma neurofeedback on memory and intelligence in the elderly.* Biological Psychology, 2014. **95**(0): p. 74–85.
28. Nan, W., et al., *Individual alpha neurofeedback training effect on short term memory.* International Journal of Psychophysiology, 2012. **86**(1): p. 83–87.
29. Sun, J.C.-Y., *Influence of polling technologies on student engagement: An analysis of student motivation, academic performance, and brainwave data.* Computers & Education, 2014. **72**(0): p. 80–89.
30. Wang, J.-R. and S. Hsieh, *Neurofeedback training improves attention and working memory performance.* Clinical Neurophysiology, 2013. **124**(12): p. 2406–2420.
31. Rashid, N.b.A., et al., *Summative EEG-based Assessment of the Relations between Learning Styles and Personality Traits of Openness.* Procedia - Social and Behavioral Sciences, 2013. **97**(0): p. 98–104.
32. Muldner, K. and W. Burleson, *Utilizing sensor data to model students' creativity in a digital environment.* Computers in Human Behavior, 2014(0).
33. Tang, K.-D., *Match and Mismatch Effects in the Study of E-Commerce Web-Based Learning System Effectiveness Analysis: Based on the Cognitive Styles and Physiological Factors, in Digital Content & Technology.* 2014, National Taichung University: Taichung, Taiwan.
34. Wu, C.-H., et al., *Integration of affective computing techniques and soft computing for developing a human affective recognition system for U-learning systems.* Int. J. of Mobile Learning and Organisation, 2014. **8**(1): p. 50–66.
35. Peters, C., S. Asteriadis, and G. Rebolledo-Mendez:, *Modelling user attention for human-agent interaction*, in *WIAMIS 2009*. 2009. p. 266–269
36. Sharma, N. and T. Gedeon, *Objective measures, sensors and computational techniques for stress recognition and classification: A survey.* Comput. Methods Prog. Biomed., 2012. **108**(3): p. 1287–1301.
37. Damaraju, E., et al., *Affective learning enhances activity and functional connectivity in early visual cortex.* Neuropsychologia, 2009. **47**(12): p. 2480–2487.

GPU Based Simulation of Collision Detection of Irregular Vessel Walls

Binbin Yong, Jun Shen, Hongyu Sun, Zijian Xu, Jingfeng Liu and Qingguo Zhou

Abstract Collision detection is a commonly used technique in the fields of computer games, physical simulation, virtual technology, computing and animation. When simulating the process of particle collision of ADS (Accelerator Driven Sub-Critical) system, complex and irregular vessel walls need to be considered. Generally, an irregular vessel wall is a curve surface, which cannot be defined as an exact mathematical function, and it is difficult to calculate the distance between particles and the wall directly. In this paper, we present an algorithm to perform collision detection between particles and irregular wall. When the number of particles reaches the level of 10^6, our algorithm implements a considerable improvement in performance if running on GPU, nearly 10 times faster than running on CPU. Results have demonstrated that our algorithm is promising.

Keywords Collision detection · Irregular vessel · Physical simulating · GPU

B. Yong · H. Sun · Z. Xu · Q. Zhou (✉)
School of Information Science and Engineering, Lanzhou University,
Lanzhou, China
e-mail: zhouqg@lzu.edu.cn

B. Yong
e-mail: yongbb14@lzu.edu.cn

H. Sun
e-mail: sunhy13@lzu.edu.cn

Z. Xu
e-mail: xuzj15@lzu.edu.cn

J. Shen
School of Computing and Information Technology, University of Wollongong,
Wollongong, Australia
e-mail: jshen@uow.edu.au

J. Liu
Linksprite, Wuhan, China
e-mail: jingfeng.liu@linksprite.com

© Springer Nature Singapore Pte Ltd. 2018

N.Y. Yen and J.C. Hung (eds.), *Frontier Computing*, Lecture Notes
in Electrical Engineering 422, DOI 10.1007/978-981-10-3187-8_42

1 Introduction

With the rapid development of GPU (Graphics Processing Unit) technology, we are more likely able to implement complex and realistic system. In our research on nuclear simulations, the accelerated particles are generally made to collide with other particles and the irregular wall. The collision detection between particles is a hot research topic [1]. Hence, we need to simulate the collision between various particles and different walls.

In this paper, we design an algorithm to implement the collision detection between the particles and irregular wall by dividing the wall into very small triangles. Next, we assign the triangles and particles into uniform grids by space subdivision, and make collision detection between these triangles and particles (these particles are considered as spheres) in parallel with the help of GPU. We also implement the algorithm with CPU and make a comparison between these two computing architectures. Consequently, experimental results demonstrate that our parallel algorithm has a good feasibility and obvious effect in accelerating the simulation.

The rest of this paper is organized as follows. Section 2 reviews the related work. We present our collision algorithm in Sect. 3. In Sect. 4, we show the experimental results, and the effects of different parameters are discussed. At last in Sect. 5, we conclude the work and address the future work.

2 Related Work

General Purpose GPU (GPGPU) is a relatively new research area. But the idea of using GPU in collision detection has been researched longer than the emergence of GPGPU.

Zheng et al. [2] showed a contact detection algorithm based on GPU and they used the uniform grid method in detection. Based on the vector relation of point, line segment and rectangle, Shen et al. [3] implemented a rapid collision detection algorithm. Li and Suo [4] researched the application of particle swarm optimization in randomly collision detection algorithm and increased real-time capacity, compared to the classic OBB (Oriented Bounding Box) bounding box algorithm. Similarly, Qu et al. used parallel ant colony optimization algorithm in randomly collision detection algorithm to improve the real-time characteristic and precision in collision detection [5]. With spatial projection transformation method, Li and Tao mapped irregular objects from three dimensional space to regular two-dimensional objects to carry on collision detection [6]. Based on MPI (Message Passing Interface) and spatial subdivision algorithm, Huiyan et al. researched an advanced algorithm to improve the performance and accuracy of collision detection [7]. Tang et al. [8] proposed a GPU-based streaming algorithm to perform collision queries between deformable models by using hierarchical culling. Zhang et al. presented a parallel collision detection algorithm with multiple-core computation by CPUs or GPUs [9]. Wang et al. proposed an image-based optimization algorithm for collision detection [10].

Although many of these GPU-based collision detection algorithms have been researched, the research on the collision of irregular walls has been seldom discussed so far.

3 Collision Detection Algorithm

3.1 Overview

In this paper, a 3D irregular wall model is represented by STL (STereo Lithograph) file format, which contains many triangle meshes. The basic idea of our algorithm is to divide the triangles of the wall into very small triangles that are in the same scale as the particles. And then we assign these triangles into different grids by their center. At last, we make collision detection between these triangles and the particles in the same grid.

We implement both CPU-based and GPU-based codes to detect collision between particles and irregular vessels. However, the main idea of the algorithm is similar. The basic process of our algorithm is as follows: Firstly, we read the model file to get the position information of the triangle meshes. Next, we divide these triangles into smaller triangles, whose sizes are limited to particular range to make sure it can be contained by the uniform grids. The next phase is to make space subdivision. In this paper, we simply use the uniform grid to divide the space. Next we put the particles and the triangles into corresponding grid, and sort them to make it easy to implement data replication between CPU and GPU memory (no sort operation and data replication in CPU code). Finally, we perform the collision detection between particles and triangles in the same grid by calculating the distance between particles and triangles.

3.2 Data Structures

When making collision detection on GPU, we firstly copy the data structures of particles and triangles from CPU memory to GPU memory. (Note: The latest CUDA versions support (UMA) uniform memory access, but our algorithm is designed not only for new CUDA versions). The data structures are organized in array to copy quickly rather than use complex data structures such as linked list. The basic arrays are particles array and triangles array. The particles array is organized every four float data for each particle, including three coordinate values and a radius. The triangles array is organized every 12 float data for each triangle, including three points (9 coordinate values) and a normal vector (3 coordinate values). There are also auxiliary arrays, such as hash arrays, index arrays, cell start arrays and cell end arrays, both for particles and triangles.

3.3 Triangles Division

We read the triangles of the irregular wall model from an STL file. In general, these triangles are much larger than the particles. Firstly, we need to divide the triangles into smaller ones. In this paper, we divide the triangles by two methods to acquire a better solution. The two methods need a different minimum value of side length to divide the triangles.

The first method is based on the center of a triangle (centre of gravity), which is denoted as triangle-center method. We assign the divided triangles into different grids by the center of a triangle, and the minimum value of side length is set as the maximum radius of the particles.

Another method is based on the circumcenter of a triangle, which is denoted as circumcenter method. In this case, we should make sure that the area of every small triangle is as large as possible to reduce the number of triangles after dividing, so as to reduce the computation complexity. When the maximum value of the side length is fixed, an equilateral triangle has the largest area.

Hence, we get the side length by equation

$$d = r \times \cos(30^\circ) \times 2 \tag{1}$$

We assign these divided triangles into different grids by their circumcenters, and the minimum value of side length is set as Eq. (1).

But there is still an issue. If the divided triangle is an obtuse triangle, the circumcenter may be out of the triangle. In this situation, it is difficult to calculate which grid the obtuse triangle belongs to. Hence, we need to divide the obtuse triangle into smaller non-obtuse triangles. We simply draw a vertical line from the vertex of the obtuse angle to the opposite edge. By this way, the obtuse triangle is divided into two right triangles.

These two methods look similar. Nevertheless, the number of divided triangles grows exponentially. With the circumcenter method, the minimum side length is $\sqrt{3}$(Eq. (1)) times of the first method. However, the total number of triangles after dividing is no more than half of the triangles divided by triangle-center method.

The basic approach to dividing a triangle is to recursively split the triangle into two smaller triangles from the midpoint of this edge. Figure 1 shows the divided model. The image in the left is the original model, and the middle is the divided model.

Fig. 1 The original, divided (wall1) and spallation target model (wall2) walls

3.4 Spatial Subdivision

We split the world space into uniform grids, and the number of grids in three dimensions is denoted as (*sizeX*, *sizeY*, *sizeZ*). Overall, there are $sizeX \times sizeY \times sizeZ$ grids. Each grid has the same scale size as the particles. For each particle we calculate the grid position (x, y, z) by its center position. And then we calculate the grid number as its hash value simply through equation

$$hash = z \times sizeY \times sizeX + y \times sizeX + x \tag{2}$$

For each triangle (after dividing) and particle, we judge the grid position by its center point. For a triangle, we calculate its triangle core or circumcenter, which is used to get the number of grid. We calculate the grid number (hash value) by Eq. (2).

3.5 Sorting and Data Replication

For a grid, the number of particles and triangles are not fixed. We herein can not store these data in fixed memory size. One way is to reorder the particles array and the triangles array by its hash value (the grid number), and then we store the particles and the triangles of each grid by their start number and end number. We sort the data via fast radix sort in the CUDPP library.

After sorting the data, we still need to fill in the hash start arrays and hash end arrays for particles and triangles. Each particle or triangle gets its start hash and end hash by comparing its cell index with the previous cell index. If there is a difference, it indicates a new grid and new start hash (the particles or triangles in one grid have a same hash value).

3.6 Collision Detection

Once the grid structures are built on GPU memory, it is used to detect particle-wall interactions. In the continuous simulation, we must consider the specific collision physical model between particles and walls to obtain the next frame. For example, a DEM (Discrete Element Method) [11, 12] may be used and the forces should be considered. In this paper, anyhow, we focus on the collision detection rather than the interaction model.

There are two methods to traverse collision detection. The first method is a wall-based method. For triangles in a grid, we get the start triangle and the end triangle, and then we loop over the neighboring grid cells of each triangle to check for collisions with each particle in these cells. The second method is particle-based method. Similarly, we find start particle and end particle in a grid, and then loop to detect neighboring collisions. There are similarities between these two methods. However, the computational efficiency is different. The detailed results can be seen in Sect. 4.

The last issue of basic collision detection is the test between a sphere and a triangle, which is equivalent to calculate the shortest distance between the sphere core and the triangle. If the distance is smaller than the sphere radius, it means a collision occurs.

4 Experiments

4.1 Experimental Environment

The experiments were basically performed on a computer with Intel Core i3 processor and 4.00 GB RAM and a GTX480 GPU. We also run the code on a computer with Tesla K80 GPU to find the improvement on different GPU unit. We used a 3D modeling software to draw a cylinder (height 100 and radius 50, the left and middle of Fig. 1, name wall1) as one simple collision wall, which was easy to verify collision detection. We also experimented on a spallation target model (the right of Fig. 1, name wall2) used in our project. The particles are different in size, but the maximum radius is set as 1. The length of uniform grid is set as 2.

Next, we mainly focus on the experiments between CPU and GPU, the differences between triangle-center and circumcenter triangulation method, wall-based and particle-based traversal methods, the influence of number of triangles and the differences on different GPUs.

4.2 CPU and GPU

In order to measure the performance of our algorithm, we choose different particle numbers as the input variable. We conduct experiments on CPU and GPU platform. As shown in the left of Fig. 2 and Table 1 (only 10^2, 10^3, 10^4, 10^5, 10^6 level of magnitudes are shown), we can see that if the number of particles is small (less than 10^3), the CPU and GPU code has a similar computation time, and CPU code may even achieve a better performance. However, with the increase of the number of particles (more than 10^4), the computation time of CPU code increases sharply. When the number of particles is 10^4, 10^5 and 10^6, compared to CPU code, the speed-up ratio of GPU code by circumcenter method is 1.89, 3.69 and 4.99.

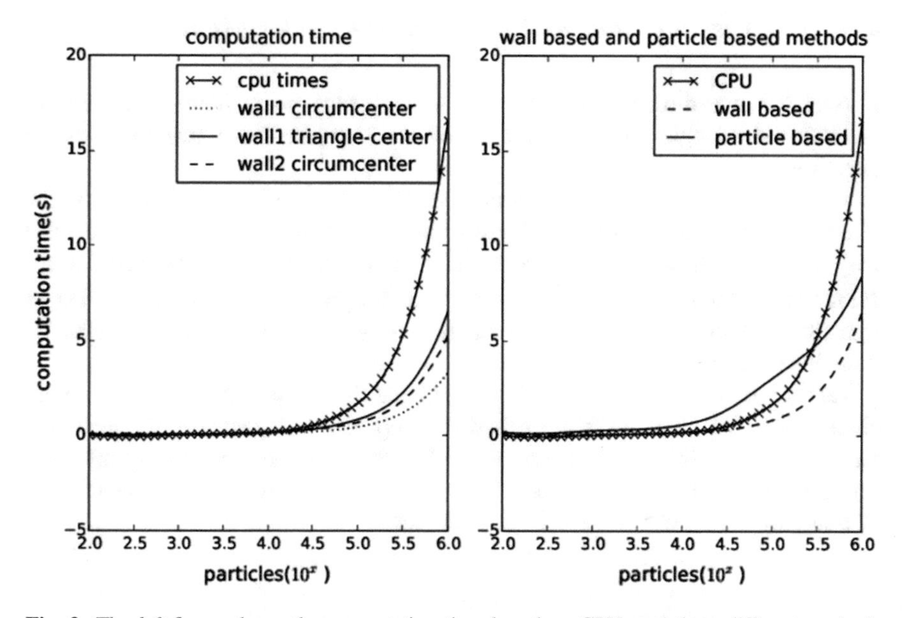

Fig. 2 The *left* figure shows the computation time based on CPU, and three different methods based on GPU: wall1 circumcenter method, wall1 triangle-center method, wall2 circumcenter method. The *right* figure illustrates the comparison between wall based method and particle based method

Table 1 Experimental results of computation time

Number	10^2	10^3	10^4	10^5	10^6
CPU time(s)	0.02577	0.03819	0.19552	1.67494	16.5695
wall1 tri-center(s)	0.05659	0.06805	0.15951	0.83642	6.51862
wall1 circumcenter(s)	0.03703	0.04499	0.10320	0.45318	3.31567
wall2 circumcenter(s)	0.04515	0.05750	0.13836	0.68392	5.26571

4.3 Two Traversal Methods

As discussed above, there are two methods to detect collision based on GPU, wall based and particle based methods. In the right of Fig. 2, the computation time of these two methods was drawn. If we use particle based method, when the number of particles is less than $10^{5.5}$, the GPU code is even worse than CPU code. When the number of particles is larger, the particle based GPU code performs better than CPU code. However, the particle based code is no better than triangle based code. In general, the number of triangles is much smaller than the number of particles. Hence, the particle based code needs more concurrent threads to finish collision detection.

4.4 The Circumcenter and the Triangle-Center Methods

In Sect. 3.3, we discussed two type of triangle division methods, circumcenter method and triangle-center method. Now we compare these two methods on experimental data.

For the model wall1, the original number of input triangles is 156. If we use the triangle-center method to divide these triangles and the minimum value of side length is set as 1, we get 226056 triangles after division. When we use the circumcenter method, the minimum value of side length is set as $\sqrt{3}$. The number of triangles is 99612 after division, which is less than half of the triangles by triangle-center method.

As shown in Fig. 2 and Table 1, we find that the circumcenter method is more efficient than the triangle-center method in computation. A plausible explanation is that the circumcenter method uses the collision space more effectively and gets fewer triangles after division. In fact, the computation time is relevant to the number of triangles, which can be seen from next section.

4.5 The Influence of Triangles Number

We tested the influence of number of triangles based on fixed number of particles. In Fig. 3, we set the number of particles as 10^5 and 10^6 and test the relationship between number of triangles and computation time. We roughly draw a conclusion that the computation time is proportional to the number of triangles. We calculate the following linear fitting function. Where t_5 and t_6 stand for the time for 10^5 and 10^6 particles, and n_{tri} stands for the number of triangles.

$$t_5 = 2.477 \times 10^{-9} \cdot n_{tri} + 0.2712 \tag{3}$$

$$t_6 = 2.078 \times 10^{-5} \cdot n_{tri} + 1.8970 \tag{4}$$

We also tested the spallation target model wall2 based on circumcenter triangle division method. The original number of triangles is 1600. The number of triangles after division is 151424. As shown in Fig. 2, it even runs faster than the simpler model wall1 based on triangle-center method. The reason is that the number of divided triangles of wall1 based on triangle-center method is 226056, larger than number of divided triangles of wall2 based on circumcenter method. Therefore,

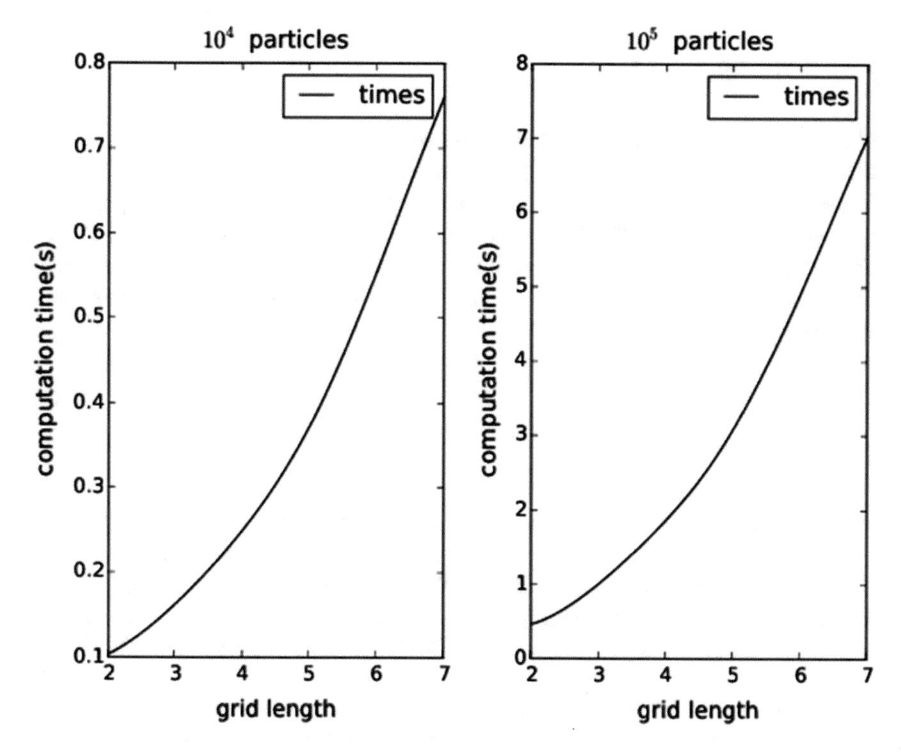

Fig. 3 Experimental results: the influence of the number of triangles

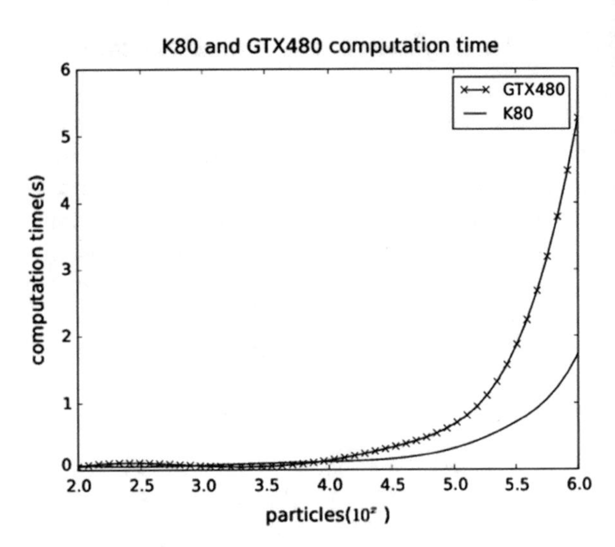

Fig. 4 Experimental results: the comparison between GTX480 and K80

using a better method to reduce the number of divided triangles is a better way to improve computational efficiency.

4.6 *On Different GPUs*

We also tested our algorithm on GTX480 and Tesla K80 GPU. As shown in Fig. 4. When the number of particles is less than 10^4, the code on GTX480 runs faster. However, once the number of particles is more than 10^4, for instance, 10^4, 10^5, 10^6, compared to code on GTX480, the speed-up ratio of code on K80 are 1.15, 2.14, 3.04. Hence, we can conclude that with the increase of the number of particles, a GPU with higher computation ability will obviously improve computation efficiency.

5 Conclusion

In our current process, it is meaningful to focus on collision detection between irregular walls and particles. In this paper, we designed and implemented an algorithm to achieve this goal. With the help of a K80 GPU, we can detect collision between a million particles and a spallation target model in 2s. The experiment results prove that the algorithm proposed in this paper is feasible and effective.

Acknowledgements This work was supported by Dongguan's Recruitment of Innovation and entrepreneurship talent program, National Natural Science Foundation of China under Grant No. 61402210 and 60973137, Program for New Century Excellent Talents in University under

Grant No. NCET-12-0250, Strategic Priority Research Program of the Chinese Academy of Sciences with Grant No. XDA03030100, Gansu Sci. and Tech. Program under Grant No. 1104GKCA049, 1204GKCA061 and 1304GKCA018, Google Research Awards and Google Faculty Award, China.

References

1. NVIDIA, *Particle Simulation using CUDA*, 1st ed., NVIDIA, 9 2013.
2. J. Zheng, X. An, and M. Huang, "Gpu-based parallel algorithm for particle contact detection and its application in self-compacting concrete flow simulations," *Computers & Structures*, vol. 112, pp. 193–204, 2012.
3. Y. Shen, Q. Jia, G. Chen, Y. Wang, and H. Sun, "Study of rapid collision detection algorithm for manipulator," in *Industrial Electronics and Applications (ICIEA), 2015 IEEE 10th Conference on*, Jun. 2015, pp. 934–938.
4. S. Xue-li and Z. Ji-suo, "Research of collision detection algorithm based on particle swarm optimization," *Computer Design and Applications (ICCDA)*, vol. 1, 2010.
5. H. Qu and W. Zhao, "Fast collision detection algorithm based on parallel ant," *Virtual Reality and Visualization (ICVRV)*, pp. 261–264, 2013.
6. S. Xue-li and L. Tao, "Fast collision detection based on projection parallel algorithm," *Future Computer and Communication (ICFCC)*, vol. 1, 2010.
7. H. Qu and W. Zhao, "Fast collision detection of space-time correlation," *Computer Science and Electronics Engineering (ICCSEE)*, vol. 3, pp. 567–571, 2012.
8. M. Tang, D. Manocha, J. Lin, and R. Tong, "Collision-streams: Fast GPU-based collision detection for deformable models," in *I3D '11: Proceedings of the 2011 ACM SIGGRAPH symposium on Interactive 3D Graphics and Games*, 2011, pp. 63–70.
9. X. Zhang and Y. J. Kim, "Scalable collision detection using p-partition fronts on many-core processors," *IEEE Transactions on Visualization and Computer Graphics*, vol. 20, no. 3, pp. 447–456, Mar. 2014.
10. L. Wang, Y. Shi, and R. Li, "An image-based collision detection optimization algorithm," in *Signal and Information Processing (ChinaSIP), 2015 IEEE China Summit and International Conference on*, Jul. 2015, pp. 220–224.
11. H. Karunasena, W. Senadeera, Y. Gu, and R. Brown, "A coupled sph-dem model for fluid and solid mechanics of apple parenchyma cells during drying," in *18th Australian Fluid Mechanics Conference*. Australasian Fluid Mechanics Society Launceston, Australia, 2012.
12. M. Rhodes, X. S. Wang, M. Nguyen, P. Stewart, and K. Liffman, "Study of mixing in gas-fluidized beds using a dem model," *Chemical Engineering Science*, vol. 56, no. 8, pp. 2859–2866, 2001.

A Logistics Privacy Protection System Based on Cloud Computing

Xin Liu, Bingmeng Hu, Qingguo Zhou and Jingfeng Liu

Abstract Generally, because the logistics business uses plain text in orders, the disclosure risk of user's privacy information widely exists in various logistics business processes. In this paper, combined with cloud computing and two-dimensional code in networking technology, we build a secure, trusted logistics system, completely protect people's privacy from the source (sender) to the recipient. The system protects us from being harassed by calls and messages. At the same time, the project uses the new security model, which greatly reduces the risk of information disclosure, to ensure the security of user's information.

Keywords Logistics management · Privacy protection · Cloud computing

1 Introduction

With the promotion of global economic integration and the refinement of social division, the logistics industry has become an important basis for the development of the national economy. The flourish of third-party logistics, the quick response and the collaborative interaction has been the subject of logistics development. But look at the current logistics management processes, we will find that the customer's personal information is at risk of being leaked. The causes of personal information

---

X. Liu · B. Hu · Q. Zhou (✉)
School of Information Science and Engineering,
Lanzhou University, Lanzhou, People's Republic of China
e-mail: Zhouqg@lzu.edu.cn

X. Liu
e-mail: liuxin13@lzu.edu.cn

B. Hu
e-mail: hubm13@lzu.edu.cn

J. Liu
LinkSprite Technologies, Inc., Wuhan, China
e-mail: Jingfeng.liu@linksprite.com

© Springer Nature Singapore Pte Ltd. 2018

N.Y. Yen and J.C. Hung (eds.), *Frontier Computing*, Lecture Notes in Electrical Engineering 422, DOI 10.1007/978-981-10-3187-8_43

disclosure are various. The first aspect is that customers discarded the delivery list they signed for. The second aspect is that some merchants and courier companies make unauthorized purchases. And the second aspect is the hidden dangers and security vulnerabilities in the e-commerce platform that may allow hackers to attack and steal the customer's information [1]. Disclosure of personal information makes customers suffer from spam, telemarketing calls and other harassment, and some even cause frauds, extortions, burglaries and other offenses. The direct display of personal information in plain text in the logistic order is the main cause of personal information leakage.

At present, more and more companies not only use courier logistics information system, and with the development of cloud computing and mobile computing, mobile electronic orders and signings systems are increasingly widespread [2, 3]. Express delivery and other logistics information systems management information is becoming the mainstream of logistics information management. Courier logistics information management system is becoming the mainstream of logistics information management. Logistics Management Information System [4, 5] is a people-oriented information systems, which uses various software, hardware, network communications equipment and other office equipment to collect, transmit, process and update information in order to improve the effectiveness and efficiency.

With the development of express industry, domestic and foreign scholars had made a lot of theoretical analyses and empirical research on these issues. In the 1980s, the US Federal Express started using GPS to design computer tracking system, which provides online real-time inquiry cargo service [6]. In the 1990s, the UPS has created the portable high-tech equipment to create digital signatures. In the recent years, with the development of delivery technology, there have been research based on RFID [7, 8] and two-dimension code [9, 10]. In contrast, the two-dimensional code has more advantages(such as large amount of information can be stored and encrypted, high recognition rate can be obtained, Chinese characters can be supported, etc.) We build a logistics privacy protection system based on cloud computing technology and the two-dimensional code. Through a third party escrow system which is trusted and access control, the system can protect privacy from malicious, illegal leak. At the same time, the system takes into account the business needs of traditional logistics to ensure the logistics companies have a smooth transition to the next generation of Internet technology.

2 Technical Routes

The system consists of two subsystems: third party escrow system and security logistics service system. The third party escrow system used to store the information of the user, which ensures that user's information can not be obtained by untrusted parties such as Taobao sellers. The security logistics service system is used to ensure the logistics enterprise employees, especially the grass-roots employees will not divulge user's privacy maliciously for profit.

Overall, our core mission is to build the RESTful API based on MVC architecture. We realize the separation of the function structure and the data exchange of each terminal device by API. As for the development process, we use Node.js/Ruby on Rails/C++ as the main function of the development of language, combined with the Java language as an auxiliary language. In the deployment level, we apply the Reverse Proxy technology in human-computer interaction. At the same time, it can release the static resources to speed up the overall system and reduce the core server load levels.

3 Third Party Escrow System

The system is designed for the user (the recipient) to fill in their own private data and logistics enterprises to obtain the recipient data. When the user has saved the information to the system, the system will generate a unique identifier, which is uniquely corresponds with personal information. After the user passes the information to the sender, the sender can use this identifier to express mail in the logistics enterprise.

The process of the third party escrow system is depicted in the Fig. 1. The main points of the system are as follows:

(1) Fill in the recipient's own information (name, mailing address, phone number, zip code, etc.) on this platform.
(2) The recipient information will be stored in the database system and the unique identifiers will be generated.

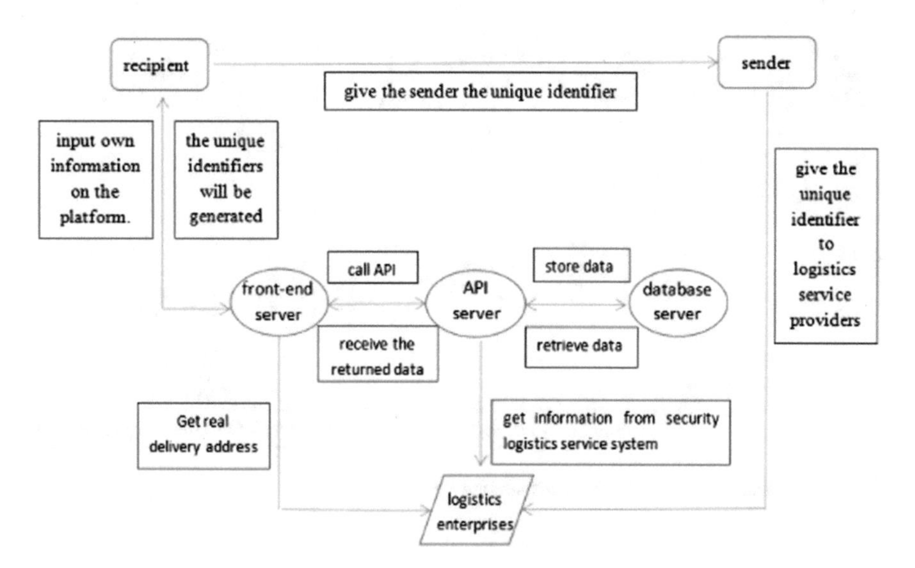

Fig. 1 The process of the third party escrow system

(3) The system will return the unique identifier to the user
(4) The user does not need to fill in the details of their delivery information(such as name, address, etc.) during online shopping, but just give logistics service providers the above unique identifier.
(5) Logistics courier service provider obtains the true information from this subsystem. (Do not use "security logistics service system")
(6) Security logistics system obtains real delivery information through this subsystem API, and directly imports it to the database of security logistics service system. (Use "security logistics service system")

According to the above process, this subsystem enables the recipient to protect their privacy from the sender, reached the recipient's privacy protection.

4 Security Logistics Service System

The work process of security logistics service system is divided into three aspects: onsite collection, transportation and receipt. The process of the security logistics service system is depicted in the Fig. 2. Specific steps are as follows:

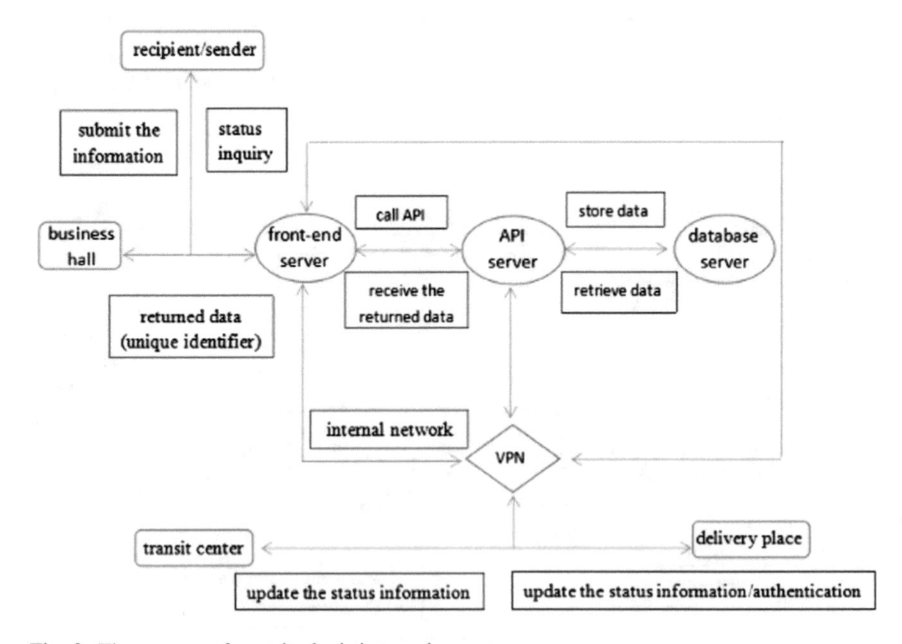

Fig. 2 The process of security logistics service system

4.1 Onsite Collection

(1) The sender needs to send express mail to the logistics companies, informing the clerk of the recipient's information and sender's contact information, then set the security password. (Do not use "the third party escrow system")
(2) The sender needs to send express mail to the logistics companies, and get the recipient's unique identifier generated by "the third party escrow system". Then the sender sets the security password. (Use "the third party escrow system")
(3) The clerk communicates with the back-end server through business clients to generate express identification number corresponding to the above one. Then, it will make a express mail ID correspond with the security password and return the information to the business clients. At the same time, a notification message included the shipment ID and security password will be sent to the sender and recipient by the sender's registration information.
(4) The clerk checks the shipment, makes express label through the business clients and pastes it on the package.
(5) After the completion of the onsite collection, the clerk updates status information and ready for mailing.
(6) Thereafter, the sender and the recipient can view the status information of the shipment with the shipment ID and security password via the server. If the sender or the recipient has not received the notification message, the query server can resend the notification message by providing the correct ID and security password.

4.2 Transportation

(1) Staffs will update the status information of the shipment via the business clients of transport vehicle and express scanning equipment before they load the shipment into transport vehicles or after they unload them.
(2) Staffs will update the status information of the shipment via the business clients of transport vehicle and express scanning equipment before the shipment is moved into or out of transit centers.
(3) When the shipment is about to be moved to the delivery place, back-end server will generate a random authentication code and send a message to the recipient, which contains the password, the shipment ID and a notification message.

4.3 Receipt

(1) The recipient goes directly to the delivery place if he receives the notification message.
(2) If the recipient has not received the notification message, the query server can resend the notification message by providing the correct ID and security password.
(3) Staffs find the shipment by the shipment ID provided by recipient. Scan this shipment and require the recipient to provide the random authentication code via the business clients and express scanning equipment;
(4) If the password is correct, the recipient can take the shipment. Then, the business clients will automatically update status and information.
(5) If the password is incorrect, the recipient can provide the security password to the staff to regenerate the random authentication code and check the password again.

5 Conclusions

In this paper, we discuss the private protection logistics platform, combined with cloud computing and two-dimensional code in networking technology. As for the authority control, the concept of the trusted third-party storage is introduced, so that the authority of processing data can be controlled efficiently and the security is also developed. The two subsystems can be independent from each other. Logistics enterprises or third-party institutions can be used by all of the system or the functionality of a subsystem. It is compatible with the existing logistics system and convenient to develop. What's more, it takes into account the business needs of traditional logistics to ensure the logistics companies have a smooth transition to the next generation of Internet technology.

Acknowledgements This work was supported by National Natural Science Foundation of China under Grant No. 61402210 and 60973137, Program for New Century Excellent Talents in University under Grant No. NCET-12-0250, "Strategic Priority Research Program" of the Chinese Academy of Sciences with Grant No. XDA03030100, Gansu Sci.&Tech. Program under Grant No. 1104GKCA049, 1204GKCA061 and 1304GKCA018. The Fundamental Research Funds for the Central Universities under Grant No. lzujbky-2014-49 and lzujbky-2013-k05, Gansu Telecom Cuiying Research Fund under Grant No. lzudxcy-2013-4, Google Research Awards and Google Faculty Award, China. It was also supported in part by Dongguan's Recruitment of Innovation and entrepreneurship talent program.

References

1. Liao X, Zhou Y. Courier service and online personal information protection [J]. Logistics Sci-Tech, 2014(1):114–115.
2. Wu L Q, Wang D X, Fang Y. Research of Mobile Delivery System of Express Logistics Based on Smartphone [C]. International Conference on Business Computing & Global Informatization. 2011:453–456.
3. Lei G P. The Study and Design of Logistics APP based on Handheld Terminal [J]. Digital Technology and Application, 2015, (8):177–178.
4. Zhang W M. Analysis and Research on Guangzhou Bao Sen Express Management System [D]. Yunnan: Yunnan University, 2015.
5. Deng Z H. On The Network Construction of Express Delivery Companies in China [J]. Logistics Engineering and Management, 2011, (5):71–73.
6. "Federal Express Corporation". trademarkia. com. Trademarkia. Retrieved 22 February 2016.
7. Zheng W L. Research of RF Tag in Automatic Identification of Freight Containers System [D]. Wuhan: WuHan University of Technology, 2008.
8. Liang Y. Possibility of Application of RFID in the Express Delivery Company [J]. Logistics Engineering and Management, 2015, (3):107–108.
9. Liu D P, Han Y J, Yin L. Application of Quick-response Code in E-receipt Process Design [J]. Logistics Technology. 2011, 30(5):96–98.
10. Zhao Y F, Yang Y, Li K. The Application of Two-dimensional Code Technology in Logistics Management. Software Engineer, 2014, 06:49–50.

Design and Implementation of the Forewarn System for College Student Management Based on Cloud Platform

Xinghua Sun, Yongfei Ye, Naidi Liu and Li Hao

Abstract It helps development of student management work that teacher obtain information of potential future problems timely, and to advance prevention and warning. This paper designs the system structure of the forewarn system for college students. The system use big data technology to evaluate and analysis warning object, scope, forewarn index and information with a large number of information data of student, issue notice to the crisis, aimed to discover and identify potential or actual risk factors. The system provides three login types, such as fixed network Web, mobile phone based on android and ios platforms to query and receive forewarn information.

Keywords Cloud platform · Forewarn system · Big data analysis · Student management

1 Introduction

In recent years, with the rapid development of social economy, university emergencies occur frequently, so it is urgent to strengthen the university emergency monitoring and responding. For the effective prevention and timely control and properly handle all kinds of unexpected events in our school, improve the rapid

Major Research Projects of Hebei North University (ZD201303); Hebei Province Population Health Information Engineering Technology Research Center.

X. Sun · Y. Ye · N. Liu · L. Hao (✉)
School of Information Science and Engineering, Hebei North University,
Zhangjiakou, Hebei, China
e-mail: 350727533@qq.com

X. Sun
e-mail: sunxinghua@189.cn

Y. Ye
e-mail: yeyongfei005@126.com

© Springer Nature Singapore Pte Ltd. 2018

N.Y. Yen and J.C. Hung (eds.), *Frontier Computing*, Lecture Notes
in Electrical Engineering 422, DOI 10.1007/978-981-10-3187-8_44

response and emergency handling capabilities, to ensure the school normal education teaching and living order, establish and sound on the cloud platform of university students' management [1–3] forewarn system (hereafter this text will be abbreviated as 'forewarn system') is imperative. Constructing university students' management forewarns system will achieve two major functions: (1) to nip in the bud, crisis nipped in the cradle, to give the related students comprehensive help and support. (2) to fast and accurate comprehensive understanding and evaluating the blown crisis, and make preliminary preparations for the next event processing, and at the same time. It can realize four main functions, such as forewarn settings, forewarn query, forewarn statistics and forewarn release.

2 System Architecture

Cloud computing, as popular software architecture, has been widely used in various fields. The core idea of cloud computing is to manage and schedule a large number of computing resources that is connected each other in the network. Calculation and other operations will be completed in the cloud and the user only need to enjoy the final results. The student's management forewarn platform based on cloud architecture can be understood as such a concept, establish a distributed forewarn system with multi terminal devices to collect and acquire forewarn data, and establish a forewarn information management center to analyze and process the upload data, and to realize crisis prediction and analysis.

Forewarn system based on cloud computing by SOA architecture uses web service to realize the communication of platform and accessed mobile terminal, and encrypt the XML and JSON data in the inter application communication, all these strategies guarantee the security of the platform very well. The platform by using a four tier architecture model with clear logic and structure has good scalability and scalability. The main architecture model of the platform is shown in Fig. 1. The layers of the four tier architecture are: application layer, cloud resources, authentication and authorization, the campus cloud layer. Application layer is a service provider for students and teachers in colleges and universities. Users can interact with the platform through the browser or mobile APP, the layer contains the application of mobile APP and WEB, which are deployed in the cloud resource layer. The cloud resource layer is the center of self-management platform, which deploy the forewarn applications with the help of the cloud server. At the same time, the layer has another role of storing and managing the application data. The authentication layer provides a unified authentication to the administrators, students, teachers and other users of the university. The campus cloud layer is mainly used by the managers of colleges and universities, including forewarn data analysis system, etc. The managers of colleges and universities analyze the forewarn data supplied by students and teachers, in order to assist the management and decision-making.

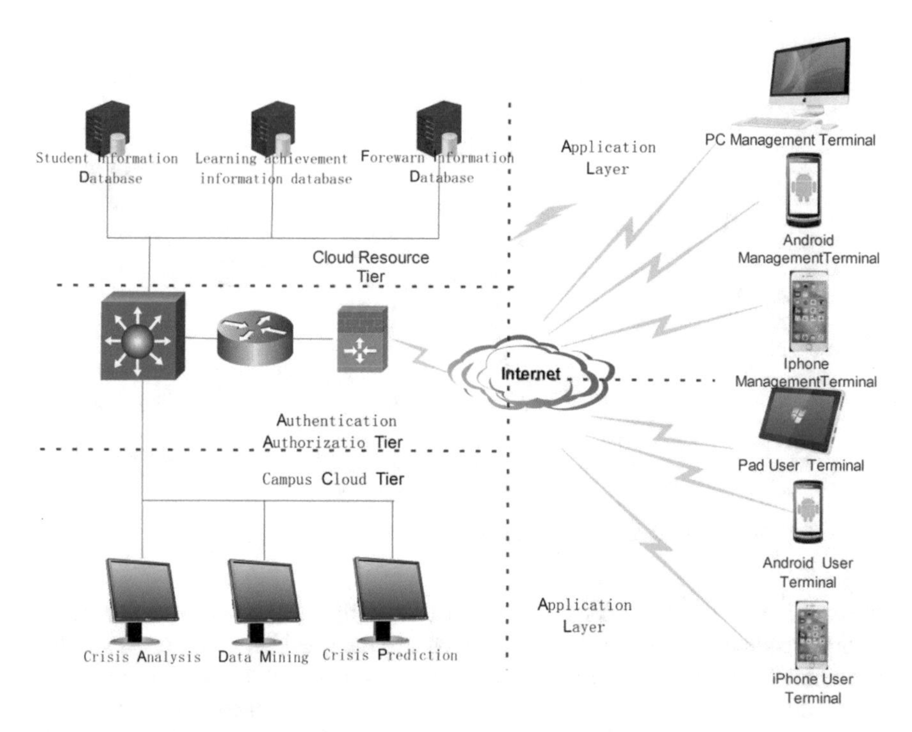

Fig. 1 The architecture model of forewarn system

3 Function Module Division and User Behavior of the Forewarn System for College Student Management

3.1 Function Module Division

The forewarn system [4, 5] for college student management include the following functions: forewarn management subsystem, crisis forewarn and alarm management subsystem, crisis forecast and analysis subsystem, user management, exchange platform, system management and maintenance module, as shown in Fig. 2.

Forewarn management subsystem includes the function of forewarn information checking, forewarn information processing, emergency location and so on. Teacher users can view and deal with forewarn information at any client, such as PC, mobile terminal, etc. Emergency forewarn and alarm subsystem is one of the most important functions of the platform, including emergency forewarn, academic record forewarn, attendance forewarn, psychological forewarn, life forewarn, disciplinary forewarn and so on. According to the situation of emergency, forewarn information is divided into red, orange and yellow three levels. When meet an emergency, emergency forewarn allows students or teachers to send forewarn

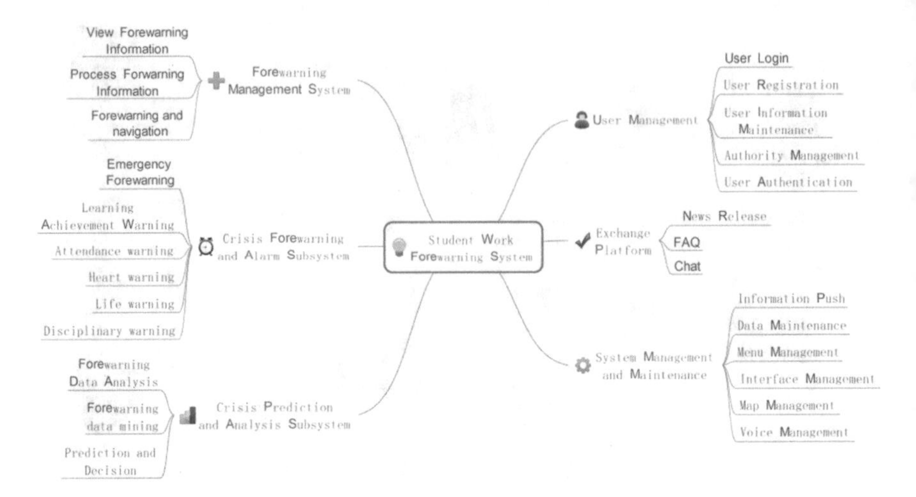

Fig. 2 The function module of the forewarn system

information by using the client. Forewarn information will be pushed to the client of the specified user, while also send message to the specified users. If the situation is extremely urgent, users can call the function—one key call to call the teacher or leader directly on the client. When the system analyze the collected data, according to the results of the analysis several other types of forewarn information will be automatically sent by the system correspondingly. By using the relevant data-mining algorithm forecasting and analysis system of the crisis analyze all kinds of collected information and extract useful data from the messy data effectively, and then make forewarn decision according to these data. Communication module includes functions such as publishing information, questions and interactive chat, this module although not the core module of the forewarn system, but the interaction module can enhance users enthusiasm to use the forewarn system and also close the relationship between teachers and students. User management module includes user login, user registration, user information maintenance, rights management, etc. System management and maintenance module includes message push, data maintenance, menu management, interface management, map management, voice management module. The function of message push is to send forewarn and system message; data maintenance to maintain various data; menu management is mainly responsible for the application menu management on web client; interface management is responsible for data sharing between all various devices that access to the cloud; map management realizes emergency positioning and navigation; voice module is mainly used to realize the voice search, voice messaging and other functions.

3.2 Users and Role Division

According to the business needs of the platform, the roles of platform user are divided into: (1) students: the main users of forewarn system; (2) the ordinary teachers: work related to attendance and academic forewarn; (3) counselors: work related to emergency forewarn, psychology forewarn, life forewarn, disciplinary forewarn; (4) leaders: work related to make decision for forewarn; (5) school administrators: maintenance the normal operation of the whole system.

4 Key Technology Realization of Forewarn System

4.1 Generation and Transmission of Forewarn Messages

The forewarn system involves a variety of forewarn information generation and transmission, but each mechanism of forewarn information generation is not the same, and this paper mainly expounds the formation mechanism of the achievements forewarn information. The main purpose of the achievements forewarn is to send a warning message to the students when the number of the students' course grade is not qualified or the number of attendance is up to a certain degree.

1. Message Generation Based on the Rules

The system in the formulation of rules with reference to the provisions of Hebei North University existing student management regulations, use the count of failing to pass the course and the number of absenteeism as main index to develop a different message transmission strategy. Rules can be express as a triple <T, U, M>, where T = {attendance, achievement} said rule type, U = {<i, j >|i, j ∈ N, i < j} said numerical range of indices, M is the string set {m1, m2, m3, …}, satisfying the condition is said to send a message. In order to make the user flexible management rules conveniently, the system design specialized rule management module for editing rules and management. After the completion of the editing rules, rule engine periodically scans the course and attendance data of students in the clock triggered. For each acquiring student data, turn out the rules to judge, if the condition is satisfied, for the students to generate a message and save in the messages table. Short message generation algorithm based on rules is described as follows:

(1) Obtain the data of each student's course grade information in the database, and generate the total failed count S_no;
(2) Get grade type rules from the rules table, and sort them according to the upper limit of the failed number from high to low, forming a regular queue $\{<S^i_{min}, S^i_{max}>\}$;

(3) Turn out from each rule S^i_{max}, if $S_{\_no} > S^i_{max}$ that will complete the rules matching, extracting the message string of the rule as the message of failed courses and returns, then go into (4), otherwise continue to (3) until the empty, then go into (4);
(4) Access to every student's attendance information from the database and form the total number of absenteeism $K_{\_no}$;
(5) Get attendance type rules from the rules table, and sort them according to the upper limit of the absenteeism number from high to low, forming a regular queue $\{<K^i_{min}, K^i_{max}>\}$;
(6) Turn out from each rule K^i_{max}, if $K_{\_no} > K^i_{max}$, that will complete the rules matching, extracting the message string of the rule as the attendance warning message and returns, then go into (7), otherwise continue to (6) until the empty, then go into (7);
(7) The end.

2. Short Message Push Based on Component

Send forewarn information is the data output part of the system, due to it's not the core part of this project, so we use the third party software component to complete the message sending. Sending information includes pushing system information and send short messages two ways.

4.2 Network Message Procession

In the Android system, due to the reason of the system itself, the network request module may occur system lag or ANR (application not Responding) anomalies, which may cause the system to collapse and abnormal. In order to solve the above problems effectively, and improve the network connection efficiency, using Volley to handle network connection is very necessary. Volley that is a network connection frame is proposed and recommends to developers by Google's official website, and has many good tools packages that are allowed developers to call directly, such as StringRequest, JsonRequest, imagerequest etc. Volley also has very high expansion to custom tools according to their own needs. To login as an example, the network request is divided into four steps: (1) instantiate the RequestQueue; (2) call the network request tool class; (3) put the network request into the message queue; (4) withdraw the network request from the message queue.

By using Volley, the message queuing problem can be solved fundamentally, while increasing the message processing process to prevent repeated accesses caused by poor network signal bug. To login as an example, when the client loads

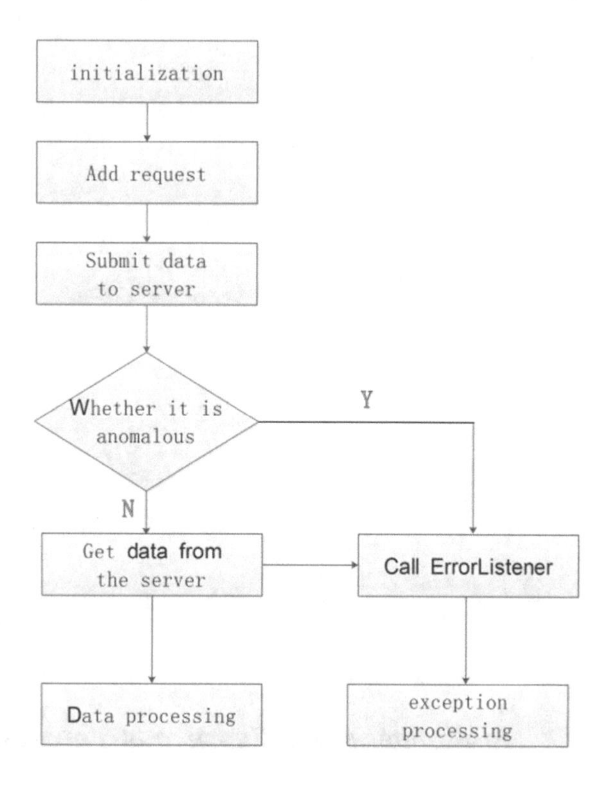

Fig. 3 Flow chart of flow chart

LoginActivity, RequestQueue is initialized. When the user clicks the login, RequestQueue will submit the data filled by user and process the corresponding data from the server. The network request can cancel on the midway, that is, when the network request is sent, by destroying Activity to cancel the network request. The Volley process is shown in Fig. 3.

4.3 Realization of Speech Recognition Function

In order to facilitate the operation for teachers and students, a large number of speech recognition technologies are used in the forewarn system; such as in the case of unexpected situations, students can quickly send out forewarn information. The system uses the third party sdk–IFLYTEK speech API to develop speech recognition module. IFLYTEK speech using complete MSP platform architecture on the Internet includes mobile client, server and client in Internet. The development steps are as follows: (1) import the jar packet of speech function; (2) calls the tool class of the speech function; (3) processing the returned data.

The workflow for speech recognition is shown in Fig. 4.

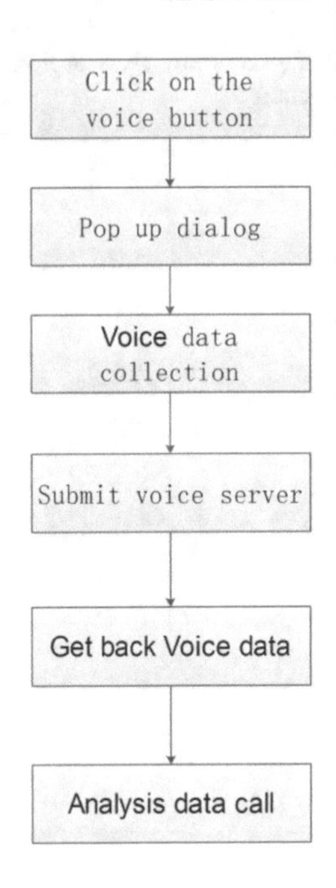

Fig. 4 Voice recognition workflow

4.4 *Realization of the Function of Forewarn Map*

In the case of emergency, the system can accurately define the location of the sudden situation and send it to the instructors and leaders. This system uses the map sdk supplied by Gao De. Implementation steps are as follows:

(1) Registration and add applications, access to Appkey. (2) Download and import the map sdk supplied by Gao De. Since the Gao De map is currently not supported gradle, so it is necessary to download and import the corresponding SDK into the project under the LIBS directory and bring. (3) Add user Key. (4) Add permissions. Add the permissions into the file "AndroidManifest.xml" in the project.

The interface of map positioning function is shown in Fig. 5.

Fig. 5 Forewarning map interface

5 Conclusion

By research the related business processes and technology of forewarn system of college students, realize the forewarn system of student management. But with the continuous progress of requirements of student management work and the improvement of information technology, student information forewarn system still have many places needs to be improved. The system can also be carried out further research in the following aspects: (1) Function: forewarn information results report output needs to be improved; (2) The interface design: user interface is not fresh, natural and beautiful; the security of the system also needs to be further enhanced.

References

1. Zhang Hai-jun, Chen Ying-hui. Student Management Research and Design of Early Warning System [J]. Computer Knowledge and Technology. 2010, (8) 6765–6766.
2. Deng Hong zhen, Yu Xiuri. A Study of Early Warning Mechanism in College Student Work [J]. Journal of College Counselors. 2010(6)95–97.
3. Wen Jian. The Design and Implementation of College Students' Study Early warning System Based on .Net Technology [D]. Changchun. Jilin University. 2014:14–20.
4. Shen Fei-fei. Am Explore of Student Crisis Forecast and Support System in Higher Education Based on Data Warehouse [J]. Journal of Chaohu College. 2007(5)155–157.
5. Ye Zhaxiong, Zhang Denghui, Dai shiqiang. Research and Exploitation of the Forewarning System for Comprehensive Evaluation of Students [J]. Journal of Zhejiang Shuren University. 2008(3)8–12.

A Decision Model of Service Learning Loyalty

Wen-Hsiang Shen and Mong-Te Wang

Abstract This study builds a quick and efficient intention model with students joining service learning loyalty. The purpose of this study the factors involved in influencing attitudes, intentions and other services for service-learning loyalty detected in University students. This study is using Autonomy questionnaire way, and the main group is the studying Service Learning Course students in central Taiwan, investigated a total of 1,500 copies were returned 1,230 effective samples, the rate was 82.3%. Regarding "Besides me, I will encourage other people to participate in volunteer service", there are 400 copies in answered "Greatest Agree", so we found out the key decision factors of learning service loyalty by the grey theoretical analysis of Decision Support System (DSS) from overall participation attitudes. Authors found there is a significant difference between overall participation attitudes and learning service correlation of approval in different gender background variables. This study builds the Students service loyalty decision model for counting the effective service loyalty by key decision variable. The results can provide other schools and following study reference, and suggests schools and government should focus learning service education function and counseling expertise ability, to offer better service to students.

Keywords Service learning · Decision Support System · Grey theory · Participation attitudes · Service loyalty

---

W.-H. Shen (✉)
Department of Information Technology, Overseas Chinese University, Taichung, Taiwan
e-mail: shen@ocu.edu.tw

M.-T. Wang
Graduate Institute of Educational Information and Measurement, National Taichung University of Education, Taichung, Taiwan
e-mail: shakathu@gmail.com

© Springer Nature Singapore Pte Ltd. 2018

N.Y. Yen and J.C. Hung (eds.), *Frontier Computing*, Lecture Notes in Electrical Engineering 422, DOI 10.1007/978-981-10-3187-8_45

1 Introduction

With the coming of the twenty-first century, the volunteer service has become an important trend of a new century, and America tread expert, Mr. Faith Popcorn predicted that 21st century will be the world of Volunteer service. “Global Volunteer declaration” of “International Volunteer year” in 2001 United Nation mentioned that now is the year of volunteer and civil society. Taiwan was public the “Volunteer service law” at the same year. As the trend, volunteer service has become the new strategy for citizen participation and practice civic duty. Howard thinks that combines professional courses and learning service not only improve learning theory but also encourage students doing society service [1].

Astin and other scholars found that the positive performance of joining the learning service includes, study performance, values, self-efficacy, leadership, career choice and plan after graduation [2]. Osborne and other scholars found that students who joining learning service increase their self-esteem and self-worth in the social area [3, 4]. Quezada and Christopherson offered that University students provide information about society attitude develop leader skills and community perceptions [5]. Learning service can improve students’ study, such as realize major knowledge deeply, increase professional knowledge, learn real life in class etc. Also they can improve various skills, such as leadership and communication, conflict and problem solving, adaptability etc. [6, 7]. The purpose of this study is developing learning service maximize by effective model to testing students loyalty, and distribute students to suitable service departments.

2 Background

The purpose of education is training students to develop their comprehensive character, include personal, knowledge and society ability [8]. The purpose of learning service is not only development of students, but also the win-win between schools and social area. (Reeb, Folger, Langsner, Ryan, & Crouse) Through the learning service promotion to establish students to develop social and civic duty, service skills, building self-confidence, and expand their human relationship and train their working attitudes [9]. The advantage of learning service is both of teachers and students work learning service to build good work partners relationship from social service. Kelley find the following [10]:

(a) “Service learning” emphasis on combining “Service” and “Learning” each other. Through the sharing of service experience to improve the learning and development of volunteer and train helpful attitude and civic duty. It will be advantaged and grow up through service learning and then value of social justice.

(b) With innovation principles, through selection, transformation and reorganization way, Learning Service improve students' new thinking, new opinions and new action, own social and civic duty, service skills personal development and learning ability of real life, strength the ability of learning and critical thinking.
(c) The main purpose of pushing learning service is encouragement and sharing. The share can help service department and those people being served to see that students' learning and improve themselves to build self-confidence and a helpful establishment of the caring culture.
(d) Involving the learning service into the teaching, not only achieve the purpose of study or strength the learning course but also promote the social balance and improve the human being welfare. It can develop the personality balanced, confirm the human respect, build completely value system, finish the professional study, combine knowledge and action, advantaged human relationship, keep justice and care society and offer service to individual and group.

3 Present Study

With the current learning service system, students will be assigned randomly to service department after finished the beginning of the questionnaire. It takes 2–3 weeks if all students finished the questionnaire and then assign them to a service department. About those students who like service learning and not like service learning are assigned to every service department random, we have to study the subjects as follows:

(a) Build the rapid testing learning service loyalty model, immediately to get the students' wish and assign them to service learning department after finish the questionnaire.
(b) Every service department has highly dedicated students to join in and encourage other students improving their learning service want.

4 Methods

4.1 Empirical Study Setting

The empirical study was taken throughout the academic year during 2015 in two 18-week the service learning obligatory course. This questionnaire is designed the first draft by the investigator to study the related works of literature and discuss with the main group, to pretest the 200 samples from the study population, and then make the formal questionnaire by Exploratory factor analysis (EFA) after deleting some relative question items [11]. For the Taiwan University freshmen to study learning service course, investigated a total of 1,500 copies were returned 1,230 effective samples, the rate was 82.3%.

4.2 Decision Support Analysis

Decision Support System often used the theory gray theory analysis method, TOPSIS method, a simple weighted method (simple additive weighting method) [12], Analytic Hierarchy Process, ELECTRE [13], and Technique for Order Preference by Similarity to Ideal Solution, (TOPSIS) [14]. The study, using gray theory analysis, the most important is to explore for both the service-learning data because the gray theory is the most important for the next ambiguity of the data model, that information is not complete, the system effectively explores and understand [15–17]. The Grey System Theory was established by Deng in 1982, it includes Grey relational analysis, Grey modeling, prediction and decision making of a system in which the model is unsure or the information is incomplete. It provides an efficient solution to the uncertainty, multi-input and discrete data problem. The relation between machining parameters and machining performance can be found out using the Grey relational analysis.

In grey system theory, Grey Relational Analysis (GRA) is an effective mathematical tool to treat the uncertain, multiple, discrete and incomplete information. This study would like to refer the localized grey relational grade which is proposed by Nagai [18]. Its procedure as follows:

4.2.1 Establishment of Original Vectors

The reference vector x_0 and inspected vectors of original data x_i are established as followed:

$$\begin{array}{c} i=1,2,\ldots,n, \quad k=1,2,\ldots,m \\ x_0=(x_0(1),x_0(2),\ldots,x_0(k),\ldots,x_0(m)) \\ x_1=(x_1(1),x_1(2),\ldots,x_1(k),\ldots,x_1(m)) \\ x_2=(x_2(1),x_2(2),\ldots,x_2(k),\ldots,x_2(m)) \\ \vdots \\ x_i=(x_i(1),x_i(2),\ldots,x_i(k),\ldots,x_i(m)) \\ \vdots \\ x_n=(x_n(1),x_n(2),\ldots,x_n(k),\ldots,x_n(m)) \end{array} \tag{1}$$

4.2.2 Calculation of GRA

In original formula, x_0 is reference sequence of local grey relational grade, and x_i are inspected sequences. The established sequences have to satisfy three conditions: non-dimension, scaling, polarization. Grey relational generation has three ways: larger-the-better (the expected goal is bigger the better), smaller-the-better (the expected goal is smaller the better), and nominal-the-better (the expected goal is between maximum and minimum), their calculations are as follows:

a. Larger-the-better

$$x_i^{\max} = \frac{x_{ij} - \min\limits_{\forall_i} x_{ij}}{\max\limits_{\forall_i} x_{ij} - \min\limits_{\forall_i} x_{ij}} \tag{2}$$

where $\max\limits_{\forall i} x_{ij}$ means the maximum number in j and $\min\limits_{\forall i} x_{ij}$ means the minimum number in j.

b. Smaller-the-better: when we expect the target to be as small as possible.

$$x_i^{\min} \equiv \frac{\max\limits_{\forall i} x_{ij} - x_{ij}}{\max\limits_{\forall i} x_{ij} - \min\limits_{\forall i} x_{ij}} \tag{3}$$

c. Nominal-the-better: when we expect the target to be between the largest and the smallest data.

$$x_i^{ob} \equiv \frac{\max\limits_{\forall i}\{e_{ij}\} - e_{ij}}{\max\limits_{\forall i}\{e_{ij}\} - \min\limits_{\forall i}\{e_{ij}\}}, \; e_{ij} \equiv \frac{|x_{obj} - x_{ij}|}{|x_{obj}|} \tag{4}$$

where $x_{obj} \neq 0$ and the target goal is zero, Nagai's equation of smaller-the-better will be used, that is, $\max\limits_{\forall i} x_{ij} \geqq x_i^{ob} \geqq \min\limits_{\forall i} x_{ij}$.

4.2.3 Localized Grey Relational Grade

The localized grey relational grade is defined as follows:

$$\Gamma_{0i} = \Gamma(x_0(j), x_i(j)) = \frac{\overline{\Delta}_{\max} - \overline{\Delta}_{0i}}{\overline{\Delta}_{\max} - \overline{\Delta}_{\min}} \tag{5}$$

where $\overline{\Delta}_{\max}$ and $\overline{\Delta}_{\min}$ are maximum value and minimum value of $\overline{\Delta}_{0i}$ respectively, $\overline{\Delta}_{0i}$ is the absolute distance between x_0 and x_i, its formula as follows:

Table 1 A correlation between each item and "Besides me, I will encourage other people to participate in volunteer service work"

"Besides me, I will encourage other people to participate in volunteer service work"					
Item	LGRA	Item	LGRA	Item	LGRA
S1	0.927316368	S8	0.854518877	S15	0.832379744
S2	0.974648714	S9	0.777243078	S16	0.841021368
S3	0.98144256	S10	0.860393595	S17	0.906618794
S4	0.990599841	S11	0	S18	0.991973166
S5	1	S12	0.090368764	S19	0.879585136
S6	0.871238574	S13	0.674103773	S20	0.923856802
S7	0.852476781	S14	0.873763253	S21	0.922648069

$$\overline{\Delta}_{0i} = \|x_{0i}\|_{\rho} = \left(\sum_{k=1}^{m} [\Delta_{0i}(k)]^{\rho} \right)^{\frac{1}{\rho}} \tag{6}$$

$\overline{\Delta}_{0i}$ is called Murkowski distance. This study applies $\rho = 2$, so $\overline{\Delta}_{0i}$ is also known as Euclidean distance.

When Γ_{0i} is close to 1, it means that x_0 and x_i are highly correlated, in contract, Γ_{0i} is close to 0, the relationship between x_0 and x_i is lower.

Grey relational ordinal: The whole decision-making is made by the comparison of the grey relation Γ_{0i}. Through the ordinal, different causes can be identified, and the most important influence can be found, becoming the relational standard in the system.

The study is using Autonomy questionnaire, the main group is the Taiwan University freshmen to study learning service course, investigated a total of 1,500 copies were returned 1,230 effective samples, the rate was 82.3%, the Cronbach's Alpha = 0.911. Regarding "Besides me, I will encourage other people to participate in volunteer service." there are 400 copies in answered "Greatest Agree", the Cronbach's Alpha = 0.967, analysis result as Table 1.

The study gets the key decision items are S5, S3, S20, S18, S4, S14, S1, and S2 from the main request "Besides me, I will encourage other people to participate in volunteer service." The relation value GAMMA $\geq$ 0.90.

5 Regression Analysis

To explore the key decision item associated with the service to learn loyalty, strength, and then analyzed via regression correlation coefficients to construct service loyalty model. On regression analyses using each later stage variable as a dependent variable and all previous stage variables as independent (or predicting)

Table 2 Significant regression model summary

Item (S*i*)	Unstandardized coefficients		Standardized coefficients	t	Sig.
	B	Std. error	Beta (*Ci*)		
Constant	1.861	0.160		11.651	0.000
S5	0.086	0.036	0.103	2.394	0.017
S18	−0.048	0.037	−0.055	−1.321	0.187
S4	0.153	0.040	0.169	3.864	0.000
S3	0.109	0.040	0.126	2.696	0.007
S2	0.013	0.036	0.015	0.376	0.707
S1	0.138	0.029	0.190	4.742	0.000
S20	−0.014	0.029	−0.018	−0.483	0.630
S17	0.100	0.029	0.128	3.453	0.001

Dependent variable: "Besides me, I will encourage other people to participate in volunteer service work"

variables, significant regressions were found and their result summary is given in Table 2.

As indicated in the table, all models are significant and that each dependent variable can be statistically accountable by its previous stage variables. That is, the dependent variables can be predicted by using previous stage independent variables.

6 Results

The meaning of how much loyalty that a student joins a service. We use action intent to calculate the percentage of loyalty. When the student self-identification can meet the defined survey items, he/she will get a high score. The loyalty value of each service practice (as $S_i score$) is evaluated by comparing the best intent (as *Bestintented*) and the time actually elect in this item (as *ActualIntented*):

$$S_i score = \frac{S_i ActualIntented}{S_i BestIntented} \times 100 \tag{7}$$

The loyalty value of a student is evaluated as the average value of all weight:

$$\text{Service Loyalty} = \frac{\sum_{i=1}^{n} C_i \cdot S_i score}{\text{n}} \tag{8}$$

where $S_i score$, Ci and n denote the ith Score, the ith regression coefficient and the total number of items in a service survey, respectively.

In the allocation of service-learning course content, the use of computing weighted score out of high to low points for each configuration services so that each service-learning course project, can be evenly distributed from the highest assigned the lowest score and makes the completion of each project You can reach the highest degree, but also make high scores to lead the low scores to complete service learning course project together.

7 Conclusion

By service learning, we train students to serve others and turn the volunteer service into whole life property, and the focus is guiding students to realize, action stage and willing to continue service by exploring stage, besides of his own as much as possible to encourage others joining the volunteer service work and creating the maximum value of whole life learning service.

This study departs from a systematic review of the literature to have identified the values in service learning. On comparing related literature approaches, a decision making model with associated service intent measures is derived as the basis for service loyalty assessment. After the grey relational analysis and regression analysis of statistical method and verification based on data collected from the empirical study, a useful model has been generated to show result through service learning survey. It is hoped that the result, along with other findings in this paper, can help to interpret the service loyalty measures and provide a useful feedback to service learning trainers as well as trainees as future civic education.

References

1. Howard, J. (2001). Service-Learning Course Design Workbook. Michigan Journal of Community Service Learning.
2. Astin, A. W., Vogelgesang, L. J., Ikeda, E. K., & Yee, J. A. (2000). *How service learning affects students*. Los Angeles: Higher Education Research Institute.
3. Osborne, R. E., Hammerich, S., & Hensley, C. (1998). Student effects of service-learning: tracking change across a semester. *Michigan Journal of Community Service Learning*, 5, pp. 5–13.
4. Furco, A., Moely, B. E., & Reed, J. (2007). *Formulating a Model of Effects of College Students' Service-Learning Experience*. Paper presented at the International Conference on Psychology. Athens Greece.
5. Quezada, R. L., & Christopherson, R. W. (2005). Adventure-based service learning: University students' self-reflection accounts of service with children. *Journal of Experiential Education*, 28(1), pp. 1–16.
6. Eyler, J.S., & Giles, D.E. (1999). Where's the learning in service learning? San Francisco: Jossey-Bass.
7. Simons, L. & Cleary, B. (2006). The influence of service learning on students' personal and social development. College Teaching, 54(4), 304–319.

8. Burns, L. (1998). Make sure it's service learning, not just community service. *Education Digest*, 64(2), pp. 38–41.
9. Reeb, R. N., Folger, S.F., Langsner, S., Ryan, C., & Crouse, J. (2010). Self-efficacy in servicelearning community action research: theory, research, and practice. American Journal of Community Psychology, 46(3), p. 459–471.
10. Lorraine, I., Kelley-Quon, L.I., Melanie, A., Crowley, M.A., Applebaum, H., Cummings, K., Kang, R.J., Tseng, C.H., Carol, M., Mangione, C.M., Shew, S.B. (2015), Academic-community partnerships improve outcomes in pediatric trauma care, Journal of Pediatric Surgery, *Journal of Pediatric Surgery*, In Press.
11. Hwang, Ching-Lai and Kwangsun Yoon, (1981) Multiple Attribute Decision Making – Methods and Applications Springer-Verlag Berlin Heidelberg, New York.
12. Rosenberg, M. J., & Hovland, C. I. (1960). *Cognitive, affective, and behavioral components of attitude*. In M. J. Rosenberg et al. (eds), Attitude organization and change: an analysis of consistency among attitude components. New Haven, CT: Yale University Press
13. Wong, K. C., Lam, Y. R., & Ho, L. M. (2002). The effects of schooling on gender differences. British Educational Research Journal, 28, 827–843.
14. Yamaguchi, D., Li, G. D., & Nagai, M. (2005). New Grey relational analysis for finding the invariable structure and its applications. *Journal of Grey System*, 8(2), pp. 167–178.
15. Deng, J. L., (1989) "Introduction of Grey System Theory," Journal of Grey System, Vol. 1, No. 1, pp. 1–24.
16. Yamaguchi, D., Li, G. D., & Nagai, M. (2005). New Grey relational analysis for finding the invariable structure and its applications. Journal of Grey System, 8(2), 167–178.
17. Yamaguchi, D., Li, G. D., Mizutani, K., Akabane, T., Nagai, M., & Kitaoka, M. (2007). A realization algorithm of grey structural modeling. Journal of Grey System, 10(1), 33–40.
18. Yamaguchi, D., Li, G. D., & Nagai, M. (2007). Verification of effectiveness for grey relational analysis models. *Journal of Grey System*, 10(3), pp. 169–181.

An Effective Model to Analyze Students' Misconceptions in Learning

Chih-Ling Chia, Wei-Hsu Chen, Mong-Te Wang and Cheng-Hsuan Li

Abstract This study has found an objective and an effective way to help teachers to find out what is the most prioritized, which should be review as remedial questions in the classroom. Teachers will teach students a unit after a formative assessment, followed by a limited time to conduct a review of the remedy for students' misconceptions. Teachers often believe that the most number of students got the wrong item, or, up to the item that was asked by the most students need to be reviewed first. However, teachers cannot review all the wrong items within the limited time. In this study, we used Failure Mode and Effects Analysis (FMEA) which is widely applied by industries and commerce. By getting rid of subjective marking from experts which had redefined Remedial Teaching Priority Number (RTPN). There are three factors in this method, which are the failure rate, correlation, and hyponymy. The study sample was a high school class of 40 students in central Taiwan. We found RTPN recommend remedial items and the most students want to ask items are basically the same, which means RTPN selected items and students want to be the true remedy is consistent with several previous items. RTPN provides direction to the teacher to remedy, and it can effectively provide students who need appropriate help.

Keywords Failure mode and effects analysis · Remedial teaching priority number · The failure rate · Correlation · Hyponym

---

C.-L. Chia (✉) · W.-H. Chen · M.-T. Wang · C.-H. Li
Graduate Institute of Educational Information and Measurement,
National Taichung University of Education, Taichung, Taiwan
e-mail: shiling1027@gmail.com

W.-H. Chen
e-mail: weisuch@gmail.com

M.-T. Wang
e-mail: shakathu@gmail.com

C.-H. Li
e-mail: ChengHsuanLi@gmail.com

© Springer Nature Singapore Pte Ltd. 2018

N.Y. Yen and J.C. Hung (eds.), *Frontier Computing*, Lecture Notes in Electrical Engineering 422, DOI 10.1007/978-981-10-3187-8_46

1 Introduction

An effective teaching achievement is that students will learn within the classroom, where they can utilize it within their means. In order to confirm whether the students have a complete right to learn or not, teachers usually test the students after teaching a unit and see how the effectiveness of learning take place [1–3].

If students have some problems and misconceptions, the teachers will review and clarify concepts immediately. Such remedial teaching is immediate and efficient. However, with the instruction immediately after the test results, how to identify problems effectively and promptly, that enable students to complete a full study of the unit is a test for the teachers.

Teachers generally will review the number of wrong answers to the questions first, or ask students what subject teachers need to solve first? These questions are the most in need of review items. However, such kind of a review of the order is correct? The most wrong item that was made by students is the most important concept in this test? Perhaps, just because the item is defective or concept guiding deviations, or even the poorest narrative due to students answer incorrectly. In order to enhance more efficient remedial teaching, this study presents an objective and efficient way to help teachers pick out the most priority; which should be to review the remedy items first in the classroom.

2 Background

The so-called remedial teaching means to use a personalized fitness instruction teaching method for low-achieving students, or students who are in low level to ensure those students have the basic academic abilities. Broadly speaking, remedial teaching can be said a part of learning tutorial. When students have difficulties in learning, they should get a diagnosis teaching [4, 5]. One of the features in complement teaching is emphasized "assessment—teaching—assessment again" in a circular course. That is, to understand the ability of difficulty in learning and degree of them, and then use different teaching strategies or methods for students' individual needs, and allows students to better understand the unit content [6]. Therefore, remedial teaching is a short-term remedy to strengthen the teaching measures, and expected students can keep up with the general progress after the implementation period.

Remedial teaching patterns are commonly used in domestic and abroad [7], for example: (1) resource room, (2) learning lab, (3) learning station, (4) learning material set, (5) computer assisted instruction.

In all subjects, the most common remedial teaching subject should be mathematics. Due to large student bodies in classes, resulting numerous students of mathematics achievement is not high in this department [8, 9]. Based on the above, when the teachers find point of difficulty among students, it is very important to give appropriate and accurate feedback immediately. Therefore, after students finish their test papers, teachers can find the problem immediately, and offer a remedial teaching for the students.

3 Present Study

In recent years, the education sector has begun to focus on students' academic backwardness, especially for disadvantaged students. Even some civil society groups have been actively involved in assisting the disadvantaged students in after-school tutoring work. Their object locks on vulnerable families (low-income families, single parents or grandparents) where academic performance of students 20–35% of the posterior segment in a class. It offers additional 3–4 weeks of lessons from school, and gives students homework guidance after school.

According to the above question we need to study the subject as follows:

To establish rapid feedback test model, so that the teachers can mark after the students finish the test, and know what kind of items should be reviewed and explained post-tests.

4 Methods

Remedial Teaching Priority Number is multiplied by failure rate, discrimination rate and influence rate. And N is the number of examinees, n is the number of items, $a_{ij} = 1$ indicates that student i answers the jth item correctly, and $a_{ij} = 0$ indicates that answers incorrectly. We define (a) failure rate, (b) discrimination rate, and (c) influence rate as follows:

(a) Failure rate:

Failure rate of the item k is

$$F_k = \frac{\sum_{i=1}^{N} (1 - a_{ik})}{N}, 0 \le F_k \le 1 \tag{1}$$

As can be seen from (1), Failure rate is the proportion of item k answered incorrectly in all examinees. The higher the value, the more representative of the need to remedy for item k.

(b) Discrimination rate:

Discrimination rate of the item k is

$$D_k = \frac{\sum_{i=1}^{N}(a_{ik} - \frac{1}{N}\sum_{i=1}^{N} a_{ik})(T_i - \frac{1}{N}\sum_{i=1}^{N} T_i)}{\sqrt{\sum_{i=1}^{N}(a_{ik} - \frac{1}{N}\sum_{i=1}^{N} a_{ik})^2}\sqrt{\sum_{i=1}^{N}(T_i - \frac{1}{N}\sum_{i=1}^{N} T_i)^2}},$$
$$T_i = \sum_{j=1}^{n} a_{ij}, \; -1 \le D_k \le 1 \tag{2}$$

Discrimination rate in this study is the use of the point-biserial correlation coefficient, this value is the same as with the Pearson correlation coefficient. The point-biserial correlation is an index of item discrimination. The value of a point-biserial discrimination index can range between −1 and 1; the closer the value is to 1, the better the discrimination, this represents the item is worth being taken seriously.

(c) Influence rate:

B is a $n \times n$ matrix, and entry b_{jk} represents the proportion of item k answered correctly and item j answered incorrectly in all examinees.

$$b_{jk} = \frac{1}{N}\sum_{i=1}^{N}(1 - a_{ij})a_{ik}, \tag{3}$$

We suppose the thresholds ε is the mean of all entries of matrix B.

$$\varepsilon = \frac{1}{n^2}\sum_{k=1}^{n}\sum_{j=1}^{n} b_{jk}, \tag{4}$$

C is a $n \times n$ matrix

$$c_{jk} = \begin{cases} 1, & b_{jk}, \varepsilon \\ 0, & b_{jk} \ge \varepsilon \end{cases}, \tag{5}$$

If $c_{jk} = 1$, then item j could be linked forwards to item k. The relation denoted as $j \to k$ means that j is a prerequisite to k [10, 11]. Influence rate focus on the causal relationship between the items, the more prerequisites of item k, the higher the influence rate is.

Remedial Teaching Priority Number of item k is

$$RTPN_k = F_k \times D_k \times S_k \tag{6}$$

5 Results

In this study, the study sample are 120 students who are from high school in central area of Taiwan. Research content is a test paper, which is "Index and the number of applications" unit of high school mathematics. After explaining this unit as well as review the idea, then do a test for the students. Test time is 50 min, after finishing the test, recovering the papers, which then will be reviewed in the classroom immediately the next day. There are total of 20 test items, this study specifically designed answer sheets in demand in a region. We use the Pareto Principle for the students to find important papers puzzled critical 20–80% interest. There are 20 items in this test paper, of which 20% is 4 items. This study set four fields, and labeled 1, 2, 3, 4 of the field. Students have to write down the most wanted ask items and fill in the fields marked 1, then in accordance with the wishes from strong to weak, labeled 2, 3, 4 respectively in the remaining three bars.

We gave 4 points to item 1; 3 points to item 2; 2 points to item 3; 1 point to item 4. We then add the final score to each points sections. Each item will have a final number, which represents interest in inquiry of the scores from students wishing to ask, we called "inquiry index."

After surveying, we aggregated results of the Survey and RTPN, sequentially order is title sequence number, the failure rate, discrimination rate, influence rate, RTPN, inquiry index, and number of queries RTPN sort and index inquiry, such as a Table 1. The study found that most students want to ask questions of the first order are 18, 19, 6, 11 item, shown in Fig. 1. The highest failure rate for the first four items in turn are 18, 19, 5, 6 item, as shown in Fig. 2. In contrast RTPN recommendations of the first four items in turn are 6, 18, 19, 15 item, shown in Fig. 3. The first three items are basically the same, which means that RTPN selected items and students want to be reviewed is consistent with several previous item, but the order is slightly different. As to item 5 ranked by RTPN is fourth, ranked by inquiry index is seventh, the gap between the two will not be too diverse.

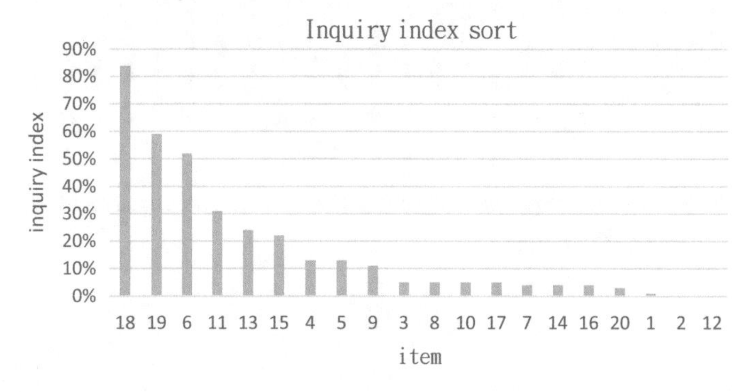

Fig. 1 The item of the current students most ask the four highly asked items about the title sequence for the first 18, 19, 6, 11

Table 1 Construction survey results and RTPN sort and inquiry index

Item	Failure rate	Discrimination rate	Influence rate (%)	RTPN (%)	Inquiry index (%)	RTPN sort	Inquiry index sort
1	0.025641026	0.113887653	0	0	1	20	18
2	0.871794872	0.230763541	95	19	0	9	19
3	0.743589744	0.525629374	85	33	5	5	10
4	0.692307692	0.491714041	60	20	13	8	7
5	0.923076923	0.393265761	95	34	13	4	7
6	0.923076923	0.581454284	95	51	52	1	3
7	0.282051282	0.235732247	10	1	4	19	14
8	0.666666667	0.372318578	50	12	5	10	10
9	0.538461538	0.708005863	30	11	11	12	9
10	0.641025641	0.032165302	40	1	5	18	10
11	0.794871795	0.015921522	95	1	31	16	4
12	0.256410256	0.450539463	10	1	0	17	19
13	0.743589744	0.429926547	70	22	24	7	5
14	0.564102564	0.596392867	35	12	4	11	14
15	0.743589744	0.525629374	85	33	22	6	6
16	0.487179487	0.333133552	20	3	4	14	14
17	0.487179487	0.249528607	25	3	5	15	10
18	0.974358974	0.520629271	95	48	84	2	1
19	0.974358974	0.520629271	95	48	59	2	2
20	0.512820513	0.469473925	25	6	3	13	17

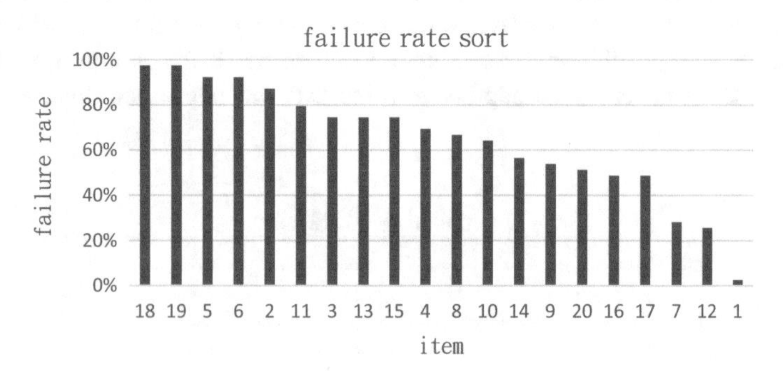

Fig. 2 Students test error rate is currently the highest in the title sequence for the first four questions 18, 19, 5, 6 item

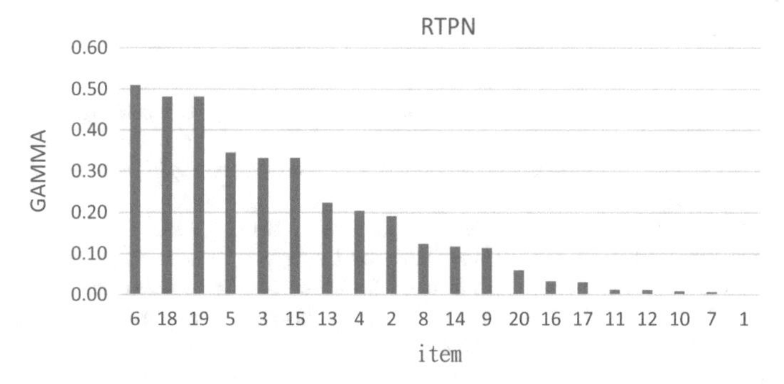

Fig. 3 RTPN recommendations of the first four questions in turn are 6, 18, 19, 5 item

6 Conclusion

In this study, we data our research based on real study samples, and referred relevant reference and literature. We executed test papers as research tools, used statistical method to analysis sample data, drew conclusions and built a better model to make recommendations for schools, students and researchers for subsequent reference.

To solve questions is one of the things that most teachers in the teaching field need to deal with. Taking mathematics at high school for example, under the limit of teaching hours and teaching schedule, it is difficult to do a comprehensive review of all the tests. Solving students' questions at the right time is very important. That is why we introduced FMEA that is commonly used in business and industry [12–15] to education field, and getting rid of experts' subjective marking methods, where we let the data speak for itself. Here we define remediation priority number RTPN, while included three factors: the failure rate, correlation, and hyponymy. Teachers in the classroom should pick out the most priority item to remedy in a formative assessment, thereby enhance the effectiveness of remedial teaching.

Based on the above analysis, we can improve that there is a certain degree of accuracy by using RTPN to know students' learning situation. Using this method to establish the model can increase a quick way to solve the problem from students. With this aspect of learning can increase students' confidence and provide the teacher to solve the problem absorbedly, select discomfort items, and focus on the completion of this learning target. In addition, the objective accuracy of this method of teaching can make up for lack of subjective, and enhance the effectiveness of teaching. Therefore, teachers can use RTPN anytime to diagnose learning content or items of relevance to enhance the effectiveness of teaching.

References

1. Johnson, D. W., Johnson, R. T., & Holubec, E. J. (1994). Cooperative learning in the classroom. Alexandria, Va: Association for Supervision and Curriculum Development.
2. Slavin, Robert E. (1995). Cooperative learning: theory, research, and practice. Boston: Allyn & Bacon.
3. Pantiz (1999). Collaborative versus Cooperative Learning: A Comparison of the Two Concepts Which Will Help Us Understand the Underlying Nature of Interactive Learning. Cape Cod Community College, peninsula, Massachusetts; USA. Retrieved 5 Nov. 2011.
4. Airasian, P.W., & Bart, W.M. (1973). Ordering theory: A new and useful measurement model. *Journal of Educational Technology*, 5, 56–60.
5. Slavin, R. E. (1989). Research on cooperative learning: An international perspective. Scandinavian Journal of Educational Research.
6. Bell, A. W. (1992). Diagnostic teaching selected lectures. The 7th International Congress on Mathematical Education (pp. 19–34). Quebec.
7. McLaughlin T. F., & Vacha, E. F. (1992). The at-risk student: A proposal for action. Journal of instructional psychology.
8. Bell, A. W. (1993). Some experiments in diagnostic teaching. Education Studies in Mathematics, 24, 115–137.
9. Slavin, R. E. (1989). Student at–risk for school failure. In R. E. Slavin, N. L. Karweit, & N. E. Madden (Eds), Effective programs for students at-risk (p. 3–19). Boston: Allyn & Bacon.
10. Chang, C. L. and Tsai, C. H. and Wei, C. C. (2000). "A New Evaluation Method for Failure Mode and Effect Analysis," Journal of the Chinese Industrial Engineers, Vol. 17, No 1, pp. 51–64, 2000.
11. Wu, H.-M., Kuo, B.-C., & Yang, J.-M. (2012). Evaluating Knowledge Structure-Based Adaptive Testing Algorithms and System Development, Educational Technology & Society, 15(2), pp 73–88.
12. TORII Kentaro /NISHIKAWA Takeichiro /HIRANO Kaho (2009). Improvement of Processes Using Risk FMEA, 東芝レビュ一Vol. 64 No. 8特集.
13. Chunsheng Yang, Yanni Zou Affiliated, Pinhua Lai, Nan Jiang (2015). Data mining-based methods for fault isolation with validated FMEA model ranking, Applied Intelligence December 2015, Volume 43, Issue 4, pp 913–923.
14. Liu, Shuo-Fang; Cheng, Jui-Hung; Lee, Yann-Long (2016). A case study on FMEA-based quality improvement of packaging designs in the TFT-LCD industry, Total Quality Management & Business Excellence, Volume 27, Issue 3–4, 2016 PP. 413–431.
15. Shamsi Ghasemi, Rahim Mahmoudv and Affiliated, Kazem Yavari (2016). Application of the FMEA in insurance of high-risk industries: a case study of Iran's gas refineries, Stochastic Environmental Research and Risk Assessment February 2016, Volume 30, Issue 2, pp 737–745.

360° Panoramic Video from a 3-Camera Rig

S. Morteza Safdarnejad, Xiaoming Liu and Jingfeng Liu

Abstract In this work, a system for creation of 360° panoramic video from a low-cost 3-Camera rig is presented. To provide 360° coverage with 3 cameras, cameras equipped with fisheye lens are utilized. Fisheye lens introduces extra challenges in stitching videos from the cameras, due to considerable distortion near the borders of field of view. First, we calibrate each camera and extract mapping parameters to correct for the fisheye effect. Then, to speed up the process, these mappings are consolidated into the mappings calculated in stitching phase, to form the composite mappings. Finally, video sequences are read from cameras, warped using the composite mappings, and corrected for color uniformity to generate the 360° panoramic video.

Keywords Panoramic video · Video stitching

1 Introduction

360° panoramic videos are gaining more and more popularity due to the rich information they provide from the captured environment. Generally, for generation of the panoramic videos, scene is captured by an appropriate number of cameras and resultant videos are stitched together to form a uniform and seamless panoramic video. For full coverage of the 360°, generally a large number of cameras are used which results into costly camera rigs and costly data acquisition modules [8]. In this paper, we introduce a low-cost system for generation of 360° panoramic videos from a 3-camera rig, as shown in Fig. 1. In this scenario, to provide the 360° coverage, small cameras equipped with fisheye lens are packed into a tight and small rig.

S. Morteza Safdarnejad · X. Liu (✉)
Michigan State University, East Lansing, USA
e-mail: liuxm@cse.msu.edu

J. Liu
LinkSprite Technologies, Inc., Wuhan, China

© Springer Nature Singapore Pte Ltd. 2018

N.Y. Yen and J.C. Hung (eds.), *Frontier Computing*, Lecture Notes in Electrical Engineering 422, DOI 10.1007/978-981-10-3187-8_47

Fig. 1 The proposed 3-camera rig for generation of 360° panoramic videos

Fisheye lens introduces considerable distortion, especially near the boundary of field of view of the camera. Although in the final panoramic video only about 140° of coverage of the fisheye camera is needed to form a seamless 360° panoramic video, in the phase of stitching mappings estimation, the distortion causes difficulties. So, effective fisheye distortion removal is required. Thus, we first calibrate each camera and extract mapping parameters to correct for the fisheye effect. After calibration, the rig is fixed is space and a calibration image is captured from each camera. These images are stitched together using a spherical compositing surface, to generate the stitching mappings. During the operational phase, videos are captured from each camera, and fisheye removal and stitching mappings are applied on each frame to map the video content to the global coordinate of the 360° videos. To speed-up the operational phase, for each camera, fisheye removal mapping and stitching mapping are consolidated into a single mapping in the calibration phase. Finally, warped videos are corrected for color uniformity and rendered into a seamless 360° panoramic video. The resultant video may be rendered in a flat surface for display on regular display devices or prepared appropriately to be displayed using 360° panoramic video services like YouTube 360°.

2 Proposed Framework

Since the proposed panoramic system is composed of only 3 cameras, fisheye lens cameras are used. Fisheye lenses provide very large wide-angle views, and thus, provide enough overlap between the cameras which is required for successful stitching of the frames from the cameras. There are algorithms for registration of frames with minimal or no overlap between the frames given jointly moving cameras, such as the seminal work of Caspi and Irani [3]. However, our experiments reveal that these methods are not very accurate. More importantly, for the case of small or no overlap between frames, it is not possible to refine the results using bundle adjustment, as will be described later, to ensure that the 360° panoramic video forms a seamless realistic videos.

For realization of this work, capabilities of OpenCV [1] libraries were found to be appropriate. Main stages of the proposed framework are discussed in the following sections.

2.1 Fisheye Lens Calibration

Each fisheye camera is separately calibrated to remove fisheye lens distortion. At this stage, a checker board pattern is used. For calibration, since 180° coverage of the fisheye lens leads to huge distortions at the boundary, different Matlab and OpenCV libraries are tried to find the best choice for calibration, with maximum usable angular coverage after calibration. OpenCV built-in calibration libraries were finally found to have superior performance. These libraries use low-order polynomial models for radial distortion correction. Figure 2 shows how the calibration removes the fisheye lens distortion. As a side effect, this distortion removal decreases field of view of the camera from about 180 to about 140, however, still these coverage provides enough overlap for estimation of stitching parameters.

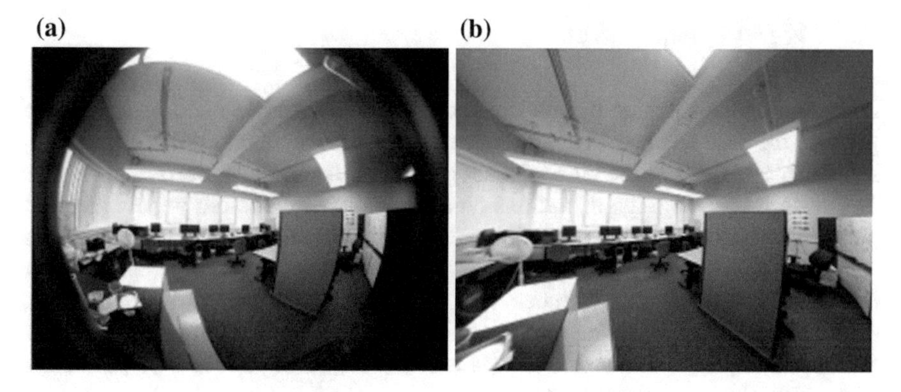

Fig. 2 **a** Input frame affected by fisheye lens distortion, **b** frame after removing fisheye lens distortion

2.2 Estimation of Stitching Parameters

To estimate the mapping from each video to the global 360° panorama coordinate, a calibrated image from each camera is used. This still image should be captured in a textured environment so that enough keypoint correspondence may be made between calibration images of all the 3 cameras. This is the most crucial step, since video stitching problem would be a generalization of this image stitching step. Also, since a single set of images are used to find the stitching parameters, the final video stitching results will lead to the best results when the environment depth is similar to the depth in the calibration images. So, if the target application involves capturing videos from objects far from the camera, the same kind of condition should be emulated for the parameter estimation stage.

For estimation of stitching mappings, first, SURF [2] keypoints are detected and matched [4] for adjacent cameras, via the calibration images. Given the matches, relative rotation and translation of cameras are estimated. To ensure that the mappings result into a seamless 360° panorama, bundle adjustment [6] is used to refine the mapping parameters.

Given the estimated parameters, in the operational phase, frames are continuously read from the 3 cameras and mapped to the global 360° panorama coordinate to form the panoramic video. To maintain synchronization of the videos from each camera compositing the final panoramic video, it is important to read 3 frames from all the 3 cameras simultaneously and store them, before starting processing of frames. Otherwise, if frames are read and processed sequentially, the delay in reading the frames caused by processing time will lead to synchronization artifacts in the panoramic video.

2.3 Video Composition

After finding the stitching parameters, a compositing surface should be selected to produce the final stitched image and video. This surface might be flat, cylindrical, spherical, etc. [5]. For the case of 360° panorama, compositing surface cannot be flat, since it leads to huge distortions on the boundaries and more importantly, the panorama cannot be wrapped to form a 360° coverage. In practice, flat panoramas start to look severely distorted if the field of view exceeds 90°. Thus, spherical or cylindrical surface is used, with better results achieved with spherical surface. Also, it is possible to composite the results on a dome to reproduce the stitched images as if they are created with a single fisheye lens camera, facing upward, perpendicular to the optical axes of the 3 cameras.

2.4 *Blending and Color Uniformity*

The frames from the 3-camera rig have some overlap, so, it is important to deal with the overlap area when composing the final stitched video. Combination of blending algorithms and exposure compensation algorithms [7] transform the stitched result into a color-uniform 360° video. For blending, multi-band blending is used [5].

2.5 *Preparation for YouTube 360°*

To enable playback of videos using YouTube 360° service, some meta data needs to be added to the resultant panoramic video to show the type of compositing surface used. For this purpose, "360° Video Metadata Tool" is used: https://github.com/google/spatial-media/releases/download/v2.0/360°.Video.Metadata.Tool.win.zip.

3 Results

In this section, we present stitching results for a sample set of frames from the 3 cameras. These frames are shown in Fig. 3. Figure 4 presents the stitched frames composited on spherical and cylindrical compositing surface. To illustrate how well the stitched frames match at the far ends to construct a full 360° panorama, we also present the stitching results using a dome as the compositing surface. As shown, the low-cost 3-camera rig equipped with fisheye lens provides an acceptable 360° panoramic system.

Fig. 3 Sample input frames

Fig. 4 Stitching results using different compositing surfaces, *top* to *bottom*: spherical, cylindrical, and dome compositing surfaces

4 Conclusion and Discussion

In this paper, a low-cost 3-camera rig equipped with fisheye lenses is proposed which enables generation of 360° panoramic videos. This low-cost system is realized via only 3 USB cameras, which omits the need for data acquisition system. Fishseye lens provides enough coverage and overlap between the cameras, so that the stitching parameters can be estimated reliably. However, it adds the fisheye lens distortion correction which is computationally expensive. To reduce this computational burden, it is possible to consolidate the distortion correction warping and the stitching mapping into a single mapping function for each camera.

Despite the reasonable performance, such a system which relies on offline and pre-calibration of the system, will be affected by parallax issue. Parallax issue will be more severe for the conditions far from the calibration condition. For the case of panoramic videos, due to computational costs and high efficiency required, parallax tolerant methods such as [9] are not feasible. Thus, for best performance, calibration should be performed in an environment with depth variations similar to the desired operational environment.

Acknowledgements This research was supported in part by Dongguan's Recruitment of Innovation and Entrepreneurship talent program.

References

1. G. Agam. Introduction to programming with openCv. Online Document, 27, 2006.
2. H. Bay, T. Tuytelaars, and L. Van Gool. SURF: Speeded up robust features. In Proc. European Conf. Computer Vision (ECCV), pages 404–417. Springer, 2006.
3. Y. Caspi and M. Irani. Aligning non-overlapping sequences. International Journal of Computer Vision, 48(1):39–51, 2002.
4. D. G. Lowe. Distinctive image features from scale-invariant keypoints. Int. J. Computer Vision, 60(2):91–110, 2004.
5. R. Szeliski. Image alignment and stitching: A tutorial. Foundations and Trends® in Computer Graphics and Vision, 2(1):1–104, 2006.
6. B. Triggs, P. F. McLauchlan, R. I. Hartley, and A. W. Fitzgibbon. Bundle adjustmental modern synthesis. In Vision algorithms: theory and practice, pages 298–372. Springer, 1999.
7. W. Xu and J. Mulligan. Performance evaluation of color correction approaches for automatic multi-view image and video stitching. In Computer Vision and Pattern Recognition (CVPR), 2010 IEEE Conference on, pages 263–270. IEEE, 2010.
8. Y. Xu, Q. Zhou, L. Gong, M. Zhu, X. Ding, and R. K. Teng. High-speed simultaneous image distortion correction transformations for a multicamera cylindrical panorama real-time video system using FPGA. Circuits and Systems for Video Technology, IEEE Transactions on, 24(6):1061–1069, 2014.
9. F. Zhang and F. Liu. Parallax-tolerant image stitching. In Proceedings of the IEEE Conference on Computer Vision and Pattern Recognition, pages 3262–3269, 2014.

Design and Implementation of a Facial Expression Response Analysis System

Kuan-Cheng Lin, Kun-Hsiang Chiang and Jason C. Hung

Abstract This study implemented a facial expression analysis system. The system can capture facial images of students via webcam and analyze facial expressions of students towards learning contents with different difficulties and a relationship between social interaction and learning effectiveness. The system selected ten face feature points for identification of 11 facial action units. Then these action units are classified into positive, neutral and negative social interactions by using a rule-based expert system.

Keywords Affective computing · Face expression · Digital learning

1 Introduction

Affective computing belongs to artificial intelligence. It uses massive computing, deduction and learning to enhance perception. The first important one is affective perception. Affective perception signals can be from objective physiological changes, such as facial expression, sound, pulse, breathe, brain waves, eyeball movement, body movement and etc. Emotion is a brain's appraisal for sensory stimuli, and a reaction resulted from emotional experiences. The reactions shown on faces are facial expressions, such as happiness, anger, sadness, fear, surprise and disgust; reactions shown on visceral nerve may change pulse, breathe, blood pressure, pupils, body temperature, brain waves, muscle and skin electric conductivity. Facial expression is a natural way to show inner emotional reactions.

K.-C. Lin · K.-H. Chiang
Department of Management Information Systems, National Chung-Hsing University, Taichung, Taiwan
e-mail: kclin@nchu.edu.tw

K.-C. Lin · K.-H. Chiang · J.C. Hung (✉)
Department of Information Technology, Overseas Chinese University, Taichung, Taiwan
e-mail: jhungc.hung@gmail.com

© Springer Nature Singapore Pte Ltd. 2018

N.Y. Yen and J.C. Hung (eds.), *Frontier Computing*, Lecture Notes in Electrical Engineering 422, DOI 10.1007/978-981-10-3187-8_48

The emotional expressions are mainly shown on mouths, eyes and eyebrows. Thus, the system must recognize and analyse the three parts.

Ekman suggested facial expression combinations of Facial Action Coding System (FACS) [1, 2]. It has one or combines several relevant facial action units, which can enhance facial expressions recognition. Ekman and Friesen defined 44 facial action units (AU) from upper face and lower face, and six basic emotions as per the facial expressions: Happiness, Anger, Sadness, Fear, Surprise, and Disgust et al. Due to complicated relationship between human facial expressions, the action units may be combined with other action units. Thus, description of facial expressions and degree may be different. Although FACS defined 44 facial action units, the operator can select action units related to the operation goal in practice according to needs, and optimal recognition effect can still be achieved.

2 The Design of Expression Response Analysis System (ERAS)

Implementation of ERAS contains three steps: facial detection, feature extraction and location, and expression classification and recognition.

We used OpenCV AdaBoost (machine learning algorithm), and Haar-Like for facial detection [3, 4]. OpenCV trained facial detection cascade classifier, including face, eyes (including left eye, right eye and eyes), nose and mouth to finish detection of face or specific locations. One static facial image contains eyes, nose and mouth. The total average detection time is around 0.6 s, and detection frequency is extremely high. The distance between eyes and brows are too close. In order to avoid mutual interference in the two extraction areas, the image sequence is duplicated to extract four areas: brows, mouth, left eye and right eye. The flow chart for facial detection and feature extraction is shown in Fig. 1.

2.1 Feature Extraction and Location

We observe changes in facial expressions mainly through three parts, brows, eyes and mouth. The feature points performed image processing and displacement measurement for the three areas. Thus, 10 feature points were selected as measurement basis (as shown in Fig. 2). Point A and Point B represent left and right corners of mouth respectively. When corners of mouth turn upwards, point A and point B raise. The upper and lower lips are linked, CC' distance can represent mouth opening and closing. Eyelids are linked. EE' and FF' distance can represent eye opening and closing. Even, eye blink or closing can be determined through closing time. Inside corners of eyebrows are also important in facial expression.

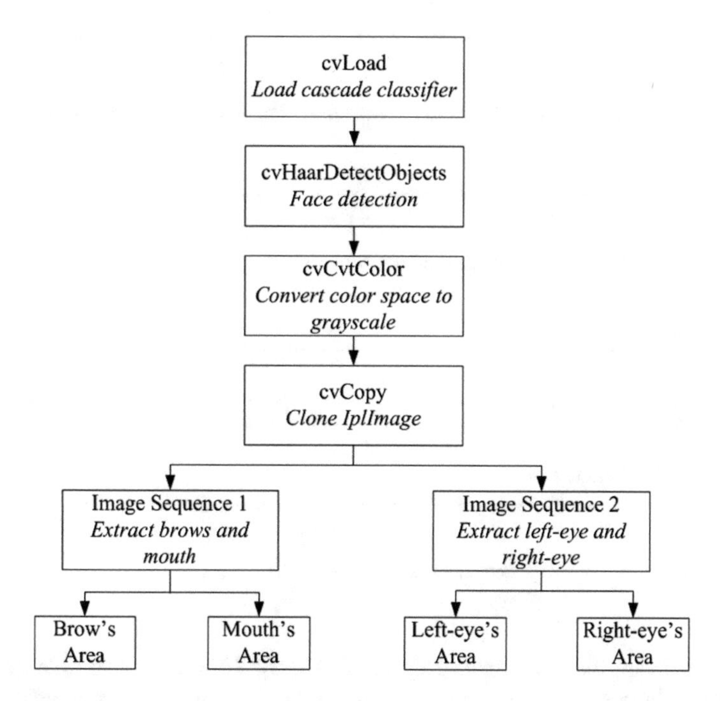

Fig. 1 Flow chart for face detection and feature extraction

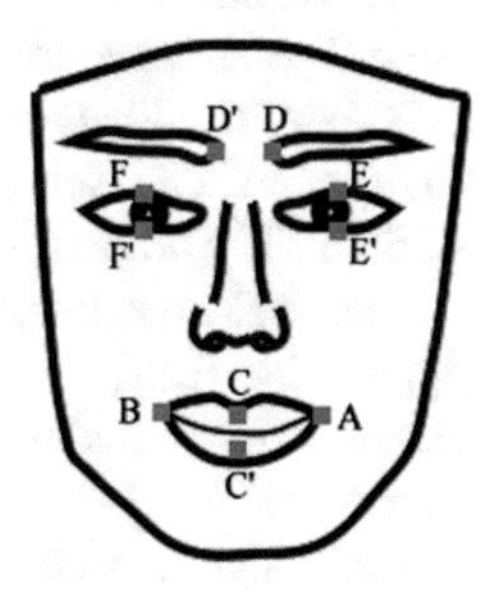

Fig. 2 Location of face feature points

This can be observed through up and down of Point D and Point D' and distance between the two points.

2.2 Feature Distance Vector

Feature vector is important basis for classification of facial expressions. The data changes have meanings. AB feature point is taken as an example. When smile is shown on face, corners of mouth turn upwards, vtAB is expanded and enlarged, and

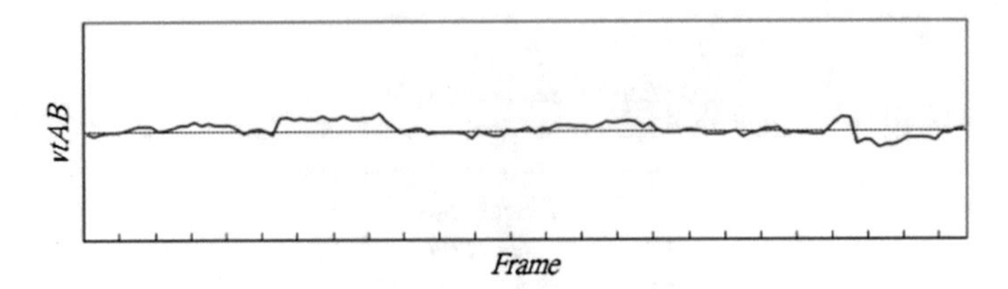

Fig. 3 Diagram of vt image sequence (before conversion)

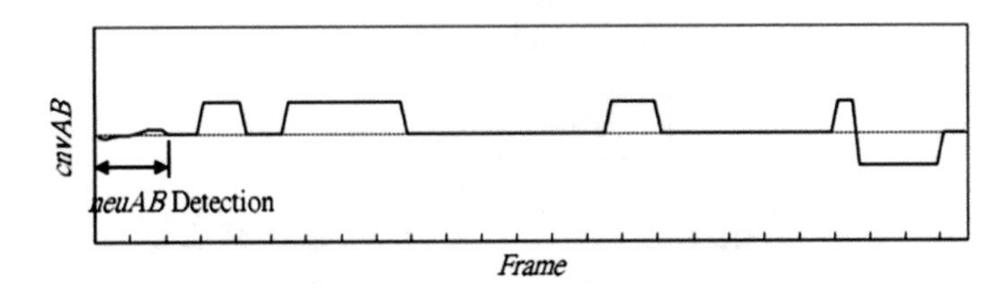

Fig. 4 Diagram of cnv image sequence (after conversion)

vtA and vtB are also enlarged. This is AU12. Swallow may also enlarge vtAB. However, corners of mouth do not turn upwards. Thus, vtA and vtB are not enlarged. This is AU20. With smile, the upper and lower lips become thin, and vtCC′ are shortened. This is AU12+24. When laugh is shown in face, the upper and lower lips are opened, and vtCC′ is enlarged. This is AU12+25,26,27 according to vtCC′ stretched length. When yawning or feeling sad, opening upper and lower lips enlarges vtCC′. However, corners of mouth are downwards. Thus, this is AU15 +25,26,27. The original data of vtAB feature vector are shown in Fig. 3.

The real-time image detection and human breathe may cause fluctuation and jiggle. The feature vector may change slightly. This is not good for subsequent classification and identification. Thus, the experimental subjects must keep no facial expression for 2–3 s. This can facilitate the system to detect neutral (neu). The neutral value is used as benchmark, −10% is lower threshold value, and +10% is upper threshold value. They are converted through the following equation.

$$cnv = \begin{cases} neu * 0.7, & if\ vt \leq neu * 0.9 \\ neu * 1.3, & if\ vt \geq neu * 1.1 \\ neu, & other \end{cases}$$

After threshold value conversion (cnv), vt vector is shown in Fig. 4.

Figures 5 and 6 represent feature vector change of mouth and upper and lower eyelids.

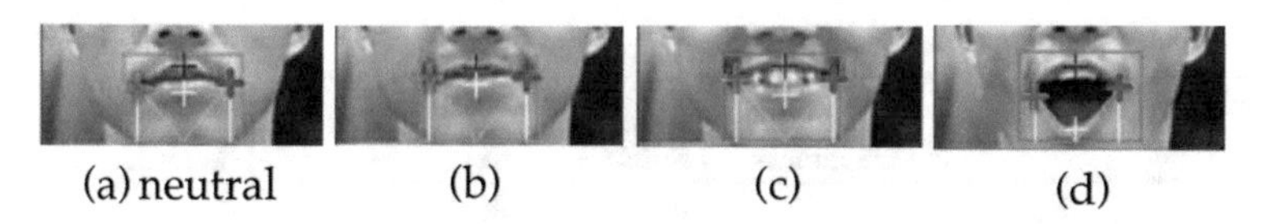

Fig. 5 Diagram of change in mouth feature vector

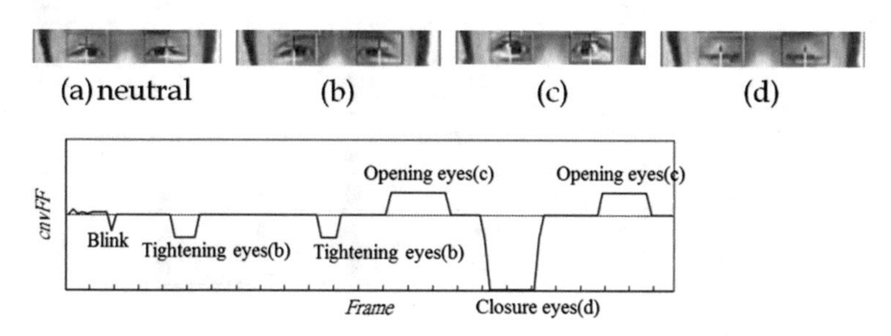

Fig. 6 Diagram of change in feature vector of *upper* and *lower* eyelids

2.3 Expression Recognition and Classification

This paper aims at social reactions represented by natural facial expressions of students after they received textbooks with different stimuli. Rule-Based Expert System was used for identification of feature vectors of each location, and FACS AU for recognition.

Brow's AUs (AU1, AU4)

Brow's AUS are used to identify up and down of points D and D′ and DD′ flexing distance. DD′ distance may not change when we consider up and down of points D and D′. Thus, the identification rule is divided into two parts for classification.

Eye's AUs (AU5, AU7, Blink, Close)

Eye action is very rapid, and action maintenance time must be considered. If eyelid closure distance (EE′ and FF′) is longer, the maintenance time is shorter; if the eyelid closure distance is shorter, the maintenance time is longer. Both are exclusive. In the same frame, when AU5 (opening eyes) is determined, AU7 (eyelids tightening, blink and closure) is impossible to occur. The webcam shooting speed is 10FPS, and the maintenance time is 0.5 s (5 frames). When the eyelid closure distance is ≤ 0.01 and the maintenance time of action is less than 0.5, the facial expression is blink. If the maintenance time exceeds 0.5, the facial expression is blink.

Mouth's AUs (AU12, AU15, AU20)

Action of corners of mouth is crucial to producing mouth expression. The basis is up and down of points A and B, and AB distance. When people feel pleased,

Table 1 AU combinations of basic emotions

Location	Emotions	Positive		Negative			
		Happiness	Surprise	Fear	Anger	Sadness	Disgust
Upper face	AU1	●	●	●		●	
	AU4			●	●	●	●
	AU5		●	●	●		
	AU7	●				●	●
Lower face	AU12	●					
	AU15		●	●	●	●	●
	AU20			●			●
	AU24	●		●	●	●	●
	AU25	●	●	●	●	●	●
	AU26	●	●	●	●	●	
	AU27	●	●	●	●	●	

corners of mouth may turn upwards, showing a smile; when people feel unpleased, there is no facial expression, or corners of mouth are downwards; when people feel bored or uninterested, corners of mouth may stretch, showing escape.

Lips's AUs (AU24, AU25, AU26, AU27)

Opening and closing of upper and lower lips can lift reaction range of mouth expression, and the main basis is CC′ distance. We can take neuCC′ as an example. If less than neuCC′, the upper and lower lips are tense, and CC′ distance is shortened; if exceeding neuCC′, the mouth is opened, and CC′ distance is enlarged. There is small open-wide AU25, medium open-wide AU26 and big open-wide AU27.

AU Combinations

Ekman and Friesen defined 44 action units from upper face and lower face, and defined six basic emotions: Happiness, Anger, Sadness, Fear, Surprise and Disgust. This paper summarizes 11 action units according to the facial expressions of the students after they received information stimuli. The AU combinations are defined according to the basic six emotions, as shown in Table 1.

3 The Implementation of Expression Response Analysis System(ERAS)

ERAS contains three modules: image extraction module, facial expression recognition module, and emotional reaction analysis module.

3.1 Image Extraction Module

Image extraction module is simple in interface operation, and has image location adjustment and warning functions. It runs on the computers of the tested students. Due to small interface and simple functions, it can reduce psychological impact on tested students, and make images and locations of tested students consistent. Its main outcome is AVI file of the facial expressions of tested students in learning. These files are convenient for future processing and application. When the green light is on (as shown in Fig. 7a), face location of one tested student is suitable, and distance to the lens is normal (60–70 cm); when the yellow light is on, face location of the tested student deviates, and is still accepted; when the red light is on, face of the tested student cannot be detected, or the system cannot perform processing and recognition. The tested students shall avoid red light. During program execution, the correction window (as shown in Fig. 7b) is in central upper edge of the screen, and below the webcam. Its function is like a mirror, and tested students can exercise facial expressions besides adjustment of face location. After pressing down "Record", the correction window is hidden to prevent interference with testing.

3.2 Facial Expression Recognition Module

The facial expression recognition module runs at back end. It mainly process facial expression images of the tested students, including image processing, feature point marking, calculation of feature vector and conversion. At last, all the feature vectors

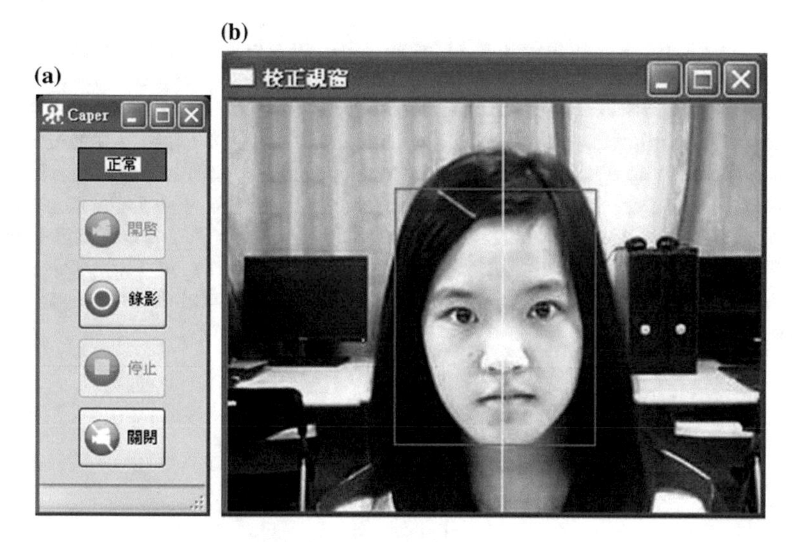

Fig. 7 Operation picture of image extraction module

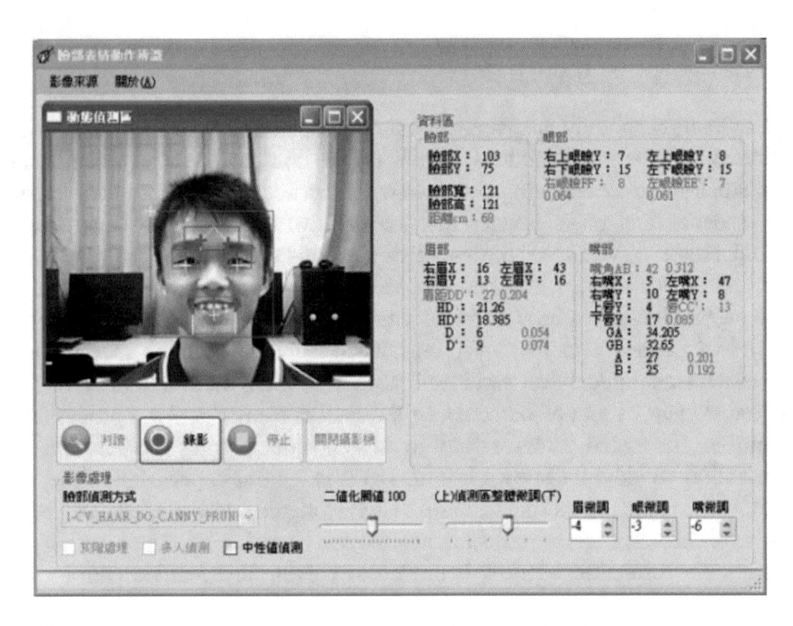

Fig. 8 Facial expression recognition module—dynamic image detection

are recognized to action units, and then text files are input. This module can be used for detection of dynamic images, as shown in Fig. 8.

3.3 Facial Expression Analysis Module

The facial expression analysis module is also back-end program. Due to close relations between facial expression and time sequence, facial expression analysis is conducted after all AU are recognized. If hold time of eye closure is ≤ 0.5 s (around 5 frames), it is eye blink. If the time exceeds 0.5 s, it is eye closure. The facial expression analysis module takes frame as analysis unit. It analyses AU from one frame. Thus, one frame produces one facial expression. The facial expression analysis rule is performed one frame by one frame as per AU combination table (Table 1). The facial expression analysis module (as shown in Fig. 9) uses AU recognition results from the facial expression recognition module to analyse facial expressions.

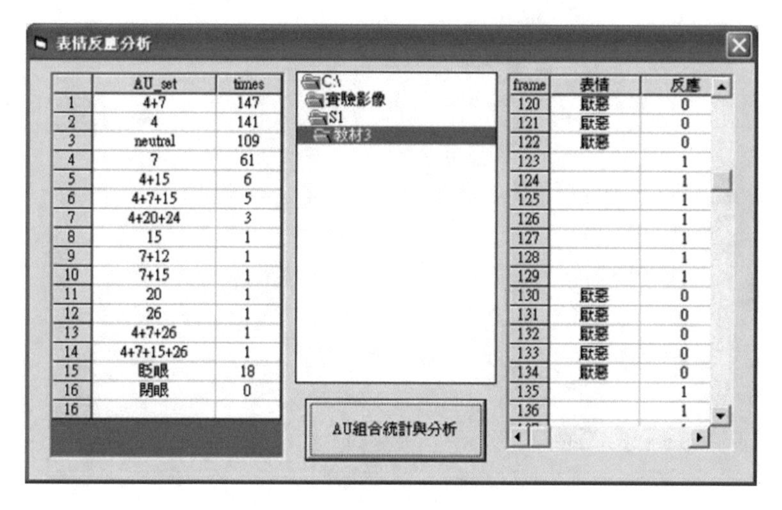

Fig. 9 Operation picture of facial expression analysis

4 Conclusions

Based on the facial action coding system, OpenCV and image processing method were used for face detection, feature extraction and location. A facial expression analysis system was implemented. This system extracted 10 face feature points, and used 9 feature vectors and neutral values as measurement benchmarks for identification and blink detection of 11 facial action units. It classified the action units into six basic facial expressions, happiness, surprise, fear, anger, sadness and disgust.

Acknowledgements The authors greatly appreciate the financial support provided by Taiwan's Ministry of Science and Technology, the Republic of China under contract No. MOST 104-2511-S-005-002.

References

1. P. Ekman, "Emotions Revealed: Understanding Faces and Feelings," Weidenfeld & Nicolson, 2003.
2. P. Ekman and W. V. Friesen, "The Facial Action Coding System: A Technique for The Measurement of Facial Movement", Consulting Psychologists Press, San Francisco, 1978.
3. D. S. Bolme, M. Strout and J. R. Beveridge, "FacePerf: Benchmarks for Face Recognition," *Proceedings of the International Symposium on Workload Characterization,* 2007, pp. 114–119.
4. J. Whitehill and C. W. Omlin, "Haar Features for FACS AU Recognition," *Proceedings of the International Conference on Automation Face and Gesture Recognition,* 2006, pp. 5–9.

pcDuino: A Friendly Open Hardware Platform for Programming

Qi Yao, Jingfeng Liu and Qingguo Zhou

Abstract Arduino and Raspberry Pi are the most popular open source hardware in the world which are invented with the hope of inspiring generation of students to learn programming and be creative. The former is a simple AVR microprocessor but powered by simple programming IDE and the complete open hardware ecosystem. The latter is a powerful ARM based mini PC but it is not very easy to get started if users has no enough programming background. To combine the advantages of both an ARM based mini PC and the complete Arduino ecosystem. LinkSprite released a powerful mini PC platform: pcDuino, which represents PC + Arduino. This paper provides a review of pcDuino technology and especially introduces the several programming methods on pcDuino.

Keywords pcDuino · Open source hardware · Scratch · Arduino

1 Introduction

pcDuino [1] is a mini but powerful PC platform compatible with Arduino, developed by LinkSprite Technologies Inc [2]. It's a powerful device especially designed for demanding users with a small budget. Combining the benefits of a high-performance ARM based mini PC, Linux, Android and Arduino ecosystem, pcDuino has become the new choice of students, teachers, makers.

Q. Yao (✉) · J. Liu
LinkSprite Inc., Wuhan, China
e-mail: qi.yao@linksprite.com

J. Liu
e-mail: jingfeng.liu@linksprite.com

Q. Zhou
Lanzhou University, Lanzhou, China
e-mail: zhouqg@lzu.edu.cn

© Springer Nature Singapore Pte Ltd. 2018

N.Y. Yen and J.C. Hung (eds.), *Frontier Computing*, Lecture Notes in Electrical Engineering 422, DOI 10.1007/978-981-10-3187-8_49

pcDuino was first launched in 2013, and the first generation of pcDuino uses Allwinner A10 SoC chip which is 1 GHz ARM Coretex A8 SoC, and it can run Ubuntu 12.04 or Android ICS 4.0.

With advances in IC manufacturing processes, many open source hardware platform start to use multi-cores SoC to meet the increasing performance requirements. Like pcDuino3, it was released in 2014 and is powered by Allwinner A20 SoC chip which takes dual-core 1 GHz ARM Cortex A7. And in the fourth quarter of 2015, LinkSprite launched a new product, called pcDuino8 Uno which is an 8 cores high performance, cost effective single board computer with H8 SoC chip up to 2.0 GHz. It runs operation systems such as Ubuntu Linux and Android. pcDuino8 Uno has HDMI interface to output its graphic desktop screen. It could support multi-format 1080 p 60 fps video decoder and 1080 p 30 fps H.265/HEVC video encoder with its built-in hardware video processing engine. It targets specially the fast growing demands from the open source community. As shown in Fig. 1, pcDuino8 Uno provides easy-to-use tool chains and is compatible with the popular Arduino ecosystem such as Arduino Shields.

Fig. 1 pcDuino8 Uno

2 Technical Specifications

The specifications of latest pcDuino are described as follows:

- CPU: AllWinner H8 8-Core Cortex-A7@2.0 GHz
- RAM: 1 GB DRAM
- GPU: Power VR SG544@720 MHz
- Storage: MicroSD card slot
- Ethernet: 10/100/1000 Mbps Ethernet port
- Arduino sockets: same as Arduino Uno 14x GPIO, 1x UART, 1x SPI 1x I2C
- 6x ADC (extra module needed to provide ADC)
- Camera: MIPI for camera
- Audio: 3.5 mm analog audio interface
- USB: 1x USB Host, 1x USB OTG
- Power: 5 V, 2000 mA Micro USB
- Size: 92 × 54 mm

Required accessories:

- Micro-USB port adapter (5 V, 2 A)
- HDMI cable and a monitor with HDMI port
- USB Hub, USB mouse and USB keyboard
- 4 GB or more micro SD card (Fig. 2)

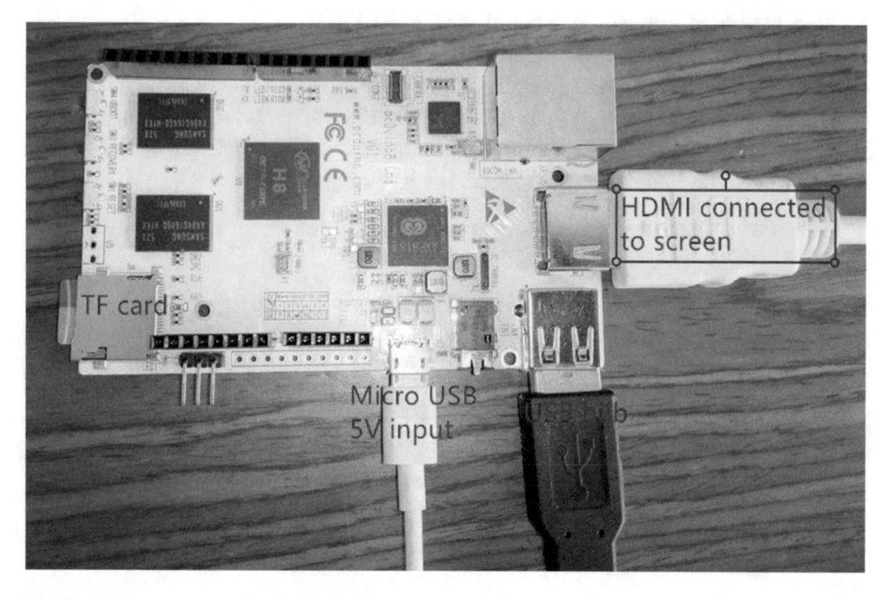

Fig. 2 Hardware setup of pcDuino8 Uno

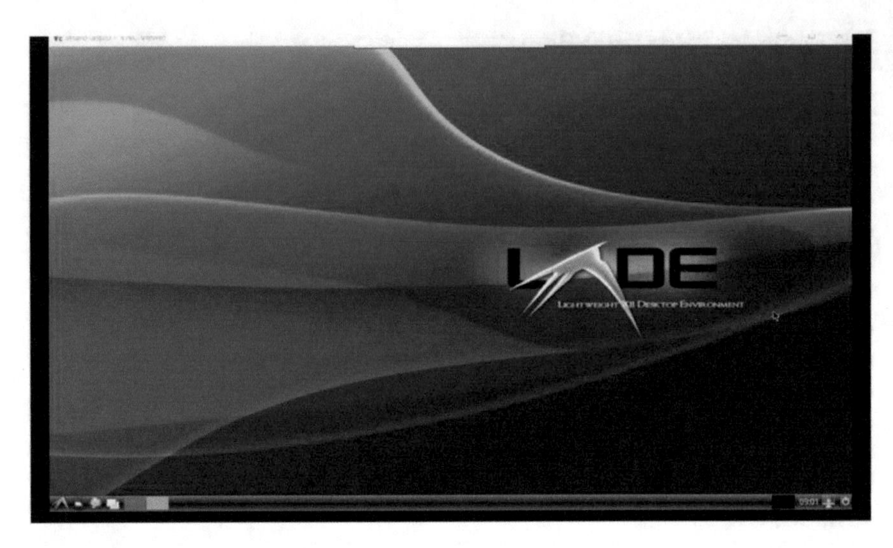

Fig. 3 Ubuntu desktop on pcDuino8 Uno

3 Hardware Setup and Booting Process

1. Download the Ubuntu system image from LinkSprite official website [3].
2. Use Win32diskImage Windows tool or dd Linux command to burn the system image onto micro SD card.
3. Plug the micro SD card into pcDuino8 Uno.
4. Plug keyboard, mouse onto board with USB HUB.
5. Plug HDMI cable onto the board and to the monitor.
6. Connect 5 V DC power input via Micro USB port.
7. For the first time boot, system initialization may take about 1–2 min.
8. System will automatically restart and enter into desktop, as shown in Fig. 3.
9. After that, the board can be utilized for any project.

Note: the default username and password are all linaro.

4 Programming Methods on PcDuino

pcDuino has not only supported GNU/C or C++, python, or JavaScript programming, but also supported Scratch or VIPLE [12] visual programming which make it very easy to access hardware resource of pcDuino.

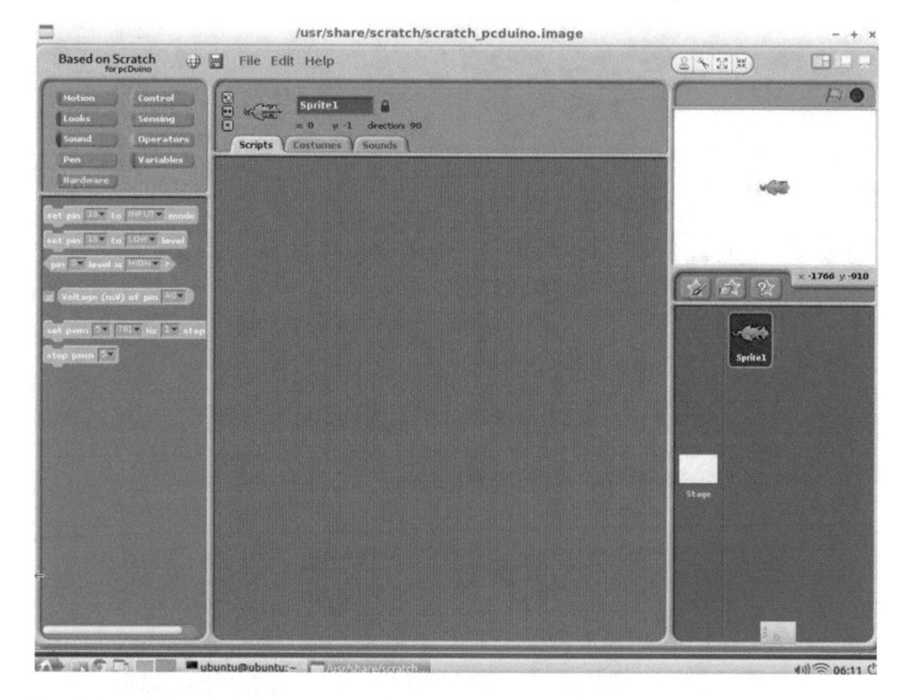

Fig. 4 Scratch on pcDuino

4.1 Scratch Programming

As shown in Fig. 4, Scratch [4] is a visual programming language for everyone and you can create your own interactive stories, games, music and art and share them online.

pcDuino team developed a customized version of Scratch for pcDuino [5], The GPIO, PWM and ADC pins can be accessed directly from Scratch panel. Users can use it easily to program, including control LED, or set up a robot.

As shown in Fig. 5, this program is used to set the GPIO3 as output port and in every second, it will output 0 or 1.

4.2 Arduino Programming

Arduino programming is another popular programming method which supplies plenty of libraries and examples. As shown in Fig. 6, with the simple APIs of libraries, users can very easily to access hardware without figuring out the details of hardware, like I2C, SPI communication protocols or configuration of some sensors' registers.

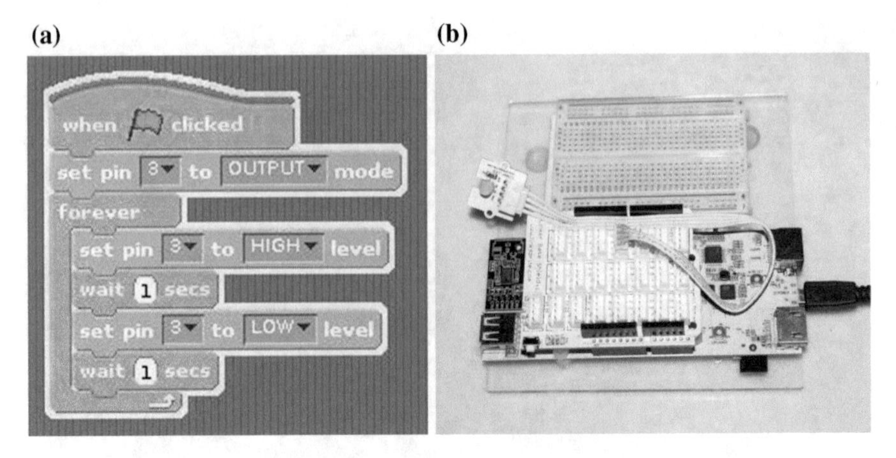

Fig. 5 **a** Scratch programming, **b** shield on pcDuino

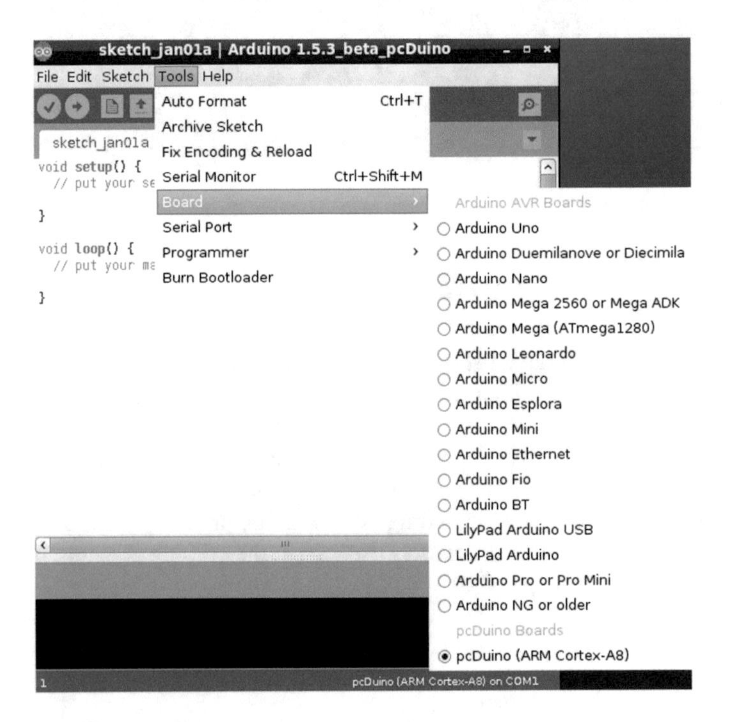

Fig. 6 Arduino IDE on pcDuino

pcDuino team especially developed a libraries for pcDuino which make it compatible with Arduino programming. The system image has installed the customized Arduino IDE.

4.3 Other Programming

Besides, C/C++, python and JavaScript are also good choices to access pcDuino's IO, for these languages, there are a developed libraries and examples which are open source on Github website [8–10].

5 Applications

pcDuino is not only for programmers but all the people who are interested in computing science or want to solve the practical problem with programming. Some applications on pcDuino have been included but not limited:

- pcDunio MongoDB Cluster [6]: Use 4 pcDuino3 Nano Lite boards to construct a cheap database cluster, as shown in Fig. 7.
- Face detection on pcDuino8 Uno: Take use of high performance of pcDuino8 Uno and run openCV to process video stream from webcam and detect the human face, as shown in Fig. 7.

Fig. 7 **a** pcDuino mongoDB cluster, **b** face detection on pcDuino8 Uno

Fig. 8 WiFi video surveillance robot

- 3D printer [7]: This product is powered by pcDuino Acadia.
- WiFi Video Surveillance Robot, as shown in Fig. 8: Remote control the robot via WiFi and get the video stream.

6 Conclusion

There is a trend that more and more students or people who want to learn programming will start with a visual programming first, including Scratch, Blockly [11] or other tools, but not with command line method. These friendly programming method hide these underlying hardware details and simplify the accessing hardware which encourage users focusing on solving a problem itself.

pcDuino is that kind of friendly platform and supplies several programming methods to users, including Scratch, Arduino IDE or python. Meanwhile, pcDuino has Arduino-compatible slot which let many Arduino shield boards mount on pcDuino directly without any change both hardware and software.

Acknowledgements This work was supported in part by Dongguans Recruitment of Innovation and entrepreneurship talent program.

References

1. http://www.linksprite.com/linksprite-pcduino/
2. http://www.linksprite.com/
3. http://www.linksprite.com/image-for-pcduino8-uno/
4. http://scratch.mit.edu/
5. Lifeng Zhao and Jingfeng Liu, Introduction to pcDuino, 2013
6. http://www.ferzkopp.net/wordpress/2016/01/04/pcdunio-mongodb-cluster/
7. http://www.linksprite.com/support-raise3d-kickstater-a-3d-printer-powered-by-pcduino-acadia/
8. https://github.com/pcduino/c_environment
9. https://github.com/pcduino/python-pcduino
10. https://github.com/rwaldron/johnny-five
11. https://developers.google.com/blockly/
12. http://venus.eas.asu.edu/WSRepository/VIPLE/

Analysis Modeling of College Students Sources

Wen-Hsing Kao, Kuo-Ping Kao, Min-Hui Ding, Jason C. Hung, Pi-Chung Wang and Bo-Shen Liou

Abstract Demographic structure of Taiwan is becoming an aging society with fewer children due to lower birth rate, which have great impacts on the teacher supply and demand, education resources and the school education. It is a currently important issue to concern the future education developments from the standpoint of demographic transition. In recent years, because of socio-economic pressures and social values change, the birth population of Taiwan is dramatically decreased, the birth rate declines, and the society of the trend of fewer children is formed. The crisis and the impact produced by the trend of fewer children directly reflect on educations. We collect date related to the source of students, including whole populations of Taiwan, birth population of Taiwan, the ratio of class number to student number from all high schools in Taiwan, enrollment number, enrollment grades and ratios of vocational schools in Taiwan, to predict the situation of the future source of students in variety collage departments based on the collecting data. By surveying the ninth grade student and the high school third grade student as subjects, their choosing intention of further studies is validated, then the ratio of ambitions and the school ranks is adjusted. In this study "Analysis model of college

W.-H. Kao · J.C. Hung · B.-S. Liou (✉)
Department of Information Technology, Overseas Chinese University,
Taichung, Taiwan
e-mail: bsliou@ocu.edu.tw

W.-H. Kao
e-mail: star@ocu.edu.tw

K.-P. Kao
Department of Tourism and Recreation Management,
Overseas Chinese University, Taichung, Taiwan
e-mail: Kuo@ocu.edu.tw

M.-H. Ding · P.-C. Wang
Department of Computer Science and Engineering,
National Chung Hsing University, Taichung, Taiwan
e-mail: christine8018458@gmail.com

P.-C. Wang
e-mail: pcwang@nchu.edu.tw

© Springer Nature Singapore Pte Ltd. 2018

N.Y. Yen and J.C. Hung (eds.), *Frontier Computing*, Lecture Notes
in Electrical Engineering 422, DOI 10.1007/978-981-10-3187-8_50

student sources”, we used the Visual Basic and C# language to write a college student source inquiring website through which the trend of application number and the birth population of the school year are obtained. By this study, we can estimate the situation of the possible low enrollment, and provide it as a consult to schools for recruiting new students or for transformation, meanwhile this study provides sufficient data analysis as a reference basis for students when deciding the school, to prevent from choosing closing departments or schools, and provide application-related information to user as the reference.

Keywords Fewer children · College student sources analysis · Population trend

1 Research Background

Taiwan’s economic development, because modern life stress, parenting high cost, enhance the level of education of people, values changed, and female gender perspectives for the rise of self-awareness and other factors, led to low fertility in Taiwan.

The world, including Japan, Germany, Italy, France and other developed countries have faced the impact of low birth rate, the earliest face this issue is Japan, on the other hand Taiwan is the country with the fastest speed of the low birth rate occurred within a short time. Crisis and the impact generated by the phenomenon of the declining birthrate, more directly and clearly reflected in the students of private universities problem.

Minister of Education Wu Se-hwa said that the number of university in Taiwan should be reduced to under 100, due to the year of 105 exactly facing the baby born in the year of Tiger. High school graduates will reduce 30,000 which is considered a “105 Doomsday” [1].

2 Research Motivation

In view of the declining birth rate has been a wave of shock to the national education system, schools at all levels are directly faced with dwindling sources of students, according to research statistics show the Vocational and Technical Schools will face the serious impact more than the University [2].

To follow the social trends, that popular course which students always take is different in technology institution. Including biomedical, catering service, tourism, design, cultural and creative departments are becoming the most wanted major course by students [3, 4]. The main reason is to attract students is class contents of these more interesting than information technology department [5, 6]. Instead of such electrical engineering departments, industrial departments, they are not taken seriously gradually student. In contracts with the popular course use to be Electrical Engineering and Industrial Management will not be the priority major anymore [7].

Taking all these factors, by inference universities will likely face the appearance of insufficient of enrollment.

3 Research Purposes

The Ministry of Education estimated the policy which in 5 years will reduce the existing 159 of large school down to the 100. Within the Ministry of Education will also start counseling large school in transition to new model or to be exit.

In the face of a shortage of students, through the system of Students source model can estimate whether there is any shortage of students in the future, provided to the school as a suggestion on enrollment or transfer into a new model system of University also provide students several options when they comes to decide their major [8].

4 Limitations of Study

In this study, analysis subject from Oversea University will not represent other Technology Information Department from general University in Taiwan.

References and data is difficult to obtain due to the current population of Taiwan and the small number of research papers relevant to students.

Through data projections students will not be able to fully perceive the future development of social trends, market-oriented industry and the future of education policy from Ministry of Education. These factors will affect the students in the choice of school or volunteer options judgment, which led to the projected inaccurate.

5 Research Methods

5.1 College Students Sources Modeling System Architecture

In this thesis, the data resource comes from University students which include: total population in Taiwan of past years, the birth rate, the proportion of students from high school in Taiwan, the population of enrollment the grade of enrollment and the number of enrollment from technology institution. To use language of Visual studio asp.net and C#, gather all the research data into a database system to write and adjust the web page for university students. Users will enter year of enrollment, which university, groups of each university major and departments through database matching system and data processing, the output will present birth rate of each year of enrollment, the proportion of the number of examination registration, the total number of examination registration, the proportion of enrollment from each group of university major to apply. Then through this system will demonstrate whether will affect the number of future enrollment of major from each university. (The system frame shown in Fig. 1).

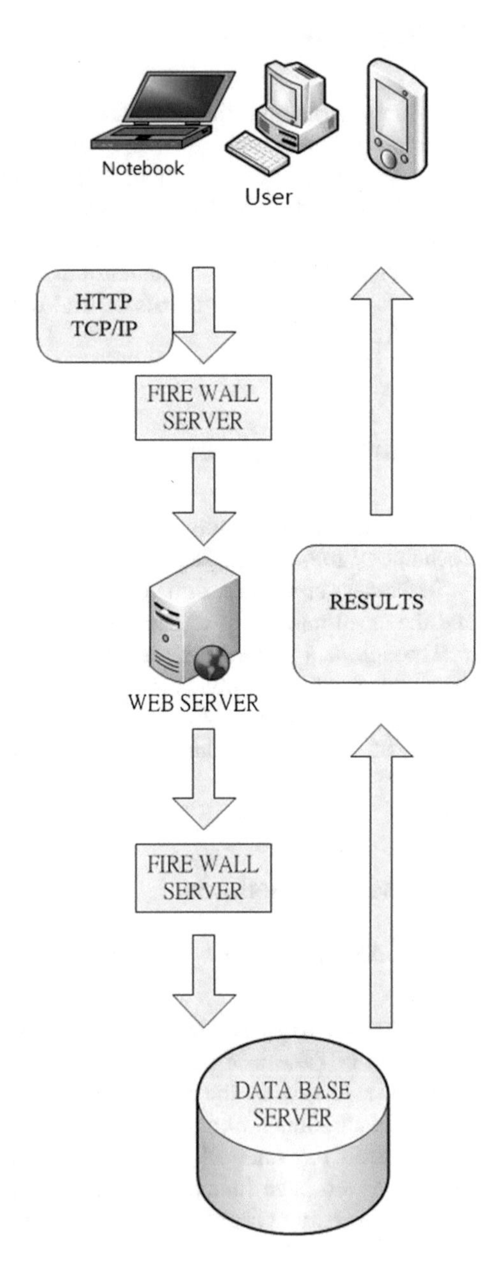

Fig. 1 System architecture

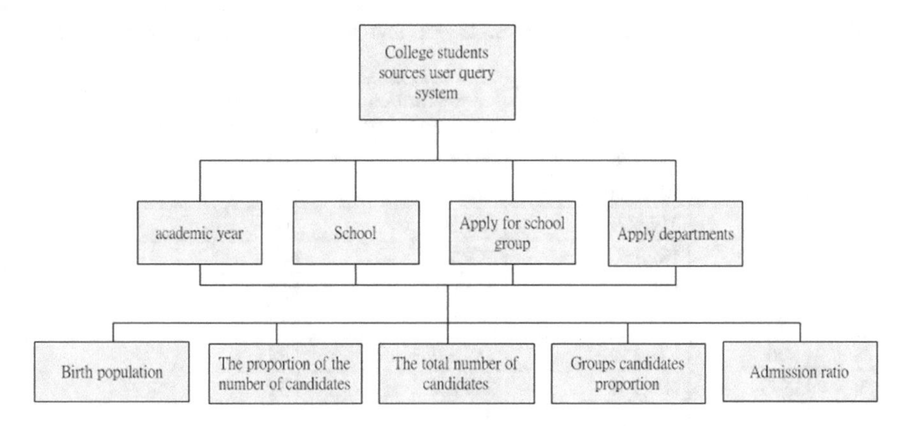

Fig. 2 System function

5.2 *College Students Sources Modeling System Function*

The functionality of student source model is users to use the drop-down menu to select the input matriculation year, the data shows that the total number of people attending school 'birth population' the proportion of the number of candidates and the total number of candidates. Apply for school groups and departments to apply for the information is directly related to the department of information directly affects the proportion of candidates in each category and the total number of candidates, the final output is the proportion of admission. (As shown, the system functions in Fig. 2).

5.3 *Student Source Estimated*

To estimate year of 103–117, a approved unified examination from high school to technology institution (electrical technology group 'electrical group' business and management group) to evaluate the calculation steps are as follows:

(1) The total number of examination candidates/total number of total birth number of Taiwanese = proportion of examination candidates 103 year 142,573/325,545 = 43.79%

To sum up the proportion of apply for technology institution from academic year of 90 through 103.

According to this data to presume the future enrollment number in electrical technology group, electrical group, and business and management group. (In Table 1 below).

Table 1 Calendar year exam registration 2-year and 4-year TVE joint college entrance examinations proportion

Birth year	Academic year	The total number of candidates	Taiwan's total population was born	The proportion of the number of candidates (%)
72	90	237,576	383,439	62
73	91	225,783	371,008	61
74	92	214,969	346,208	62
75	93	189,100	309,230	61
76	94	182,591	314,024	58
77	95	181,530	342,031	53
78	96	172,530	315,299	55
79	97	169,974	335,618	51
80	98	158,154	321,932	49
81	99	158,025	321,632	49
82	100	159,243	325,613	49
83	101	155,733	322,938	48
84	102	146,047	329, 581	44
85	103	142,573	325,545	43.79

(2) And then to apply the proportion of 43.79% for the academic year of 103 a fixed number, multiplied by the number of known birth number of Taiwan, will comes up a total estimate of the number of candidates

Republic of China 86 years born population 326 002 * 103 school candidates proportion 0.43795234452 (43.79%) = reckoning the 104 years the total number of applicants 140,000 2,773.

(3) Estimated total number of candidates academic year of 104 will be 140,000 2,773 * (Electrical technology candidates proportion of 8.98% shown in Tables 1, 2 and 3) = academic year of 104 estimated number of people enrolled 4,798 people owned Electricity

Estimated total number of candidates in academic year 104 will be 140,000 2,773 * (electrical group candidates proportion of 3.35% shown in Tables 1, 2 and 3) = 104 Estimated number of people enrolled Motor Class 10,000 2,824 people

Estimated total number of candidates academic year of 104 will be 140,000 2,773 * (Business management group candidates proportion 22.89% Tables 1, 2 and 3) = 104 Estimated number of people enrolled Business management group 30,000 2,683 people.

(4) by inference in the years of Republic of China 99 is minimal year students, the number of births 160,000 6,886 people, reckoning in the academic year of 117 electrical group will have 2,453 of student and Electrical technology group will

Table 2 Estimates of future total number of candidates

Birth year	Academic year	The total number of candidates	Taiwan's total population was born	Electrical group 3.35%	Electrical technology group 8.98%	Business management group 22.89%	The proportion of the number of candidates (%)
85	103	142,573	325,545	4786	12,806	32,637	43.79[1]
86	104	142,773[2]	326,002	4,793[3]	12,824[3]	32,683[3]	43.79
87	105	118,882	271,450	3,991	10,678	27,214	43.79
88	106	124,230	283,661	4,170	11,158	28,438	43.79
89	107	133,712	305,312	4,489	12,010	30,609	43.79
90	108	114,022	260,354	3,828	10,242	26,101	43.79
91	109	108,406	247,530	3,639	9,737	24,816	43.79
92	110	99,446	227,070	3,338	8,932	22,765	43.79
93	111	94,781	216,419	3,182	8,513	21,697	43.79
94	112	90,154	205,854	3,026	8,098	20,638	43.79
95	113	89,543	204,459	3,006	8,043	20,498	43.79
96	114	89,523	204,414	3, 005	8,041	20,493	43.79
97	115	87,035	198,733	2,922	7,818	19,924	43.79
98	116	83,785	191,310	2,813	7,526	19,179	43.79
99	117	73,088	166,886	2,453	6,565	16,731	

Table 3 103 The number of all types of cluster registration

103 academic year	Electrical technology group	Electrical group	Business management group
Groups enrollment	12,806	4,786	32,636

Table 4 103 The total number of school enrollment distribution

103 academic year	Electrical technology group	Electrical group	Business management group
Distribute the total number of admissions	3,934	1,804	10,088

Table 5 Overseas Chinese University Information Technology Department of the cumulative total number of admissions rankings

Distribution admission Overseas Chinese University Information Technology Department of the cumulative total number of admissions rankings			
Academic year	Electrical group	Electrical technology group	Business management group
101	1,538	*	*
102	995	1,712	4,591
103	861	1,386	3,784

have 6,565 students also business management group will have 16,731 students. (As shown in Table 2)

(5) The enrollment of academic year of 103 will be 12,806 (Table 3)/the Electrical technology total number of admissions is 3,949 in academic year of 103 (Table 4) = Rate of Electrical technology is 3.24.

The enrollment of academic year of 103 will be 4,786 (Table 3)/the electrical group total number of admissions is 1,804 in academic year of 103 (Table 4) = Rate of Electrical technology is 2.65.

The enrollment of academic year of 103 will be 32,636 (Table 3)/the business management group total number of admissions is 10,088 in academic year of 103 (Table 4) = Rate of Electrical technology is 3.23.

Distribution rate representation of the distribution number in registered groups, on behalf of the higher proportion of distribution the less of enrollment students; on the other hand the lower proportion of distribution the more of enrollment students.

(6) The students whose enroll to Information Technology Department of Oversea University from electrical group at ranking 861 (Table 5) *the rate of electrical group 2.65 = to estimate the number of candidate will be 2,282 in electrical group of academic year of 117.

The students whose enroll to Information Technology Department of Oversea University from electrical technology group at ranking 1386 (Table 5) *the rate of electrical group 3.24 = to estimate the number of candidate will be 4,490 in electrical technology group of academic year of 117.

The students whose enroll to Information Technology Department of Oversea University from business management group at ranking 3,784 (Table 5) *the rate of electrical group 3.23 = to estimate the number of candidate will be 12,0222 in business management group of academic year of 117.

(7) To estimates of future enrollment number between independent recruitment from electrical group and applying examination in electrical group.

Independent recruitment number in electrical group is 2,453 > estimates of applying examination 2,282.

Independent recruitment number in electrical technology group is 6,565 > estimates of applying examination 4,490.

Independent recruitment number in business management group is 16,731 > estimates of applying examination 12,222.

In conclusion, those department of electrical technology information information technology and business management will step in the dilemma of lack enrollment in the year 117 of Republic of China.

6 Conclusion

In this thesis, the data resource comes from University students which include: total population in Taiwan of past years, the birth rate, the proportion of students from high school in Taiwan, the population of enrollment the grade of enrollment and the number of enrollment from technology institution. To use language of Visual studio asp.net and C#, gather all the research data into a database system to write and adjust the web page for university students. Users will enter year of enrollment, which university, groups of each university major and departments through database matching system and data processing, the output will present birth rate of each year of enrollment, the proportion of the number of examination registration, the total number of examination registration, the proportion of enrollment from each group of university major to apply. Then through this system will demonstrate whether will affect the number of future enrollment of major from each university.

The aspect of student, this query page may provide applicants ratio and the information of proportion, so that students will estimate the major which they are taking has low number on enrollment or to be suspend. In order to avoid the option on suspension of the major department or due to be closed from poor management, which provide information for students to consider the reputation of school. Getting the diploma was printed defunct department or school, the situation will affect students' achievement and interests of future job.

In school terms, the query system provided with number of examination registration and school enrollment ratio. These data will provide information on course schedule, registration. Schools can also plan well schedule of course which involve with industry or marketing. In order to avoid the situation on poor number of enrollment also eliminate the low profit department which will help school management quality of education and improve reputation school.

7 Future Research

(1) Students analyze system's future direction will be toward more accurate calculation of improvements, trying to change the relevant variables added factor of students, such as: regional variables, the gap between urban and rural education and political policies and other factors.
(2) Through better design of the questionnaire in the future, such as: increasing the national high school level the number of classes in each category, so that the statistical analysis of the results from the data can be applied in a more precise adjustment.
(3) The contribution of this paper will be people can pay attention of the whole trend of enrollment from University students, or to provide more complete data to researchers who concerned with the future education of Taiwan.

References

1. Tsai-Wen, Wu. "The Financial Crisis Warning Model Combined with Population Structure factors of Taiwan Private Universities.'' Soochow University. 2014.
2. Hossler, Don, John Braxton, and Georgia Coopersmith. "Understanding student college choice." Higher education: Handbook of theory and research 5 (1989): 231–288.
3. Brown, Steven D., et al. "Social cognitive predictors of college students' academic performance and persistence: A meta-analytic path analysis." Journal of Vocational Behavior 72.3 (2008): 298–308.
4. Christophel, Diane M., and Joan Gorham. "A test-retest analysis of student motivation, teacher immediacy, and perceived sources of motivation and demotivation in college classes." Communication education 44.4 (1995): 292–306.
5. Kuh, George D., et al. "Unmasking the effects of student engagement on first-year college grades and persistence." The Journal of Higher Education 79.5 (2008): 540–563.
6. Bound, John, Michael F. Lovenheim, and Sarah Turner. "Why have college completion rates declined? An analysis of changing student preparation and collegiate resources." American economic journal. Applied economics 2.3 (2010): 129.
7. Wang, Ru-Jer. "From elitism to mass higher education in Taiwan: The problems faced." Higher Education 46.3 (2003): 261–287.
8. Becker, Gary S. Human capital: A theoretical and empirical analysis, with special reference to education. University of Chicago Press, 2009.

Extracting New Opinion Elements in the Semi-automatic Chinese Opinion-Mining System from Internet Forums

Shih-Jung Wu, Rui-Dong Chiang and Wei-Ting Chang

Abstract Articles posted on a forum often contain new Internet terms related to opinion elements. Consequently, existing Chinese opinion-mining systems may exhibit low recall and precision because they cannot recognize these new Internet terms. Therefore, we designed an algorithm to elaborate on the opinion elements of such articles by extracting the new terms. By ignoring any uncommon opinion element that appears only once, we determine whether the new term identified through manual judgment is a useful opinion element for a specific domain and add it to the thesaurus. In comparison with semi-automatic annotation methods, our approach can save considerable labor. The same Chinese word may have different meanings depending on the context, and this fact is prone to cause difficulties by changing the polarity or meaning of certain opinion elements, leading to errors in the analysis results of many Chinese systems. We designed appropriate algorithms to address this problem. Meanwhile, this system extracts the opinion elements from an article based on its established thesaurus and simultaneously considers various sentence patterns, the default topic, and clause priority to determine the opinion tendency of the author.

Keywords Opinion mining system · Sentiment analysis · Customer review

S.-J. Wu (✉)
Department of Innovative Information and Technology, Tamkang University, New Taipei City, Taiwan
e-mail: wushihjung@mail.tku.edu.tw

R.-D. Chiang · W.-T. Chang
Department of Computer Science and Information Engineering, Tamkang University, New Taipei City, Taiwan

© Springer Nature Singapore Pte Ltd. 2018

N.Y. Yen and J.C. Hung (eds.), *Frontier Computing*, Lecture Notes in Electrical Engineering 422, DOI 10.1007/978-981-10-3187-8_51

1 Introduction

Most of Chinese opinion-mining systems use part-of-speech analysis (POS) or syntactic parsing (Parser) to first analyze each sentence and identify its specific pattern. These systems use a nearby approach to identify the relevant opinion elements and synthesize them to express the opinion tendency of the author or authors [1, 2]. However, when POS and Parser are used to analyze Chinese articles, mistakes can be easily generated [3, 4]. Therefore, our system does not use POS and Parser. Instead, the opinion elements of articles are extracted based on a thesaurus and combined with various sentence patterns, the default topic and clause priority to analyze the authors' opinions. Because of the difficulty of designing a universal Chinese opinion-mining system for all domains, our system focused on a single domain for each analysis. Addressing only a single domain allows us to ignore feature words or opinion words that belong to other domains to reduce errors. Moreover, we applied the method of a "default topic" to address the problem of missing topics for unformatted articles posted on an online forum and applied "clause priority" to correct the combination error that can arise when using the nearby approach. In our system, we use four types of opinion elements (topic, feature, item, and opinion word) to form an opinion sentence from sentences that express subjective opinions. Moreover, in our opinion sentences, we use (topic, feature, item, opinion word), (topic, feature, ∅, opinion word), (topic, ∅, item, opinion word) or (topic, ∅, ∅, opinion word) to extract the opinions of the authors. ∅ means null.

Because articles posted on a forum often contain new Internet terms that are not included in the default Chinese thesaurus, the system will not be able to recognize these new Internet terms. This lack of recognition will further affect the precision and recall of the system in the extraction of opinion elements. Most studies regarding the automatic extraction of features or opinion words use POS or Parser to analyze an article and then extract the opinion elements. The systematic precision of using POS or Parser is between 60% and 70%, and the recall is between 50% and 80% [5, 6]. To improve the precision and recall of the system, we must use a manual annotation approach to extract the opinion elements and new Internet terms that lie outside the scope of the thesaurus. Because manual annotation requires substantial labor, some studies have proposed the use of Semi-Automated Tagging to reduce labor [7]. However, the application of Semi-Automated Tagging still requires the manual inspection of all sentences, which requires considerable labor. The method of opinion-element extraction and expansion developed for our system can process more than 2,000 articles (more than 20,000 sentences) each month on average and takes the reasonable approach of ignoring any opinion word that appears only once (with frequency 1); it requires only approximately two hours to inspect more than 400 terms. By contrast, the inspection of the same number of sentences using the manual annotation approach requires approximately four to seven days; therefore, our method can save a considerable amount of manual inspection time.

To evaluate the polarity of an opinion word, content dependency must be considered [1, 8–10]. However, for a Chinese system, it is not sufficient to consider only the content of each word independently. The same Chinese word may have different meanings when used in different terms. When an opinion word is combined with a specific word or concatenated with other opinion word, it may cause a change in the polarity or meaning of the opinion. It is also possible that a non-opinion word may become an opinion word and that the original opinion word may become a non-opinion word. Therefore, we designed three algorithms to improve the capability of our system to cope with situations.

The primary contributions of this paper are as follows: (1) We present a semi-automatic Chinese opinion-mining system that can be applied to an Internet forum to analyze the opinion tendency expressed by articles on the forum in a specific domain (2) We extract the opinion elements from articles based on a thesaurus; in the system, we implement related algorithms for the extraction and expansion of opinion elements to increase the precision and recall for the identification of opinion elements and effectively reduce the manual labor required for this task. (3) We propose the simultaneous consideration of various sentence patterns, a default topic, and clause priority to combine the extracted opinion elements into complete output opinion sentence.

2 Related Work

Some studies of the extraction of opinion elements have used a manual approach to establish a thesaurus of features; others have used natural-language technologies, such as POS and Parser, to search for features or opinion words. Using manual annotation, it is possible to establish a complete thesaurus of features and opinion words. Although the precision of this approach may be very high, it involves considerable labor [11]. To save time, most studies use a natural-language technique combined with certain specific patterns to extract features or opinion words. However, the drawback of this approach is that many terms may not belong to the thesaurus of features or opinion words [5, 7, 12]. Some researchers have proposed different approaches to expanding the thesaurus and corpora. For the study of Chinese, the major thesauruses that are available on the Internet include HowNet and NTUSD. HowNet is a large thesaurus available on the Internet. NTUSD is a dictionary of opinion words that was developed by Lun-Wei Ku and Hsin-Hsi Chen. However, at present, these resources for elaboration cannot effectively recognize newly emerging Internet terms. At present, there are few analogous traditional Chinese systems. Ku et al. [13] have developed CopeOpi, this system analysis articles and their opinion tendencies based on the previously established NTUSD dictionary. Liu et al. [14] have established a system in the field of Chinese film. However, these systems do not address the question of thesaurus expansion to include new Internet terms.

3 System Architecture

The thesaurus of the system includes a general thesaurus, a general thesaurus of opinion words, and a thesaurus related to a specific domain. The Chinese universal dictionary [15] is used as the general thesaurus for normal Chinese terms. The general thesaurus of opinion words contains a collection of many traditional, normal opinion words, and the domain-specific thesaurus contains only relevant opinion elements that belong to the specific domain.

During system operation, after selecting forum data for a specific domain, (1) during the pre-processing of the data, we first convert all English letters in the article to lower case, and interrogative sentences and hypothetical sentences are deleted. (2) We use the thesaurus to annotate the opinion elements in the article. (3) We execute the algorithms to extract and expand upon opinion elements and store the results in the thesaurus. Moreover, we can edit the properties of the opinion elements, such as whether they belong to a specific domain, their polarity, and the content dependency among opinion elements, during stages (2) and (3). (4) We use various sentence patterns, a default topic and clause priority to combine the relevant opinion elements into an output opinion sentence. Finally, manufacturers and consumers can rapidly acquire the information.

4 Algorithm for the Extraction and Expansion of Opinion Elements

On Internet forums, new Internet terms periodically emerge, and these new terms may become new opinion elements. We propose a word-hyphenation algorithm to extract new Internet terms that may become new opinion elements from an article posted on an Internet forum. Moreover, we designed three algorithms to address the possible change in the polarity or meaning of an opinion word when that opinion word is combined with a specific word or concatenated with other opinion words. Because most identified opinion elements are opinion words, in the experiment and the following description, we annotate and discuss the role of only opinion words. Moreover, because the word-hyphenation algorithm is the only algorithm discussed here that is relatively complicated; we provide a detailed discussion of this algorithm in this section.

4.1 The Word-Hyphenation Algorithm

As depicted in Fig. 1, the word-hyphenation algorithm involves a four-step process.

Step I: Because a sentence that expresses an opinion requires at least a topic and an opinion word to be an opinion sentence, we process the article to extract segments that may be missing a topic or an opinion word.

Step II: The text of an article on an Internet forum often contains new Internet terms or incorrect terms that are not defined in the thesaurus. After the word hyphenation of the article, any text that cannot be recognized by the thesaurus will treated as n individual words. Let $P\{t_1, t_2, \ldots, t_n\}$ be a contiguous set of individual words, $t_1, t_2, \ldots, t_n$, in the article; then, for each $P\{t_1, t_2, \ldots, t_n\}$, we can combine m contiguous single words to form a new term, $t_{s+1} \ldots t_{s+m}$, where $2 \leqq m \leqq n$ and $0 \leqq s \leqq n-m$.

Step III: Some people tend to use abbreviations to replace relatively long terms that appear particularly often in features, products, or projects. If a particular short term is often used as an abbreviation or if the short term has another meaning, its frequency of occurrence should be higher than the corresponding long term that it abbreviates; we can directly replace the short term with the long term.

Step IV: There are many terms with a frequency of occurrence of *1*, but there is no significant difference between the number of opinion words with a frequency of *1* and above and the number of opinion words with a frequency of *2* and above. For example, for the experimental data considered in this paper, in the articles from July of 2011 on the Mobile

Word-hyphenation algorithm

Step I: Preprocess the article and extract from the article any paragraphs that may be missing a topic or an opinion word.
Step II: Perform word hyphenation on the paragraphs $P\{t_1,t_2,\ldots,t_n\}$.
For each $P\{t_1,t_2,\ldots,t_n\}$ do
Consider n individual, sequential words t in paragraph $P\{t_1,t_2,\ldots,t_n\}$.
Valid shift s in P
for $m \leftarrow 2$ to n
for $s \leftarrow 0$ to $n\text{-}m$
generate a new term, $t_{s+1} \ldots t_{s+m}$
Calculate the frequency of new terms after combination.
Step III: If the frequency of short terms is smaller than or equal to the frequency of long terms, delete the short terms contained within the long terms.
Step IV: Output the terms with a frequency greater than 1 for manual inspection and retain the terms with a frequency of 1 for the next round.

Fig. 1 Word-hyphenation algorithm

01[1] forum, there are *2,849* terms with a frequency of *1* and above, and *10* of these terms have been manually determined to be opinion words; there are *292* terms with a frequency of *2* and above, and six of these terms have been manually determined to be opinion words. Therefore, to reduce the time required for manual inspection, we do not manually inspect terms with a frequency of *1*, and they are carried forward to the next iteration. If a term is subsequently used again by the originator, it will be manually inspected when the accumulated frequency becomes greater than *1*, and therefore, this new opinion word will be missed once at most.

4.2 Processing the Changes in the Polarity and Meaning of an Opinion Word

In Chinese, when an opinion word (referred to as *OP*) is combined with a specific word or concatenated with other opinion words, it may generate a change in the polarity or meaning of the expressed opinion. Consequently, to process this change, we use three different algorithms to handle terms in the following three forms: "*OP + OP*," "*OP +* 不*+OP*," and "*OP +* 了."

- Algorithm—"OP + OP"
  When an opinion word OP_1 is concatenated with another opinion word OP_2, it might generate a new opinion word OP_1OP_2, and the polarity of this new opinion word could be different from that of OP_1. Hence, an algorithm to decipher this type of opinion word is required and manual intervention is required to determine whether the new opinion word belongs to the domain.
- Algorithm—"OP + 不+OP"
  For an opinion word *OP*, the term "*OP +* 不*+OP*" is always an interrogative term in the Chinese language. For example, "好*/good*" is an opinion word, but "好不好*/good or bad?*" is an interrogative term. Therefore, to avoid such misinterpretation, this algorithm identifies terms with the "*OP +* 不*+OP*" form and adds them to the general thesaurus.
- Algorithm—"OP + 了"
  When an opinion word is followed by a "了," its tone will sometimes be reversed. In the telecommunication domain, "好*/good*" is an opinion word, whereas "好多*/a lot of*" is not an opinion word. However, "好多了*/much better*" is once again an opinion word.

[1]http://www.mobile01.com/.

5 Sentence Patterns and the Combination of Opinion Elements

Because there is no restriction on the length of articles that may be posted on Mobile 01, if the author mentions a topic in the article, multiple sentences are typically used to thoroughly describe the topic. The author often uses subsequent clauses for supplementary description. The consideration of only clauses that contain topics would result in some clauses relevant to the topic at hand being discarded because some clauses do not mention a topic. As a result, the system would achieve only incomplete opinion expression, and the subsequent analysis would suffer from information loss. Therefore, we propose the concept of a default topic. If a topic is mentioned in the text of an article but not in all clauses, we apply the topic mentioned in the previous sentence to subsequent clauses, thereby avoiding the problem of incomplete opinions when the subsequent clauses do not mention a topic. The combination of opinion elements primarily utilizes the nearby approach and clause priority. When a feature is mentioned in an earlier clause and another feature and an opinion word are also contained in the next clause, the relation between the feature and opinion word in the later clause is considered to have higher priority. The prior feature is replaced by the subsequent feature.

In the following, we introduce three sentence patterns: a general sentence, an equative sentence, and a comparative sentence. In fact, any sentence may contain two or more basic sentence patterns.

- General sentence pattern
  When pairing opinion elements in general sentences, the default topic, i.e., the topic of the previous sentence, will be used when there is no topic mentioned in the sentence under consideration. When analyzing an article, its features, items, and opinion words will be paired according to the pairing method for opinion elements described above in the subsection pertaining to clause priority.
- Equative sentence pattern
  The equative sentence pattern refers to the case in which opinion elements of the same type are connected with conjunctions. The standard pattern for an equative sentence is "A conjunction B conjunction C conjunction…D" (where A, B, C, and D are the same type of topic, feature, item, or opinion word). Because only one type of opinion element can be placed in the tuple, when opinion elements on the same level are connected with conjunctions, there must be a process for separation. A new tuple is added for each level of opinion elements. Then, the opinion elements are paired according to clause priority, as described above.
- Comparative sentence pattern
  In articles posted to an online forum, comparisons are often presented between products or companies, and such comparisons use two types of comparative sentences.

(1) "A…比較 (more)…opinion word" (where A could be any combination of topic, feature, and item).
This is a relatively simple sentence that describes only a unilateral good (or bad) opinion, and the pairing processing is the same as for general sentences. However, if the topic does not appear in the clause of such a comparative sentence, the default topic must be used.
(2) "A…比 (than)…B…opinion word" (where A and B can be any combination of topic, feature, and item).
For this type of comparative sentence, if there is no topic in front of "比/*than*," the topic that appears in the nearest previous sentence will be used as the default topic; the feature portion of the sentence is also centered around "比/*than*." If the topics of the two sentences are different and a feature is mentioned in the previous sentence but not the current one, then the feature referenced in the previous sentence is used for the sentence under consideration. We use a comparison of sentence properties to assign the opposite polarity to opinion words in the later sentence (add a negative word "*not*/不" before the opinion word that belongs to the later sentence).

6 Experimental Data for Extracting and Expanding Opinion Elements in the System

Here, we demonstrate how the system effectively reduces labor requirements and identifies relevant new opinion elements using various algorithms. The data source is articles from the comprehensive discussion forum on Mobile 01 mobile communication, and the range of data is from July of 2011 to February of 2013, a total of *20* months. Table 1 presents the data from each month in the telecommunication domain, which contains *2,613* articles and *22,230* sentences per month on average.

Table 1 The amount of data posted each month for the telecommunication domain

Month	Number of articles	Number of sentences	Month	Number of articles	Number of sentences
2011/07	*2275*	*23071*	2012/05	*2250*	*18368*
2011/08	*2055*	*21889*	2012/06	*2175*	*19242*
2011/09	*2020*	*20812*	2012/07	*2301*	*16894*
2011/10	*2025*	*20945*	2012/08	*2522*	*18215*
2011/11	*2377*	*21313*	2012/09	*3185*	*27139*
2011/12	*2683*	*22397*	2012/10	*3790*	*31892*
2012/01	*1973*	*17024*	2012/11	*3623*	*32739*
2012/02	*2427*	*20797*	2012/12	*3985*	*28626*
2012/03	*2904*	*23587*	2013/01	*3066*	*23386*
2012/04	*2133*	*17246*	2013/02	*2475*	*19013*

The system first uses the thesaurus to automatically flag opinion elements. It then manually determines which of these opinion elements belongs to the telecommunication domain, and they are added to the thesaurus for the telecommunication domain. The experimental data are presented in Fig. 2. For the first several months, the manual inspection required approximately one hour per month on average; for subsequent months, because the most commonly used opinion words had already been identified, the manual inspection required only approximately 20 min. In the following, we will analyze and discuss the performance and effects of the word-hyphenation algorithm.

Table 2 presents the results of analyzing an article using the word-hyphenation algorithm. In the table, the time costs are listed by date for the manual inspection of terms with a frequency of 1 and above and with a frequency of 2 and above. Some terms must be checked against the text that appears before and after them in the article, and such determinations usually take a long time. The "number of judgments based on context" represents the number of such determinations. The time required for the manual inspection of terms with a frequency of 1 and above is approximately 3–5 times higher than that required for the manual inspection of terms with a frequency of 2 and above.

From September of 2012 to December of 2012, because a relatively large number of articles were posted during that time, a large number of manual judgments were required. However, most involved meaningless terms, and there were few opinion words to be found. Because of the word-hyphenation algorithm, we needed to spend only approximately 1 h per month to evaluate approximately 400 terms (approximately 2 h if the time required for manual inspection during automatic annotation is considered). In comparison with previous approaches without the assistance of algorithms, which required a manual search for opinion elements that would typically require four to 7 days to process 1 month of data, our method offers a considerable reduction in labor and time costs.

Table 3 presents the experimental results obtained using the algorithms that process the changes in the meanings of opinion words. The number of opinion

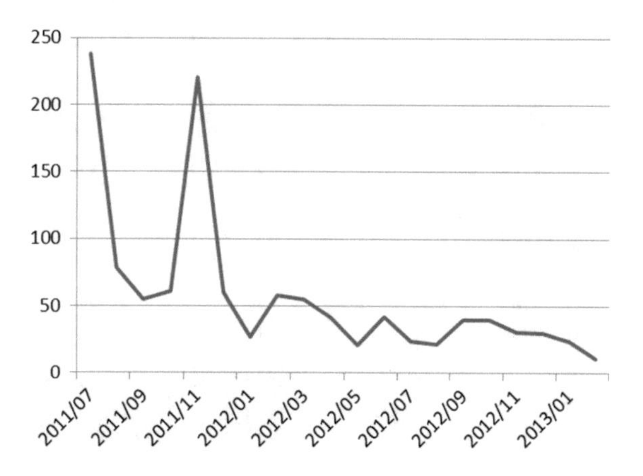

Fig. 2 Number of manual evaluations required for the automatic annotation of opinion elements in each month

Table 2 Experimental data for word-hyphenation algorithm

Date	Frequency of 1 and above				Frequency of 2 and above			
	Number of opinion words	Number of non-opinion words	Number of judgments based on context	Time cost (minutes)	Number of opinion words	Number of non-opinion words	Number of judgments based on context	Time cost (minutes)
2011/07	*10*	*2,839*	*15*	*187*	*6*	*286*	*10*	*44*
2011/08	*9*	*4,438*	*15*	*125*	2	*315*	*8*	*40*
2011/09	*8*	*5,796*	*13*	*107*	*1*	*274*	*5*	*29*
2011/10	*8*	*7,132*	*14*	*109*	2	*295*	*8*	*39*
2011/11	*0*	*3,420*	*20*	*160*	*0*	*367*	*6*	*36*
2011/12	*0*	*5,830*	*14*	*163*	*0*	*302*	*5*	*30*
2012/01	*0*	*2,686*	*16*	*148*	*0*	*335*	*9*	*44*
2012/02	*0*	*5,583*	*13*	*184*	*0*	*287*	*7*	*35*
2012/03	*0*	*7,968*	*13*	*158*	*0*	*397*	*6*	*38*
2012/04	*1*	*9,602*	*21*	*145*	*0*	*327*	*11*	*49*
2012/05	*1*	*11,166*	*16*	*126*	*1*	*344*	*12*	*53*
2012/06	*0*	*12,768*	*18*	*134*	*0*	*361*	*11*	*51*
2012/07	*0*	*14,138*	*17*	*120*	*0*	*347*	*9*	*44*
2012/08	*1*	*15,565*	*15*	*116*	*0*	*393*	*8*	*44*
2012/09	*1*	*18,034*	*15*	*168*	*0*	*630*	*8*	*56*
2012/10	2	*20,240*	*19*	*167*	*0*	*661*	*8*	*57*
2012/11	2	*22,122*	*7*	*115*	*0*	*621*	*7*	*52*
2012/12	2	*24,400*	*20*	*174*	*0*	*672*	*6*	*52*
2013/01	3	*26,327*	*20*	*156*	*1*	*635*	*6*	*50*
2013/02	3	*26,997*	*21*	*97*	*0*	*337*	*7*	*38*

Table 3 Experimental results of the individual algorithms for processing changes in the meanings of opinion words

Date	Algorithm—"OP + OP"			Algorithm —"OP + 不+OP"	Algorithm—"OP + 了"		
	Number of opinion words	Number of non-opinion words	Time cost (minutes)	Number	Number of opinion word	Number of non-opinion word	Time cost (minutes)
2011/07	3	4	9	1	2	12	7
2011/08	5	5	15	0	1	4	1
2011/09	1	3	3	0	0	10	1
2011/10	0	7	0	1	0	2	0
2011/11	1	7	3	0	0	0	0
2011/12	0	0	0	0	0	0	0
2012/01	0	0	0	0	0	0	0
2012/02	2	3	6	0	0	0	0
2012/03	1	3	3	0	0	0	0
2012/04	0	5	0	0	0	0	0
2012/05	0	0	0	0	0	0	0
2012/06	1	1	3	0	0	0	0
2012/07	0	1	0	0	0	0	0
2012/08	1	2	3	0	0	0	0
2012/09	0	3	0	0	0	0	0
2012/10	1	3	3	0	0	0	0
2012/11	0	1	0	0	0	0	0
2012/12	1	1	3	0	0	0	0
2013/01	0	1	0	0	0	0	0
2013/02	0	2	0	0	0	0	0

words generated from the data by each algorithm was small, and for each month, only 5–10 min was required for the manual evaluation of these new opinion words.

7 Conclusions

In this paper, we established a Chinese opinion-mining system that can be applied to an Internet forum and analyzed the opinion trends expressed by the articles on one such forum. The experimental results confirm that our method for the extraction and expansion of opinion elements can not only identify new terms emerging in Internet usage but also effectively reduce labor costs. In future, we will work toward continuous improvement of this system.

References

1. M. Chen and T. Yao, "Combining dependency parsing with shallow semantic analysis for Chinese opinion-element relation identification," in Universal Communication Symposium (IUCS), 2010 4th International (pp. 299–305), 2010, pp. 299–305.
2. P. Ting-Chun and S. Chia-Chun, "Using Chinese part-of-speech patterns for sentiment phrase identification and opinion extraction in user generated reviews," in Digital Information Management (ICDIM), 2010 Fifth International Conference on (pp. 120–127), 2010, pp. 120–127.
3. L. Li, "Opinion Mining for Chinese Sentence in a Specific Domain," Master Master, School of Electronic Information and Electrical Engineering, Shanghai Jiao Tong University, 2008.
4. L. Wu, Y. Zhou, F. Tan, F. Yang, and J. Li, "Generating syntactic tree templates for feature-based opinion mining," presented at the Proceedings of the 7th international conference on Advanced Data Mining and Applications - Volume Part II, Beijing, China, 2011.
5. Z. Shanzong, L. Yuanchao, L. Ming, and T. Peiliang, "Research on Feature Extraction from Chinese Text for Opinion Mining," in Asian Language Processing, 2009. IALP '09. International Conference on, 2009, pp. 7–10.
6. G. Xu, C.-R. Huang, and H. Wang, "Extracting Chinese Product Features: Representing a Sequence by a Set of Skip-Bigrams," in Chinese Lexical Semantics. vol. 7717, D. Ji and G. Xiao, Eds., ed: Springer Berlin Heidelberg, 2013, pp. 72–83.
7. B. Liu, M. Hu, and J. Cheng, "Opinion observer: analyzing and comparing opinions on the Web," presented at the Proceedings of the 14th international conference on World Wide Web, Chiba, Japan, 2005.
8. G. Qiu, B. Liu, J. Bu, and C. Chen, "Expanding domain sentiment lexicon through double propagation," presented at the Proceedings of the 21st international joint conference on Artifical intelligence (pp. 1199–1204), Pasadena, California, USA, 2009.
9. L. Garcia-Moya, H. Anaya-Sanchez, and R. Berlanga-Llavori, "Retrieving Product Features and Opinions from Customer Reviews," Intelligent Systems, IEEE, vol. 28, pp. 19–27, 2013.
10. E. Cambria, B. Schuller, X. Yunqing, and C. Havasi, "New Avenues in Opinion Mining and Sentiment Analysis," Intelligent Systems, IEEE, vol. 28, pp. 15–21, 2013.
11. N. Kobayashi, K. Inui, and Y. Matsumoto, "Opinion Mining from Web Documents: Extraction and Structurization," Information and Media Technologies, vol. 2, pp. 326–337, 2007.

12. C. Bosco, V. Patti, and A. Bolioli, "Developing Corpora for Sentiment Analysis: The Case of Irony and Senti-TUT," Intelligent Systems, IEEE, vol. 28, pp. 55–63, 2013.
13. L.-W. Ku, H.-W. Ho, and H.-H. Chen, "Opinion mining and relationship discovery using CopeOpi opinion analysis system," Journal of the American Society for Information Science and Technology, vol. 60, pp. 1486–1503, 2009.
14. L. Chien-Liang, H. Wen-Hoar, L. Chia-Hoang, L. Gen-Chi, and E. Jou, "Movie Rating and Review Summarization in Mobile Environment," Systems, Man, and Cybernetics, Part C: Applications and Reviews, IEEE Transactions on, vol. 42, pp. 397–407, 2012.
15. C.-H. Tsai, "Tsai's List of Chinese Words," University of Illinois at Urbana-Champaign, 1996.

Computer Simulation of Illustrating the Impact of Uneven Saline Distribution on Thermal Lesion During Radiofrequency Ablation Treatments

Huang-Wen Huang and Lin Hui

Abstract Radiofrequency ablation (RFA) is a technique by which deposition of electromagnetic energy is used to thermally heat tissues. And incomplete treatment with heat-based therapy alone may sometimes occur. To improve the efficacy of thermal therapies, many attempts have been used by modifying the tumor's underlying physiologic characteristics. The objective of this study is to determine whether fluid saline injection during radiofrequency ablation (RFA) can increase the coagulation area, and how parameters of both electrical conductivity and blood perfusion would have impact on thermal lesion formation. Although the heat generated by this high-frequency current arises in all the conducting tissue pathways, high temperatures only develop in tissues near the electrode, where the current density is high. Barely any heat arises in tissues further from the electrode because of the fall in current density and the cooling (or radiator) effect of blood flow. Continuous saline infusion is assumed at several locations to investigate current density and temperatures reaction. A simple 2D geometry was used to illustrate electrical current pathways and temperature field. Finite-element numerical simulations are performed to solve Laplace equation of electric field calculation, and Pennes bio-heat transfer equation of calculation temperature field. Results showed that injected saline regions which could raise higher temperatures due to increasing electrical conductivity and thus enlarge the thermal lesion.

Keywords Radiofrequency ablation (RFA) · Saline injection · Thermal lesion · Pennes Bio-heat transfer · Electrical conductivity

---

H.-W. Huang · L. Hui (✉)
Department of Innovative Information and Technology, Tamkang University, Yilan County 26247, Taiwan
e-mail: amar0627@gmail.com

H.-W. Huang
e-mail: hhw402@mail.tku.edu.tw

© Springer Nature Singapore Pte Ltd. 2018

N.Y. Yen and J.C. Hung (eds.), *Frontier Computing*, Lecture Notes in Electrical Engineering 422, DOI 10.1007/978-981-10-3187-8_52

1 Introduction

Radiofrequency ablation (RFA) is a technique by which deposition of electromagnetic energy is used to thermally heat tissues and is now the standard method for local treatment of hepatic malignancy. Radiofrequency ablation (RFA) is a minimum invasive and localized treatment designed to destroy tumors by heating tissue to temperatures that exceed 55. It has received increased attention as an effective minimally invasive approach for the treatment of patients with a variety of primary and secondary malignant neoplasms [1]. One potential strategy to increase the efficacy of RF ablation is to modulate the biologic environment of treated tissues [2]. Along these lines, several investigators [3–5] have demonstrated the possibility of increasing RF tissue heating and coagulation during RF ablation by injecting saline or other NaCl solutions into the tissues during RF application. With advance computer technology availability at present, information technology and management are used to analyze the RFA tumor treatment through computer simulation and visualization. The analyzed results could assist physicians in cancer treatment planning and managing useful data through information technology. Thus, the objective of this study is to analyze the impact of NaCl solution on thermal lesion formation in liver through computer simulation and 3-D visualization.

In liver ablation treatments which is very popular in Taiwan, radiofrequency ablation is a minimally invasive procedure widely used in destruction of primary and metastatic liver tumors, as well as tumors in kidney, lung, and other solid organs [6–9]. The ablation is less expensive than other treatment options and it provides excellent alternative means of treatment which is hard to perform by surgical resection. Typical setting of RFA includes a RF generator and two electrodes: one needle electrode directly inserted into the tumor and one dissipative electrode (ground pad) [10]. During RFA procedure, the needle electrode is inserted under image guidance and an alternating current (about 500 kHz) is then applied across the inserted electrode and the ground pad for around minutes. When electric current travels through tissue, Joule heat is generated due to the movement of ions in cellular fluids against friction, resulting in temperature elevation. As tissue temperature is raised beyond 50 °C, irreversible cellular damage occurs, and then an area of coagulation necrosis is formed around the electrode [10, 11]. The cool-tip RFA system platform makes the straight-needle maximum energy delivery possible to create larger ablation zones. Through this paper, the comprehension of electromagnetic heating will be visually delivered and the uneven saline areas cases are explained.

2 Methods

2.1 Formulation of the Problem

The computer model was set up to investigate the present study. Figure 1 showed a square geometry with dimensions: 100-mm by 100-mm cross-section illustrating the computational domain of the study in a very long parallelepiped which the longitudinal impact s on the domain are neglected.

Figure 1 showed a square computational domain of 0.1 m by 0.1 m (10 cm by 10 cm) in (x and y-dimensions) indicating four (4) small square regions for saline fluid injection in tissues. At *top* boundary, a segment region, x >= 0.04 and x <= 0.06, is placed at 23 V electrical potential and other segments are insulated. Grounds are placed at the other three boundary faces.

2.2 Mathematical Equations

Pennes Bio-Heat transfer equation used as solved in 2D plane is the following Eq. (1) with assumption of a thickness $d_z = 1$ m. Q_{met} is relatively small thus neglect.

$$\begin{aligned} d_z \rho C_p \frac{\partial T}{\partial t} + d_z \rho C_p \mathbf{u} \cdot \nabla T + \nabla \cdot \mathbf{q} = d_z Q + q_0 + d_z Q_{bio} \\ Q_{bio} = \rho_b C_{p,b} w_b (T_b - T) + Q_{met} \end{aligned} \tag{1}$$

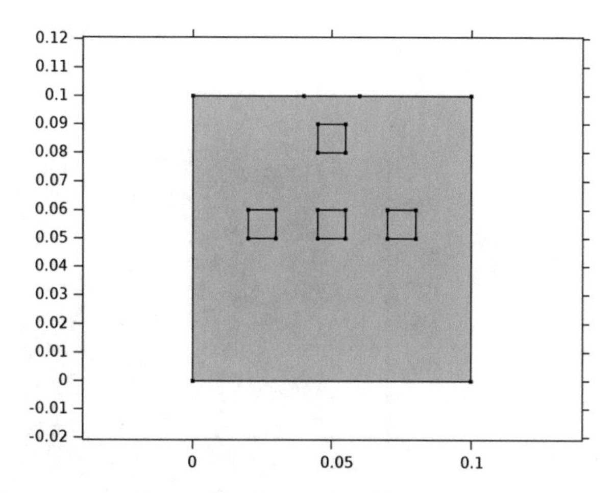

Fig. 1 A square computational domain of 0.1 m by 0.1 m (10 cm by 10 cm) in (x and y-dimensions) indicating four (4) small square regions for saline fluid injection in tissues. At *top* boundary, a segment region, x > = 0.04 and x <= 0.06, is placed at 23 V electrical potential and other segments are insulated. Grounds are placed at the other three boundary faces

T_b is 37 °C, C_p is 4000 j/(kg.C), density ρ is 1000 kg/(m3) and blood perfusion is 1 kg/(m3.s) (or 0.001 1/s). As to conservation of electrical current, the Laplace equation is used

$$\nabla \cdot (\sigma(x,y)\nabla V(x,y) = 0 \quad (2)$$

For uneven electrical conductivity by injected NaCl solutions, it can be increased by different concentration of NaCl solutions as shown in Table 1. The electromagnetic heat source Q is the following,

$$Q = J \cdot E = \frac{|J|^2}{\sigma} \quad (3)$$

2.3 Numerical Modeling

COMSOL finite-element application program is used to solved the geometry shown in Fig. 1. For finite element meshes in this study, to capture well electrical current density near saline injected area. Finer triangular meshes are used in the area near four squares. Figure 2 showed structured and unstructured meshes in computational domain (cross-sectional 2D) of 0.1 m by 0.1 m (10 cm by 10 cm). Four square

Table 1 The electrical conductivity of NaCl Solutions

	0.1% NaCl	0.13% NaCl	0.2% NaCl	0.3% NaCl	1.0% NaCl	5.0% NaCl
Estimated Actual σ (siemen/m)	0.33	0.66	1.0	1.7	4.5	25.0

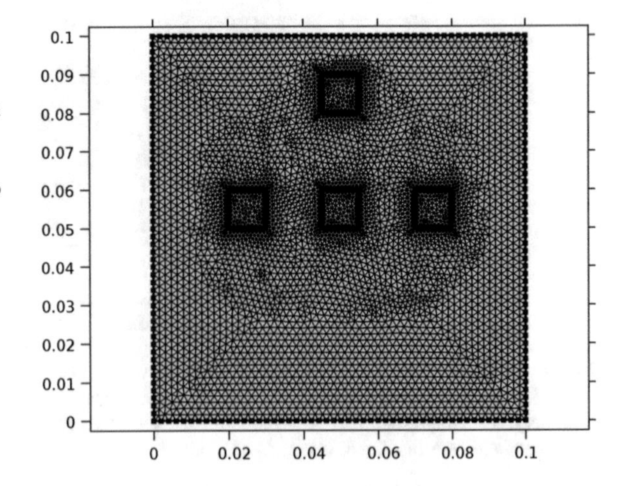

Fig. 2 Structured and unstructured meshes in computational domain (cross-sectional 2D) of 0.1 m by 0.1 m (10 cm by 10 cm). Four square spots are with more dense finite elements to capture parameters of both variable electrical conductivity and blood perfusion within spots. Complete mesh consists of 8880 elements

spots are with more dense finite elements to capture parameters of both variable electrical conductivity and blood perfusion within spots. Complete mesh consists of 8880 elements.

3 Results

To capture thermal lesion impacts, due to different electrical conductivities of injected regions of saline concentration. Several important quantities are important for analysis.

3.1 Electrical Potential, Streamline and Current Density Arrow Line Distributions

The driving electrical potential (volts) at the top boundary is 23 V which is the maximum volts that will generate maximum temperatures above 50 °C. Figure 3 showed electrical potential (volts), streamline and current density arrow line distributions for uniform electrical conductivity and uniform blood perfusion value in whole treated area. Higher current density is observed in top small square area and lesser one is located at the corner ones. Maximum volts of 23 is located at 2 cm segment boundary and gradually diffuse out to three ground boundary surfaces. Arrow lines indicated electrical current path direction. Figure 3 legend showed electrical potential.

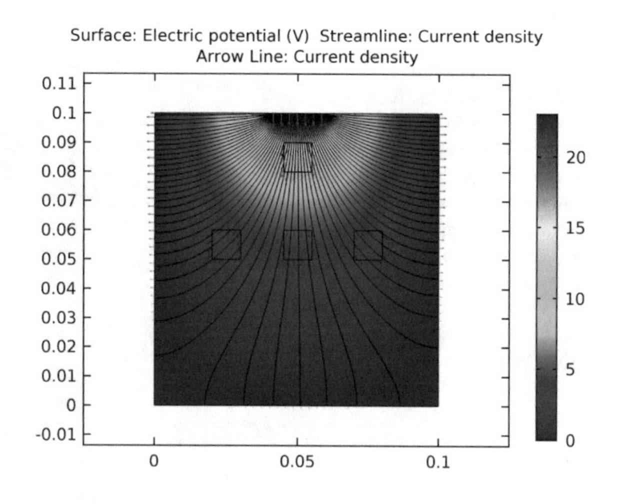

Fig. 3 Electrical potential (Volts), streamline and current density arrow line distributions for uniform electrical conductivity and uniform blood perfusion value in treated area

3.2 Temperature Field

Figure 4 showed temperature distribution with uniform electrical conductivity and uniform blood perfusion value in treated area. The boundary temperatures are set to 37 °C and initial temperatures in the domain are also set to 37 °C. The blood perfusion is 1.0 kg/(m^3.s) or (0.001 1/s) and the electrical conductivity is 0.33 S/m.

3.3 Temperature Distributions with Varied Electrical Conductivities

To illustrate temperatures and current density and streamlines changes due to varied electrical conductivities, several sequential figures are demonstrated here. Figure 5 showed temperature and current density streamline distributions with electrical conductivity ratios of (a) 1, (b) 2, (c) 5 and (d) 10, respectively for uneven saline injections. The ratio of 1 indicated the electrical conductivity is 0.33 S/m is used in all squares which are identical to all other area. The ratio of 2 indicated the electrical conductivity is 0.66 S/m is used in all squares but all other area is 0.33 S/m.

Figure 5a legend showed temperature field for electrical conductivity ratio of 1. It revealed uniform electrical conductivity of 0.33 S/m in entire domain. Streamlines originated from the top segment boundary with max. voltage of 23 (V) and a top square has higher current density as the square is near the source of electrical power and streamlines are dense within the square.

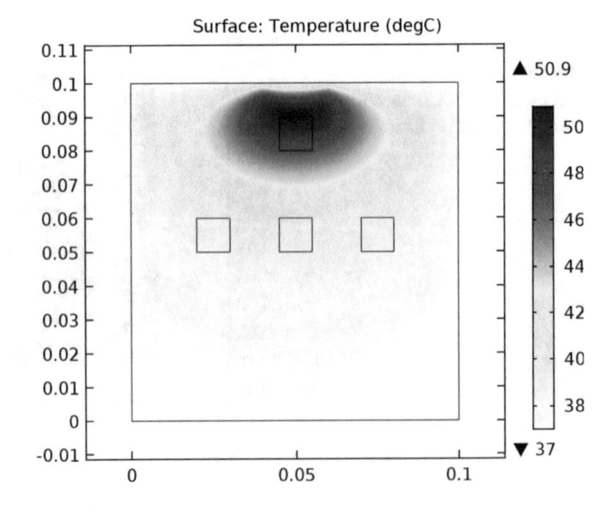

Fig. 4 Temperature distribution with uniform electrical conductivity and uniform blood perfusion value in treated area. The boundary temperatures are set to 37 °C and initial temperatures in the domain are also set to 37 °C. The blood perfusion is 1.0 kg/(m^3.s) or (0.001 1/s) and the electrical conductivity is 0.33 S/m

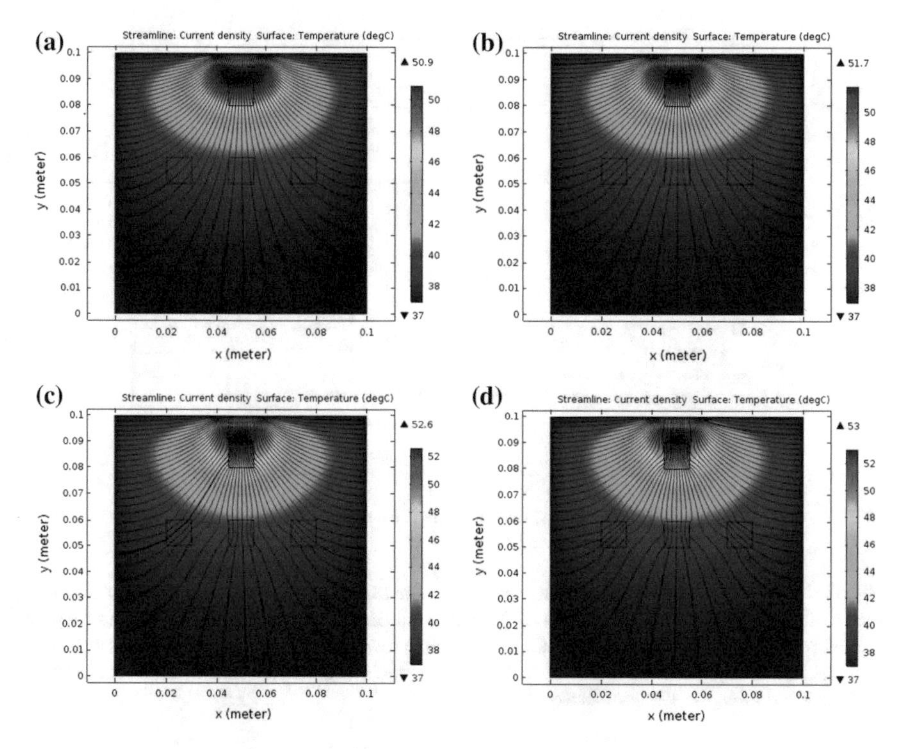

Fig. 5 Temperature and current density streamline distributions with electrical conductivity ratios of **a** 1, **b** 2, **c** 5 and **d** 10, respectively for uneven saline injections

3.4 Maximum Temperatures Within Top and Corner Squares with Varied Electrical Conductivities

To illustrate temperature raised due to varied electrical conductivities, the electrical conductivity ratio is used. Figure 6 showed maximum temperatures within top and corner uneven saline square regions at different electrical conductivity ratio (the ratio of 1 means uniform eletrical conductivity in the whole treated region). (For example, if the ratio is not 1, then there are uneven saline fluid in four square regions. Let us say, the ratio is 2, that means the electrical conductivity within those four square regions are double the value of those outside region.)

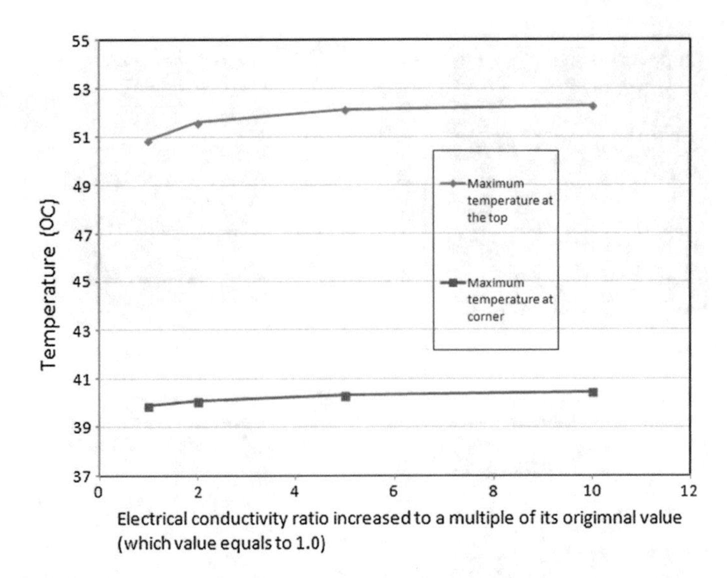

Fig. 6 Maximum temperatures within *top* and corner uneven saline square regions at different electrical conductivity ratio (the ratio of 1 means uniform electrical conductivity in the whole treated region). (For example, if the ratio is not 1, then there are uneven saline fluid in four square regions. Let us say, the ratio is 2, that means the electrical conductivity within those four square regions are double the value of those outside region.)

4 Discussion

Electrical conducting paths or streamlines are very important information in radiofrequency ablation treatments as many components (or organs) exist in bio-systems. As Fig. 5 illustrated that the top square increased dense streamlines as electrical conductivity increased from 1 to 10. The streamlines were distorted near the top square for higher electrical conductivity (shown in Fig. 5d). At the same, the other three squares below also showed dense streamlines, however they are not as severe as the top square which is closer to the emitting power source. This could also be seen in Fig. 6 that showed maximum temperatures raised more for the top square relatively as compared to those below squares (corners and center), when electrical conductivity increased within the squares.

There is some perturbation existed in higher electrical conductivity ratios as shown in Fig. 5c that a rather thicker streamlines appeared in left-hand part originated from the top square. That is due to automatic mesh generation did not perform exact identical meshes on both right and left sides as shown in Fig. 2. Thus it may cause perturbation in computational results.

In current radiofrequency heating, the highest temperatures (above 50 °C) are located approximate 1 cm below the driving electrical potential. The heated region where the temperatures were above 50 °C becomes larger as electrical conductivity increased. For the heated region to move deeper in the vertical direction, the cooling is needed at the top boundary surface and electrical driving power has to be increased as for the electrical potential.

5 Conclusions

The paper addressed the important visual information on radiofrequency thermal lesion formation impact by electrical conductivity. COMSOL finite element program is used to capture streamlines and current density, as well as their relationship with temperature distributions. The uneven saline injection at the area during radiofrequency ablation treatments could produce distorted streamlines and higher temperatures which could increase enlarged thermal lesion area.

Acknowledgements The author would like to thank the Ministry of Science and Technology for partially supporting this research under no. MOST104-2221-E-032-015.

References

1. G. Scott Gazelle, S. Nahum Goldberg, Luigi Solbiati, Tito Livraghi, Tumor Ablation with Radio-frequency Energy, Radiology 217 (2000) 633–646.
2. Goldberg SN, Gazelle GS, Mueller PR. Thermal ablation therapy for focal malignancy: a unified approach to underlying principles, techniques, and diagnostic imaging guidance. AJR Am J Roentgenol 2000; 174:323–331.
3. Curley MG, Hamilton PS. Creation of large thermal lesions in liver using saline-enhanced RF ablation. In: Proceedings of the 19th International Conference IEEE/EMBS. Chicago, Ill, October 30–November 2, 1997.
4. Livraghi T, Goldberg SN, Lazzaroni S, et al. Saline-enhanced radio-frequency tissue ablation in the treatment of liver metastases. Radiology 1997; 202:205–210.
5. Miao Y, Ni Y, Mulier S, et al. Ex vivo experiment on radiofrequency liver ablation with saline infusion through a screw-tip cannulated electrode. J Surg Res 1997; 71:19–24.
6. D. A. Gervais, F. J. McGovern, R. S. Arellano, W. S. McDougal, and P. R. Mueller, "Renal cell carcinoma: Clinical experience and technical success with radio-frequency ablation of 42 tumors," Radiology **226**, 417–424 (2003).
7. M. Ahmed, C. L. Brace, F. T. Lee, Jr., and S. N. Goldberg, "Principles of and advances in percutaneous ablation," Radiology **258**, 351–369 (2011).
8. D. E. Dupuy, T. DiPetrillo, S. Gandhi, N. Ready, T. Ng, W. Donat, and W. W. Mayo-Smith, "Radiofrequency ablation followed by conventional radiotherapy for medically inoperable stage I non-small cell lung cancer," Chest **129**, 738–745 (2006).
9. T. Livraghi, L. Solbiati, M. F. Meloni, G. S. Gazelle, E. F. Halpern, and S. N. Goldberg, "Treatment of focal liver tumors with percutaneous radiofrequency ablation: Complications encountered in a multicenter study," Radiology **226**, 441–451 (2003).
10. S. N. Goldberg, "Radiofrequency tumor ablation: Principles and techniques," Eur. J. Ultrasound **13**, 129–147 (2001).
11. D. Haemmerich, "Biophysics of radiofrequency ablation," Crit. Rev. Biomed. Eng. **38**, 53–63 (2010).

Clinical Application of Decision Support System for Treatment of Migraine

Lin Hui, Huan Chao Keh, Meng Chu Chiang and ZhenYao Liu

Abstract In this modern society, migraine belongs to a kind of common disease. The hospitals begin to use clinical decision support system a lot to improve the accuracy of diagnosis, this kind of system can help the physicians make the decisions and give a treatment by entering the information of the patients in advance. In the treatment of migraine, the patients need to keep diaries of headache, the physicians can make a diagnosis and trace according to the diaries. In this paper, we come up with constructing a clinical decision support system for treatment of migraine. We use the headache diaries as data, in order to store the data easily, we transform paper diaries to electronic diaries first, store the data in the databases of the servers or the mobile platforms (e.g. smartphones, tablet computers), and the data can be shown on the front-end interface. We also make use of data mining to analyze the factors, medicines and associations of migraine, and the result will be recorded in the system to improve the efficiency of the physicians' inspection.

Keywords Migraine · Clinical decision support system · Data mining

1 Introduction

1.1 Definition

Migraine is a common disease in the world, it is not only caused by psychological factors, it is a physiological disease, for example, the change of brain chemicals; The HIS (International Headache Society) put forward the standard of the diagnosis

L. Hui (✉) · Z. Liu
Department of Innovative Information and Technology,
Tamkang University, Yilan County 26247, Taiwan
e-mail: amar0627@gmail.com

H.C. Keh · M.C. Chiang
Department of Computer Science and Information Engineering,
Tamkang University, New Taipei City, Taiwan

© Springer Nature Singapore Pte Ltd. 2018

N.Y. Yen and J.C. Hung (eds.), *Frontier Computing*, Lecture Notes
in Electrical Engineering 422, DOI 10.1007/978-981-10-3187-8_53

of migraine, and make more detail description and definition of this symptom. According to the report of The World Health Organization, that migraine is one of the four major chronic diseases lead to disability, although migraine does not lead to life-threatening, but will seriously affect the patient's life, work and the family status. Migraine and its sequel will not only increase unemployment, disability benefits and medical costs, also lead to the economic loss of the USA for 17 billion dollars a year [1]. In Taiwan, Taipei Veterans General Hospital had the statistics of migraine epidemiology [2], the prevalence that Taiwanese suffering from migraine is about 9%, it means in Taiwan, there are about 1.5 million to 2 million people suffering from migraine. And the degree of severity of migraine of each patient is different. Now it is not sure that which kind of reason lead to migraine, but according to the present studies show that [2], because of the stimulation of the diet or the environment, the patients may be leaded to migraine, just like the change of the neurotransmitters in the brain, such as the change of concentration of serotonin, will lead to a series of pain. In order to track the status of the patients more clearly, doctors advise that the patients can get into the habit of keeping a diary, record the time of the onset of the headache every time and the time the headache lasts, if there is a sign or not before the onset and the seizure frequency, next time the diary can help the doctor do the diagnosis and it can also be the reference of the follow-up treatment.

1.2 Objective

The purpose of this research project is mainly that we expect to design a system that can help the physicians track the migraine, the type of the system is Clinical decision support system, using this system can put forward the record of the patients' headache to provide the doctors with the reference of diagnosis and help them make a decision. We hope we can help hospital transform the paper record of the headache diaries they provide to digital information, and transform it into a mobile application, avoid the disadvantage of the paper diaries that they occupy space and they are not easy to store, the paper diaries are suitable to use for a short record, but most of the patients need to do a record are suffering from chronic migraine, the onset of them are mainly for years, this will make the data difficult to store and analyze, so we hope the patients do the record of headache on the mobile platform, but based on that some of the patients don't used to use mobile platform and some of them are too old, we will still keep the part of the paper record, although it will improve the difficulty of storage and statistic.

2 Related Work

2.1 *The Classification of Migraine*

IHS in 1988 put forward the criteria for the diagnosis of migraine, and gave clear definitions and descriptions of migraine, they have divided the headache into 7 types (Table 1), but we only talk about two kinds of Standard of Diagnosis of migraine, migraine without warning and ominous migraine (Table 2). So-called omen is defined as from changing gradually to more than 5 min at least and fewer than 60 min, and neurological symptoms can completely restore. Common are: Omen visual symptoms 'Unilateral paresthesia' one side weakness and inarticulate, and the time from a warning to the migraine cannot be more than 1 h, moreover there may be premonitory symptoms and postdrome symptoms during a few hours to 1–2 days of the onset of the migraine, including the combination of different symptoms, such as fatigue, hard to concentrate, neck stiffness, sensitive to light or sound, nausea, blurred vision, yawning, pale [3].

Table 1 The classification of migraine

The classification of migraine	
1	Migraine without aura
2	Migraine with aura
3	Ophthalmoplegia migraine
4	Retinal migraine
5	May be the precursor of migraine related symptoms of Periodic syndrome in children
6	Complicated migraine
7	The diagnosis of migraine is not in conformity with the above standards

Table 2 The criteria of diagnosis of migraine without warning and ominous migraine

Migraine without warning	Ominous migraine
A. At least five times can match the onsets of B–D B. The onset of the migraine lasts 4 to 72 h C. Headache with at least two of the following features: 1. Unilateral 2. Pulsation 3. The moderate or severe degree (daily activities by limiting or banning) 4. Up and down the stairs, or similar daily activities will make the headache worse D. When a headache attacks there may be at least one of the following circumstances: 1. Nausea or vomiting 2. Photophobia and afraid of noisy	A. At least twice can match the onset of B B. With three of the following four features: 1. More than once ominous symptoms can restore fully, shows local cerebral cortex and (or) brainstem dysfunction 2. At least one warning symptoms gradually appears in more than 4 min, or more than two kinds of symptoms occurred 3. Warning symptoms last less than 60 min, if there are more than one warning symptom, the duration also increases with the increase of proportion 4. Headache attacks about 60 min after warning (also can be before the warning or together with the warning)

2.2 Chronic Migraine

The US Food and Drug Administration (FDA) defined the chronic migraine as the headache's time is 15 days each month and more than 4 h every day. Various symptoms such as pain, afraid of light, afraid of noisy, nausea, often aggravate along with the intensification of activity, and affect the daily life, but want to use drugs to maintain normal daily activities and work productivity, may lead to drug dependence. The early stage of the course of disease of chronic migraine is usually episodic migraine, at the beginning, seizure frequency of the headache is less than two days each week, then the headache will frequent seizures gradually (more than 15 days each month), 8 days of 15 days performing as migraine conforms to the diagnosis of chronic migraine. It is estimated that each year, about 3% of patients with paroxysmal migraine will progress into chronic migraine [4].

2.3 Common Migraine Treatment

According to the treatment of migraine, International SOS provided that the migraine can be divided into two parts: acute headache and chronic recurrent headaches.

Acute Headache

Most of the patients need most is quiet and rest, noise and strong light will aggravate the headache. It will help a lot if the patents can lie down in the quiet and dark room, if they can sleep, when they wake up they will have an obvious relief on their headache. If they have a severe headache, painkillers such as aspirin, paracetamol can also relieve the pain. Other prescriptions doctors commonly used to treat migraine are Cafergot, Non-steroidal anti-inflammatory agent and a new generation of migraine pain killers—Imigran. The drugs of acute attack should be taken as soon as possible, the best time to take the medicine is the time the warning appears or the beginning of the headache, if patients take medicine at the peak of the migraine, usually slow action cannot save a critical situation, and the effect is limited.

Imigran and other Selective serotonin catalysts (this kind of medicine is collectively called as 'Triptans'), can restrain meningeal vascular and brainstem serum tension element receptors, provide more choices for the treatment of migraine, it is best to take medicine in 40 min of the headache attacks or before the warning of the headache, the early to take medicine, the better the results [5]. Although this kind of drugs is expensive, there is a pay limit of the health insurance, the effect is good, currently the medicine that Taiwan headache medical association suggested include Propranolol, Topiramate and Valproic acid, are the first line drugs in the hospitals [2]. Besides drugs, recuperate in life, the good life habit, sufficient sleep is not too much, avoid overfatigue, etc., can also relief the condition of headaches. Some patients may have special inducing factors, such as food, alcohol, temperature changes, if it is found that these conditions, the patient can remind themselves, avoid similar status, can prevent migraine. In addition, regular exercise, especially aerobic exercise, such as jogging, swimming, riding a bicycle, etc., can improve health and prevent headaches. Patients also need to record their headache condition at any time, the project of record can refer to the headache diary that designed by Taiwan headache medical association (Fig. 1), next time the physician can make diagnosis by referring to the diary.

Headache Diary - VGH

Date : ______Y______M ID : ____________ Name : ____________

Intensity (0= no headache 1=mild 2=moderate 3=severe)															
	1	2	3	4	5	6	7	8	9	10	11	12	13	14	15
Morning															
Afternoon															
Night															
Sleep															
If you have the following symptoms, please tick															
Nausea ?															
Vomiting ?															
Sensitive to light ?															
Sensitive to sound ?															
Throbbing ?															
Unilateral onset ?															
Worsened by physical activities ?															
Aura before headache ?															
Visual aura															
Other aura															
Headache duration hours/day ?															
Preventive medications															
Acute treatment, working or not ? (0=no 1=a little 2= effective 3= very effective)															
Menstrual period, please tick															

Fig. 1 Headache clinic special paper headache diary

Chronic Recurrent Headaches

Effective treatment goal is to reduce the frequency of headache to the standard of paroxysmal headache and accept preventive treatments. If migraine sufferers appear the following symptoms, you may need to take preventive medicine: (1) Headache attacks every month more than twice (2) The onset time is long, more than 48 h (3) The degree of headache is very serious (4) acute treatment often cannot completely relieve pain (5) The time of omen of headache is too long. The main medicine to prevent headache in the hospitals are: β-blockers, Calcium blockers, antidepressants, 5HT2 receptor antagonist, antiepileptic drugs and so on, according to the condition of patients, physicians can prescription for different types of medicine [6]. These drug prescribing appropriate, can get the result of treatment and prevention, but to determine the effect of suspension and no rebound effect, may need treatment for more than two to three months, also need to combine with cognitive behavior therapy, psychological treatment, drug therapy and physical therapy to achieve control of headache.

OnabotulinumtoxinA (commodity name Botox), is the only one drugs that the United-States Food and Drug Administration (FDA) approves to use for chronic migraine. Courses include 155-unit injection to specific areas in head and neck, and through clinical trials confirm therapy (PREEMPT) injection once every three months. Botox injection effect will be decreasing over time, so usually require multiple doses. After treatment, the patents can do an assessment that if they can stop injection or extend the injection interval according to the frequency of onset of the migraine [7].

2.4 Clinical Decision Support System

Clinical decision support system (CDSS or CDS), Its purpose is to help physicians and other health care professionals provide decision-making tasks. For example, when you need to perform a specific diagnosis, CDSS can do further specific testing or treatment. And the concept of it is updating constantly. The definition of the current mainstream job is proposed by Robert Hayward: "Clinical Decision Support systems link health observations with health knowledge to influence health choices by clinicians for improved health care" [8]. Although in the society of information explosion, the physician can get all kinds of useful health information on the Internet, but most of all on the Internet is the electronic format, lead to that physicians lack of information skills and time to find and evaluate the information they need, CDS may be complement of the part for the physician. In terms of strategic purpose, CDS may allow doctors began to use computers, handheld devices and other electronic devices to contact related clinical knowledge, to confirm prescription drugs, refer to all kinds of medical practice, communication with the other physician and interview patients to provide more high-quality medical education [9]. CDS covers a variety of tools, in order to improve the decision-making process of clinical work. These tools include computerized warning to remind nurses and patients, clinical guidelines, the order of the specific conditions, the key patient data reports and the summaries, document templates, help diagnosis and provide the relevant reference information, etc. [10]. CDS is designed to be a kind of application that the physicians can use it in the clinical operation, input the data of the patients to CDS, CDS will provide some relevant information, such as: possible diagnosis, the alarm of drug interaction or prompt list for preventing disease, etc., Table 3 [11] for a variety of types of CDS, from the table, as shown in the CDS can be applied to solve a variety of different conditions and has the potential to improve the quality of medical treatment. From literature, using CDS to do prevention and management of chronic obstructive pulmonary disease (COPD), they finally worked out by the CDS can solve about the discovery of COPD diagnosis and layered, case management, and send high accuracy diagnosis suggestion to the doctors can also be used in helping to discover new cases of COPD [7] (Table 4).

2.5 *Data Mining in the Field of Medical*

In any advanced diagnosis or treatment process contains some diagnosis and intervention, sometimes even more. In any such data collection medical institutions or the useful lists are necessary. Because of the appropriate database between medical institutions and the network system, all appropriate data or images are easy to reach [12]. Data mining in medical research are: Chen Shiyuan [19] from the medical records, looked for the relationship between cases and drug use, and hope that through the data mining technology, prevent drug abusive problems in the

Table 3 A variety of cases of CDS

Types of CDS	Cases
At least five times can match the onsets of B–D	At least twice can match the onset of B
prevention and healthcare	Immunization, screening, disease management guidelines, secondary prevention
Diagnosis	Put forward the proposal according to the symptoms and diagnosis of physical characteristics of the patients
Planning and implementation of treatment	Treatment principle, the recommended dose of drug, drug interaction warning
The follow-up management	Order inference, reminding monitoring of drug adverse event
Improve the efficiency of the hospitals	The minimum residence time of health plan
Reduce the cost and convenience to improve the patients	The caution of repeating the test, prescribing guidelines

Table 4 Cases of data mining in the field of medicine

Name	Data mining algorithms	Mining content
Heart disease	Association Rule and decision tree	Use a single data to excavate standard of diagnosis of heart disease and the treatment of heart disease, using mixed data mining to compare the relationship between the heart disease diagnosis and treatment [14]
Cardiovascular disease	C4.5 algorithm	Use data mining to find out the cause of the cause cardiovascular disease, the results can be used in clinical medicine [15]
Diabetes	Apriori and FP-Growth	Use data mining to classify the correlation between different types of diabetes from diabetes database [16]
Lung cancer	Decision tree	Use mining biomarkers NSCLC to predict the type of lung cancer is a new type of lung cancer or other unknown [17]
Seasonal flu	SOM (Self-Organizing Maps)	Use SOM algorithms to analyze the data of infectious diseases, mainly mining under a certain time the spread of infectious diseases in different parts of the mainland China features [18]
Stroke	K-mean and decision tree	Analyze links with interesting degree of correlation analysis model in the database of Chinese and western medicine stroke patients to establish a stroke patient's condition prediction model

health insurance system; Wu Suying [20] used data mining technology to build a hospital disease classification of knowledge management system; Tang Shousheng [21] used data mining technology in tuberculosis patients' medical prediction, etc. We list some cases of data mining in the field of medicine.

3 Research Method

This study constructs a special migraine clinical decision support system, provide doctors tracking and diagnosis of the patients with headache. And together with a hospital in Taipei, this study uses the paper headache diary used by hospital neurology clinic of migraine, convert them to comply with all kinds of action platform software 'electronic headache diary', the patients do the record by themselves on the mobile platform, then update to the database, physicians and patients can bring up the records from the front-end interface by themselves. Therefore, this study of Fig. 2 step as the research method, research steps in order to complete in accordance with the picture stage target, finally construct a complete system.

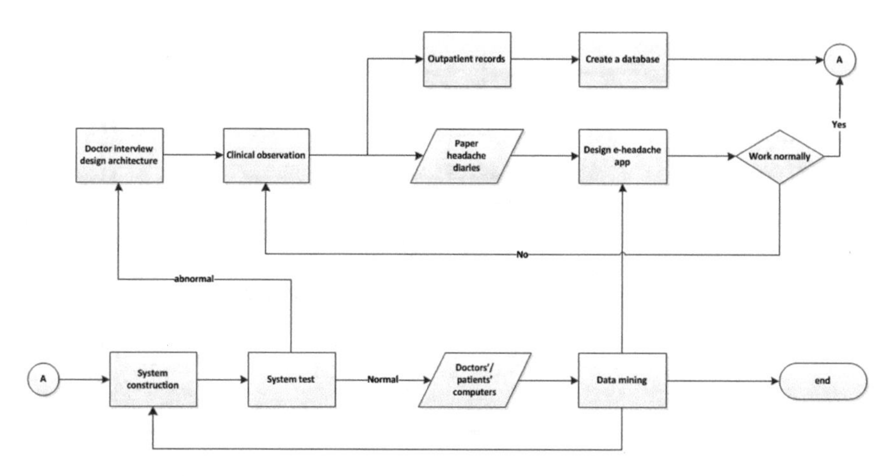

Fig. 2 Research steps

3.1 Research Steps

Figure 2 is the research steps of the study, in accordance with the steps to carry out and complete, to construct a clinical decision support system that can supply physicians with headache track and diagnosis of patients with migraine. Studies on Fig. 2 steps show as follows:

The Physician Interviews and Design Architecture

This research cooperation with a hospital in Taipei, the study focuses on the headache clinic neurology, in order to design the system's main structure, our clinicians discuss with neurology, to satisfy physicians' need to do system backbone, we also give advice to physician to determine whether to incorporate this recommendation system, after many meetings, we design a complete architecture diagram of the system. This research design the system architecture (Fig. 3), in terms of patients, patients use electronic headache diary to record his headache, and upload it to the database of our system, and the patients can download their dairy to their own smartphones; On examining the diary, patients can choose directly that watch diary over the mobile phone or from the front-end interface, this study will set the initial account as patient's id card number or medical record number, account shall be set by patients after login, the patients can see their record of the diaries that they uploaded to the system in the front-end interface. In terms of physician, the physician will have their own account and password, using the group account and password can enter the physician exclusive interface, physicians can enter a patient case number in the page or id number to bring up the headache diary, you can choose directly to examine the diary from the patient's action platform. The design of the system is aim to be simple and not complex, for the purpose of not only easy to use and can save time, is convenient for physicians and patients to review and examine the contents of the diaries.

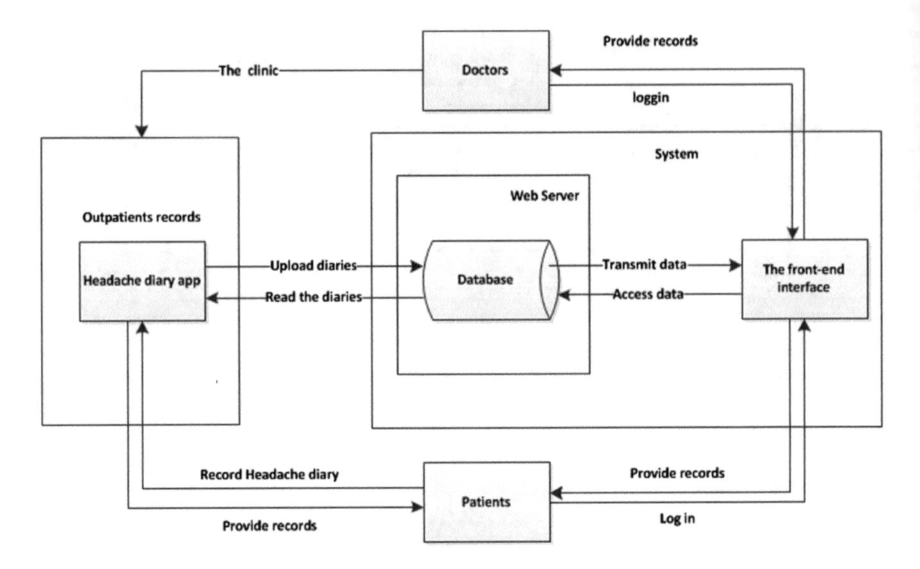

Fig. 3 System architecture

With Clinical and Observation

In order to improve the system accuracy, we will directly get into the outpatient service examines internal nerve headache clinic, to direct face-to-face contact patients with diagnosis way and detailed record the process of physician visits, to find out the advantages and disadvantages to be the reference of modification of the system. In the process of diagnosis, we can observe patients with migraine disease from time of the onset to find out what similarities or special relationship as a reference, moreover can also from a conversation with the patient to get some advice about the system and other ideas.

Design Electronic Headache Diary

In this study, we design the e-diary for Android, we use Android as a priority because Android is a free open source system, and it is suitable for use in testing the APP development stage, the development will meet some unexpected bugs, so it needs to test online and improve, and can use online test to get use suggestions, to improve and complete the application. When the Android App is done, we will design other version applications, such as IOS and Windows phone, to provide different mobile platform with the application; we will according to a paper diary as a reference provided by the hospital paper diaries ask the patients to record the projects: the degree of headache, symptoms of headache, omen of the headache, how many hours of headache, drug name and dosage, the effect of pain relievers, menstruation, etc., we will make development projects included in the above functions, in addition to design a function of statistic of headache situation can be used by the clinicians directly, this can help the doctors to know the condition of the patients. System developers will go directly to a hospital in Taipei, help the patient install electronic headache diary and update version and we will transform the website that where can download the application into QR code, and stick it outside the clinic, teach the patients who are the first time to use the application how to use it correctly, and record the doctors' and patients' situation of using the application, etc., the patients who are the first time to install the application, we will ask them to leave their information to do the track of the system and contact the patients.

The Clinic Record

Because in this study we will go to a hospital in Taipei, we will record all the process and content of the clinical physician visits, for example: the question that the clinicians ask the patients, including patient's basic information, the patient's medical history, and the patient's family history, the drugs the patients take and the inspection of patients, the habit of life of patients and some other data, we will organize the data and transform it into files, for the follow-up data mining.

Set Up the Database

The database in this study is mainly used to store the data of the patients' headache diaries, the hospital can use the patients' migraine data to do further study. This study uses the relational database SQL Server, launched by the Microsoft, the database interface is simple and it is easy to operate. We will divide the data into five data table: user account data, personal data, the degree of headache, symptoms with the headache and drugs to match the e-headache diary. We used advanced programming Language like Java, ASP.NET, C#, and SQL Language for this part (Table 5).

System Construction

This study constructed the front-end user operation interface and back-end database for the system operation, the front-end user interface constructed the interface of website by web-based method, as most of the headaches associated with personal physique, the headache can be treated but is difficult to cure, according to clinical observation, migraine sufferers rarely pain for a lifetime, most major attack at the age of 15, 6 to 30 or 40 years old, when the age is 50 or 60 years old, the diseases will relief or disappear [13], in order to satisfy the age, we mainly design the system

Table 5 Data table

Data table name	Terms
User account data	ID, name, account, password
Personal data	Gender, birth, case number, job, marriage, the degree of education, phone number, address…
The degree of headache	Time, headache time(morning, noon, afternoon, night)
Symptoms with the headache	Concomitant symptom
Drugs	Date, drug name, dose, significant degree

for easy use and it will not bring other burdens on the patients, and the doctors and patients can easily get start in the application. We will use such as net platform and some other web-based to construct the system interface, such as web interface main settings: user login and registration, personal information, such as a headache diary shows.

System Testing

System of this research is mainly to assist physician to track the patients' headache situation, using electronic headache diary can reduce the trouble to carry, also can achieve the effect of immediately record and update, moreover physician can also use it to track the drugs of the patients, to provide physicians improve treatment method or replacing prescription drugs. After completion of the system construction, therefore, must pass a physician verification, to test if the system meets the needs of the doctors or not, whether can actually upload the situation of the patients, if we can't meet the requirements, we will return to the first step in with the physician to discuss again, review and find out the causes of the system construction of incorrect, and correct programs, repeat the step until complete our system.

Data Mining

In this study, we hope we can find out the factors lead to migraine in the record of the headache of the patients and the outpatient records, we will also use the tracking of the situation that the patients take medicine to do mining, to find out which kinds of drugs affect migraines, the effect is good or bad for migraine, whether to take some drugs can cause conflict or reduced, what types of patients and fit to eat what drugs, etc., we will also design the result of data mining into questions and add them to the system and e-headache diaries to improve the correct rate of the diagnosis of migraine.

4 Conclusion

This research will interview with the doctors, with the diagnosis and observation for a long time, complete the electronic version of the APP migraine headache diary and tracking system in the early time. And then on the academic and medical to construct a new system architecture clinical decision support system and organize huge amounts of data into database. And the contributions in medicine, including: this design can make use of the APP in the clinical medicine, and this app has a professional medical background; System can be used in clinical medicine; Can completely store and save for a long time headache patients data; Information can let a hospital to do more in-depth study of migraine.

When after the migraine tracking system finishes online using, provide the doctors to track and transform the patients' headache diaries into information, save the extra space and improve the service efficiency, we will sort and save the huge amounts of data that got from the patients, allow clinicians to save and view patient's headaches data through the system, the patients' headache data shown in the system cam supply the doctors as the second reference of the diagnosis of migraine to improve accuracy of physicians' diagnosis of diseases, the doctors can also do the track of drugs from the data to provide physicians with the reference of improving treatment or replacing prescription drugs, in addition the hospital can also use this data to do more in-depth research and analysis of migraine, increase the understanding of the chronic migraine symptoms.

Acknowledgements The author would like to thank the Ministry of Science and Technology for partially supporting this research under no. MOST-105-2221-E-032-061.

References

1. Dodick, D. W., & Gargus, J. J. (2008). Why migraines strike. Scientific American, 299(2), 56–63.
2. Wu, Pei-Lin, Huang, Chiu-Ku, & Chen, Chun-Hsien. (2012). Comparison of Clinical efficacy and safety of migraine prevention drugs, 20(4), 285–300. (in Chinese).
3. Chen, Chien-Chih, Chuang, Yu-Min, Yang, Ching-Hua, & Chen, Teng-Lang. (2009). prodromal symptoms of dizziness of migraine. Cheng-Ching Health care management magazine, 5(2), 41–46.
4. Teri Robert. (2011). Chronic Migraine - The Basics. The HealthCentral. Retrieved 22th Nov, 2014, from: http://www.healthcentral.com/migraine/c/123/144625/chronic-migraine.
5. Effective management of migraine (2)(2006). Taiwan headache society. Retrieved 11th Dec, 2014 from: http://www.taiwanheadache.com.tw/epaper_34.asp.
6. Wang, Shu-Chun (2012). The latest research and treatment of migraine. Medical coverage, 3–7. Retrieved 12th Dec, 2014 from: http://www.sle.org.tw/ezcatfiles/nw19/img/img/389/8001.pdf.
7. Velickovski, F., Ceccaroni, L., Roca, J., Burgos, F., Galdiz, J. B., Marina, N., & Lluch-Ariet, M. (2014). Clinical Decision Support Systems (CDSS) for preventive management of COPD patients. Journal of translational medicine,12(Suppl 2), S9.
8. Hayward, R. S., El-Hajj, M., Voth, T. K., & Deis, K. (2006). Patterns of use of decision support tools by clinicians. In AMIA Annual Symposium Proceedings (Vol. 2006, p. 329). American Medical Informatics Association.
9. Hayward, R. (2004). Clinical decision support tools: Do they support clinicians?. Canadian Medical Association Journal, 170(10; SUPP), 66–85.
10. HealthuIT. (n.d.). CDS. Retrieved 12th Dec, 2014 from the World Wide Web: http://www.healthit.gov/policy-researchers-implementers/clinical-decision-support-cds
11. Berner, E. S. (2007). Clinical Decision Support Systems (pp. 3–22). New York: Springer Science + Business Media, LLC.
12. Groselj, C. (2002). Data mining problems in medicine. In Computer-Based Medical Systems, 2002. (CBMS 2002). Proceedings of the 15th IEEE Symposium on (pp. 377–380). IEEE.
13. Su, Yi-Ching(2011). Annoying migraine. A healthy life, 8. Retrieved 11th Dec, 2014 from: http://www.health.ntpc.gov.tw/web66/_file/1459/upload/ehealth/10008/pages/index-01-01.html.
14. Shouman, M., Turner, T., & Stocker, R. (2012, March). Using data mining techniques in heart disease diagnosis and treatment. In Electronics, Communications and Computers (JEC-ECC), 2012 Japan-Egypt Conference on (pp. 173–177). IEEE.

15. Rajmohan, K., Paramasivam, I., & Sathyanarayan, S. (2014, February). Prediction and Diagnosis of Cardio Vascular Disease–A Critical Survey. In Computing and Communication Technologies (WCCCT), 2014 World Congress on (pp. 246–251). IEEE.
16. Sankaranarayanan, S., Dr Pramananda Perumal, T. (2014). Diabetic Prognosis through Data Mining Methods and Techniques. International Conference on Intelligent Computing Applications. IEEE.
17. Osmania Univ., Dass, M. V., Rasheed, M. A., & Ali, M. M. (2014, January). Classification of lung cancer subtypes by data mining technique. In Control, Instrumentation, Energy and Communication (CIEC), 2014 International Conference on (pp. 558–562). IEEE.
18. Xu, T., Zhou, J., Gong, J., Sun, W., Fang, L., & Li, Y. (2012, May). Improved SOM based data mining of seasonal flu in mainland China. In Natural Computation (ICNC), 2012 Eighth International Conference on (pp. 252–255). IEEE.
19. Chen Shiyuan (2000). Data Mining in Acquiring Association Knowledge between Diseases and Medicine Treatments. Master Thesis in the department of information management, National Sun Yat - sen University.
20. Wu Suying (2004). Applying Data Mining Technique to Construct Knowledge Management System for the Management of Disease Classification in the Hospital. Master Thesis in the department of information management, National Chung Cheng University.
21. Tang Shousheng (2004). Application of Data Survey Technology in Predicting Pulmonary Tuberculosis. Master Thesis in the department of information management, National Chung Cheng University.

Exploring the Social Influence Mechanism of Applying Social Network Technology in a Classroom

Fang-Ling Lin

Abstract This study examines the impacts of social influence on the academic performance of college students by applying a social network technology to assist instruction in a classroom. Social influence can push systems toward uniformity of behavior while selection can lead to fragmentation. Hence, understanding the effects of these two forces in group or organization behaviors help instruction development. This research applies the stochastic actor-based model to discover these two social forces of knowledge sharing among peer-mediated learners by social media. Drawing on the literature of the knowledge construction and social influence, it aims to have a better understanding of social network technology in two directions: describing how relationships are creating and identifying individual and contextual attribute which facilitate its spontaneous knowledge construction.

Keywords Social network analysis · Social network technology · Approach of learning

1 Introduction

Social network technology have been deeply embedded in modern life and alter learning behaviors of college students [1, 2]. Varied types of social network technology, such as blogs, wikis, and chat rooms support new learning activities [3]. These types of social network technology give students a platform to exhibit work and to share experiences. These activities enhance students' self-confidence or satisfy their curiosity [4]. The self-disclosure activities help students be aware the existence of audience as well as to strengthen motivations [5].

Hwang et al. [6] have addressed the participation in social network technology is an important learning practice for modern students. Social network technology,

F.-L. Lin (✉)
Department of Information Management, Lunghwa University of Science and Technology, Taoyuan, Taiwan
e-mail: fangling@mail.lhu.edu.tw

© Springer Nature Singapore Pte Ltd. 2018

N.Y. Yen and J.C. Hung (eds.), *Frontier Computing*, Lecture Notes in Electrical Engineering 422, DOI 10.1007/978-981-10-3187-8_54

connecting participants, expanding their relations and accelerating information exchange form a learner-centered interactive media [7]. Social interaction supports knowledge construction and academic performance for students [7, 8]. Tacit knowledge is not private but social [9].

Also, learning processes related to learning strategies included student motivations, awareness of responsibility, method of assessments, and classmates [10]. Biggs et al. [11] came up with the concept of an 'approach to learning' for how students perceived and went about learning a learning task. They argued that students have three motives: to keep out of trouble with minimal effort, to engage the task appropriately, and to maximize grade, respectively. Moon [12] had considered deep or surface learning as a possible result of a learning process.

Inspired by the above researchers, this study suggests an instructional design with the social network technology. A SlideShare-like social software: OpenSlide has been installed for the pedagogy of an operating system course for undergraduates of the information management department. Functions of OpenSlide system (Fig. 1) include:

(1) Homework visualization: reformat the demonstrated homework to web slides and embed a player for browsing (Fig. 2).
(2) Personal directory: provide personal account and space for documents.

Fig. 1 Functions of OpenSlide system

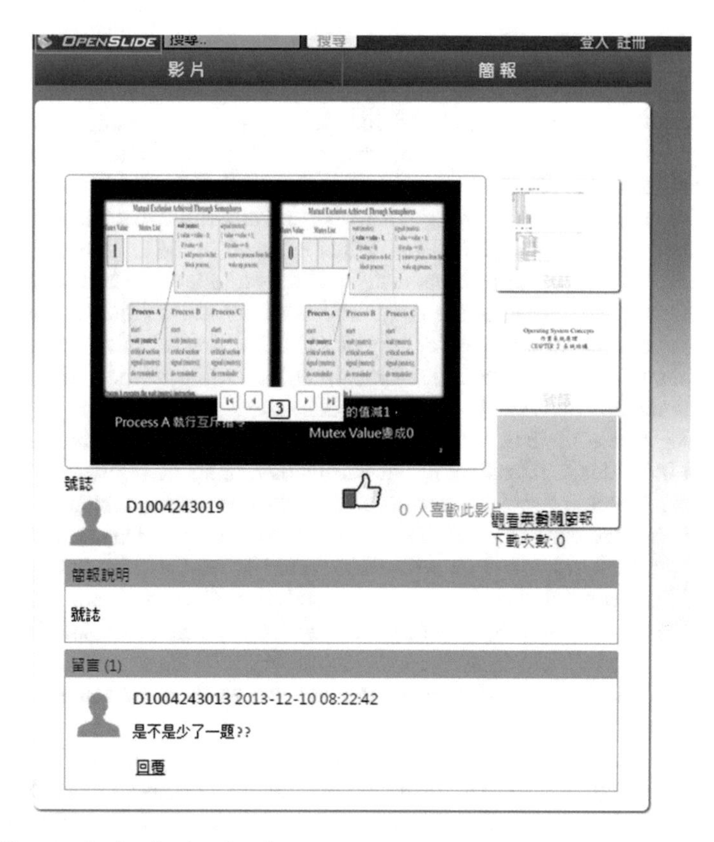

Fig. 2 Homework visualization function

(3) Documents upload and download: let users upload personal documents for demonstration. The documents can be set to be downloadable for other users.
(4) Comment replication: let users post comments for demonstrations of others.
(5) Tag cloud: tag to let documents found easily. Popular tags have larger font sizes.
(6) Topic searching: let users search related documents with the keyword.

The course design of instructional requires students to upload their files of homework to the OpenSlide system and share with classmates for discussion. Students were encouraged to learn from each other's homework, to discuss issues, and to give comments. Through using the OpenSlide system, this study assumed the facilitating of students' interactions and applied the stochastic actor-based methods [1] to model the forces of social influence for students with different learning approach.

2 Stochastic Actor-Based Model

The process of making a friend and maintaining the relationship related to the person's behavior and personality [13]. In a study of students' attitude toward using a computer, Erickson [14] had argued that in the confused stage of learning a new tool, individuals will obtain the guidance of actions from matching attitudes of similar others in the referenced group. When engaged in the referenced group, the similar attitude will be enhanced and the various attitudes will be altered. Hence, similar actions of actors can be seen as a process of influence. Finally, actors behave as the connected others with similar attitude [15]. The process of selecting involves the individuals alter their behaviors or attitude according to the intimated peers. Socially connected individuals assimilate each others gradually.

Matching and assimilation cause the same empirical evidence: the similarity of connected individuals. Investigating the mechanism of social influence should also studying the process of social selection. If the selection causes similarity, it implies individuals behave similarly but their relations are changed. On the contrary, the influence mechanism implies individuals keep stable relationships but behave differently. The serial changes of relations and behaviors generate interactively dependent relationships [15]. Investigating the co-occurrence of social networks and behavior needs to obtain all the information of individuals and their peers. The difficulty of collecting data challenges the research methods and interpretation of similar underlying behavior among peers.

To overcome the above limitations, Snijders [16] has proposed a method of simulation investigation for empirical network analyses (SIENA) to implement the stochastic actor-based model. In this system, simulation can evaluate the change of behavior and social network. Therefore, it allows researchers to test the effects of selection or influence by controlling one another at the same time.

A simple example introduces the stochastic actor-based model. In Fig. 3, supposed the actor i thinks actor j is a friend but without similarity in the beginning. In the second observation, actor i still thinks actor j is a friend and behaves as actor j, i.e., in the (a) process of assimilation. Many researchers have classified this change as an influence because actor i adjust behavior with actor j. Because this conclusion assumed their relationship keep stable during the period between two observation. This judge might be wrong. Actually, it might go through a serial process of (b)–(d) before we obtain the second evidence of observation. The friendship might disappear after the beginning stage like (b) process. The actor i change his behavior alone like process (c). The actor i and actor j with similar behavior or attitude start their friendship with homophily. The social selection rather than the social influence becomes the critical factor in the process (b) and (d). Traditional statistical approach ignores the unseen changes that make biased conclusions. A real social influence mechanism needs to explain the underlying changes.

Therefore, the stochastic actor-based model needs data with at least two observations of the continual time serials. Using t_1, t_2, …, t_o represents the time of observation. Tie represents a direction relationship from the sender. x_{ij} is a binary

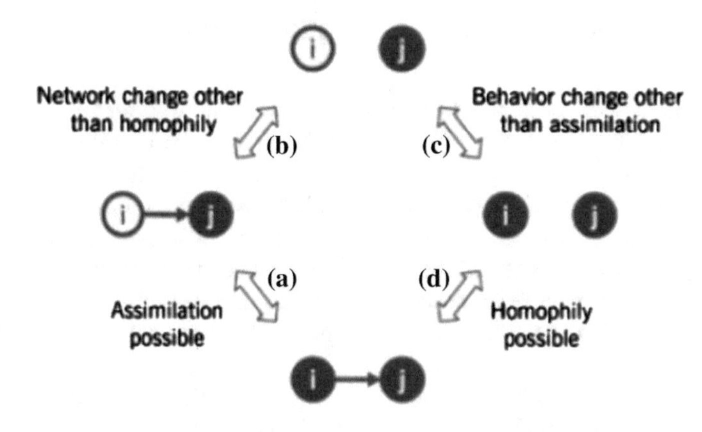

Fig. 3 The basic process of change among similarity and friendship [15]

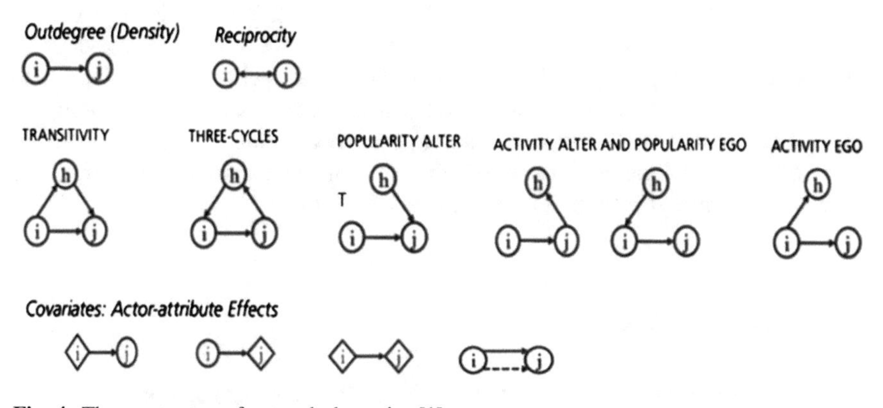

Fig. 4 The parameters of network dynamics [1]

variable which represents actor i has a relation to j with value 1 and no relation with value 0. An adjacency matrix with n × n dimension represents a whole network of n actors. Interdependent behavioral variables are assumed to be discrete, ordinal and represented by vector z = (z_i). In the stochastic actor-based model, dependent variables are time serials with observed adjacency matrix and behavioral vectors: $(x, z)(t_1)$, $(x, z)(t_2)$, …, $(x, z)(t_m)$.

The kernel of the stochastic actor-based model is a computation of the objective function to determine the probability of tie changes among actors. The objective function contains all the parameters of network dynamics, such as density, reciprocity, several triangular structures, popular, popular alters and so on (see Fig. 4). The objective function also measures the covariant effects and behavior effects [1].

3 Social Network Benefits of Knowledge Sharing

One important stage of knowledge acquiring is to know who own the knowledge and to connect with the owners [17–19]. Knowing the experts is the critical factor of information efficiency [20]. Hence, the most often correspondents of knowledge exchanges in a group might be prestigious [21]. The prestige one has many connections to others, they become the primary channel of knowledge sharing [22]. Greenberg [23] has claimed that the individual positioned in the central location can discuss with others with fewer costs and can be acquired easily also. Lai and Wong [24] believes that the participants in the population centers are active and transfer information efficiently.

Scholars also proposed the characteristic of network flow [21] and several other connection properties such as frequent contacts, emotional and reciprocal benefits [25]. The strength of exchange network of Knowledge workers has direct impacts on their competitiveness [26].

Using the OpenSlide system to assist the learning activities in the classroom is aimed to conduct a process of facilitating changes in the whole collectivities of students by encouraging peer interactivities. The social network technology help students have new connections with classmates and extend their social networks of knowledge sharing. With different learning approach, students gain the different achievement. Investigating the network dynamics in the classroom can reveal the collective system dynamic of a course in the classroom.

4 Research Design

This study analyzed the network dynamics of classroom interactivities and academic achievement of different learning approaches with the stochastic actor-based model. The author proposed a process of relational changes from friendship to knowledge sharing according to work of Veenstra and Dijkstra [15]. The social influence mechanism of knowledge sharing in the classroom assumes the following hypotheses.

H1: The relationships of knowledge sharing among students were influenced by using the OpenSlide system.

H2: The relationships of knowledge sharing among students were influenced by their new friendships in the course.

H3: The academic achievements of students with deep learning approaches were growing.

H4: The academic achievement of students with surface learning approach were growing.

A semester course syllabus includes a midterm test and a final examination, assignments and group activities are given. Students uploaded their homework for each assignment and group leaders upload group work to slide share system. Students were encouraged to learn from others' homework. Instructors didn't involve in online discussion.

Students have completed paper-based social network questionnaires in the beginning and the end of semester respectively which examine their social relationships of knowledge sharing. The instructor guided the students to select correspondents according to questions in name list of classmates. The items can be seen in Table 4 and Table 5 of appendix A. They also completed a questionnaire about the study process [11] which includes items of deep learning motivation, deep learning strategy, surface learning motivation, and surface learning strategy.

The progress from midterm to final examination was computed by the rank of the whole class and analyzed using statistical methods by removing the covariance.

5 Results and Discussion

5.1 Demographics

Of 39 students of this course, 36 participated in the activities in the ensuing 4 months. Table 1 is the demographic of the behavior variables. Table 2 illustrates the demographics of social networks. The average connections were 10 correspondents. It implies each student know 1/3 classmates in this course. Through the OpenSlide system, each student knows 1.5 new friends averagely.

5.2 Network Dynamics Analysis

Table 2 illustrates the fit model of network dynamics which was analyzed by RSiena software [27]. RSiena is a program for the statistical analysis of network

Table 1 The demographic of the behavior variables and network variables

N = 36	Min	Max	Mean	Std.
DA (deep learning approach)	20.00	50.00	29.28	6.36
SA (surface learning approach)	15.00	38.00	26.39	4.83
ds1 (network size of knowledge discussion before the course)	3.00	16.00	9.11	3.10
Friendship (network size)	2.00	20.00	13.14	4.49
ds2 (network size of knowledge discussion after the course)	6.00	16.00	10.70	2.42
OS (new friends by the OpenSlide system)	0.00	9.00	1.56	2.30

Table 2 The fit model of network dynamics

Effect name	Parameter value	Standard error	p-value	t-statistic
Basic rate parameter network	11.079	0.941		
Outdegree (density)	−1.192	0.265	<0.001	0.096
Reciprocity	−0.89	0.234	<0.001	0.002
Transitive triplets	0.143	0.031	<0.001	0.055
Number of actors at distance 2	−0.362	0.119	0.002	0.104
OS	2.465	0.347	<0.001	0.105
Class	0.406	0.233	0.082	−0.05
Same gender	−0.012	0.150	0.938	0.034
SA alter	−0.092	0.021	<0.001	−0.048
SA ego	0.006	0.015	0.703	0.053
SA similarity	−0.486	0.546	0.374	−0.023
DA alter	0.003	0.012	0.817	0.004
DA ego	0.003	0.011	0.778	0.061
DA similarity	−0.287	0.440	0.515	−0.093

Table 3 The predictability of behavior variables

Model	Beta coefficient	T	Sig.
Deep approach	0.393	2.213	0.034
Surface approach	−0.147	−0.858	0.398

data, with the focus on social networks. It includes the parameters of reciprocity, transitive triples, popular alters (receiver) and egos (sender) The influence of actors includes egos with learning approach (senders), alters with learning approach (receivers). The estimated parameters demonstrate the significant effects of reciprocity and transitive triples. Using the OpenSlide system had significantly positive impacts on developing a network of knowledge sharing.

The actors with surface learning approach have significantly negative receiving relationships. It implies that the actors with surface learning approach were unpopular. The actors with deep learning approach were expected to be popular alters. Unfortunately, this assumption was not verified.

5.3 *The Learning Approach and the Academic Achievement*

Table 3 illustrates the predictability to academic achievement. Students with deep learning approach were significantly positive predicted to have higher academic achievement. However, the small interpretation power of R square ($R^2 = 0.192$) implied that some other underlying influential factors were needed for investigating.

6 Conclusion

This study had explored the efficiency of applying a social network technology in a classroom by verifies the measurable network variables and behavior variables by using the stochastic actor-based model. It also intended to develop a model of network dynamics of knowledge sharing in a course. Some hypothesis had been supported. There are still a number of unresolved issues that need further study. This study provides a better understanding of social network technology in two directions: describing how relationships are created and identifying individual and contextual attribute which facilitate its spontaneous knowledge construction.

Appendix

See Tables 4 and 5.

Table 4 The first social network questionnaires of Knowledge sharing

Please identify the people from whom			
ID	Name	You have ever discussed with to solve problems related to homework	You are good friends
xxxxx	xxxxx	☐	☐
xxxxx	xxxxx	☐	☐
———(omitted)			

Table 5 The second social network questionnaires of Knowledge sharing

Please identify the people from whom			
ID	Name	You have ever discussed with to solve problems related to homework	You are new friends from the OpenSlide system
xxxxx	xxxxx	☐	☐
xxxxx	xxxxx	☐	☐
———(omitted)			

References

1. Madge, C., Meek, J., Wellens, J., & Hooley, T. (2009). Facebook, social integration and informal learning at university: it is more for socializing and talking to friends about work than for actually doing work. Learning, Media, and Technology, 34(2), 141–155.
2. Hung, H. T., and Yuen, S. C-Y (2010) 'Educational use of social networking technology in higher education', *Teaching in Higher Education*, 15: 6, 703–714.
3. Subrahmanyam, K., Reich, S.M., Waechter, N., & Espinoza, G. (2008). Online and offline social networks: use of social networking sites by emerging adults. Journal of Applied Developmental Psychology, 29(6), 420–433.
4. Crook, C & Harrison, C (2008), *Web 2.0 Technologies for Learning at Key Stages 3 and 4: Summary Report*, available at http://dera.ioe.ac.uk/1480/1/becta_2008_web2_summary.pdf, accessed 06/10/2011.
5. Hemmi A., Bayne S. & Land R. (2009) The appropriation and repurposing of social technologies in higher education. *Journal of Computer Assisted Learning* 25, 19–30.
6. Hwang, A., Kessler, E.H., & Francesco, A.M. 2004. Student Networking Behavior, Culture, and Grade Performance: An Empirical Study and Pedagogical Recommendations. *Academy of Management Learning and Education*, 3 (2): 139–150.
7. Dewey, J. (1963). *Experience and education*. New York: Macmillan.
8. Vygotsky, L. (1978). The interaction between Learning and Development (pp. 79–91). In Mind in Society. (Trans. M. Cole). Cambridge, MA: Harvard University Press.
9. Polanyi, M., (1966). *The Tacit Dimension.* Doubleday and Co., Garden City, NY.
10. Xie, Y., Ke F., & Sharma, P, (2008), The effect of peer feedback for blogging on college students' reflective learning processes, *Internet and Higher Education*, 11 (2008) 18–25.
11. Biggs, J.B., Kember, D., & Leung, D.Y.P. (2001) The Revised Two Factor Study Process Questionnaire: R-SPQ-2F. *British Journal of Educational Psychology. 71,* 133–149.
12. Moon J (1999) Reflection in Learning and Professional Development: theory and practice London; Kogan Page.
13. Friedkin, N.E. (1998). *A structural theory of social influence*. Cambridge: Cambridge University Press.
14. Erickson, T.E. (1987). Sex differences in student attitude towards computers. Paper presented at the Annual Meeting of the American Educational Research Association, Portland. Oregon. November.
15. Veenstra, R. & Dijkstra, J.K. (2011). Transformations in Adolescent Peer Networks. In B. Laursen & W. A. Collins (eds.) *Relationship Pathways: From Adolescence to Young Adulthood* (pp.). New York: Sage.
16. Snijders TAB, van de Bunt G, Steglich C (2010) Introduction to stochastic actor-based models for network dynamics. Soc Networks 32:44–60.
17. McDermott, R., (1999). Why Information Technology Inspired but Cannot Deliver Knowledge Management. California Management Review, *41*(4).
18. Cross, R., Parker, A., Prusak, L. & Borgatti, S.P. (2001). Knowing what we know: supporting knowledge creation and sharing in social networks. *Organizational Dynamics, 30* (2), 100–20.
19. Borgatti, S.P. & Cross, R. (2003). A relational view of information seeking and learning in social networks. *Management Science, 49*, 432–445.
20. Lin, F.L. & Chiou, G.F. (2008). Support-seeking and support-giving relationships of school technology coordinators. *British Journal of Educational Technology, 39*(5), 922–7.
21. Lin, N. (2001). Social networks and status attainment. In N. Lin (Ed.), *Social Capital: A Theory of Social Structure and Action.* Cambridge University Press, Chapter 6.
22. Burt, R.S. (1992). *Structural Holes: The Social Structure of Competition,* Cambridge: Harvard University Press.
23. Greenberg, B.S. (1964). Diffusion of news of the Kennedy assassination. *Public Opinion Quarterly, 28*, 225–32.

24. Lai, G. & Wong, O. (2002). The tie effect on information dissemination: the spread of a commercial rumor in Hong Kong. *Social Networks, 24* (1), 49–75.
25. Binder, J.F., Roberts, S.G.B, & Sutcliffe, A.G. (2009). Closeness, loneliness, support: core ties and significant ties in personal communities. *Social Networks*, 34. pp 206–214.
26. Lin, F.L. & Chiou, G.F. (2012). Network of Practices: A Case Study of Knowledge Competition of School Technology Coordinators, *Proceedings of Advances in Social Networks Analysis and Mining (ASONAM) 2012, Aug. 26–29, Istanbul, Turkey.*
27. R Development Core Team (2015) *R: A Language and Environment for Statistical Computing*. R Foundation for Statistical Computing, Vienna, Austria. http://www.R-project.org.

Improvement of Power-Spectral-Subtraction Algorithm Using Cross-Term Compensation for Speech Enhancement

Ching-Ta Lu, Yung-Yue Chen, Jun-Hong Shen, Ling-Ling Wang and Chung-Lin Lei

Abstract Although the power-spectral-subtraction (PSS) algorithm is widely used in speech enhancement, this method suffers from musical residual noise. So the enhanced speech sounds annoying to the human ear. This study proposes using the cross term between the spectrum of speech and noise signals to be additionally subtracted from the power spectrum of noisy speech, enabling background noise to be efficiently removed. Experimental results show that the proposed method can significantly improve the performance of the PSS algorithm by the consideration on the cross term. The quantity of musical residual noise can be efficiently removed, while speech components are well preserved in the enhanced speech.

Keywords Speech enhancement · Cross-term · Power-spectral-subtraction · Harmonic adaptation

C.-T. Lu (✉) · J.-H. Shen · L.-L. Wang · C.-L. Lei
Department of Information Communication, Asia University, Taichung, Taiwan, ROC
e-mail: Lucas1@ms26.hinet.net

J.-H. Shen
e-mail: shenjh@asia.edu.tw

L.-L. Wang
e-mail: ling@asia.edu.tw

C.-L. Lei
e-mail: BligsLcl@gmail.com

Y.-Y. Chen
Department of Systems and Naval Mechatronics Engineering,
National Cheng Kung University, Tainan, Taiwan, ROC
e-mail: yungyuchen@mail.ncku.edu.tw

© Springer Nature Singapore Pte Ltd. 2018

N.Y. Yen and J.C. Hung (eds.), *Frontier Computing*, Lecture Notes
in Electrical Engineering 422, DOI 10.1007/978-981-10-3187-8_55

1 Introduction

Single channel speech enhancement is one of the most widely used methods for the enhancement of noisy speech. The spectral subtraction method is a popular single channel speech enhancement technique. This method is simple and substantially reduces the noise level in noisy speech. However, this method suffers from musical residual noise in enhanced speech.

Recently, many novel schemes [1–12] have been proposed to enhance speech signals corrupted by additive noise. A single-channel speech enhancement algorithm based on non-linear and multi-band adaptive gain control (AGC) was proposed [1]. This method reduced the background noise in the temporal domain using a non-linear and automatically adjustable gain function for multi-band AGC. The gain function varies in time and is deduced from the temporal envelope of each frequency band to highly suppress the frequency regions where noise is present and slightly suppress the frequency regions where speech is present. So the background noise can be efficiently reduced. Lee et al. [2] proposed a new approach to estimating the a priori signal-to-noise ratio (SNR) based on a multiple linear regression (MLR) technique. They found the a priori SNR based on the MLR technique by incorporating regression parameters such as the ratio between the local energy of the noisy speech and its derived minimum along with the a posteriori SNR. The regression coefficients obtained using the MLR were assigned according to various noise types. Experimental results showed that this method outperformed the conventional methods. An improved multi-band spectral subtraction (I-MBSS) method [3] was proposed to cope with noisy speech degraded by real-world noises. The I-MBSS method used an adaptive noise estimation approach to estimating the noise for each subband. The subtraction parameters were adjusted according to noise masking threshold. Experimental results showed that this method performed better than the classical multi-band spectral subtraction algorithm. A new geometric approach to spectral subtraction was presented in [4]. This paper addressed the shortcomings of the spectral subtraction algorithm. A method for estimating the cross terms involving the phase differences between the noisy (and clean) signals and noise was proposed. Objective evaluation of this algorithm showed that it performed significantly better than the traditional spectral subtractive algorithm. A dynamic spectral subtraction algorithm [5] was proposed to enhance noisy speech. An over-subtraction factor was employed to improve the ability for background noise removal. Huang et al. [6] investigated a parametric gain approach to noise reduction in the frequency domain. In comparison with the traditional parametric Wiener gain, the major novelty of this method is that the parametric gain is formulated to estimate the noise by using the mean-squared error between the noise and the noise estimate. The enhanced signal was then obtained by subtracting the noise estimate from the noisy observation signal. Virag [7] proposed using a psychoacoustical model to adjust the over-subtraction parameters so as to render the residual noise inaudible [7]. Although traditional methods which utilize the power-spectral-subtraction (PSS) algorithm to roughly estimate the speech spectra can provide an acceptable performance, the estimated speech spectra can be further improved for computing the noise masking threshold. Lu [8] proposed using a two-step-decision-directed method to obtain speech estimate. In turn, this estimate was employed to compute the noise masking threshold of a perceptual gain factor. Experimental results showed that the amounts of residual noise could be efficiently suppressed by embedding the two-step-decision-directed algorithm in the perceptual gain factor.

The traditional PSS algorithm is computationally simple to implement but suffers from musical residual noise. In addition, the subtractive rules are based on incorrect assumptions about the cross term being zero. In order to improve the performance of the PSS algorithm for the noise removal, a cross term between speech and noise is considered in this study. The PSS algorithm assumes that speech and noise signals are uncorrelated, and the noise is zero mean. The cross term between speech and noise does not exist. However, speech and noise signals are correlated in real environments. The assumption of the PSS algorithm is not really correct. This is a major reason why the performance of the PSS algorithm is not satisfied. Since clean speech cannot be observed, we employ the PSS algorithm to obtain the spectral estimate of the speech signal. The output signal is named as a pre-processed signal. Hence, the cross term between the pre-processed signal and noise is calculated. The spectrum of enhanced speech is obtained by subtracting the cross term from the spectrum of the pre-processed signal. A great quantity of background noise can be efficiently removed. So the enhanced speech sounds less annoying than that produced by the PSS algorithm; meanwhile, speech quality can be maintained at an acceptable level. Accordingly, the performance of the PSS algorithm can be significantly improved by the proposed method.

The rest of this paper is organized as follows. Section 2 briefly introduces the PSS algorithm. Section 3 describes the proposed cross term compensation approach for the PSS algorithm. Section 4 demonstrates the experimental results. Conclusions are finally drawn in Sect. 5.

2 Brief Review of Power-Spectral-Subtraction Algorithm

A noisy speech signal $y(m,n)$ can be modeled as the sum of clean speech $s(m,n)$ and additive noise $d(m,n)$ in the frame m of the time domain, given as

$$y(m,n) = s(m,n) + d(m,n) \tag{1}$$

where n is the sample index in a frame.

The observed signal $y(m,n)$ is analyzed and transformed to the frequency domain using the short-term Fourier transform (STFT), given as

$$Y(m,k) = \sum_{n=0}^{N-1} y(n+mM) \cdot h(n) \cdot e^{-j(2\pi/N)nk} \tag{2}$$

where k is the frequency bin index, h is an analysis window of size N, and M is the frame update step in time.

The short-term power spectrum of the noisy speech can be expressed as

$$\begin{aligned} |Y(m,k)|^2 &= Y(m,k) \cdot Y^*(m,k) \\ &= |S(m,k)|^2 + |D(m,k)|^2 + S(m,k) \cdot D^*(m,k) + S^*(m,k) \cdot D(m,k) \\ &= |S(m,k)|^2 + |D(m,k)|^2 + 2\,\mathrm{Re}\{S(m,k) \cdot D^*(m,k)\} \end{aligned} \tag{3}$$

where * denotes the operator of complex conjugate. $\mathrm{Re}\{\cdot\}$ represents the real part of a complex variable.

In Eq. (3), the terms $|D(m,k)|^2$, $S(m,k) \cdot D^*(m,k)$ can be obtained by

$$|D(m,k)|^2 = E\left\{|D(m,k)|^2\right\} \tag{4}$$

$$S(m,k) \cdot D^*(m,k) = E\{S(m,k) \cdot D^*(m,k)\} \tag{5}$$

where $E\{\cdot\}$ denotes the expectation operator.

If the additive noise $d(n)$ is assumed to be zero mean and uncorrelated with the clean speech signal $s(n)$, the cross term $E\{S(m,k) \cdot D^*(m,k)\}$ is zero. The estimated power of clean speech $\left|\tilde{S}(m,k)\right|^2$ can be represented as

$$\left|\tilde{S}(m,k)\right|^2 = |Y(m,k)|^2 - \left|\tilde{D}(m,k)\right|^2 \tag{6}$$

where $\left|\tilde{D}(m,k)\right|^2$ denotes the estimated power of noise.

Equation (6) is known as the PSS algorithm. Adequately estimating the noise magnitude of noise $\left|\tilde{D}(m,k)\right|^2$ can obtain the estimated power of clean speech $\left|\tilde{S}(m,k)\right|^2$.

3 Proposed Cross Term Compensation for the PSS Algorithm

In real environments, speech and noise signals are correlated. The cross term $E\{S(m,k) \cdot D^*(m,k)\}$ in Eq. (5) is not zero. Figure 1 shows an example of the power spectra for speech and noise (Fig. 1b), and the corresponding magnitude of

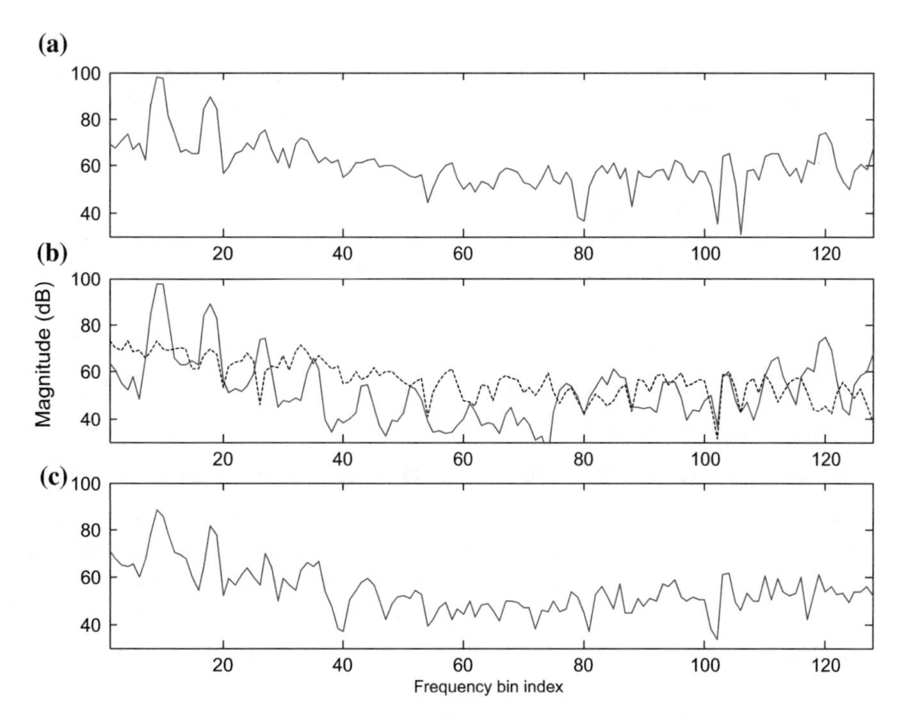

Fig. 1 An example of power spectrum. (From *top* to *bottom*) **a** noisy speech corrupted by factory noise with an average SegSNR = 10 dB, **b** clean speech spoken by a female speaker (*solid*) and factory noise (*dotted*), **c** the magnitude of the cross term between the spectrum of clean speech and factory noise

cross term (Fig. 1c). A speech signal is corrupted by factory noise with an average SegSNR = 10 dB shown in Fig. 1a. Ideally, the magnitude of the cross term should be zero. In the real case, the magnitude of the cross term of short-term spectra is far from zero as shown in Fig. 1c. Therefore, the assumption that the magnitude of the cross term should be zero is not adequate. This fact would cause the performance of the PSS algorithm given in Eq. (6) to be deteriorated. Consequently, the magnitude of the cross term between the spectrum of speech and noise signals should be considered in Eq. (6) to improve the performance of the PSS algorithm.

In order to improve the performance of the PSS algorithm, we propose using a compensation factor to compensate the effects caused by the cross term of the spectra for speech and noise signals. According to Eq. (3), the power spectrum of enhanced speech $\left|\widehat{S}(m,k)\right|^2$ can be expressed by

$$\left|\widehat{S}(m,k)\right|^2 = |Y(m,k)|^2 - \left|\widetilde{D}(m,k)\right|^2 - 2\,\mathrm{Re}\left\{S(m,k)\cdot\widetilde{D}^*(m,k)\right\} \tag{7}$$

Because the power spectra of clean speech and noise are unknown, a modified version of minima-controlled-recursive-averaging algorithm [13] can be employed to estimate the spectral magnitude of noise. The power spectrum of enhanced speech given in Eq. (7) can be rewritten as

$$\begin{aligned}\left|\widehat{S}(m,k)\right|^2 &= |Y(m,k)|^2 - \left|\widetilde{D}(m,k)\right|^2 \\ &\quad - 2\cdot\left|\widetilde{S}(m,k)\right|\cdot\left|\widetilde{D}(m,k)\right|\cdot\cos\left\{\theta\left[\widetilde{S}(m,k)\cdot\widetilde{D}(m,k)\right]\right\}\end{aligned} \tag{8}$$

where $\theta(\cdot)$ represents the phase of the complex vector $\widetilde{S}(m,k)\cdot\widetilde{D}(m,k)$.

In Eq. (8), the value of the cross term $\cos\left\{\theta\left[\widetilde{S}(m,k)\cdot\widetilde{D}^*(m,k)\right]\right\}$ lies between -1 and 1. A negative value of the cross term would boost the spectral magnitude of enhanced speech $\left|\widehat{S}(m,k)\right|$. But the spectra of speech and noise are unknown and estimated. Inaccurate boost on the spectrum of enhanced speech may introduce serious speech distortion. Therefore, we constrain the value of the cross term to be non-negative. The power spectrum of enhanced speech can be approximated by

$$\left|\widehat{S}(m,k)\right|^2 \approx |Y(m,k)|^2 - \left|\widetilde{D}(m,k)\right|^2 - 2\cdot\lambda(m,k)\cdot\left|\widetilde{S}(m,k)\right|\cdot\left|\widetilde{D}(m,k)\right| \tag{9}$$

where $\lambda(m,k)$ $(0 \le \lambda(m,k) \le 1)$ is a weighting factor for the cross term.

In Eq. (9), the magnitude of speech $\left|\widetilde{S}(m,k)\right|$ can be roughly estimated using the PSS algorithm given in Eq. (6). If the correlation level between speech and noise signals increases, the magnitude of the cross term is increased. Since the cross term is further subtracted from the power spectrum of noisy speech, the magnitude of enhanced speech decreases. This enables background noise to be efficiently removed. Conversely, the magnitude of the cross term is decreased when the correlation level between speech and noise decreases. In the case of a vowel region, the weighting factor of the cross term should be decreased. A greater quantity of speech components can be preserved, enabling the quality of enhanced speech to be maintained. Conversely, the weighting factor of the cross term should be increased in noise-dominated region. A greater quantity of the cross term is subtracted from the power spectrum of noisy speech. So the background noise can be efficiently removed by additional consideration on the cross term. The performance of the PSS algorithm is accordingly improved. The value of the weighting factor $\lambda(m,k)$ can be defined as

$$\lambda(m,k) = \begin{cases}\lambda_v, & \text{if } Y(m,k) \in \text{owel frames} \\ \lambda_{v\_neighbor}, & \text{if } Y(m,k) \in \text{the neighbors of vowel frames} \\ \lambda_N, & \text{otherwise}\end{cases} \tag{10}$$

where λ_v, $\lambda_{v\_neighbor}$, and λ_N are empirically chosen to be 0.6, 0.8, and 0.9 respectively.

4 Experimental Results

In the experiments, speech signals were Mandarin Chinese spoken by five female and five male speakers. These speech signals were corrupted by various kinds of additive noise, such as white, F16-cockpit, factory, helicopter-cockpit, and car noise signals which were extracted from the Noisex-92 database. Three input average segmental SNR levels, including 0, 5 and 10 dB, were used to evaluate the performance of a speech enhancement system. A minima-controlled-recursive-averaging algorithm with variable length was performed to estimate the power of noise for each frequency bin [13]. In order to evaluate the performance of the proposed method, the PSS algorithm and the dynamic spectral subtraction (DSS) algorithm [5] were also conducted for performance comparisons.

Table 1 presents the performance comparisons in terms of the average segmental SNR (Avg_SegSNR) improvement which is computed by subtracting the Avg_-SegSNR of noisy speech from that of enhanced speech. The maximal Avg_-SegSNR improvement corresponds to the best speech quality. It can be found that all of the speech enhancement algorithms provide significant Avg_SegSNR improvement in low-SNR environments. The proposed approach can significantly improve the performance of the PSS algorithm. The major reason is attributed to

Table 1 Comparison of SegSNR improvement for the enhanced speech in various noise corruptions

Noise type	SNR (dB)	Average SegSNR improvement		
		PSS	DSS	Proposed
White	0	3.55	6.23	8.05
	5	3.26	5.21	6.09
	10	2.83	4.11	4.13
F16	0	3.64	5.43	6.59
	5	3.03	4.54	5.06
	10	2.56	3.60	3.58
Factory	0	3.72	5.26	6.26
	5	3.16	4.40	4.72
	10	2.60	3.46	3.11
Helicopter	0	3.62	5.50	6.97
	5	3.06	4.75	5.72
	10	2.54	3.77	3.99
Car	0	1.99	3.35	11.31
	5	1.53	2.67	10.48
	10	1.26	2.10	8.34

Table 2 Comparison of perceptual evaluation of speech quality (PESQ) for the enhanced speech in various noise corruptions

Noise type	SNR (dB)	PESQ			
		Noisy	PSS	DSS	Proposed
White	0	1.64	1.85	2.00	2.20
	5	1.94	2.21	2.39	2.59
	10	2.28	2.60	2.77	2.97
F16	0	1.86	2.11	2.22	2.35
	5	2.20	2.48	2.61	2.74
	10	2.56	2.86	2.99	3.13
Factory	0	1.84	2.07	2.17	2.25
	5	2.18	2.47	2.57	2.65
	10	2.55	2.87	2.97	3.02
Helicopter	0	2.05	2.27	2.37	2.45
	5	2.39	2.66	2.78	2.88
	10	2.75	3.04	3.14	3.24
Car	0	3.48	3.33	3.64	3.37
	5	3.86	3.77	3.96	3.78
	10	4.14	3.98	4.14	3.96

that the proposed method can efficiently remove background noise by additional consideration on the cross term between the spectrum of speech and noise; while speech components are well preserved. In addition, the proposed approach outperforms the DSS algorithm, which is a kind of over-spectral subtraction algorithms, in the conditions of heavy noise corruptions (input Avg_SegSNR is less than 5 dB). Table 2 presents the performance comparisons in terms of the perceptual evaluation of speech quality (PESQ). The maximal PESQ score corresponds to the best speech quality. Except the condition of car noise corruption with input Avg_SegSNR = 10 dB, the proposed method outperforms the PSS algorithm. These results ensure that the proposed method can efficiently remove background noise; meanwhile, the speech quality can be maintained at a high level. In addition, the proposed method also outperforms the DSS algorithm in most conditions.

Figure 2 demonstrates an example of waveform plots for performance comparisons. A speech signal uttered by a male speaker was corrupted by white noise with Avg_ SegSNR = 0 dB. By comparing the waveform plots of enhanced speech shown in Fig. 2c, d, the proposed method can significantly improve the performance of the PSS algorithm in the removal of background noise, in particular during speech-pause regions. The proposed method also slightly outperforms the DSS algorithm for the removal of background noise. Although the proposed method can efficiently reduces a quantity of background noise in speech-pause regions, the enhanced speech does not been severely deteriorated during speech-dominant regions. Therefore, the quality of enhanced speech can be maintained at an acceptable level.

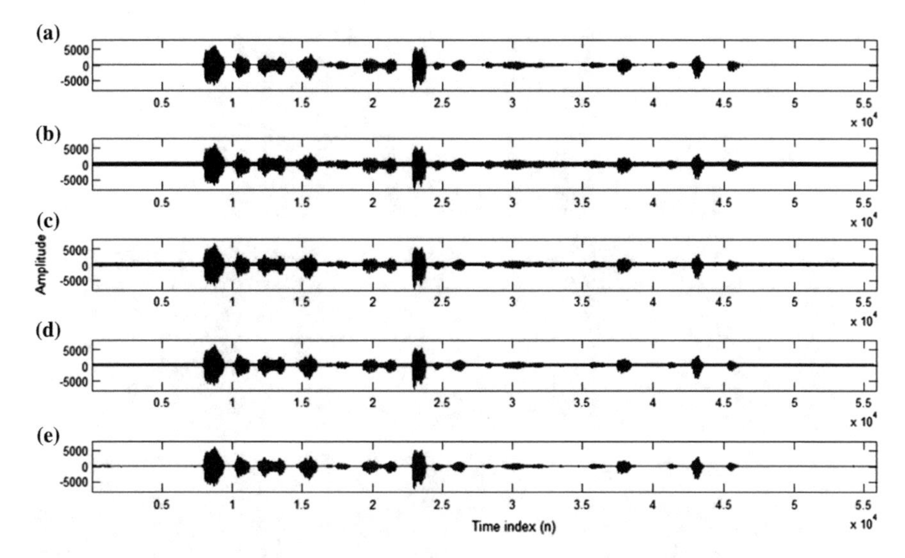

Fig. 2 Example of speech waveform plots. (From *top* to *bottom*) **a** clean speech spoken in Mandarin Chinese by a male speaker, **b** noisy speech corrupted by white noise with an average SegSNR = 0 dB, **c** enhanced speech using the PSS method, **d** enhanced speech using the DSS method, **e** enhanced speech using the proposed method

Figure 3 presents spectrogram comparisons for various speech enhancement methods. A speech signal is corrupted by factory (non-stationary) noise with Avg_SegSNR = 0 dB. By observing the spectrograms of enhanced speech shown in Fig. 3c, e, plenty of musical residual noise exists in the enhanced speech of the PSS algorithm. This noise is very annoying to the human ear. Conversely, the proposed method (Fig. 3e) can significantly improve the PSS algorithm not only in the removal of background noise, but also in the reduction of musical residual noise. Accordingly, the enhanced speech of the proposed method sounds less annoying than that produced by the PSS algorithm (Fig. 3c). The proposed method is superior to the DSS method (Fig. 3d) in the removal of background noise. The quantity of residual noise in the proposed method is less than the DSS method. In addition, the spectrograms also reveal the fine structure of spectra in speech-activity regions. A muffled signal is absent at the output of enhanced each speech enhancement method. By comparing the spectrograms of enhanced speech during speech-pause regions, the proposed method shown in Fig. 3e is better able to remove background and residual noise than the other methods shown in Fig. 3c, d. Moreover, isolated spectral patches are also absent at the enhanced speech of proposed method (Fig. 3e). The proposed method does not suffer from musical effect of residual noise, thus the enhanced speech sounds comfortable.

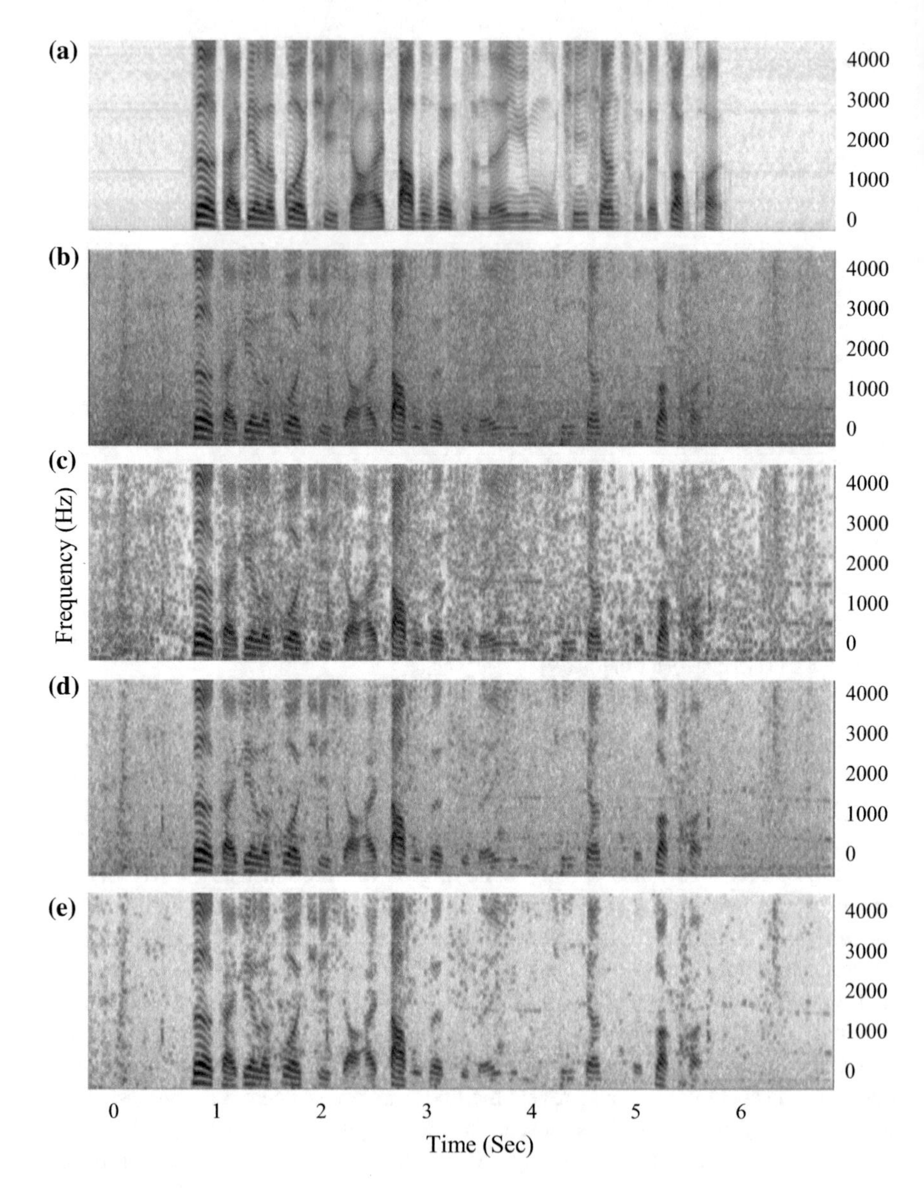

Fig. 3 Spectrograms of speech spoken by a male speaker, **a** clean speech, **b** noisy speech (corrupted by factory noise with average segmental SNR = 0 dB), **c** enhanced speech using the PSS method, **d** enhanced speech using the DSS method, **e** enhanced speech using the proposed method

5 Conclusions

This study proposes using the cross term between the spectrum of speech and noise to be additionally subtracted from the power of noisy speech, enabling background noise to be efficiently removed. Initially, the PSS algorithm is employed to obtain the power spectrum of the pre-processed speech signal. Hence, the cross term between the spectrum of pre-processed speech and estimated noise is computed and subtracted from the power spectrum of noisy speech. A great quantity of residual noise can be efficiently removed. Experimental results show that the proposed method can significantly improve the performance of the PSS algorithm by additionally subtracting the cross term between the spectrum of speech and noise. The quantity of musical residual noise can be efficiently removed, while speech components are well preserved. Concequently, the performance of the PSS algorithm can be significantly improved by the proposed approach.

Acknowledgements This research was supported by the Ministry of Science and Technology, Taiwan, under contract numbers MOST 104-2221-E-468-007, and MOST 104-2628-E-006-012-MY3.

References

1. Lezzoum, N., Gagnon, G., VoixLu, J.: Noise reduction of speech signals using time-varying and multi-band adaptive gain control for smart digital hearing protectors. Applied Acoust., vol. 109, pp. 37–43 (2016)
2. Lee, S., Lim, C., Chang, J. H.: A New a Priori SNR Estimator Based on Multiple Linear Regression Technique for Speech Enhancement. Digital Signal Process., vol. 30, pp. 154–164 (2014)
3. Upadhyay, N.: An Improved Multi-band Speech Enhancement Utilizing Masking Properties of Human Hearing System. In: IEEE International Symposium on Electronic System Design, pp. 150–155 (2014)
4. Lu, Y., Loizou, P. C.: A Geometric Approach to Spectral Subtraction. Speech Commun., vol. 50, pp. 453–466 (2008)
5. Sukanya, S. M., Ananthakrishna, T: Dynamic Spectral Subtraction on AWGN Speech. In: IEEE International Conference on Signal Processing and Integrated Networks, pp. 92–97, 2015
6. Huang, G., Chen, J., Benesty, J.: Investigation of A Parametric Gain Approach to Single-Channel Speech Enhancement. In: IEEE International Conference on Acoustics, Speech, and Signal Processing, pp. 206–210, 2015
7. Virag, N.: Single Channel Speech Enhancement Based on Masking Properties of the Human Auditory System. IEEE Trans Speech Audio Process., vol. 7, no. 2, pp. 126–137 (1999)
8. Lu, C. –T.: Enhancement Of Single Channel Speech Using Perceptual-Decision-Directed Approach. Speech Commun., vol. 53, pp. 495–507 (2011)
9. Griffin, A., Zorila, T. C., Stylianou, Y.: Improved Face-to-Face Communication Using Noise-Reduction and Speech Intelligibility Enhancement. In: IEEE International Conference on Acoustics, Speech, and Signal Processing, pp. 5103–5107, 2015
10. Sun, C., Zhu, Q., Wan, M.: A Novel Speech Enhancement Method Based on Constrained Low-Rank and Sparse Matrix Decomposition. Speech Commun., vol. 60, pp. 44–55(2014)

11. Djendi, M., Scalart, P.: Reducing Over-and-under-Estimation of the a Priori SNR in Speech Enhancement Techniques. Digital Signal Process., vol. 32, pp. 124–136 (2014)
12. Momeni, H., Abutalebi, H. R., Tadaion, A.: Joint Detection and Estimation of Speech Spectral Amplitude Using Noncontinuous Gain Functions. IEEE/ACM Trans. Audio Speech Language Process., vol. 23, no. 8, pp. 1249–1258 (2015)
13. Lu, C. –T., Lei, C. –L., Tseng, K. –F., Chen, C. –T.: Estimation of noise Magnitude Using Minima-Controlled-Recursive-Averaging Algorithm with Variable Length for Speech Enhancement. In: National Symposium on Telecommunications (2015)

An Efficient Three-Dimensional Data Query System to Digital Archive Using API Proxy

Wei-Tsung Su, Ching-Sheng Wang, Wei-Cheng Chen, Siou-Dian Li and Yu-Lin Lin

Abstract The data presentation in digital archive is too complicated to understand for general tourists. Although tourists can easily acquire some information from digital archiving systems (DAS) through location-based services (LBS), they are unable to have an in-deep tour without understanding the data represented in three dimensions as time, space and topic. In recent years, spatial computing is proposed to transform the use of DAS from location-aware into location-understanding. Unfortunately, the application programming interface (API) of most DAS is not intuitive and inefficient to query the data stored in DAS. In this paper, we will present a three-dimensional data query system (3DQS) using the proposed API proxy mechanism. The contributions of proposed 3DQS include (1) providing a friendly API to query three-dimensional data in DAS, (2) caching data being requested, and (3) prefetching data according to the time, space, and topic localities predicted from users' interests.

Keywords Digital archive · Location-based service · Application programming interface · API proxy

1 Introduction

Digital archive is widely employed to preserve the historical data of a cultural heritage in various types of format, such as text, image, video, and even 3D model [1, 2]. Thus, many digital archiving systems (DAS), such as Archivematica, CollectiveAccess, and DSpace [3], are developed to store digitized data objects for long-term access. Researchers can digitize and store the tangible data objects or

W.-T. Su · C.-S. Wang (✉) · W.-C. Chen · S.-D. Li · Y.-L. Lin
Department of Computer Science and Information Engineering,
Aletheia University, New Taipei City, Taiwan, ROC
e-mail: cswang@mail.au.edu.tw

W.-T. Su
e-mail: au4451@au.edu.tw

© Springer Nature Singapore Pte Ltd. 2018

N.Y. Yen and J.C. Hung (eds.), *Frontier Computing*, Lecture Notes in Electrical Engineering 422, DOI 10.1007/978-981-10-3187-8_56

other cultural relics they found into DAS to systematically trace the original face of history back. In the case of our research project, the related cultural relics of George Leslie Mackay, who was the first Presbyterian missionary to Taiwan, has been archived in DSpace by computer scientists. Geographical scientists can add geographical information into archived data objects. Then, cultural scientists can easily query these archived data objects to trace the footprint of George L. Mackay in Taiwan back.

A lots of museums or historical heritage zone provide tourists with web-based or mobile applications interfaced with DAS. Under these applications, the tourists can search some data from DAS according to their interests or to their geographical locations if location-based service (LBS) is supported. However, because the data presentation in DAS is generally too tangled and complicated to understand, the tourists cannot easily acquire the implied information directly from archived data objects. Consequently, they are unable to have an in-deep tour without understanding these archived data objects from various dimensions in time, space and topic. In order to enrich the sematic meaning of archived data objects, metadata [4] is typically applied to describe these archived data objects with supplementary information. Moreover, some DAS provides API to ease the development of applications to query archived data objects. For example, DSpace provides REST API [5] to assist in querying archived data objects. Unfortunately, the native API is not intuitive to application developers. In addition, it is inefficient to query a large number of archived data objects.

In this paper, we will present a three-dimensional data query system (3DQS) to DAS. The contributions of proposed 3DQS include (1) providing a data query interface to query archived data objects in time, space and topic dimensions; (2) supporting data caching to reduce the response time of data query; and (3) exploring the performance improvement of data prefetching according to the time, space, and topic localities predicted from user's interests. The experimental results show that the average response time can be significantly reduced by 60% by data caching. The performance should be further improved if data prefetching can be implemented in 3DQS in the future.

2 Related Work

2.1 Metadata

Metadata is the data which provides information about other data. There are many metadata standards designed and developed for different purposes. For example, IEEE LOM specifies the learning objects, ISO 19115 is designed to describe geographical information, and CDWA is the framework for describing the works of art, and so on.

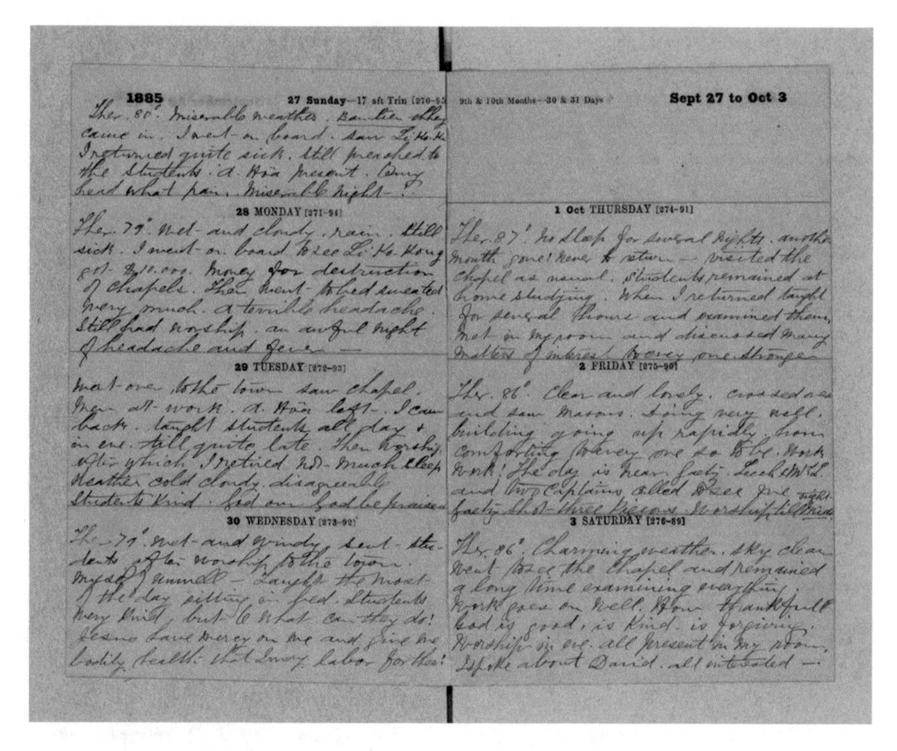

Fig. 1 Diary of George L. Mackay

In our project, Dublin Core [6], which focus on the interoperability of network resources, is utilized to describe the diary of George L. Mackay in Taiwan. As shown in Fig. 1, the tangible diary (Sept. 27 to Oct. 3, 1885) can be easily digitized and stored in DAS. However, the semantic meaning of this tangible diary must be further represented as different fields, such as title, abstract, subject, type, date, and so on, in metadata as shown in Fig. 2. Therefore, metadata can greatly help to filter the archived data objects more precisely using their semantic meaning.

2.2 API Management

Web API is the most common way to access data via Internet. DSpace provides its REST API to increase the possible usage of archived data objects. The application developers can query data using a simple HTTP request. API manager is typically employed to manage web API in an efficient way. The followings are the benefits of using API manager to manage web API. First, data access can be secured by authentication and data encryption mechanism provided by API manager. Second, the usage of web API can be monitored and analyzed for improving next revision of

```xml
<?xml version="1.0" encoding="utf-8" standalone="no" ?>
<dublin_core schema="dc">
<dcvalue element="contributor" qualifier="author">Mackay</dcvalue>
<dcvalue element="identifier" qualifier="other">18851003</dcvalue>
<dcvalue element="description" qualifier="abstract">[abstrat]</dcvalue>
<dcvalue element="subject" qualifier="none">Education</dcvalue>
<dcvalue element="title" qualifier="none">Mackay Diary</dcvalue>
<dcvalue element="type" qualifier="none">Image</dcvalue>
<dcvalue element="date" qualifier="issued">1885-10-03</dcvalue>
</dublin_core>
```

Fig. 2 The semantic meaning of diary of George L. Mackay represented in Dublin Core metadata

Table 1 Comparison of API managers

	WSO2 API manager	API umbrella
Ease of development	Low	Medium
Extensibility	High	Low
User-friendly	High	Low
API analysis	More	Few
Data caching	Content-aware caching	Standard HTTP caching

API. Third, the average response time can be reduced by API manger which supports data caching.

There are several open source API manager projects, such as WSO2 API Manager [7], API Umbrella [8], and so on. The comparison of WSO2 API Manager and API Umbrella is shown in Table 1. Data caching mechanism used in these two API managers is particularly emphasized in this paper. Compared with API Umbrella which provides standard HTTP caching, WSO2 API Manager provides content-aware caching with its own cache mediator which can significantly reduce the average response time.

3 Three-Dimensional Data Query System

3.1 Three-Dimensional Data Query

Spatial computing [9] is an emerging technology to create a new understanding of locations. Under this paradigm, 3DQS attempts to transform the use of DAS from location-aware into location-understanding by providing three-dimensional data query. As shown in Fig. 3, three-dimensional data query allows tourists to freely traverse a historical heritage zone across time, space, and topic. For example, in our project, the use case of using three-dimensional data query is to traverse the

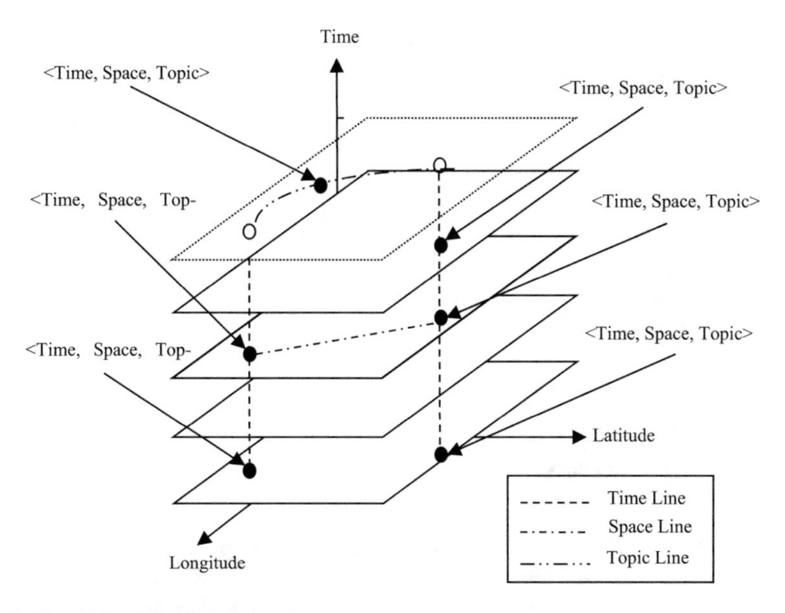

Fig. 3 Three-dimensional data query

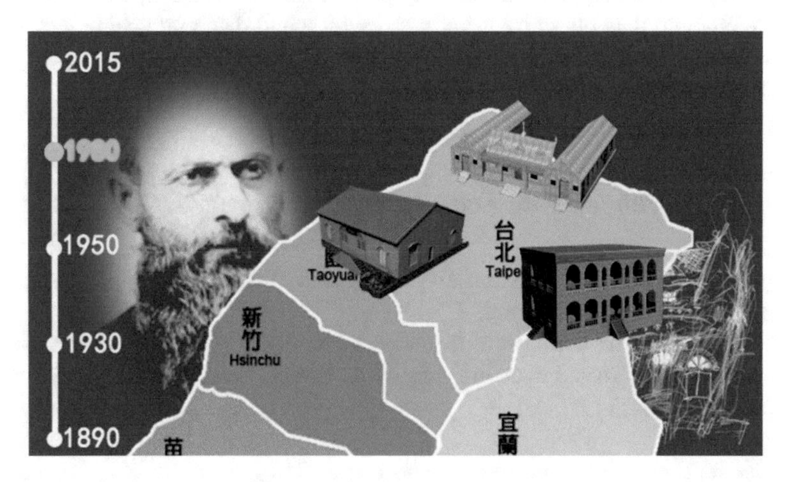

Fig. 4 Use case of three-dimensional data query

footprint of George L. Mackay. As shown in Fig. 4, the tourists can visually see the years in which different buildings, related to George L. Mackay, was completely constructed in Taiwan.

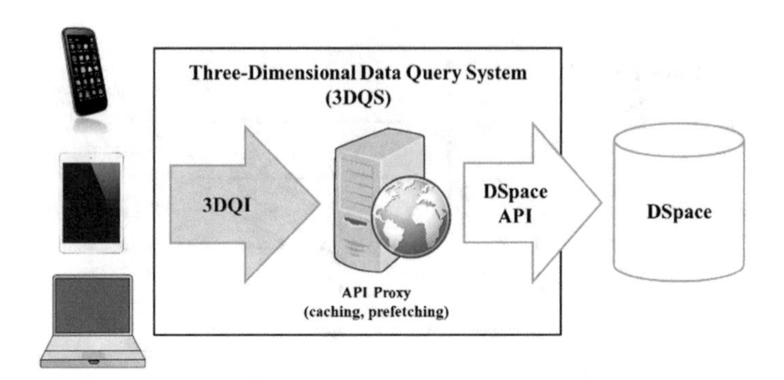

Fig. 5 System overview of 3DQS

3.2 System Overview

As shown in Fig. 5, the proposed 3DQS is consisted of three-dimensional data query interface (3DQI) and API proxy. In 3DQI, a set of API, supporting three-dimensional data query, is provided to application developers. With 3DQI, the developers can intuitively query archived data objects using their semantic meaning without understanding the metadata used in DAS. In addition, API proxy can provide efficient data query by using data caching and data prefetching techniques based on API manager as described above. The details of 3DQI and API proxy will be introduced in the following paragraphs.

3.3 Three-Dimensional Data Query Interface (3DQI)

The proposed 3DQI has two advantages, (1) data hiding and (2) process encapsulation, as described as follows.

Data Hiding. The application developers can query the archived data objects with specific topic (e.g. Education) via DSpace API directly using the following request.

```
POST /rest/items/find-by-metadata-field
[{"key":"dc.subject","value":"Education"}]
```

Prior to submit this request, the developers must know that the metadata standard is Dublin Core and the field 'subject' indicates the topic of this archived data object.

With 3DQI, the developers can submit the same query using the following request.

```
POST /Mackay/items
[{"key":"topic","value":"Education"}]
```

Thus, the developers do not need to know that the metadata standard used in DSpace is Dublin Core. In addition, the same 3DQI request can be used to query the archived data objects from other data sources instead of DSpace.

Process Encapsulation. In the case that the developers attempts to query the archived data objects with date between 20-Feb-1900 and 12-Jan-1985. Actually, the developers cannot finish this data query in a single request via DSpace API. In order to finish this data query, all data objects, with expanded metadata, must be first obtained. Then, the filter must be performed to retrieve the data objects with metadata filed 'date' in the specified range. This complicated process will be a burden to applications developers. With 3DQI, the developers can finish the same query using a single request as follows.

```
POST /Mackay/items
[{"key":"start_date","value":"20-Feb-1900"},
{"key":"end_date","value":"12-Jan-1985"}]
```

Once receiving this 3DQI request, the proposed 3DQS will execute the above process and return the final results to the developers directly.

As a result, 3DQI can help the developers to easily query the archived data objects in time, space, and topic dimensions. Thus, the developers can focus on the application itself without understanding the specific data query techniques for different data sources.

3.4 API Proxy

In 3DQS, API proxy is responsible for data caching and data prefetching. Data caching is a passive approach to cache data on the basis of API request [10]. In contrast, data prefetching is an active approach to cache data on the basis of prediction of user's interests [11]. The system architecture of API proxy is shown in Fig. 6. API proxy is consisted of two components, data cache manager and data locality analyzer, which are responsible for data caching and data prefetching, respectively.

Data cache manager applies the data caching mechanism provided by the API manager. According to our survey, API Umbrella and WSO2 API Manager both provide data caching. In API Umbrella, standard HTTP caching layer is provided in front of API. However, WSO2 API Manager caches data using its content-aware cache mediator. In addition, the in-memory implementation is supported to significantly reduce the response time once data hits in WSO2. Therefore, in 3DQS,

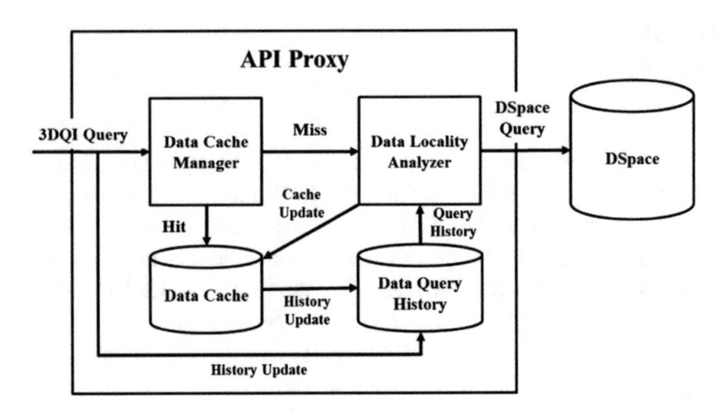

Fig. 6 System architecture of API proxy

WSO2 API Manager is employed to be the basis of API proxy. As shown in Fig. 6, when the developers submit a 3DQI request, data cache manager will first check if the data exists in data cache. If cache hit, the data will be acquired from data cache directly to reduce the response time. Otherwise, the request will be passed to data locality analyzer.

Data locality analyzer will pre-fetch the data according to the time, space, and topic localities determined by user's data query history. For example, if a tourist queries the data in 1980, there is high probability that this tourist may query the data in 1981 in the next query. Thus, data locality analyzer can pre-fetch the data in 1981 in this query. If the data localities can be well predicted using soft computing techniques [12], the cache hit rate can be increased. Consequently, the average response time can be further reduced.

4 Performance Evaluation

In this section, we will show that the performance improvement by using data caching in 3DQS. Each API request was performed 100 times to obtain the average response time. In Fig. 7, the average response time to access different numbers of data objects using DSpace, API Umbrella, and WSO2 API Manager without data caching is compared. Since API Umbrella has built-in standard HTTP caching, the experimental result shows that the average response time can be even shorter than directly using DSpace API. In contrast, the average response time of WSO2 API Manager is higher because of the indirect data query.

After enabling the data caching in WSO2 API Manager, the average response time can be significantly reduced by 60%, as shown in Table 2, if the same API request is performed. However, because the data caching is on the basis of API request, the average response time is not obviously improved if different API requests are performed in the experiments.

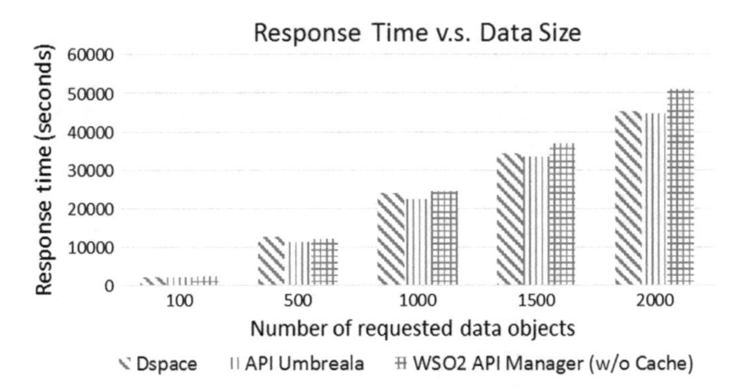

Fig. 7 Performance comparison among different API managers

Table 2 WSO2 API manager with/without caching

Number of data objects	100	500	1000	1500	2000
WSO2 API manager	2,562	12,194	24,440	36,833	51,009
WSO2 API manager (cache)	110	179	404	698	34,837

Acknowledgements This paper is based on the work supported by the Ministry of Science and Technology, Taiwan, R.O.C., under grants MOST104-2632-M-156-001 and MOST104-2221-E-156-007.

References

1. Q. L. Nguyen, A. Lake and M. Huber, “Evolvable and scalable system of content servers for a large digital preservation archives,” in 4th Annual IEEE Systems Conference, San Diego, CA, 2010.
2. D. Sacher, D. Biella and W. Luther, “Towards a versatile metadata exchange format for digital museum collections,” in Digital Heritage International Congress, Marseille, 2013.
3. “DSpace,” [Online]. Available: http://www.dspace.org/.
4. “Meradata,” [Online]. Available: https://en.wikipedia.org/wiki/Metadata.
5. “DSpace REST API,” [Online]. Available: https://wiki.duraspace.org/display/DSDOC5x/REST+API.
6. “Dublin Core,” [Online]. Available: http://dublincore.org/.
7. “WSO2 API Manager,” [Online]. Available: http://wso2.com/api-management/try-it/.
8. “API Umbrella,” [Online]. Available: http://apiumbrella.io/.
9. S. Shekhar, S. K. Feiner and W. G. Aref, “Spatial Computing,” Communications of the ACM, vol. 59, no. 1, pp. 73–81, 2016.
10. I. Ollite and N. Mohamudally, “Performance analysis of a 2-tier caching proxy system for mobile RESTful services,” in IEEE International Conference on Computer as a Tool (EUROCON), Salamanca, 2015.

11. Y. Chen, S. Byna and X. H. Sun, "Data access history cache and associated data prefetching mechanisms," in ACM/IEEE Conference on Supercomputing, Reno, NV, USA, 2007.
12. D. Guttman, M. T. Kandemir, M. Arunacgalam and R. Khanna, "Machine learning techniques for improved data prefetching," in International Conference on Energy Aware Computing Systems & Applications (ICEAC), Cairo, 2015.

Effects of Design Group Students' Thinking Style on Team Collaboration

Weiting Hsiao and Wenlung Chang

Abstract In adolescence adolescents are facing a big transformation both in psychology and physiology, they need time to adapt during the course of transformation, and they are prone to generate negative thinking disposition if not well adapted, thereby affecting subsequent life. Based on the above critical reason, this study aims at describing and explaining effects of present situation and characteristics of design group students' positive, negative thinking style on team collaboration disposition. This study adopted questionnaire method to collect data from design group students of a senior vocational school in central region, and a total of 157 valid student questionnaires were collected; used SPSS 18.0 to conduct Pearson correlation analysis and multiple regression empirical analysis, and carried out estimation and verification for structural relationships between each group of variables. The study results show that: (1) design group students' positive thinking and negative thinking have a significant negative correlation, (2) design group students' positive thinking and negative thinking respectively has a significant relationship with team collaboration disposition; (3) positive, negative thinking style has a significant predictability for team collaboration disposition.

Keywords Positive, negative thinking style · Design group *students*' · Team collaboration

W. Hsiao
Overseas Chinese University, Taichung, Taiwan
e-mail: weiting@ocu.edu.tw

W. Chang (✉)
Changhua, Taiwan
e-mail: uenlung@chsc.tw

© Springer Nature Singapore Pte Ltd. 2018

N.Y. Yen and J.C. Hung (eds.), *Frontier Computing*, Lecture Notes in Electrical Engineering 422, DOI 10.1007/978-981-10-3187-8_57

1 Study Motivation and Purpose

Because in adolescence adolescents are facing a big transformation both in psychology and physiology, they need time to adapt during the course of transformation, and they are prone to generate negative thinking disposition if not well adapted, thereby affecting subsequent life. Therefore, this study would like to understand the present situation and characteristics of design group students' positive, negative thinking style, and this is the first purpose of this study.

In addition, during the course of team collaboration and project promotion, emotional control ability of members is also an important factor affecting team performance [1], therefore, this study intends to analyze the relationships between design group students' positive, negative thinking style and team collaboration disposition, and this is the second purpose of this study.

Especially when an individual joins a collaborative team, both positive emotions generated through optimism and negative emotion generated through conflict would affect team performance [2]. In short, team members' positive and negative thinking disposition affect the team's collaboration disposition. Therefore, this study expects to explore the connection between both, and this is the third purpose of this study. In conclusion, our main study purposes are as follows:

1. Understanding present situation and characteristics of design group students' positive, negative thinking style.
2. Analyzing relationship between positive, negative thinking style and team collaboration disposition.
3. Exploring effects of team members' positive, negative thinking style on the team's collaboration disposition.

2 Literature Review

Whether in terms of general psychology or educational psychology, study for "self" has always been one of the important study topics since ancient times. However, the study for self is quite difficult. The reason is that self can not be measured from external observation, and can only rely on oneself experience. Therefore, this chapter will respectively explain the meanings of "definition and course of thinking", "thinking style", "Internal and external factors affecting thinking" and "collaboration disposition", and use them to construct study framework.

3 Definition and Course of Thinking

What is thinking? "Thinking" is using our mind to feel, to study and uptake knowledge or understand knowledge. Thinking is also the course and activity of an individual in using intelligence, basing on current information and existing knowledge and experience, to engage in problem solving and new knowledge exploring. Simply speaking, when an individual received any message, separate independent messages will be processed to bring out correlation and meaning among them, and in the course of processing messages, the individual is in a thinking state.

In summary of various arguments, the researcher of this study attempted to make the following definition for thinking and thinking course: "the course of thinking is —after an individual discovered a problem—a process of using existing knowledge and experience as a basis and applies assistance from understanding ability and existing information to integrate sub-thinking systems of viewpoints and knowledge which may mutually conflicting or contrasting to engage in root cause exploring and problem solving."

4 Thinking Style

4.1 Characteristics and Effect of Positive Thinking

Positive thinking refers to that an individual holds a positive, optimistic and aggressive belief for the interpretation of self and surrounding people and things. This allows individuals to transform irrational belief triggered by negative events to rational belief, adjust cognitive model and flexibly think solving method when encountered a setback, and is beneficial to adaptation of personal life. Positive thinking tendency is helpful to positive emotional feeling, allows a happier and healthier feeling in the face of all things, and easier to explain things with uncertainty from a positive viewpoint and a more optimistic thinking. In addition, they are also more enthusiastic, vital, active and alert, have more self-efficacy, and would become involved with more pleasant emotion while in the course of interpersonal interaction and goal achieving [3, 4].

4.2 Characteristics and Effect of Negative Thinking

Positive thinking refers to that an individual holds a negative, pessimistic and passive belief for the interpretation of self and surrounding people and things. Crandell and Chambless (1986) pointed out that negative thinking (including separation, low self-assessment, helplessness and hopelessness) is a bias of cognitive content, and can be used as an identifying factor for depressive and non-depressive populations. Norem and Chang (2002) expressed that negative thinking is pessimistic, and pessimism is a quality linking to more negative results; negative thinking people prefer to make unrealistically low expectation for the situation they face, and then reduce their activity through psychological play or reflection. Therefore, negative thinking is a process of responding activity, it is prone to generate negative thinking (Sugiura 2002) and also easy to cause occurrence of depression if using negative coping strategy in the face of stress source.

Summing up the above literature and study results, the researcher believes that thinking disposition is subject to personal intrinsic disposition, relevant experiences and external stimulus affairs and thus has different tendencies. Generally speaking, adolescents' thinking disposition is more sensitive and makes them having positive thinking dispositions of curiosity, enthusiasm and vitality, and courage to ask question or propose dissatisfaction when facing various affairs, and critical attitude when facing challenges.

4.3 Internal and External Factors Affecting Thinking (Positive and Negative Orientation)

As mentioned in previous section, positive, negative thinking refers to explaining belief for self and surrounding people and things of an individual, and can also be referred to as attitude. Many scholars believe that attitude is composed by three factors of cognition, affection and behavior. The so-called cognition factor refers to an individual's knowledge, belief, values and imagery for situation and things, but is not involved with personal subjective feelings. While affection factor refers to emotional reaction to object of attitude, is personal emotional views on things, including emotion and feeling for specific object of attitude; and behavior factor refers to adopted action or behavior for object of attitude. Therefore, this study attempts to divide internal and external factors affecting thinking into internal physiological factor inherently born and external emotional qualities learned. In other words, the positive thinking disposition can be developed.

4.4 Collaboration Disposition

Research shows that collaboration is voluntary, interactive, ongoing, inclusive, and requires commitment to a common goal (Friend and Cook). Friend and Cook assert collaboration is a means for solving problems and obtaining goals through a voluntary process whereby two or more stakeholders come together as equally valued participants to work on a mutual goal.

The purposes of collaborating are many and varied, but concerns about learning through collaboration have become increasingly prominent. Assimakopoulos and Macdonald (2003), focusing on networks of innovation, examine informal, social learning processes but do not explicitly consider the attitudes that influence these processes. Much of the literature addresses attitudes to learning in either competitive or collaborative frames.

Attitudes have so far been described as though they operate at a macro-level, throughout the life of a collaboration. Given the inherently dynamic nature of all aspects of collaboration, attitudes may also change over time at the macro level, as, for example, individuals get to know each other, early successes lead to trust development, or changes of personnel undermine it. Attitude refers to the positive or negative feelings derived from knowledge sharing.

4.5 *Study Method and Implementation Course*

This study adopted questionnaire method to collect data, the objects of study is design group students of a senior vocational school in Taiwan central region, it was expected to issue 168 student questionnaires, and survey time is from January 2016 to February 2016. The preparation of the study tool is based on and modified from Wu Ming-Chang and Wang Xiao-Tien (2011): "Present College Student's Idea and Life Experience Scale"; after adopted SPSS 18.0 to conduct statistical analysis and validation on survey data and ensured correctness of original data, used Pearson correlation analysis and multiple regression empirical analysis to carry out statistics for linear relationships between each group of variables in the model of this study, and to present structural relationships between positive, negative thinking style and team collaboration disposition. In the followings, this chapter will focus on explaining: study framework and hypotheses, study object and data collection, study tool and modification as well as data analysis method.

4.6 *Study Framework and Hypotheses*

Based on the abovementioned motivation and purpose, this study proposed the linear relationship model between three variables, namely "positive thinking", "negative thinking" and "team collaboration disposition" (Fig. 1).

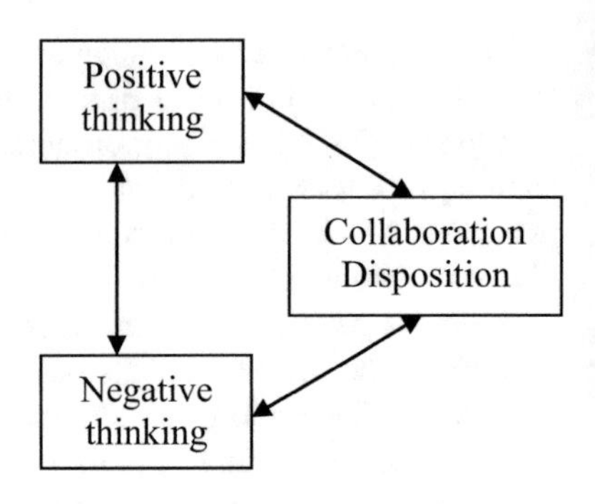

Fig. 1 Study framework

For exploring the affecting relationships between "positive thinking", "negative thinking" and "team collaboration disposition" of design group students' team members, this study set up the following hypotheses:

Hypothesis 1: Design group students' "positive thinking" and "negative thinking" is relevant.
Hypothesis 2: Design group students' "positive, negative thinking" and "team collaboration disposition" is relevant.
Hypothesis 3: Design group students' "positive, negative thinking" would directly affect "team collaboration disposition".

4.7 Study Object and Data Collection

This study adopted purposive sampling approach to conduct measuring work for design group students in a senior vocational school of Taiwan central region by class, collected 168 questionnaires from four classes (42 students per class) of second and third year design group students in the school, and removed invalid questionnaires with regular answers and too many missing questions, and a total of 157 valid questionnaires were collected, the validity rate is 93.5%, of which number of boy samples is 33 and number of girl samples is 124. The survey was conducted from January 2016 to February 2016 (Table 1).

Table 1 Basic information of samples

Variable	Category	Number of samples	Percentage (%)
Age	16	20	12.7
	17	85	54.8
	18	52	33.1
Gender	Male	33	21.0
	Female	124	79.0

4.8 Study Tool and Modification

This study utilized questionnaire survey method, and since the purpose of this study is for describing and explaining present situation and characteristics of design group students' positive, negative thinking, so adopted self-report inventory as main tool. The preparation of the study tool is based on and modified from Wu Ming-Chang and Wang Xiao-Tien (2011): "Present College Student's Idea and Life Experience Scale". All questions utilize Likert Scale as measuring scale, the subjects can tick a answer from "strongly disagree" to "strongly agree" for each question, and a score of 1–5 is give to each answer in sequence. The basic meaning and structure of this scale consists of three main dimensions, respectively is positive thinking, negative thinking, and exploring of effect on team collaboration disposition. The preparation of various dimension questionnaires had made reference to review and assessment of foreign and domestic relevant scales, and modified for this study.

(1) Positive thinking dimension adopted study scale of Ingram and Wisnicki (1988), a total of 22 questions; constructed four content dimensions, respectively are 7 questions for "positive expectation for future development", 6 questions for "positive self-assessment", 5 questions for "positive thinking for daily life", and 4 questions for "comprehensive self-assessment", mainly for exploring whether design group students hold positive idea for these four content dimensions.
(2) Negative thinking dimension adopted study scale of Hollon and Kendall (1983), a total of 25 questions; constructed four content dimensions, respectively are 10 questions for "negative self-expectation", 7 questions for "powerlessness of give-up, hopelessness", 5 questions for "not adaptive to present situation and want to change", and 3 questions for "low self-esteem", mainly for exploring whether design group students hold negative idea for these four content dimensions.
(3) Team collaboration disposition dimension adopted study scale of Jian Zhong Wang (2001), a total of 11 questions, to explore whether design group students' positive, negative thinking will affect team collaboration disposition.

The reliability of this study questionnaire was assessed by using Cronbach's α coefficient. The reliability of three scales of positive thinking, negative thinking, and team collaboration disposition are respectively 0.905, 0.968, and 0.937, all achieve good level, indicating that the measuring tool (questionnaire) of this study has trusted reliability level.

4.9 Data Analysis Method

After utilized SPSS 18.0 to carry out statistical analysis and validation on survey data and ensured correctness of original data, this study used descriptive statistical procedure to conduct frequency distribution analysis for basic information of study objects. This study then used Pearson correlation analysis and multiple regression empirical analysis to carry out statistics for linear relationships between each group of variables in the model of this study, and to present structural relationships between positive, negative thinking and team collaboration disposition, and carried out statistical test for each hypothesis proposed in this study.

5 Study Results

5.1 Results of Correlation Analysis

This study adopted Pearson correlation analysis to verify the relationships between dimensions, and used standardized coefficients to estimate correlation values (Table 2) between various dimensions, the final model paths are shown in Fig. 2. The path analysis coefficients of the study model reveal: (1) positive thinking and negative thinking have a negative and significant relationship ($\gamma = -0.726$, ** $p < 0.01$); verified that study hypothesis 1 is valid, showing that when an individual shows positive thinking, will also show negative thinking. (2) Positive thinking and self-perceived team collaboration disposition have a positive and significant relationship ($\gamma = 0.600$, ** $p < 0.01$), while negative thinking and team collaboration disposition have a negative and significant relationship ($\gamma = -0.432$, ** $p < 0.01$); verified that study hypothesis 2 is valid.

Table 2 The correlation matrix of latent variables

	Positive thinking	Negative thinking	Collaboration disposition
Positive thinking	1		
Negative thinking	-0.726^{**}	1	
Collaboration disposition	0.600^{**}	-0.432^{**}	1

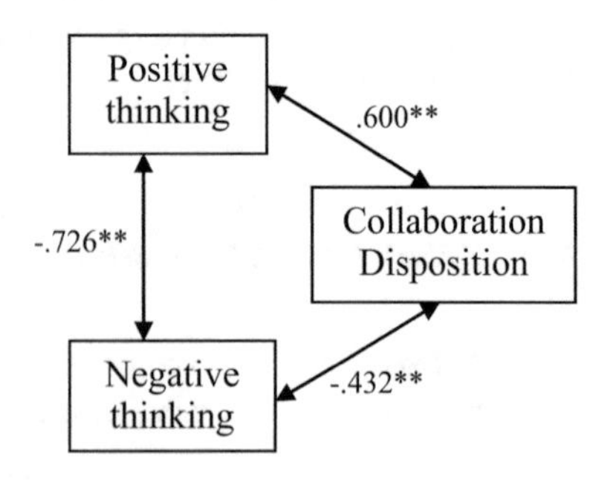

Fig. 2 Relevant structure model of positive thinking and negative thinking versus team collaboration disposition

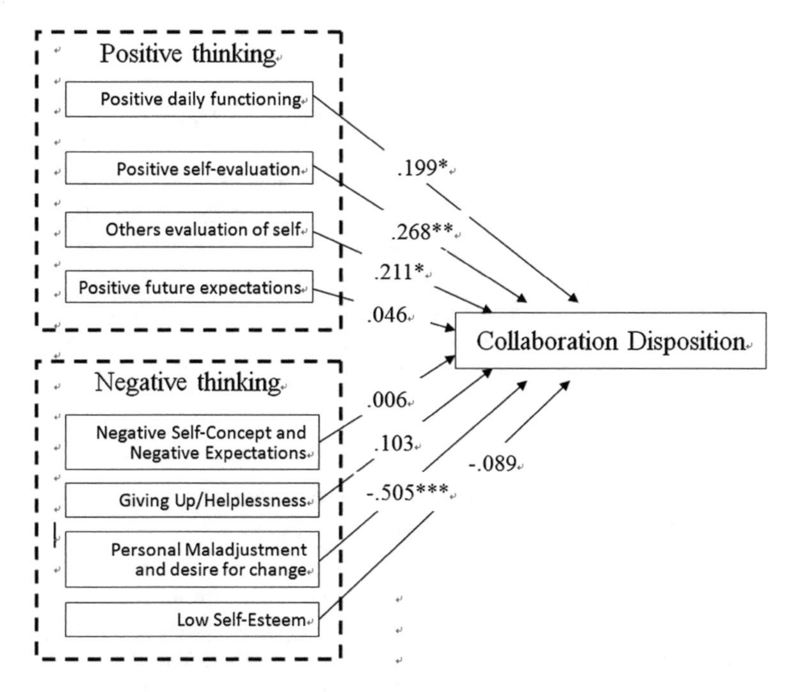

Fig. 3 Study structure model

5.2 Empirical Results of Multiple Regression Analysis

This study adopted multiple regression analysis to verify effects of various content dimensions of positive, negative thinking on team collaboration disposition, and used standardized coefficients to estimate effect values of various content dimensions, the final model paths are shown in Fig. 3. From the above results, verified that study hypothesis 3 is valid.

6 Discussions, Conclusions and Recommendations

This study examined affecting factors of design group students' positive, negative thinking on team collaboration disposition, and used the results to analyze the effectiveness and structure model of the affecting factors. The analysis results in this study lead to some interesting issues for discussion.

6.1 Discussions

6.1.1 Coexistence of Positive Thinking and Negative Thinking

The results of this study are consistent with the views of past scholars, that is, positive, negative affection disposition are independent personality dispositions, and a person may have both high positive affection disposition and negative affection disposition at same time, or have both low positive, negative affection disposition at same time [5]. Therefore, this study suggests that design group students should actively understand and accept the fact of having both positive and negative thinking, and then try to play more positive thinking pattern to replace negative thinking pattern.

6.1.2 Correlation of Positive Thinking and Negative Thinking Versus Team Collaboration Disposition

According to the study results, it can be seen that design group students' positive thinking and self-perceived team collaboration disposition have significant positive relationship. While negative thinking and team collaboration disposition have significant negative relationship. In other words, the more the design group students' positive thinking, the more the design group students' team collaboration disposition; the more the design group students' negative thinking, the less the design group students' team collaboration disposition. As mentioned above in this study, because positive thinking individuals mostly hold a positive, optimistic and aggressive belief for the interpretation of self and surrounding people, things and objects, and thus also would be involved with more pleasant emotion while in the course of interpersonal interaction and goal achieving. Therefore, compared to negative thinking individuals holding negative, pessimistic, passive beliefs, their team collaboration disposition can be better predicted.

6.1.3 Effects of Content Dimensions of Positive Thinking and Negative Thinking on Team Collaboration Disposition

This study aims at exploring affecting factors of team collaboration disposition, the study results reveal that three variables ("positive thinking for daily life", "positive self-assessment" and "comprehensive self-assessment") in various content dimensions have positive and significant direct effect on "team collaboration disposition". It is noteworthy that only the variable "not adaptive to present situation and want to change" in four content dimensions has negative and significant direct effect on "team collaboration disposition". This finding seems to indicate that:

i. A person with "positive thinking" always thinks in the direction of problem-solving, is committed to solve problems for him-/herself and help others, understand his/her and the other party's needs, provide his/her own knowledge and experience and invite the other party to jointly conceive method for solving problem, these approaches form team collaboration and thus achieve a win-win outcome, naturally will have a significant effect on "team collaboration disposition".

ii. In the content dimensions of "negative thinking", only the variable "not adaptive to present situation and want to change" has negative and significant direct effect on "team collaboration disposition". This finding is in line with the conclusion of Zing Zhou (2007), reveals that a negative attitude in the workplace may be an effective catalyst for improvement and progress. Which even points out: "the dissatisfaction for maintaining status quo and present situation of things would encourage people to develop ideas and find solutions." That is, performance of negative thinkers may be better than positive thinkers, because they would expect problems and avoid mistakes in order to more likely make plan to do things right and do a good job rather than proceeds with relying on feel good about themselves.

6.2 Conclusions

Through exploring design group students' positive, negative thinking, this study examined eight factors affecting team collaboration disposition. With regard to the affecting effectiveness of these content dimensions, this study finally obtained the following three conclusions:

(1) Design group students' positive thinking and negative thinking have a negative and significant relationship, reveals that when an individual shows positive thinking, he/she may also show negative thinking at the same time;
(2) There exist significant relationships among positive thinking, team collaboration disposition and negative thinking, reveal that the more the design group students' positive thinking, the more the design group students' team collaboration disposition; the more the design group students' negative thinking, the less the design group students' team collaboration disposition.
(3) Negative thinking has no significant predictability for team collaboration disposition, and only the "not adaptive to present situation and want to change" in its four content dimensions reaches negative and significant direct effect on "team collaboration disposition". This symbolizes that negative thinking may be an effective catalyst for improvement and progress, it would make an individual take into account the community honor and group performance and thus improve team collaboration disposition.

6.3 Recommendations

The main purpose of this study expects to understand whether positive, negative thinking disposition would affect design group students' performance of team involvement attitude. Since in the past most studies regarding thinking tend to explore single dimension thinking disposition, however, positive, negative thinking are independent personality dispositions, so the help for present teaching situation is actually quite limited if explores students' team involvement attitude only from a single dimension of thinking style, therefore this study adopted two thinking style, not only can compensate for the deficiencies of the past studies in this field, but also try to increase the accumulation of adolescent thinking behavior theory. Based on the conclusions of this study, thereby proposed the following recommendations:

(1) In psychology, the most common method used by researchers of self concept is usually self-report inventory. The content includes a number of questions describing characters, and a subject is only required to answer those questions suitable to personal circumstances when used. In the future, more variables of cooperative behavior aspects can be increased for exploring, and qualitative study methods such as interview can be added to provide more information in order to achieve a deeper understanding.
(2) For a theoretical study targeting on positive, negative thinking and team collaboration disposition, in the course of this study there are certain limitations and expectations for future study; for the convenience of sampling and issuance of questionnaires, the samples of this study are only limited to design group students of a senior vocational school in central region, so there may be environmental and cultural characteristics and limitations, in order to get more diversified understanding and discovery on this study topic, future study subjects should be able to expand to northern or southern schools and adolescent groups of different learning backgrounds.
(3) In addition, number of possible variables affecting team collaboration disposition are many, this study only explored effects of positive, negative thinking on design group students' team collaboration disposition from individual perspective, therefore, it is recommended that follow-up studies may include different contextual variables, and even can add school teacher's perspective and adopt field case observation method to explore more deeply.
(4) This study has confirmed that positive, negative thinking of team members serve as antecedents in team collaboration; like other workplaces, in order to develop positive thinking and transform negative thinking, schools should be committed to create a positive atmosphere. If school curriculum design can cultivate positive emotions and create a positive thinking environment to help design group students in establishing positive emotional intelligence, mutual trust among members would be enhanced and team collaboration could be recognized, thereby members would be willing to provide better decision-making and more creative solutions to achieve good collaboration performance.

References

1. Birx, E., Lasala, K.B., & Edd, W.M. (2011). Evaluation of a team building: Retreat to promote nursing faculty cohesion and job satisfaction. *Journal of Professional Nursing*, *27*, 174–178.
2. Jordan, M. H., Field, H. S., & Armenakis, A. A. (2002). The relationship of group process variables and team performance: A team-level analysis in a field setting. *Small Group Research*, *33*, 121–150.
3. Watson, D., & Clark, L. A. (1984). Negative affectivity: The disposition to experience aversive emotional states. *Psychological Bulletin, 96*(3), 465–490.
4. Williams, S., Zainuba, M., & Jackson, R. (2003). Affective influences on risk perceptions and risk intention. *Journal of Managerial Psychology, 18*(1/2), 126–137.
5. George, J.M., & Zhou, J., 2007. Dual tuning in a supportive context: joint contribution of positive mood, negative mood, and supervisory behaviors to employee creativity, *Academy of Management Journal, 50*(3), 605–622.

An Augmented Reality Mobile Navigation System Integrating Indoor Localization and Recommendation Mechanism

Ching-Sheng Wang, Chien-Liang Chen and Shih-Han Chen

Abstract Several well-known museums and researchers have developed mobile navigation systems to improve the navigation effect. But most of those systems are lack of efficient recommendation and indoor localization function. This situation causes the difficulty for visitors to have an overview and a flexible navigating experience. In this paper, an augmented reality mobile navigation system that supports indoor localization and real-time recommendation service is proposed. It aims at providing personalized suggestions corresponded to user's context through the analysis of interest and location information of visitors. In addition, accompanied by the augmented reality display function, this system provides diversified 3D and multimedia navigation information, so as to enrich and improve user's navigation experience effectively.

Keywords Augmented reality · Mobile navigation · Indoor localization · Recommendation · Positioning · iBeacon

1 Introduction

When visiting a museum, as most navigation services nowadays lack an accurate Indoor localization function, in large exhibition places with spatial crossing, visitors frequently find it hard to know where they are and have to shuttle back and forth among the exhibition spaces and exhibits aimlessly, which is a waste of time. Besides, as the introduction and contents about general exhibits are limited, visitors often have to interpret the connotations of the exhibits by relying on their knowl-

---

C.-S. Wang (✉) · C.-L. Chen (✉) · S.-H. Chen
Department of Computer Science and Information Engineering,
Aletheia University, New Taipei City, Taiwan, ROC
e-mail: cswang@mail.au.edu.tw

C.-L. Chen
e-mail: au5199@mail.au.edu.tw

© Springer Nature Singapore Pte Ltd. 2018
N.Y. Yen and J.C. Hung (eds.), *Frontier Computing*, Lecture Notes
in Electrical Engineering 422, DOI 10.1007/978-981-10-3187-8_58

edge in order to self-determine the relevance between works. As a result, museum tours tend to be superficial.

Over the past few years, with the prevalence of smart mobile devices and progress of positioning technologies, several well-known museums and researchers have developed mobile navigation systems to improve the navigation experience [1–4]. The Brighton Museums App provides navigation information of five museums in Brighton. In addition to providing region maps, historical pictures, special activities and the updated news about the museums [1]. The American Museum of Natural History App provides pictures and textual introduction about the exhibits by using the visitor's location via the Wi-Fi positioning function and the navigation of the move lines inside the museum [2]. The Giuseppe Castiglione App of Taiwan's National Palace Museum provides a rich amount of navigation information and supports a positioning function dominated by GPS and passive RFID positioning [3]. Besides, the Street Museum App rolled out by the Museum of London can superimpose the museum's historical pictures onto the corresponding streetscapes to present a special effect of interlacement of history, which is presented through the GPS positioning function [4]. But, the positioning mechanisms mentioned therein are all relatively simple, and the functions provided are still limited.

Moreover, although the above-mentioned mobile navigation systems can provide diversified multi-media navigation information, the users often cannot effectively absorb such a large amount of navigation information. Therefore, using a recommendation system to provide practical and useful information can be very helpful to the user [5]. A number of scholars have developed practical recommendation systems for mobile devices. For example, Willem et al. used the collaborative recommendation mechanism to recommend the artifact information and routes that may interest the users by using the user preferences for artifacts as the recommendation indicator [6]. A restaurant recommendation system that can operate on mobile devices is proposed, which can recommend restaurants in the user's area after the analysis of the user's personal data and GPS location, using a Bayes network as the main structure [7]. Another mobile restaurant recommend system using collaborative filtering is designed to recommend suitable restaurants to a user and provide the navigation routes to the recommended restaurant according to the user's GPS location [8].

The above studies have all proved the feasibility of indoor localization and recommendation function for mobile navigation system, but the functions provided are still limited. Compared with those positioning systems at well-known museums, the system developed herein is able to achieve the accurate Indoor localization, efficient recommendation and augmented reality display function, thus improving the navigation effect and experience. The remainder of this paper is organized as follows: Sect. 2 introduces the main framework of the system; Sect. 3 presents of the proposed iBeacon Indoor localization mechanism; Sect. 4 describes the specification of recommendation mechanism; Sect. 5 shows the implementation and demonstration of system; Sect. 6 provides the conclusion of this paper.

2 System Overview

Figure 1 illustrates the system's architecture, which mainly composed of the four parts: AR navigation interface, navigation information management system, the iBeacon Indoor localization system, and the recommendation system. When using this system for the first time, the users need to fill in their basic personal information and interests, to be used as the basis for the analysis of the explicit behaviors by the recommendation system. Meanwhile, the system can preliminarily recommend exhibits and tour routes that are consistent with the interests of the user. In addition, the system can continuously record the location and time of the users by the positioning function, as the basis for the analysis of the implicit behaviors by the recommendation system before updating the recommended exhibits and tour routes through integration with the original interests.

After the navigation system receives the recommendation results, it can summarize and send back relevant navigation information to the user's mobile device. The users can continuously assess the preferences of the exhibits during the visit, allowing for more detailed analyses of the exhibits and interests, as well as real-time updating of follow-up exhibits and tour routes. Hence, with the increasing amount of assessment information, the accuracy of the recommendation system will be gradually improved.

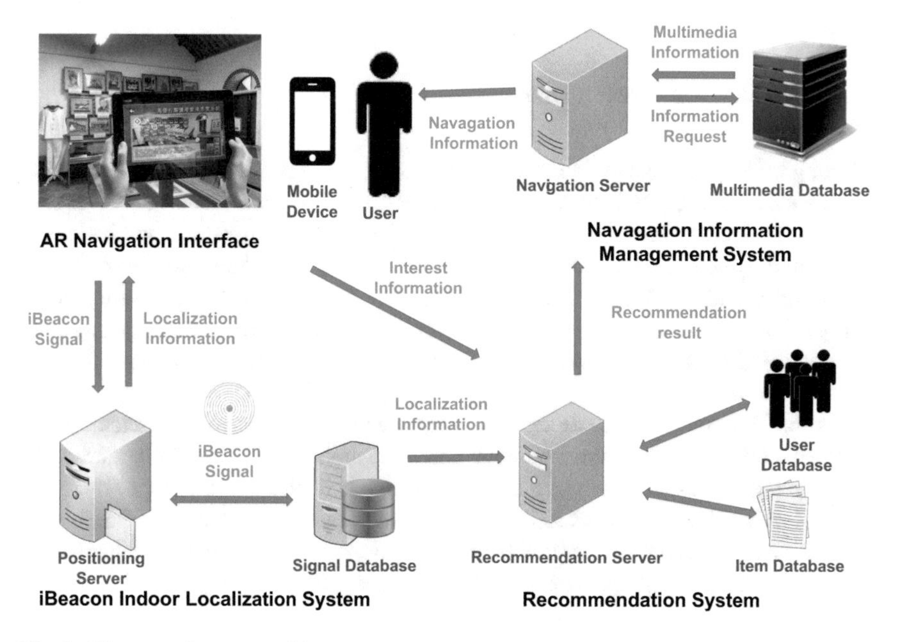

Fig. 1 Diagram of system architecture

The Indoor localization system is responsible for the real-time positioning of the user. It can help the users identify their current positions and find related exhibits faster and more easily. Moreover, the localization system can identify specific users through the user identification function, allowing the navigation server to provide specific data to the specific user for personalized navigation. The recommendation system provides information about the exhibits that are consistent with the interests of the user. The recommendation system will analyze the questionnaire and score table of the user, as well as the navigation records by referring to the user database and exhibit database records, and will then recommend exhibits and tour routes that may interest the users. Meanwhile, it can record new recommendation routes to establish more reference data, which will further improve the accuracy of the recommendation system.

The navigation information management system is mainly responsible for the transmission, reception and integration of various navigation needs and information. The navigation server is responsible for receiving and integrating the requests of the recommendation system and for obtaining relevant information from the multimedia database. Then, through the user identification function, the specific data are sent to the mobile device at the user end to display the navigation information. Besides, through the mark-less AR image recognition technology, the system can identify the pictures displayed in the exhibition area through the mobile device and then display the corresponding 3D and multimedia navigation information on the AR navigation interface.

3 Localization Mechanism

Although GPS is increasingly more popular today, but it is only applicable outdoors and cannot offer an Indoor localization function. The main methods of Indoor localization include infrared ray, ultrasonic, UWB, Wi-Fi, ultrasonic, passive/active RFID and Bluetooth technologies. Of these, based on Bluetooth Low Energy (BLE), also called Bluetooth 4.0 or Bluetooth Smart, iBeacon positioning technology has advantages of low cost, high popularity, and feasible positioning accuracy and range [9], which is very suitable to realize Indoor localization and location-aware service.

As the iBeacon's signal strength decreases with an increase in distance, on the basis of considering the environment and iBeacon's individual differences, and according to the distance between a mobile device and iBeacon, this paper develops signal levels of different iBeacons for different distances. Through the matrix formed by these signal levels, this paper then adopts a fault-tolerant mechanism as assistance for realizing accurate and stable positioning [10].

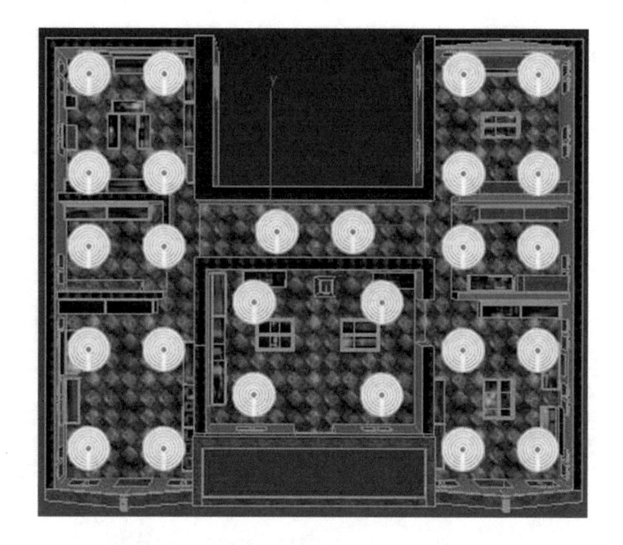

Fig. 2 Schematic diagram of Tamsui Oxford College localization model

The localization model is as shown in Fig. 2, in which the circle denotes the iBeacon tag. The system implementation contained the initial setting and real-time positing stages. At the initial setting stage, iBeacon were placed at proper locations according to the actual environment of Tamsui Oxford College. Afterwards, the signal intensity reference values of the individual iBeacons were set after analyzing the signal strength of the individual iBeacons at different distances, in order to establish the matrix of signal level containing all of the localization areas in the database. At the real-time positioning stage, when the visitor holds mobile device and moves through the site, the mobile device will scan the iBeacon signal, and transmit the collected iBeacon-related information to the positioning server for analysis.

Figure 3 presents the localization algorithm of the system. Based on the signal strength of the individual iBeacon and signal level correspondence table, the server will first change the received iBeacon signal strength to the signal level and save it in the signal level matrix S; next, based on the signal level in the level matrix, the signals will be ranked again from strong to weak and saved in matrix Ssort; finally, based on the strongest Ssort [0] signal and location in Ssort [0] and accompanied by following Ssort [1], Ssort [2], and Ssort [3] signals and locations, positioning can be completed rapidly and accurately.

```
=================================================================
S is the signal level matrix composed of all iBeacon signals.
Ssort is the signal level matrix after ranking based on signal levels in S.
Ssort [0] is the strongest signal; Ssort [1] is second strongest; and so on.
=================================================================
begin
1.
2.   while IsOnline then
3.        switch Ssort[0] then
4.             case L0
5.                  show Location (Ssort[0])
6.             case L1
7.                  if Ssort[1] = L2 and Ssort[2] = L2 then
8.                       show Location (Ssort[0], Ssort[1], Ssort[2])
9.                  else if Ssort[1] = L2 then
10.                      show Location (Ssort[0], Ssort[1])
11.            case L2
12.                 if Ssort[1] = L2 and Ssort[2] = L2 and Ssort[3] = L2 then
13.                      show Location (Ssort[0], Ssort[1], Ssort[2], Ssort[3])
14.            default
15.                 no_Show()
16.       end
17.  end
end
=================================================================
```

Fig. 3 Localization algorithm

4 Recommendation Mechanism

Although positioning technology is useful in indoor navigation, the users still face the problem of an overwhelming amount of exhibit information, which makes the users unable to determine the exhibits and tour routes. Hence, it is practical and important to add a recommendation system to the indoor navigation system, which could recommend suitable exhibits and tour routes after analysis according to users' behavior or active assessment information. Based on the concept of a recommendation system, this paper presented a mobile navigation system that could analyze the explicit behaviors of different users, including items such as personal interests and evaluations of the exhibits, as well as implicit behaviors, such as location and time, in order to recommend the exhibits that may interest the user. Coupled with the positioning function, it could proactively recommend suitable navigation routes and timeframes for navigation to the user. In this work, we employ a method that incorporates with hybrid filtering to achieve the recommendation. Three filtering methods are described as following:

1. **Content-based filtering** is based on a user profile to recommend exhibits to user by comparing the content of exhibits. The content of each exhibit is a set of identity attributes such as the keywords which is used to describe the exhibit. The user profile is built up according to user's preference and analysis the content of exhibits which have been seen by the user. Hence, the user's preference and the history of the user's behavior are gathered in our system. Besides, a key issue with content-based filtering is whether the user's profile is able to gather regarding as user's explicit behavior and implicit behavior. To overcome this problem, the visit times of each exhibit and the number of times seen exhibit for each user are collected in our system. Overall, although content-based filtering has certain recommendation effects, it cannot determine the potential interests of the user. The lack of an evaluation mechanism, in the case of the user's implicit behavior, may affect the accuracy of the recommendation mechanism. Therefore, we gather user's preference, the visit times of each exhibit and the number of times seen exhibit for each user to achieve more accuracy of recommendation mechanism.
2. **Collaborative filtering** is based on multiple users' experiences to recommend the exhibit to the user if they have the same experiences. It collects user feedback by rating scores for seen exhibits and take advantage of similarities and difference among users to achieve recommend mechanism. There are two approaches can be used in the collaborative filtering. One is User-based collaborative filtering and the other is Item-based filtering. User-based filtering attempts to recommend an exhibit to the users if most people regard the exhibit as being interesting. It mainly analyzes exhibits according to the rating scores of questionnaires filled in by users. Item-based filtering recommends suitable exhibits to the users according to the similarities between exhibits. However, the collaborative filtering also suffers a number of problems, e.g. sparsity and cold start problem. Most users do not rate the most exhibits after seeing the exhibits. Because of sparsity, it is possible that the similarity between two users cannot be obtained. Cold start refers to an exhibit cannot be recommended unless it has been rated by enormous users. Beside, a new user also has to rate sufficient exhibits so it can be to provide more accuracy of recommendation mechanism.
3. **Hybrid filtering** takes the advantage from the content-based and the collaborative filtering methods, and avoids their disadvantages, resulting in relatively better recommendations. It's simplicity, effectiveness and the ability to collect implicit behaviors. It also has the ability to identify the potential interests of the user, and avoid the problems of newly added data and the cold start issue, based on the user profile obtained from content-based filtering, allowing it to obtain diversified and accurate recommendation results.

Thus, in our system, we adopted the hybrid filtering for the implementation of the recommendation mechanism. First, a questionnaire survey was conducted to understand users' interests, as a basis for a preliminary recommendation of exhibits and tour routes. Then, during the visit, the system continuously recorded the user's preferences for various exhibits and predicted the exhibits that could possibly interest the users by combining Eqs. 1 and 2 to analyze the user's interests and update the tour routes.

In Eq. 1, Sim (x, y) denotes the level of similarity between exhibit x and exhibit y, N represent the users who both rated exhibit x and exhibit y, r_{ux} and r_{uy} represent the scores which is rated by user u regarding exhibits x and y respectively, and $\bar{r}_u$ denotes the average score of the exhibits that have been rated by user u. In Eq. 2, w denotes the unseen exhibit which we want to predict rating whether the user is interested, z denotes the seen exhibit similar to w, M denotes the number of seen exhibits with a higher level of similarity to exhibit w, and R_{uw} denotes the predicted rating value of user u in exhibit w. A higher rating value indicates that the user is more interested in the exhibit; therefore, the recommendation system would recommend the exhibits that have a higher predicted rating value of interest to the user. In other words, we take the seen exhibits similar to the unseen exhibit, and from those similar seen exhibits, we pick exhibits which are rated by the users. We weight the user's rating for each of these seen exhibits by the similarity between the seen exhibits and unseen exhibit. Finally, we scale the prediction by sum of similarities to obtain a value for the predicted rating.

$$Sim(x,y) = \frac{\sum_{u \in N} (r_{ux} - \bar{r}_u)(r_{uy} - \bar{r}_u)}{\sqrt{\sum_{u \in N} (r_{ux} - \bar{r}_u)^2 \sum_{u \in N} (r_{uy} - \bar{r}_u)^2}} \tag{1}$$

$$R_{uw} = \frac{\sum_{z \in M} Sim(z,w) * r_{uz}}{\sum_{z \in M} Sim(z,w)} \tag{2}$$

5 Implementation and Demonstration

This paper used Tamsui Oxford College, a Level 2 historical site in Taiwan, as the experimental site. The system implementation was divided into three parts, including the navigation system, the localization system, and the recommendation system. The software 3DS Max was used in the navigation system to create 3D virtual scenarios and 3D objects, and Unity was used for the integration of navigation scenes and dynamic tour routes. In the localization system, iBeacon was used with the matrix of signal level localization mechanism to complete the positioning of the user. In the recommendation system, hybrid filtering was used for the analysis and recommendation of navigation information consistent with the needs of the user.

Figure 4 shows the on-site test result in Oxford College and navigation interface of the system. The navigation interface mainly can be divided into three parts. The first part is the (AR information display area) in the middle of the interface: aiming at the specific pictures and texts of Oxford College, the users may detect and present the related 3D and video navigation information; the second part is the (text information display area) in the lower middle of the interface: it can display the related text navigation information of exhibition area and exhibits where the user is located in; and the third part is the (position information display area) in the right lower corner of the interface: the user's location can be timely displayed on a small map of Oxford College, so that the user may easily know where he/she is. When using the system for the first time, the users have to fill in basic information and select the interest options via user interfaces (as shown in Fig. 5), in order to produce the preliminary recommended tour routes (yellow lines).

During the visit, if the users were particularly interested in certain exhibit items, the users could use the camera of mobile devices to scan the pictures of exhibits to browse the detail augmented reality 3D and multimedia navigation information (as shown in Fig. 6), thus deepening the depth and enjoyment of the navigation. With the movements of the user, the visited routes would be automatically changed into

Fig. 4 Schematic diagram of on-site testing result and navigation interface

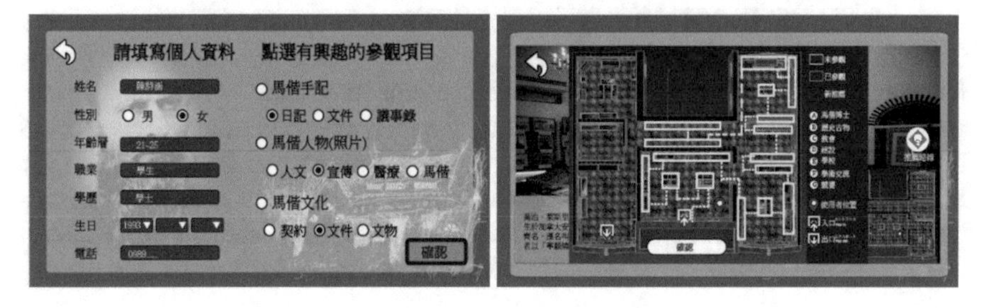

Fig. 5 Schematic diagram of fill-in interface of user interest option and the corresponding tour route (*yellow lines*)

Fig. 6 Schematic diagram of demonstrating the detailed 3D navigation information

Fig. 7 Schematic diagram of updating the visited tour route (*green lines*) and recommended new tour routes (*red lines*)

green color to facilitate identification (as shown in Fig. 7). Moreover, the system continuously recorded the visited locations and time as the basis for the analysis of the implicit behaviors of the user, and updating the recommended new tour routes (red lines) individually (as shown in Fig. 7).

6 Conclusion

This paper proposed a mobile navigation system that combines iBeacon Indoor localization and hybrid filtering recommendation mechanisms. Based on the iBeacon Indoor localization function, the system could accurately record the location and time of the visitor as a reference for navigation and exhibit recommendations. And, the proposed hybrid filtering recommendation mechanism could determine the potential interests of the users and avoid problems such as false behavior and cold starts. It could further actively recommend exhibits and navigation routes consistent with the interests of the user, thus realizing personalized and diversified mobile navigation. In addition, accompanied by the AR display function, this system provides diversified 3D and multimedia navigation information, so as to improve the navigation effect and enrich user's navigation experience effectively.

Acknowledgements We are grateful to the support of Ministry of Science and Technology, Taiwan, R.O.C., under contract number MOST 104-2221-E-156-007.

References

1. Brighton Museums, "Brighton Museums App," Available at: https://play.google.com/store/apps/details?id=com.surfaceimpression.brightonmuseums [Accessed: March 2016].
2. American Museum of Natural History, "Explorer App," Available at: https://play.google.com/store/apps/details?id=org.amnh.explorer [Accessed: March 2016].
3. National Palace Museum, "Giuseppe Castiglione App," Available at: https://play.google.com/store/apps/details?id=iiiGuide.taipei.npm.castiglione.italy [Accessed: March 2016].
4. Museum of London App, "Streetmuseum App," Available at: http://www.museumoflondon.org.uk/discover/museum-london-apps [Accessed: March 2016].
5. V. Viswanathan, Ilango Krisnamurthi, "Finding relevant semantic association paths through user-specific intermediate entities," Human-centric Computing and Information Sciences, vol. 2, no. 9, 2012.
6. Willem Robert van Hage, Natalia Stash, Yiwen Wang, Lora Aroyo "Finding Your Way through the Rijksmuseum with an Adaptive Mobile Museum Guide," Lecture Notes in Computer Science, Vol. 6088, 2010, pp. 46–59.
7. Moon-Hee Park, Jin-Hyuk Hong, Sung-Bae Cho, "Location-Based Recommendation System Using Bayesian," Ubiquitous Intelligence and Computing, Lecture Notes in Computer Science Vol. 4611, 2007, pp. 1130–1139.
8. Fan Yang, Zhi-Mei Wang, "An Optimized Mobile Restaurant Recommend System," Proceedings of the 10th WSEAS international conference on Automation and information, 2009, pp. 411–417.
9. Ching-Sheng Wang, Wei-Tsung Su and Yu-Cheng Guo, "An Augmented Reality Mobile Navigation System Supporting iBeacon Assisted Location-Aware Service," 2016 International Conference on Applied System Innovation, May 2016.
10. Ching-Sheng Wang, Xin-Mao Huang and Ming-Yu Hung, "Adaptive RFID Positioning System Using Signal Level Matrix," World Academy of Science, Engineering and Technology, Volume 46, November 2010, pp. 746–7 adfa, p. 10, 2011.© -Verlag Berlin Heidelberg 2011.

Constructing a Home-Based Knee Replacement Exercise Monitoring System with G Sensor

Lun-Ping Hung, Yuan-Hung Chao, Yu-Ling Tseng and Yi-Lun Chung

Abstract As Taiwan enters an aging society, osteoarthritis (OA) has become a common chronic disease among the elderly population. Knee is the most predilection site of OA. In the later stages of OA, knee replacement surgery is considered the best treatment of OA. A complete course of treatment, however, does not end once the operation is over; instead, it includes the follow-up rehabilitation exercise, tracks and observation. Rehabilitation exercise requires patient cooperation to continue with the rehabilitation at home. Due to various factors, however, many patients are unable to carry out the rehabilitation adequately to ensure satisfactory results. Hence, it is important to have a home-based rehabilitation system that could enable the communication between the patients and their doctors, and record the information about their rehabilitation. The system developed in this study uses a 3-Axis G Sensor to measure the knee angles during the rehabilitation, and uses Bluetooth to transmit the rehabilitation data to the computer. The data are then stored in a cloud database and are presented visually in the back-end. The physiatrists in the hospital have real-time access to and monitor every patient in their course of rehabilitation, thus giving them prompt feedbacks. The proposed remote rehabilitation platform makes it possible for patients and doctors to improve the quality of home rehabilitation after knee replacement surgery.

Keywords Osteoarthritis · Knee replacement · Home telerehabilitation · G sensor

L.-P. Hung (✉) · Y.-L. Tseng · Y.-L. Chung
Department of Information Management, National Taipei University of Nursing and Health Sciences, Taipei, Taiwan
e-mail: lunping@ntunhs.edu.tw

Y.-H. Chao
School and Graduate Institute of Physical Therapy, College of Medicine, National Taiwan University, Taipei, Taiwan

© Springer Nature Singapore Pte Ltd. 2018

N.Y. Yen and J.C. Hung (eds.), *Frontier Computing*, Lecture Notes in Electrical Engineering 422, DOI 10.1007/978-981-10-3187-8_59

1 Introduction

Despite the rapid development of modern technology, advances in medical technology in particular, aging society has become a serious problem in many countries. As the UN DESA report on World Population Prospects shows, in 2015, people aged over 65 comprise 8.28% of the global population. In 2025, this number is projected to increase to 10.42% [1]. According to the WHO, a country is defined as an aging society when the elderly population exceeds 7% of the total population. With the demographic aging, health issues relating to chronic diseases have emerged, and the number of people with chronic diseases will grow year by year. According to the "(6th) Survey of Health and Living Status of the Middle Aged and the Elderly in Taiwan" published by the National Health Insurance Administration, Ministry of Health and Welfare, 87.3% of the aged population is diagnosed with more than one chronic diseases [2]. In the past several decades, Spain has shown a drastic surge in the number of people with chronic diseases, resulting in 70% of the country's medical budget spent on 15 million elderly patients.

Among all the chronic diseases prevalent among the elderly, osteoarthritis (OA) is a chronic condition especially common for those aged 65 years old or above. It can affect one's sleep, normal daily activities and even life expectancy. About 74% of the people aged 65 or above in the United States are suffering from OA [3]. OA is believed to be mainly caused by the overuse of joints, when there is insufficient chondrocyte and glial cells to maintain the cartilage, causing the cartilage to gradually wear away. In the worst cases, bone spurs might even appear. The knee, which carries most of the weight of the body, is one of the joints most likely affected by OA. In the U.S., 27 million people are affected by OA, leading to a total of $185.5 billion in treatment costs every year [4].

When OA occurs, the cartilage breaks down gradually, which is non-repairable and irreversible once it begins. For the severely affected patients, the symptoms may include stiffness, pain, or even deformation in the knees, causing inconveniences in daily activities. Depending on the different circumstances, various treatments are available, namely. non-surgical methods, such as taking painkillers or anti-inflammatory drugs or doing exercises, for those with mild symptoms to improve their condition, and knee replacement for those with severe symptoms, which is considered an effective treatment for OA.

After the knee replacement, however, the patient still needs to face some functional deficiencies in the knee. Rehabilitation is therefore needed after the operation to facilitate the recovery. Effective and efficient post-operative rehabilitation care is important to the patients. Under the current medical procedures, patients are instructed to perform post-operative rehabilitation exercises at home, and monitor the progress by themselves. It is, however, difficult to ensure the rehabilitation compliance of all patients, and the patients are uncertain about the exercise result. When they encounter difficulties in the process, there is no real-time assistance from the physicians. The physicians do not know about the rehabilitation process until the next follow-up visit. The fact is, however, when the patients are not doing the exercises correctly, no satisfactory rehabilitation results can be reached. In some cases, injuries may occur.

As suggested by the American Academy of Orthopaedic Surgeons (AAOS) in the U.S., rehabilitation is the most important treatment after a knee replacement and the greatest concern is the maximum knee bending angle and training of the muscle around the knee in the process of the rehabilitation. Continuous post-operative rehabilitation can reduce the stiffness on the joint after the operation, and relieve the abnormal muscle force distribution, which might otherwise lead to knee instability and affect the patient's daily activities. It could also result in a decline in the knee replacement rate [5].

Most of the rehabilitation is conducted by the patients at home. The effectiveness of the rehabilitation, therefore, largely depends on the consciousness and self-control of the patients. Since rehabilitation is a long process, patients may easily quit during the course, thus missing the best timing and opportunity for their recovery. Home-based rehabilitation is becoming more important in the modern society. For the elderly in need of long-term rehabilitation, the lack of necessary resources in the traditional settings for home-based rehabilitation is a serious problem.

To promote home-based rehabilitation, information technology has been adopted, among which the Internet of Things (IoT) has been widely used in the medical field. Many studies used home as the experimental site for patients in need of medical support, and tested various body sensors, in order to examine the patients' rehabilitation progress. The data are transmitted to the cloud. Those studies had promoted the concept of telerehabilitation [6]. The advance in technology has led to intelligent rehabilitation systems. Telerehabilitation, in this sense, refers to the use of various electronic and communication technology to facilitate the patients' rehabilitation at home. Using those systems, patients can send their data to the physicians via the internet, and the physicians can conduct remote diagnosis and treatment. As medical information technology and communication technology mature in recent years, the resources and devices needed for telerehabilitation are becoming available. The concept of telerehabilitation has been integrated into various medical studies focusing on rehabilitation. The development of rehabilitation technology and telerehabilitation platforms is on the way to become an indicator for the medical technological level. Currently, telerehabilitation systems are roughly divided into three categories:

Image-based rehabilitation. It is the most widely applied technology in terms of rehabilitation. Kinect, a motion sensor developed by Microsoft, uses a camera to capture the gesture and motion of the human body. The interaction is enabled by the recognition of body language. A real-time interface is provided as a display for the users. Su et al. [7] used Kinect to develop and analyze systems through Dynamic Time Warping (DTW) and fuzzy logic. However, the judgment and analyses were merely based on the images taken by the camera. When a person is lying on his or her back or move beyond the camera's coverage, the system is unable to transmit images entirely, thus cannot determine the user's motion and gesture.

Virtual reality rehabilitation. Virtual Reality (VR) is a new technology which works by simulating a high-fidelity 3D space with the computer. When a user wears a special camera or display device, they could enter this space with their vision, hearing and touch. In this way, even if the user is in a virtual world, they have the illusion that they are in the real world. With the help of controllers, it is even possible to make interactions. Kairy et al. [8] applied VR on patients suffering from upper extremity complications after a stroke. The subjects were divided into two groups for comparison. The results showed that the integration of VR in rehabilitation could enhance the consistency of the rehabilitation, making rehabilitation more interesting to the patients. The patients feel that they were in the real world so that they make their best efforts to attempt to make the movements. For VR, high image definition is required, thus demanding for advanced computer and image technology.

Sensor rehabilitation. In tele-rehabilitation, various sensors are needed to sense the precision of the movements in the process, namely the patients' motion or physiological conditions, which are then transmitted via the Internet. Currently, the popular sensors include inertial measurement units such as gyroscopes and 3-axis accelerometers. Those measuring devices could provide complete records of patients' information concerning their precise movements, tilt and gravity during the rehabilitation. Bae and Tomizuka [9] applied an inertial measurement unit (IMU) and a boot-shaped ground reaction force (GRF) sensor on patients in need of gait rehabilitation. A physical therapist was present at each gait rehabilitation session to propose appropriate rehabilitation plans for the patients. The sensors were used to help the patients adjust their gait and record the data, which were transmitted to the cloud platform. The physical therapist could thus determine the rehabilitation of the patients, and provide the patients with real-time customized rehabilitation services.

In the present medical environment, there are already some successful cases of the application of technology in tele-rehabilitation. As the elderly patients tend to be affected by OA after a knee replacement, the physicians need to know the patients' rehabilitation progress, and facilitate the success of the rehabilitation. Therefore, based on the concept of IoT, this study aims to develop a device using G sensors so that patients can ensure the precision of movements and knee bending angles during their rehabilitation at home. The data captured by the sensors are sent to a cloud platform wirelessly to be viewed by physicians. The proposed home-based tele-rehabilitation system intends to help patients suffering from OA after the knee replacement.

2 System Environment and Configuration

The system environment in this study is divided into three sections: the patient end, the cloud database, and the physician end. After a patient is discharged from the hospital, the rehabilitation physician issues a prescription for home rehabilitation. The content of the prescriptions is uploaded along with the health education videos provided by the medical teams. The cloud database stores related information, and

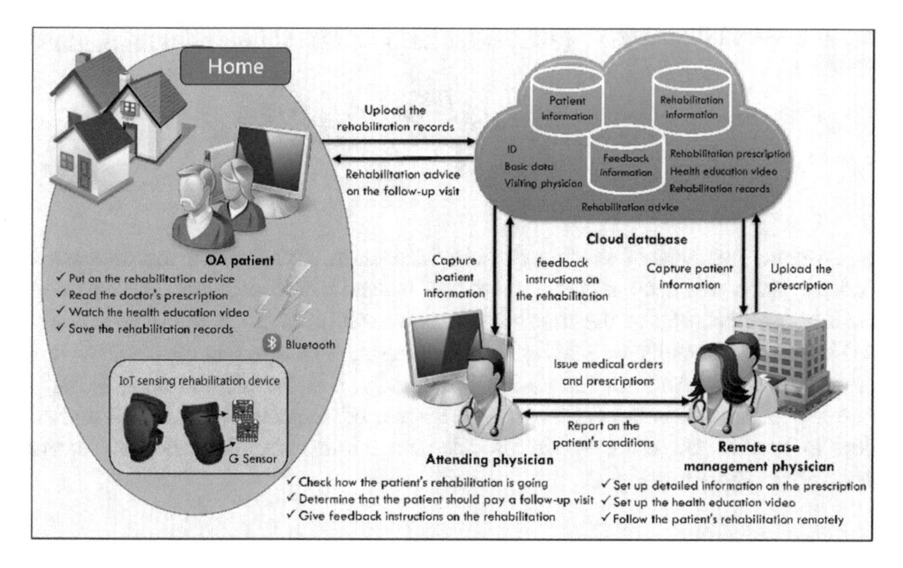

Fig. 1 Diagram of system environment and configuration

is accessible for both the physicians and patients, including the patient profile, instructions on the rehabilitation movements, health education videos, home rehabilitation records, and feedbacks from the physicians. At the patient end, patients can download the rehabilitation information from the cloud to their computers. After viewing the rehabilitation movements, the patients wear an IoT knee rehabilitation device during rehabilitation. The data are then sent to the cloud database and update the existing records.

The physicians can track the patients' rehabilitation progress at anytime, and are alerted when the rehabilitation movements are too difficult to the patients, or the patients are unable to bend their knees to a certain angle required by the physicians, or the current rehabilitation targets are too easy to the patients. The physicians then report the information to the attending physicians, who then determine whether the patients need a follow-up visit. Based on the information in the system, the physicians can provide feedbacks or suggestions to the patients. The proposed system can support telerehabilitation using the IoT technology. The system environment and configuration is shown in Fig. 1.

3 Rehabilitation Monitoring

3.1 IoT Sensing Device for the Knee

The sensor in the IoT sensing device for the knee used in this study is comprised of a 3-axis G sensor and a bluetooth module on the basis of Arduino. In the

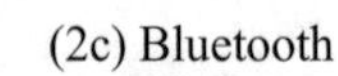

Fig. 2 Relevant IoT sensing devices

(2a) Arduino nano (2b) G sensor (2c) Bluetooth

experiment, two sets of sensing devices were used. Considering the dimension, Arduino nano was used as the motherboard instead of the Arduino. The bluetooth module uses HC-05 for the transmission and Arduino nano 5 V for power. The ADXL-345, a popular and affordable G Sensor, was used in this study. It is powered through Arduino nano 3.3 V. Given that it is the I2C protocol, the SDA Port and SCL Port on the G Sensor were connected respectively to the A4 and A5 Port to control the transmission of data and commands. The relevant sensing devices are shown in Fig. 2.

3.2 Transmission Technology

This study used Bluetooth technology for data transmission. Bluetooth uses frequency-hopping spread spectrum (FHSS), a technology for frequency modulation in a small range. With Bluetooth, data could be transmitted and communicated between devices freely without using cable. It is therefore a technology specially designed for small network over short distances.

With features of compact size, low power consumption, high level of interference resistance, fast connection speed, and energy-saving, Bluetooth can provide convenience, usability and fitness for the device worn on the knee. A Bluetooth module can transmit the data captured by the 3-axis G Sensors to the computer for follow-up computation. The Bluetooth technical specifications are shown in Table 1.

3.3 Measurement of the Knee Bending Angle

During the knee rehabilitation, the patients were asked to wear two rehabilitation knee braces, which contain two sets of sensors, which are affixed to the upper and

Table 1 Bluetooth technical specifications table

Item	Operating frequency	Transmission distance	Mobility	Directionality	Power
Specifications	2.4 GHz	10–100 m	Extremely good	Omni direction	No more than 100 mW

Fig. 3 The schematic diagram of the 3-axis G Sensor concerning the tilt angle of the x axis

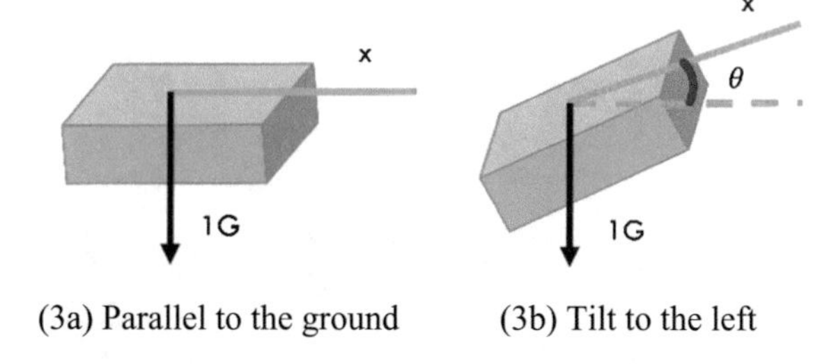

lower ends of the knee joint with a 10-cm distance between each other. The sensors were in proper alignment with each other for further measurement and computation.

The patients' knee bending angles were measured via the 3-axis G Sensors, which calculate the tilt angle by trigonometric functions and based on the gravitational acceleration. For example, when the G Sensor is parallel to the ground, the output value of the x axis is 0 g; when the G Sensor begins to tilt, the output value of the x axis changes accordingly. Figure 3 shows the tilt angle of the x axis (among the three axes).

Based on the above schematic diagram, when the 3-axis G Sensor tilts to the left [10] and is expressed by a trigonometric function, the equation for the output value of the x axis is written as follows:

$$G^X = 1\text{G} \times \sin\theta_X \tag{1}$$

After the tilt of the x axis is derived from Eq. 1, we can obtain Eq. 2:

$$\theta_X = \sin^{-1}\frac{G^X}{1G} \tag{2}$$

Based on the tilt angle of the 3-axis G Sensor, the knee bending angle can be calculated. Figure 4 is the schematic diagram of the patient's knee bending during the knee rehabilitation. In this study, two 3-axis G Sensors were used to measure the knee bending angle, which are G_A and G_B in the following figure. When the knee bends, the tilt angle of 3-axis G Sensor G_A as opposed to the ground is θ_a and the tilt angle of G_B as opposed to the ground is θ_b. By using $\theta_a + \theta_b$, the knee bending angle θ_k can be calculated, as shown in Eq. 3.

$$\theta_k = \theta_a + \theta_b \tag{3}$$

In the home-based rehabilitation experiment of this study, the patient was required to bend the knee at 90° for three times. If the knee bending angle reached 90° in one of these movements, according to the above diagram, the tilt angle of G_A as opposed to the ground θ_a is 50° and the tilt angle of G_B as opposed to the ground is θ_b 40°. Equation 3 puts the two values together, and the patient's knee bending angle is shown as θ_k 90°.

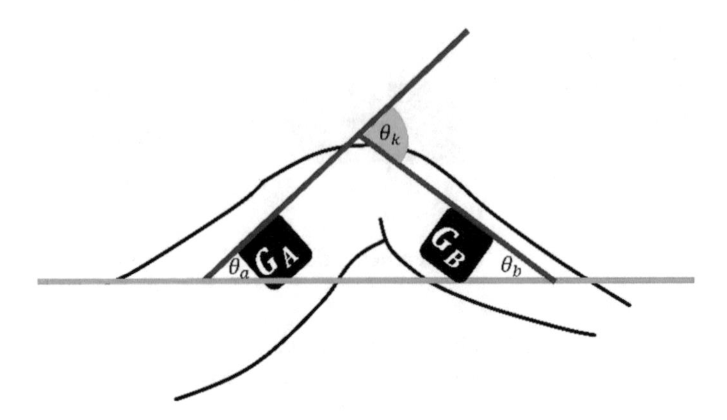

Fig. 4 The schematic diagram for the calculation of the knee bedding angle

3.4 Determine the Number of Rehabilitation

The proposed system allows the physicians to monitor the knee bending angles, as well as the number and time of home rehabilitations. The number of rehabilitation exercises includes the times of successful and failed attempts to make the movement, namely reaching the target knee bending angle required by the physicians.

The determination of the number of rehabilitation exercises is calculated by the knee bending angle of the patient each time. The angles are recorded every 0.5 s. When it is less than 10°, it is not regarded as the beginning of the rehabilitation and the indicator shows red; when it is greater than 10°, it means that the patient has started the rehabilitation and the indicator shows green. When the bending indicator changes from green to red, it means that the patient has made one rehabilitation movement. The largest angle detected when the indicator is green is recorded as the knee bending angle. As shown in Eqs. 4 and 5, Ca (current angle) denotes the current bending angle sensed by the sensor. When s is 0, it means that the rehabilitation has not started; when s is 1, it means that there is an ongoing rehabilitation. The number of rehabilitation increases one when s changes from 0 to 1 and back to 0 again. The maximum angle in each rehabilitation exercise, defined as Maxangle, has a default value of 0. This maximum angle changes in each rehabilitation exercise.

$$\mathrm{f}(s) = \begin{cases} 0, & Ca < 10 \\ 1, & Ca \geq 10 \end{cases} \tag{4}$$

$$\sum_{\mathrm{i}=Ca_1}^{Ca_n} \mathrm{f}(\mathrm{maxangle}) = \begin{cases} 0, & s=0 \\ i, & s=1, \quad i > maxangle \\ maxangle, & s=1, \quad i \leq maxangle \end{cases} \tag{5}$$

The maximum angle is compared with the target angle set by the physician, in order to determine the number of successful and failed attempts. Considering the possible errors in the process of rehabilitation, a threshold value is set for the determination of the number of successful and failed attempts. For example, when the target angle for a patient is 90°, the movement is considered successful as long as the knee bending angle reaches 80°; otherwise, it is deemed as a failed attempt.

4 System Functions

4.1 Physician End

To monitor the progress of the home-based rehabilitation, the physicians can access the system on a computer and browse the information stored in the back-end database. They can manage the data of the patients and view the rehabilitation records. Based on the information, they make suggestions to the patients on their home-based rehabilitation movements. A message board is provided to allow the physicians to answer patients' questions in a timely manner, thus facilitating the interaction and communication between the physicians and patients. The system interface for the physicians is divided into the several sections featuring patient information, rehabilitation records, and medical records.

4.2 Patient End

After patients undergo knee replacements, they need to start knee rehabilitation. The proposed system can help the patients monitor their knee rehabilitation at home and ensure the precision of their rehabilitation movements. When the patients log into the system, the rehabilitation prescription is displayed automatically, and reminds the patient to perform rehabilitation exercises at the designated time. Before the exercise, the patient reads the rehabilitation instructions and watches the health education video to understand how to do the exercise. After they put on the IoT rehabilitation sensors and start the exercise, the sensors make alerts when the patients reach the target angle. When the rehabilitation exercise is completed, the data are uploaded to the back-end database. If the patients have any questions about the rehabilitation or encounter any problems, they can send messages to the physician via the message board.

5 Conclusion

In the current aging society, many elderly suffer from OA, thus more patients perform knee replacement surgery. Hence, there is a larger demand for home-based rehabilitation. This study developed a home-based rehabilitation system that can help physicians to monitor the patients' rehabilitation progress and provide feedbacks to the patients. For patients, they can access rehabilitation information and view health education videos. The proposed system is comprised of a low-cost 3-Axis G Sensor, which measures the knee angles during the rehabilitation. The rehabilitation information is sent to the cloud database and accessed by the physicians, so as to ensure the rehabilitation effect and provide timely suggestions.

References

1. United Nations Population Division Department of Economic and Social Affairs.: World Population Prospects: The 2015 Revision. (2015)
2. Bureau of Health Promotion, Department of Health, Taiwan: 2007 Taiwan Longitudinal Study on Aging Survey Report, Taiwan Aging Study Series. 11 (2014)
3. Baird C. L. Schmeiser D. Yehle K. T: Self -care women with osteoarthritis living at different levels of independence. Health Care for Woman International. 24, 617–634 (2003)
4. Pao-Feng Tsai, Jason Y. Chang, Cornelia Beck, Jody Hagen, K. J. S Anand, Yong-Fang Kuo, Paula K. Roberson, Karl Rosengren, Linda Beuscher: The feasibility of implementing Tai Chi for nursing home residents with knee osteoarthritis and cognitive impairment. Activities Directors Q. Alzheimers Other Dementia Patients. 10, 9–17(2009)
5. KP I, JB N, N I, A. M.: Targeted early rehabilitation at home after total hip and knee joint replacement: Does it work?. Disability and Rehabilitation. 29(6), 495–502 (2007)
6. Nagender Suryadevara, Anuroop Gaddam, Ramesh Rayudu, S.C. Mukhopadhyay: Wireless Sensors Network Based Safe Home to Care Elderly People: Behaviour Detection. Sensors and Actuators A Physical. 186, 277–283 (2012)
7. Su C-J, Chiang C-Y, Huang J-Y: Kinect-enabled home-based rehabilitation system using Dynamic Time Warping and fuzzy logic. Applied Soft Computing. 22, 652–666 (2014)
8. Dahlia Kairy, Mirella Veras, Philippe Archambault, Alejandro Hernandez, Johanne Higgins, Mindy F. Levin, Lise Poissant, Amir Raz, Franceen Kaizer: Maximizing post-stroke upper limb rehabilitation using a novel telerehabilitation interactive virtual reality system in the patient's home: study protocol of a randomized clinical trial. Contemporary Clinical Trials. 47, 49–53 (2016)
9. Joonbum Bae, Masayoshi Tomizuka: A tele-monitoring system for gait rehabilitation with an inertial measurement unit and a shoe-type ground reaction force sensor. Mechatronics. 23(6), 646–651 (2013)
10. C. J. Fisher: Using an Accelerometer for Inclination Sensing. Analog Devices (2010)

Improved Face Recognition from Weighed Face Representations for Deepcam

Wei Wan, Jingfeng Liu and Donald Knasel

Abstract Recent advances in big data and deep convolutional neural network (CNN) have pushed the performance of face recognition significantly and becoming comparable to that of human being. At this moment, data is more important than algorithm when it comes to the contribution to the performance of face recognition. At the same time, video surveillance is becoming increasingly popular in consumer market thanks to the wide adoption of smartphones. This gives a new way to collect more data. This paper shows a way to integrate face recognition to Deepcam, a peer to peer WiFi security camera. By applying real time face recognition on the real time video stream from Deepcam, this paper shows how to improve the performance of face recognition by weighted combination of the face representations of the faces from consecutive frames.

Keywords Face recognition · Deep learning · Convolutional neural network (CNN) · Triplet-based loss function

1 Introduction

In recent two years, the performance of face recognition algorithms increased impressively and becoming comparable to that of human being. A good example is that the accuracy on LFW has been increased from 95 to 99% [1]. The improvement is mainly because of the large sets of faces available from Internet, such as the boom of social network. On the other hand, the advent of deep convolution neural network (CNN) [2], and the triplet-based loss function training scheme [3], can

W. Wan · J. Liu (✉)
LinkSprite Inc., Wuhan, China
e-mail: jingfeng.liu@linksprite.com

W. Wan
e-mail: wei.wan@linksprite.com

D. Knasel
Mysnapcam LLC, Atlanta, USA

© Springer Nature Singapore Pte Ltd. 2018
N.Y. Yen and J.C. Hung (eds.), *Frontier Computing*, Lecture Notes in Electrical Engineering 422, DOI 10.1007/978-981-10-3187-8_60

Fig. 1 Deepcam

accommodate a complicated model with large number of parameters to effectively reduce to dimensionality of the face from tens of thousands to a hundred.

At this moment, the bottleneck of the performance of the face recognition is the availability of face data, rather than the algorithm [4]. Thanks to the wide adoption of smartphones, video surveillance is becoming increasingly popular in consumer market. Users can do live view of their properties easily and get push notification on smart phone if certain events happen, like motion or noise. Deepcam wifi camera as shown in Fig. 1 [5] is one of the examples. These features were only available with enterprise level security systems. With the deployment of the Deepcam, it can detect motion and capture a large amount of faces each day. By allowing the user to assist the unknown or low confidence faces, Authface learns on to go and increases its accuracy.

An overview of the reset of the paper is as follows: in Sect. 2, we review the architecture of Deepcam; Sect. 3 describes the system of Authface, the face recognition system we used for Deepcam; In Sect. 4, we propose the novel weighted face representations method we used in Authface, together with some results. Finally Sect. 5 concludes the work.

2 Deepcam

Deepcam is a peer to peer WiFi security camera that offers cloud recording and face recognition [5]. The specifications of latest Deepcam is described as follows:

- 1280720 HD Camera with IR Night Vision and two way audio

- View live, Playback and Manage by Local/Internet PC or Mobile Devices
- Image sensor: 1/4 in. CMOS
- Compression: H.264
- Image resolution: 1280 * 720
- Extended Interface: Built-in MIC and speaker and audio output socket
- Min. Illumination: 0.5 lx
- White Balance: Auto
- Frame Rate: 25 fps (720P), 30 fps (VGA)
- Wi-Fi: Wi-Fi (IEEE802.11b/g/n)
- Day/night Switch: IR-Cut Light Filter Switch

Deepcam uses peer to peer platform to transverse the NAT device so that consumer doesn't need to configure the router to view the video from WAN. When motion detection or noise is detected by Deepcam, Deepcam will upload a video clip that contains the events to the cloud recording, and at the same time, pushed to Authface, which is a face recognition server.

The overall system diagram is shown in Fig. 2. When Deepcam powers on, it will check in with the checkin server first, and it will also check in with the firmware server to see if there is any updated firmware for it. After that, it will send keep alive signal every tens of seconds. When the mobile client wants to live view the particular Deepcam, the mobile client will check with the checkin server to see the status of the deepcam, and the checkin server will use STUN to see if a peer to

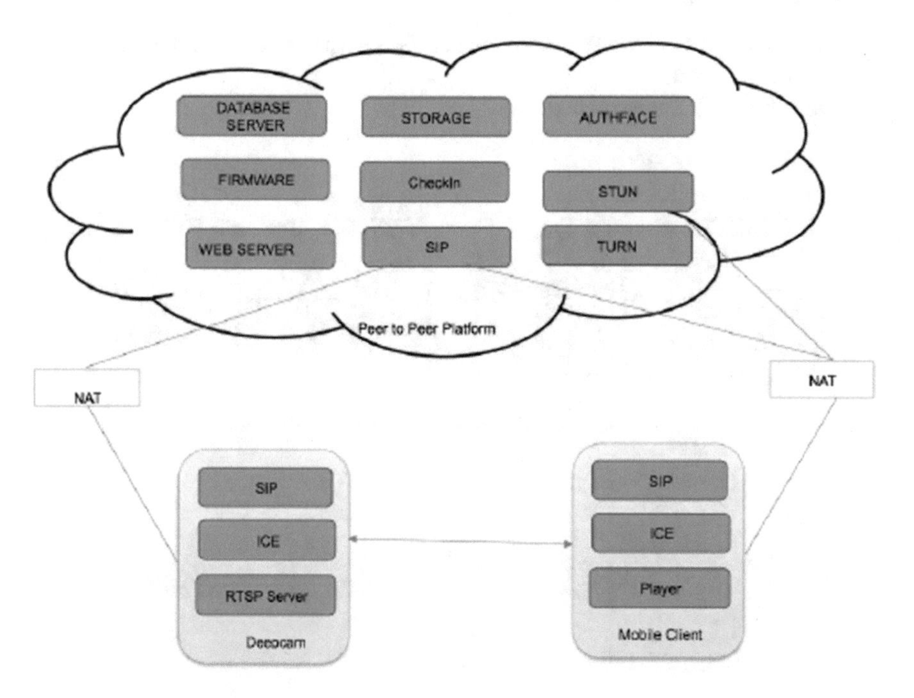

Fig. 2 Deepcam architecture

peer connection can be setup between the mobile client and the camera. If not, it will try to relay the packets using TURN server. Deepcam detects motion locally by doing pixel analysis. If motion is detected, it will push a short video clip that contains the motion events to the cloud storage, and at the same time, push the video clip to Authface server for face recognition.

3 Authface

Authface is a face recognition server that communicates with the clients by web socket. The deepcam's storage server pushes the video clip that contains motion to the Authface server by websocket. Once the video clip has been received by Authface, it will process the video in the following steps:

1. Grab frame from the video clip using OpenCV [5].
2. Detect the largest face in the frame using dlib.
3. Detect eyes and nose using dlib.
4. Align the detected face to be a 96 × 96 image.
5. Feed to aligned face to a trained deep CNN to obtain the face representation.
6. Label the face representation using a trained SVM classifier.
7. If the confidence level is lower than a threshold, report unknown face, and send the face back to the client

Client can also send the 96 × 96 aligned face to the server with the proper label to train the classifier.

Figure 3 shows the overall steps.

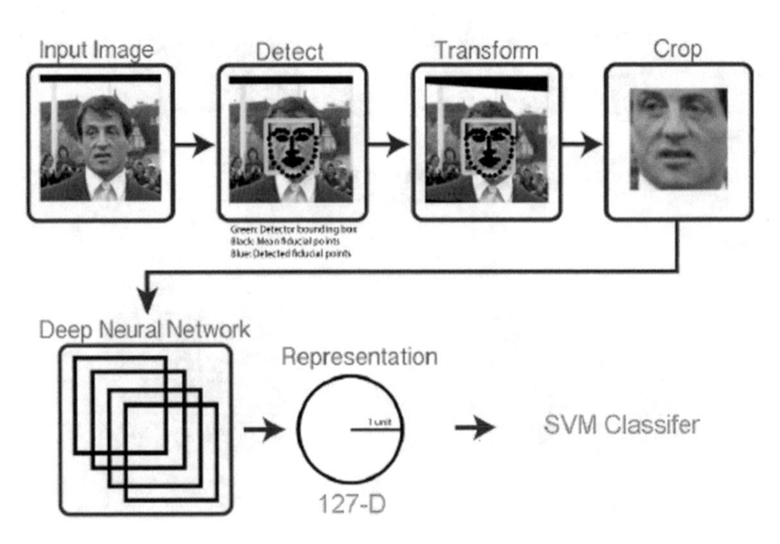

Fig. 3 Authface architecture

4 Weighted Face Representations

After the face got detected and processed into a 96 by 96 RGB aligned picture, it will be feed into a trained CNN to get the face representation, which is of a much reduced dimensionality. The resultant face representation is then classified by a trained classifier according to the nearest label from the Euclidean distance point of view.

In the video clip, the face of the same person can appear in consecutive frames. After the face is first detected, it can be tracked using KLT algorithm [6, 7]. The faces on different frames are shown in Fig. 4.

In our experiment, the CNN model is of 17 layers, and has been trained by combining the two largest (of April 2016) publicly-available face recognition datasets based on names: FaceScrub [8] and CASIA-WebFace [9]. Besides the model, the classifier we use to classify the face representation is a SVM. It has been trained for 40 persons.

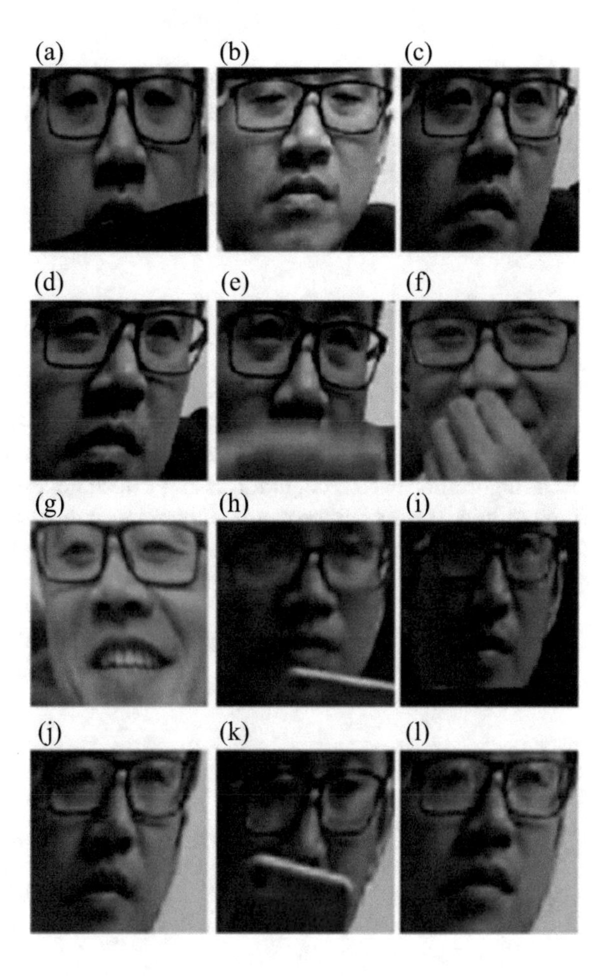

Fig. 4 Faces for different frames

If we feed directly the faces shown Fig. 4 to CNN and then the resultant face representation to a trained classifier, we will obtain the following results (the correct label should be 'Jingfeng'):

Face: a	Detected Person: Sky	Confidence level: 0.21
Face: b	Detected Person: Messi	Confidence level: 0.42
Face: c	Detected Person: Aimee	Confidence level: 0.25
Face: d	Detected Person: Messi	Confidence level: 0.34
Face: e	Detected Person: Jingfeng	Confidence level: 0.37
Face: f	Detected Person: Wangfei	Confidence level: 0.22
Face: g	Detected Person: ZhangLi	Confidence level: 0.28
Face: h	Detected Person: Jingfeng	Confidence level: 0.37
Face: i	Detected Person: Messi	Confidence level: 0.41
Face: j	Detected Person: Jingfeng	Confidence level: 0.28
Face: k	Detected Person: Jingfeng	Confidence level: 0.20
Face: l	Detected Person: ZhengSanQi	Confidence level: 0.27

Now instead of feeding individual face representation to the classifier and got the label, we average the face representations and then feed to the classifier. This is possible as all the faces are detected from the same person using KLT algorithm [6, 7]. The classifier predicts Jingfeng with confidence level of 0.42. By comparing the results to the above table, we can see dramatical improvement.

5 Conclusions

We described an improvement for deep learning based face recognition by weighting the face representations from the same person. This is possible that once face has been detected from the video, it can be tracked in the following frames by KLT algorithm. Face recognition is not a technique alone, in this paper, we also present how this technique can be integrated into a consumer video surveil- lance camera: Deepcam. The architecture of both the peer to peer camera and the face recognition server: Authface have been discussed.

Acknowledgements This work was supported in part by Dongguan Recruitment of Innovation and entrepreneurship talent program.

References

1. G. B. Huang, M. Mattar, T. Berg, and E. Learned-Miller Labeled faces in the wild: A database for studying face recognition in unconstrained environments, 2007. 1, 2, 5.
2. Y. LeCun, L. Bottou, Y. Bengio, and P. Haffner, Gradient-based learning applied to document recognition, Proceedings of the IEEE, 86(11):2278–2324, 1998.

3. F. Schroff, D. Kalenichenko and J. Philbin, FaceNet: A Unified Embedding for Face Recognition and Clustering, The IEEE Conference on Computer Vision and Pattern Recognition (CVPR), 2015, pp. 815–823.
4. Dong Yi, Zhen Lei, Shengcai Liao and Stan Z. Li, Learning Face Representation from Scratch, arXiv preprint arXiv:1411.7923. 2014.
5. http://www.deepcam.com/.
6. Carlo Tomasi and Takeo Kanade Detection and Tracking of Point Features, Carnegie Mellon University Technical Report CMU-CS-91-132, 1991.
7. http://www.mathworks.com/help/vision/examples/face-detection-and-tracking-using-live-video-acquisition.html?requestedDomain=www.mathworks.com.
8. http://vintage.winklerbros.net/facescrub.html.
9. http://arxiv.org/abs/1411.7923.
10. http://opencv.org/.

A GIS-Based Learning System to Support General Education—Using Cultural Heritage Course as an Example

Kai-Yi Chin, Ko-Fong Lee and Hsiang-Chin Hsieh

Abstract In present day general education classrooms, teachers often must turn to traditional teaching methods to help students acquire knowledge and understand content. However, the use of conventional teaching models rooted in the past may result in limited one-way knowledge transmission in the real world, rendering the overall process less effective. The continual use of older teaching techniques in our modern age of learning results in a potential loss of meaningful interactions with students during the knowledge transfer phase. To cope with this problem, we developed a GIS-based learning system designed around the curriculum of a MacKay Cultural Heritage course to deliver relevant information to students in classroom learning environments. The proposed system consists of two main components: the "Multimedia Database of MacKay Cultural Heritage" and the "Learning Platform of MacKay Culture". Both components of this course were used to demonstrate the significance of the added educational value imparted by Dr. MacKay's contributions and his impact on the state of education, medicine and religion in Taiwan. Our system also allowed for the incorporation of geographic information technologies into a web-style map to improve the delivery of information to students as a complement to conventional teaching practices. In addition, the proposed system served as a medium for providing personalized learning support for each student. Thus, we believe that our GIS-based learning system

K.-Y. Chin
Department of Digital Humanities, Aletheia University, New Taipei City, Taiwan
e-mail: au0292@mail.au.edu.tw

K.-F. Lee (✉)
Department of Information Computer Science and Information Engineering,
National Taipei University of Technology, Taipei City, Taiwan
e-mail: kookyrational@hotmail.com

H.-C. Hsieh
Institute for Information Industry, Taipei City, Taiwan
e-mail: palapala@iii.org.tw

© Springer Nature Singapore Pte Ltd. 2018

N.Y. Yen and J.C. Hung (eds.), *Frontier Computing*, Lecture Notes
in Electrical Engineering 422, DOI 10.1007/978-981-10-3187-8_61

contained the necessary features to effortlessly facilitate students' knowledge acquisition and promote student motivation to achieve the learning aims of the MacKay Cultural Heritage course.

Keywords General education · GIS · MacKay · Interactive learning environments

1 Introduction

Significant importance has been placed on the development of a well-rounded general education that is based on the concept of holistic education. In Taiwan, it is mandatory for university level students to enroll in a general education course intended to help students build a solid base of academic skills, covering both humanities and science literacy. The aim was to create opportunities for students to develop a mastery of different aspects of basic education, to gain broader knowledge scope in their studies. Unfortunately, this theoretical holistic approach to education is difficult to put into practice. Many studies indicated that the general education course usually is implemented by the traditional methods of teaching [1]. Such methods may only provide one-way knowledge transmission in the real world, and is thus less effective as it lacks meaningful interaction with students during the knowledge transfer phase. Therefore, the issue of how to promote holistic curriculum planning and finding modern ways to foster the development of general education has become a very important area of educational research.

Most colleges in Taiwan have decided to create general education courses based on individual lectures as a way to standardize the curriculum across several different schools and disciplines. However, many studies evident that curriculums involving the broad scope of a general course should not be limited by the traditional methods of teaching, and it is especially true for courses that allow students to partake in activities either inside or outside of the classroom [2]. For example, content describing important historical facts about Tamsui, MacKay's significant influence on the region, the unique aboriginal history, and efforts of heritage preservation or reconstruction would be more impactful for students who could physically participate in field visits, especially if students could study and explore these concepts outside of a traditional campus setting. These experiences can increase interactions with the learning environment and allow learners to connect time and space with static learning content, particularly for courses in humanities that are meant to build and develop knowledge.

Within present day institutions, educators have started to explore digital learning systems as a tool for teaching. As described in Fig. 1, Hwang et al. [3], proposed a system that combined PDAs (Personal Digital Assistants) and multimedia resources to facilitate student-learning activities in plant identification and botany instruction. In another study, Chen and Huang [4] also provided students with an integrated system that utilized PDA and multimedia resources to conduct an interactive

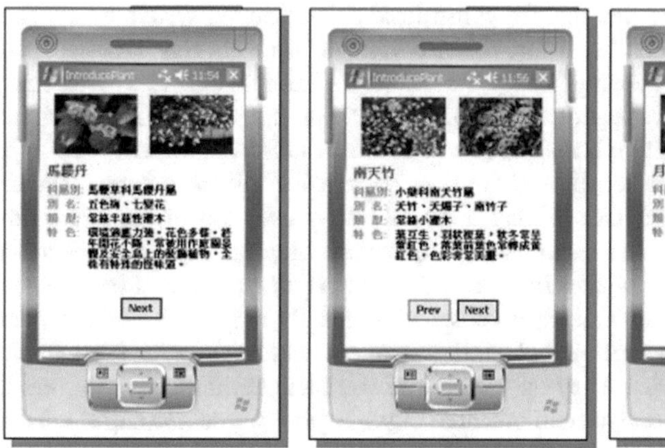

Fig. 1 Using PDAs in plant gardens for identifying specimens [3]

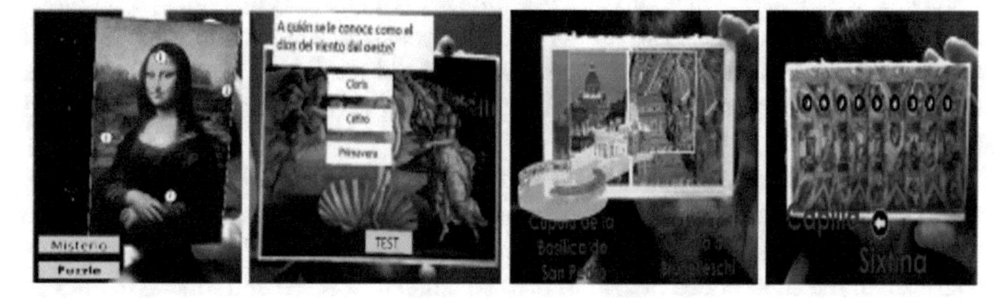

Fig. 2 Use of augmented reality and 3D objects in a museum setting [5]

learning activity within a museum setting. Furthermore, Fig. 2 shows how Di Serio et al. [5] combined augmented reality and 3D-objects using smart phones that could display painting and historic information to students taking part in the activity. With the limitations of traditional teaching approaches, many educators must supplement the delivery of learning content with additional text, voice, video, and image resources. These additional resources integrated into the existing learning system can further support learning activity in fields that are rooted in humanities or social sciences. However, instructors who implement these learning systems often focus on creating a platform for simply showing multimedia resources and they overlook the more complex process of integrating time and space concepts into knowledge acquisition. This means that students may have trouble with constructing learning contexts and their learning trends are stunted without tools to adapt or modify their learning process.

The GIS (Geographic Information System) is a remarkable synthesis system developed over the past 50 years, which combines a wide range of information systems with identifiable geographic locational materials and allows users to synergistically obtain information located in the real world. For example, as shown

Fig. 3a, Benz and his colleagues [6] constructed an eCognition system by utilizing satellites that had the ability to provide environmental monitoring, disaster management, and explore geographies to obtain valuable information. This system may be applied to a wide variety of fields because GIS can insert, store, inquire about, and analyze geographic information [7]. In addition, GIS contains mechanisms to store any spatial characteristics of the data collected in different image layouts. Most interestingly, GIS can overlay a digital world over top of the real world landscape and allows the user to manage the layout and organization of the space through different image layouts in a way that suits the ultimate aim of the end goal [8]. The use of GIS for research purposes was recently employed by Clarke and Gaydos [9] to analyze geographic information in various locations within the United States of America. They selected Washington and California as two areas to study in detail as a reference and potential city blueprint for developing a suitable model city structure of the future. Figure 3b demonstrates how Professor Yo and team [10] applied GIS to protect, examine, and execute plans for management of precious water resources. Our survey of the related publications over the last several years has demonstrated that GIS as a teaching tool truly opens up many possibilities in the area of education. The ability of this system to collect information about existing landscapes coupled with the functionality of modifying and adding custom layouts means more can be accomplished through meta-analysis with space and time.

Therefore, our study proposed an integrated learning platform that uses information systems, cartography, and additional multimedia elements to expand the scope of teaching within general education curriculums. We have based our research on the capabilities of GIS to take advantage of space and time dynamics to

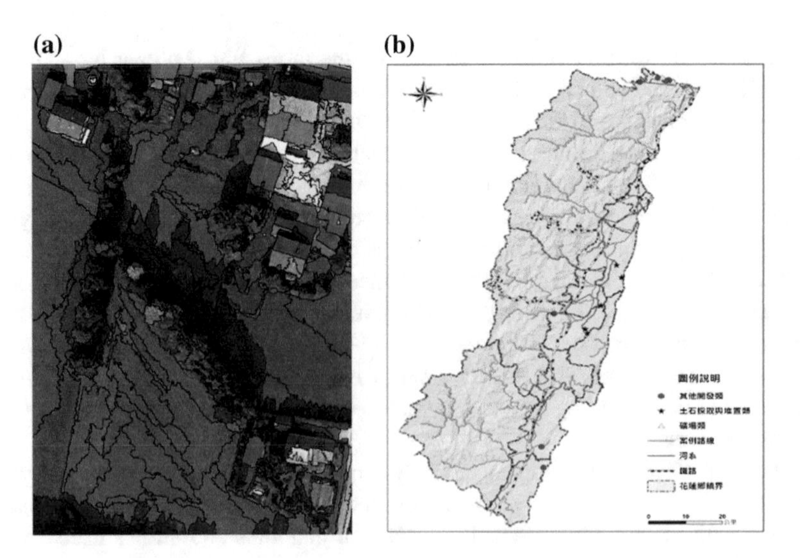

Fig. 3 **a** Cut results of terrain by using eCognition [6]. **b** Case position and roadmap [10]

enhance the selected course curriculum surrounding the historical and cultural contributions of Dr. MacKay. In the year 1872, Dr. MacKay arrived in Tamsui to preach but he ended up profoundly affecting the region with his contributions. He was the founding force behind Oxford College at Aletheia University, the first western-style institution that opened in 1882. Even after 132 years, Oxford College continues to propagate a strong spirit of educational inquisitiveness and maintains its status as a hub of precious cultural information that holds a special place in Taiwanese modern history. Dr. MacKay has left a lasting impression on the people, culture and landscape of Tamsui and some of the buildings he resided and worked in are still standing today, including the Reverend MacKay Residence, Oxford College, Missionary House, The House of Reverends, and The House of Maidens, to name a few. The relics and personal effects that remain have been used as curriculum material for courses at the college general education level.

Furthermore, this study developed novel educational content using information from MacKay's life and legacy. Our study constructed "The Multimedia Database of MacKay Cultural Heritage" and "The Learning Platform of MacKay Culture" to show the spatial distribution of the locations in Taiwan where MacKay lived, as well as where he conducted his educational and medical pursuits. This study provided students with an understanding of MacKay's life story and showed what it was like for a pioneer of the nineteenth century to navigate life in northern Taiwan. We included cartography and multimedia resources to provide students with self-directed controls and operations to easily explore and observe Dr. MacKay's educational spirit and the resulting impact he had on Taiwanese culture.

2 The Multimedia Database of MacKay Cultural Heritage

As shown in Fig. 4, the first component of our system was the "Multimedia Database of MacKay Cultural Heritage", composed of five major modules: User Manager Module, Building Manager Module, Multimedia Manager Module, User Course Manager Module, and the Message Manager Module. These modules provide users with the course content and information necessary for integrating and sharing knowledge with others regarding Dr. MacKay's effect on Taiwanese cultural heritage. We designed the system so that the "Multimedia Database of MacKay Cultural Heritage" could increase the availability of tourism resources through a web interface, while the "Multimedia Database of Mackay Cultural Heritage" could promote the acquisition of in-depth cultural knowledge. In addition, associated peripheral facts that could be used to fill gaps in understanding for participants to make connections and draw conclusions from the information presented.

The User Manager Module interface is shown in Fig. 5a, this module provides access to a main manager to manage, create, and modify user information. Figure 5b is a screenshot of the Building Manager Module interface, which is used to create and manage information associated with the buildings of cultural

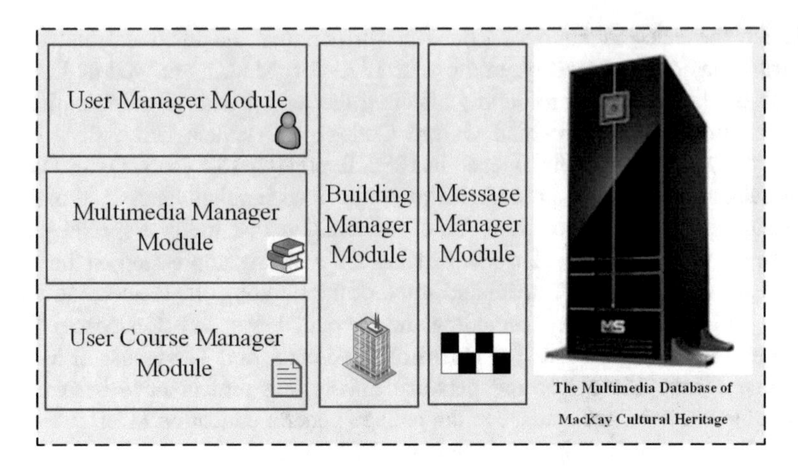

Fig. 4 The architecture of The Multimedia Database of MacKay Cultural Heritage

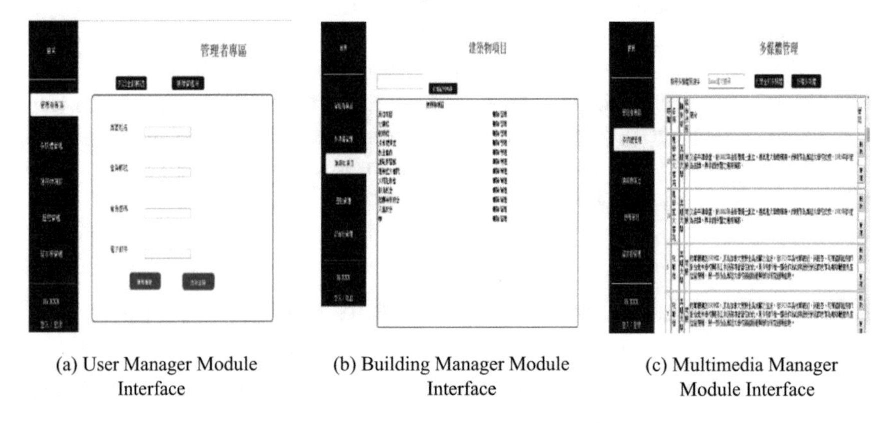

(a) User Manager Module Interface (b) Building Manager Module Interface (c) Multimedia Manager Module Interface

Fig. 5 The Multimedia Database of MacKay Cultural Heritage

significance in MacKay's legacy. Next, Fig. 5c shows the Multimedia Manager Module, which has the function of managing multimedia files associated with Mackay's cultural heritage and this module is expandable to accommodate the addition of new files into the database. The Multimedia Manager Module contains features that can edit, manage, and share with others via the Internet for those who are active on social media and the World Wide Web.

The User Course Manager Module can record the number of clicks on a page, track the number of multimedia sources that are being accessed and identify user information related to specific multimedia projects. The interface of the User Course Manager Module is easy to navigate and the data display window is able to chart user information in graph form, as shown in Fig. 6a. The final module in the system is captured in Fig. 6b; this Message Manager Module is the communication

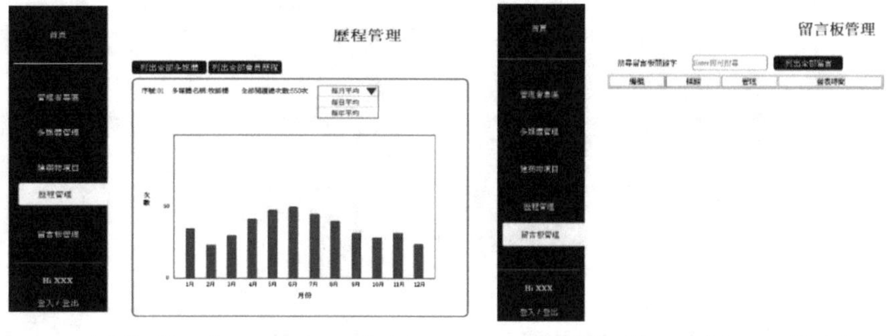

(a) The User Course Manager Module (b) The Message Manager Module

Fig. 6 The Multimedia Database of MacKay Cultural Heritage

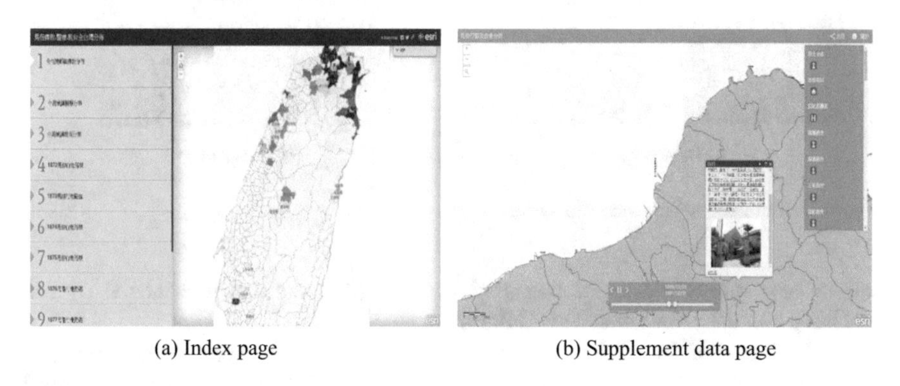

(a) Index page (b) Supplement data page

Fig. 7 GIS content sub-system

center of the system, which allows for management of messages that are directed towards the system administrator. Students and other users of the system with inquiries have the ability to contact the administrator and responses can be managed and sent from this interface.

3 The Learning Platform of MacKay Culture

This study was developed with the intention of creating The Learning Platform of MacKay Culture for university level students to learn more about the contributions and impact of Dr. MacKay. The Learning Platform of MacKay Culture consists of two sub-systems that are available to students in their learning process. The GIS content sub-system is shown in Fig. 7, which displays map information using GIS-style interface; while the multimedia appearance sub-system is shown in Fig. 8, which presents multimedia content using Web-style interface. Through the

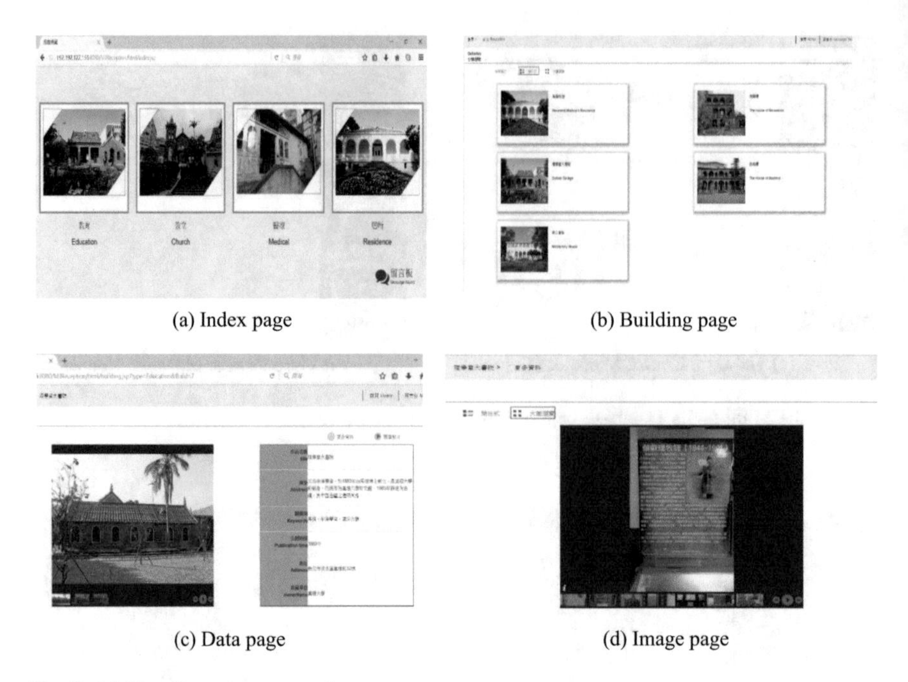

(a) Index page

(b) Building page

(c) Data page

(d) Image page

Fig. 8 Multimedia appearance sub-system

use of two sub-systems, our proposed learning platform can present the overall teaching content about MacKay Culture.

In the index of the GIS content sub-system, users have access to geographical information that details the locations where Dr. MacKay preached, practiced medicine, and taught in Taiwan, as seen in Fig. 7a. The GIS content sub-system was designed to guide students in exploration of Dr. MacKay's movements throughout Taiwan's north, central, south, and east regions and allow students to uncover the processes and historical challenges that were present at that time. Students can better understand Dr. MacKay's movements process, and synthesize information based on his contributions and experiences to construct a MacKay culture knowledge map during their learning endeavors. Figure 7b is a screenshot of the GIS content sub-system with supplemental data displayed on the screen. If users click on a pre-defined target on the map, a supplemental data dialog box will appear that contains hyperlinks to additional multimedia content and resources. This function allows students to obtain more knowledge about the selected topic and provides a degree of control over the learning process.

The index page of the multimedia appearance sub-system is shown in Fig. 8a. This page categorizes information about Dr. MacKay into four general areas: education, church, medical, and residence. These general categories are intended for the student to choose learning content that is the most interesting and relevant to their learning needs. Figure 8b is the building page of the multimedia appearance sub-system where students can browse and obtain detailed building information by clicking into specific building pages. For example, selecting the "Education" option from the index page will lead the user to a new interface with a selection of five buildings on the building page. This design layout was selected so students could search for information related to MacKay quickly and efficiently. From the data page of the multimedia appearance sub-system, students can read about special features of different buildings, including information containing the official name, a brief introduction to the property, keywords, publication time, address and history, as shown in Fig. 8c. In addition, this page enables students to watch videos and browse through related images. The final page is the image page of the multimedia appearance sub-system, as shown in Fig. 8d. This page shows historically and culturally significant information that is relevant to MacKay through both images and text features.

4 Conclusion and Future

This system was developed based on existing educational content related to Dr. MacKay and we wanted to bring to light the impact that he made on Taiwanese culture and society. Our study has made Dr. MacKay's contribution to Taiwanese history and culture more accessible to the public by incorporating these important records into a digital-style learning system. This study provided a platform for the "Multimedia Database of MacKay Cultural Heritage" and the "Learning Platform of MacKay Culture" where information on Dr. MacKay could be collected and aggregated into one place. This learning system has the potential to guide students for generations to come in their learning about Dr. MacKay's educational spirit and the cultural contribution that was a part of his legacy. Our proposed system allows students to learn in a new and different way, where self-direction and self-motivated learning can happen for students who are interested and more engaged in the learning process. This study and the novel application of technologies to the field of teaching can expand the reach of how general education courses are delivered to the student to improve knowledge retention and significantly change the way students learn new information.

Acknowledgements This work was supported by the National Science Council of the Republic of China under Contract No. MOST 104-2632-M-156-001.

References

1. J. Langone, D. M. Malone, P. M. Stecker, and E. Greene. "A Comparison of Traditional Classroom Instruction and Anchored Instruction with University General Education Students". *Journal of Special Education Technology*, Vol. 13, no. 4, pp. 99–109, 1998.
2. S. Springer, L. Colins. "Interacting inside and outside of the language classroom". *Language Teaching Research*, Vol. 12, no. 1, pp. 39–60, 2008.

3. G. J. Hwang, C. C. Tsai, and S. J. Yang. "Criteria, Strategies and Research Issues of Context-Aware Ubiquitous Learning". Educational *Technology & Society*, Vol. 11, no. 2, pp. 81–91, 2008.
4. C. C. Chen, and T. C. "Huang. Learning in a u-Museum: Developing a context-aware ubiquitous learning environment". *Computers & Education*, Vol. 59, no. 3, pp. 873–883, 2012.
5. Á. Di Serio, M. B. Ibáñez, and C. D. Kloos. "Impact of an augmented reality system on students' motivation for a visual art course". *Computers & Education*, vol. 68, pp. 586–596, 2013.
6. U. C. Benz, P. Hofmann, G. Willhauck, I. Lingenfelder, and M. Heyne. "Multi-resolution, object-oriented fuzzy analysis of remote sensing data for GIS-ready information". I*SPRS Journal of photogrammetry and remote sensing*, vol. 58, no. 3, pp. 239–258, 2004.
7. 許文國、張智傑、白璧玲、廖泫銘(民100)。GIS 與BIM 之整合應用。台灣地理資訊學會年會暨學術研討會。台大醫院國際會議中心。
8. 趙忠明、周天穎、嚴泰來(民100)。空間資訊技術原理及其應用理論基礎篇。新北市: 儒林圖書。
9. K. C. Clarke, and L. J. Gaydos. "Loose-coupling a cellular automaton model and GIS: long-term urban growth prediction for San Francisco and Washington/Baltimore". *International journal of geographical information science*, vol. 12, no. 7, pp. 699–714, 1998.
10. 游冠秀(民103)。GIS軟體應用於水保檢查作業之研究。國立宜蘭大學,土木工程學系碩士班,碩士論文。

SBAR: A Framework to Support Learning Path Adaptation in Mobile Learning

Alva Muhammad, Jun Shen, Ghassan Beydoun and Dongming Xu

Abstract Most of the previous studies in mobile learning focused on pedagogical or technical implementation. However, very little attention was paid to address the abstract model of the general view of adaptation particularly in ubiquitous environments. The main aim of this study is to propose a comprehensive framework for supporting adaptation, more precisely the learning path adaptation, in mobile and ubiquitous learning environment. We introduce Situation, Background, Assessment, and Recommendation (SBAR) framework to define the conceptual model of the context and adaptation in the mobile learning environment. The conceptual model is explored by a scenario of daily activities in ubiquitous environments.

Keywords Framework · SBAR · Adaptation · Learning path · Learning path adaptation · Mobile learning · Ubiquitous learning

1 Introduction

The rapid development of mobile technology in recent years significantly promoted the development of mobile-based learning. The mobile devices were used by nearly all students, particularly in the universities, to support their daily activities, such as instant messaging, sharing files, social networks, or blogging. It is becoming impossible to ignore the existence of a mobile device as a valuable tool for sharing,

A. Muhammad (✉) · J. Shen · G. Beydoun
University of Wollongong, Wollongong, Australia
e-mail: ahm145@uowmail.edu.au

J. Shen
e-mail: jshen@uow.edu.au

G. Beydoun
e-mail: beydoun@uow.edu.au

D. Xu
The University of Queensland, Brisbane, Australia
e-mail: d.xu@business.uq.edu.au

© Springer Nature Singapore Pte Ltd. 2018

N.Y. Yen and J.C. Hung (eds.), *Frontier Computing*, Lecture Notes in Electrical Engineering 422, DOI 10.1007/978-981-10-3187-8_62

collaborating, and communicating with peers. Mobile learning also closes the gap when the students do not attend class but want to take advantage of its gadget support for learning on the sidelines of activity. These facts have led to a popularization of mobile learning as a learning tool for students, which is characterized by the proliferation of mobile learning variations, such as context-based learning [1], collaborative learning [2], ubiquitous learning [3], and seamless learning [4].

We will use the term mobile learning and ubiquitous learning interchangeably in this paper, although the term ubiquitous learning is often described as a mobile learning form, which enhanced with an intelligent environment and context awareness [5]. In contrast to other learning activity, the activity in mobile learning occurred when a learner is on the move to a location and still connected to learning resources with their mobile devices. So with the support of mobile devices, the learners can learn not only "anytime, anywhere", but also learn the right thing, at the right place and the right time [3, 6]. When the learners are in motion with their mobile device, the ubiquitous system dynamically computes to support learners by communicating with embedded devices in the environment. However, the idea of learning on the move results in a series of follow-up challenges. The challenges are ranging from technological (connectivity, learning object design, bandwidth) to pedagogical (determining what is right material, how to learn, the effective learning styles) or a mixed one (personalizing learning under certain situation).

Through a lengthy discussion on mobile learning since a decade ago, the problems arise are still of great importance nowadays, especially in the context of personalization or adaptation for improving learning outcome [7]. The literature suggests that an online learning system should personalize or adapt learning contents to match their learner's needs and characteristics [8, 9]. However, the discussions in the literature are mainly focused on adapting the learners with various characteristics, such as learning style, prior knowledge, or motivation. This discussion leaves a gap in adapting learning path on mobile learning based on situation identification, so that the system can provide learners with the right learning materials related to the given situation.

Thus, in this paper, we present the framework of an adaptive system for mobile and ubiquitous learning that takes into account situational aspects of the environment and the background of the learner to recommend sequences of learning activities in a certain order. Specifically, this paper offers a new model for understanding the design and implementation of learning path adaptation based on situational identification in the ubiquitous mobile learning environment. Finding learning path in online learning is a challenging issue while a large number of resources are established in a system [10]. In our view, the adaptation of learning path within mobile learning environment can be seen as a way to improve lifelong learning and to contribute in creating great opportunities for ubiquitous and seamless learning.

The rest of this paper is structured as follows. The next section discusses the related works and research issues in mobile learning. Section 3 introduces the framework of this study. In Sect. 4 develops an adaptation model and methods of framework. The Sect. 5 is discussion and analysis to show that the proposed framework is effective for the adaptive learning path in ubiquitous learning. Finally, the last section concludes this paper and also provides pointers for further work.

2 Related Works

This section discusses several works related to pedagogical and technical design in mobile learning and also the adaptation issues in the mobile learning environment.

2.1 Research Issues in Mobile Learning

Previous studies reported that the potential usefulness of mobile devices to improve learning activities are not just because they are mobile [11, 12]. However, the main pedagogical issue is about how to utilize mobile learning activities to seamlessly support the regular curriculum [11]. As a consequence to address this issue, the *mobility of learning* should be placed as the object of analysis [6]. In this way, it is easy to understand how knowledge can be transferred across contexts, how learning can be managed across life transitions, and how new technologies can be designed to support learner on the move in daily life. Some researchers have emphasized the pedagogical principles based on the attributes of the mobile learning environment to improve the mobile learning process. For example, the inquiry-based learning in science using mobile devices have proven could increase the learners' activities [13, 14]. Some evidence were also supporting the claim that combining mobile learning with appropriate pedagogical approaches can create special educational value for students' science learning [15].

From the technological perspective, most research on mobile learning has been focused on design, integration, and their evaluation. Additionally, most research on mobile learning indicate the trend to "support adaptation or personalized contextual, easily retrievable, auto-updated, and intelligent pushed content" [16]. As the learning content became more personalized, the mobile learning system must provide adaptation of learning content, organization of learning form, presentation of learning materials and interaction with devices according to learners' profiles or learning purposes. To provide adaptation, the mobile learning system should collect and analyze the data from relevant students' learning information, such as learning style, learning behaviors, and learning progress. There are several frameworks for adapting mobile learning that has been proposed in the literature. For instances, a machine learning based framework using ANFIS has been proposed as adaptive engine in mobile learning [7]. The foundation of the framework is to adapt learning content, location and presentation of the users. Another framework based on context-aware also provide adaptation in ubiquitous learning, an adaptation engine senses the context and produces adapted educational activity and infrastructure

[17]. Most previous studies on adaptive mobile learning proposed to adapt the interface, the content, or the learning style [7, 18]. In this paper, we propose the adaptation based on learning path in mobile learning.

2.2 Learning Path Adaptation

The term 'learning path' in this paper will be used to refer to the *sequence* of learning material which is designated to improve the student's knowledge in mobile learning activities. The purpose of sequencing learning path is to provide learners with the most suitable individual learning object to learn according to the learner characteristics. The concept of learning path adaptation is developed from the field of curriculum sequencing in adaptive hypermedia system [19], which is usually driven by sequencing rules produced in the courseware. The choice and sequence of knowledge elements can be determined by several characteristics of the student, for instance by their learning styles, preferences, abilities or by some constraints, such prerequisites of the course and length of study.

Most research in learning path adaptation formulates the learning path where the objective is to minimize the path or route of learning [20]. Furthermore, the problem formulation in learning path adaptation can be modeled as Traveling Salesman Problem (TSP), Constraint Satisfaction Problem, Multi-Objective Optimization Problem and Weighted Directed Graph [10]. Once the model is constructed, the learning path generation algorithm searches the best possible match between each student and the learning objects or pedagogical requirements on the contrary. The bio-inspired algorithm, such as Genetic Algorithm, Ant Colony Optimization, and Particle Swarm Optimization have successfully demonstrated to solve the problem in constructing learning path [19, 21, 22, 23, 24].

3 Situation, Background, Assessment, and Recommendation (SBAR) Framework

This section discusses the overall framework which proposed in this research. The first part provides a basic foundation of the framework. The implementation of the model and adaptation process is discussed in the second part.

3.1 Basic Foundation

The process of mobile learning at the very basic level, like the process in traditional e-learning, can be identified as an interaction among the following components:

(i) pedagogy-driven activities, (ii) technology-driven processes, and (iii) knowledge transfer with the actors involved [25]. Furthermore, the whole cycle will include the following operations: design, search, adapt and use the content. To effectively manage the entire learning in ubiquitous environment, we introduce the Situation, Background, Assessment and Recommendation (SBAR) model as a conceptual framework. The SBAR technique has been widely implemented at healthcare systems to communicate the patient's condition from team members to another [26, 27].

The idea of implementing SBAR framework in this research is to present the considered challenges from several complexities when adapting learning path in mobile learning. In this work, we modeled the framework into two main domains: Context and Adaptation. Then, the framework consists of four main layers: Situation, Background, Assessment, and Recommendation. Furthermore, in each layer consist of several dimensions and variables. We use the idea from [17] to define each variable and attributes in the context. Figure 1 summarizes the overall framework which has four main layers:

1. *Situation*, that focus on handling the situation when learning occurs. This first layer is characterized by the following variables: environment state (e.g. activities, location, time, availability and current status) and infrastructure state (e.g. network connectivity, bandwidth, devices, user interface, input and output).
2. *Background*, that focus on defining the state of the learner history. This second layer is characterized by the following variables: learner state (e.g. level of knowledge, learning style, motivation, cognitive abilities and styles, personal preferences) and educational state. (e.g. subject, learning objectives, requirements and prerequisites, material types, instructional design, content, presentation and media).

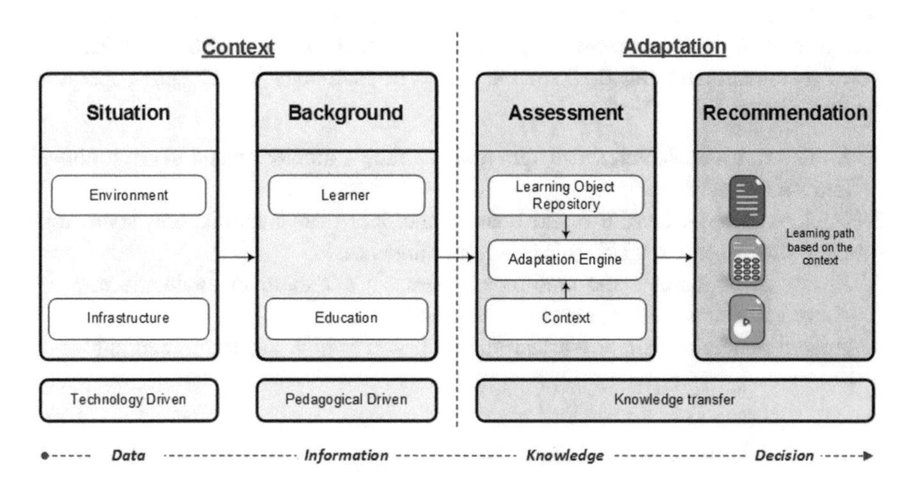

Fig. 1 The SBAR framework

3. *Assessment*, which summarizes the facts and gives an assessment to provide a decision. This layer is characterized by the adaptive engine to select the most appropriate content that reinforces the good decision for the learner.
4. *Recommendation*, that present the decision related to learning activities that should be taken by the learner. This domain is characterized by the presentation layer to display the result for the learner.

From the description above, the framework allows mapping the core features of the learner and their learning activity on a set of operational items which are exploited for adaptation purpose. This approach can be used to improve the effectiveness of adaptation for the learner. In the next section, the analysis of the adaptation process is presented.

3.2 Adaptation Process

In general, the adaptation in e-learning aims to adjust the educational environment by accommodating learners' needs or personalizing the learning content. However, the adaptation in mobile and ubiquitous learning is different from adaptation in other e-learning activities, which is usually emphasizing in learner's profile, learning style or content. Adaptation in ubiquitous learning is closely related to the context in which the learner is doing an activity. For example, when the learner is on the motion, then the assessment will focus on adapting the situation over the background. On the contrary, when the learner in idle situation, the adaptation focus on the background rather than the situation.

From the framework, the context domain will sense the data related to situation and background, then manage it and use the data to predict the future state. The adaptation process lies in assessment layer, which responsible for interpreting the situation and background. After the adaptation engine assesses the situation and background of the learner, the adaptation engine will recommend the learning path based on situational context. Thus, the phases of each layer in adaptation operation are as follow:

1. Situation: track, detect, sense, and monitor the parameter related to environment and infrastructures.
2. Background: retrieve, use and manage the data from identification layer, estimate the context related to learner and education.
3. Assessment: manage the context of learner's situation, evaluate the context, choose relevant content relate to the context.
4. Recommendation: adapt the learning path and predict the future content.

The adaptation model above may be applied generally in any ubiquitous environment. In the next section will discuss the adaptive learning path based on the context of the situation.

4 Enrichment of Learning Path Adaptation

The main aim of the paper is to investigate how this model could identify the activities done by learner and predict learning activities in the model. The adaptation engine in assessment layer will read the context and match with the learning object repository to produces the recommendation module for the learner. The main challenge in adapting learning path relates with pairing the massive learning object (LO) with the context of learning. In general, the adaptation engine makes recommendation according to following steps [28]: the process begins with manipulating learning objects by labeling the metadata to support adaptation. Then, the adaptation engine selects the LO and modeling the course to make any adjustment relate to the context of learning.

The primary process is sequencing the LO and automatically build the course structure through adaptation aspects. The course is then created by mapping the learning objects with the context. The mapping process may be modeled in various ways, such as constraint satisfaction problem [19] or multi-objective optimization problem [21]. After the model has been formulated, several methods based on heuristic algorithm could be chosen to solve the problem. Some authors have considered the evolutionary algorithm, such as genetic algorithm, particle swarm optimization, or ant colony optimization to solve the learning path adaptation [23, 24, 29]. After the sequencing stages, the result of sequencing process is a tailored learning path or a sequence of learning objects that suit the context of learning. Lastly, the system needs to translate the sequencing learning object and present it so that the student could navigate module through the contents. Figure 2 illustrates the entire process of adaptation engine in assessment layer.

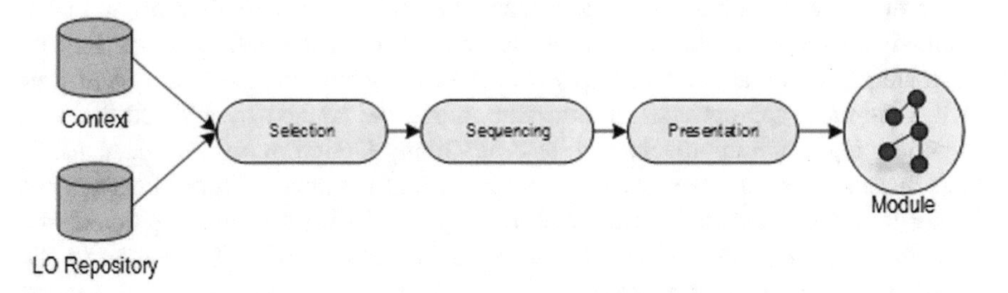

Fig. 2 The adaptation engine

5 Discussion and Analysis of the Framework

This section discusses the specific situation by which the proposed framework can provide detail analysis of the adaptation process in mobile learning. A portrait of a student's daily activities can demonstrate how to utilize the mobile device to support ubiquitous learning.

Assume that John is a second years' student who studies Bachelor of Computer Science. He is currently studying Data Mining this session (week 4). John is an active student who enjoys working with a picture and also trying a new thing. Today at 9 AM, on his way to campus by bus, he has 15 min times to learn using his mobile device. He uses headphone in his Android phone (OS Lollipop; 1136 × 640 pixels), with 4G internet connection. John learns classification technique using a decision tree. The system provided learning materials in video, presentation file and example of program algorithm. John is watching the video, however when the internet connection is weak, the system recommends the presentation file.

From the above situation, we may identify the initial condition using SBAR framework to predict the future learning path. In the context domain, the scenario above can be identified as follow:

- Situation layer:
  - *Environment* (*En*): sitting on the bus, 9 AM, 15 min times.
  - *Infrastructure* (*In*): headphone, Android phone, OS Lollipop, 1136 × 640 pixels, 4G connection.
- Background layer:
  - *Learner* (*Ln*): John, student Bachelor of Computer Science, second year, learning style active and visual.
  - *Education* (*Ed*): data mining, week 4.

Thus, the process in the adaptation domain can be described as follow:

- Assessment layer: first, the engine identifies the En, In, Ln, and Ed. If En is on idle and In is good, then the video format is provided to the learner.
- Recommendation layer: display video that appropriate to the device.

In this case, we assume that the presented path is the material that has not been studied, so the path that has been studied is not taken into account for the assessment. When adaptation is performed in the given context, the system proposes the most appropriate learning path among the context by adopted the evolutionary algorithm approach, such as Ant Colony Optimization (ACO). In ACO, ants seek the shortest path from their colony and a source of food. Pheromones along the route (arcs) are represented as weights. For the adaptation purpose, the system pairs the context with the learning object and calculate the weight as the heuristic value to the ant colony. The fitness of an arc can be set on several objective functions in mobile learning, such as the location, time available for study, or the network condition. The most important points that are emphasized in ubiquitous learning environments are mainly in the intelligent environment and context awareness. The above situation shows that learners with their personal background and education setting can take learning opportunities directly from the situation where they occurred, as the system support him by mobile devices at hand.

6 Conclusion

Though many adaptation framework approaches are proposed in recent years, only a few of them are suitable for situational adaptation, to result in an efficient tool for mobile and ubiquitous environments. In this paper, we discuss the furthering method of the adaptation in mobile learning, making it efficient in predicting the future learning path. The overall framework of Situation, Background, Assessment, and Recommendation (SBAR) have been described. This framework offers a new approach for understanding the context and providing learning path adaptation to improve active learning. The framework consists of two dimensions: context and adaptation. Each dimension consists of several attributes and variables. The presented framework may help researchers of mobile and ubiquitous learning to identify the situation and opportunities to provide further adaptation for the learner. In the future investigation might be possible to use the real situational identification and to optimize the provided learning path to enhance learning outcome in mobile learning.

Acknowledgements The first author gratefully acknowledges support from the Indonesia Endowment Fund for Education (LPDP), Ministry of Finance, the Republic of Indonesia.

References

1. Wen-Chung, S. and S.-S. Tseng, *A Knowledge-based Approach to Retrieving Teaching Materials for Context-aware Learning*. Journal of Educational Technology & Society, 2009. **12**(1): p. 82–106.
2. Huang, J.J.S., et al., *Social Learning Networks: Build Mobile Learning Networks Based on Collaborative Services*. Journal of Educational Technology & Society, 2010. **13**(3): p. 78–92.
3. Peng, H., et al., *Ubiquitous knowledge construction: mobile learning re-defined and a conceptual framework*. Innovations in Education and Teaching International, 2009. **46**(2): p. 171–183.
4. Looi, C.-K., et al., *Leveraging mobile technology for sustainable seamless learning: a research agenda*. British Journal of Educational Technology, 2010. **41**(2): p. 154–169.
5. Vinu, P.V., P.C. Sherimon, and R. Krishnan, *Towards pervasive mobile learning – the vision of 21st century*. Procedia - Social and Behavioral Sciences, 2011. **15**: p. 3067–3073.
6. Sharples, M., J. Taylor, and G. Vavoula, *A theory of learning for the mobile age*, in *Medienbildung in neuen Kulturräumen*. 2010, Springer. p. 87–99.
7. Al-Hmouz, A., et al. *Modeling Mobile Learning System Using ANFIS*. in *Advanced Learning Technologies (ICALT), 2011 11th IEEE International Conference on*. 2011.

8. Xu, D., et al., *Enhancing e-learning effectiveness using an intelligent agent-supported personalized virtual learning environment: An empirical investigation.* Information & Management, 2014. **51**(4): p. 430–440.
9. Premlatha, K.R. and T.V. Geetha, *Learning content design and learner adaptation for adaptive e-learning environment: a survey.* Artificial Intelligence Review, 2015: p. 1–23.
10. Muhammad, A., et al., *Learning Path Adaptation in Online Learning Systems*, in *Computer Supported Cooperative Work in Design (CSCWD), 2016 IEEE 20th International Conference on.* in-press, IEEE: Nanchang.
11. Malandrino, D., et al., *A Tailorable Infrastructure to Enhance Mobile Seamless Learning.* Learning Technologies, IEEE Transactions on, 2015. **8**(1): p. 18–30.
12. Motiwalla, L.F., *Mobile learning: A framework and evaluation.* Computers & Education, 2007. **49**(3): p. 581–596.
13. Song, Y., L.-H. Wong, and C.-K. Looi, *Fostering personalized learning in science inquiry supported by mobile technologies.* Educational Technology Research and Development, 2012. **60**(4): p. 679–701.
14. Jones, A.C., E. Scanlon, and G. Clough, *Mobile learning: Two case studies of supporting inquiry learning in informal and semiformal settings.* Computers & Education, 2013. **61**: p. 21–32.
15. Chee-Kit, L., S. Daner, and X. Wenting, *Exploring Students' Progression in an Inquiry Science Curriculum Enabled by Mobile Learning.* Learning Technologies, IEEE Transactions on, 2015. **8**(1): p. 43–54.
16. Chiang, F.-K., et al., *Research and trends in mobile learning from 1976 to 2013: A content analysis of patents in selected databases.* British Journal of Educational Technology, 2015.
17. Economides, A.A., *Context-Aware Mobile Learning*, in *The Open Knowledge Society. A Computer Science and Information Systems Manifesto: First World Summit on the Knowledge Society, WSKS 2008, Athens, Greece, September 24–26, 2008. Proceedings*, M.D. Lytras, et al., Editors. 2008, Springer Berlin Heidelberg: Berlin, Heidelberg. p. 213–220.
18. Tabuenca, B., et al., *Time will tell: The role of mobile learning analytics in self-regulated learning.* Computers & Education, 2015. **89**: p. 53–74.
19. de-Marcos, L., et al. *Competency-Based Intelligent Curriculum Sequencing Using Particle Swarms.* in *Advanced Learning Technologies, 2008. ICALT 2008. Eighth IEEE International Conference on.* 2008.
20. Al-Muhaideb, S. and M. Menai, *Evolutionary computation approaches to the Curriculum Sequencing problem.* Natural Computing, 2011. **10**(2): p. 891–920.
21. Seki, K., T. Matsui, and T. Okamoto, *An adaptive sequencing method of the learning objects for the e-learning environment.* Electronics and Communications in Japan (Part III: Fundamental Electronic Science), 2005. **88**(3): p. 54–71.
22. Dharani, B. and T.V. Geetha. *Adaptive learning path generation using colored Petri nets based on behavioral aspects.* in *Recent Trends in Information Technology (ICRTIT), 2013 International Conference on.* 2013.
23. Dharshini, A.P., S. Chandrakumarmangalam, and G. Arthi, *Ant colony optimization for competency based learning objects sequencing in e-learning.* Applied Mathematics and Computation, 2015. **263**: p. 332–341.
24. Chu, C.-P., Y.-C. Chang, and C.-C. Tsai, *PC2PSO: personalized e-course composition based on Particle Swarm Optimization.* Applied Intelligence, 2009. **34**(1): p. 141–154.
25. Štuikys, V., *Model-Driven Specification in Designing Smart LOs*, in *Smart Learning Objects for Smart Education in Computer Science: Theory, Methodology and Robot-Based Implementation.* 2015, Springer International Publishing: Cham. p. 103–122.
26. Haig, K.M., S. Sutton, and J. Whittington, *SBAR: a shared mental model for improving communication between clinicians.* The joint commission journal on quality and patient safety, 2006. **32**(3): p. 167–175.
27. Beckett, C.D. and G. Kipnis, *Collaborative Communication: Integrating SBAR to Improve Quality/Patient Safety Outcomes.* Journal for Healthcare Quality, 2009. **31**(5): p. 19–28.

28. Garrido, A. and E. Onaindia, *Assembling Learning Objects for Personalized Learning: An AI Planning Perspective.* Intelligent Systems, IEEE, 2013. **28**(2): p. 64–73.
29. Afsarmanesh, H. and J. Tanha, *A High Level Architecture for Personalized Learning in Collaborative Networks*, in *Collaborative Networks for a Sustainable World*, L. Camarinha-Matos, X. Boucher, and H. Afsarmanesh, Editors. 2010, Springer Berlin Heidelberg. p. 601–608.

Dynamic Data Representations for Spatio-temporal Data Visualization

Anthony Y. Chang and Chang Yu-Wen

Abstract Spatio-temporal computing can be viewed as a specific computation to link spatial features in physical or virtual spaces with simultaneously in discrete or continuous and asynchronous time steps. With the increase of temporal and spatial digital data sets, the importance of temporal, spatial, and spatio-temporal aggregation computation has been reflected in a significant number of disciplines. We develop a graph grammar based visual system with spatio-temporal relational grammars. The theoretical foundations are developed for reasoning tasks by the artificial intelligence and dynamic spatial interaction. We apply the spatio-temporal computing theory to process dynamic data visualization with interactions. The dynamic visual analytics designed to transform the data into some combination terms that we can understand more easily. The presentation algorithms may produce an interactive streaming media on the screen. We can also use motion as a display technique to represent data that is either static or dynamic.

Keywords Spatio-temporal computing · Data representation · Data visualization · Big data

1 Introduction

Representing spatio-temporal knowledge is an essential part of many computer applications. Researchers of artificial intelligence, linguistics, knowledge systems, and information science require a time model for knowledge representation and analysis.

Spatio-temporal computing can be viewed as a specific computation to link spatial features in physical or virtual spaces with simultaneously in discrete or continuous and asynchronous time steps. With the increase of temporal and spatial

A.Y. Chang (✉) · C. Yu-Wen
Department of Information Technology, Overseas Chinese University, Taichung, Taiwan, ROC
e-mail: achang@ocu.edu.tw

© Springer Nature Singapore Pte Ltd. 2018

N.Y. Yen and J.C. Hung (eds.), *Frontier Computing*, Lecture Notes in Electrical Engineering 422, DOI 10.1007/978-981-10-3187-8_63

digital data sets, the importance of temporal, spatial, and spatio-temporal aggregation computation has been reflected in a significant number of disciplines.

There are many possibilities for representing time and techniques to make interesting temporal patterns perceptible. Allen [1] has proposed an interval framework and Vilain and Kautz [2] have proposed a point framework for representing indefinite qualitative temporal information. These frameworks are influential and have been applied in diverse areas such as multimedia database [3], composite services [4], multimodal analysis of human behavior [5], semantic web [6], and data mining [7]. In this paper, two fundamental temporal frameworks are integrated and extended for spatial representations.

This paper proposes a graph grammar based visual system with spatio-temporal relational grammars. The theoretical foundations are developed for reasoning tasks by the artificial intelligence and dynamic spatial interaction.

2 Data Model About Temporal and Spatial Relations

Representation and Reasoning about temporal/spatial information play an important role in current computer science.

Based on qualitative point relations, we use an encoding method to generalize and prove the 13 interval exclusion relations. Suppose As and Ae are the starting and ending points of the line segment A. And, Bs and Be are those of B. We define a binary relation, ⊗, (either $<$, $=$, or $>$ for "A is before B", "A is the same as B", or "A is after B") of two points. The 13 interval relations (Fig. 1) introduced by Allen 0.

Based on qualitative point relations, we use an encoding method to generalize and prove the 13 interval exclusion relations. Suppose A_s and A_e are the starting and ending points of the line segment *A*. And, B_s and B_e are those of *B*. We define a binary relation, ⊗, (either $<$, $=$, or $>$ for "*A* is before *B*", "*A* is the same as *B*", or

X equal (e) Y:
X: Y:
X before (<) Y:
X before^{-1} (>) Y:
X meets (m) Y:
X meets^{-1} (mi) Y:
X overlaps (o) Y:
X overlaps^{-1} (oi) Y:
X during (d) Y:
X during^{-1} (di) Y:
X starts (s) Y:
X starts^{-1} (si) Y:
X finishes (f) Y:
X finishes^{-1} (fi) Y:

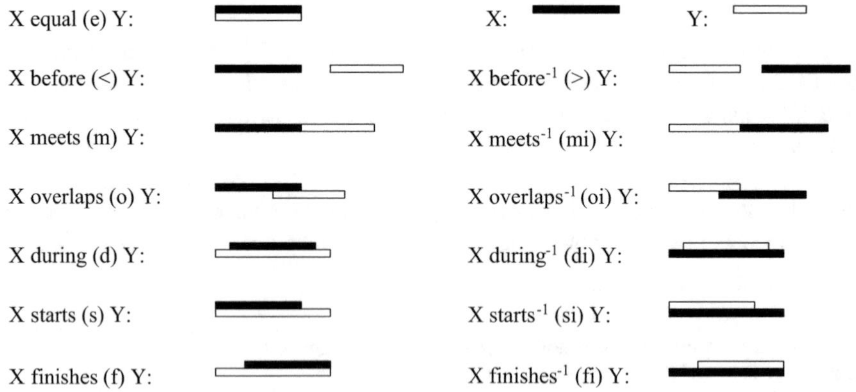

Fig. 1 The 13 interval relations

"*A* is after *B*") of two points. The 13 *interval relations* introduced by Allen [1] make the binary relations hold in the first part of the following table:

To analyze a generalized model of spatio-temporal relations, we consider the following situations. Conclusively, the generalized model of temporal/spatial relations include six cases:

- two points on a line
- two points on a plane
- two line segments on a line
- two line segments on a plane
- two 2-D objects on a plane
- two 3-D object project onto x-y, z-x and y-z planes

For an arbitrary pair of points, A and B, located on a 1-dimensional line, there are three point relations: A < B, A = B, or A > B, for A is before B, A is at the same position as B, and A is after B, respectively. If these two points are located on a 2-dimensional plane, there exists nine (i.e., 3 * 3) cases. The X and the Y coordinates of these two points on the plane are independent. The possible relations between these two points on a plane can be denoted as A (<, <) B, A (<, =) B, A (<, >) B, A (=, <) B, A (=, =) B, A (=, >) B, A (>, <) B, A (>, =) B, and A (>, >) B, where the first element in the pair representing a point relation denotes the order on the X coordinate while the second is for the Y coordinate. Considering two line segments located on a 1-dimensional line, the situation becomes complicated. Since each line segment has a starting point and an ending point, we analyze the spatio-temporal relations of two line segments according to these points. Allen's research is the special case of two line segments on the 1-dimensional line, with each line segment of length greater than zero.

Considering two line segments on a 2-dimensional plane, according to the above table and since the position of these two line segments are independent at the X and the Y coordinates, there exists 182 = 324 possible relations between these two line segments on a plane. These relations, similar to those of two points on a plane, are denotes by pairs as: (<, <), (<, >), (<, d), (<, di),…, (00, 01e), and (00, 00). We use these 324 binary relations to model spatial point-interval relations of two lines on a plane. Suppose X_A and X_B are the projection of two segments A and B (Fig. 2). Y_A and Y_B are the projection on Y. X_A is {start} to X_B and Y_A is {before} to Y_B. We represent the spatial relation between A and B as (A, (s, <), B).

Relations of n-D objects can be used in object representation and recognition. A object in 3-D space can be projected onto y-x, z-x, and y-z planes. The projections correspond to surfaces generated from 3D objects. Similarly, a 2-D object is projected to x and y axes (Fig. 3). If we look at two objects in the n-dimensional space, we can project the positional relation between these two objects from n directions to n 1-D space. The projections of 2-D object are x-interval and y-interval, but not the point. Thus, an n-dimensional relation can be formularized by a conjunction of n 1-D interval relations. A conjunction of two 1-D relations, which

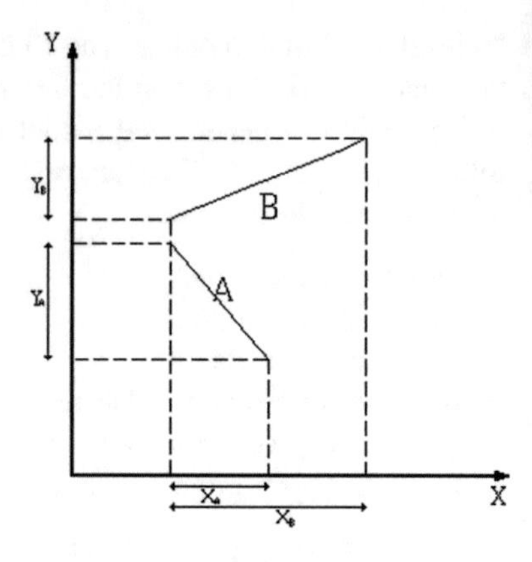

Fig. 2 Projection of two line segments on a 2-D plane

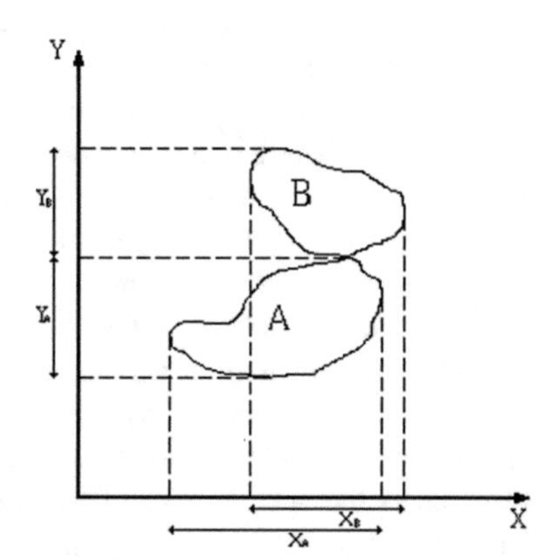

Fig. 3 Relations between two 2-D objects project onto x-y plane

denotes a 2-D relation, has 13^2 variations, i.e. {(<, <), (<, >), (<, d),…, (=, fi), (=, =)} where the first element in the pair representing a interval relation denotes the order on the X coordinate while the second is for the Y coordinate. Similarly, there are 13^3 3-D relations.

3 Similarity Distance of Relations

Relations are similar to each other in certain degree. For example, "during" and "starts" are similar since the only difference is the starting points of the two intervals are different. However, "before" and the inverse of "meets" are not quite the same.

In Table 1, each of the 13 interval relations and 5 point-interval relations are defined by four "⊗" relations. These relations can be used as a base of our evaluation criterion. A *relational-distance* of two "⊗" relations belong to two different temporal relations occurs if those two temporal relations hold different relations in the same column of Table 1.

Definition 3.1 *Point relation states* (PRS) defined with respect to a point relation r of index n have n incompatible differences from r. The following Table 2 gives a definition of point relation distance.

Definition 3.2 An *extended point-interval relation states* (EPIRS) defined with respect to a point-interval or interval relation r of index n have n incompatible differences from r. Let R and R′ are two interval relations or point-interval relations. The encoding point relation of R (see Table 1) is R_{AsBs}, R_{AsBe}, R_{AeBs}, R_{AeBe}, and the encoding point relation of R′ is R'_{AsBs}, R'_{AsBe}, R'_{AeBs}, R'_{AeBe}. We have a *EPIRS* formula:

$$EPIRS(\mathrm{R},\mathrm{R}') = PRS(\mathrm{R}_{As\otimes Bs}, \mathrm{R}'_{As\otimes Bs}) + PRS(\mathrm{R}_{As\otimes Be}, \mathrm{R}'_{As\otimes Be}) + PRS(\mathrm{R}_{As\otimes Bs}, \mathrm{R}'_{As\otimes Bs}) + PRS(\mathrm{R}_{As\otimes Be}, \mathrm{R}'_{As\otimes Be})$$

The index of EPIRS with respect to each temporal relation in Table 1 can also be retrieved from the length of a shortest path in a *distance graph* (see Fig. 4).

We have developed a fast computation mechanism to representation the varying of temporal relations. Using a bit-slicing representation of the relations, the index of image similarity can be computed in a few operations.

Table 1 Starting and ending point relation

$A_s \otimes B_s$	$A_s \otimes B_e$	$A_e \otimes B_s$	$A_e \otimes B_e$	ID	Point-interval relations	Simplified conditions
$<$	$<$	$<$	$<$	1	{$<$}	$A_e < B_b$
$>$	$>$	$>$	$>$	2	{$>$}	$A_b > B_e$
$>$	$<$	$>$	$<$	3	{d}	$A_b > B_b \wedge A_e < B_e$
$<$	$<$	$>$	$>$	4	{di}	$A_b < B_b \wedge A_e > B_e$
$<$	$<$	$>$	$<$	5	{o}	$A_b < B_b \wedge A_e < B_e$
$>$	$<$	$>$	$>$	6	{oi}	$A_b > B_b \wedge A_e > B_e$
$<$	$<$	$=$	$<$	7	{m}	$Ae = B_b$
$>$	$=$	$>$	$>$	8	{mi}	$A_b = B_e$
$=$	$<$	$>$	$<$	9	{s}	$A_b = B_b \wedge A_e < B_e$
$=$	$<$	$>$	$>$	10	{si}	$A_b = B_b \wedge A_e > B_e$
$>$	$<$	$>$	$=$	11	{f}	$A_b > B_b \wedge A_e = B_e$
$<$	$<$	$>$	$=$	12	{fi}	$A_b < B_b \wedge A_e = B_e$
$=$	$<$	$>$	$=$	13	{e}	$A_b = B_b \wedge A_e = B_e$
$=$	$=$	$>$	$>$	14	{los}	$A_b = B_b = B_e < A_e$
$<$	$<$	$=$	$=$	15	{loe}	$A_b > A_e = B_b = B_e$
$=$	$<$	$=$	$<$	16	{ols}	$A_b = A_e = B_b < B_e$
$>$	$=$	$>$	$=$	17	{ole}	$B_b > A_b = A_e = B_e$
$=$	$=$	$=$	$=$	18	{oo}	$A_b = A_e = B_b = B_e$

Table 2 Point relation states (PRS)

PRS	>	=	<
>	0	1	2
=	1	0	1
<	2	1	0

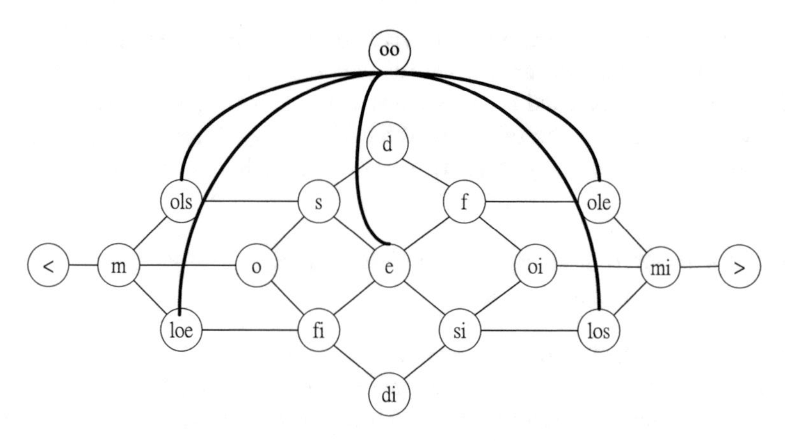

Fig. 4 States of well-defined interval relations

We use 10, 00, and 01 to represent the comparisons of end points of a line segment, >, =, and <, respectively. Eight bits are used to represent a relation of two line segments on a line. And 16 bits are used to represent a relation of two line segments on a plane. For instance, the two relations, d and mi, in the starting and ending point relation table in Table 1 can be represented as (10 01 10 01)2 and (10 00 10 10)2, respectively. An exclusive or logical operation is used to compute the similarity syndrome, which represents the difference with (00 01 00 11)2.

An integer array, A, of 2^8 or 2^{16} elements (for either temporal or spatial relation, respectively) is used to simulate the EPIRS index function. The similarity syndromes (represented as unsigned integers) are indices to the array. For instance, A (00 01 00 11) = 3, which represents that the two relations, {d} and {mi}, have three incompatible error bits at the second and the fourth "⊗" relation. Therefore, one logical operator and one direct table lookup compute the relational-distance. The efficiency is very important in computing polygon similarity.

According to Raghavan [8] representation, the Binary temporal relations, OCPN (object composition Petri nets) and PACFG (probabilistic, attributed context free grammar) could be translated each other. Agents may indicate their interest in one or more resources and time specifications in computational setting. We could derive the qualitative relations and quantitative values by closure table efficiently, then translate to Petri nets representations as basis of designs of user interface. Also, we can maintain the states constraints and resources in varying time series.

4 Resource Allocating for Actions with Time

Allen's work 0 discusses relations for 1-D objects (i.e. time intervals). This Section introduces a mechanism to extend the relations of objects to an n-dimensional space. Relations of 2-D objects can be used in screen layout designs. Relation of 3-D objects can be used in 3-D graphics, such as Virtual Reality applications. A cube in 3-D space can be projected onto a 2-D plane. Similarly, a square is projected to a line segment. If we look at two objects in the n-dimensional space, we can project the positional relation between these two objects from n directions to n 1-D space. Thus, an n-dimensional relation can be formularized by a conjunction of n 1-D interval relations.

Let rs denote a set of 1-D temporal interval relations (i.e., rs $\in$ 29Relset). The relation composition table discussed in 0 can be refined (e.q., make each relation as an atomic set of that relation) to a function maps from the Cartesian product of two rs to a rs. Assuming that f 1 is the mapping function interpreting Allen′s table, we can compute f2, the relation composition function of 2-D objects, and f3, the one for 3-D objects, from f 1. There are 13 relations for 1-D objects. A conjunction of two 1-D relations, which denotes a 2-D relation, has 13^2 variations. Similarly, there are 13^3 3-D relations. Fortunately, 4-D relations are not quite applicable and the memory space required for 2-D and 3-D relation tables is manageable by nowadays computers.

Since a 2-D relation is conjunction of two 1-D relations, we use the notation, rs1 $\times$ rs2, to denote a 2-D relation set where rs1 and rs2 are two 1-D relation sets. Thus f2 is a mapping from Cartesian product of two rs $\times$ rss to a rs $\times$ rs. Similarly f3 is obtained. The following are signatures of these functions:

f1 = 29RelSet o 29RelSet $\rightarrow$ 29RelSet
f2 = (29RelSet $\times$ 29RelSet) o (29RelSet $\times$ 29RelSet) $\rightarrow$ 29RelSet $\times$ 29RelSet
f3 = (29RelSet $\times$ 29RelSet $\times$ 29RelSet) o (29RelSet $\times$ 29RelSet $\times$ 29RelSet) $\rightarrow$ 29RelSet $\times$ 29RelSet $\times$ 29RelSet
where 29RelSet $\times$ 29RelSet $\in$ $\{\{<\} \times \{<\}, \{<\} \times \{>\}, \ldots, \{=\} \times \{=\}\}$
29RelSet $\times$ 29RelSet $\times$ 29RelSet $\in$ $\{\{<\}\times\{<\}\times\{<\}, \{<\}\times\{<\}\times\{>\},\ldots, \{=\}\times\{=\}\times\{=\}\}$.

Functions f 2 and f 3 are computed according to the following formulas:

$\forall$ i1 $\times$ j1, i2 $\times$ j2 $\in$ P (29RelSet $\times$ 29RelSet)
f2 (i1 $\times$ j1, i2 $\times$ j2) = $\prod$ f1 (i1, i2) $\times$ f1 (j1, j2)
$\forall$ i1 $\times$ j1 $\times$ k1, i2 $\times$ j2$\times$ k2 $\in$ P (29RelSet $\times$ 29RelSet $\times$ 29RelSet)
f3 (i1 $\times$ j1$\times$ k1, i2 $\times$ j2 $\times$ k2) = $\prod$ f1 (i1, i2) $\times$ f1 (j1, j2) $\times$ f1 (k1, k2)

where $\prod A \times B = \{a \times b \mid \forall a \in A, b \in B\}$

$\prod A \times B \times C = \{a \times b \times c \mid \forall a \in A, b \in B, c \in C\}$

The functions are implemented as table mappings. Table generated by the above formulas is stored in memory to reduce run-time computation load.

2-D projection representation is the simplest kind of spatial information of practical relevance. Image retrieval algorithms could be constructed based on projection relations. The system is able to handle rotated and reflected images. The matching mechanism focuses on the relative positions among a set of objects instead of searching a single object of a particular shape.

5 Conclusions

Constraint satisfaction techniques play an important role in current computer science. Many difficult problems involving search from areas such as ma-chine vision, scheduling, graph algorithms, machine design, and manufacturing can be considered to be the cases of the constraint satisfaction problem. If we can represent data values as visual features and group these features into visual objects, we will have a powerful tool for organizing related data.

This paper constructs an algebra system of spatio-temporal interval relations and the set of enhanced mechanism for spatio-temporal relation composition. The spatio-temporal approximate measuring algorithm proposed in this paper can be used in other computer applications for comparing temporal relations and spatial objects. We hope the spatio-temporal computing theory and the interactive dynamic data visualization could be used in many related applications such as intelligence lives, Internet of things technology, human computer interaction, scientific visualizations, cultural and historical convergence, information service technology, social networking and interaction of mobile applications…etc. [9].

References

1. James F. Allen: Maintaining Knowledge about Temporal Intervals. Communications of the ACM, vol. 26, No. 11 (1983)
2. M. Vilain and H. Kautz: Constraint Propagation Algorithms for Temporal Reasoning. In: Proceedings AAAI-86, Philadelphia, 1986, pp. 377–382. (1986)
3. Sujal Subhash Wattamwar and Hiranmay Ghosh: Spatio-Temporal Query for Multimedia Databases. Proceeding of the 2nd ACM Workshop on Multimedia Semantics, 2008, pp. 48–55. (2008)
4. Azlan Ismail, Jun Yan and Jun Shen: Verification of Composite Services with Temporal Consistency Checking and Temporal Satisfaction Estimation. Web Information Systems Engineering - WISE 2009. LNCS vol. 5802, pp. 343–350. (2009)
5. Chreston Miller and Francis Quek: Toward Multimodal Situated Analysis. Proceedings of the 13th International Conference on Multimodal Interfaces, 2011, pp. 239–246. ACM Press. (2011)
6. Jerry R. Hobbs and Feng Pan: An Ontology of Time for the Semantic Web. Transactions on Asian Language Information Processing (TALIP), vol. 3, Issue 1, 2004, pp. 66–85. (2004)
7. Haval Patel, Wynne Hsu and Mong Li Lee: Mining Relationships among Interval-Based Events for Classification. Proceedings of the 2008 ACM SIGMOD International Conference on Management of Data, pp. 393–404. (2008)

8. S.V. Raghavan, B. Prabhakaran and S.K. Tripathi: Synchronization Representation and Traffic Source Modeling in Orchestrated Presentation, Selected Areas in Communications, IEEE Journal on Volume: 14, No. 1, pp. 104–113. (1996)
9. Charles D. Hansen, Chris Johnson: The Visualization Handbook. Academic Press. (2004)

Commodities Selection of Supermarket Email-Flyers by Recommender Systems

Shih-Jung Wu

Abstract Traditional physical retail stores often attract customers by sending email-flyers (or promotional electronic Direct Mail, DM). However, most customers will treat electronic DM as junk mail if it includes too many types of commodities or is sent too frequently. To avoid this problem, we propose a recommendation algorithm to select a small and fixed number of commodities for preparing customized electronic DM suitable for a well-known Taiwanese supermarket. Our method first selects a small and fixed number of commodity types that customers will most likely purchase; then, at product level, the method selects one commodity of each commodity type to promotion. Considering the habits of customers have already purchased commodities will buy again, when the recommendation success rate for recommending commodity types (or commodities) reached a certain value, we could replace some commodity types (or commodities) those the customers had already purchased in the last month. Compared to two item-based collaborative filtering approaches, Cosine and Bigram, the experimental results show that our approach has higher recommendation success rate and then increases the possibility of customers back to physical retail stores to purchase.

Keywords Retailing · Recommender system · Collaborative filtering

1 Introduction

Due to the emergence of electronic commerce, consumption behavior is gradually changing from shopping at traditional, physical stores to shopping online, which is more convenient. Because of this change in consumption behavior, traditional physical retail stores are encouraged to change their sales and promotion modes. The primary sales modes of traditional physical retail stores include television

S.-J. Wu (✉)
Department of Innovative Information and Technology, Tamkang University, Yilan County 26247, Taiwan
e-mail: wushihjung@mail.tku.edu.tw

© Springer Nature Singapore Pte Ltd. 2018

N.Y. Yen and J.C. Hung (eds.), *Frontier Computing*, Lecture Notes in Electrical Engineering 422, DOI 10.1007/978-981-10-3187-8_64

advertising and sending both physical and electronic DM. Due to the relatively low cost of electronic DM, many traditional physical retail stores draw customers back to their physical retail stores by sending a large amount of electronic promotional DM, which includes all the latest promotional commodities.

The main purpose of sending electronic DM is to draw customers back to physical retail stores. However, not all of the commodities included in electronic DM are necessarily liked by customers. If electronic DM includes too many commodities or is sent too frequently, customers will require a copious amount of time to find the commodities that they like, which may give them a negative impression of electronic DM and cause them to treat it as junk mail. Recommender systems can be used to analyze and select the commodities recommended in DM and can be used to produce customized electronic DM for recommendations. Our recommendation algorithm considers both customer preferences and the commodity combination under the condition that customized electronic DM can only include a small fixed amount of commodity types. Thus, customers will only need to spend a short time reading electronic DM to find the commodities that they like. They then will become more interested in reading electronic DM, and the probability of their return to physical retail stores will be increased.

In this study, the customer transaction records of a well-known Taiwanese supermarket were used as test data. Unlike traditional supermarket, there are many imported goods, which cannot find in other supermarkets, in this supermarket that is located in a well-known department store. Most of the commodities sold in this supermarket are consumables. Food is a commodity that has repurchasing characteristics; therefore, unlike computer, communication and consumer electronics (3C products) and books, the probability that customers will repurchase the same food commodity is very high. Hence, our recommender system will first recommend the commodities that customers have already purchased. When customers return to physical retail stores to shop, we hope that they will also purchase commodities that they have never purchased in the last month. Therefore, when the recommended commodities have already drawn the customer to return to the physical retail stores, we can replace some of the commodity types that have never purchased by the specified customer in the last month, such that different commodities will be recommended.

The primary contributions of this study are: (1) a recommender system is applied to supermarkets; (2) the recommendation algorithm proposed in this study can be used not only to uncover the order of the commodities that the customers will most likely purchase but also to produce the most ideal commodity combination for recommendation, and thus the recommendation algorithm is suitable for customizing electronic DM; (3) to draw customers and increase the number of customers back to physical retail stores, our recommendation algorithm can also recommend commodity combinations that include both commodities those customers have already purchased and commodities those they have not purchased in last month. This is a new concept that is different from traditional recommender system which is applied in e-commerce. Utilizing customized electronic DM recommends commodities to increase the number of customers back to physical stores.

There are five sections in this paper. The motive and objective of the study are introduced in Sect. 1. Related studies on recommender systems are introduced in Sect. 2. The proposed recommender system is discussed in Sect. 3. Related experimental data are shown in Sect. 4. Conclusions are made and future work is outlined in Sect. 5.

2 Related Work

The study of recommender systems is a research area that focuses on practical guidance. Recommender systems facilitate the quick retrieval of information or commodities that are useful or interesting to us. In recent years, a number of online shops have built recommender systems. Collaborative filtering is used for analysis by most of the common recommender systems. In addition, the preferences of groups that share the same interest or have the same experience are used to recommend information that is appealing to customers [1–11].

The calculation methods for the correlation coefficient, which is most often mentioned in the collaborative filtering system, include the nearest neighbor algorithm, the Pearson correlation coefficient, cosine similarity, adjusted cosine similarity, the vector approach, the Euclidean distance and the Jaccard coefficient [6, 12–17]. The general calculation methods for the correlation coefficient focus exclusively on recommending commodities that customers have not purchased. However, customers of the large do-it-yourself home decoration store La Boîte à Outils display the consumption behavior of repurchasing the same commodity. Therefore, whether customers will repurchase the same commodity must be taken into consideration. Thus, a new calculation method has been proposed [10] to correct the correlation for repurchasing the same commodity type. First, similar commodities have been classified into one type, and the Bigram concept [18, 19] has been used to calculate the correlation among commodity types. In this study, the Bigram concept was also used for correlation calculations.

Collaborative filtering is generally classified into two types—user based and item based. The preferences of a group of people similar to the target user are primarily used to recommend new commodities in user-based collaborative filtering. The commodities that have been purchased by the target user are used to recommend similar commodities in item-based collaborative filtering. Generally speaking, because the number of commodities that the customers have already purchased is far smaller than the number of customers, the calculation execution efficiency of the item-based algorithm is superior to the user-based algorithm [13]. The item-based algorithm was used in this study.

3 Proposed Method

The objective of this study is to predict the commodities that customers will most likely purchase next month through an analysis using the recommender system. The closer the occurrence of consumption behavior to the recommendation month, the more likely it is that the consumption behavior will become the consumption behavior in the following month. If the period included in the test data is too long, the correlation may be inaccurate. To avoid this situation, the transaction records of the customers who had shopped at the store in the month prior to the recommendation month were used as our test data. The correlation among commodities was calculated. The commodities that had been purchased by all of the customers in the month prior to the recommendation month were used to customize the recommendations to all of the customers. The proposed Collaborative Filtering for the Supermarket (CFS) consisted of three parts: (1) calculating the correlation between commodity types, (2) finding the most suitable commodity type combination, and (3) recommending the commodities that the customers would most likely purchase.

3.1 Calculation of the Correlation Among Commodity Types

The customer transaction records for the month prior to the recommendation month were used as the test data. Equations (1)–(4) were used to calculate each customer's preference score on each commodity type. The result was then sorted, and the commodity type that the customers would most likely purchase was recommended. Due to the simplicity of the Bigram calculation and its low time complexity, that calculation was used to determine the correlation among different commodity types. Equation (1) is the Bigram calculation formula. In terms of the consumption behavior during the period close to the recommendation month, the commodity types that the customers would most likely purchase were those that had already been purchased in this supermarket. Thus, our recommender system would first recommend commodity types that had already purchased. Therefore, as shown in Eq. (2), due to the customers' high probability of purchasing commodity types that they had already purchased, we defined the correlation between commodity types as one.

Equation (3) shows the commodity types that the customers purchased in the last month that they had made a purchase. To first recommend the commodity types that the customers had already purchased, we calculated the customers' preference scores on each commodity type (Eq. 4) based on the commodity types that they had purchased in the last month. The higher the preference scores, the higher the probability for the customers to make a purchase. The correlation of commodity types that the customers had not purchased in last month could be calculated using Eq. (1). If commodity types that the customers had already purchased were recommended, then the correlation could be calculated using Eq. (2). Finally, after the results were sorted, the commodity types that the customers would most likely purchase were recommended. If there were two or more commodity types with the same highest score, then the results were sorted based on the number of customers who had purchased that commodity type, and the sorted commodity types were recommended (from highest to lowest). In other words, after we used Eq. (4) to calculate each customer's preference score on each commodity type, we produced a customized recommendation list for each customer based on his preference scores. If the preference scores were the same, then the commodity types were sorted based on their popularity during the prior month.

$$\mathrm{b}_{a,k} = \frac{\sum_{i=1}^{i=m} C_{i,a} * C_{i,k}}{\sum_{i=1}^{i=m} C_{i,a}} (C_a \neq C_k) \tag{1}$$

$$\mathrm{b}_{k,k} = 1 \tag{2}$$

$$Buy\_List_j = [C_{j,a}]_{a=1\ldots N}\ with\ C_{j,a} = \begin{cases} 1\ if\ j\ buy\ C_a \\ 0\ if\ j\ not\ buy\ C_a \end{cases} \tag{3}$$

$$Purchase_{j,k} = \max_{1 \leq a \leq N} (\mathrm{C}_{j,a} * \mathrm{b}_{a,k}) \tag{4}$$

In the above formulae, $\mathrm{b}_{a,k}$ represents the correlation between commodity type C_a and commodity type C_k, m represents the total number of customers who had made a purchase at the store during the prior month, and $i \in m$. $C_{i,a}$ and $C_{i,k}$ indicate whether customer i purchased commodity type a and commodity type k in the month prior to the recommendation month. If the customer purchased these two commodities, then $C_{i,a}$ and $C_{i,k}$ are one; if the customer did not purchase them, then $C_{i,a}$ and $C_{i,k}$ are zero. Buy_List_j represents the commodity types that the customer j purchased during the last month. Moreover, M represents the total number of customers who made a purchase. $j \in M$, N represents the total number of commodity types, and $Purchase_{j,k}$ represents customer j's preference score on commodity type k.

3.2 Commodity Type Combination

To avoid the situation in which all of the recommended commodity types were those that the customers had already purchased. Therefore, if the recommended commodity types alone could draw the customers back to the physical retail stores, and then we could replace some of the commodities that the customers had not already purchased in last month. In our algorithm, a combination mode was used for recommendation.

Assuming one piece of DM could recommend D types of commodities, we could set in our algorithm that the DM includes P types of commodities already purchased by the customers. Therefore, when $P = D$, we would use the first part to calculate the order of the commodity types that each customer would most likely purchase and select the recommended commodity types based on the order. However, when $P < D$, we recommend P types of commodities, that the customer had already purchased, by the order, then the insufficient commodity types would be supplemented by the commodity types that the customers had not purchased in last month by the correlation order. If the number of the commodity types that the customers had purchased was less than P, then commodity types that the customers had not purchased in last month would be used for supplementation. Because the primary commodities sold in each supermarket may be different, recommendation combination modes were produced by adjusting parameters D and P. Through experiments and analysis, we could find the most suitable recommendation combination mode for each supermarket.

3.3 Commodity Combination

For each customer, the recommended commodity type combination was determined in the previous stage. Under the condition that only a small fixed amount of commodities could be recommended, at this stage, the primary task is to select one commodity of each commodity type that the customer would most likely purchase. When selecting commodities for recommendation, we must consider not only the customers' preferences for the commodities but also whether the recommended commodities are sales items. However, because we could not obtain the related sales promotion data, we only considered the customers' preferences for the commodities recommended in this study. We also used Eqs. (1)–(4) and replaced commodity type by commodity to calculate the correlation among commodities. Afterwards, based on each recommended commodity type for each customer, we used Eq. (4) to calculate each customer's preference score for each commodity. After the results were sorted, the commodity that each customer would most likely purchase would be recommended. For the same commodity type, if there were two or more commodities with the same highest score, then the commodity with the highest popularity would be selected for recommendation.

4 Experiments and Discussion

The customer transaction records from January to August 2011 of a well-known Taiwanese supermarket were used as the test data. The test data included 110,538 members. The records included 282,704 transactions involving 163 commodity types and 25,956 commodities (Table 1). The number of customers would increase with increasing time. However, not every customer would return to the retail store and shop during the recommendation month. Therefore, it was difficult to analyze the supermarket customers' behavior. Thus, to further study the supermarket

Table 1 Customer transaction records from January through August 2011

Code	Month	Jan.	Feb.	Mar.	Apr.	May	Jun.	Jul.	Aug.
A	Total number of transactions	37962	69649	104320	135945	171690	209159	246085	282704
B	Monthly number of transactions		31687	34671	31625	35745	37469	36926	36619
C	Total number of customers	26785	43407	57394	68463	79895	91005	101180	110583
D	Monthly number of customers	26785	23405	23844	22194	24385	25687	25273	24800
E	Ratio % (D/C)	100.0	53.9	41.5	32.4	30.5	28.2	25.0	22.4
F	Number of customers who shopped at the supermarket during two consecutive months		6783	6454	6772	6928	7701	7787	7622
G	Ratio % (F/D)		29.0	27.1	30.5	28.4	30.0	30.8	30.7

customers' behavior, we analyzed the customers who had purchased commodities during two consecutive months. Table 1 (F) shows the number of such customers each time that sampling was conducted. Because this supermarket's electronic DM often includes four to six commodity types and more than 30 commodities, we also selected four to six commodity types for recommendation. Because the results obtained from recommending four, five, or six commodities were similar, we only use the experiment in which six commodity types were recommended for explanation. In Sect. 4.1, we use the success rates for commodity type recommendation to compare the cosine correlation coefficient (COS), the Bigram (BIG) and the recommendation efficiency of the CFS algorithm. In Sect. 4.2, we use the success rates for commodity recommendation to compare the COS, the BIG and the recommendation efficiency of the CFS algorithm.

4.1 Comparison Among the Category Level

For the convenience of explanation, in this section we use CFS_D_P to represent the situation of the parameters used in this algorithm. D represents the number of recommended commodity types, whereas P represents the largest number of the commodity types that the customers has already purchased (If the number of the commodity types that the customers had purchased was less than P, then commodity types the customer had not purchased in last month would be used for supplementation.). We treat the issue of commodity recommendation as a prediction of the commodities that the customers will most likely purchase during the following month, with which we can increase the probability that the customers will return to the physical retail store to shop. Therefore, provided that the customer purchased one or more recommended commodity types or commodities, the recommendation for that customer was deemed a success; otherwise, the recommendation was deemed a failure.

Figure 1 shows the recommendation success rates for customers for each month; the x-axis represents the months recommended by the customized electronic DM, whereas the y-axis represents the recommendation success rates for recommended commodity types. We can clearly see that the recommendation success rates of the CFS (CFS_6_6) algorithm for target customers were approximately 12% higher than those of the BIG and 20% higher than those of the COS. CFS_6_6 means that there were six commodity types that could be recommended and there were up to six commodity types that the customer had already purchased. The main reason is the CFS that will consider the habits of customers have already purchased commodity categories will buy again. Recommended product categories are usually popular commodity categories those customers have already purchased. However, COS and BIG recommend the commodities that the customers did not purchase in last month.

In Fig. 2, customers' behavior of repurchasing the same commodity is considered in the CFS algorithm, the commodity types recommended by the CFS algorithm will tend to be the popular commodity types that the customers have already purchased. Only when the number of commodity types that the customer has already purchased is less than the number of recommended commodity types, our algorithm will recommend the commodity types that the customers did not purchase in last month to them. On the other hand, in Fig. 3, the commodity types recommended by the COS and the BIG will tend to be popular commodity types that the customers did not purchase during their last visits. But our method CFS_6_6 recommends the commodity type that the customer has already purchased. So the success rate of COS and BIG is higher than CFS_6_6.

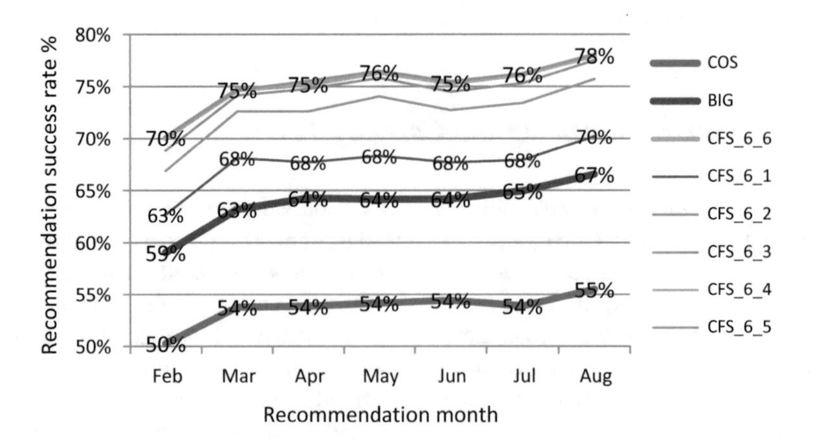

Fig. 1 Recommendation success rates for recommendation

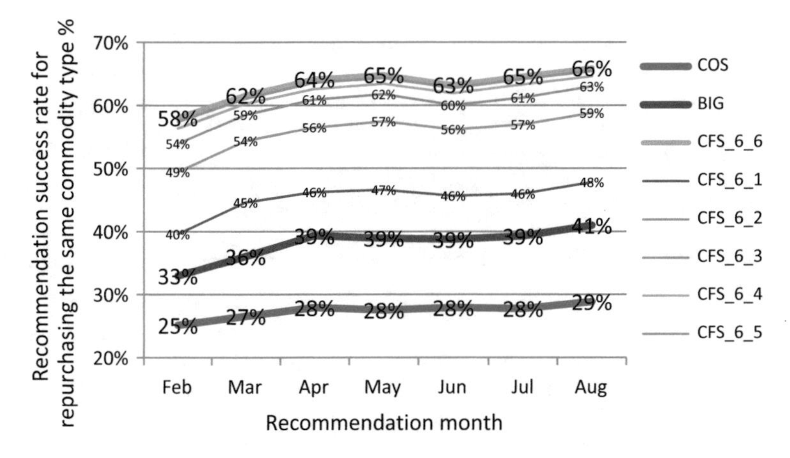

Fig. 2 Recommendation success rates for repurchasing the same commodity type when recommending six commodity types

According to the cross-selling concept, to increase supermarket profit we also hoped that customers would purchase commodity types that they had not purchased in the last month. Therefore, when the recommendation success rate for recommending commodity types that the customers had already purchased reached a certain value, we could replace some commodity types those the customers had already purchased. In Fig. 2, CFS_6_4 and the CFS_6_5 algorithms were used for making recommendations, the overall recommendation success rate and the recommendation success rate for commodity types that had been already purchased were both close to the recommendation success rate calculated using the CFS_6_6 algorithm. However, Fig. 3 show that the recommendation success rate for the commodity types that the customers did not purchase during their last visits calculated using the CFS_6_4 algorithm was approximately 5% higher than that of using the CFS_6_5 algorithm, and it was approximately 10% higher than that of using the CFS_6_6 algorithm. Therefore, we will use the CFS_6_4 algorithm to find the suitable combination in this study. In other words, when we selected six commodity types for recommendation, we only needed to recommend four (at most) commodity types that the customer had already purchased to interest him.

4.2 Comparison Among Product Level

In this study, one commodity of each commodity type that the customer would most likely purchase would be recommended. Figures 4, 5 and 6 show that the results of commodity recommendation were similar to those of commodity type recommendation. Figure 4 shows that the recommendation success rate for each month calculated using the CFS (CFS_6_6) algorithm was approximately 18%

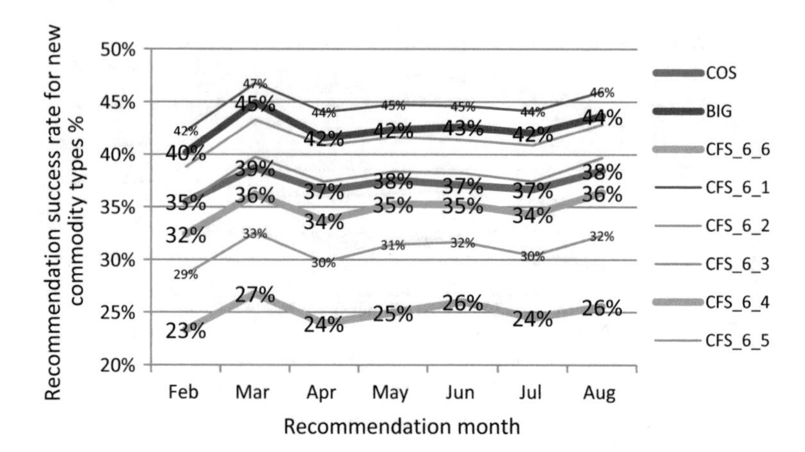

Fig. 3 Recommendation success rates for the commodity types that the customers did not purchase during their last visits when recommending six commodity types

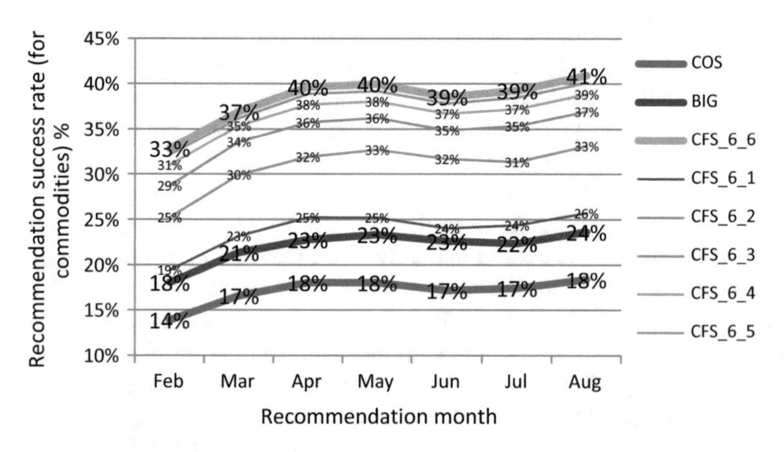

Fig. 4 Recommendation success rates for commodities when recommending six commodity types

higher than the recommendation success rate calculated using the BIG, and it was approximately 25% higher than the recommendation success rate calculated using the COS.

As shown in Fig. 5, when the CFS_6_4 and the CFS_6_5 algorithms were used for recommendation, the overall recommendation success rate and the recommendation success rate for commodities that had been already purchased were both close to the recommendation success rate calculated using the CFS_6_6 algorithm. In addition, as shown in Fig. 6, the recommendation success rate for commodity types that customers had not purchased in last month calculated using the CFS_6_4

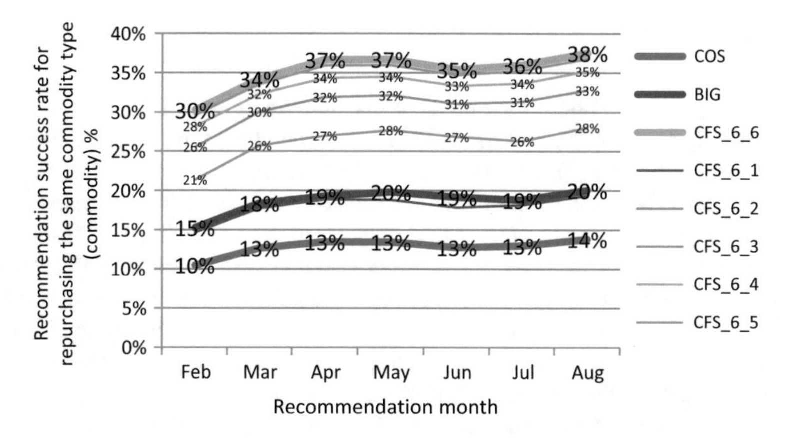

Fig. 5 Recommendation success rates for repurchasing the same commodity when recommending six commodity types

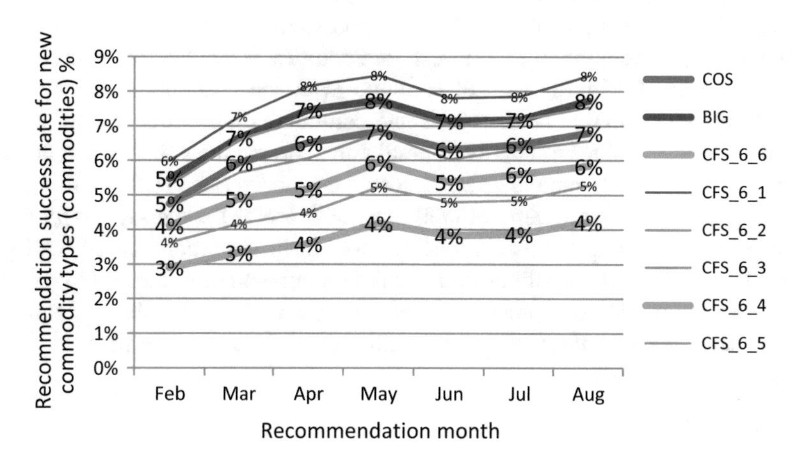

Fig. 6 Recommendation success rates for commodities those customers did not purchase during their last visits when recommending six commodity types

algorithm was approximately 2% higher than that of using the CFS_6_5 algorithm, and approximately 4% higher than that of using the CFS_6_6 algorithm. Therefore, we will use the CFS_6_4 algorithm to find the suitable combination in the product level.

5 Conclusions

In our study, we provide a new concept that utilizing customized electronic DM recommends commodities which have already purchased in the last month to increase the number of customers who will back to physical stores. The proposed algorithm could be applied not only to customized electronic DM recommendations but also to marketing with coupons. Currently, numerous Taiwanese stores send coupons to customers; however, the commodities that are the subject of the coupons are not always those that customers like. So, those sales commodities do not interest customers. If coupons can be provided based on the customers' preferences and are available for subsequent purchases, customers will be more interested and more willing to return to the physical retail stores to shop.

This study focused on a supermarket. Therefore, most of the commodity types studied were food. However, for different commodity types, the characteristics will be different, resulting in different consumption behavior. For instance, for 3C electronics, it is necessary to evaluate whether there exists a possibility of repurchasing based on the time factor and the past purchasing records; therefore, there will be no repurchasing behavior within a short time. In addition, the experiments did not consider the issue of discounted commodities; instead, we only considered their popularity. In future, we will continue to conduct studies focusing on the characteristics of different commodities and commodity types; in addition, we will consider different reference factors to predict the commodities that customers like. With the popularity of mobile devices, location-based service (LBS) will be considered in recommender system. Mobile users will receive the message of promoted commodities according the user's location near or in supermarket. CFS will generate recommended commodities of a small fixed amount of commodity types that is suitable for mobile equipment. We will merge LBS in mobile application to spread the promoted messages quickly and effectively to mobile users in further research.

References

1. E. Rich, "User modeling via stereotypes," in Readings in intelligent user interfaces, T. M. Mark and W. Wolfgang, Eds., ed: Morgan Kaufmann Publishers Inc., 1998, pp. 329–342.
2. J. Bobadilla, F. Ortega, A. Hernando, and A. Gutiérrez, "Recommender systems survey," Knowledge-Based Systems, vol. 46, pp. 109–132, 2013.
3. X. Luo, Y. Xia, and Q. Zhu, "Incremental Collaborative Filtering recommender based on Regularized Matrix Factorization," Knowledge-Based Systems, vol. 27, pp. 271–280, 2012.
4. D. Goldberg, D. Nichols, B. M. Oki, and D. Terry, "Using collaborative filtering to weave an information tapestry," Commun. ACM, vol. 35, pp. 61–70, 1992.
5. U. Shardanand and P. Maes, "Social information filtering: algorithms for automating & ldquo; word of mouth & rdquo," presented at the Proceedings of the SIGCHI Conference on Human Factors in Computing Systems, Denver, Colorado, USA, 1995.

6. J. A. Konstan, B. N. Miller, D. Maltz, J. L. Herlocker, L. R. Gordon, and J. Riedl, "GroupLens: applying collaborative filtering to Usenet news," Commun. ACM, vol. 40, pp. 77–87, 1997.
7. L. Terveen and W. Hill, "Beyond Recommender Systems: Helping People Help Each Other," ed, 2001.
8. G. Linden, B. Smith, and J. York, "Amazon.com Recommendations: Item-to-Item Collaborative Filtering," IEEE Internet Computing, vol. 7, pp. 76–80, 2003.
9. R. M. Bell and Y. Koren, "Lessons from the Netflix prize challenge," SIGKDD Explor. Newsl., vol. 9, pp. 75–79, 2007.
10. B. Pradel, S. Sean, J. Delporte, #233, b. Gu, rif, l. Rouveirol, N. Usunier, Fran, #231, o. Fogelman-Souli, Fr, and r. Dufau-Joel, "A case study in a recommender system based on purchase data," presented at the Proceedings of the 17th ACM SIGKDD international conference on Knowledge discovery and data mining, San Diego, California, USA, 2011.
11. J. Wang, B. Sarwar, and N. Sundaresan, "Utilizing related products for post-purchase recommendation in e-commerce," presented at the Proceedings of the fifth ACM conference on Recommender systems, Chicago, Illinois, USA, 2011.
12. B. Sarwar, G. Karypis, J. Konstan, and J. Riedl, "Item-based collaborative filtering recommendation algorithms," presented at the Proceedings of the 10th international conference on World Wide Web, Hong Kong, Hong Kong, 2001.
13. M. Deshpande and G. Karypis, "Item-based top-<i>N</i>recommendation algorithms," ACM Trans. Inf. Syst., vol. 22, pp. 143–177, 2004.
14. G. Adomavicius and A. Tuzhilin, "Toward the next generation of recommender systems: a survey of the state-of-the-art and possible extensions," Knowledge and Data Engineering, IEEE Transactions on, vol. 17, pp. 734–749, 2005.
15. L. Candillier, F. Meyer, and M. Boull, "Comparing State-of-the-Art Collaborative Filtering Systems," presented at the Proceedings of the 5th international conference on Machine Learning and Data Mining in Pattern Recognition, Leipzig, Germany, 2007.
16. C. G. González, W. Bonventi, Jr., and A. L. V. Rodrigues, "Density of Closed Balls in Real-Valued and Autometrized Boolean Spaces for Clustering Applications," in Advances in Artificial Intelligence—SBIA 2008. vol. 5249, G. Zaverucha and A. Costa, Eds., ed: Springer Berlin Heidelberg, 2008, pp. 8–22.
17. X. Su and T. M. Khoshgoftaar, "A survey of collaborative filtering techniques," Adv. in Artif. Intell., vol. 2009, pp. 2–2, 2009.
18. B. Sarwar, G. Karypis, J. Konstan, and J. Riedl, "Analysis of recommendation algorithms for e-commerce," presented at the Proceedings of the 2nd ACM conference on Electronic commerce, Minneapolis, Minnesota, USA, 2000.
19. K. Choonho and K. Juntae, "A recommendation algorithm using multi-level association rules," in Web Intelligence, 2003. WI 2003. Proceedings. IEEE/WIC International Conference on, 2003, pp. 524–527.

Enhancement of Salt-and-Pepper Noise Corrupted Images Using Fuzzy Filter Design

Yung-Yue Chen, Ching-Ta Lu and Pei-Yu Chang

Abstract This investigation presents a fuzzy filtering method for the removal of salt-and-pepper noise of a corrupted image which is caused by the corruption of impulse noise in the record or transmission process. This fuzzy filtering method comprise with a size adjustable local window which is used to analyze each extreme pixel (0 or 255 for an 8 bits gray-level image) and a fuzzy smoother which can interpolate non-extreme values inside the local window to construct a noiseless center pixel. By the help of the proposed fuzzy filtering method, the center pixel with an extreme value is replaced by the fuzzy interpolation value and enables noisy pixels to be restored smoothly and continuously. From the tough tests, experimental results reveal the fact that salt-and-pepper noises (only for known extreme values 0 and 255) of a corrupted image for different noise corruption densities (from 10 to 90%) can the efficiently removed via the universal interpolation ability of the proposed fuzzy filtering method; meanwhile, the denoised image is free from the blurred effect.

Keywords Image denoising ability · Salt-and-pepper noise · Size adjustable window · Fuzzy smoother · Universal interpolation

1 Introduction

Inevitably, the impulse noises corrupt images due to the following reasons: malfunctioning pixels in camera sensors, fault memory locations in hardware, transmission in a noisy channel, and bit errors in transmission. In real applications,

Y.-Y. Chen (✉) · P.-Y. Chang
Department of Systems and Naval Mechatronics Engineering, National Cheng Kung University, Tainan, ROC
e-mail: yungyuchen@mail.ncku.edu.tw

C.-T. Lu
Department of Information Communication, Asia University, Taichung, Taiwan, ROC
e-mail: Lucas1@ms26.hinet.net; shenjh@asia.edu.tw; ling@asia.edu.tw; BligsLcl@gmail.com

© Springer Nature Singapore Pte Ltd. 2018

N.Y. Yen and J.C. Hung (eds.), *Frontier Computing*, Lecture Notes in Electrical Engineering 422, DOI 10.1007/978-981-10-3187-8_65

there are two types of impulse noises: the random valued noise and the salt-and-pepper noise. Salt-and-pepper noises which appear with the maximum or the minimum gray levels always seriously corrupt images and significantly deteriorate the quality of an image. How to efficiently remove this kind of impulse noise for the corrupted image becomes an important research task.

Recently, many investigations have been contributed for the reconstruction of images contaminated by salt-and-pepper noise [1–10]. In most of these designs, the median and the mean filters were popular for the removal of salt-and-pepper noise because of their good denoising power and computational simplicity. However, some details and edges of the original image cannot be well-restored by the mentioned algorithms when the noise density is over 60%.

An adaptive median filter which enabled noisy pixels to be removed by choosing the median value in an adaptive window for each pixel is proposed by Hwang and Hadded [11]. This method performed well at low noise densities but is awful at high noise densities because a large window size which may lead to a blur in the denoised image has to be adopted. Zhang and Karim [12] and Bovik, [13] proposed the modified switching based median filters; in these studies, the decision of an impulse noise pixel is based on a pre-defined threshold value. The major drawback of these switching based methods is: a robust decision threshold is difficult to be defined practically. Also details and edges could not be recovered satisfactorily by these filters because they usually did not take the local features into account, in particular, when noise density was high. Dong and Xu [14] proposed a directional-weighted-median (DWM) filter for the elimination of random-valued impulse noises. This method analyzed the neighborhood information of the center pixel on four directions by weighting the pixels in a local window. By following the denoising process, a noise-corrupted pixel could be detected, and denoised optimally by the weighted median filter. Lu and Chou [15] proposed a modified DWM (MDWM) filter which can give an additional constraint on performing switching median filtering and find a better edge direction. Experimental results of this method indicate that it could enhance the performance of the DWM filter (Dong and Xu [14]) significantly by removing a greater quantity of background noise and preserving the details of the original image simultaneously. Ahmed and Das [16] contributed an iterative adaptive fuzzy filter via using an alpha-trimmed mean method. This method developed an adaptive fuzzy filter with an iterative process which needs a huge computation cost for optimally denoising impulse noise corrupted images. For adaptively restoring the current noise pixel, Li et al. [17] proposed a method to estimate the noise density of a non-recursive local window for a noise corrupted pixel. Recently, Zhang and Li [18] proposed an adaptive weighted mean filter for the removal of salt-and-pepper noises with heavy densities. This method firstly determined the adaptive window size with using the maximum and the minimum values of two successive windows. In this method, noise-free pixels would be left and unchanged, and noise candidate pixels were replaced by the weighted mean of the window. Experimental results of this method reveal that a noisy image with high noise density could be well-restored. Li et al. [19] developed an image block-based noise detection rectification method to the remove impulse noises by estimating noise density of an image and a global image information-based method. Experimental results of this investigation indicate a feature of this method: the denoising performance can be dramatically improved. For reserving edges of images and improving image denoising ability, Liu et al. [20] proposed a weighted mean filter which is with a two-phase noise detector. In the first phase of this method, a rand-ordered difference method was adopted for detecting noise candidates, and a minimum difference of edge pixels was then utilized as the second phase to identify edge pixels from noise candidates. Maybe some of these methods represent good denoising performance to noised images, however, huge computation burdens or the real-time calculation ability is usually a main drawback for the above mentioned methods. Based on the above findings, how to real-time improve the performance for denoising heavy-noise corrupted images (the noise density exceeds 60% or higher) is still a tough research task currently. In this investigation, a new derived fuzzy filtering method which is combined with a size adjustable local window and a fuzzy smoother is proposed to real-time solve the removal problem of salt-and-pepper noises with heavy densities of images. In the beginning of the denoising process, the size adjustable local window is employed to analyze each noisy pixel which has either the maximum or the minimum pixel value at the center of the window. Based on the extreme character of a pixel: 0 or 2^N for an N bits gray-level image, a noisy pixel can be selected as a noise candidate, and this noise candidate pixel will be rejected for the purpose of reconstructing a new center pixel in a local window. If there is no non-extreme pixel in the local window, the window size will be expanded until the local window contains non-extreme pixels. These non-extreme pixels are then collected as the input pixels of the fuzzy smoother of the proposed fuzzy filtering method. Besides, one set of fuzzy rules which are interfered by the membership functions defined with the size of the local windows is given for the purpose of optimally interpolating a center pixel. From the above arrangement, a set of optimal weighting factors can be mathematically and quickly calculated based on the interference values of membership functions and a new noiseless center pixel can be obtained via interpolating non-extreme pixels with optimal weighting factors which are inferred by fuzzy membership functions. Old noisy center pixels in a noise corrupted image are then replaced by these new noiseless center pixels, and the whole image which is corrupted by the salt-and-pepper noise can be restored. Satisfactory contributions are revealed from experimental results for the proposed method. These contributions show us that this proposed method can effectively remove salt-and-pepper noise from a corrupted image with respect to different noise densities (from 10 to 90%); meanwhile the denoised image prevents from the blurred effect. Besides, this proposed method doesn't suffer the computation burden as the conventional designs and possesses the real-time calculation ability.

The rest of this investigation is formulated as the following: Sect. 2 describes the proposed fuzzy filtering method, Sect. 3 demonstrates the experimental results, and Conclusions are made finally in Sect. 4.

2 Fuzzy Filtering Method

This fuzzy filtering method comprise with a size adjustable local window which is used to analyze each extreme pixel (0 or 255 for an 8 bits gray-level image), and a fuzzy smoother which can interpolate non-extreme values inside the local window to construct a noiseless center pixel for a noise corrupted image. As to how to determine the size of a local window, the window size is expanded until non-extreme pixels to be included in the window. Based on the input messages of non-extreme pixels of the local window, a set of optimal fuzzy weighting factors can be mathematically found by a fuzzy smoother, and then a new noiseless center pixel can be obtained via multiplying the non-extreme pixels with the optimal fuzzy weighting factors to replace old noisy center pixels in the noise corrupted image. Finally, the denoising image can obtained.

2.1 *Adjustable Window Size*

A local window which slides from left to right and from top to bottom in an image is employed to analyze the neighbor properties for each extreme pixel. Non-extreme pixels in the local window will be used to restore the noise-corrupted center pixel. The local window $W_{2s+1}(i,j)$ can be expressed as

$$W_{2s+1}(i,j) = \{X_{i+m,j+n} | \text{where } m, n \in (-s \sim s)\} \tag{1}$$

where s controls the size of the window, and it is a positive integer. $X_{i,j}$ denotes the input pixel at the location ith row and jth column of the local window. The window size is equal to $(2s+1) \times (2s+1)$.

The number of non-extreme pixels will be utilized to determine the weighting factor. A non-extreme flag $F_{i,j}^{non-extreme}$ which is used to determine whether a pixel is non-extreme can be formulated as

$$F_{i,j}^{non-extreme} = \begin{cases} 1, & \text{if } X_{i,j} \in \text{non} - \text{extreme pixels} \\ 0, & \text{otherwise} \end{cases} \tag{2}$$

From (2), it is easy to check over: the pixel $X_{i,j}$ belongs to non-extreme pixels, and the non-extreme flag $F_{i,j}^{non-extreme}$ is set to unity, and this represents that the pixel $X_{i,j}$ is noise-free.

As the above expression, $F_{i,j}^{non-extreme}$ can be used as a tool to compute the number of non-extreme pixels by summing up the number of non-extreme flags in a local window and is given as

$$N_{i,j}^{non-extreme} = \sum_{m=-s}^{s} \sum_{n=-s}^{s} F_{i+m,j+n}^{non-extreme} \tag{3}$$

where $N_{i,j}^{non-extreme}$ equals the number of non-extreme pixels in a local window.

The number of non-extreme pixels computed by (3) can be regarded as an index to determine whether the size of a local window should be expanded. If this index is greater than zero $(N_{i,j}^{non-extreme} > 0)$, the size of the used local window will not be expanded. When the window does not contain none of non-extreme pixels (i.e., $N_{i,j}^{non-extreme} = 0$), the size of the local window should be expanded by one pixel at each of its four sides (i.e., replace s in (1) by s + 1). The window expansion procedure is repeated until at least one non-extreme pixel appears. In practical, the window size can be started up with a small matrix and is usually initialized to 3×3.

Remark 1 The size of the local window will not be further expanded when the window size reaches a dimension of 7×7 because a window size which is larger than this dimension will produce blurred effect to the denoised image and the whole denoising process based on this window size will suffer a huge computation burden.

2.2 Fuzzy Smoother

The proposed fuzzy filtering method of this study reconstructs a noiseless pixel to replace noise-corrupted center pixel based on messages of the pixels inside the adjustable window, and only non-extreme pixels are mathematically selected to reconstruct the center pixels of a local window. A pixel will be reserved as a non-extreme pixel by the following criterion, and a pixel is with extreme values 0 or 255 will be excluded

$$\tilde{X}_{i,j} = \{X_{i,j},\ X_{i,j} \neq 0 \text{ and } X_{i,j} \neq 255\} \tag{4}$$

where $\tilde{X}_{i,j}$ represents the non-extreme pixel.

The proposed fuzzy smoother $\hat{S}_{i,j}$ in this investigation will used to reconstruct the center pixels of the local window, and this smoother can be mathematically expressed by adopting the interpolation property of the fuzzy universal approximation. The center pixels $\hat{S}_{i,j}$ can be described by the following rules:

$$\begin{aligned} &R^{\gamma}:\ \text{IF}\, m \text{ is } \mu_i^m \text{ and } n \text{ is } \mu_j^n \\ &\text{Then}\, \hat{S}_{i,j} = \tilde{X}_{i,j} \end{aligned} \tag{5}$$

where $R^{\gamma}(\gamma = 1, \ldots, 2s+1 \times 2s+1)$ are fuzzy rules, m and n are the indexes of input pixels to the proposed fuzzy smoother. $\hat{S}_{i,j}$ is center value proposed by the γth rule, and $\mu_i^m(i = 1, \ldots, 2s+1)$ and $\mu_j^n(j = 1, \ldots, 2s+1)$ are the membership functions which characterize the ith and jth fuzzy sets defined in the m and n variables' coordinates, respectively. These membership functions for the proposed fuzzy smoother are expressed as the following for continuously interpolation purpose:

$$\mu_i^m = \exp\left[-\frac{1}{2}\left(\frac{x-c_i^m}{\sigma_i^m}\right)\right]^2 \tag{6}$$

$$\mu_j^n = \exp\left[-\frac{1}{2}\left(\frac{y-c_j^n}{\sigma_j^n}\right)\right]^2 \tag{7}$$

where c_i^m, c_j^n, σ_i^m and σ_l^n are respectively the mean values and the standard deviations of the membership functions, and they are nonlinear function in C^∞.

Based on the arrangement above, a center pixel with an extreme value is replaced mathematically as below for the denoising purpose by the interpolation of the pixels with non-extreme values inside the local window.

$$\hat{S}_{i,j} = \sum_{i=-s}^{s} \sum_{j=-s}^{s} \kappa_{ij} \tilde{X}_{ij} \tag{8}$$

where

$$\kappa_{ij} = \frac{\mu_i^m \mu_j^n}{\sum_{i=-s}^{s} \sum_{j=-s}^{s} \mu_i^m \mu_j^n} \tag{9}$$

In (8), the restored center pixel $\hat{S}_{i,j}$ can be obtained by optimally interpolating non-extreme values inside the local window, i.e., a noise-corrupted image can be enhanced by (8) because the restored pixel $\hat{S}_{i,j}$ is a "non-extreme pixel" due to it is constructed by all non-extreme values.

3 Experimental Results

For verifying the denoising performance of this proposed method, A desktop with a Intel® Core™ i7 4600 M CPU (2.9 GHz) and the Microsoft Windows 7 (32 bits) operation system is adopted. In the desktop, the size of the main memory is 4 GB. This proposed algorithm is programmed and executed in the famous software: MATLAB. A standard gray-level test image "Lena" which is with size 512 × 512 is utilized to assess the noise reduction performance of this proposed fuzzy-based denoising algorithm in the real experiments. The test image was corrupted by doping salt-and-pepper impulse noises with different noise densities (50 and 80%).

The DWM filter (Dong, and Xu [14]), the MDWM (Lu and Chou [15]) filter, and the LGII (Li et al. [19]) were implemented for the comparison purpose. Restoration results were measured by the peak signal-to-noise ratio (PSNR) quantitatively.

A restored image can be quantitatively assessed by the PSNR and the PSNR can be expressed as (Bovik [9])

$$\mathrm{PSNR(dB)} = 10 \cdot \log_{10}\left(\frac{\mathrm{MAX}}{\mathrm{MSE}}\right) \tag{10}$$

where **MAX** denotes the largest value of gray-level of images, and **MAX** equals 255 for an 8-bits gray-level image. The MSE as below represents the mean-square-error between original and restored images

$$\mathrm{MSE} = \frac{1}{M \cdot N} \sum_{i=0}^{M-1} \sum_{j=0}^{N-1} \left| S_{i,j} - \hat{S}_{i,j} \right|^2 \tag{11}$$

where $S_{i,j}$ and $\hat{S}_{i,j}$ represent the original and the restored pixels. M and N which is equal to 512 are the width and the height size, respectively.

Remark 2 (10) means: the larger the value of the PSNR is, the better quality of the restored image is. □

In the case of middle noise, a corruption as Fig. 1 which is 50% noise density for the Lena image is used. The LGII and the proposed methods reveal the superior denoising performance to the DWM and the MDWM filters. The LGII filter slightly outperforms the proposed method for the cases with a 50% noise density. As to the cases of heavy noise corruptions (noise density is greater than 80%) as Fig. 2, the MDWM filter and the proposed method provide better results in the removal of heavy density impulse noises than the other two methods. From the point of view with PSNR, this proposed method manifests the exceptional denoising performance than the others. Existing literatures indicated a conclusion that the denoising performance of the DWM filter can be significantly improved with a modified version: the MDWM filter, under images corrupted with heavy noise corruptions, such as noise density greater than 80% in this case. The main reason for MDWM filter being capable of improving the denoising performance is: the MDWM filter excludes the pixels with extreme values (0 or 255) on the optimum direction before the operation of median filtering, and yields an effective removal of impulse noises. As the MDWM filter, the proposed method also excludes the neighbor pixels with extreme values for the restoration of noise-corrupted image, and most of salt-and-pepper noises are efficiently removed.

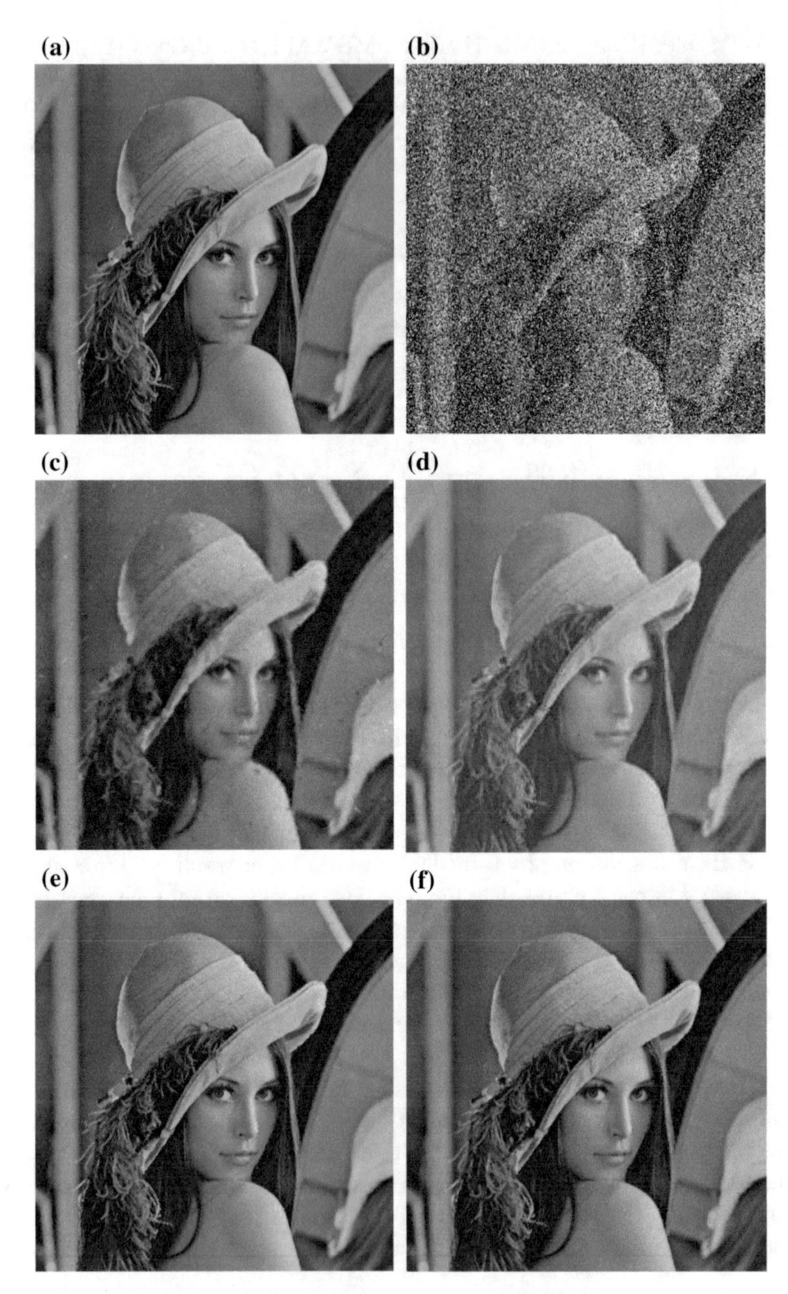

Fig. 1 Output results for Lena image with 50% noise density. **a** Original image; **b** noisy image; **c** output image of the DWM filter; **d** output image of the modified DWM filter; **e** output image of the LGII method; **f** output image of the proposed method

Fig. 2 Output results for Lena image with 80% noise density. **a** Original image; **b** noisy image; **c** output image of the DWM filter; **d** output image of the modified DWM filter; **e** output image of the LGII method; **f** output image of the proposed method

4 Conclusions

A new filtering method based on concepts of the size adjustable local window and the fuzzy interpolation ability for the removal of salt-and-pepper noises is proposed in this study. Based on the filtering of size adjustable local window and the optimal weighting factors design, pixels with salt-and-pepper noises which are with extreme value 0 or 2^N can be effectively removed by interpolating non-extreme pixels inside the local window. Besides, this proposed method possesses the promising real-time calculation ability due to its simple and no iteration structure hence it will not suffer the computation burden as the conventional designs.

The noise reduction performances of this proposed method are verified by the tough standard: peak signal-to-noise ratio (PSNR) for restoration of a denoised image. Comparisons of this proposed method are made with respect to several existing methods such as DWM filter, MDWM filter, and the LGII filter for "Lena image". Verification results indicate that this proposed method has a superior ability in effectively removing salt-and-pepper noise of a corrupted image with respect to different noise densities (from 10 to 90%) than these three filters; meanwhile the denoised image can get free from the blurred effect.

Acknowledgements This research was supported by the Ministry of Science and Technology, Taiwan, under contract number MOST 104-2628-E-006-012-MY3, and MOST 104-2221-E-468-007.

References

1. Akkoul, S., Ledee, R., Leconge, R., Harba, R.: A new adaptive switching median filter. *IEEE Signal Process. Lett.*, 17(6), 587–590 (2010)
2. Chen, T., Ma, K. K., Chen, H.: Tri-state median filter for image denoising. *IEEE Trans. Image Process.*, 8(12), 1834–1838 (1999)
3. Chan, R. H., Ho, C. W., ikolova, M.: Salt-and-pepper noise removal by median-type noise detectors and detail-preserving regularization. *IEEE Trans. Image Process.*, 14(10), 1479–1485 (2005)
4. Chen, P. Y., Lien, C. Y.: An efficient edge-preserving algorithm for removal of salt-and-pepper noise. *IEEE Signal Process. Lett.*, 15, 833–836 (2008)
5. Duan, F., Zhang, Y. J.: A highly effective impulse noise detection algorithm for switching median filters. *IEEE Signal Process. Lett.*, 17(7), 647–650 (2010)
6. Mial, Z., Jiang, X.: Weighted iterative truncated mean filter. *IEEE Trans. Signal Process.*, 61(16), 4149–4160 (2013)
7. Sun, T., Neuvo, Y.: Detail-preserving median based filters in image processing. *Pattern Recog. Lett.*, 15(4), 341–347 (1994)
8. Toh, K. K. V., Isa, N. A. M.: Noise adaptive fuzzy switching median filter for salt-and-pepper noise reduction. *IEEE Signal Process. Lett.*, 17(3), 281–284 (2010)
9. Wang, Z, Bovik, A. C., Sheikh, H. R., Simoncelli, E. P.: Image quality assessment: from error visibility to structural similarity. IEEE Trans. Image Process., 13(4), 600–612 (2004)
10. Wan, Y., Chen, Q. Q., Yang, Y.: Robust impulse noise variance estimation based on image histogram. *IEEE Signal Process. Lett.*, 17(5), 485–488 (2010)

11. Hwang, H., Hadded, R.: Adaptive median filter: New algorithms and results. *IEEE Trans. Image Process.*, 4(4), 499–502 (1995)
12. Zhang, S., Karim, M.A.: A new impulse detector for switching median filters. *IEEE Signal Process. Lett.*, 9(11), 360–363 (2002)
13. Bovik, A.: *Handbook of Image and Video Processing,* New York, Academic Press (2000)
14. Dong, Y.Q., Xu, S.F.: A new directional weighted median filter for removal of random-valued impulse noise. *IEEE Signal Process. Lett.*, 14(3), 31–34 (2007)
15. Lu, C.-T, Chou, T.-C.: Denoising of salt-and-pepper noise corrupted image using modified directional-weighted-median filter. Pattern Recog. Lett., 33(10), 1287–1295 (2012)
16. Ahmed, F., Das, S.: Removal of high density salt-and-pepper noise in images with an iterative adaptive fuzzy filter using alpha-trimmed mean. *IEEE Trans. Fuzzy Syst.*, 22(5), 1352–1358 (2014)
17. Li, Z., Liu, G., Xu, Y., Cheng, Y.: Modified directional weighted filter for removal of salt & pepper noise. *Pattern Recog. Lett.*, 40, 113–120 (2014)
18. Zhang, P., Li, F.: A new adaptive weighted mean filter for removing salt-and-pepper noise. *IEEE Signal Process. Lett.*, 21(10), 1280–1283 (2014)
19. Li, Z., Cheng, Y., Tang, K., Xu, Y., Zhang, D.: A salt & pepper noise filter based on local and global image information. Neurocomputing. 159, 172–185 (2015)
20. Liu, L., Chen, C.L.P., Zhou, Y., You, X.: A new weighted mean filter with a two-phase detector for removing impulse noise. Inform. Sci., 315, 1–16 (2015)

Efficient Processing of Spatial Window Queries for Non-flat Wireless Broadcast

Jun-Hong Shen, Ching-Ta Lu and Chien-Tang Mai

Abstract Consider skewed access patterns of mobile clients, popular data are broadcast more times than regular ones via non-flat wireless broadcast, resulting in the decrease of the client waiting time for mobile devices to retrieve popular data. Window queries are one of the fundamental spatial queries for location-based services. Such queries retrieve spatial objects in a fixed window region according to clients' current location. In this paper, considering the skewed access patterns, we propose an efficient processing method for the window queries via non-flat broadcast in the wireless environments. From the experimental results, we have verified that our proposed method performs better than the existing methods.

Keywords Skewed access patterns · Spatial index · Window queries · Wireless data broadcast

1 Introduction

In pervasive computing, location-based services (LBSs) via wireless data broadcast are efficient to provide an enormous amount of mobile clients with applications according to their current locations [10]. Window queries are one representative type of spatial queries for LBSs. Such queries retrieve spatial objects in a fixed window region according to clients' current location. For example, a tourist issues a query asking restaurants within one mile according to his current location. Considering skewed access patterns of mobile clients, a few spatial objects are more frequently retrieved than the others. Therefore, these popular objects should be

---

J.-H. Shen · C.-T. Lu (✉) · C.-T. Mai
Department of Information Communication, Asia University, Taichung, Taiwan
e-mail: lucas1@ms26.hinet.net

J.-H. Shen
e-mail: shenjh@asia.edu.tw

C.-T. Mai
e-mail: purelydinkum@gmail.com

© Springer Nature Singapore Pte Ltd. 2018

N.Y. Yen and J.C. Hung (eds.), *Frontier Computing*, Lecture Notes in Electrical Engineering 422, DOI 10.1007/978-981-10-3187-8_66

appeared more times in a broadcast cycle than the others, i.e., non-flat broadcast. As a result, the average client waiting time of retrieving data from the wireless channel is reduced. In this paper, we address on the design of the spatial index for supporting window queries via non-flat wireless broadcast. In the architecture of non-flat wireless broadcast, servers with the antenna cyclically broadcast a set of selected spatial objects, call a broadcast cycle, on a wireless channel to generate a broadcast stream, and mobile clients tune into the channel to retrieve the desired data.

Since mobile devices have scarce energy resource, they should stay in the doze mode for most of the time, and switch into the active mode when necessary. Interleaving spatial indexes with spatial objects could help the mobile devices know the arrival times of the related indexes and spatial objects. As a result, the mobile devices can skip examining the unwanted data and retrieve only the desired data, resulting in the decrease of energy consumption [7, 9]. Two measures, the access time and tuning time, are concerned in evaluating the performance of spatial indexes over wireless broadcast. The access time is the client waiting time from the time issuing a query to the time when the desired data are retrieved. The tuning time is the amount of time that the mobile device actually stays in the active mode to retrieve the related indexes and data for processing a query.

Considering skewed access patterns of mobile clients, in the literature, many researchers have proposed spatial indexes to handle the spatial processing for non-flat wireless broadcast. Im et al. proposed a grid-based distributed index (*GDIN*) [3], which is developed for window queries over non-flat broadcast by using a regular space partition. Since regular data located near popular data are repeated as the same time as the popular data in *GDIN*, a two-tier spatial index (*TTSI*) [6] is proposed to solve this problem by using a two-tier data space to distinguish popular and regular data. Multi-leveled air index scheme (*MLAIN*) [4] is developed to improve the problem occurred in *GDIN* by using a multi-level space partition. Since all popular data are broadcast as the same frequency within a broadcast cycle in *MLAIN*, the length of the broadcast cycle is lengthened, resulting in the increase of the access time. To solve the problem occurred in *MLAIN*, a skewed spatial index (*SSI*) [8] is proposed to classify the popular data into multiple groups, and repeat these groups in the broadcast cycle according to their broadcast frequencies. The above methods were designed for broadcasting data on the one wireless channel. A grid-based indexing scheme on multiple channels (*GRIM*) [5] was proposed to support window queries considering skewed access patterns on the multiple-channel broadcast.

The above cell-based methods have a severe problem: all the cells overlapped with the query region should be examined, even though some of the cells do not contain the answered objects. This problem leads to unnecessary examination of the cells, resulting in the increase of the access time and tuning time. Therefore, in this paper, considering skewed access patterns of the mobile clients, we propose a largest-empty-rectangle based spatial index, *LER*, for efficiently processing window queries via non-flat wireless broadcast. The largest empty rectangle is the rectangle of maximum size extending from the minimum bounding rectangle containing the objects in the hot cell, without touching any spatial object. If the query region of a window query is contained in the largest empty rectangle of the cell, the answered objects are definitely located in this cell, and no more cells should be examined. As a result, the access time and tuning time can be reduced. From our simulation study, we have verified that our proposed method has a better performance than *SSI* and *MLAIN*.

The rest of this paper is organized as follows. In Sect. 2, we present our proposed method. In Sect. 3, we describe the experimental results. Finally, a conclusion is presented in Sect. 4.

2 Proposed Method

In this section, we present our proposed method that includes the following functions.

1. Map spatial objects from the two-dimensional space to the one-dimensional space.
2. Generate a non-flat broadcast allocation for spatial objects.
3. Interleave spatial index information into the non-flat broadcast allocation.
4. Provide the access mechanism of the query processing.

2.1 Mapping

To preserve locality of a mapping from 2D to 1D for spatial objects, we apply the Hilbert curve. For the Hilbert curve of order n, a space is recursively divided into four equal-sized cells until there are 2^{n*2} cells in the space, and each cell is assigned a sequence number from 0 to $(2^{n*2} - 1)$, which is called the Hilbert curve value. Take the map containing 13 restaurants of interest in Fig. 1a for the running example. In Fig. 1a, spatial objects d_2 and d_3 are the most popular ones, spatial objects d_8 and d_{12} are popular ones, and the rest are regular ones. Assume that the capacity, η, of each cell is 2 spatial objects. At first, the space is divided into four equal-sized cells according to the Hilbert curve of order 1, as shown in Fig. 1b, where each cell is labeled by $C(i, j)$, i is the order of the Hilbert curve, and j is the sequence number of the Hilbert curve of order i. Since the number of spatial objects in one of cells exceeds 2, these cells are further divided according to the Hilbert curve of order 2. The sequence numbers of cells of the Hilbert curve of order 2 are assigned by following the same rotation and reflection pattern at each cell of the Hilbert curve of order 1 such that each cell is continuously visited. The division result of the Hilbert curve of order 2 is shown in Fig. 1c, where each cell contains at most 2 spatial objects. In Fig. 1c, the cell containing popular objects is called the hot cell, which is labeled by $H(i, j)$, e.g., $H(2, 2)$.

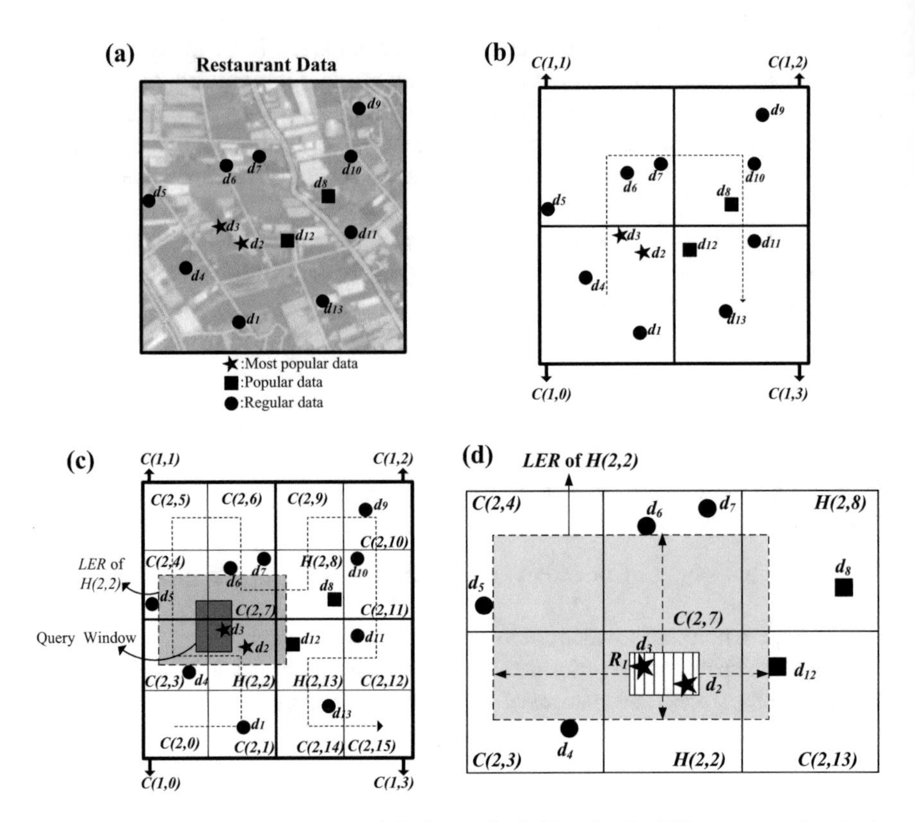

Fig. 1 The running example: **a** the original map; **b** division by the Hilbert curve of order 1; **c** division by the Hilbert curve of order 2; **d** the large empty rectangle

2.2 Non-flat Broadcast

Considering skewed access patterns of the mobile clients, Acharya et al.'s broadcast disks method [1] is an efficient way to organize a popularity hierarchy of broadcast objects for quick access of popular ones. Therefore, after the space is divided into cells according the Hilbert curve of order *n*, we apply Acharya et al.'s broadcast disks to generate the non-flat broadcast program of the spatial objects on the wireless channel. Assume that there are three disks, D_1–D_3, and their corresponding relative frequencies are $\lambda_1 = 4$, $\lambda_2 = 2$ and $\lambda_3 = 1$, respectively. D_1 has the smallest number of cells and D_3 has the largest one. We take the spatial objects in Fig. 1c as the example. The cells are sorted in the descending order of the access probabilities of the spatial objects inside them, and classified into disks D_1–D_3, as shown in Fig. 2. The cells in each disk have the similar access probability, and the sequence of the cells in each disk is according to the Hilbert curve. Each disk is further split to chunks. The least common multiple of the relative frequencies, *L*, of the three disks is 4. Then, the cell in D_1 are put into $L/\lambda_1 = 4/4 = 1$ chunk, the cells in D_2 are

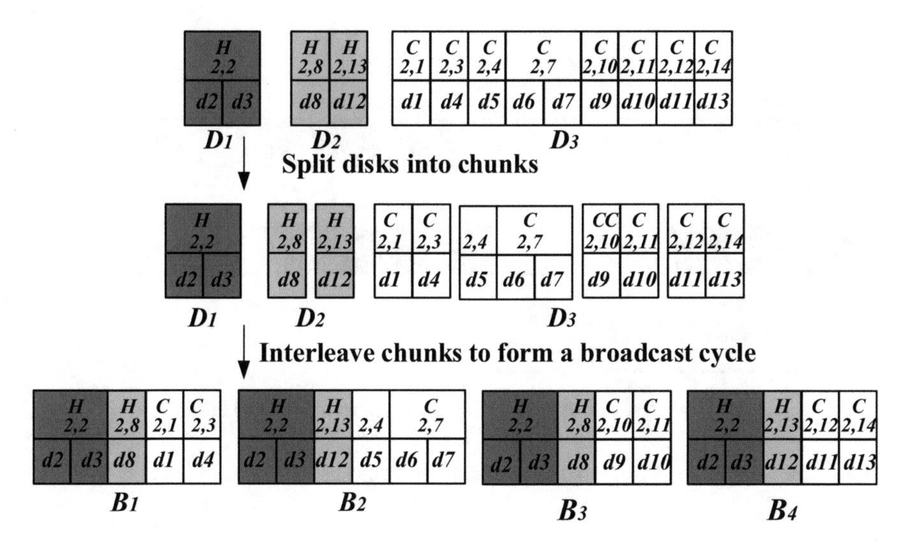

Fig. 2 Using broadcast disks to form a broadcast stream of non-flat broadcast

put into $L/\lambda_2 = 4/2 = 2$ chunks, and those in D_3 are put into $L/\lambda_2 = 4/1 = 4$ chunks, as shown in Fig. 2. One chunk in each disk is cyclically interleaved to form a minor cycle. One broadcast cycle consists of $L = 4$ minor cycles, B_1–B_4.

2.3 Index Structure

Consider that a window query is issued in Fig. 1c. Although the answered object d_3 is contained in cell $H(2, 2)$, all the cells overlapped with the query region should be examined, resulting in the increase of the access time and tuning time, i.e., $H(2, 2)$, $C(2, 3)$, $C(2, 4)$, and $C(2, 7)$. To avoid the above unnecessary examination, in this paper, we use the largest empty rectangle [2] to solve this problem. The largest empty rectangle, *LER*, is a rectangle of maximal size without touching any object. Take cell $H(2, 2)$ in Fig. 1c and d for example. Spatial objects d_2 and d_3 are contained in the minimum bounding rectangle R_1 in Fig. 1d. The largest rectangle extending from the edges of R_1 without touching any other spatial object is *LER* of cell $H(2, 2)$. In Fig. 1c, since the query region is totally contained in *LER* of cell H $(2, 2)$, the answered object for the query is limited in this cell, and no other cells should be examined. Since the hot cells are more frequently retrieved than the regular ones, *LER* of the hot cells are provided in the index structure of the proposed method.

Our proposed method interleaves three kinds of indexes with spatial objects on a broadcast cycle: the global index, the hot cell index and the cell index. The global index is interleaved before each minor cycle. The hot cell index is interleaved

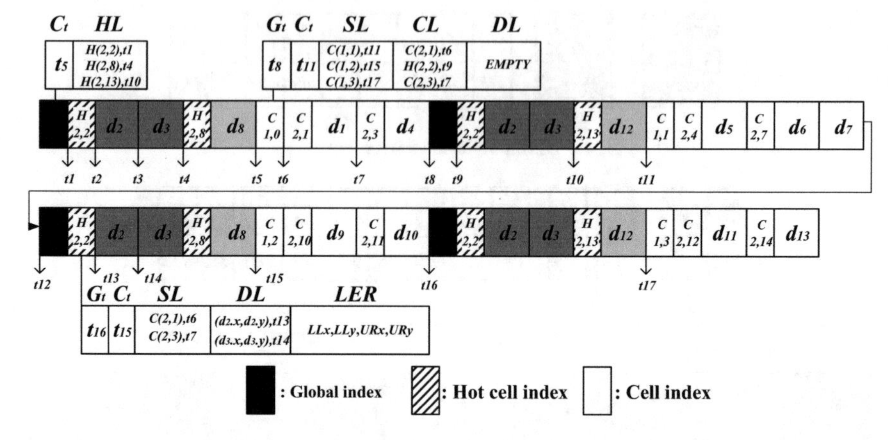

Fig. 3 The non-flat broadcast program interleaved with spatial indexes

before each hot cell. The cell indexes of the Hilbert curve from order 1 to order $(n - 1)$ are interleaved before the first element of the Hilbert curve of the next order. After that, the cell index of the Hilbert curve of order n is interleaved before each cell of the Hilbert curve of order n. Figure 3 shows the non-flat broadcast program interleaved with spatial indexes for the spatial objects shown in Fig. 1c. In Fig. 3, since cell $C(2, 4)$ containing object d_5 is the first element inside cell $C(1, 1)$, the cell index for $C(1, 1)$ is interleaved before the cell index for $C(2, 4)$.

The details of index structure are described as follows.

1. Global index: $G = <Ct, HL>$

   - C_t is the arrival time of the nearest upcoming cell index of the Hilbert curve of order 1 on the wireless channel. For example, C_t of the first global index in Fig. 3 is pointing to the beginning of cell index $C(1, 0)$, $t5$.
   - HL contains the arrival time of the cell indexes for all hot cells. For example, HL of the first global index in Fig. 3 contains the information about all the hot cells, $H(2, 2)$, $H(2, 8)$ and $H(2, 13)$.

2. Hot cell index: $H(i, j) = <G_t, C_t, SL, DL, LER>$

   - G_t is the arrival time of the nearest upcoming global index. For example, in Fig. 3, G_t of the hot cell index of $H(2, 2)$ in the second row is pointing to the beginning of the upcoming global index, $t16$.
   - SL contains the information about the sibling cells of the Hilbert curve of the same order.
   - DL contains the coordinates of the objects in that cell, and their corresponding arrival time. The mobile device can examine DL to decide whether the spatial objects are contained within the query region before accessing them.

- *LER* is the largest empty rectangle of this cell. The lower-left and upper-right coordinates of the largest empty rectangle are recorded in *LER*.

3. Cell index: $C(i, j)$ = <G_t, C_t, *SL*, *CL*, *DL*>

- *CL* contains the information about the child cells of the Hilbert curve of the next order. For example, in Fig. 3, *CL* of the cell index of $C(1, 0)$ contains the information about its child cells of the Hilbert curve of order 2, $C(2, 1)$, $H(2, 2)$ and $C(2, 3)$.

2.4 Window Query Processing

Similar to *SSI* [8], the cells of the Hilbert curve of order *n* overlapped with the query region for processing a window query should be checked. Such cells are put into checking set *CS*, and their corresponding cell indexes are further examined to check whether the spatial objects are within the query region via their coordinates stored in *DL* before accessing them. The processing procedures for the window query running in the mobile device are handled as follows.

1. Tune into the wireless channel to get the arrival time of the nearest upcoming global index, and switch into the doze mode to wait for the coming of the global index.
2. Examine *HL* of the global index to get the arrival times of the hot cells in *CS*, put them into checking queue *CQ*, and remove those cells from *CS*. Note that *CQ* is a sorted queue recording the arrival time of the next visited index or object in the ascending order.

   a. If *CS* is empty, follow the arrival time in *CQ* to retrieve the cell indexes to get the corresponding objects. That is, the query region is overlapped with only the hot cells. Otherwise, put the arrival time in C_t into *CQ*.

3. Follow the arrival time in *CQ* to retrieve the cell indexes to get the corresponding objects until *CS* is empty. Note that the visited element in *CQ* is removed after visited.

   a. If the visited index is the hot cell index, examine whether the query region is contained in *LER*. If yes, the answered objects are restricted to this hot cell. Follow *DL* in this hot cell index to retrieve the answered objects and end the processing. Otherwise, examine *SL* in this hot cell with *CS*, put the corresponding arrival time into *CQ*, and decide whether C_t should be put into *CQ*.
   b. If the visited index is the cell index, examine *SL*, *CL* and *DL* of this cell index with *CS*, put the corresponding arrival time into *CQ*, and decide whether C_t should be put into *CQ*. Follow *DL* in this cell index to retrieve the answered objects if necessary.

Take the window query in Fig. 1c for example. Cells *H*(2, 2), *C*(2, 3), *C*(2, 4) and *C*(2, 7) overlapped with the query region should be examined, i.e., *CS* = {*H*(2, 2), *C*(2, 3), *C*(2, 4), *C*(2, 7)}. Consider that the mobile device tunes in at the beginning of the first global index in Fig. 3 and retrieves this global index. After examining *HL*, *CS* = {*C*(2, 3), *C*(2, 4), *C*(2, 7)}. Since *CS* is not empty, C_t is put into *CQ*, i.e., *CQ* = [*t*1, t5]. The mobile device then retrieves hot cell index *H*(2, 2) at *t*1. After examining *H*(2, 2), the mobile device verifies that the query region is contained in *LER* of *H*(2, 2), and the query processing can be finished after retrieving the answered object in this hot cell. Then, the mobile device checks *DL* of *H*(2, 2) with the query region, and decides that only spatial object d_3 should be retrieved, i.e., *CQ* = [*t*3]. The mobile device switches into the doze mode and wakes up at *t*3 to retrieve spatial object d_3.

3 Simulation Study

To evaluate the effectiveness of our proposed method, we compare the proposed method with *SSI* [8] and *MLAIN* [4] on the average access time, tuning time and energy consumption.

3.1 Simulation Model

A real data set containing 5,922 points of interest in Greece (http://www.rtreeportal.org) is used to evaluate the performance among the methods. On a wireless broadcast channel, a logical transmission unit is a bucket, which occupies 64 bytes in our simulation. A spatial object occupies 1024 bytes, and two one-byte integers represent the coordinate of a spatial object.

To simulate the skewed access patterns, the access probability, *Pr*(*i*), of each spatial object *i* is generated by the *Zipf* distribution [1], $\Pr(i) = \frac{(1/i)^\theta}{\sum_{j=1}^{N} (1/j)^\theta}$, $1 \leq i$ *N*, where *N* is the total number of spatial objects and θ is the *Zipf* factor. When the value of θ is increasing, the access probabilities are becoming increasingly skewed. Assuming that the access probability of any spatial object in the same disk is uniform, the demand access probability for each disk from mobile clients is also generated by the *Zipf* distribution with *Zipf* factor ϕ. The relative frequency, R_j, of disk *j* is determined by $R_j/R_S = (S - j)\Delta + 1$, and $R_S = 1$, $1 \leq j \leq S$, where Δ is the factor for relative frequencies, and *S* is the number of disks.

3.2 Experimental Results

In our experimental results, 10,000 window queries are issued. The average access time and tuning time are measured in terms of buckets. For evaluating energy consumption, we assume that each mobile device consumes 1,150 mW/s in the active mode and 24.16 mW/s in the doze mode according to the simulations in [4, 8]. Moreover, the downlink channel with the 1 Mbps bandwidth is used. The average energy consumption of the query processing is calculated by 1150 * *Avg_TT* + 25.16 * (*Avg_AT* – *Avg_TT*), where *Avg_AT* and *Avg_TT* are the average access time and tuning time in seconds, respectively.

The search region of a window query is a square with the side length controlled by *WinLengthRatio*, which is the ratio of the side length of the query region to that of the total space. We evaluate the performance on different values of *WinLengthRatio*, 0.08, 0.095, 0.11, 0.125 and 0.14%. The number of disks, *S*, is 3, and the value of the factor for relative frequencies, Δ, is set to 1. The *Zipf* values of θ and ϕ are set to 2.5 and 2, respectively.

Figure 4a shows the average access time with the increase of the value of *WinLengthRatio*. Among the three methods, our proposed method, *LER*, has the shortest average access time. Moreover, our proposed method has the average improvement of 25.4 and 57% on the average access time over *SSI* and *MLAIN*, respectively. In Fig. 4a, we can observe that with the increase of the value of *WinLengthRatio*, the average access times for the three method remain stable. The reason is as follows. The span of the answer objects in a query region affects strongly the access time. Since the span of the answer objects for the settings in this simulation are quite similar, the average access times of each method with the different values of *WinLengthRatio* are stable.

Table 1 lists the corresponding average tuning time. In Table 1, we can observe that the average tuning times for the three methods are increasing while the value of *WinLengthRatio* is increasing. With the increase of the value of *WinLengthRatio*, the search region becomes wide and the number of the answered objects in this region increases. This leads to the increase of retrieving more index information and

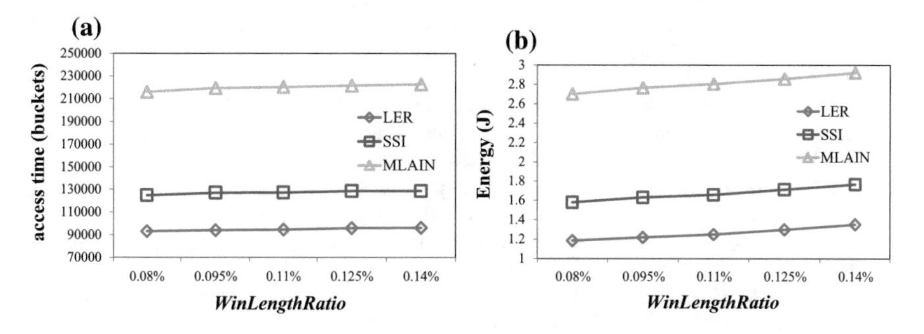

Fig. 4 Experimental results: **a** the average access time; **b** the average energy consumption

Table 1 The average tuning time

WinLengthRatio (%)	LER	SSI	MLAIN
0.08	77. 2	87.64	87.77
0.095	117.68	130.36	130.52
0.11	161.68	177.67	178.03
0.125	224.4	245.5	246.95
1.14	310.58	337.48	339.58

the targeted objects, resulting in the increase of the average tuning time. Our proposed method has the shortest average tuning time among the three methods. Figure 4b shows the corresponding average energy consumption, and our proposed method has the average improvement of 24.4% and 55.2% on the average energy consumption over *SSI* and *MLAIN*, respectively.

4 Conclusion

In this paper, we have addressed the problem of processing window queries for skewed access patterns via non-flat wireless broadcast. We have proposed an efficient processing method that applies the concept of the largest empty rectangle to avoid the unnecessary examination of the broadcast stream while processing such queries. From our simulation study, the proposed method has a better performance on the average access time, tuning time and energy consumption than the existing methods.

References

1. Acharya, S., Franklin, M., Zdonik, S., Alongso, R.: Broadcast Disks: Data Management for Asymmetric Communications Environments. In: Proc. of the 1995 ACM SIGMOD Int. Conf. on Management of Data, pp. 199–210 (1995)
2. Aggarwal, A., Suri, S.: Fast Algorithms for Computing the Largest Empty Rectangle. In Proc. of the third Annual Symp. on Computational Geometry, pp. 278–290 (1987)
3. Im, S., Youn, H.Y., Choi, J., Ouyang, J.: A Novel Air Indexing Scheme for Window Query in Non-Flat Wireless Spatial Data Broadcast. J. of Commun. and Networks, vol. 13, no. 4, pp. 400–407 (2011)
4. Im, S., Choi, J.: MLAIN: Multi-leveled Air Indexing Scheme in Non-flat Wireless Data Broadcast for Efficient Window Query Processing. Computers and Mathematics with Applications, vol. 64, pp. 1242–1251 (2012)
5. Im, S., Choi, J.: Quick Data Access on Multiple Channels in Non-flat Wireless Spatial Data Broadcasting. IEICE Trans. Commun., vol. E95-B, no. 9, pp. 3042–3046 (2012)
6. Im, S., Hwang, H.: A Two-Tier Spatial Index for Non-flat Spatial Data Broadcasting on Air. IEICE Trans. Commun., vol. E97-B, no. 12, pp. 2809–2818 (2014)
7. Jung, S., Mo, H.: A Spatial Indexing Scheme for Location based Service Queries in a Single Wireless Broadcast Channel. Journal of Information Science and Engineering, vol. 30, pp. 1945–1963 (2014)

8. Shen, J.H., Jian, M.S.: Spatial Query Processing for Skewed Access Patterns in Nonuniform Wireless Data Broadcast Environments. International Journal of Ad Hoc and Ubiquitous Computing, vol. 25, no. 1/2, pp. 4–16 (2017)
9. Shen, J.H., Chang, Y.I.: An Efficient Nonuniform Index in the Wireless Broadcast Environments. The Journal of Systems and Software, vol. 81, no. 11, pp. 2091–2103 (2008)
10. Song, D., Park, K.: A Partial Index for Distributed Broadcasting in Wireless Mobile Networks. Information Sciences, vol. 348, pp. 142–152 (2016)

A Research on Multi-attribute Sequence Query Processing Techniques for Motion Databases

Edgar Chia-Han Lin

Abstract Due to the development of computer technology and the mature development of motion capture technology, the applications of motion databases become more and more important. How to analysis the huge data stored in the database and efficiently retrieved the matched data is an important research issue. 3D animation design is one of the important applications of motion databases. Based on our teaching experience, the bottleneck of the students' learning of animation is the motion animation of the 3D characters. Therefore, the motion database can be used to assist the design of the motions for 3D characters. However, it is still a difficult problem because of the high complexity of the matching mechanism and the difficult of user interface design. In this paper, the motion data can be represented as multi-attribute multi-dimensional sequences while the corresponding index structures and query processing mechanism are proposed for efficiently processing the motion queries. Moreover, Microsoft Kinect is used in this paper as the user interface. The captured data can be used as the user query and the further comparison will be performed to find the matched motion data.

Keywords Motion database · Index structure · Query processing · Kinect

1 Introduction

Due to the great improvement of technology and the mature development of motion capture technology, the applications of motion database become more important in recent years. How to analyze the large amount of motions recorded in the databases to efficiently retrieve the desired motions become an important research issue. Motion data is a sequential data which is formed by the series of reference coordinate values of different parts of human body. Due to the large amount of reference coordinate values and the characteristics of sequential data, the analysis and

---

E.C.-H. Lin (✉)
Department of Information Communication, Asia University, Taichung, Taiwan
e-mail: edgarlin@asia.edu.tw

© Springer Nature Singapore Pte Ltd. 2018

N.Y. Yen and J.C. Hung (eds.), *Frontier Computing*, Lecture Notes
in Electrical Engineering 422, DOI 10.1007/978-981-10-3187-8_67

retrieval of motion data is time consuming. There are many researches which focus on reduction the complexity of the motion data or designing the index structure for the motion data to enhance the efficiency of data retrieval.

In [1], the complex multidimensional time series data is separated into several smaller segments to reduce the complexity of data. Furthermore, the segments are used to reduce the difficulty of the matching mechanism. In these years, many researches focus on the representation of the motion data [2, 3]. In [4, 5], the features of the motion poses will be extracted for each motion frame which map to a multi-dimension vector. The motion content is represented by those vectors while the index structure is also constructed. In [6, 7], the motion poses are represented by a hierarchal structure and the key frames are extracted to find the motion information. However, the extraction of key frames is time consuming. In [8–10], various motion database representation mechanisms are proposed.

In [11], a method to find the similar motions is proposed. Based on the index structure, the partial queries, i.e., the motions of some specific body parts, can be processed. Five index structures are constructed for the motions performed by different body parts. Moreover, a hierarchical structure of the human body is used to integrate the index structures such that the partial motion queries can be processed. A new similarity measurement is proposed in [12] to enhance the efficiency and effective of similar matching mechanism.

Since the motion data can be represented as multiple attribute (multiple reference points) time series data, many researches focus on the similarity search of multiple attributes databases. In [13], the Dynamic Time Warping (DTW) and Longest Common Subsequence (LCSS) approaches are extended as the similarity measurement of the multiple attribute data. In [14], a pivot-based index structure for combination of feature vectors is proposed. The query processing problem is transformed to a searching problem of multiple attributes data set. In [15], an aggregate nearest-neighbors retrieval algorithm is proposed for multi-points query problem. iDistance [16] is an index structure based on distance. In [17], a preprocessing procedure is proposed to construct the neighbor graph for motion database such that the nearest neighbor search can be efficiently performed. However, the preprocessing procedure is time consuming.

In our previous works [18–24], the index structures and corresponding query processing mechanisms for multiple attribute time sequence are proposed for exact or approximate matching problem.

In this paper, a multi-dimensional multi-attribute index structure for motion data and a multi-resolution query processing technique are proposed to find the matched motions form the motion database based on users' query and query resolution.

The queries will be specified by using Kinect device and the queries will be represented by a multi-attribute multi-dimensional string which will be used in the query processing procedure. Since the matched motions will be used to construct the motions in a 3D animation, the motion data of different body parts can be applied individually based on a semantic model. Therefore, we don't need to find the exactly matched motions from the database. The partial matched results will be useful for users to construct their own 3D animations. That is, the user can specify different query resolution to find the different matched results.

2 Multi-attribute Multi-dimensional Sequence Model for Motion Data

In this section, the data model of motion data is proposed to represents the content of motion data in the database. Since the motion data can be represented by the movement of multiple reference points. The movement of each reference point can be represented by the movement in 3 coordinates. That is, the movement of each reference point can be represent as a multiple 3-dimension sequence. Moreover, each reference point can be considered as an attribute of the motion data. Therefore, the motion data can be represented by a multi-attribute 3-dimension sequence.

There are two different data models which are proposed to represent the semantic and geometric meanings of motions. The semantic model is constructed based on the content information of the motions while the geometric data model is constructed based on the coordinates of the reference points in 3D spaces. The multi-resolution queries are based on the hierarchy of human body which is used in the semantic model. Only the interested body part will be used to find the matched data.

To construct the semantic model, the hierarchy of human body will be defined first. In the proposed semantic model, the semantic meanings of human body are defined as the following different types: Whole Body, Half Body, Hand and Foot. The type of motions corresponding to each type of human body will be further defined as: Stand, Squat, Walk, Run, Jump, Wave and Kick. Each motion in the database will be analyzed and hierarchically classified into some particular semantic class such as Hand -> Wave. Therefore, each motion in the database will be structurally annotated based on the semantic data model. The annotated metadata can be used as the query criteria specified by users.

The geometric data model is constructed based on the motion properties of the reference points in the 3D spaces. The movement of each reference point can be represented by the feature values of 3 dimensions. That is, the feature values are used to represent the movement characteristic of three different dimensions. The feature values used to represent the motion data are shown as follows.

X-dimension	Left	Steady	Right
Y-dimension	Up	Steady	Down
Z-dimension	Forward	Steady	Backward

For example, the movement of the reference point corresponding to right wrist in Fig. 1 can be represented as the following 3-dimension string.

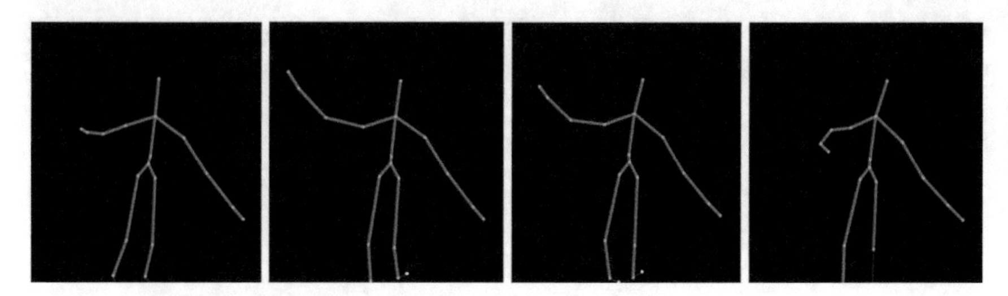

Fig. 1 Example of a motion data

$$\begin{matrix} L & R & R \\ U & D & D \\ F & F & B \end{matrix}$$

Therefore, the motion data can be represented as a set of 3-dimensional strings which can be used to construct the index structure for further query processing mechanism.

Since the dimensions of each attribute is the same, therefore, the index structure can be constructed for each dimension individually. However, it is time consuming to combine the query results, which comes from different index structure, to find the final results. Moreover, the number of feature values of each dimension is few such that the number of data recorded in the data node of the index structure may be very large. Therefore, the multi-dimensional data sequences can be combined into a single multi-dimensional symbol sequence. That is, only one index structure will be constructed and the efficiency of the query processing procedure can be enhanced. The 3-dimensional string in the previous example can be represented as the following 3-dimensional symbol string.

$$(L, U, F)(R, D, F)(R, D, B)$$

The motion data recorded in the database will be transformed into multiple 3-symbol strings according to different attributes. These 3-dimensional symbol will be used to construct the index structure for efficient query processing.

3 Index Structure and Query Processing

3.1 Index Structure Construction

Since the motion data is represented as a multiple 3-dimensional symbol strings, the suffix tree based index structure proposed in our previous work [18] can be modified and applied to efficiently find the matched motion data. On the other hand, the semantic hierarchy can be used to record the semantic meaning of each motion which can be used to find the matched results by traverse the hierarchy.

In order to index the motion data in the database, a 2-level index structure is proposed in this paper. The multiple strings of 3-dimensional symbols, which corresponding to multiple attributes, will be used to represent a clip of motion data. Since the 3-dimensional symbols used to represent the movement of each attribute are in the same set, only one index tree will be constructed. However, since only one index tree is constructed in the proposed approach, a second-level index structure will be constructed for the data recorded in the data node to find the matched data for different attribute efficiently.

In each data node of the index tree, the second-level index structure is constructed. The attribute table will be used to record the data vector which is a bit-vector to indicate the matched data via a mapping table.

Definition *Mapping Table*
Mapping table is a table which records the data number corresponding to each bit of the data vector. The length of mapping table is the amount of motion data recorded in the database. When a new motion data is inserted into the database, the length of mapping table will be increased and the inserted data number will be recorded.

Definition *Attribute Table*
Attribute table is the second level index structure in the proposed approach. In each data node of the index tree, the corresponding data will be recorded. However, since only one index tree is constructed, the 3-dimensional symbol string of each attribute will be indexed by the same index tree. Therefore, the data recorded in the data node should be further indexed by the attribute table to distinguish the matched data for the different attributes.

Definition *Data Vector*
In the proposed approach, a data vector is used to record the corresponding data in the data node of the index tree. The data vector is recorded in the corresponding element of the attribute table. Each bit of the data vector corresponds to one particular data recorded in the database, which can be mapped via the mapping table. The data vector will be updated only when a new motion data is inserted to the corresponding data node.

The following data update procedure is applied when a new data is inserted to the data node. The insertion vector is a bit-vector which is used to identify the inserted data.

Data Update Procedure (*data number*, *data vector*)

Step 1. An *insertion vector* will be generated based on the *data number*.
Step 2. New *data vector* = OR (*data vector*, *insertion vector*).

The update procedure will be applied only when a new data inserted into the data node. The data vector of the corresponding attribute will be retrieved and updated.

Since only the data node related to the inserted data will be updated, the length of the data vectors recorded in the data nodes may be different. Therefore, some preprocessing steps should be applied in the query processing procedure.

3.2 Query Processing Mechanism

A kinect based query interface is also designed for user to specify queries. The feature values of the query motion will be extracted and represented as a query string. The query processing mechanism will be applied to process the query string and the match results will be efficiently found via the index structure.

The semantic query can be specified by users via the interface to find the corresponding motions. The geometric queries will be specified by using the Kinect device. The query motion can be acted by user and the Kinect device will capture the user motion and transform into the multiple 3-attribute strings as the query string. The multiple 3-attribute strings will be decomposed and traverse the corresponding index tree to find the matched data. At last, the matched data will be compared and the matched results will be found. Since only the movement of each reference point is considered, the absolute location of the human body captured by Kinect device won't affect the matching results. The experiment results show the flexibility of specifying queries.

The proposed query procedure can be shown as follows. Note that since the length of data vector may be different, the matched data vectors will be updated in Step 5 before the further comparison.

Query Processing Procedure

Step 1. Decompose the query motion into different query strings according to the attributes which is chosen based on user resolution.
Step 2. Traverse the index tree and process the query string one by one.
Step 3. All the matched *data vectors* are retrieved form the data node.
Step 4. The *mapping key* is generated by mapping table which is used to update all the matched *data vectors*.
Step 5. For all matched data vectors, the updated data vectors will be obtained by using logical OR operation as shown in follows.
updated *data vector* = OR (*mapping key*, *data vector*)
Step 6. Applied logical AND operation to all updated matched *data vectors* and then the matched data will be found via the *mapping table*.

4 Experimental Results

In this paper, a 3D data management query system is proposed. The desired motions can be specified semantically or geometrically by users. A semi-automatically annotation system is developed to annotate the semantic metadata of each motion recorded in the database. Moreover, the multiple 3-dimensional symbol strings used to represent each motion are automatically generated by considering the relationship of the coordinates between reference points. Then, a 2-level index structures is constructed for all 3-dimensional symbol strings corresponding to different attributes. The query interface is designed for user to specified desired motions as queries. The semantic queries can be manually specified via the interface. Moreover, the geometric queries can be captured via the Kinect device and the corresponding multiple 3-dimensional symbol strings will be transformed and specified as query strings. Therefore, the index structures are traversed and the matched results can be found.

Since the data vectors are used to indicate the motion data recorded the data node of the index structure, it is time consuming to update all the vectors while a new motion data coming. In the proposed approach, we allow the length of data vectors to be different. The data vector will be updated only when a new data with particular attribute is inserted into the data node or it matches a user query. Therefore, no extra scanning of index tree will be performed. Based on the proposed approach, the efficiency of motion data insertion will be greatly enhanced.

The experiments show that the desired motions can be easily specified by users and the matched results can efficiently found.

5 Conclusions

In this paper, a motion data management query system based on Kinect is proposed. The semantic meaning of motions will be hierarchically annotated. Moreover, the geometric meanings of the motions will be represented as multiple 3-dimensional symbol strings and the corresponding index structure are proposed to efficiently find the matched motions. Moreover, the Kinect device is used to capture the user queries and represented as a query string for further query processing mechanism. The experimental results show that the desired motions can be easily specified and the matched results can be found efficiently. Since the user queries may not exactly describe the desired motions, the approximate results should be further considered. We are currently working on extending the proposed methodology to find the approximate results. The similarity measurement and the corresponding matching algorithm are currently under development.

References

1. C. Lu. and N. J. Ferrier, "Automated analysis of repetitive joint motion," IEEE Trans. Inf. Technol. Biomed., vol. 7, no. 4, pp. 263–273, Dec. 2003.
2. M. Muller, T. Roder, and M. Clausen, "Efficient content-based retrieval of motion capture data," ACM Trans. Graphic. (TOG), Vol. 24, pp. 667–685, 2005.
3. G. Liu, J. Zhang, W. Wang, and L. McMillan, "A system for analyzing and indexing human-motion databases," in Proc. 2005 ACM SIGMOD Int. Conf. Manage. Data, pp. 924–926.
4. S.-P S.-P. Chao, C.-Y. Chiu, J.-H. Chao, Y.-C. Ruan, and S.-N. Yang, "Motion retrieval and synthesis based on posture features indexing," in Proc. 5th Int. Conf. Comput. Intell. Multimedia Appl., Sep. 2003, pp. 266–271.
5. C.-Y. Chiu, S.-P. Chao, M.-Y. Wu, S.-N. Yang, and H.-C. Lin, "Contentbased retrieval for human motion data," J. Vis. Commun. Image Representation, vol. 15, pp. 446–466, Apr. 2004.
6. F. Liu, Y. Zhuang, F. Wu, and Y. Pan, "3D motion retrieval with motion index tree," Comput. Vis. Image Understanding, vol. 92, pp. 265–284, Jun. 2003.
7. Q. Gu, J. Peng, and Z. Deng, "Compression of human motion capture data using motion pattern indexing," Comput. Graph. Forum, vol. 28, no. 1, pp. 1–12, 2009.
8. J. Wang, D. Fleet, and A.Hertzmann, "Gaussian process dynamical models for human motion," IEEE Trans. Pattern Anal. Mach. Intell., vol. 30, no. 2, pp. 283–298, Feb. 2008.
9. K. Tang, H. Leung, T. Komura, and H. P. H. Shum, "Finding repetitive patterns in 3D human motion captured data," in Proc. 2nd Int. Conf. Ubiquitous Inf. Manage. Commun., Suwon, Korea, 2008, pp. 396–403.
10. X. Wang, Z. Yu, and H. Wong, "3D motion sequence retrieval based on data distribution," in Proc. 2008 IEEE Int. Conf. Multimedia Expo., 2009, pp. 1229–1232.
11. Gaurav N. P. and Balakrishnan P., "Indexing 3-D Human Motion Repositories for Content-Based Retrieval," IEEE Transactions of Information Technology in Biomedicine, vol. 13, no. 5, Sep. 2009.
12. B. Kruger, J. Tautges, A. Weber and A. Zinke, "Fast Local and Global Similarity Searches in Large Motion Capture Databases," Proc. 2010 ACM SIGGRAPH/Eurographics Symposium on Computer Animation, 2010.
13. M. Vlachos, M. Hadjieleftheriou, D. Gunopulos, and E. Keogh, "Indexing multi-dimensional time-series with support for multiple distance measures," in Proc. SIGMOD, Aug. 2003, pp. 216–225.
14. B. Bustos, D. Keim, and T. Schreck, "A pivot-based index structure for combination of feature vectors," in Proc. SAC 2005, New York, pp. 1180–1184.
15. D. Papadias, Y. Tao, K. Mouratidis, and C. K. Hui, "Aggregate nearest neighbor queries in spatial databases," ACM Trans. Database Syst., vol. 30, no. 2, pp. 529–576, 2005.
16. C. Yu, B. C. Ooi, K.-L. Tan, and H. V. Jagadish, "Indexing the distance: An efficient method to KNN processing," in Proc. VLDB, San Francisco, CA, 2001, pp. 421–430.
17. Chai J., Hodgins J. K., "Performance animation from low-dimensional control signals," ACM Trans. Graph, vol. 24, no. 3 (2005), pp. 686–696. SIGGRAPH 2005.
18. Chia-Han Lin and Arbee L.P. Chen, "Indexing and Matching Multiple-Attribute Strings for Efficient Multimedia Query Processing," in IEEE Transactions on Multimedia Vol. 8, No. 2, pp. 408–411, Apr. 2006.
19. Chia-Han Lin and Arbee L. P. Chen, "Approximate Video Search Based on Spatio-Temporal Information of Video Objects," in The First IEEE International Workshop on Multimedia Databases and Data Management, 2006.
20. Edgar Chia-Han Lin, "Research on Sequence Query Processing Techniques over Data Streams," in Applied Mechanics and Materials, Vol. 284–287, 3507–3511, 2013.

21. Edgar Chia-Han Lin, " Research on Multi-Attribute Sequence Query Processing Techniques over Data Streams," in 2nd International Conference on Advanced Computer Science Applications and Technologies, 2013.
22. Edgar Chia-Han Lin, "Research on Multi-Attribute Sequence Query Processing Techniques over Data Streams," in Applied Mechanics and Materials, Vol. 513–517, 2014.
23. Edgar Chia-Han Lin, "A research on 3D motion database management and query system based on Kinect," to appear in Lecture Notes in Electrical Engineering, Vol. 329, 2015.
24. Edgar Chia-Han Lin, "A Research on Multi-dimensional Multi-attribute String Matching Mechanism for 3D Motion Databases," in Lecture Note in Electrical Engineering, Vol. 375, pp. 575–582, 2016.

A Research on Integrating AR and Multimedia Technology for Teaching and Learning System Design

Edgar Chia-Han Lin, Yang-Chia Shih and Rong-Chi Chang

Abstract Due to the great progress of information technology and the mature development of the multimedia technology, the computer and multimedia elements is involved in the learning materials. Recently, the due to the popularity of mobile devices, augmented reality development has become more sophisticated, more and more applications using augmented reality technology to attract the attention of the user or consumer. In this Paper, an interactive multimedia learning system with augmented reality technology is proposed to improve the effectiveness and efficiency of teaching and learning behavior.

Keywords Augmented reality • Interactivity multimedia

1 Introduction

In these years, due to the improvement of information technology, the digitalized learning material become more popular. The computer, which including the mobile devices, and multimedia data are included in the material of teaching assistant systems. For student, the new type of teaching material not only increase the impression of the content but also enhance the learning interest and learning motivation. Furthermore, the game elements are included in the design of teaching material. The main purples of learning by game is the reinforcement of learning motivation. The students would like to play the game and learning while solving the

E.C.-H. Lin (✉)
Department of Information Communication, Asia University, Taichung, Taiwan
e-mail: edgarlin@asia.edu.tw

Y.-C. Shih
Department of Biotechnology, Asia University, Taichung, Taiwan
e-mail: angelashih@asia.edu.tw

R.-C. Chang
Department of Digital Media Design, Asia University, Taichung, Taiwan
e-mail: roger@gm.asia.edu.tw

© Springer Nature Singapore Pte Ltd. 2018

N.Y. Yen and J.C. Hung (eds.), *Frontier Computing*, Lecture Notes in Electrical Engineering 422, DOI 10.1007/978-981-10-3187-8_68

problems designed in the game. Due to the population of mobile devices, they can be used as a virtual classroom in which the learning activities can be applied on. Moreover, the multimedia technology can be integrated into the mobile devices which make the student can learn anywhere anytime.

AR (Augmented Reality) technology combines the virtual 2D or 3D images with the actual shooting images. By using mobile devices and identification technology, image, text or audio-visual can be presented in front of the user. AR can be applied on many different fields such as medical, business, industry, entertainment and so on.

To design an AR application, the images should be analyzed and the corresponding 3D animation or the multimedia content will be shown on the sceen. That is, the virtual objects can be shown in the real 3D space.

In this paper, we develop a learning system for the course "General Biology Experiment" which integrate the AR and multimedia technology to enhance the effectiveness of teaching and learning.

2 Related Works

Rendering 3D objects in a VR application can be found in many applications. VisionLens develop an APP which shows the 3D model via AR technology. The 3D model can be combined with the product which will be helpful to attract the customer. There is another AR application in Japan which is used to promote a 3D cartoon. The App provides interesting interactions between the 3D cartoon characters and the users.

In UK, there is an App called Aurasma which apply the AR technology on newspapers and posters. Based on the identification and AR technology, the pictures in newspapers or posters will become videos which are easier for the reader to realize the content. In 2014, IKEA publish an APP of "IKEA Catalog" which applies the AR technology to put the 3D model of the product in a real environment such that the customer will image the situation after buying the product and put in a real environment. In addition to the commercial usage of AR technology, some researchers focus on different types of AR applications such as interactive tennis game [3], a desktop card game [6], Chinese phonetic alphabet learning materials, nine planets teaching system [5] and so on [1, 2, 4].

Since the AR technology can be used to combine the virtual information with the real environment, it can be used in various domain such as learning system. It can help the students to "virtual practice" by using the mobile device without go to the real classroom.

3 The Design of AR Integrated Multimedia Learning System

In this paper, a teaching and learning system with three steps for the course "General Biology Experiment" is proposed. Through the system, students can learn from the multimedia based learning material before the class. Moreover, they can interact with the applications including AR applications designed in the system to understand the detailed information of the experiments. At last, when the students actually experiment in the classroom, the system will provide some guideline for them to complete the experiment correctly. The system architecture of the proposed approach is shown as follows (Fig. 1).

3.1 Development of Learning System Based on AR and Multimedia Technologies

In this section, we focus on how the biological experiment teaching activities with AR technology and multimedia content can help students' learning. Different topic of the course may need different way to integrate the multimedia content. Table 1 shows the topic of the course "General Biology Experiment." In this section, we develop a multimedia assistant learning system including the interactive multimedia and AR applications corresponding to different topics.

3.1.1 The Establishment of the Virtual Learning Objects

For each course unit of "General Biology Experiment," we have to find the appropriate way to interactive with the students. Moreover, we will design the

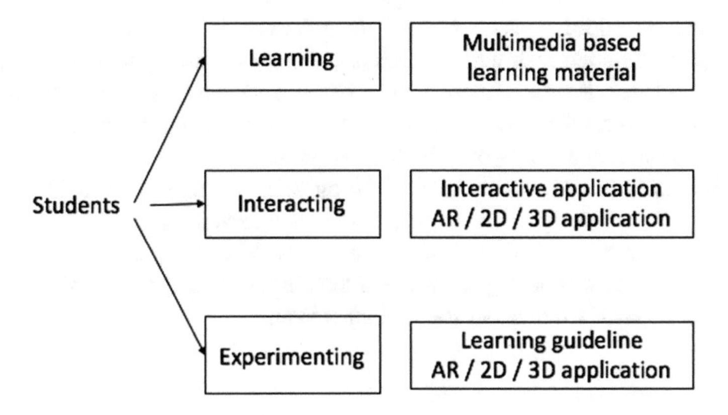

Fig. 1 The architecture of the 3 step learning system

Table 1 Course content of "General Biology Experiment"

Course content
DNA model
Microscope: animal and plant cell
Microbial culture
Bacteria stain and counting
Experimental observation of pool plankton pool
Experimental observation of fungi and lichens
Vegetative organs of angiosperms
Observed linear animals—and roundworm C. elegans
Observed annelid—earthworms and Nereid
Fish bones taxidermy
Frog anatomy and bone taxidermy
Fermentation
The plant cell mitosis
Water quality and freshwater biological field survey
Meiosis of plants
Coevolution of plants and insects
Purification of bacterial DNA
Electrophoretic analysis of DNA
Photosynthesis and plant pigment analysis
ELISA

2D/3D virtual objects which can be used in the multimedia and AR learning applications.

For example, in the course unit "Microscope: Animal and Plant Cell," we design a 3D virtual object of microscope for student to observe and understand the usage and function of microscope before they use it in the real classroom. Moreover, with AR technology, when students hold a mobile device and facing the microscope in the classroom, a 3D microscope with explanations will be shown in the screen. The students can rotate or move 3D objects to observe and learn the structure of microscope. In addition, each part of the microscope can also be separated to provide more detailed instructions. Unlike ordinary textbooks with static 2D images, the students' learning interest may be enhanced by using the 3D dynamic objects. Moreover, since the 3D objects can be manipulated by students, the students' learning effectiveness may be also improved.

Furthermore, for the course unit of "Frog anatomy and bone taxidermy." 2D virtual objects are designed and integrated in the system for student to learn the structure of frog bone taxidermy before the class. Moreover, we create 3D virtual objects using AR technology to allow students to learn more with the detailed description of each structure in the real classroom.

3.1.2 The Design of the Identification Patterns

In AR applications, the design of identification pattern will affect the timing of AR object appears. It becomes an important research issue to design an appropriate identification pattern which can fit the real environment. In this section, the identification patterns are designed based on the real class environment or the characteristics of the topics. The designed patterns are used in the AR applications to invoke the 3D virtual objects in the screen while the students learning by using the application.

3.2 The Development of AR Integrated Multimedia Learning System

This section focus on how the teaching activities of the course "General Biology Experiment" with the multimedia integrated application can be used to enhance the students' learning effectiveness. Since there are different characteristics for different course topics, we design different multimedia interaction applications for each topic. The interaction style for each course topic is shown in the following table.

The multimedia based teaching and learning system integrates different interaction styles for different topics in which the student can learn the content by himself before the class and also can learn the content during the class (Table 2).

Table 2 Interaction style of each course topic

Course content	Interaction Style
DNA model	AR with 3D objects
Microscope: animal and plant cell	AR with 3D objects
Microbial culture	Interactive 2D applications
Bacteria stain and counting	Interactive 2D applications
Experimental observation of pool plankton pool	Video with 2D animation
Experimental observation of fungi and lichens	
Vegetative organs of angiosperms	Interactive 2D applications
Observed linear animals—and roundworm C. elegans	Video with 2D animation
Observed annelid—earthworms and Nereid	
Fish bones taxidermy	AR with 3D objects
	Interactive 2D applications
Frog anatomy and bone taxidermy	AR with 3D objects
	Interactive 2D applications
Fermentation	Video with 2D animation

(continued)

Table 2 (continued)

Course content	Interaction Style
The plant cell mitosis	Video with 2D animation Interactive 2D applications
Water quality and freshwater biological field survey	Video with 2D animation
Meiosis of plants	Video with 2D animation
	Interactive 2D applications
Coevolution of plants and insects	Video with 2D animation
Purification of bacterial DNA	Video with 2D animation
Electrophoretic analysis of DNA	
Photosynthesis and plant pigment analysis	Video with 2D animation
ELISA	Video with 2D animation

4 Experimental Results

The proposed system is used in the classes of Asia University to help the students to learn the course "General Biology Experiment." 100% students indicate that the proposed system is interesting and would like to use it while learning the course. Although more than 80% students indicate that the proposed system is helpful while they learn the course, but more than 60% students indicate that they learn more in the real class. It is acceptable since the virtual objects is used in the proposed system which are differ from the real objects the students used in real class. The experimental results shows that the proposed system is helpful while the student learn the course. However, it can also be improved to integrate more learning activities such that the student can interact and practice with the system and learn something which may not be obtained easily in the real class.

5 Conclusions

Due to the great progress of information technology and the mature development of the multimedia technology, the computer and multimedia elements is involved in the learning materials. Recently, the due to the popularity of mobile devices, augmented reality development has become more sophisticated, more and more applications using augmented reality technology to attract the attention of the user or consumer. In this Paper, an interactive multimedia learning system with AR technology is proposed to improve the effectiveness and efficiency of teaching and learning behavior. The experimental results shows that the proposed system is helpful while students learn the course. We will improve the proposed system by integrate more learning activities to enhance the learning effectiveness of the students.

Acknowledgments This work was partially supported by Asia University (Project No. 104-asia-07).

References

1. Carl, L. E. and Anderson, T., The virtual reality casebook. NY: Van Nostrand Reinbold, 1994.
2. Crosbie, J. H., Lennon, S., McNeill, M. D. and McDonough, S. M., "Virtual reality in the rehabilitation of the upper limb after stroke: the user's perspective," *Cyber Psychology and Behavior*, 9(2): 137–141, 2006.
3. Henrysson, A., Billinghurst, M., Ollila, M., "Face to face collaborative AR on Mobile phones," *IEEE Computer Society*, 80–89, 2005.
4. Holden, M. K., "Virtual environments for motor rehabilitation: Review," *Cyber Psychology and Behavior*, 8(3): 187–219, 2005.
5. Kondo, T., "Augmented learning environment using mixed reality technology," *World conference on E-learning in corporate, government, healthcare, and Higher education,* Vol. 2006, 83.
6. Sony, The eye of judgment, from https://www.playstation.com/en-us/games/the-eye-of-judgment-ps3/, 2009.

Storyboard-Based Automatic Summary Video Editing System

Shih-Nung Chen

Abstract The recent popularity of smart mobile devices has led to a significant increase in the needs of multimedia services. Finding new more efficient methods for automatic classification and retrieval of a large number of multimedia files will significantly reduce manpower costs. However, most current video content analysis methods adopt low-level features to analyze video frame by frame, and need to improve high-level semantic analysis on a number of issues. Hence, this study presents a storyboard-based automatic video editing system that uses storyboard information, such as character dialogue, narration, caption, background music and shot changes, to enable accurate video content retrieval and automatic render summary videos. The proposed system can be applied to the course video trailer and the commercial video trailer for quick preview video content or suitable viewing configuration for smart mobile devices. Consequently, the audience can quickly understand the whole video story and the video editors can substantially reduce the time taken to publish videos.

Keywords Video search · Storyboard · Video content retrieval · Automatic video editing · Summary video

1 Introduction

The rapid development of Internet and multimedia technologies has significantly increased the ability to present and communicate diverse information that not only provides visual and immersive enjoyment, but also raises awareness of the needs of life, thereby accelerating multimedia development and applications in learning and commerce. The high efficiency technology of image processing and production enhances learning motivation and interests for people. Researchers have recently

S.-N. Chen (✉)
Department of Information Communication, Asia University, Taichung, Taiwan, ROC
e-mail: nung@asia.edu.tw

© Springer Nature Singapore Pte Ltd. 2018

N.Y. Yen and J.C. Hung (eds.), *Frontier Computing*, Lecture Notes in Electrical Engineering 422, DOI 10.1007/978-981-10-3187-8_69

performed many studies in digital images, digital videos and other related digital media technologies, which mostly for digital compression, pattern recognition, image analysis, image processing, automatic indexing and searching. However, few studies have focused on computer-assisted digital video editing technology.

Moreover, with the popularity of wireless broadband networks and increase in total information processing capacity for multimedia information, searching and access to knowledge through the Internet is no longer confined to traditional text-based retrieval methods. Since presentation of image, sound, video and other multimedia content often conveys accurate information more efficiently than pure text mode. Therefore, encapsulating the knowledge of digital multimedia packages for dissemination will become increasingly important in the future. Effective management and retrieval of large numbers of digitized multimedia files is a very important research topic. Therefore, researchers continue to develop the core technology of multimedia content management and retrieval, such as keyword extraction, speech recognition, image recognition and indexing terms, in order to manage the large amounts of multimedia files with minimum labor and time costs.

2 Related Work

Early digital video research mostly used a single frame or a single pixel as a unit of analysis [1, 2], or considered all the information of the frame [3]. Digital video research has recently turned to object-based video processing [4]. It is mainly aimed at extraction of features, such as background, shot change, color, texture, motion, for video data classification [5–7]. Restated, recent studies have focused not on a single pixel changes, but instead on object-based processing. In video retrieval research, an object-based index structure can more accurately mark the differences between the videos [8, 9]. For motion tracking, the processing algorithm detects and tracks an entire moving object rather than a single pixel [10]. These object-based analysis are more to our human visual perceptual experience, and also makes the object moving information more meaningful.

Photographic technology has been available for over 100 years, and the camera phones also became a commodity around 2000. The number of photographic equipment and smart mobile devices is now too large to count. Since the market of mobile devices has grown significantly, and the functions of mobile devices is rapidly improving, recording video clips is rapidly becoming very popular. Although the production of short videos has substantially risen, the video editing and production process has not changed much over the years. People still need to spend a lot of time to learn editing and post production skills. Most people can convey the overall emotion of their videos, but cannot easily handle visual impressions due to the additional sensitivity and skill required. Research on automatic video editing in computer vision is currently rare. Previous studies mostly focus on video special effects, such as fades, camera shot change, bluescreen and compositing, but the needs of research and related technology of automatic video editing will increase substantially in the future.

The objective of a notice film is to briefly summarize the film, which is the result of creative and technical work, and to attract an audience to the film. The vital content of a notice film is mostly selected by the editor based on subjective experience. However, literature on editing of notice films and the correlation between film content and audience in terms of semantic awareness is rarely seen. Yang presented an approach based on semantic networking, and conducted literature review to collect, analyze and conclude suggestions on the editing of educational notice films. Yang applied content analysis to discover the existing semantic content of the film Taiwan Treasure: Rice, released by Discovery. Moreover, Yang collected inspired vocabularies experimentally, and established the semantic network structure using the Kawakita Jiro Method (KJ method). The Yang approach developed three perspectives of semantic word groups, namely rice cuisine, historical background and Taiwanese spirit, and edited the notice films according to each perspective. Yang then re-collected the inspired vocabularies of the notice films experimentally method alongside a narrative interview for the semantic network analysis of KJ method. Eventually, Yang completed the semantic network map, and presented the similarities and differences in semantic networks by transactional analysis [11].

3 Design and Implementation

Technological developments and approaches have been developed to improve streaming media service quality. People use sophisticated digital media players or Internet streaming services to watch digital videos, which may be digitized analog films or pure digital videos. Most traditional video publishers store digital video files directly through the digitization process, but this does not include information for future publication in various forms. A video publisher needs to find desired content, or even select a desired piece of content that must from the huge video libraries to filter out, browse and edit the designated videos. Most of these jobs are still processed manually, wasting time and labor costs.

Furthermore, the increasing popularity of smart mobile devices has significantly increased the needs of multimedia services. Improving the efficiency of automatic classification and retrieval of a large number of multimedia files can significantly reduce the cost of manpower. However, most previous video content analysis methods use low-level features to analyze video frame by frame, but high-level semantic analysis still needs improvement on a number of issues. Therefore, this study develops a storyboard-based automatic video editing system that applies storyboard information, such as the character dialogue, narration, caption, background music and shot changes, to retrieve video content accurately and automatically render summary videos. The proposed system can be applied to the course video trailer and the commercial video trailer for quick preview of video

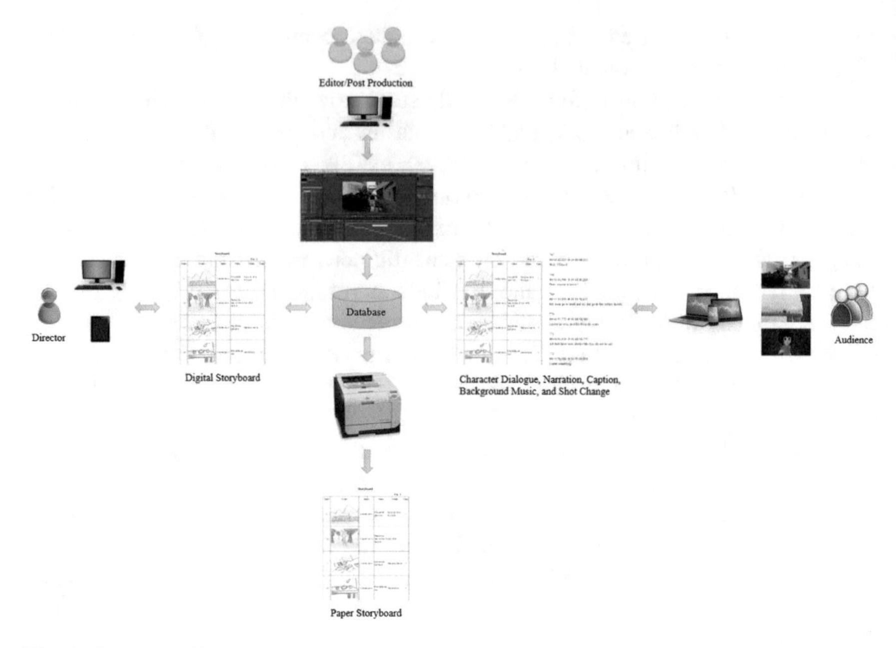

Fig. 1 System architecture

content, or generate a suitable viewing configuration for smart mobile devices. Figure 1 shows the system architecture.

3.1 Script

A script is a work written by screenwriters for a film, video game or television program. It can be an original works or an adaptation from existing pieces of writing, and it narrates the movement, actions, expression and dialogues of the characters. Scripts can be classified into many types. Users may select scripts in a way that they can easily understand for themselves. However, for all others involved in filmmaking, the script must be arranged in a clear and easy to read style. Because people need to understand clearly the meaning of the script, they have developed certain conventions to convey the following required information:

- The scene location to tell the story.
- The time of day in which the scene happens.
- The scene number, indicating the starting a new scene.
- A specific camera angle or perspective.
- Who is in the scene?
- The action description of the characters and the cameras.
- The dialogue between the characters.

- The effects of different camera shots.
- A transition from one scene/camera position to the next.
- Background music for the scene.

A script brings initial ideas to life by capturing what the director wants the audience to see and hear. Screenwriters adopt script conventions to effectively communicate their ideas to the cast and crew of a production in an established and universally understood format [12].

3.2 *Storyboard*

A script is a document written in the language that outlines every aural, visual, behavioral, and lingual element required to tell a story. However, a storyboard is a graphic representation of how the video will unfold, shot by shot. It is made up of a number of squares with illustrations or pictures representing each shot, with notes about what is going on in the scene and what is being said in the script during that shot. The aim of a storyboard is to provide the crew who are making the project with sequential images, related to the script, that describe action, location and camera movement. This enables the crew to fully understand the director's vision of the script. This important working document allows discussion and planning of various aspects of the story, such as technical preparation, mood, timing, location, set design, stunts, special effects (SFX) and visual effects (VFX) [12]. The aims of a storyboard are listed below.

- Making each scene easy to understand.
- Explaining the director's concept.
- Ensuring that the film can be visualized.

Storyboards can be used for both animation and live action productions. The drawings can be simple, but each scene should be worked through until the planned sequence of events runs smoothly. Each sketch should fill the frame in exactly the same way as the shot will fill the screen in the finished film. In a film production, the heads of department in the production company (producers, director, 1st assistant director, cinematographer, production designer, art director, stunt coordinator, script supervisor, special effects, visual effects, hair and makeup, wardrobe, locations manager) then meet to read and discuss the storyboards. The storyboards become the "blueprints" for shooting the film, video, animation or television commercial, and thus play a crucial role that storyboards play [12].

3.3 Development Environment

The system was implemented by Microsoft Visual Studio Professional 2012 and OpenCV 2.4.11 (Open Source Computer Vision Library 2.4.11). The API (Application Programming Interface) can read and decode various video formats into the video memory. Therefore, the file format is not an issue in real-time reading and decoding. The development environment was as follows:

Intel Core i7-4790 CPU 3.60 GHz, 16.0 GB RAM
Microsoft Windows 7 Professional 64bit Service Pack 1
Microsoft Visual Studio Professional 2012
Microsoft Access 2013
OpenCV 2.4.11

3.4 Conversational Storyboard Editing System

This study develops a conversational storyboard editing system to replace the traditional paper storyboard drawn to fill frames on paper, as shown in Fig. 2. Figure 3 shows the proposed electronic editing system using a conversational fill-in-the-blank form. The electronic storyboard can be saved to a database and printed in a traditional paper table format. Figure 4 shows the print and export function of the proposed conversational storyboard editing system. The print

Storyboard

Page: 1

Scene	Visuals	Audio	Video	Effects	Time
1-1		Country music	A beautiful farm view	Extra long shot, Pan right	5
1-2		Country music	Farmers are busy working in the field	Long shot	5
1-3		Country music	Pear flowers and leaves	Medium close up	5
1-4		Country music	A tea table and trees	Medium shot	5

Storyboard

Page: 2

Scene	Visuals	Audio	Video	Effects	Time
2-1		Natural sound	A little girl walks in a street near a traditional courtyard houses	Long shot	5
2-2		Natural sound	The little girl walks forward, an old man is talking to an old lady	Medium close up	4
2-3		Natural sound	The little girl walks on a road near fruit trees	Long shot	4

Fig. 2 Example of storyboard

Fig. 3 Conversational storyboard editing system

function can print the electronic form to the traditional paper table format, and the export function can export the electronic form to PDF, Excel and Word file formats. The Excel and Word files can be re-edited for accurate video content retrieval.

3.5 *Accurate Video Content Retrieval System*

This study differs from other similar researches in the methodology and the type of information extracted, including low-level features. The main purpose of this study is to identify the most used keywords or content in the storyboard to make improve the accuracy of video content retrieval. The following method is adopted by other researches.

- Using a single frame or a single pixel as a unit of analysis.
- Object-based video processing.
- Video shot change detection.

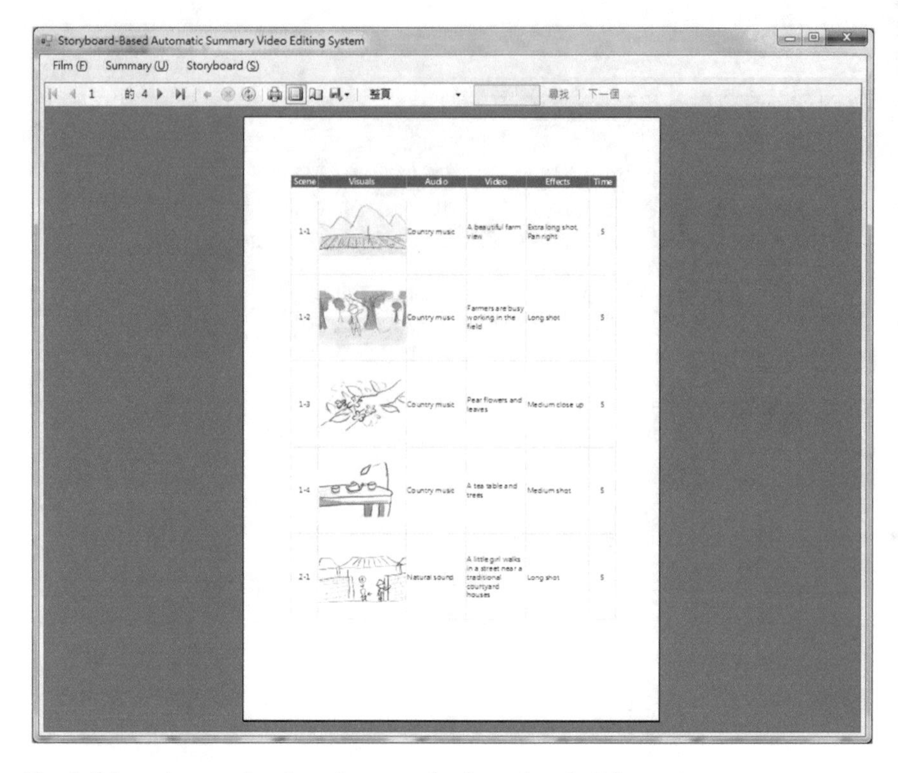

Scene	Visuals	Audio	Video	Effects	Time
1-1		Country music	A beautiful farm view	Extra long shot, Pan right	5
1-2		Country music	Farmers are busy working in the field	Long shot	5
1-3		Country music	Pear flowers and leaves	Medium close up	5
1-4		Country music	A tea table and trees	Medium shot	5
2-1		Natural sound	A little girl walks in a street near a traditional courtyard houses	Long shot	5

Fig. 4 Print and export function of conversational storyboard editing system

- Feature extraction of video shots.
- Using semantic network to analysis video connotation in manual processing and narrative interview.

The above methods were unable to determine accurately the exact meaning of the video frame, or to retrieve important content and keywords of the video. This study develops an accurate video content retrieval system, as shown in Fig. 5. The proposed system uses storyboard information, such as the character dialogue, narration, caption, background music and shot changes, to ensure accurate video content retrieval.

3.6 Automatic Summary Video Editing System

The early approach to video content retrieval was to add text to video filenames or description metadata. It is obviously not based on retrieving the video content.

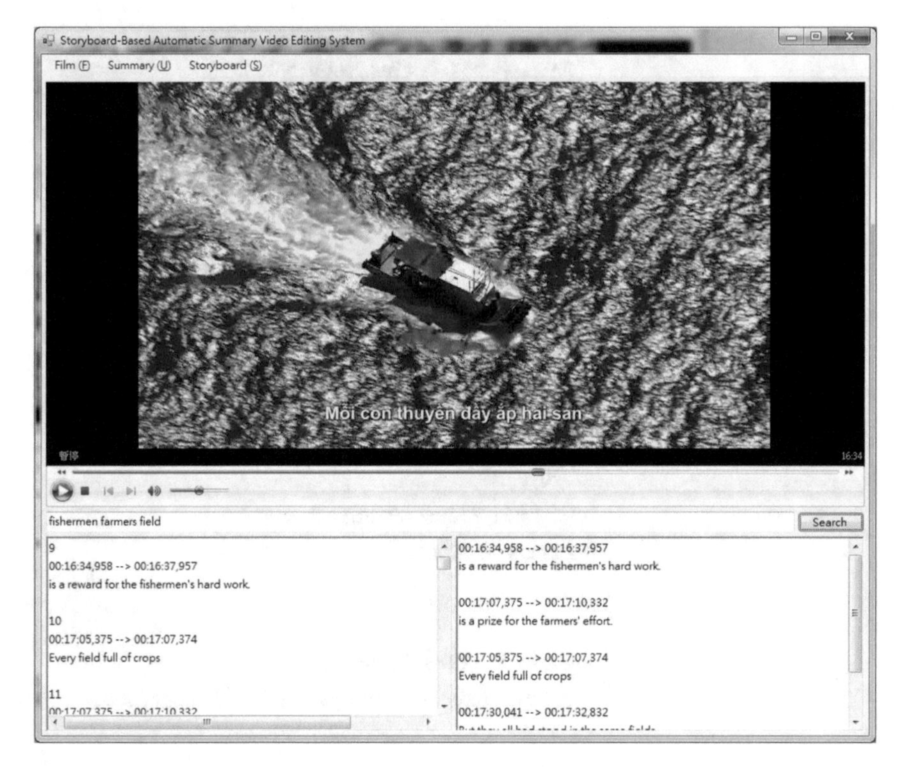

Fig. 5 Accurate video content retrieval system (video source: beyond beauty taiwan from above 2013 intro)

Adding exact text to describe the huge information within the video content is not practical; therefore, retrieving content-based multimedia information has become a key database technology. Content-based video retrieval system automatically extracts the frame features, such as color, shape, texture, motion and audio, for automatic retrieval and processing. However, these methods are too complicated and time-consuming, and do not process much meaningful information in each frame.

This study applies the character dialogue, narration, captions, background music and shot changes in the storyboard as keywords to ensure accurate video content retrieval and obtain the time code, as shown in Fig. 6. The video clips are retrieved and stored based on the time slice index determined by the user (by default, the time slice is same as in the original planning in storyboard), as shown in Fig. 7. The first frame of the video clip is captured as a key frame to display in the user interface. The user can click the key frame in the index table of the user interface to preview

Fig. 6 Dialogue time code

767
00:44:42.221 → 00:44:44.212
Well. I'll fix it.

768
00:44:44.599 → 00:44:46.260
Does anyone at home?

769
00:44:46.434 → 00:44:50.643
My mom go to work and my dad go to buy lottery tickets.

770
00:44:51.772 → 00:44:54.388
I guess he won. as it has been six years.

771
00:44:56.819 → 00:44:58.775
All dads have been always like this. do not be sad.

772
00:44:58.988 → 00:45:00.899
I need something.

the video clip. The proposed system uses a summarization algorithm that automatically combines selected video clips into a single summary video. The automatic summarization algorithm arranges the large number of video clips chronologically, and according to the summary video length given by the user, to automatically render a summary video. The audience can quickly understand the schematic structure and content of whole video in a limited time. The summary video can thus meet the demand and selection purposes to regroup, render and reuse video content to make effective knowledge management and retrieval. Therefore, it can be applied to the course and commercial video trailers for quick preview of video content, or suitable viewing configuration for smart mobile devices.

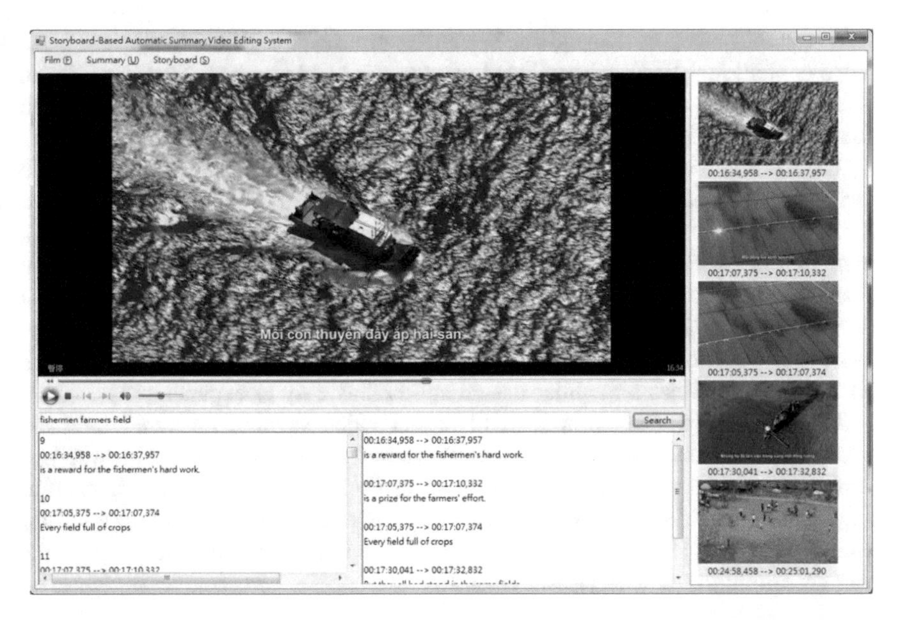

Fig. 7 Automatic summary video editing system (video source: beyond beauty taiwan from above 2013 intro)

4 Conclusions

This study presents a storyboard-based automatic summary video editing system comprising three subsystems, conversational storyboard editing system, accurate video content retrieval system, and automatic summary video editing system. The conversational storyboard editing system replaces the original storyboard paper drawn to fill the frames in the paper with an electronic editing system using a conversational fill-in-the-blank form. This form can be saved to the database, and printed in the traditional paper table format. The accurate video content retrieval system applies the character dialogue, narration, caption, background music and shot changes in the storyboard as keywords to ensure accurate video content retrieval and obtain the time code. The automatic summary video editing system retrieves and stores the video clips according to the time slice index decided by the user (by default, the time slice is the same as that in the original storyboard). The first frame of the video clip is captured as a key frame to display in the user interface. The user can click the key frame in the index table of the user interface to preview the video clip. The proposed system applies a summarization algorithm that automatically combines specified video clips into a single summary video. The automatic summarization algorithm arranges the large number of video clips chronologically, and according to the summary video length given by user, to automatically generate a summary video. The audience can quickly understand the schematic structure and content of the entire video in a limited time. The summary video can thus meet the demand and selection purposes to regroup, render and reuse video content for effective knowledge management and retrieval. Therefore, it can be applied to the course and commercial video trailers to quickly preview video content or select an appropriate viewing configuration for smart mobile devices.

The contribution of this study is to propose a storyboard-based automatic summary video editing technology. The benefits of this technology are as follows: (1) High Precision and Low Cost: the method uses information from the storyboard, such as the character dialogue, narration, caption, background music and shot changes, to ensure accurate video content retrieval and automatically generate a summary video. (2) Practicability: the method adopts smart mobile devices combined with easy-to-use features to increase the popularity of the system. (3) Flexibility: the method has a modular design and is thus easy to expand.

This technology can be applied to digital video services delivered over the network. On the receiving side, the video player users can real-time browse and retrieve the rendered summary video clips, then select the video clip of interest to play. Digital video service providers simply transfer the searchable information for users to search and browse, then render and transfer the summary video clips to the user according to the user video demands.

References

1. Lai-Man, P., Wing-Chung, M.: A Novel Four-Step Search Algorithm for Fast Block Motion Estimation. IEEE Transactions on Circuits and Systems for Video Technology, vol. 6, no. 3, pp. 313–317 (1996)
2. Hussain, M., Chen, D., Cheng, A., Wei, H., Stanley, D.: Change Detection from Remotely Sensed Images: From Pixel-Based to Object-Based Approaches. ISPRS Journal of Photogrammetry and Remote Sensing 80, pp. 91–106 (2013)
3. Deng, Y., Manjunath, B.S.: Content-Based Search of Video Using Color, Texture, and Motion. In: IEEE International Conference on Image Processing, vol. 2, pp. 534–537 (1997)
4. Belongie, S., Carson, C., Greenspan, H., Malik, J.: Color- and Texture-Based Image Segmentation Using EM and Its Application to Content-Based Image Retrieval. In: Sixth International Conference on Computer Vision, pp. 675–682 (1998)
5. Eric, B., Stéphane, M.: Nonlinear Temporal Modeling for Motion-Based Video Overviewing. In: Third International Workshop on Content-Based Multimedia Indexing (2003)
6. Truong, B.T., Venkatesh, S.: Video Abstraction: A Systematic Review and Classification. The ACM Transactions on Multimedia Computing, Communications, and Applications, vol. 3, no. 1, pp. 1–37 (2007)
7. Money, A.G., Agius, H.: Video Summarisation: A Conceptual Framework and Survey of the State of the Art. Journal of Visual Communication and Image Representation, vol. 19, pp. 121–143, (2008)
8. Ianeva, T., de Vries, A.P., Westerveld, T.: A Dynamic Probabilistic Retrieval Model. In: IEEE International Conference on Multimedia and Expo, pp. 1607–1610 (2004)
9. Bhat, S.A., Sardessai, O.V., Kunde, P.P., Shirodkar, S.S.: Overview of Existing Content Based Video Retrieval Systems. International Journal of Advanced Engineering and Global Technology, vol. 2, no. 2, pp. 476–483 (2014)
10. Bregler, C.: Learning and Recognizing Human Dynamics in Video Sequences. In: IEEE International Conference on Computer Vision and Pattern Recognition, pp. 568–574 (1997)

11. Chun-Wei, Y.: Investigation on the Methods of Educational Films Editing via Semantic Network Concepts. National Hsinchu University of Education (2012)
12. Scriptwriting and Storyboards, http://generator.acmi.net.au/resources/scriptwriting-and-storyboards

Optimal Thrust Allocation for Ship Dynamic Positioning System

Wenjuan Li, Yifei Yang, Wei Yuan and Zhen Su

Abstract The thrust allocation of a ship dynamic positioning system allocates the desired forces to thrusters by calculating thrust and angle of each thruster. Due to the problem of allocation failure arising from the thrust optimization allocation algorithm in the situation of the desired forces caused mutation, this paper puts forward the method based on the truncation of multiobjective optimization, which aims to maximize the ability of the propellers. The simulation results prove that the proposed method can effectively solve the problem.

Keywords Thrust allocation · Dynamic positioning · Multiobjective optimization · Truncation

1 Introduction

Dynamic positioning system (DPS) is defined as an autonomous control system that acts as to maintain the ship position and heading at a reference point by means of thrusters. The control system block contains a feedback control law and a thrust allocation system [1–3].

The effect of the thrust allocation system is to map the desired generalized forces determined by the motion controller into individual thruster forces and directions. It is important that a thrust allocator always providesa solution and provides the solution in time. Further the thrust allocator should try it's best to produce the desired generalized forces. Application overviews of thrust allocation can be found in [4–6]. Solutions to the thrust allocation problem usually contain pseudo-inverse,

W. Li (✉) · Y. Yang · Z. Su
Marine Equipment and Technology Institute, Jiangsu University
of Science and Technology, Zhenjiang, China
e-mail: wenjuanli03211017@126.com

W. Yuan
Department of Electronic and Information, Jiangsu University
of Science and Technology, Zhenjiang, China

© Springer Nature Singapore Pte Ltd. 2018

N.Y. Yen and J.C. Hung (eds.), *Frontier Computing*, Lecture Notes
in Electrical Engineering 422, DOI 10.1007/978-981-10-3187-8_70

quadratic programming (QP), sequential quadratic programming (SQP), augmented Lagrange and so on.

When the control force is under the ability of thrusters, all those methods aforementioned are applicable. Actually, the output value of control force may be beyond the ability of thrusters due to the randomness of sea environment such as steady sea encountered mutation. Heading priority is proposed [7]. However, when the condition does not last a long time, heading priority is not suitable. The time of finding the optimum heading may be longer than the lasting time of the sea environment changing. To deal with this problem, Ruth [8] proposed the method based on power constraints to get the scale factor, but the factoris difficult to be obtained.

In this paper, the method based on the truncation of multiobjective optimization [9] is usedto solve the problem of allocation failure arising from the thrust optimization allocation algorithm in the situation of the desired forces caused mutation, taking account of the amplitude and rate saturation of thrusters, the angle change rate of azimuth thrusters, prohibited area of thrusters and so on, which aims to maximize the ability of the propellers. After that, the problem is converted to a general optimization problem which can be resolved by augmented Lagrange method.

2 Generally Optimal Thrust Allocation

It is common to use a 3-degree-of-freedom model in DPS which only takes account of surge, sway and yaw motions. Considering a ship equipped with n thrusters. The generalized force vector $F = [F_x, F_y, M_z]^T \in R^3$ produced jointly by the actuators is given by

$$F = B(\alpha)T \tag{1}$$

F_x is surge force, F_y is sway force, M_z is yaw moment. $T = [T_1, T_2, \ldots, T_n]^T$, $\alpha = [\alpha_1, \alpha_2, \ldots, \alpha_n]^T$, T_i is the generalized force of the i-th thruster, α_i is the azimuth angle of the i-th thruster. The i-th column of the $3 \times n$ matrix $B(\alpha)$ is given by

$$B_i(\alpha_i) = \begin{pmatrix} \cos\alpha_i \\ \sin\alpha_i \\ -l_{yi}\cos\alpha_i + l_{xi}\sin\alpha_i \end{pmatrix} \tag{2}$$

where (l_{xi}, l_{yi}) is the location of i-th thruster in a coordinate system with origin in the center of rotation, positive x-axis forward, and positive y-axis towards starboard.

Considering the following optimization problem

$$\min J = T^T W T + s^T Q s \tag{3}$$

subject to

$$\begin{aligned} & F = B(\alpha)T + s \\ & T_{\min} \leq T \leq T_{\max} \\ & \Delta T_{\min} \leq T - T_0 \leq \Delta T_{\max} \\ & \alpha_{\min} \leq \alpha \leq \alpha_{\max} \\ & \Delta\alpha_{\min} \leq \alpha - \alpha_0 \leq \Delta\alpha_{\max} \end{aligned} \quad (4)$$

where W is a positive definite matrix, s is a vector of slack variables, Q is the weight matrix which is chosen large. $T_{\max} \in R^n$ and $T_{\min} \in R^n$ are the maximum and minimum of the thruster force, $\Delta T_{\max} \in R^n$ and $\Delta T_{\min} \in R^n$ are the maximum and minimum change rate, respectively. The azimuth, α, is required to belong to the given sectors defined by the vectors of lower and upper bounds, $\alpha_{\min}$ and $\alpha_{\max}$.$\Delta\alpha_{\min}$ and $\Delta\alpha_{\max}$ are the rate-of-change of the azimuth angles. In general, azimuth angles do not allowed to have a large change during the sampling time. For the tunnel thruster, the constraint of the azimuth angle can be represented as $\alpha_i = \alpha_i^{\min} = \alpha_i^{\max}$.

Additionally, an azimuth thruster can rarely produce thrust in any direction without disturbing other thrusters or submerged equipment on the vessel. Therefore, limitations on thrust directions, such as forbidden zones, should be considered.

We can see that, the general thrust allocation is a nonlinear optimization problem with constrains.

3 Solving the Problem of Allocation Failure

In this section, we introduce the multiobjective optimization method which is used to calculate the maximums and minimums of the surge/sway thrust and yaw moment. And then we give the details of the proposed thrust allocation logic.

3.1 Multiobjective Optimization

A generic multiobjective optimization problem may be formulated as:

$$\min F(X) = (f_1(X), f_2(X), \ldots, f_m(X))^T$$
$$s.t. \begin{cases} g_i(X) \leq 0, \; i = 1, \ldots, p \\ h_j(X) = 0, \; j = 1, \ldots, q \\ X \in S \\ X = (x_1, x_2, \ldots, x_c)^T \end{cases} \quad (5)$$

where $f_1(\mathrm{X}), f_2(\mathrm{X}), \ldots, f_m(\mathrm{X})$ are the m objectives functions, $(x_1, x_2, \ldots, x_c)$ are the c optimization parameters, and $S \in R^n$ is the solution or parameter space. $g_i(\mathrm{X})$ and $h_j(\mathrm{X})$ are the inequality constraints and equality constraints. The number of inequality constraints and equality constraints are p and q.

There is general consensus that multiobjective optimization methods can be broadly decomposed into two categories: Scalarization approaches and Pareto approaches. While different names are used for these categories, the fundamental discriminator is always the same. In the first group of methods the multiobjective problem is solved by translating it back to a single objective, scalar problem. The Pareto methodskeep the elements of the objective vector F separate throughout the optimization process and typically use the concept of dominance to distinguish between inferior and non-inferior solutions. The goal of all these methods is to provide designers and decision makers with a set of 'optimal' alternatives to choose from.

3.2 Truncation of Multiobjective Optimization

The method based on the truncation of multiobjective optimization taking account of the amplitude and rate saturation of thrusters, the angle change rate of azimuth thrusters, prohibited area of thrusters and so on, aims to maximize the ability of the propellers.

Two models are contained of this method. One is the maximums of the surge/sway force and yaw moment given by (6), the other is the minimums of the surge/sway force and yaw moment formulated as (7).

$$\max \quad F = [F_x, F_y, M_z]^T$$
$$s.t. \begin{cases} inequ.: & T_{min} \leq T \leq T_{max} \\ & \alpha_{min} \leq \alpha \leq \alpha_{max} \\ & \Delta T_{\min} \leq T - T_0 \leq \Delta T_{\max} \\ & \Delta\alpha_{\min} \leq \alpha - \alpha_0 \leq \Delta\alpha_{\max} \end{cases} \tag{6}$$

$$\min \quad F = [F_x, F_y, M_z]^T$$
$$s.t. \begin{cases} inequ.: & T_{min} \leq T \leq T_{max} \\ & \alpha_{min} \leq \alpha \leq \alpha_{max} \\ & \Delta T_{\min} \leq T - T_0 \leq \Delta T_{\max} \\ & \Delta\alpha_{\min} \leq \alpha - \alpha_0 \leq \Delta\alpha_{\max} \end{cases} \tag{7}$$

The most widely used method, Weighted Sum (WS), is adopted to solve the multiobjective optimization problem.

3.3 Thrust Allocation Logic

Step 1. Obtain the generalized forces determined by the controller.
Step 2. Usegeneral optimization method to solve the problem of thrust allocation.
Step 3. If the allocation is successful, get the result.
Step 4. If not, usemultiobjective optimizationMultiobjective optimization method to calculate the maximums and minimums of the surge/sway force and yaw moment for the current sampling instant based on the previous sampling instant, and then truncate the generalized forces based on the maximums and minimums. The thrust allocation problem is converted to a general thrust allocation problem.
Step 5. Return to Step 2.

4 Simulation

A simulation described in this paper is performed with a model vessel. This is an approximately 1:20 scale replica of an supply vessel, with the length of 3.75 m. The vessel is equipped with two azimuth thrusters at stern and two tunnel thrusters in the bow. Tunnel thruster is a propeller which is mounted inside the hull and can provide thrust in transverse direction only. Azimuth thrusters are attractive in dynamic positioning system since they can produce forces in different directions. And, it should be noted that the azimuth thrusters are only allowed to supply positive force.

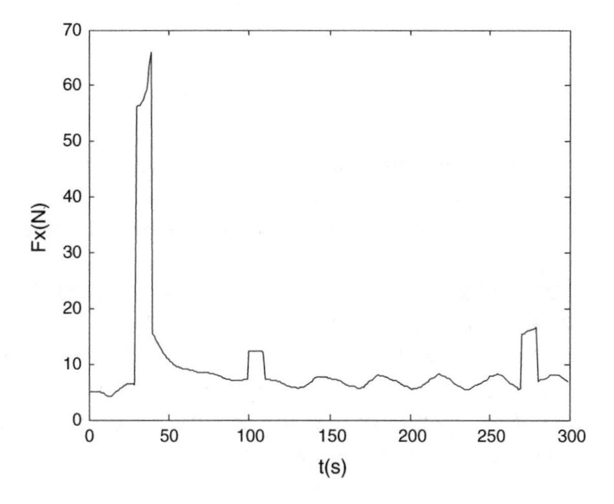

Fig. 1 Surge force

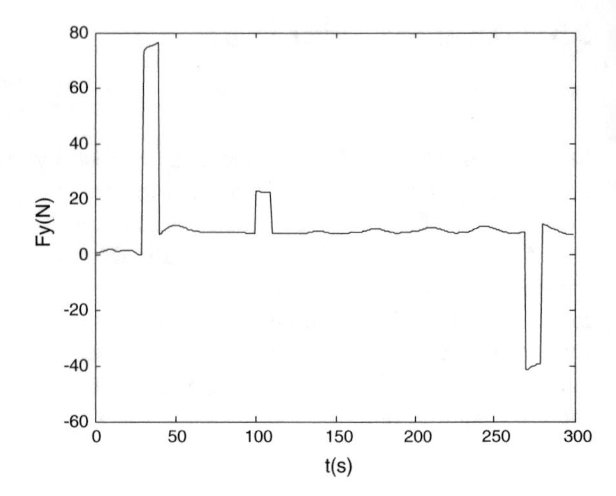

Fig. 2 Sway force

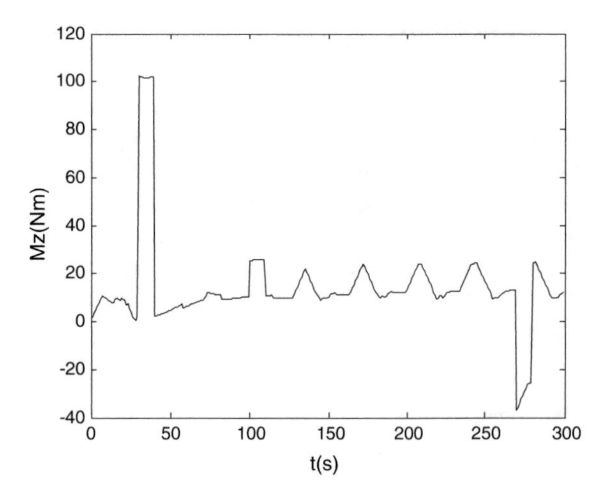

Fig. 3 Yaw moment

In the simulation, assuming steady sea encountered mutation, the surge/sway force and yaw moment increase rapidly as shown in Figs. 1, 2 and 3. To demonstrate the allocating algorithm, three cases are considered. Each case lasts ten sampling periods.

As the first mutation for example, the generalized force vector is [56.23, 73.11, 102.49]. The last thrust allocation results are shown that the thrust of azimuth thruster 1 is 2.96 N, the angle is 354°, the thrust of azimuth thruster 2 is 3.27 N, the angle is 354°, the thrusts of two tunnel thrusters are 0.37 N and 0.34 N. Considering the properties of the thrusters, the current ability of the thrusters cannot reach to the desired forces and moment as [56.23, 73.11, 102.49]. So, the thrust allocation data cannot be obtained by using general thrust allocation method. The new method is proposed.

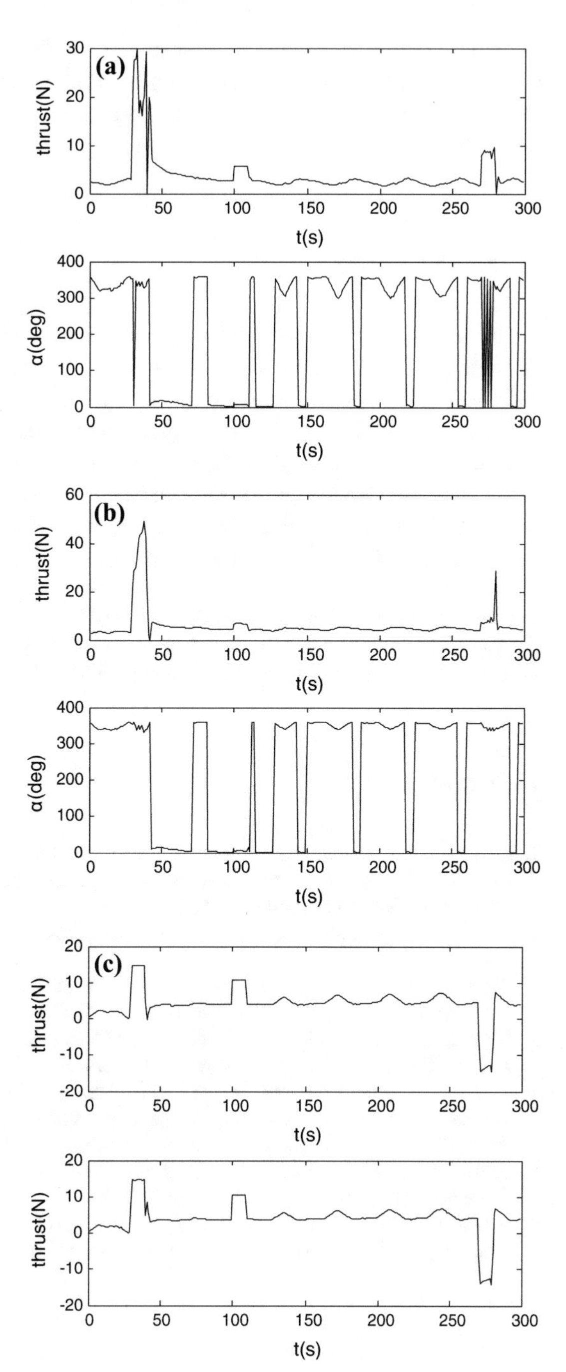

Fig. 4 The thrust and the angle of the azimuth thruster and the thrust of tunnel thruster: **a** azimuth thruster 1, **b** azimuth thruster 2, **c** tunnel thruster 3 and 4

Figure 4a–c give the thrust allocation data. From the simulation results, we can see that the proposed thrust allocation method is capable of handling the allocation problem when the steady sea is encounteredmutation, though the form of mutation is varied.

5 Conclusions

Under the mutation of sea environment, the generalized forces are usually beyond the ability of thrusters lasting a short time. The existing thrust allocation methods and heading priority scheme are not suitable when the limitations of the amplitude and rate saturation of the thrusters, prohibited area, the angle change rate of azimuth thrusters and so on are all taken into account. In this paper, a new thrust allocation method is proposedbased on the truncation of multiobjective optimizationto avoid the fault of thrust allocation. The simulation results show that the proposed method can efficiently deal with the above problem.

Acknowledgements This work was supported by Collaborative innovation center of JiangsuColleges and Universities high-tech shipand Marine Equipment and Technology Institute Jiangsu University of Science and Technology (No. HZ2016006).

References

1. Fossen, T. I.: Guidance and control of ocean vehicles. John Wiley & Sons Ltd. (1994)
2. Fossen, T. I.: Marine Control Systems: Guidance, Navigation and Control of Ships, Rigs and UnderwaterVehicles. Trondheim, Norway: Marine Cybernetics (2002)
3. Fossen, T. I.: Handbook of marine craft hydrodynamics and motion control. John Wiley & Sons Ltd, (2011)
4. Johanson, T. A.: Optimizing nonlinear control allocation. 43rd IEEE Conference on Decision and Control, vol. 4, pp. 3435–3440. Bahamas (2004)
5. Millan, J.: Thrust Allocation Techniques for Dynamically Positioned Vessels. Canada (2008)
6. Fossen, T. I, Johansen, T. A.: Perez T. Underwater vehicles: A survey of control allocation methods for underwater vehicles. InTech, Australia (2009)
7. Fossen, T. I.: Strand, J.P.:Nonlinear passive weather optimal positioning control (WOPC) system for ships and rigs: experimental results vol. 37, pp. 701–715. Automatica (2001)
8. Ruth, E.: Propulsion control and thrust allocation on marine vessels. Trondheim: Norwegian University of Science and Technology (2008)
9. Altannar, C., Panos, M. P.: A survey of recent developments in multiobjective optimization. Vol. 154, pp. 29–50. Annals of Operations Research (2007)

Modeling and Simulation of Draghead on Trailing Suction Hopper Dredger

Su Zhen, Diandian Cao and Sun Jian

Abstract In this paper, particle filter algorithm is applied to the trailing suction hopper dredger for analyzing drag head parameters. A particle filter is designed in the process of trailing suction hopper dredger for predicting the drag head density. Using the filtering results and measured results to evaluate its performance based on multiple sets of measured data, and the results performed well. The filter has certain help for the development of trailing suction hopper dredger decision support system.

Keywords Trailing suction hopper dredger · Drag head density · Particle filter · Modeling · Simulation

1 Introduction

Drag head is the most front-end equipment of TSDH. In the process of dredging, the drag head is close to the bottom of the sea. The mixture of water and sand is sucked into the drag head through the centrifugal pump. In the dynamic modeling of the TSDH, in order to simulate the production of mixture in the drag head, the model should forecast density of mixture entering the drag head. The density can be measured by the density instrument, but the density of mixture has about 10 s of delay due to the pipe transportation. Therefore, it is necessary to use the model to estimate the accurate density of mixture to avoid the phenomenon of cavitation caused by the high density of the mixture. In this paper, the particle filter algorithm

S. Zhen
School of Mechatronic Engineering and Automation,
Shanghai University, Shanghai, China

S. Zhen (✉) · D. Cao · S. Jian
Marine Equipment and Technology Institute,
Jiangsu University of Science and Technology, Zhenjiang, China
e-mail: suzhen415@126.com

© Springer Nature Singapore Pte Ltd. 2018

N.Y. Yen and J.C. Hung (eds.), *Frontier Computing*, Lecture Notes
in Electrical Engineering 422, DOI 10.1007/978-981-10-3187-8_71

is used to predict the density of the drag head, which can provide the operator with numerical reference. The operator can take measures to avoid pump cavitation [1].

1.1 Excavation Process

The modern drag head is provided with water jets, a double row of teeth and visor that can be adjusted in real time, as shown in Fig. 1. Ship speed, water jets and visor angle size will affect flow and density of mixture, which determines the production of drag head. Formation of the mixture in the drag head is a rather complicated process, which can be found little in the literature. The physical model depends on the physical phenomena in the drag head and a lot of field data and is related to soil-dependent parameters, which are unknown or can't be directly measured. The interaction of water jets, cutting and erosion with each other is unclear. In the case, this paper will use a black-box model to predict the density of the drag head.

According to the literature [2], the following models are obtained:

$$\rho_i = -a_{dh} Q_i^2 + b_{dh} V_{sh} + c_{dh} \tag{1}$$

with

ρ_i The density of mixture
Q_i The flow of mixture
V_{sh} Ship speed.

These parameters can be obtained by sensor measurement and a_{dh}, b_{dh}, c_{dh} are positive coefficients.

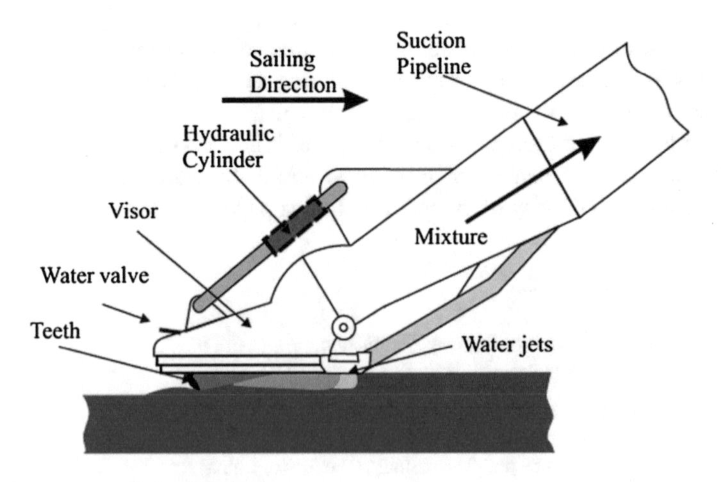

Fig. 1 The structure of drag head

2 Prediction Model for Density of Drag Head Based on Particle Filter

2.1 Particle Filter Principle

Particle filter is a approximate Bayesian filtering algorithm based on Monte-Carlo simulation. The core idea is to approximate the probability density function of the random variables of the system by some discrete random sampling points. In order to obtain the minimum variance of the state, integral operation is replaced by the sample mean value. The particle set is appropriately weighted and delivered according to Bias criteria [3].

Monte-Carlo method is used to collect the weight of the particle set from the posterior probability, which is expressed by the particle set, and then the integral operation is transformed into sum. The posterior probability function is:

$$\hat{p}(X_{0:k}|Z_{1:k}) = \frac{1}{N}\sum_{i=1}^{N} \delta_{X_{0:k}}(dX_{0:k}) \tag{2}$$

where $\left\{X_{0:k}^{(i)}: i = 1, 2, \ldots, N\right\}$ is a random sample set of the posterior probability distribution, and δ is Dirac-delta's function. The function of state sequence $g_t: R^{(t+1)n} \to R^n$ of Mathematical expectation is:

$$E[g_t(X_{0:t})] = \int g_t(X_{0:t})p(X_{0:k}|Z_{1:k})dX_{0:k} \tag{3}$$

Approximately:

$$\overline{E[g_t(X_{0:t})]} = \frac{1}{N}\sum_{i=1}^{N} g_t(X_{0:t}^{(i)}) \tag{4}$$

According to the Sequential importance sampling principle [5]:

$$\overline{E[g_t(X_{0:t})]} = \sum_{i=1}^{N} g_t\left(X_{0:t}^{(i)}\right)\tilde{w}_t(X_{0:t}^{(i)}) \tag{5}$$

$$w_t(X_{0:t}) = \frac{p(Z_{1:t}|X_{0:t})p(X_{0:t})}{q(X_{0:t}|Z_{1:t})} \tag{6}$$

$$w_{t-1}(X_{0:t-1}) = \frac{p(Z_{1:t-1}|X_{0:t-1})p(X_{0:t-1})}{q(X_{0:t-1}|Z_{1:t-1})} \tag{7}$$

Combining (6) and (7) gives:

$$w_t(X_{0:t}) = w_{t-1}(X_{0:t-1}) \frac{p(Z_t|X_t)p(X_t|X_{t-1})}{q(X_t|X_{0:t-1}, Z_{1:t})} \tag{8}$$

where $\tilde{w}_t(X_{0:t}^{(i)})$ is $w_t(X_{0:t})$ normalized weight. $q(X_t|X_{0:t-1}, Z_{1:t})$ is a known and easily sampled reference distribution. $X_{0:t}^{(i)}$ is a sample obtained from $q(X_t|X_{0:t-1}, Z_{1:t})$. Thus, the true distribution $p(Z_t|X_t)$ which is difficult to get is converted into a known reference distribution $q(X_t|X_{0:t-1}, Z_{1:t})$:

$$q(X_t|X_{0:t-1}, Z_{1:t}) = p(X_t|X_{t-1}) \tag{9}$$

Substituting (8) into (7) leads to:

$$w_t(X_{0:t}) = w_{t-1}(X_{0:t-1})p(Z_t|X_{0:t}) \tag{10}$$

2.2 Particle Filter Steps

(1) Extracting N random samples $\left\{X_{0:k}^{(i)}: i = 1, 2, \ldots, N\right\}$ from the reference distribution function $q(X)$, and the weight is set to 1/N
(2) Calculating the importance weights for $X_t^{(i)}$ making each sample meet with $w_t^{(i)} \propto p(X_t)/q(X_t)$
(3) Normalizing importance weight $\tilde{w}_t\left(X_{0:t}^{(i)}\right) = \frac{w_t\left(X_{0:k}^{(i)}\right)}{\sum_{i=1}^{N} w_t\left(X_{0:k}^{(i)}\right)}$
(4) Re-sampling N times in discrete set $\left\{X_{0:k}^{(i)}\right\}_{i=1}^{N}$, and the re-sampling probability of every $X_t^{(i)}$ is proportional to $\tilde{w}_t^{(i)}$ [4].

2.3 Prediction Model

Historical data is measured from the TSDH, as shown in Fig. 2. It can be seen that the TSDH's historical data is shown in the form of cycle. That corresponds to the dredging operation of TSDH per cycle.

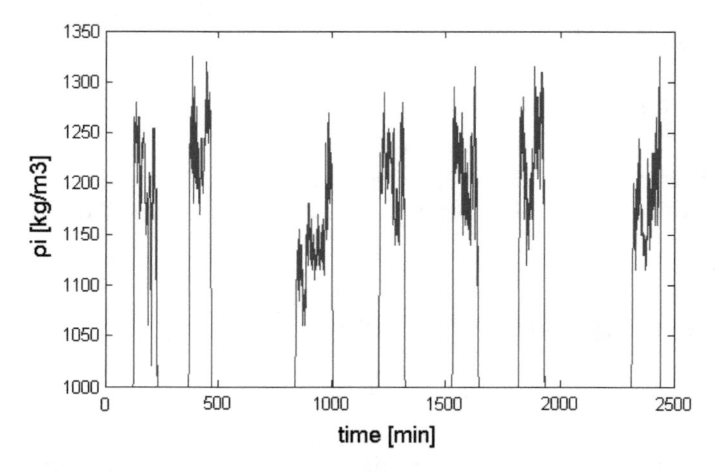

Fig. 2 Density of drag head of every cycle

According to the measured data of ρ_i to establish the prediction curve, predicting density of mixture when the flow Q_i is known. For example, it can predict ρ_i by Q_i of 80 min ahead. Prediction model's state is presented by (1):

$$X(k) = \left[a_{dh}(k) \quad b_{dh}(k) \quad c_{dh}(k) \right]^T \tag{11}$$

The state equation is:

$$\begin{cases} a_{dh}(k+1) = a_{dh}(k) + w_a(k) & w_a \sim N(0, \sigma_a) \\ b_{dh}(k+1) = b_{dh}(k) + w_b(k) & w_b \sim N(0, \sigma_b) \\ c_{dh}(k+1) = c_{dh}(k) + w_c(k) & w_c \sim N(0, \sigma_c) \end{cases} \tag{12}$$

The observation equation is:

$$\rho_k = -a_{dh}(k) Q_i^2 + b_{dh}(k) V_{sh} + c_{dh}(k) + v(k), v(k) \sim N(0, \sigma_v) \tag{13}$$

where a_{dh}, b_{dh}, c_{dh} and ρ_k contains noise, which is zero-mean Gaussian white noise and variance is σ_a, σ_b, σ_c, σ_v. Initial value of drag head model is a_{dh}, b_{dh}, c_{dh}, which is the average value obtained by least square curve fitting (see Table 1).

Table 1 Initial value of drag head model

Cycle	a_{dh}	b_{dh}	c_{dh}
1	−1.16	57.31	894.78
2	−0.8315	20.5	1178.8
3	0.3	54.9	1038

3 Prediction Result Analysis

3.1 Dredging Conditions

For validating particle filter effect, here use data of three cycles. The data is obtained from Yangtze River Estuary. Parameters are shown in Table 2.

3.2 Density of Mixture Prediction

By using particle filtering algorithm, the value of a_{dh}, b_{dh}, c_{dh} can be calculated according to ρ_i that is observed data. Then, ρ_i can be predicted by using value of a_{dh}, b_{dh}, c_{dh} obtained from particle filtering. The value of a_{dh}, b_{dh}, c_{dh} obtained from particle filtering and their average value are shown in Fig. 3.

Through the prediction model, the prediction curve of ρ_i (see Fig. 4) is obtained by using the data of cycle 1. Solid line is the measured value of ρ_i. The dashed line is the filtered value of ρ_i, and "+ + + +" line is the predicted value of ρ_i.

Table 2 Parameters of Yangtze River Estuary

Operation site	Yangtze River Estuary
Soil property	Silt
Saturated mud density	1.67 t/m^3
Diameter of grains	0.024 mm

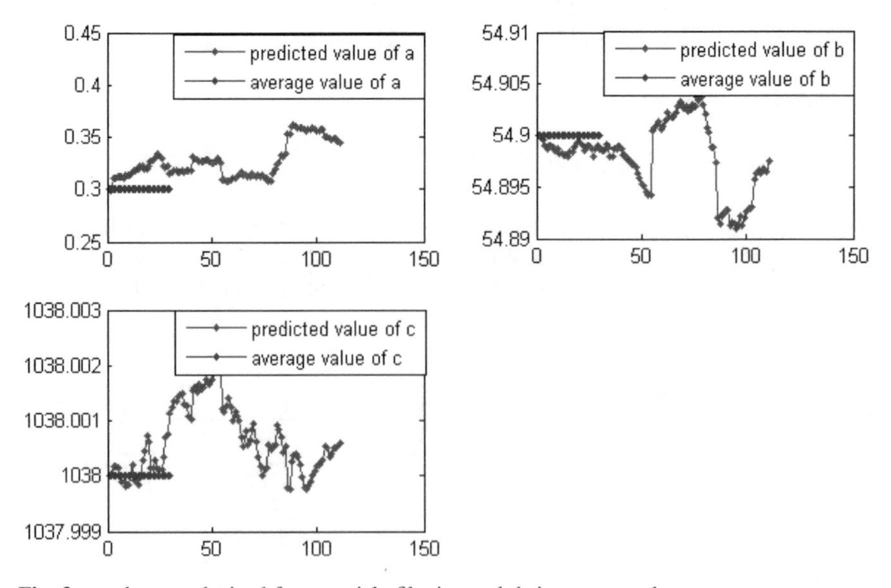

Fig. 3 a_{dh}, b_{dh}, c_{dh} obtained from particle filtering and their average value

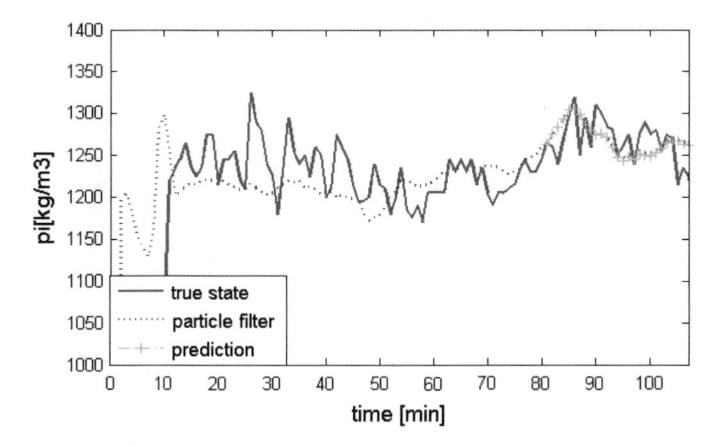

Fig. 4 Prediction of density of cycle 1

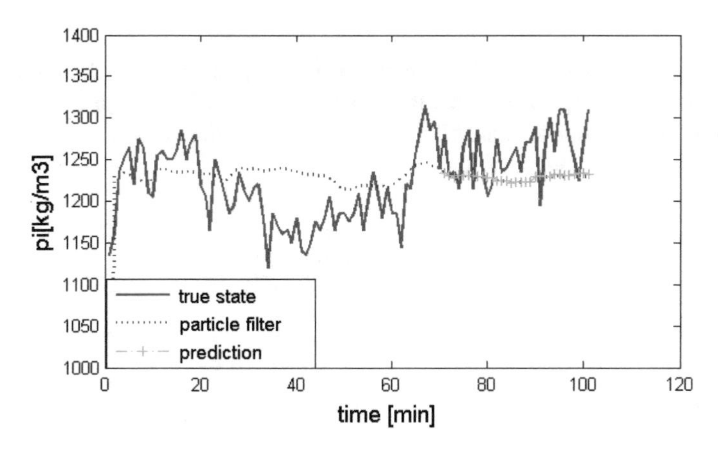

Fig. 5 Prediction of density of cycle 2

It can be seen that particle filter filtered the noise caused by the density sensor measurement, and making the results more smooth. The predicted result can fit the actual measured value well. In order to verify the accuracy of the prediction model, the prediction curves of cycle 2 and cycle 3 are given (Figs. 5 and 6).

3.3 Prediction Model Performance

Two performance indexes are identified below to evaluate the performance of the prediction model [5]:

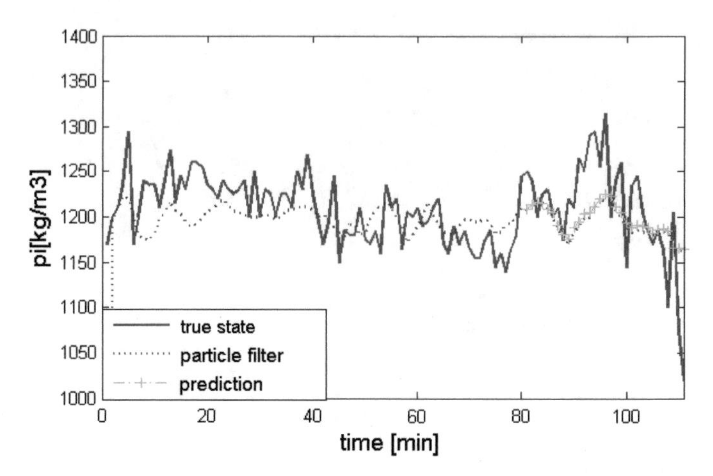

Fig. 6 Prediction of density of cycle 3

(1) Variance accounted-for (VAF), which is not very sensitive to the noise signal, even the input signal is noise, it can still give a reasonable model assessment result. The maximum value of VAF is 1. However, the value may be negative if the model is not accurate [5].

$$VAF = 1 - \frac{\mathrm{var}(y - \hat{y})}{\mathrm{var}(y)} \tag{14}$$

with

y measured data
$\hat{y}$ predicted data.

(2) Sum of Squares is applied as performance evaluation for the minimum prediction error in the nonlinear calibration procedure [5]. The sum of squares is defined as follows:

$$J(\theta) = \frac{1}{N}\sum_{k=1}^{N} (\hat{y}(k,\theta) - y(k))^2 \tag{15}$$

with

k The time index of the sampled data
θ The parameter to be estimated

Table 3 Evaluation results of prediction model

	$\sqrt{J(\theta)}$	*VAF*
Cycle 1	1215	0.20
Cycle 2	1116	0.62
Cycle 3	1208	0.22
Average	1180	0.35

$\hat{y}(k,\theta)$ The predicted data
$y(k)$ The measured data
N The number of samples.

According to the data measured in Yangtze River Estuary, using above two performance indexes assesses density of mixture obtained from predictive model. The results are given in Table 3.

It can be seen that the predicted results and measured values fits well, which shows that the particle filter is accurate. In the case of density that can't be measured, the density of drag head can be predicted according to the relevant parameters and the model. Then, production can be calculated, which can provide effective operation parameters for the operator. When the density becomes very large, Operator should take measures to reduce the density of mixture to avoid pump cavitation. Due to the limited measured data, the accuracy of the model need to be validated under different operation conditions.

4 Conclusions

(1) This paper combined particle filter with drag head model to establish the prediction model of density of drag head. The a_{dh}, b_{dh}, c_{dh} is least squares fitting parameters, which is substituted prediction model to get filtering parameters. Then, the density of mixture can be predicted by using the filter parameters and measured data.
(2) Comparing the predicted curve with the measured data, it is concluded that the particle filter has a good filtering effect on noise. At the same time, the prediction results also have a good accuracy.
(3) Due to the limited measured data, the accuracy of the model need to be validated under different operation conditions.
(4) Prediction results can be used to estimate the production of drag head and adjusting drag head control parameters, such as visor angle, water jets. It also can help to develop intelligent system of TSDH.

Acknowledgements Project supported by Collaborative innovation center of Jiangsu Colleges and Universities high-tech ship and Marine Equipment and Technology Institute Jiangsu University of science and Technology (Grant No. HZ2016011).

References

1. J. Osnabrugge, P. M.: Optimizing Manpower and Reducing Fuel Consumption While Maintaining Maximum Dredging Production. The Art of Dredging, Brussels, Belgium. (2013).
2. D. Wangli.: Intelligent Data Analysis and Control of a Hopper Dredger. Master's Thesis, Delft University of Technology. 42, 79, 169, (2006).
3. Paweł Stano, Zsofia Lendek, Robert Babuska, Jelmer Braaksma.: Particle Filters for Estimating Average Grain Diameter of Material Excavated by Hopper Dredger. Part of 2010 IEEE Multi-Conference on Systems and Control, Yokohama, Japan, September 8–10, (2010).
4. Pawe Stano.: Nonlinear State and Parameter Estimation for Hopper Dredgers. PhD Thesis, Delft University of Technology. 141–156, (2013).
5. J. Braaksma.: Model-Based Control of Hopper Dredgers. PhD Thesis, Delft University of Technology. 20, 21, 25 (2008).

Control Strategy of Three Phase PWM Rectifier in Ship Power Grid

Wenhua Yuan, Xu Zhang, Zhengang Wan, Liang Qi and Qiang Zhao

Abstract In this paper, a brief introduction of ship power grid system has been introduced. In the light of the positive and negative sequence voltage theory, an unbalanced control strategy of three-phase PWM rectifier (VSR) has been adopted for restraining the negative sequence current at the AC side. In the strategy, an algorithm of decomposing the positive and negative sequence voltage has been introduced and employed, and it has been compared with the traditional way of notch filter. It has been indicated that this algorithm can decompose the voltage correctly. A three-phase VSR simulation platform was built with Simulink software, and the correctness of the control strategy has been proved by simulation results.

Keywords PWM VSR · Unbalanced power grids · Instantaneous power

1 Introduction

The onshore grid is of infinite capacity system, while the ship grid is independent and of finite capacity system. Ship power capacity is finite compared with onshore grid, and some single phase equipments are of large capacity in ship power system. Unbalanced network voltage will be caused when these single phase equipment capacity is started. Therefore, in addition to land-based grid power factor and harmonic problem, the grid imbalance is more prone to rise in ship power system.

W. Yuan (✉) · X. Zhang · Z. Wan · L. Qi · Q. Zhao
School of Electronics and Information, Jiangsu University of Science and Technology, Zhenjiang 212003, Jiangsu, China
e-mail: yuanwhvs@foxmail.com

W. Yuan
School of Electrical and Information Engineering, Jiangsu University, Zhenjiang 212013, Jiangsu, China

© Springer Nature Singapore Pte Ltd. 2018

N.Y. Yen and J.C. Hung (eds.), *Frontier Computing*, Lecture Notes in Electrical Engineering 422, DOI 10.1007/978-981-10-3187-8_72

In many applications the ship's equipment need to use a rectifier. Therefore, it is necessary to consider the problem of unbalanced grid. In the conventional design of VSR, three-phase power grid generally assumed to be balanced. While with unbalanced gird, abnormal operating state will emerge in VSR which is designed on the basis of balanced grid. For example, negative sequence component will appear in VSR, which will cause severe asymmetry in AC current. A series of issues will be caused, such as non-characteristic harmonic components emerging in DC voltage and AC current, serious distortion in DC voltage and AC current waveform, as well as VSR absorbing unbalanced instantaneous power from grid system etc. [1].

In order to ensure three-phase PWM rectifier still run normally under conditions of unbalanced grid, appropriate control strategies must be proposed. Luis Morán et al. brought forward research reports in 1992, analyzed and derived the time domain expressions of AC current and DC voltage of VSR under unbalanced power grid [2]. Luis Morán et al. believed that negative sequence components of power grid the power grid is the main cause to AC current distortion; they also pointed out that, under unbalanced grid conditions, the conventional control scheme will generate even harmonic components in DC side and odd harmonics in AC side, resulting in AC current distortion. However, Luis Morán failed to improve the design of control strategies. H.S. Kim et al. systematically presented feedforward control strategy of negative sequence voltage in d-q coordinate system [3], making sinusoidal input current in AC side. But the negative sequence fundamental voltage of 100 Hz in d-q coordinate system was added to AC side in this program, making even harmonics components included in DC side output voltage. In order to obtain better DC output voltage characteristics, Pascal Rioual and other scholars adopted the method of restraining the input active harmonics and reactive components. Although output harmonic voltage had been restrained to some extent, negative sequence components were contained in AC current, leading to AC current asymmetry, While non-static error control can not be obtained by PI regulating [4]. In 1998, Hong-seok Song et al. proposed an independent control strategy by using two synchronously rotating coordinate systems with positive and negative sequence [5]. Based on this theory, Yong Sung Suh et al. studied the instantaneous balance of input and output power with synchronous control model of double inner current loops [6]. The strategy received a better output voltage characteristic, and non-static error control could also be achieved, except for complex algorithm, a large amount of online operating, more implementation costs and double digital signal processor (DSP) control. In this paper, algorithm of decomposing positive and negative sequence voltage has been introduced and applied based on positive and negative sequence EMF theory when three phase grid is unbalanced. The control strategy of suppressing negative-sequence current of AC side has been focused combined with analysis of instantaneous active and reactive power.

2 The Control System Analysis of VSR with Unbalance Grid

With unbalanced grid, the negative sequence components are included in grid EMF such as e_d and e_q, which are the variables in synchronous d-q axes. Then the current reference value of d-q axis can be defined by the following formula [3]:

$$i_d^* = \frac{2}{3}\frac{\left(e_{dp} + e_{dn}\sin(2\omega t) - e_{qn}\cos(2\omega t)\right)p^*}{\sqrt{\left(e_{dp} + e_{dn}\cos(2\omega t) - e_{qn}\sin(2\omega t)\right)^2 + \left(e_{qp} + e_{dn}\sin(2\omega t) + e_{qn}\cos(2\omega t)\right)^2}} \tag{1}$$

$$i_q^* = \frac{2}{3}\frac{\left(e_{qp} + e_{dn}\cos(2\omega t) + e_{qn}\sin(2\omega t)\right)p^*}{\sqrt{\left(e_{dp} + e_{dn}\cos(2\omega t) - e_{qn}\sin(2\omega t)\right)^2 + \left(e_{qp} + e_{dn}\sin(2\omega t) + e_{qn}\cos(2\omega t)\right)^2}} \tag{2}$$

From the formulas (1) and (2), it can be seen that the impact of voltage negative sequence second harmonic lead to distortion of current reference value, which will ultimately distort input current of AC side. However, in order to obtain the desired sinusoidal current control of AC side, current command value must be ideal. Accordingly, current reference value is very important for sinusoidal of rectifier AC side current.

When input voltage is unbalanced, reactive power and active power can be described as follows:

$$\begin{aligned} p = {} & \frac{3}{2}\left\{e_{dp}i_{dp} + e_{qp}i_{qp} + e_{dn}i_{dn} + e_{qn}i_{qn}\right\} + \frac{3}{2}\cos 2\omega t\left\{e_{dn}i_{dp} + e_{qn}i_{qp} + e_{dp}i_{dn} + e_{qp}i_{qn}\right\} \\ & + \frac{3}{2}\sin 2\omega t\left\{-e_{qn}i_{dp} + e_{dn}i_{qp} + e_{qp}i_{dn} + e_{dp}i_{qn}\right\} \end{aligned} \tag{3}$$

$$\begin{aligned} q = {} & \frac{3}{2}\left\{-e_{qp}i_{dp} + e_{dp}i_{qp} + e_{dn}i_{qn} - e_{qn}i_{dn}\right\} + \frac{3}{2}\cos 2\omega t\left\{e_{dn}i_{qp} - e_{qn}i_{dp} - e_{qp}i_{dn} + e_{dp}i_{qn}\right\} \\ & + \frac{3}{2}\sin 2\omega t\left\{e_{qp}i_{qn} + e_{dp}i_{dn} - e_{qn}i_{qp} - e_{dn}i_{dp}\right\} \end{aligned} \tag{4}$$

As can be seen from these equations, both reactive power and active power are composed by DC component and second harmonic (100 Hz) components. Therefore, harmonic content of reactive power and active power should be taken into consideration by current control about ideal reference value. However, in order to obtain the desired reference current very complicated calculation is needed. To suppress negative sequence current of rectifier AC side under unbalanced grid condition, resulting in ideal reference current, the control method of this paper is to control the negative sequence current component to zero, that is: $i_{dn} = i_{qn} = 0$.

The harmonic power is the product of positive sequence voltage and negative sequence current [3, 7]. If the negative sequence reference current is zero, then the reference value of positive sequence current can be represented as following equation:

$$\begin{bmatrix} I_{dp}^{*} \\ I_{qp}^{*} \end{bmatrix} = \frac{2}{3\left(e_{qp}^{2} + e_{dp}^{2}\right)} \begin{bmatrix} e_{dp} & -e_{qp} \\ e_{qp} & e_{dp} \end{bmatrix} \begin{bmatrix} p_{0}^{*} \\ q_{0}^{*}(=0) \end{bmatrix} \tag{5}$$

In above equation, p_0^* is average value of active power command, it is related with its DC side voltage average value. When DC voltage regulator using PI regulator, output of its regulator corresponds with DC current command. So it can be get as follow:

$$p_{0}^{*} = \left[\left(K_{vP} + \frac{K_{vI}}{s}\right)\left(v_{0}^{*} - v_{0}\right) + i_{L}\right] \cdot v_{0} \tag{6}$$

At the same time, the control algorithm of positive sequence current regulator can be achieved as follows:

$$\begin{cases} v_{dp}^{*} = -\left(K_{iP} + \frac{K_{iI}}{s}\right)\left(i_{dp}^{*} - i_{d}\right) - \omega L i_{q} + e_{dp} \\ v_{qp}^{*} = -\left(K_{iP} + \frac{K_{iI}}{s}\right)\left(i_{qp}^{*} - i_{q}\right) + \omega L i_{d} + e_{qp} \end{cases} \tag{7}$$

According to above description, based on mathematical models of unbalanced case, the specific control system is described as Fig. 1.

Based on above analysis, various abnormal operational statuses appearing in VSR with unbalance grid can be improved by control strategy of rectifier, eliminating effect on operation by unbalanced grid. For example, to eliminate negative sequence current in rectifier AC side when grid is unbalanced, by controlling the output of negative sequence voltage which is exactly the same with grid negative sequence EMF amplitude and phase in AC side. Because of their cancellation effect, negative sequence voltage component will no longer be produced in AC inductor, by which negative sequence current will no longer be produced. However, due to the presence of negative sequence voltage of the AC side, second harmonic component which is the product of negative sequence voltage and the positive sequence current will be included in input power. Therefore, second harmonic voltage components will be produce in DC voltage. It can also be inferred from DC current transient expression that a certain amount of negative sequence must be contained in AC side, if second harmonic current eliminated in DC side. It can be clearly seen that the control of negative sequence AC and DC voltage harmonics are contradictory, therefore, only one can by chosen. For this reason, only select different control methods to meet control requirements on different occasions.

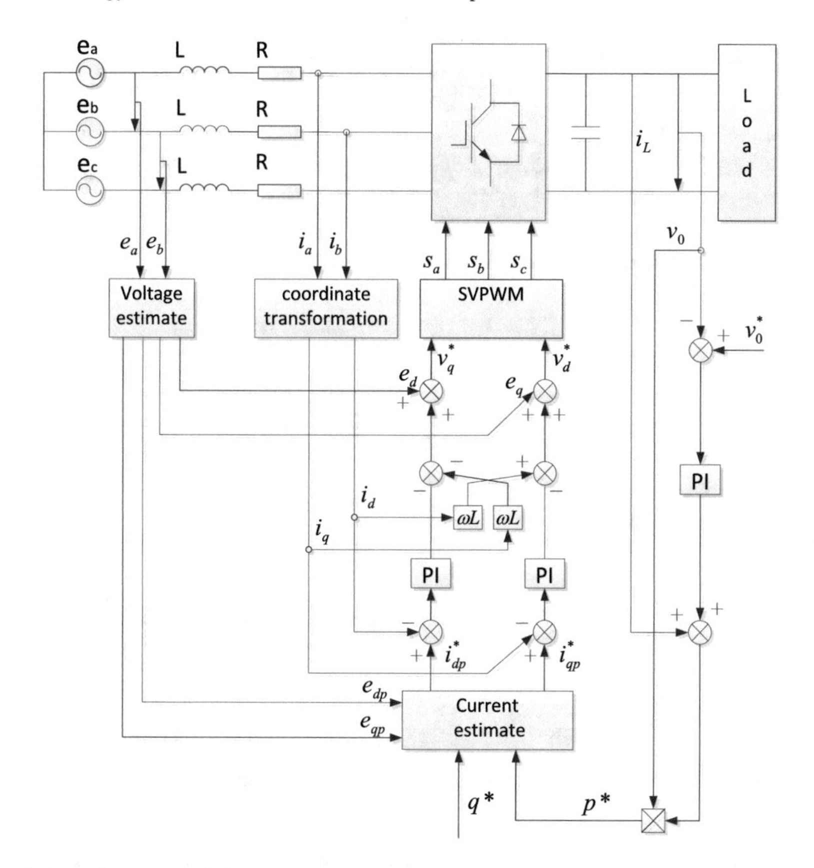

Fig. 1 Unbalance control diagram with restraining negative sequence

3 Deconstruction of Positive and Negative Sequence Voltage

In unbalance VSR control system, unbalanced EMF needed to be detected. Deconstruction of symmetrical component of the input voltage indicates that negative sequence voltage is the second harmonic component of synchronous model. Therefore, a key of unbalanced control is to decompose three-phase unbalanced voltage into positive sequence and negative sequence. The traditional approach is to be detected by band pass filter with center frequency of 100 Hz, which is shown in Fig. 2. So negative sequence voltage may also be represented by the formula (8) as follows:

$$e_{na} = e_a - e_{pa}, e_{nb} = e_b - e_{pb}, e_{nc} = e_c - e_{pc}, e_{nc} = e_c - e_{pc} \tag{8}$$

However, the application of notch filter will give rise to measurement delay or phase lag resulting in poor transient characteristics. A new kind of algorithm [7] is

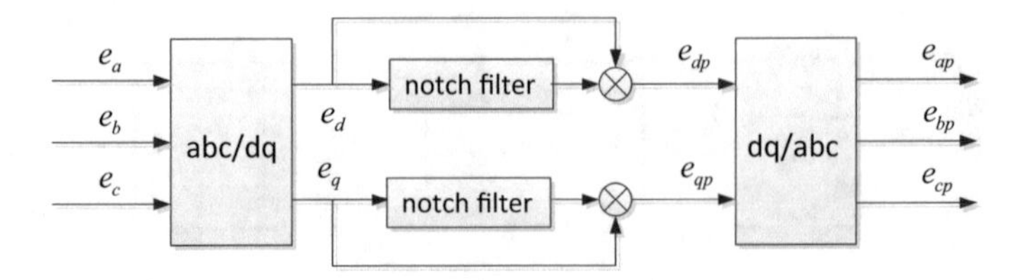

Fig. 2 Traditional detection method to positive sequence voltage

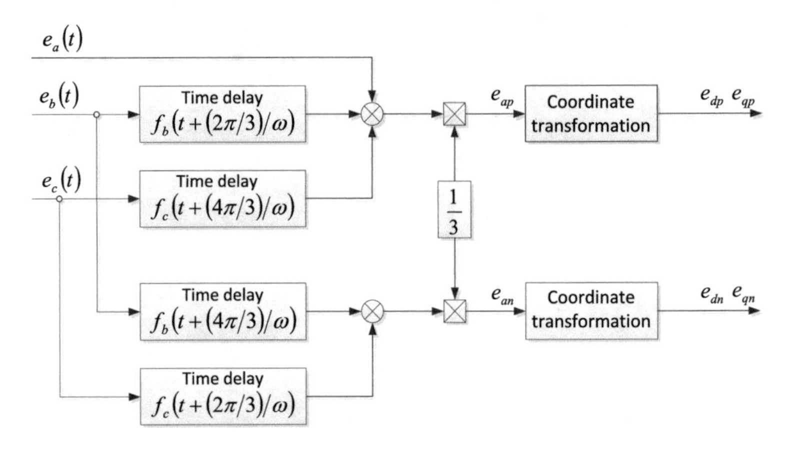

Fig. 3 New algorithm of calculating the positive and negative voltage

introduced: calculating positive and negative sequence voltage in synchronous rotating coordinate through the three-phase input voltage.

$$\begin{cases} e_{ap}(t) = \frac{1}{3}\left(e_a(t) + e_b\left(t + \frac{1}{3}T\right) + e_c\left(t + \frac{2}{3}T\right)\right) \\ e_{an}(t) = \frac{1}{3}\left(e_a(t) + e_b\left(t + \frac{2}{3}T\right) + e_c\left(t + \frac{1}{3}T\right)\right) \end{cases} \tag{9}$$

where "*T*" is the cycle of grid voltage. The specific control block diagram is shown in Fig. 3.

Simulation results comparison in Matlab of two methods of decomposing positive and negative sequence voltage is represented in Fig. 4, when three phase input voltage $e_a = 0\,V$, $e_b = e_c = 110\,V$.

Figure 4 shows that positive sequence voltage fluctuates in a stable state which is decomposed by notch filter in Fig. 4b, but the delay method can obtain very precise positive sequence voltage in Fig. 4a. The effect of decomposing positive and negative sequence voltage by delay method is obviously better than that of traditional application of notch filter. Whether decomposed positive and negative sequence voltage is accurate or not will directly affect unbalanced control system results. As a result, the precise delay decomposition method is applied to decompose positive and negative sequence voltage.

4 Unbalanced Control System Simulation

Simulation parameters

Before simulating, system simulation parameters also need to determine for the simulation model of system simulation. Basic parameters of VSR's main circuit are: rated output power: 10 kW; AC input voltage RMS: $e_a = e_b = e_c = 110\,V$; DC rated output voltage: $v_0 = 360\,V$; system rated load resistance: $R = 13\,\Omega$; AC line resistance: $R = 0.1\,\Omega$; AC side input inductance: $L = 3\,mH$; DC side output filter capacitor: $C = 3400\,\mu F$; input voltage frequency: $50\,Hz$; Switching frequency: $2\ kHz$.

System simulation

According to the analysis of unbalance control strategy, simulation model of unit power factor is designed with MATLAB/simulink.

The simulation waveforms when the grid voltage is seriously imbalanced, that is $e_a = 0\,V$; $e_b = e_c = 110\,V$. The simulation results waveform as follows (Figs. 5, 6 and 7).

Above simulation waveform shows that when grid A phase voltage is zero, control strategy applied to suppress AC side current harmonics can not only control positive sequence voltage and current to closer same phase (power factor close to 1) but also control three phase current close to sine, as shown in Figs. 8 and 9. If phase

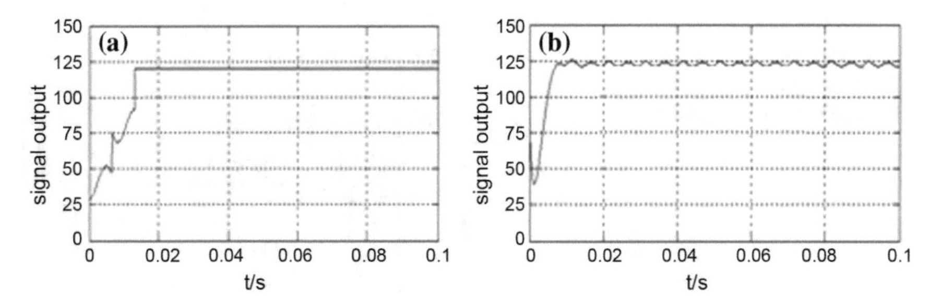

Fig. 4 Comparison between the two methods when a-phase voltage value is zero

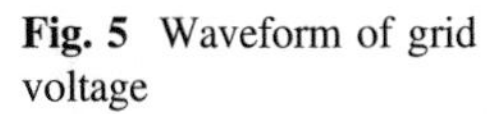

Fig. 5 Waveform of grid voltage

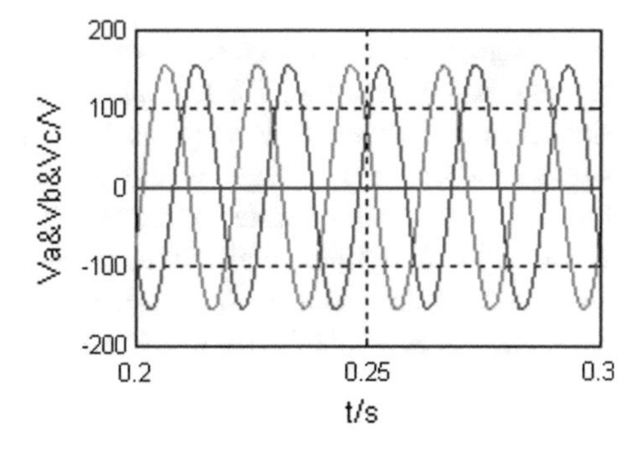

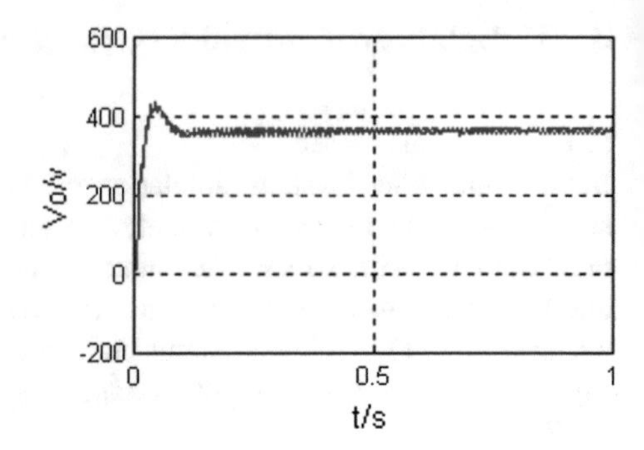

Fig. 6 Waveform of v_0

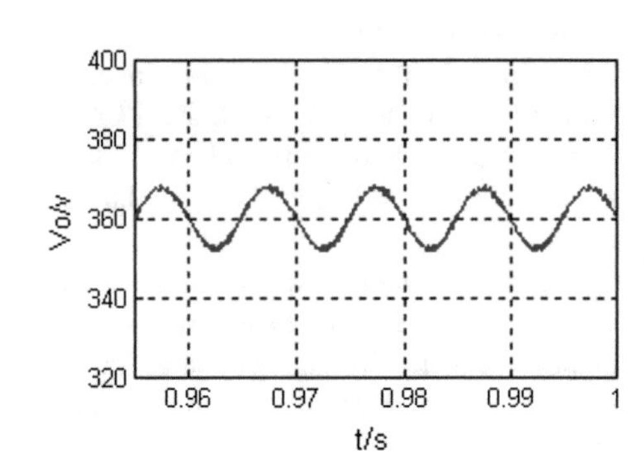

Fig. 7 Amplified waveform on steady-state of v_0

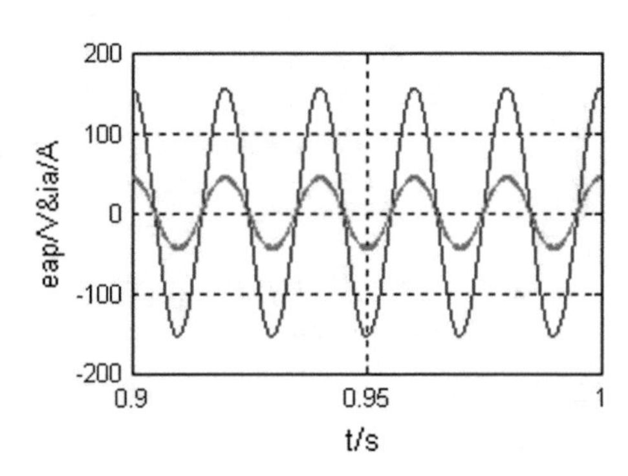

Fig. 8 Contrast waveform between A-phase positive sequence voltage and current

A voltage is zero, AC current waveform adopted by traditional VOC control strategy is shown in Fig. 10. The superiority of the control strategy suppressing AC side current harmonics is obvious by comparing with traditional VOC control strategy when grid voltage is unbalanced. At the same time, spectrogram of Figs. 11 and 12 can validate the above analysis: second harmonics will be produced at DC side of rectifier, while third harmonics will be produced at AC side. But unbalance control strategy introduced in this paper can significantly reduce third harmonics at AC side (THD < 2.61%).

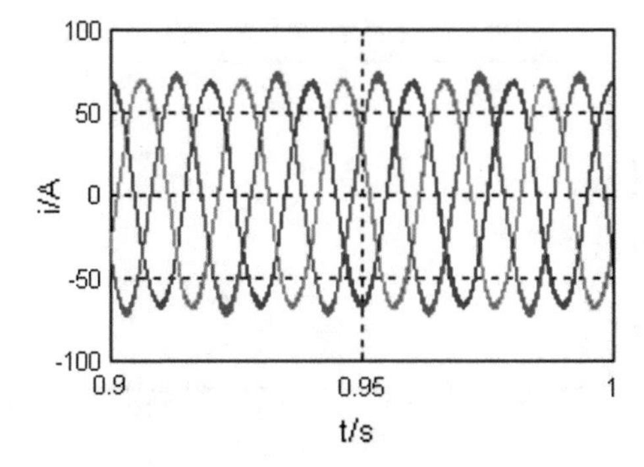

Fig. 9 3-phase current waveform at AC-side

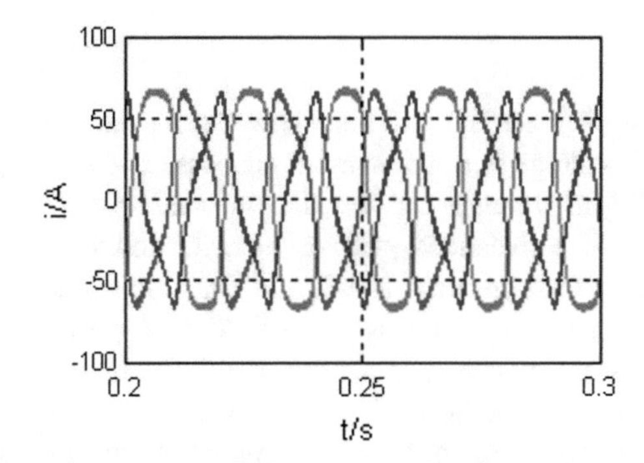

Fig. 10 Grid currents waveform of conventional VOC strategy

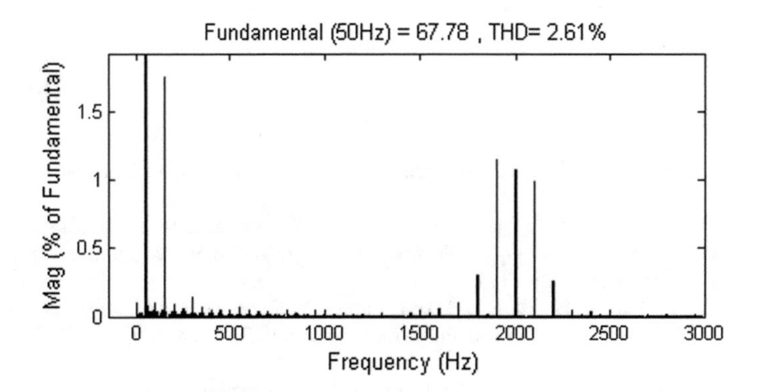

Fig. 11 Current spectrum at AC-side

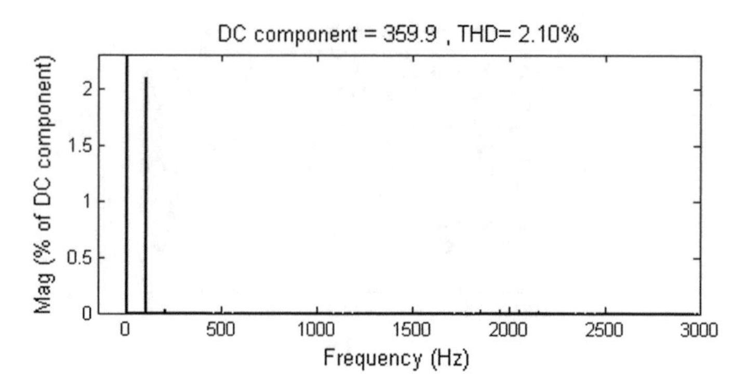

Fig. 12 Voltage spectrum at DC-side

5 Conclusion

Combined with the characteristics of ship power grid, the control strategy of VSR when the ship grid voltage is unbalanced is discussed in this paper. According to the principle that grid voltage can be decomposed into positive and negative sequence voltage with unbalanced grid voltage, the influence of the existence of negative sequence voltage on rectifier AC and DC side is discussed. Based on the analysis of instantaneous active power and reactive power, unbalance control strategy suppressing negative sequence current at AC side is proposed. Although a preliminary study on the method of optimization control and power analysis is made in this paper, an accurate transfer function model of controlling what is between input and output has not yet established, and the contradiction between three phase input current sine and DC output current containing second harmonics has not been solved, which need further research.

References

1. Chongwei Zhang, Xin Zhang. PWM rectifier and control [M]. China Machine Press. 2003, 1–55/296–306.
2. Luis Morán, Phoivos D. Ziogas, and Geza Joos. Design aspects of synchronous PWM rectifier-inverter systems under unbalanced input voltage conditions [J]. IEEE Trans. Ind. Applicat. 1992(11/12), 1286–1293.
3. H.S. Kim, H.S. Mok, G.H. Choe, D.S. Hyun, and S.Y. Choe. Design of current controller for 3-phase PWM converter with unbalanced input voltage [C]. In Proc. IEEE PESC'98. 1998, 503–509.
4. P. Rioual, H. Pouliquen, J. Louis. Regulation of a PWM rectifier in the unbalanced network state using a generalized model [J]. IEEE Trans. Power Electron. 1996(5), 495–502.
5. H. Song, K. Nam. Dual current control scheme for PWM converter under unbalanced input voltage conditions [J]. IEEE Trans. Ind. Electron. 1999(10), 953–959.

6. Y.S. Suh, and T.A. Lipo. Modeling and analysis of instantaneous active and reactive power for PWM ac/dc converter under generalized unbalanced network [J]. IEEE Tran. Power Electron. 2006(7), 1530–1540.
7. Y.S. Suh, V. Tijeras, and T.A. Lipo. A nonlinear control of the instantaneous power in d-q synchronous frame for PWM ac/dc converter under generalized unbalanced operating conditions [C]. IEEE Ind. Appl. Soc. Annu. Meeting, Chicago, IL. 2002(10), 1189–1196.

Adaptive Double-Resampling Particle Filter Algorithm for Target Tracking

Qiang Zhao, Chen Wei, Liang Qi and Wenhua Yuan

Abstract Based on the traditional particle degradation and depleted of particle filter and the number of particle set, which cannot be adaptive to change brought by the filtering accuracy and convergence rate of decline. A new methods of Innovation and resampling particle filter was applied to the paper. This approach can solve the problems mentioned above. The algorithm first uses the observation information to establish the particle distribution program of the resampling. Then to conduct a resampling on the basis of the initial resampling. The second resampling used the particle cross aggregation algorithm. This can improve efficiency of the particles, and avoid the increase of the calculation when using too many particles. The simulation result based on the DR/GPS shows that compared with the traditional PF algorithm, the algorithm can improve the accuracy and stability of the filter.

Keywords Double-resampling · Particle filter · Innovation · Adaptive · Target tracking

1 Introduction

The particle filter uses a weighted particle set to describe the probability distribution of state variable, using Monte Carlo method to realize Recursive Bayesian filter, it is an effective nonlinear non-Gaussian system suboptimal prediction method, and has been successfully applied in the field of target tracking [1–3]. But in order to correctly approximate the posterior probability density, it needs a large quantity of frequency samples. However, a large number of update operations of particle optimization will lead to lower efficiency of the algorithm. It is the main defect of particle filter.

Q. Zhao (✉) · C. Wei · L. Qi · W. Yuan
School of Electronics and Information, Jiangsu University of Science and Technology, Zhenjiang 212003, Jiangsu, China
e-mail: zhaoqiang_200@163.com

© Springer Nature Singapore Pte Ltd. 2018

N.Y. Yen and J.C. Hung (eds.), *Frontier Computing*, Lecture Notes in Electrical Engineering 422, DOI 10.1007/978-981-10-3187-8_73

In order to improve the efficiency of particle filter, the most direct and effective way to improve the efficiency of particle filter is to adjust the size of the particle size. It can achieve the purpose of filtering accuracy and reasonable operation speed. In fact, the particle filter has the problems of particle degeneracy and impoverishment, it leads to filter failure because too few effective particles, the most fundamental reason is that the probability distribution description ability is insufficient, the filter is premature convergence. FOX [4] proposed a Kullback-Leibler Distance (KLD) sampling method which realize the adaptive adjustment of particle sets by computing the K-L distance between posterior probability distribution and maximum likelihood probability distribution of particle sets. In addition, some researches about adaptive adjustment method [5] of particle set were proposed to reduce the number of sampled particles [5], but the rationality of particle distribution is not considered.

At the present, the main research means is to improve the micro-ability and adaptability of particle set [4]. Therefore, on the basis of analyzing a large number of literature, this paper puts forward a kind of double sampling adaptive particle filter algorithm. First, the proposed algorithm takes the new observed value of the target state and the residual error of the prediction result as the current observation. Then, according to the new information to reflect the relationship between the accuracy of the target prediction and the uncertainty of the system, the noise variance of the new information estimation system is determined, and use it to determine the sampling of the proposed distribution [6–9]. If predication is accurate, this method can obtain accurate density estimation only using a small amount of particles. When target motion changes greatly, the method uses more samples to ensure accurate tracking. The actual operation is to use a double-resampling when the sample set has more particles. First step, according to the new information of observation to control the number of particles double sampling, it can enhance the ability to predict the particle set. The second step, according to the particle space distribution to control particle size, it can ensure the consistency of particle size and particle spatial scale. This method has the rationality of spatial distribution. The Monte Carlo simulation experiment shows that this method can balance the particle diversity in prediction stage and particle size in update phase. Under the premise of keeping the accuracy of the particle filter, it can effectively reduce the calculation of the particle filter.

2 The Resampling Distribution Scheme Based on the Observed Innovation

Considering the following nonlinear discrete dynamic system is:

$$\begin{cases} x_{t+1} = f_t(x_t, w_t) \\ z_t = h_t(x_t, v_t) \end{cases} \tag{1}$$

In this formula, $x_t \in R^n$ is n dimensional vector state vector at time t. $z_t \in R^m$ is m dimensional observation vector, w_t is system noise, v_t is observation noise, they variance obey $\begin{bmatrix} (\sigma_x^t)^2 & \\ & (\sigma_z^t)^2 \end{bmatrix}$ distribution.

Assume $\hat{x}_t$ and x_t^p are the estimation and predication status value in step t respectively. And $e^t = \hat{x}_t - x_t^p$ is observation innovation at current time. Assume that state vector x_k predicts more accurately, then σ_x^t should be small, the new sampling should be located in a small neighborhood prediction area. However, when x_t predicts not accurate, σ_x^t should be taken larger value. Enable the sampling distribution range is large enough to contain the real target state. Thus, the system noise variance can be express as:

$$\boldsymbol{\sigma}_x^t = \begin{cases} \min(e^t, \boldsymbol{\sigma}_{\max}) & if \quad e^t \geq \boldsymbol{\sigma}_x^{t-1} \\ \max(\boldsymbol{\alpha} e^{t-1}, e^t, \boldsymbol{\sigma}_{\min}) & if \quad e^t < \boldsymbol{\sigma}_x^{t-1} \end{cases} \tag{2}$$

In order to avoid poor particle phenomena caused by poor system noise, setting the lower limit of the system noise variance is $\boldsymbol{\sigma}_{min}$, and the $\boldsymbol{\sigma}_{min}$ is the system maximum noise variance. In order to avoid the estimation error makes the system noise decrease quickly, the algorithm uses the attenuation factor $\boldsymbol{\alpha}$ to control the reduction speed of noise variance, in the experiment $\boldsymbol{\alpha} = 0.9$.

The number of particle is related to system noise, when the system noise is small, using a small amount of particles can approximate the target distribution; and when the system noise is bigger, the particle sampling range will be expanded, so should increase the number of particles. Using Sigmoid function to represents the relationship between the number of particles and uncertainty measure.

$$N_t = N_{min} + (N_{max} - N_{min})\left(\frac{2}{1 + exp(-\boldsymbol{\beta}(r_k - r_{min}))} - 1\right) \tag{3}$$

In the formula, N_{min} expresses the min particles, and N_{max} expresses the max particles. The uncertainty measure $r_t = \boldsymbol{\sigma}_x^t \boldsymbol{\sigma}_y^t$ of the target state in the t moment is estimated according to the state covariance. The lower limit of uncertainty measure is $r_{\min} = \boldsymbol{\sigma}_{x\min} \boldsymbol{\sigma}_{y\min}$, $\boldsymbol{\beta} = 0.01$ is control coefficient.

3 Adaptive Particle Filter Algorithm Based on Double-Resampling

3.1 Particles Sparse Polymerization Double-Resampling

When the target motion changes greatly, or the carrier surrounding environment changes, the algorithm needs more samples to ensure the effectiveness of the filter, and at this time, the spatial distribution of particles also shows clustering trend. At this time, the particles' space differentiation appears redundant causing that updating particles centralization values consumes too much computing resources. In this paper, before updating the weights of particle set, the algorithm weighted aggregation for particles based on spatial scale mesh of particles, to reduce the size of the particle set. So the algorithm is called particles sparse polymerization resampling [10].

Firstly, this paper gives the definition of a grid dividing the state space:

Definition 1 if dividing the i dimension of n dimensional space S into m_i closed left and right open intervals which are equal length. Thus, it can divide the whole space S into $m_1 \times m_2 \times \cdots \times m_n$ disjoint n dimensional grid g_n.

Definition 2 using the sample particles which attach to grid cell g_i on the space to express grid density of g_i, denoted as $\mathrm{den}(g_i)$. As $\mathrm{den}(g_i) = 0$, it says g_i is empty; otherwise, it says g_i is not empty. The current grids and the adjacent grids in its $l(l < n)$ dimension direction compose a corresponding mesh grid set of 3^l scale. This paper uses "#" to mark the grid set relevant variables. The k dimensional grid set corresponding to grid g_i is marked $\#g_i^l$. When $l = n$, $\#g_i^l$ is called the whole dimensional grid set. When $l < n$, $\#g_i^l$ is called the non-whole dimensional grid set. Based on the weights of the particles, taking weighted average with all the particles within the unit space to obtain a polymeric particle, and the unit space is called the polymerization unit. The particle aggregation method with a grid set as polymerization unit implementation is called cross polymerization.

3.2 Particle Cross Polymerization

Assuming that the particles are distributed in a k non-empty grid cell, the grid cell containing N_k particles can be described as g_k: $\left\{\omega_k^i, x_k^i \middle| i = 1, 2, \ldots N_k\right\}$, in the formula $k = 1, 2, \ldots, K$, x_k^i denotes as the state of particle, ω_k^i denotes as the weight of particle, the corresponding l dimension grid set is $\#\mathrm{g}_k^i$: $\left\{x_k^i, \omega_k^i \middle| i = 1, 2, \ldots \#N_k\right\}$, $\#\mathrm{N}_k^l$ is the number of particles including in the polymerization unit $\#\mathrm{g}_k^l$. Taking the grid set as the polymerization unit, all particles within the polymerization unit weight combination to get the aggregated particles of the central grid. Polymerization equation is:

$$\begin{cases} \hat{x}_t^k = \dfrac{\sum_{i=1}^{\#N_k^l} x_k^i \omega_k^i}{\sum_{i=1}^{\#N_k^l} \omega_k^i} \\ \hat{\omega}_t^k = \dfrac{\sum_{i=1}^{\#N_k^l} \omega_k^i}{3^l} \end{cases} \tag{4}$$

In the formula, $(\hat{x}_t^k, \hat{\omega}_t^k)$ expresses polymeric particle corresponding to the k grid. Polymerization achieves consistent with the particle collection size and particle space size. Polymeric particle shows that higher weight of the particle has a greater impact on spatial distribution, which has a reasonable distribution of computing resources according to the weight of the particle. Each particle average weights to $3l$ polymeric particle (l is the dimension of space), averaging the distribution of particle on the certain extent, it can effectively relieve the weight degradation problem caused by reason that weight is too concentrated to a single particle. Thus, the small weight particles will not be abandoned, and it can avoid the phenomenon of particle diversity scarcity caused by the particle weight sampling, which effectively reduces the weight concentration.

3.3 Steps of Adaptive Double-Resampling Particle Filter Algorithm

Combining particle cross aggregation algorithm, the following gives the specific steps of adaptive double sampling particle filter.

(1) Initialization

According prior distribution $p(x_0)$ to initialize sampling set $\{x_0^i\}_{i=1}^{N_0}$, and ordering the particle weight is $\omega_0^i = \frac{1}{N_n}, i = 1, \ldots, N_0, t = 1$.

(2) Sampling

According to the sampling weight ω_{t-1}^i, choosing N_t samples from sampling set $\{x_{t-1}^i\}_{i=1}^{N_{t-1}}$, then a group new sampling set $\{\hat{x}_{t-1}^i\}_{i=1}^{N_t}$ is obtained.

(3) Updating the weights

Each sampling value calculates the new sampling according to the state transition equation $x_t^i = \hat{x}_{t-1}^i + \omega_t, i = 1, \ldots N_t$ in the sampling set $\{\hat{x}_{t-1}^i\}_{i=1}^{N_t}$.

Ordering the weight of sampling i for the confidence coefficient of its corresponding candidate item O_i, the normalization weights is $\omega_t^i = g(O_i, \Omega)$.

(4) Double-resampling

According to the grid scale L, dividing the State grid lattice space $\{g_k | k = 1, 2, \ldots, K_t\}$ and calculating the grid density;

(1) According formula (4), taking particles sparse polymerization double-resampling to generate polymeric particles set $\hat{S}_t = \{x_t^k, \omega_t^k\}_{k=1,2,\ldots,N_t}$.
(2) According to the Update Model $p(o_t|x_t)$, updating the weights $\omega_t^k|_{k=1,2,\ldots,N_t}$ of polymeric particles.
(3) Normalize the particle weights ω_t^i.

(5) Estimate the target state: $\hat{x}_t = \sum_{i=1}^{N} \omega_t^i x_t^i$.

(6) Calculate predictive state: $x_t^p = \sum_{i=1}^{N_t} s_t^i$.
(7) Calculate observation innovation: $e^t = \hat{x}_t - x_t^p$.
(8) According step 3 to update the number of samples N_{t+1}.
(9) Return to the step 2, going on the steps until the end of tracking.

The following figure is the algorithm structure (Fig. 1):

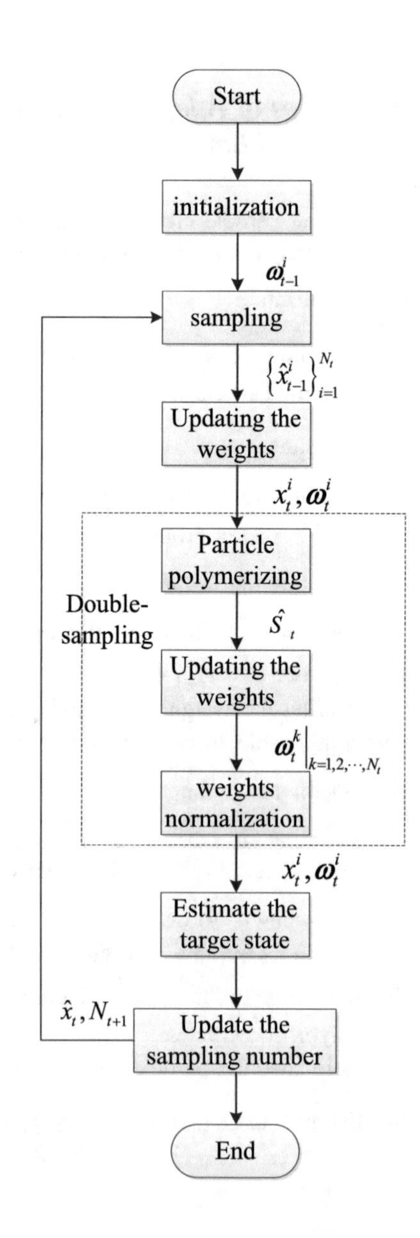

Fig. 1 The algorithm structure

4 Simulation Experiment and Analysis

4.1 Establish DR/GPS Integrating System Model

Taking the DR/GPS integrated navigation system as an example, the model of DR/GPS integrated navigation system is established as follows:

$$\begin{cases} X_t = \phi_{t,t-1} X_{t-1} + W_t \\ Z_t = h(X_t) + V_t \end{cases} \tag{5}$$

Here, the state transition matrix, the observation matrix and the noise matrix are respectively such as:

$$\phi_{t,t-1} = diag[F_e, F_n]$$

$$F_e = \begin{bmatrix} 1 & T & (-1+\alpha_e T + e^{-\alpha_e T})\alpha_e^{-2} \\ 0 & 1 & (1+e^{-\alpha_e T})\alpha_e^{-1} \\ 0 & 0 & e^{-\alpha_e T} \end{bmatrix}, F_n = \begin{bmatrix} 1 & T & (-1+\alpha_n T + e^{-\alpha_n T})\alpha_n^{-2} \\ 0 & 1 & (1+e^{-\alpha_n T})\alpha_n^{-1} \\ 0 & 0 & e^{-\alpha_n T} \end{bmatrix}; \tag{6}$$

$$h(X_t) = \begin{bmatrix} x_e & x_n & \frac{v_n a_e - v_e a_n}{v_e^2 + v_n^2} & T\sqrt{v_e^2 + v_n^2} \end{bmatrix}^T \tag{7}$$

$$W_t = [0\,0\,w_e\,0\,0\,w_n]^T; \quad V_t = [v_1 \quad v_2 \quad v_3 \quad v_4]^T.$$

Here, $\alpha_e = 1/\tau_e$, $\alpha_n = 1/\tau_n$, τ_e, τ_n is time constant of vehicle about Eastward and Northward maneuver acceleration respectively. w_e, w_n is white noise sequence which is meeting the $N(0, \sigma_e^2)$ and $N(0, \sigma_n^2)$ distribution respectively. v_1 and v_2 is the observed noise from the output of the GPS receiver in the East position and North position, respectively. They are both approximated as $(0, \sigma_1^2)$ and $(0, \sigma_2^2)$ Gaussian white noise. ε_ω is gyro drift, it is approximated as $(0,\sigma_\varepsilon^2)$ Gaussian white noise. ε_s is the observation noise of the mileage meter, according to the actual situation, it is assumed to be non-Gauss, and the corresponding noise distribution is shown in Fig. 2, and the likelihood probability density function is approximate to the formula (8).

$$\begin{aligned} p(z_t|x_t) = & \frac{1}{2}\exp\left[-\frac{1}{2}(z_t - h(x_{t|t-1}) - \bar{n})^T R^{-1}(z_t - h(x_{t|t-1}) - \bar{n})\right] \\ & + \frac{1}{2}\exp\left[-\frac{1}{2}(z_t - h(x_{t|t-1}) + \bar{n})^T R^{-1}(z_t - h(x_{t|t-1}) + \bar{n})\right] \end{aligned} \tag{8}$$

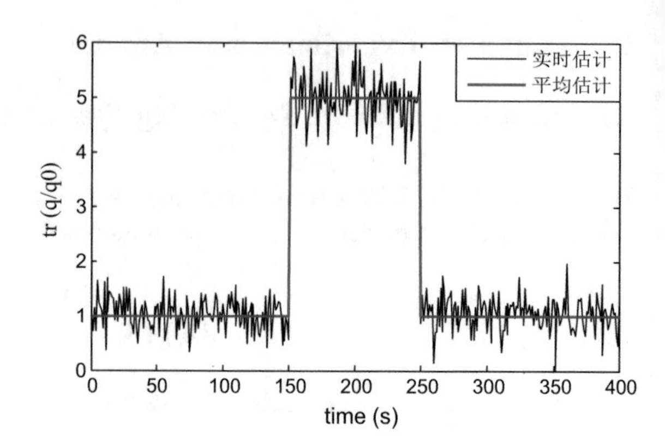

Fig. 2 Process noise mean trace

4.2 Simulation Experiment and Analysis

In order to analyze the performance of the proposed algorithm, this paper uses Matlab as the development language, simulation experiment is carried out on the platform of DR/GPS integrated navigation system.

The initial conditions of simulation and the related parameters are:

$$T = 1s; \sigma_e^2 = \sigma_n^2 = (0.3m/s^2)^2; a_e = a_n = 1;$$

$$\sigma_1^2 = (15m)^2; \sigma_2^2 = (16m)^2; \sigma_s^2 = (0.3m/s^2)^2;$$

Respectively, initial state value, initial state error variance matrix, initial process noise and measurement noise variance matrix is:

$$x_0 = [0 \quad 0.1 \quad 0 \quad 0 \quad 0.1 \quad 0]^T;$$
$$P_0 = diag[100, 1, 0.04, 100, 1, 0.04]; R_0 = diag(15^2, 16^2, 0.005^2, 0.7^2);$$
$$Q_0 = diag(0.1^2, 0.5^2, 0.3^2, 0.1^2, 0.5^2, 0.3^2)$$

The simulation time is set to 400 s. In order to verify the effectiveness of the proposed algorithm under different noise conditions. In the process of simulation, the vector is assumed to be a variable acceleration motion toward east and north in 150–250 s whose acceleration is 3sin(t/10)m/s2. In this process, the noise variance is constant, and the mean value is enlarged to 5 times. Within the 250–400 s, the carrier is doing the uniformly accelerated motion whose acceleration is 2 m/s2. At this point, the observation noise covariance is enlarged to 3 times of the initial value to increase the observation noise.

According to the simulation conditions, we use PF and APF algorithm proposed in this paper to simulate, the simulation results are shown in Figs. 2, 3, 4 and 5 and Table 1.

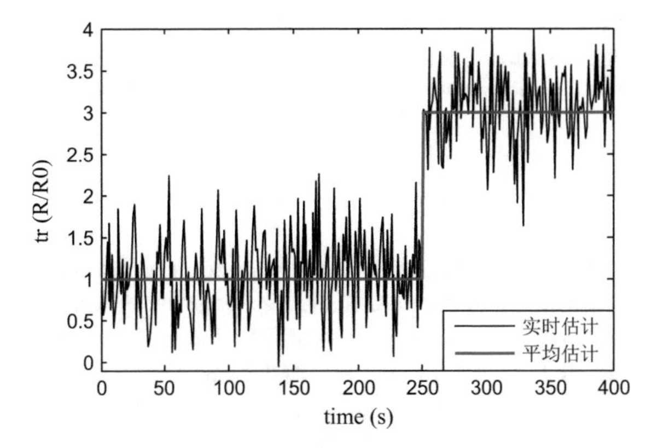

Fig. 3 Measurement noise covariance trace

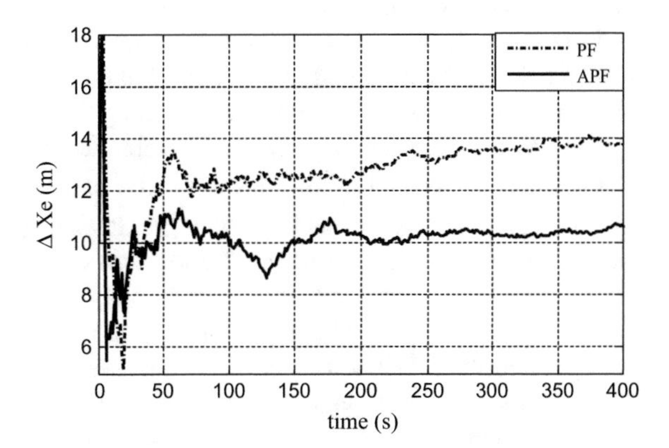

Fig. 4 Comparison of east position errors between PF and APF

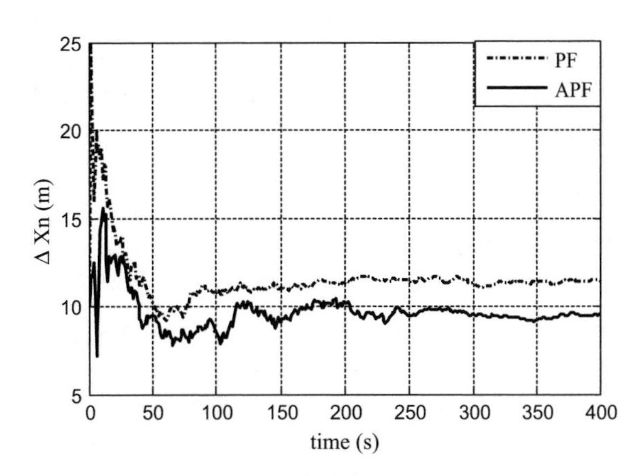

Fig. 5 Comparison of north position errors between PF and APF

Table 1 Comparison of simulation results under different particle numbers

Algorithm	N	$\eta/\%$	E_{ARMS}	Estimated time/s
PF	100	2.563	0.3901	0.305
	500	1.004	0.1993	0.556
APF	100	6.715	0.1489	0.198
	500	8.811	0.0227	0.287

Figures 2 and 3 describe the variation of the noise in the simulation process. The comparison of the position error about the Eastward and the northward is given in Figs. 4 and 5. From Figs. 4 and 5, due to the irregular change of the noise, we can see that the traditional PF algorithm is less than the APF algorithm in the convergence speed and stability. It is mainly due to the traditional PF algorithm has more or less the particle degeneracy and impoverishment, the number of particles can't be changed adaptively. In this paper, the proposed algorithm can effectively overcome the above problems by changing the particle size and distribution.

In order to further compare the estimation accuracy of the two algorithms, we selected different particle numbers. The PF algorithm and APF algorithm are used to investigate the mean square root mean square error (ARMSE), the filter efficiency and the state estimation time. The simulation results are shown in Table 1.

Filtering efficiency in the Table 1 can be expressed as $\eta = \frac{1}{E_{ARMS}N}$. The higher filtering efficiency shows that the higher filtering accuracy can be obtained with less number of N particles. From the Table 1 we can see that when the number of particles are in the same time, we can choose N = 100, so APF algorithm has higher estimation accuracy and filtering efficiency, and needs shorter state estimation time than PF algorithm. When the number of particles are the same, choosing N = 500, and the E_{ARMS} of the PF algorithm is 0.1993, it shows that the filtering accuracy is improved and the filtering efficiency η and the state estimation time are decreased. This shows that the filtering precision is improved at the expense of filtering efficiency and state estimation time. While under the premise of the E_{ARMS} of APF algorithm is smaller than PF (i.e. filtering precision is high), the filtering efficiency is higher than PF times, the state estimation time is instead smaller than PF. This shows that the APF algorithm can achieve greater efficiency with shorter time and higher accuracy.

5 Conclusion

In this paper, an adaptive particle filter based on adaptive particle filter is proposed. It is aiming at these problems which are occurred in traditional particle filter. These problems are such as particle degradation problem, dilution problem, and the number of particle sets can't be self-changed to bring about the decline of filtering precision and convergence rate. The algorithm can determine the number of particles by real-time detection of the observation information. In the double sampling process, this particle cross polymerization method is effective to improve the particle degeneracy and impoverishment problem and more effective improve the filtering precision. The validity of the algorithm is proved by DR/GPS simulation experiment.

Acknowledgements This work is supported by National Natural Science Foundation of China (Grant no. 61503162, 51505193), Natural Science Foundation of Jiangsu Province in China (Grant no. BK20150473).

References

1. Brasnett P, Mihaylova L, Bull D, et al. Sequential Monte Carlo tracking by fusing multiple cues in video sequences [J]. *Image and Vision Computing* (S0262-8856), 2007, 25 (8):1217–1227.
2. Zhou S K, Chellappa R, Moghaddam B. Visual tracking and recognition using appearance-adaptive models in particle filters [J]. *IEEE Trans. on Image Process* (S1057-7149), 2004, 13(11):1491–1506.
3. Arulampalam M.S. Maskell S., Gordon N, et al. A tutorial on particle filters for online nonlinear/non-gaussian Bayesian tracking[J]. *IEEE Transactions on Signal* (S1053-587X), 2002, 50(2):174–188.
4. Fox D. Adapting the sample size in particle filters through KLD-sampling[J]. *The International Journal of Robotics Research*, 2003, 22(12):985−1003.
5. Pan P, Schonfeld D. Dynamic proposal variance and optimal particle allocation in particle filtering for video tracking[J]. *IEEE Transaction on Circuits and Systems for Video Technology,* 2008, 18(9):1268−1279.
6. Zheng Zuohu, Wang Shouyong, Wan Yang. Model of Maneuvering Target Track With Adaptive Parameter Adjustment[J]. *Journal of Air Force Radar Academy*, 2011, 25 (2):113–116.
7. Zhao Xuemei, Chen Ken, Li Dong. Application of strong tracking Kalman filter in video target tracking[J]. *Computer Engineering and Applications*, 2011, 47(11):128–131.
8. Ba hongxin, He xinyi, Fang Zheng, Li Chunfang. A New Variance Adaptive Filtering Algorithm for Maneuvering Target Tracking[J]. *Journal of Wuhan University of Technology*, 2011, 35(3):448–452.
9. Zhang Kai, Shan Ganlin, Ji Bing, Chen Hai. Improved Variable Dimension Filtering for Attitude Tracking of Target[J]. *Electronics Optics & Control,* 2012, 19(3):40–43.
10. Li Tian-Cheng, Sun Shu-Dong. Double-resampling Based Monte Carlo Localization for Mobile Robot[J]. *Acta Automatica Sinica,* 2010, 36(9):1279–1286.

Research on Expert System for Line Heating

Liang Qi, Xiaochun Ma, Feng Yu and Xian Zhao

Abstract Line Heating (LH) is the main method for forming ship-hull plates in most of shipyards all over the world, which is called "heated bending" usually. The accuracy of final shapes and productivity of plates solely depend on skilled workers' experiences for LH process is mainly operated manually. A new expert system is developed in order to determine the processing parameters which can effectively improve the productivity.

Keyword Ship-hull plate · Line Heating (LH) · Expert system

1 Introduction

The ship-hull's surface mainly consists three-dimensional, complex and undevelopable curved plates. The technique of Line Heating (LH) is an efficient and economic process for mouldless deformation of curved surface shapes. Traditionally, LH process is mainly applied to form double-curved hull plates and depends on abundant experiences of skilled workers who usually accomplish LH process manually. However, the manual and experiential producing pattern restricts the shipbuilding's cycle and quality and becomes a bottleneck in modern production line [1]. Over the past decades, shipbuilding industry has strived for automation

L. Qi (✉) · X. Ma · X. Zhao
School of Electronics and Information, Jiangsu University of Science and Technology, Zhenjiang, China
e-mail: alfred_02030210@163.com

X. Ma
e-mail: mxc19932@163.com

X. Zhao
e-mail: zhao37082@163.com

F. Yu
Jiangnan Shipyard (Group) Co., LTD, Shanghai, China
e-mail: yufeng200807@sina.com

© Springer Nature Singapore Pte Ltd. 2018

N.Y. Yen and J.C. Hung (eds.), *Frontier Computing*, Lecture Notes in Electrical Engineering 422, DOI 10.1007/978-981-10-3187-8_74

realization of LH process. Many scholars and technicians have carried on fruitful research on LH automation as important research content.

As unceasing development and improvement of artificial intelligence, the efficient approach may be presented to determine processing parameters precisely. Expert system based on artificial intelligence is composed of knowledge base and inference engine, which can provide operators with processing strategy.

2 Establishment of Knowledge Base

Information of LH process can be divided into two categories: (1) it contains information of plate shape, material and the desired deformation which pertain to the basic features of plates; (2) it contains information of gas flow, the moving speed of the heat source and the cooling water flow which pertains to processing features. Therefore, the paper divides the information of LH process into two levels: feature cell and processing cell.

2.1 Feature Cell's Constitution

The most basic features such as the length and width of plates, the material are the most basic features. These feature cells constitute the basic feature of line heating set F, and can be expressed as:

$$F = \{f_1, f_2, \ldots, f_i, \ldots, f_n\}$$

where f_i is the i-th feature cell of steel plate, and n represents the number of cells.

Feature cell is the basic information of plates and includes the calculation and decision of the process needed all the basic information. If the f is used to represent a feature cell of the steel plate, it can be expressed as a 5-dimensional feature vector according to the properties of the plate:

$$f = (f\_ID,\ f\_S,\ f\_M,\ f\_E,\ f\_O)$$

f_ID is the feature cell number for a specific plate. All feature cells are the only number

f_S is the shape of the feature cell, which is the shape of the steel plate, including the length, width and thickness of the plate

f_S can be expressed as a 3-dimensional vector:

$$f\_S = (f\_S\_L, f\_S\_W, f\_S\_H)$$

where f_S_L is the length of the plate; f_S_W is the width of the steel plate; f_S_H is the thickness of the plate.

f_M is the material characteristics of the feature cell, including carbon content, Young's modulus, yield stress, Poisson's coefficient, thermal expansion coefficient. Therefore, f_M can be represented as a quintuple:

$$f\_M = (f\_M\_C, f\_M\_E, f\_M\_Y, f\_M\_C_P, f\_M\_C_T)$$

where f_M_C is the carbon content of the steel; f_M_E is Young's modulus of the steel; f_M_Y is yield stress steel; $f\_M\_C_P$ is Poisson coefficient steel; $f\_M\_C_T$ is the thermal expansion coefficient of the steel.

f_E represents the desired amount of deformation of steel, including steel line deformation and angular deformation, then f_E can also be expressed as a 2-tuple:

$$f\_E = (f\_E\_D_W, f\_E\_D_\alpha)$$

2.2 *Processing Cell's Constitution*

For each of the feature cell of plate, LH process generally needs to go through the multiple steps. That is, the processing chain O, which can be expressed as:

$$O_i = \{o_{i1}, o_{i2}, \ldots, o_{ii}, \ldots, o_{im}\}$$

O is used to represent a processing cell of the steel plate, which can be expressed as a 4-dimensional feature vector according to the properties of the steel plate:

$$o = \{o\_ID, o\_F, o\_H, o\_C\}$$

where o_ID represents processing cell number;

o_F indicates the feature cell of the processing cell;

o_H represents heating element of the processing cell, including gas flow and the moving speed of the heat source. o_H can be expressed as a 2-tuple:

$$o\_H = (o\_H\_F_G, o\_H\_V_H)$$

where $o\_H\_F_G$ the gas flow; $o\_H\_V_H$ represents the moving speed of the heat source.

o_C means that the cooling conditions of the processing cell, including cooling water flow and cooling apparatus moving speed, the o_C also be represented as a 2-tuple:

$$o\_C = (o\_C\_F_W, o\_C\_V_C)$$

where $o\_C\_F_W$ denotes a cooling water flow; $o\_C\_V_C$ represents the moving speed of the cooling device.

3 Establishment of Inference Engine

3.1 Design of Line Heating Expert System Inference Engine

In actual LH process, workers may design processing parameters relying on the own experiences. Due to the complexity of LH process and information, the expert system using hybrid positive and negative reasoning method. Inference engine implementation of forward reasoning is mainly based on the basic parameters of the ship's information, such as the ship number, plate number, material, and the final target curve of the plate. The expert system's inference engine implementation of reverse reasoning is mainly based on the angular variables of the target curve shape. And it can be calculated by the LSSVM [2]. In the processing parameters determination, the output quantity is acetylene flow $Q_{C_2H_2}$ and heating rate v_{HL}, the input quantity is the linear variable ΔL and the angle variable θ, as shown in Fig. 1.

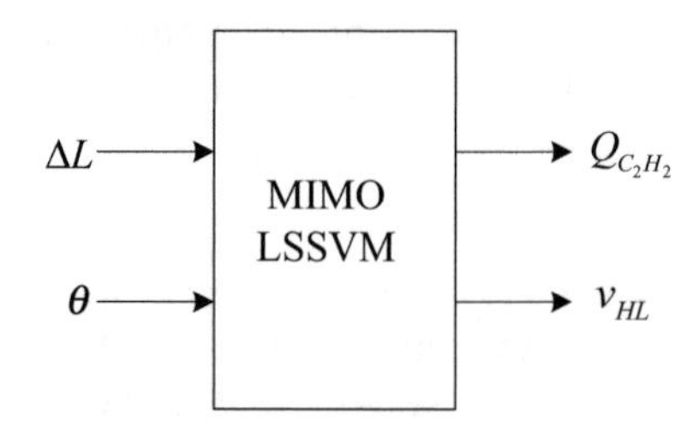

Fig. 1 The diagram of processing parameter prediction

4 Application Example of Expert System

Take a plate (1000 * 1000 * 20 mm) as an example, introduces the knowledge base and inference engine representation method.

4.1 Knowledge Base

① Feature cell (Table 1)
② Processing cell (Table 2)

Table 1 The feature cell

Feature cell	Content	Value
f1	ID	001
f2	Length	1000 mm
f3	Width	1000 mm
f4	Thickness	18 mm
f5	C%	0.2%
f6	Young's modulus	206 Gpa
f7	Yield stress	245 Mpa
f8	Poisson's coefficient	0.3
f9	Thermal expansion coefficient	12.1–13.5
f10	Desired line deformation	1.52
f11	Desired angular deformation	0.18

Table 2 Processing cell

Processing cell	Content	Value
o1	ID	001
o2	Acetylene flow	1505 L/h
o3	Heating rate	3.0 mm/s
o4	Heating times	3
o5	Water flow	1.6 m^3/h
o6	Length of heating line	1000 mm

4.2 Inference Engine

The paper provides processing plan by LSSVM prediction method, shown as Table 3.

Table 3 The sailed plate's processing parameters

Number	Acetylene flow (L/h)	Heating rate (mm/s)	Length of heating line (mm)	Heating times	Water flow (m3/h)
1	1720	3.7	769	4	1.6
2	1500	2.9	769	3	1.6
3	1912	4.3	739	5	1.6
4	1580	3.3	611	4	1.6
5	1930	4.4	685	4	1.6
6	1930	4.4	682	4	1.6
7	1508	3.1	595	6	1.6
8	1930	4.4	273	4	1.6
9	1508	2.9	838	2	1.6
10	1930	2.9	533	3	1.6
11	1520	3.3	593	3	1.6
12	1520	2.9	727	3	1.6
13	1520	3.1	610	3	1.6
14	1650	2.9	816	3	1.6
15	1755	3.8	677	4	1.6
16	1550	3.2	706	4	1.6
17	1755	3.8	721	5	1.6
18	1930	4.4	756	4	1.6
19	1520	3.1	1022	3	1.6
20	1695	3.6	679	3	1.6
21	1520	3.1	640	3	1.6
22	1520	2.9	1003	3	1.6
23	1670	3.4	961	3	1.6
24	1580	3.3	831	3	1.6
25	1495	3.1	1002	3	1.6
26	1520	2.9	1047	3	1.6

5 Conclusion

The paper establishes knowledge base of expert system by feature modeling and designs inference engine by LSSVM method. Application example shows expert system of LH process can provide processing strategy well.

Acknowledgements The authors would like to acknowledge the support of Industry-Academia Cooperation Innovation Fund Projects of Jiangsu Province (BY2011143), the support of the Special Natural Science Foundation for Innovative Group of Jiangsu University during the course of this work.

References

1. Qi L, Zhang CL. Effect of forming factors on surface temperature and residual deformation of the plate in line heating [J]. International Journal of Materials and Structural Integrity, 2013, 7 (1):171–181.
2. Smits GF, Jordan EM. Improved SVM regression using mixtures of kernels [C]. Proceedings of the 2002 Int Joint Conf on Neural Networks, Honolulu, USA, 2002, 3(3):2785–2790.

Part II
Frontier Computing Workshop

Multi-server Authentication Scheme with Hash Function

Wei-Chen Wu

Abstract In this paper, we propose an efficient remote user authentication scheme for multi-server environments with hash function. Compared with the Lin et al. scheme that takes several modular operations, our scheme achieves results more efficiently yet keeps the merits. The important merits include the following: the scheme can be used in multi-server environments; the system does not need to maintain any verification table; the users who have registered with various servers do not need to remember different login passwords for each. Our scheme can also withstand replay and modification attacks. Therefore, it allows users to choose their passwords freely. Although the Lin et al. scheme proposed a new remote user authentication scheme for multi-server architecture, a security problem exists where the password change protocol is vulnerable to a denial of service attack. Thus, we shall alter the Lin et al. scheme to avoid these problems.

Keywords Password • Authentication • Hash function • Security

1 Introduction

Computer systems and their interconnections via networks have increased the dependence of both organizations and individuals on stored information. This dependence, in turn, has led to a heightened awareness of the need for information security and the protection of data and resources from electronic eavesdropping, electronic fraud, and network-based attacks. Consequently, cryptography and security have matured, leading to the development of improved network security. In 1981, a remote password authentication scheme over an insecure channel was

W.-C. Wu (✉)
Computer Center, Hsin Sheng College of Medical Care and Management,
No.418, Jhongfeng sec 3 Gaoping Rd., Longtan Township,
Taoyuan County, Taiwan
e-mail: wwu@hsc.edu.tw

© Springer Nature Singapore Pte Ltd. 2018

N.Y. Yen and J.C. Hung (eds.), *Frontier Computing*, Lecture Notes
in Electrical Engineering 422, DOI 10.1007/978-981-10-3187-8_75

proposed by Lamport [1]. Since then, several schemes have been proposed to address this problem to achieve more functionality and efficiency [2, 3–12].

In a traditional password scheme, each user has an identity (*ID*) and a secret password (*PW*). If a person wants to log into the network system, he/she must submit his/her *ID* and its corresponding *PW*. Then, through a dictionary of verification tables, a network system authenticates its users. To avoid the plain password table being stored in a public network system, the scheme proposed a dictionary of verification tables to store each user and the corresponding one-way hashing value of passwords in the remote system [5]. Such a verification table has some risks of modification because the passwords are stored in a remote system. Therefore, several schemes have addressed this problem and attempted to improve it [2, 3–12]. These new remote password schemes still do not work well in the multi-server environment. In 2003, Lin et al. proposed a new remote user authentication scheme for a multi-server environment [6]. However, it spends too much time on several modular operations and exists where the password change protocol is vulnerable to a denial of service attack. For this reason, we propose a multi-server environments scheme with hash function to solve the above problems.

The organization of this paper is as follows: In Sect. 2, we analyze the security of the Lin et al.'s scheme. In Sect. 3, we propose a remote user authentication scheme for multi-server environments with hash function. In Sect. 4, we evaluate the security and efficiency of our scheme and make a comparison. In the final section, we state our conclusions.

2 Security of Lin et al.'s Scheme

Lin et al. proposed a new remote user authentication scheme based on simple geometric properties of the Euclidean plane [6]. There are three kinds of participants in this scheme: login users, various servers, and a central manager (*CM*).

The scheme of Lin et al. has a security problem: The password change protocol is vulnerable to a denial of service attack. Suppose the attacker obtains the legal user identity *ID* and chooses a random PW_A as the legal user's old password. Taking PW_A' as the legal user's new password, the steps are as follows:

1. The attacker should type the old password PW_A and the new password PW_A'.
2. The system calculates Q_i from the line *LS* and K_i, where *LS*: $g(K_i) = Q_i$ and then computes (D_i, W_A) from Eq. (2.1) according to *ID* and PW_A as below:

$$\begin{aligned} D_i &= e_i^{ID} \bmod p \\ W_A &= e_i^{PWA} \bmod p \end{aligned} \tag{2.1}$$

3. The system reconstructs the line L_A according to (K_i, Q_i) and (D_i, W_A). Here L_A: $Y = f(X) = a_A X + b_A \bmod p$, where $a_A = (W_A - Q_i)/(D_i - K_i) \bmod p$ and

$b_A = Q_i - (K_i(W_A - Q_i)/(D_i - K_i)) \bmod p$. Thus, L_A: $Y = f(X) = (W_A - Q_i)/(D_i - K_i)X + (Q_i - (K_i(W_A - Q_i)/(D_i - K_i))) \bmod p$.

4. The system can calculate the following value Y_A from L_A: $f(X_i) = Y_A$, where X_i can be calculated through Eq. (2.2).

$$Y_A = f(X_i) = (W_A - Q_i)/(D_i - K_i)X_i + (Q_i - (K_i(W_A - Q_i)/(D_i - K_i))) \bmod p \tag{2.2}$$

5. The system obtains the new point (D_i, W_A') via Eq. (2.2) according to the new password PW_A' and ID as below:

$$D_i = e_i^{ID} \bmod p$$
$$W_A' = e_i^{PWA'} \bmod p$$

6. The new line L_A' is reconstructed by the two points (D_i, W_A') and (X_i, Y_i). Here L_A': $Y = f(X) = a_A'X + b_A' \bmod p$, where $a_A' = (Y_i - W_A')/(X_i - D_i) \bmod p$ and $b_A' = W_A' - (D_i(Y_i - W_A')/(X_i - D_i)) \bmod p$. Thus, L_A': $Y = f(X) = (Y_i - W_A')/(X_i - D_i)X + (W_A' - (D_i(Y_i - W_A')/(X_i - D_i))) \bmod p$

The new intersection point (K_i', Q_i') can be obtained from the lines L_A' and LS as follows:

$$LS: g(X) = a'X + b' \bmod p.$$
$$L_A': Y = f(X) = (Y_i - W_A')/(X_i - D_i)X + \left(W_A' - \left(D_i(Y_i - W_A')/(X_i - D_i)\right)\right) \bmod p$$

Thus,

$$\Rightarrow (Y_i - W_A')/(X_i - D_i)X + ((W_A' - (D_i(Y_i - W_A')/(X_i - D_i))) - b') = 0$$
$$\Rightarrow (Y_i - W_A')/(X_i - D_i)X = (b' - (W_A' - (D_i(Y_i - W_A')/(X_i - D_i))))$$
$$\Rightarrow K_i' = X = (b' - (W_A' - (D_i(Y_i - W_A')/(X_i - D_i))))/(Y_i - W_A')/(X_i - D_i)$$

7. The system updates the value K_i into K_i'. Finally, the attacker finishes the denial of service attack. The concept of the attack is shown in Fig. 1.

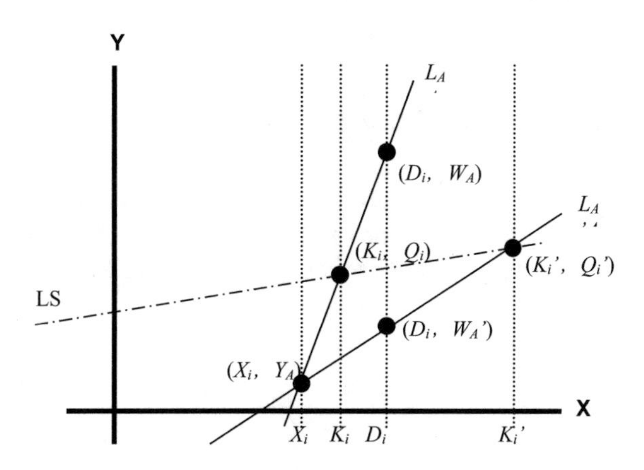

Fig. 1 The concept of the process of the attack

The Lin et al. scheme proposed a remote user authentication scheme with a multi-server environment. However, it spends too much time on several modular operations and has a security problem where the password change protocol is vulnerable to a denial of service attack.

3 Our Proposed Scheme

In this section, we further correct the Lin et al. scheme to avoid the denial of service attack and propose a remote user authentication scheme for multi-server environments with hash function. Our scheme is also divided into four phases: (1) the initialization phase, (2) the registration phase, (3) the login phase, and (4) the authentication phase.

The *CM* denotes the control manager, and $H(\bullet)$ denotes a strong one-way hash function that has the following properties: (1) a condensing property that inputs an arbitrary length message and outputs a fix-length digest message; (2) it is computationally infeasible to find an input from output that can easily compute $H(x) = y$; (3) it is collision-free—given a message x, it is computationally infeasible to find $H(x) = H(x')$ if $x \neq x'$. In the initialization phase, the *CM* sets up each server's secret d_i and in the registration phase, a new user chooses his/her own *ID* and *PW*. In the login phase, when a legitimate user wants to log into the server, he/she must enter the *ID*. Then the server will verify the legitimacy of the login user in the authentication phase. The details of the scheme for multi-server architectures are described as follows:

3.1 The Initialization Phase

Assume that S_m is a set of servers in a multi-server environment, where S_m:{Ser_1, Ser_2, Ser_3...Ser_m}. Suppose that S_n is a set of servers, with which the new user can register, where $S_n \subseteq S_m$. Initially, the *CM* chooses a random secret key d_i for each server Ser_i, where $1 \leq i \leq m$.

3.2 The Registration Phase

When a new user wants to log into the multi-server computer system, the user must register with the server. First, the new user chooses his/her own identity *ID* and password *PW*. The *CM* then performs the following operations *i* times:

1. Compute $X_i = H(ID, d_i)$ and $Y_i = H(PW, d_i)$. (3.1)

2. Compute $q_i = X_i \oplus Y_i \oplus PW$, $D_i = H(ID, q_i)$ and $W_i = H(PW, q_i)$. (3.2)
3. Construct a line L_i according to the two points (X_i, Y_i) and (D_i, W_i). Here L_i: $Y = f(X) = aX + b$, where $a = (W_i - Y_i)/(D_i - X_i)$ and $b = Y_i - X_i ((W_i - Y_i)/(D_i - X_i))$.
4. Choose a line LS randomly, where LS: $g(X) = a'X + b'$, and obtain an intersection point (K_i, Q_i) according to the two lines L_i and LS.

Assume that the server Ser_i provides service for the new user and that SP_i is the service period for each server Ser_i. SP includes the user's identity and the service expiration date. If the server does not provide service for the new user, the service period is zero. Next, the following procedure is performed:

1. The CM calculates $C_i = H(SP_i, d_i)$.
2. The CM delivers the public parameters $\{SP_i, C_i, q_i, K_i, LS\}$ to the registered user. These parameters can be stored in a smart card or other storage devices.

3.3 The Login Phase

In this phase, users are authorized to use multiple servers once they have made their way through the password authentication. Assume that S_n is a set of servers that a registered user wants to log into. The user first keys in his/her ID and PW, and then the system will perform the following steps:

1. Generate a random number r and encrypt the random number $E_{qi \oplus PW}(r)$ with $q_i \oplus PW$.
2. Obtain a time sequence T, which is like a timestamp from the terminal.
3. Compute $A_i = H(r, q_i)$ and $B_i = H(A_i, T)$ according to r and T.
4. Calculate Q_i from the line LS and K_i, where LS: $g(K_i) = Q_i$.
5. Calculate $D_i = H(ID, q_i)$, $W_i = H(PW, q_i)$ and reconstruct the line L_i according to the two points (K_i, Q_i) and (D_i, W_i).
6. Calculate Z_i from the line L_i and B_i, where L_i: $f(B_i) = Z_i$ and send the message M which includes $\{ID, (K_i, Q_i), Z_i, SP_i, C_i, q_i, T, E_{qi \oplus PW}(r)\}$ to Ser_i.

The user can also log into another registered server at the same time.

3.4 The Authentication Phase

In the authentication phase, the server receives the message M at T' and performs the following tasks to authenticate the user's login request.

1. The server shall check whether the time interval between T and T' is greater than ΔT, where ΔT denotes the expected legal time interval for transmission delay between the login terminal and the server Ser*i*. If $(T' - T) \geq \Delta T$, the message M might have been replayed, and the server will reject the login request.
2. Next, the server checks the correctness of *ID*. If the format of *ID* is incorrect, the server will reject the login request.
3. Ser_i checks the validity of the service period SP_i. If the service period has expired, the server will reject the login request. Otherwise, the server computes $C_i' = H(SP_i, d_i)$ and checks whether $C_i' = C_i$. If not, the server rejects the login request.
4. Ser_i decrypts the message $D_{Xi' \oplus Yi'}(E_{qi \oplus PW}(r))$ with $X_i' \oplus Y_i'$ to derive r'.

   Proof:

$$q_i = X_i \oplus Y_i \oplus PW$$

$$\Rightarrow q_i \oplus PW = X_i \oplus Y_i \oplus PW \oplus PW$$

$$\Rightarrow q_i \oplus PW = X_i \oplus Y_i$$

   Thus, $D_{Xi' \oplus Yi'}(E_{qi \oplus PW}(r)) = r$

5. Ser_i calculates $A_i' = H(r', q_i)$ and then computes $B_i' = H(A_i', T)$ according A_i'.
6. Ser_i can reconstruct the original line L_i' according to the two points (K_i, Q_i), (B_i', Z_i) and calculates the point (X_i', Y_i') following $X_i' = H(ID, d_i)$ and L_i': $f(X_i') = Y_i'$. If the point is on the line L_i: $f(X_i') = Y_i'$, the server will accept the login request; otherwise, it will reject the login request.

3.5 To Change the Password

In our proposed scheme, users can choose and change their passwords freely. When a user wants to change his password at server Ser_i, the steps are as follows:

1. The user should type the old password *PW* and the new password *PW'*.
2. The system calculates Q_i from the line *LS* and K_i, where *LS*: $g(K_i) = Q_i$ and computes (D_i, W_i) from Eq. (3.1) according to *ID* and *PW*.
3. The system reconstructs the line L_i according to (K_i, Q_i), (D_i, W_i) and compares the value Y_i from L_i: $f(X_i) = Y_i$ with the value Y_i from Eq. (3.1). If equal, the user will be authenticated and allowed to update the password and continue with the following tasks. Then, the system obtains the new point (D_i', W_i') via Eq. (3.2) according to the new password *PW'*.
4. The new line L_i' is reconstructed by the two points (D_i', W_i') and (X_i, Y_i'). The new intersection point (K_i', Q_i') can be obtained from the lines L_i' and *LS*. Finally, the system updates the value K_i into K_i'. The process is shown in Fig. 2.

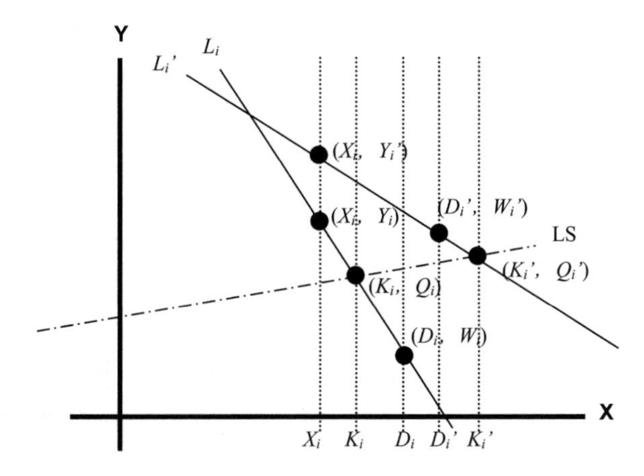

Fig. 2 The concept of the process of changing passwords

4 Analysis

In this section, we will analyze the security of our proposed scheme. We will also evaluate the efficiency of our scheme and make a comparison.

4.1 Security Analysis

In our proposed scheme, the attacker attempts a denial of service attack and changes L_A *into* L_i according to the password change protocol. Because the system can compare the value Y_i from L_i: $f(X_i) = Y_i$ with the value Y_i from Eq. (3.1), and discover it is not equal, the attempt will fail. The simulated steps are as follows:

1. The attacker should type the old password PW_A and the new password PW_A'.
2. The system calculates Q_i from the line LS and K_i, where LS: $g(K_i) = Q_i$ and calculates (D_i, W_A) from Eq. (3.2) according to ID and PW_A as follows:

$$D_i = H(ID, q_i)$$
$$W_A = H(PW_A, q_i)$$

4. The system reconstructs the line L_A according to (K_i, Q_i) and (D_i, W_A). Here L_A: $Y = f(X) = a_A X + b_A$, where a_A and b_A are as below:

$$a_A = (W_A - Q_i)/(D_i - K_i)$$
$$b_A = Q_i - (K_i(W_A - Q_i)/(D_i - K_i))$$

Thus, $L_A{:}Y = f(X) = (W_A - Q_i)/(D_i - K_i)X + (Q_i - (K_i(W_A - Q_i)/(D_i - K_i)))$.

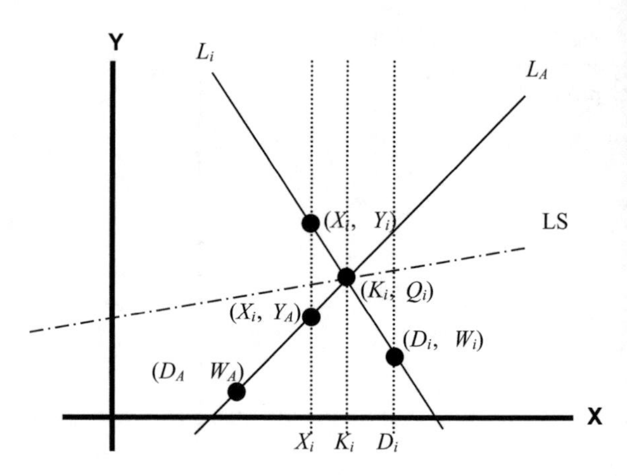

Fig. 3 The concept of the process of the attack

5. The system can compare the following value Y_i from L_i:$f(X_i) = Y_i$ with the value Y_i from Eq. (3.1). It will not be equal, and the user will not be authenticated to update the password. The concept of the process of the attack is shown in Fig. 3.

$$Y_i = (W_A - Q_i)/(D_i - K_i)X_i + (Q_i - (K_i(W_A - Q_i)/(D_i - K_i)))$$

$$\Rightarrow Y_i \neq H(PW, d_i)$$

4.2 Efficiency Analysis

Table 1 gives several comparisons among various methods in the initialization, registration, login, and authentication phases, along with the process to change the password. It is obvious that our scheme is more efficient than the Lin et al. scheme, where *T*( ) is computation time; *r* is a random number; ⊕ is a bitwise exclusive-or operation; *S* is symmetric encryption; *ME* is a modular exponentiation operation; *MM* is a modular multiplication operation; *E* is an exponentiation operation; and *f* is a one-way hash operation.

Furthermore, a symmetric encryption/decryption is at least 100 times faster than an asymmetric encryption/decryption in software. An exponential operation is approximately equal to 60 symmetric en-/decryptions [13, 14]. As listed in Table 1, Lin et al.'s scheme needs nearly 60 symmetric operations in the Initialization Phase (ours takes about zero), 300 symmetric operations in the Registration Phase (ours takes about zero), 240 symmetric operations in the Login/Authentication Phase (ours takes about one operation), and 300 symmetric operations in the Change the Password (ours takes about zero). Moreover, it takes 0.0087 s to finish a symmetric

Table 1 Comparisons of computation costs

	Initialization phase	Registration phase	Login phase	Authentication phase	Change the password
Our proposed scheme	$T(r)$	$5T(f) + T(\oplus) + T(r)$	$4T(f) + T(S) + T(\oplus) + T(r)$	$4T(f) + T(S) + T(\oplus)$	$7T(f) + 2T(\oplus)$
	≈ 0	≈ 0	$\approx T(S)$ ≈ 0.0087 s	$\approx T(S)$ ≈ 0.0087 s	≈ 0
Lin et al.'s scheme [7]	$2T(r) + T(ME)$	$5T(ME) + T(MM) + T(r)$	$4T(ME) + T(r)$	$4T(ME) + 2T(E)$	$5T(ME)$
	$\approx 60\ T(S)$ ≈ 0.522 s	$\approx 300\ T(S)$ ≈ 2.61 s	$\approx 240\ T(S)$ ≈ 2.09 s	$\approx 240\ T(S)$ ≈ 2.09 s	$\approx 300\ T(S)$ ≈ 2.61 s

en-/decryption. We thus ignore the computation load of the $T(r)$, $T(\oplus)$, $T(MM)$, and $T(f)$ because it is lighter than that of a symmetric en-/decryption [15].

5 Conclusion

In this paper, we correct Lin et al.'s scheme to avoid the denial of a service attack. Our proposed remote user authentication scheme for multi-server environments with hash function instead of several modular operations also increases efficiency with respect to cost. In our scheme, servers are not required to have a dictionary of verification tables, and users are allowed to choose their passwords freely. Moreover, it can be used in multi-server environments, and users who have registered with various servers do not need to remember different login passwords for each.

References

1. Lamport, L., Password authentication with insecure communication. *Communications of the ACM 24* (Nov. 1981), 770–772.
2. Chien, H. Y., Jan J. K. and Tseng, Y. M, An efficient and practical solution to remote authentication: Smart card. *Computers and Security*, 21 (4) (2002), 372–375.
3. Hwang, M. S., Lee, C. C. and Tang, Y. L., A simple remote user authentication scheme. *Mathematical and Computer Modelling*, 36 (2002), 103–107.
4. Liaw, H. T., Lin J. F., Wu, W. C., An efficient and complete remote user authentication scheme using smart cards. *Mathematical and Computer Modelling*, 44(1–2) (2006), 223–228.
5. Lennon, R. E., Matyas, S.M. and Mayer, C.H., Cryptographic authentication of time-invariant quantities. *IEEE Trans. on Communications*, 29 (6) (1981), 773–777.
6. Lin, I. C., Hwang, M. S. and Li, L. H., A new remote user authentication scheme for multi-server architecture. *Future Generation Computer Systems*, 19 (1) (Jan. 2003), 13–22.
7. Sun, H. M., An efficient remote user authentication scheme using smart cards. *IEEE Trans. on Consumer Electronics*, 46 (4) (Nov. 2000), 958–961.
8. Sun, H. M., Cryptanalysis of password authentication schemes with smart cards. *Information Security Conference 2001*, (May 2001), 221–223.

9. Wang, S. J. and Chang, J. F., Smart card based secure password authentication scheme. *Computers and Security*, 15 (3) (1996), 231–237.
10. Wu, W. C. and Chen, Y. M., A simple and efficient multi-authentication scheme on space geometry. *The 3rd International Conference on Mechanical and Electrical Technology*, (2011).
11. Wu, W. C., A novel authentication scheme using polynomial for multi-server environments. *International Conference on Computer, Communication, Control and Automation,* (2011).
12. Yang, W. H. and Shieh, S. P., Password authentication schemes with smart cards. *Computers and Security*, 18 (8) (1999), 727–733.
13. Lee J, Chang C., Secure communications for cluster-based ad hoc networks using node identities. *Journal of Network and Computer Applications*, 30(4) (2007), 1377–1396.
14. Schneier B, Applied cryptography: Protocols, algorithms, and source code in C. Second ed., John Wiley and Sons Inc., New York, NY, USA (1996).
15. Chen H, Hsueh S., Light-weight authentication and billing in mobile communications. *IEEE 37th Annual 2003 International Carnahan Conference on Security Technology* (2003), 245–252.

Performance Evaluation of Information Security Risk Identification

Yu-Chih Wei, Wei-Chen Wu and Ya-Chi Chu

Abstract In recent decade, information security becomes a crucial issue on protecting the benefits of business operation. Many organizations perform information security risk management in order to analysis their weakness, and ensure the security of the business processes. However, identifying the threat-vulnerability pairs for each asset during the processes of risk assessment is both difficult and time-consuming for the risk assessor. Furthermore, if the identified results diverged from the real situation, the organization may put emphasis on the unnecessary controls to prevent the non-existing risk. In order to resolve the problem mentioned above, we utilize the data mining approach to discover the relationship between asset and threat-vulnerability pair. And then, we propose a risk recommendation system for assisting user identifying threat and vulnerability. The experiment result shows that the risk recommendation system can improve the performance of efficiency and accuracy of the risk assessment. We also develop a risk assessment system in order to collect the historical selection records and measure the elapsed time for further research.

Keywords Threat · Vulnerability · Risk recommendation · Security

Y.-C. Wei (✉) · Y.-C. Chu
Telecommunication Laboratories, Chunghwa Telecom Co., Ltd,
Taoyuan, Taiwan, ROC
e-mail: vickrey@cht.com.tw

Y.-C. Chu
e-mail: gyh2211@cht.com.tw

W.-C. Wu
Computer Center, Hsin Sheng Junior College of Medical Care and Management,
Taoyuan, Taiwan, ROC
e-mail: wwu@hsc.edu.tw

© Springer Nature Singapore Pte Ltd. 2018

N.Y. Yen and J.C. Hung (eds.), *Frontier Computing*, Lecture Notes
in Electrical Engineering 422, DOI 10.1007/978-981-10-3187-8_76

1 Introduction

In recent years, more and more corporations rely on information technology to assistant them achieving their business goals. However, by the ease of use oriented mentality on system configuration and operation, making their systems more vulnerable and easy to be compromised. Consequently, a systematic approach for information security risk management is necessary to help them identify information security requirements and to create an effective management system. Risk is the effect of uncertainty on objectives, and information security risk often expressed in terms of a combination of the consequences of an information security event and the associated likelihood of occurrence [1]. Risk identification is an important step of the risk assessment, which to determine what could cause a potential loss, and to gain insight into how and why the loss might happen. Thus, if a corporation expects performing risk assessment successfully, finding appropriate threat-vulnerability pair of each asset is a crucial step. However, in the process of identifying threat-vulnerability pair, it is difficult to recognize the feasible combination for the risk assessor, especially for the one who lacks of information security competence.

Without the support of a recommendation system, a risk assessor may encounter at least three challenges: First, in spite of the threat and vulnerability lists are provided as a candidate list for risk assessors, it is still time-consuming for choosing the appropriate one from more than a hundred combinations. Second, the threat-vulnerability pairs may irrational if without considering the root cause discreetly. For example, physical server appliance may have some vulnerability on the lack of physical protection. Theoretically, environmental damage and physical breakage are reasonable threats, however, the mistake may happen when people choose another irrational threat such as "insufficient software testing". Third, not all the users have the ability to find the security issue for the information asset, and may choose non-exist risk finally. Non-exist threat-vulnerability pairs may make organization spend unnecessary time and cost on preventing the risk which may not happen. And this may mislead the manager to neglect the real weaknesses, or invest in improper security measures.

There are number of information risk assessment approaches have been proposed today. These methods of identify threats and vulnerabilities based on the International Organization for Standardization (ISO) such as ISO 31000:2009 [2], ISO/IEC 27001:2013 [3] and ISO/IEC 27005:2011 [1]. Stølen presents a risk assessment model called CORAS [4], it use threat diagram and structured brainstorming to analysis risk. These methods always performed by the leading of experts and may take too much time for the whole process. Some researches identify threats and vulnerabilities according to the security requirements, such as OCTAVE [5] which address on the security requirements of information asset only, but lack of comprehensive consideration. Other mechanism uses business process to complete risk assessment [6]. But in their work, it is hard for common users to depict each asset's risk on their lifecycle. In addition, other researches [7] identify risk by building security ontology. However, it is complicated for users who lack of security concept and also impossible build by their own at all. There still some researches recommend threat or vulnerabilities for users, but it is only suitable for specific domain, such as cloud computing [8].

Due to the deficiencies mentioned above, we propose a recommendation approach for risk identification iteration to resolve the problem mentioned above. After the asset identification step, in this paper, the asset category has been classified. By the use of data mining mechanism, the threat-vulnerability pairs for each category have been learned and provided as the recommendation list. The risk assessor can choose the appropriate pairs correspondence to the real encounter problems from the recommendation list. The main contribution of our proposed approach is to improve the efficiency and accuracy on identifying the threat-vulnerability pairs. In order to evaluate the performance of the efficiency improvement, we first invite information experts for evaluating the accuracy of the threat-vulnerability pairs on the recommendation list. In addition, we design a risk assessment system which can provide the recommendation of threat-vulnerability pairs for the risk assessor, and can measure the elapsed time of the risk assessor's selections.

The remainder of the paper is organized as follows: Sect. 2 describes relevant research on risk assessment and their problems in the past. Section 3 presents our research model which recommends vulnerabilities-threat pair for different categories of asset. Section 4 contains experimental design and results. Conclusions and future directions are given in Sect. 5.

2 Related Work

The international standards on information security risk assessment, such as ISO/IEC 27005:2011 [1] and NIST SP800-30 [9], not only form the basis of general information security risk assessment standard framework, but also developed risk assessment approaches. However, they may not explicit provide suggestions of the potential treat and vulnerability for each asset. In risk identification phase, the treat and vulnerability must be identified by risk assessor with brainstorming, questionnaire or other technical tools. This may take too much time and not intuitive for users.

Other Existing risk management mechanism such as CORAS [4] and OCTAVE [5], also propose their own methods in risk assessment based on standards. In CORAS, a Platform for Risk Analysis of Security Critical Systems is proposed, it uses threat and vulnerability modeling alongside with threat diagrams and structured brainstorming to identify risks. These approaches suggest some common security principles or security best practices. However, they do not determine and evaluate specific security needs of assets to identify their risks. The other approaches, such as OCTAVE, determine the criticality and impact of vulnerabilities from the review of security requirement. It considers the possible condition or situations that can threaten an organization's information assets by existing security checklists, standards or brainstorming. Although OCTAVE uses security requirements, they determine only the impact of vulnerabilities by security requirements or only the requirements of a product; or using asset unspecific security standards.

In addition to the above we mentioned, there are some models use different ways to identify risks. AURUM is an ontology-based method [10]. It supports the decision maker to answer the following questions: Which threats threaten critical assets? Which threat is a multiplier? Which vulnerabilities have to be exploited by a threat to become effective? And it shows the potential threat of selected asset by threat tree. For each threat highly granular vulnerabilities, which a threat could exploit, have been modeled in the ontology. All the threats AURUM provided are recommended from standard, and may lack of flexibility to adapt the new information technology. Furthermore, it is complicated for users to build new ontology depend on themselves. Webb et al. [11] presents a situation aware information security risk management (SA-ISRM) process model that can be used to facilitate improved situation awareness in information security risk management. By intelligence-driven process, it provides accurate, relevant, and complete information in a timely manner to enable quality decision-making. However, the whole risk management process is complicated, it is not easy to consider all the elements without system implementation.

3 Methodology

In this paper, the processes to find the threat-vulnerability pairs for each asset are shown in Fig. 1. We will describe each step in the remainder subsections.

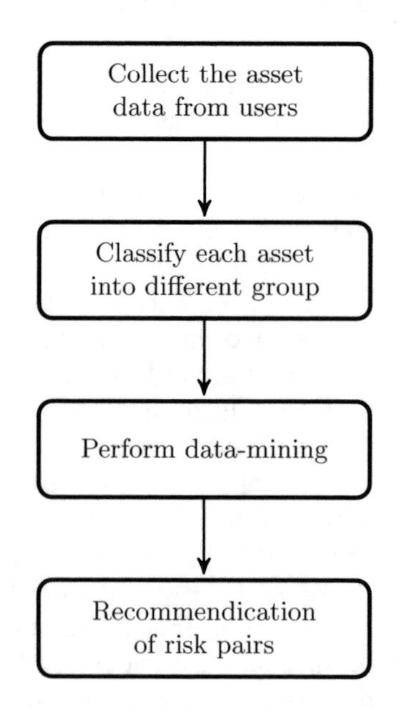

Fig. 1 The procedure of the proposed risk pair recommendation mechanism

3.1 Asset Collection and Classification

In this paper, in order to produce high qualify threat-vulnerability recommendation list, the original data source is collected from the business units in the same organization, which had been certified compliant with ISO/IEC 27001:2013. The threat- and vulnerability- list, which provided for risk assessor choosing, contain hundred of items. All the threats and vulnerabilities are extended from ISO/IEC 27005:2011. We collect many information assets and each asset has several threat-vulnerability pairs.

It is noting that information assets can be classified into number of categories, such as hardware, software, people, etc. [1]. Therefore, we classify the information asset into five categories: hardware, software, people, information and service. In addition,we also classify several groups which functioning similar for each category of information asset. For example, Ubuntu and Open Office are both belong to software category. But, they play different role in system operation. Ubuntu is a kind of operation system software, and Open Office is a kind of package software. Due to the different function of them, we establish operation system group and package software group respectively. Ubuntu belong to the operation system and Open Office belong to package software. In this paper, we create many groups in each category, as shown in Table 1.

Table 1 Illustration of the groups of asset categories

Hardware	Software	Information	People	Service
Web server	Operation system	SOP	System manager	Electric service
Application server	Application program	Installation manual	Network manager	Air conditioning
Database server	Application system	User manual	Security manager	Network service
Log server	Development tool	Operation manual	Help desk	MIS
Personal computer	Package	Planning	Operator	Email service
Laptop	Network tool	Design document	DBA	VPN service
Network device	Compression tool	Testing report	System analyst	LDAP service
End point device	Audit tool	Contract	Quality manager	TeleCom
Printer	Analysis tool	Confidential consent	End point user	Security service
Scanner	Statistics tool	Audit report	Auditor	VPN
Storage equipment	Execution file	System log file	File manager	Maintenance service
Server room	Utility program	Parameter file	Supplier	LDAP
Office	Execution files	Database	Safeguard	Login service
Control room	Self-developed tool	Source code	Administrative	System operation

3.2 *Mining the Recommendation List*

After classifying each information asset into different groups, each group will contain many assets and each asset contains several threat-vulnerability pairs which have been chosen by risk assessor. Then, we perform mining to learn the association between threats and vulnerabilities of each group. In this paper, in order to learn the threat-vulnerability pair in the recommendation list, association rule mining has been chosen for performing the learning task. Association rule mining is a popular and well researched methodology for discovering the interesting relations between variables. A typical and widely-used example of association rule mining is market basket analysis. The Waikato Environment for Knowledge Analysis (Weka) [12] is one of the most popular tool for performing data mining. It provides a general-purpose environment for data preprocessing, classification, regression, clustering, association rules, and visualization. Using Weka can assist users in extracting useful information from data and enable them to easily identify a suitable algorithm for generating an accurate predictive model [13]. In this paper, three association rule mining algorithms including Apriori, Predict Aprori and Tertius had been performed for evaluation, eventually, Predictive Aprori had chosen due to the outstanding results.

It is nothing that Predictive Aprori can generate n best associate rule depends on n selected by user. It combines the standard confidence and support statistics into a single measure called Predictive Accuracy [14]. The *support* is used to measure the accuracy, and the *confidence* can be count by the number of transactions which match the rule. Essentially, the algorithm successively increases the support threshold because the value of predictive accuracy depends on it. Predictive Aprori can mining the potential association rules to fulfill the better performance even though the support value is not big enough.

4 Experiment and Evaluation

The data source of the historical threat-vulnerability selection record is collected from three business units in the same organization including billing operation, system management, and network management, which had been certified compliant with ISO/IEC 27001:2013. It is noting that business unit which certified compliant with ISO/IEC 27001:2013 illustrated that the risk management of the business unit is more mature and completeness than others not been certified.

In order to evaluate the the threat-vulnerability pairs of the recommendation list, firstly, we invite two risk assessment experts to determine the recommendation list whether actually improves the efficiency and accuracy in identifying threat-vulnerability pairs. One of the two experts is major in risk management, and has many experiences in consulting the organization about risk assessment approach and get the ISO/IEC 27001 lead auditor certification. Another one is a professional auditor in ISO/IEC 27001, who with a great deal of maturity. After reviewing the threat-

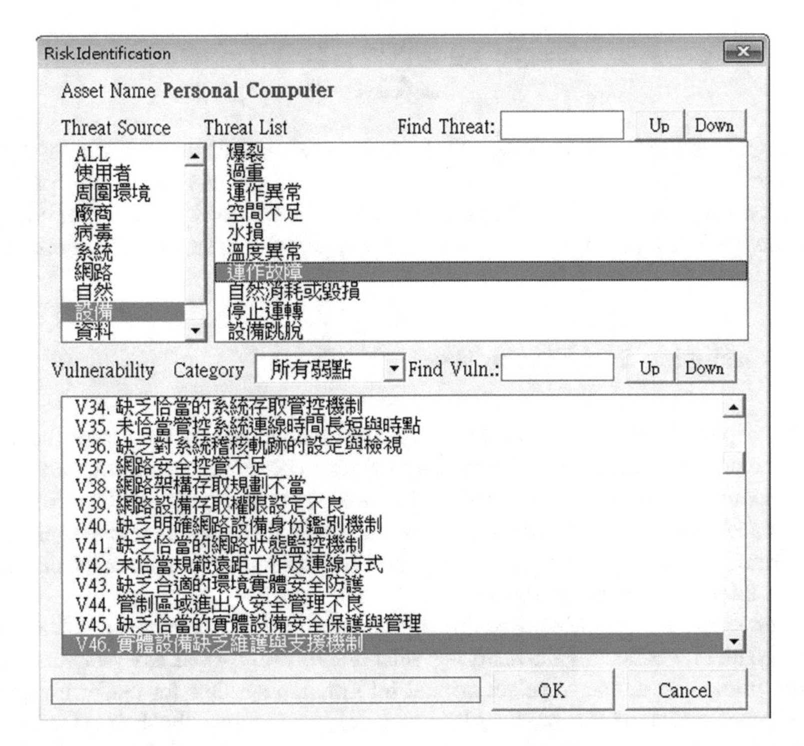

Fig. 2 The risk identification form for selecting threat-vulnerability pair which provided by the risk management system

vulnerability pairs, the domain expert has confirmed that the recommendation list can help risk assessor to filter the appropriate risk pair.

After the accuracy of the threat-vulnerability pairs have been confirmed by domain expert, we then put the list in the risk assessment system. When risk assessor perform risk identification process, a pop-up window, as shown in Fig. 2, provided for risk assessor selection. The threat-vulnerability pairs have been ranked according to the belonging group of the asset. When risk assessor choose the risk identification function which has not ever choose threat-vulnerability pairs, the system first provided and ranked by the accuracy learned by the preditive aprori ming. After that, when risk assessor choose either threat or vulnerability from the selection list, the system automatic provide the recommendation list for increasing the efficiency. If the risk assessor disagrees with the provided recommendation list, they can choose the other suitable risk pair from the list.

In order to evaluate the contribution of the recommendation list, we will select more than one hundred critical information systems for performing risk assessment. Each of the risk assessor's selection elapsed time MT_{sel} has been measured according to Eq. 1. t_s is the timestamps of the decision making, and t_p is the timestamps of the selection providing.

$$MT_{sel} = \sum_{sel \in \{self, rec\}} \frac{t_s - t_p}{|N_{sel}|} \quad (1)$$

For preventing the risk assessors rely on the recommendation list, they will not informed that the system will provide recommended threat-vulnerability pairs or selection suggestion while system training. Furthermore, the system will randomly provide the recommendation threat-vulnerability pair within the opportunity of 50%.

5 Conclusion and Future Work

In this paper, we propose a mechanism that can recommend risk assessor the suitable threat-vulnerability pairs when they identify risk for each information asset. The recommendation list is learned by the use of predictive apriori from the historical selection data of certified business unit. From the result of prior experiment performed by security experts, they confirmed that the recommendation list can help user to filter the appropriate risk item.

In order to evaluate the elapsed time of the selection, we implement a risk assessment system for collecting the historical selection records for risk assessor selection. In the future, we intend to select critical information systems for performing risk assessment, and provide more detail about the elapsed time on the their selection no matter the selection is according to the recommendation list or choosing by them self. Eventually, we can introduce the proposed risk assessment system for supporting business in information security planning.

References

1. Information technology - security techniques - information security risk management. ISO/IEC 27005:2011 pp. 1–68 (June 2011)
2. Risk management – principles and guidelines. ISO 31000:2009 pp. 1–24 (November 2009)
3. Information technology - security techniques - information security management systems – requirements. ISO/IEC 27001:2013 pp. 1–23 (September 2013)
4. Lund, M.S., Solhaug, B., Stølen, K.: Model-Driven Risk Analysis - The CORAS Approach. Springer Berlin Heidelberg (2011), http://dx.doi.org/10.1007/978-3-642-12323-8
5. Alberts, C., Dorofee, A., Stevens, J., Woody, C.: Introduction to the octave approach (2003), https://resources.sei.cmu.edu/asset_files/UsersGuide/2003_012_001_51556.pdf
6. Taubenberger, S., Jürjens, J., Yu, Y., Nuseibeh, B.: Resolving vulnerability identification errors using security requirements on business process models. Information Management & Computer Security 21(3), 202–223 (2013)
7. Ekelhart, A., Fenz, S., Neubauer, T.: Ontology-based decision support for information security risk management. In: Systems, 2009. ICONS '09. Fourth International Conference on. pp. 80–85 (March 2009)
8. Almorsy, M., Grundy, J., Ibrahim, A.S.: Collaboration-based cloud computing security management framework. In: Cloud Computing (CLOUD), 2011 IEEE International Conference on. pp. 364–371 (July 2011)

9. Guide for conducting risk assessments. Tech. rep. (2012), http://dx.doi.org/10.6028/NIST.SP.800-30r1
10. Ekelhart, A., Fenz, S., Neubauer, T.: Aurum: A framework for information security risk management. In: System Sciences, 2009. HICSS '09. 42nd Hawaii International Conference on. pp. 1–10 (Jan 2009)
11. Webb, J., Ahmad, A., Maynard, S.B., Shanks, G.: A situation awareness model for information security risk management. Computers & Security 44, 1 – 15 (2014), http://www.sciencedirect.com/science/article/pii/S0167404814000571
12. Weka 3: Data mining software in java, http://www.cs.waikato.ac.nz/ml/weka/
13. Frank, E., Hall, M., Trigg, L., Holmes, G., Witten, I.H.: Data mining in bioinformatics using weka. Bioinformatics 20(15), 2479–2481 (2004)
14. Frank, E., Hall, M., Holmes, G., Kirkby, R., Pfahringer, B., Witten, I.H., Trigg, L.: Data Mining and Knowledge Discovery Handbook, chap. Weka-A Machine Learning Workbench for Data Mining, pp. 1269–1277. Springer US, Boston, MA (2010), http://dx.doi.org/10.1007/978-0-387-09823-4_66

The Discussion of Potential Care Needs for Handicapped Citizens in Taipei City by Using Spatial Analysis

Jui-Hung Kao, Hsiao-Hsien Rau, Chien-Yeh Hsu, Li-Min Hsu and Chien-Ta Cheng

Abstract Physically and mentally disabled citizens are just like the general population who have independent human rights, but outside assistance is frequently needed to achieve independency due to the disabilities. An independent life for people with disabilities has gradually become the goal of social promotion. Under the circumstances, the support strategies and facilities for the independency of physically and mentally disabled citizens should be focusing on the design and the goal of exploring the areas needing resources of social welfare by introducing spatial analysis. According to the demographic analysis on the population sizes of physically and mentally disabled citizens in Taiwan between 2008 and 2015, the highest percentage of pathogenesis in different counties/cities was 71.95% caused by illness (organ defects), and the proportion and average life expectancy of people

J.-H. Kao (✉)
Graduate Institute of Biomedical Electronics and Bioinformatics, College of Electrical Engineering and Computer Science, National Taiwan University, Taipei, Taiwan
e-mail: kao.jui.hung@gmail.com

J.-H. Kao
Center for GIS RCHSS, Research Center for Humanities and Social Sciences, Academia Sinica, Taipei, Taiwan

H.-H. Rau
Graduate Institute of Biomedical Informatics, Taipei Medical University, Taipei, Taiwan

C.-Y. Hsu
Department of Information Management, National Taipei University of Nursing and Health Science, Taipei, Taiwan

C.-Y. Hsu
Master Program in Global Health and Development, Taipei Medical University, Taipei, Taiwan

L.-M. Hsu
National Taiwan University Hospital, Taipei, Taiwan

L.-M. Hsu
Department of Social Welfare, Taipei City Government, Taipei, Taiwan

C.-T. Cheng
Department of Information Management, Tunghai University, Taichung, Taiwan

© Springer Nature Singapore Pte Ltd. 2018

N.Y. Yen and J.C. Hung (eds.), *Frontier Computing*, Lecture Notes in Electrical Engineering 422, DOI 10.1007/978-981-10-3187-8_77

with disabilities in the individual county/city were significantly correlated. That is, the longer average life expectancy, the higher proportion of physically and mentally disabled citizens. Because the demands on healthcare and service are gradually increasing, the establishment of support measures and facilities has become an important topic at present to achieve of the goal of well life independency. The study used a database of disabled persons published by the Taipei City Government in 2015, which was based on the statistics analyzed by the Department of Social Welfare, Taipei City Government. The description of the distribution of physically and mentally disabled citizens and resources was presented by using geographic informatics, and the coverage areas of supportive institutions were calculated based on the information network of route map combining with the statistics of physically and mentally disabled citizens made by Taipei City Government. Hop spots that cannot be covered were explored by applying the Kriging Method. Furthermore, the adequate candidate locations for setting social welfare institutions to support physically and mentally disabled citizens were calculated based on hot spots, minimal statistical areas as well as weighted population center points. Due to the lack of statistical analysis on physically and mentally disabled citizens, the characteristics of demographics were hard to be clarified. This study emphasized the relationships between physically and mentally disabled citizens and their support institutions as well as the analysis on conformity between demands and the resource of social welfare. The aim was to discuss the impacts of policy implementation on the changes in behavior or survival of physically and mentally disabled citizens.

Keywords Social welfare institution · Physically and mentally disabled citizens · Kriging method · The weighted central spot of population

1 Introduction

In early years, the word “crippled” refers to people with physical loss, also having the meaning of despise. Nowadays, this word has been replaced by “physical disabilities”.

According to the laws and regulations for “Handicapped Citizens Protection Act” legislated by the Legislative Yuan in 1997, people with certain disabilities will receive handicap manuals. However, limbs disability has occupied the largest group and become the most complicated [1].

Recently, the Government and social community have increased the concern for underprivileged groups. In addition to the amendments to laws and orders, the education, medical care and rights for handicapped citizens are far more guaranteed than past [2].

This study tends to use geographic tools to analyze the certain areas of the staff, the distribution of resources, the assessment of demands on welfare promotion or implementation, the circulation pipeline of social welfare and services linked from individual hot spot in order to construct a well-connected welfare service information network for handicapped citizens. Hopefully, handicapped citizens will be benefited by reallocating overall resources and receive better services.

2 Study Area

2.1 Spatial Evolution of Life Among Handicapped Citizens

After the industrial revolution, handicapped people were provided job opportunity based on their skills or special medical treatment if needed. In late nineteenth century, education issue of handicapped youth was solved by implementation of compulsory education. All sorts of institutions for deaf, mute or blind people were gradually established, which became the foundations of special schools [3].

Before the reflection of human rights, institutional settlement was still important measurement for handicapped citizens in the United Kingdom and the United States in early twentieth century. In 1960 s, the discussion about settlements in large institutions gradually reverberated with the ideas and practicing strategies of deinstitutionalization and feedback to the community. In 1980 s, institutional services switched to community-based healthcare, so the tendency toward privatization became the direction of administration in UK and USA. The direction of policy in Taiwan regarding community-based healthcare is similar. The aforementioned experience in communization indicated that many handicapped citizens needed to leave original institutions and "return" to community. But the community is hard to be defined? The American experience told us that the results of deinstitutionalization may inversely increase the risks of turning handicapped citizens into homeless. Or they often lived in a dilapidated community. [4]. The experience in the UK also showed the disadvantage of privatization against users in remote areas. In this case, community-based services may display a great improvement between places [5, 6].

3 Results

3.1 Geographic Information

A map is a tool describing surface landscapes on the earth. However, geography, conventionally focused on natural landscapes, has covered human activities. A novel concept has thereby been transformed into a complex and rich technique and tool gradually by applying technologies accordingly.

The key elements of geographic information system usually refer to computer software, hardware, geographic data, professionals and operation procedures, and the establishment of geographic system often indicates effective data collection, storage, update, process and analysis in order to reveal all kinds of geographic information. In summary, geographic information system is a graphical interface, which can be a single layer or a result stacked from several related layers, to present geographic information in a manner of visualization [7].

This study used several spatial analyses' software to describe the demographics of handicapped citizens in Taipei City. The descriptions of analysis are as follows:

1. Choropleth Map

It is a method that is plotted with gradient colors according to the features, providing an easy way to visualize the change of same attribute data from different administration unit.

2. Symbol Classification Map

It refers to the distribution of spatial data on the spatial system after the addresses of resource/institutions for handicapped citizens are summarized into dot features using spatial analytical software. The area of the symbol represents the attributes of dot features, which is proportional to the changes of attributes.

3. Density Analysis

Density analysis is used to analyze the Kernel Density of resource/institutions for handicapped citizens to observe the clustered/dispersion tendency of institutions.

4. Overlay Analysis

A choropleth map based on demographics of handicapped citizens is combined with a distribution map with the locations of resources for handicapped citizens. Overlay analysis is a novel analysis to create correlation between spatial and attribute data in order to observe whether the distribution pattern of resources in Taipei City is in accordance with required population size.

5. Network Analyst

A Network Analyst can calculate the commands in service areas and thus analyze the service scope of resource spots for scope. In comparison with Buffer Analysis, which sets distances between regions based on the events, Network Analyst can accurately plot the scope defined by the researchers using network.

3.2 Spatial Analysis

1. Moran's I

Spatial Autocorrelation refers to the phenomenon of distributing a previous spatial unit to a place and the similarity between surrounding spatial units. The attributes of spatial distribution on geographic statistics can be divided into "Dispersed", "Random" and "Clustered". "Dispersed" indicates that a geographic phenomenon is separated from similar geographic features, when "clustered" indicates that a geographic phenomenon is assembled with similar geographic features. However, "random" refers to that similar geographic phenomenon does not show any regularities.

I. Global Moran'I

The calculation formula of Global Moran'I is as follows where N is the total number of spatial units indexed by, i is the spatial unit and i and j are the "neighbors" of the spatial unit i.

where

$$I = \frac{n \sum_i \sum_j W_{iJ} (X_i - \overline{X}) (X_j - \overline{X})}{\left(\sum_i \sum_j W_{iJ}\right) \sum_i (X_i - \overline{X})^2} \tag{1}$$

n represents the number of spatial units indexed by and;
X_i, X_j represent the spatial measures of certain attribute between feature i, j;
W_{ij} represents an element of a matrix of spatial weights;
$\overline{X}$: the mean of variable of interest;

There are two definitions of adjacent spatial units: one is defined by distance and another is defined by the connectivity of administrative areas. If it is defined by distance, it means that other spatial units within a certain distance are adjacent spatial units; however, if it is defined by connectivity, it means that the adjacent administrative area (spatial unit) is the neighbor. W is the spatial weight between feature i and j to define the impact in-between. Usually the impact of individual geographic attribute is dispersed; that is, the further away, the less influence. Therefore, common calculation that shows the average measure of all spatial units within that geographic area is the "inverse distance method".

Values of Moran's I range from −1 to +1; and the closer to −1 or +1, the higher autocorrelation is. Negative values indicate negative spatial autocorrelation (tendency to "dispersion") and positive values indicate positive spatial autocorrelation (tendency to "clustered"). The higher the absolute value is, the higher the correlation. A zero value indicates a random spatial pattern (or indicates no spatial autocorrelation).

II. Local Moran'I

The calculation formula of Local Moran'I is as follows where X_i is the measure of spatial unit i, which also is the mean of variable of interest. X_j is the measure of spatial unit j which is next to spatial unit i; W is the spatial weight, and n is the total number of spatial units indexed by.

$$I_i = \frac{X_i - \overline{X}}{\sum_{i=1}^{n} (X_i - \overline{X})^2} \sum_{j=1}^{n} W_{ij} (X_i - \overline{X}) (X_j - \overline{X}) \quad (2)$$

where

n represents the number of spatial units indexed by;
X_i, X_j represent the spatial measures of certain attribute between feature;
W_{ij} represents an element of a matrix of spatial weights;
$\overline{X}$: the mean of variable of interest;

The analysis of Local Moran'I will help us define whether the measure of each spatial unit is correlated with the adjacent spatial unit. If the result is "HH", it indicates that the measures of the spatial unit as well as the measures of adjacent spatial units are very high; if the result is "HL", it indicates that the measure of the spatial unit is very high but the measures of adjacent spatial units are very low. On the contrary, if the result is "LL", it indicates that the measures of the spatial unit as well as the measures of adjacent spatial units are very low; if the result is "LH", it indicates that the measure of the spatial unit is very low but the measures of adjacent spatial units are all very high. "No Significant" indicates that the correlation is non-significant.

2. Kernel Density Estimation

This method is designed to calculate the events within the radius based on the center point of s_1 and s_2 of the core area of k_1 and k_2. Most of the event will be allocated to different event groups to generate density surface, and the radius τ_1 and τ_2 of the core area will affect the accuracy of estimation. This spot pattern analysis is helpful to explore the locations of hot spots lacking social welfare resources. That is, the greater the density is, the greater the size of handicapped citizens needing early intervention.

$$\hat{\lambda}_\tau(s) = \frac{1}{\delta_\tau(s)} \sum_{i=1}^{n} \frac{1}{\tau^2} k\left(\frac{s - s_i}{\tau}\right)$$

where

$\hat{\lambda}_\tau(s)$: means the estimated function based on kernel method and;
τ: means the searching radius;
s: means coordinate vectors;
s_i: means the coordinate vector at location i.

4 Discussion

4.1 Demographic Analysis of Handicapped Citizens in Taipei City

The population density in Taipei City is highest and the population is 2,704,810 people (up to the end of December in 2015). As to the population of handicapped citizens, the total number of 118,596 persons (up to the end of December in 2015) (Fig. 1).

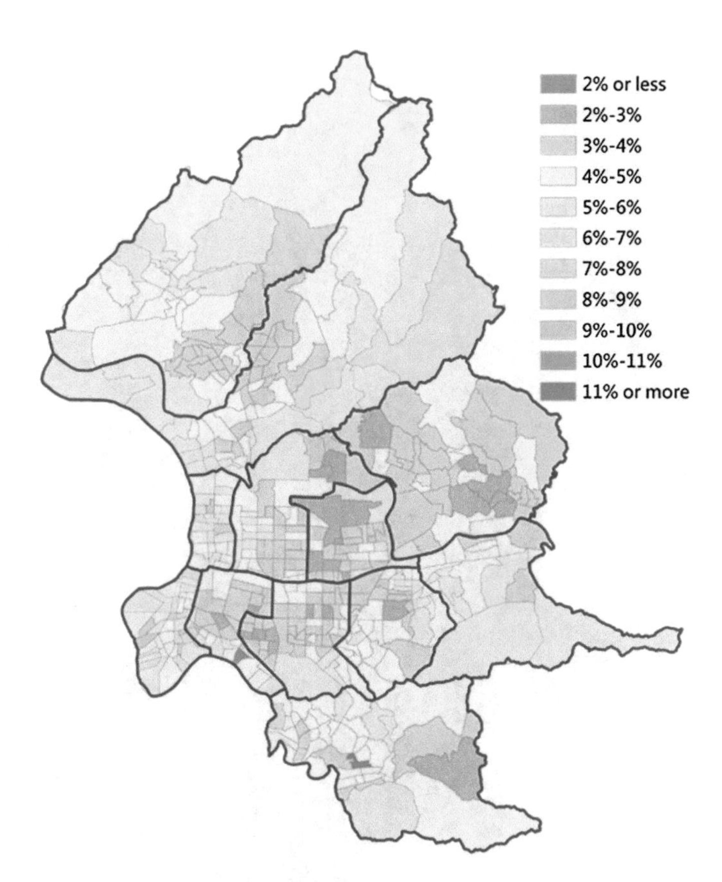

Fig. 1 Distribution of population density of handicapped citizens in Taipei City

4.2 The Analysis of Population Density and Social Welfare Institutions for Handicapped Citizens in Taipei City

Registered institutions in Taipei City are total of 43 (Fig. 2). there are 23 residential institutions (full-day care) (total of 1054 persons, including 952 persons in full-day care and 102 persons in night-staying care) (Fig. 3a); there are 22 day-care institutions (Fig. 3b), including 10 early intervention day-care institutions (1277 persons, including 351 persons receiving early intervention) (Fig. 3c), total service capacities are 2354 person. There are 2 open service institutions for various handicapped citizens (Fig. 3d).

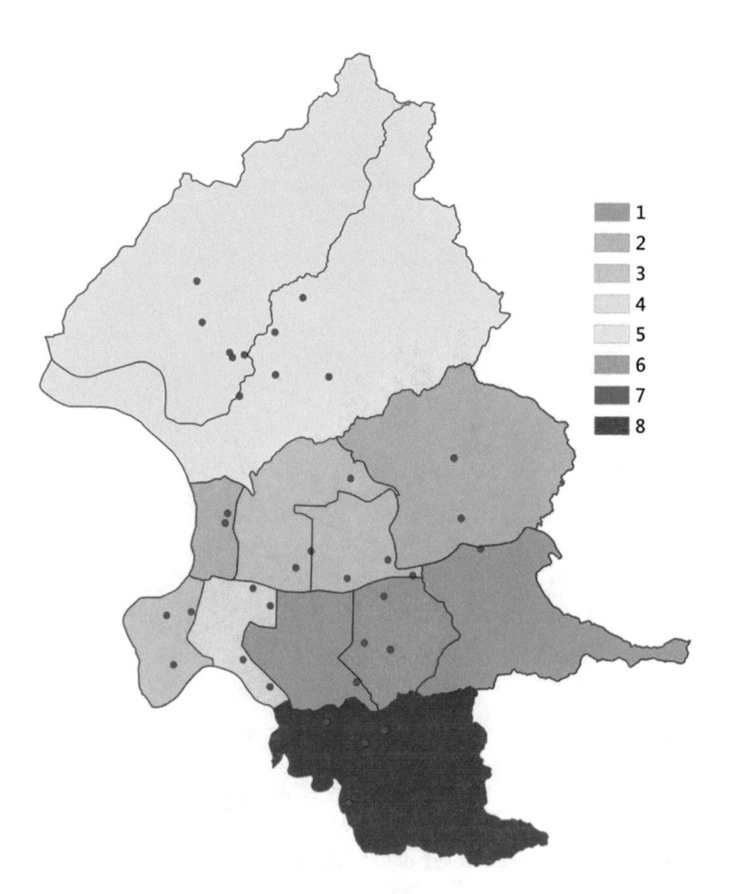

Fig. 2 Distribution of social welfare institutions in Taipei City

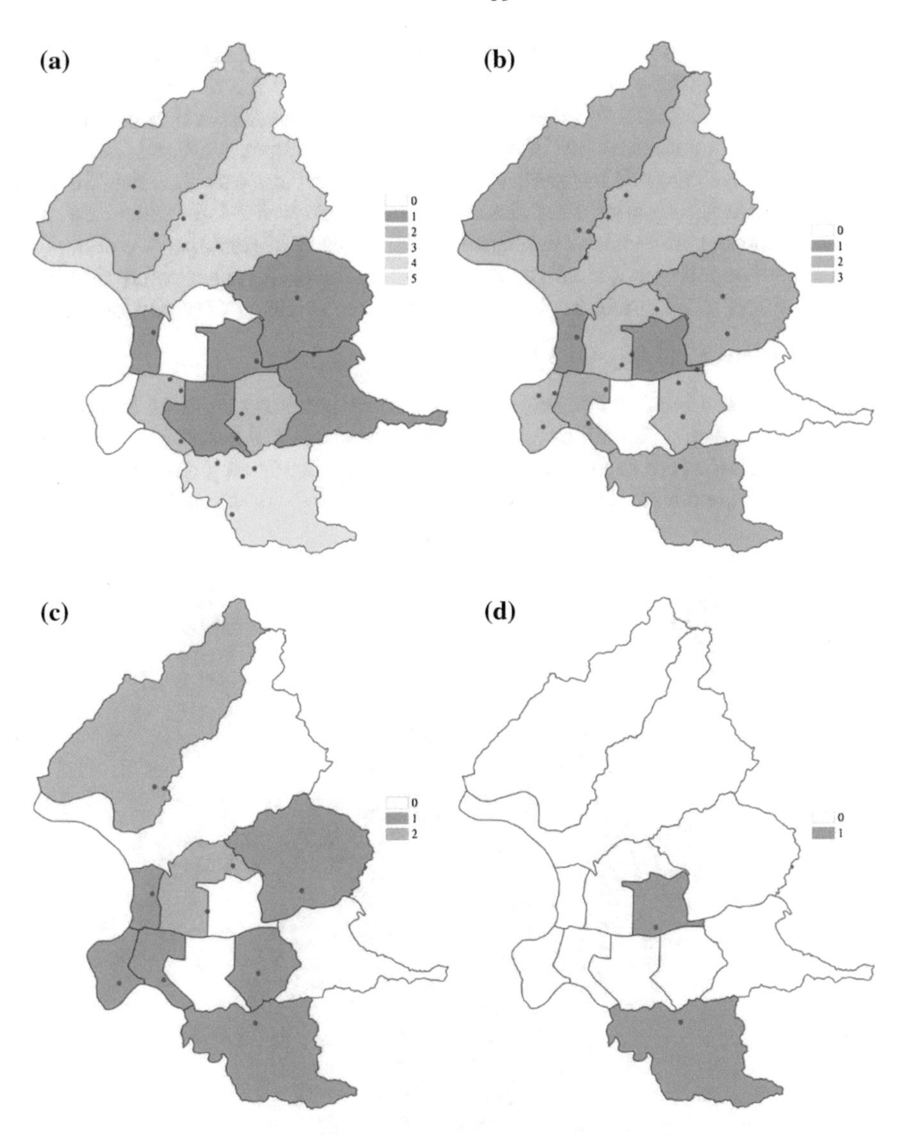

Fig. 3 Distributions of social welfare institutions in Taipei City based on the nature of provided services

4.3 Analysis on Demands of Handicapped Citizens in Taipei City

The resources of social welfare are closely related to the population, and the accessibility of social welfare has great impact on the ease of receiving assistance for people with mental disability or people needing long-term intervention or rehabilitation. Generally speaking, the comfortable pace of walking speed is about 1.22 m/sec [8], thereby it takes approximately 13 min to travel 1 km. As to the accessibility of services and related factors concerning the satisfaction of services, the study results published by Tseng, showed that the distance between residence and institutions to provide the greatest satisfaction of services was "less than 30-min walk" [9], that is, the most ideal distance in this study was 2 km (around 26-min walk). According to the report published by the Ministry of Health and Welfare [8] and the total population of handicapped citizens in Taipei City, the overall disability rate in Taiwan in 2016 will be 3.45% while the rate in Taipei City will be 2.45%.

1. Early Intervention Day-Care Institutions

There are 9 early intervention day-care institutions providing services to 318 persons. Approximately 30.64% of disabled and anthropogeny delayed children in the city have the access to such services. By conducting GeoCoding analysis with the network of route map and 9 early intervention day-care institutions in Taipei City,

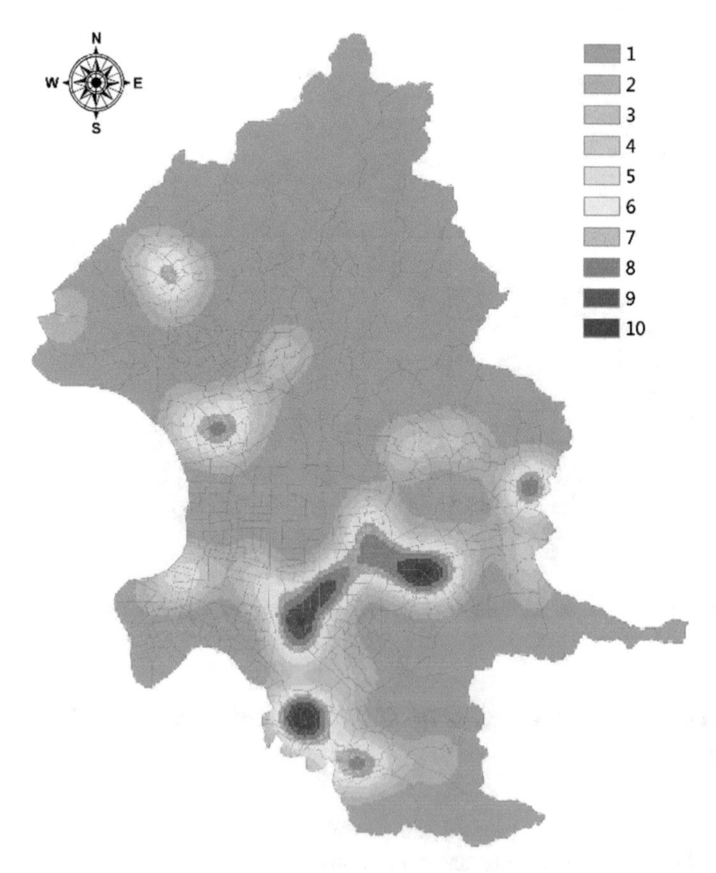

Fig. 4 Diagram of Kernel density regarding early intervention institutions lacking of social welfare resources in Taipei City

the number needing early intervention within 2 km coverage areas of the institutions are 519 people. However, there are 519 people needing such services and not living within the coverage areas, and 201 people living in the areas without sufficient services. Especially in districts Da'an, Wenshan and Xinyi, these are districts with limited resources of early intervention (Fig. 4).

One early intervention institution provides services to 33 persons with hearing disability daily, hence total of 660 persons can receive the service monthly. There are total 13289 persons with hearing disability in Taipei City; 2335 of them have severe disabilities, and 1675 persons with severe hearing disability are offered priority services.

2. Day-Care Institutions

Adult Day Services (ADS) is one of the long-term community-based care services, offering daytime care. The target subjects of day-care institutions are mostly

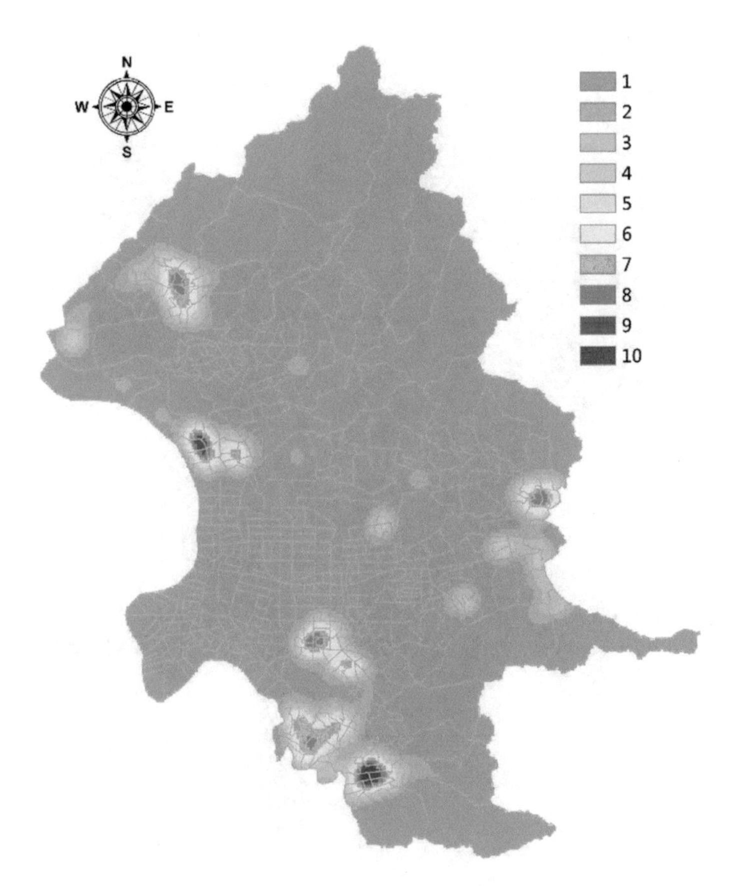

Fig. 5 Diagram of Kernel density regarding day-care institutions lacking of social welfare resources in Taipei City

disabled persons with dementia or disabilities (e.g., limbs disability, mental disability, multi-disability, senile dementia or autism).

With the GeoCoding analysis, the persons needing day-care within 2 km coverage areas of the institutions are 1822 people. However, total of 1875 people who need day-care services are not covered by sufficient service, including 979 persons needing such services are not living within the coverage areas (Fig. 5).

3. Full-Day and Night-Staying Residential Institutions

The main purpose of institutional development is to reduce the difficulties for home-care. "Residential institutions" consists of full-day (24-7 service) and night-staying services. "Placement service" refers to a complete package service provided by the institutions such as preparations before entering the institution, healthcare services during the stay, counseling services before the end of placement and follow-ups after placement.

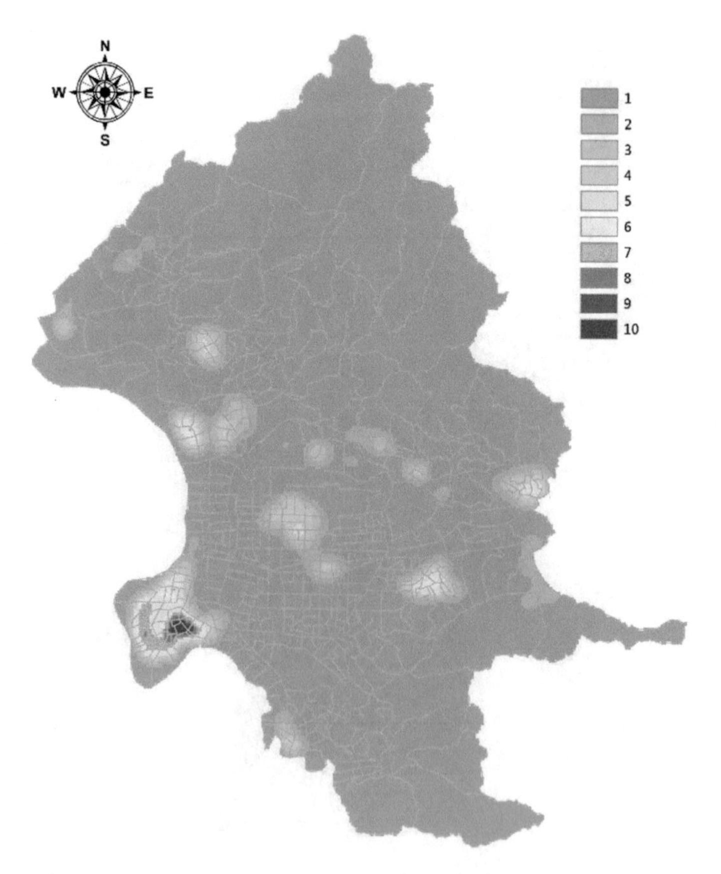

Fig. 6 Diagram of Kernel density regarding residential institutions lacking of social welfare resources in Taipei City

The Taipei City Government lists 15 full-day residential institutions for the services of 835 persons and 5 night-staying institutions for the service of 102 persons. With GeoCoding analysis, the persons needing residential institutions within 2 km coverage areas are 23324 people. However, total of 36123 people in Taipei City who need residential services are not covered, including 13736 persons needing such services are not living within the coverage areas (Fig. 6).

5 Conclusion

Most of the suggestions of welfare services for handicapped citizens are focusing on providing more day-care services. The technology of spatial analysis not only can localize the location of service resources, but also can explore proper locations.

Currently the primary target subjects of day-care services for handicapped citizens in Taipei City are those with chronic psychosis. Therefore, decision making or suggestions for the expansion of hot spots of institutions are based on the population distribution of persons with limb disability. Moreover, the residential conditions of handicapped citizens can be shown by localizing populations. Hypothetically individual spots represent individual psychosis, and the clustered rate can be calculated using density analysis. There are two factors involving density analysis-grid and searching radius. When enlarging the search radius and expanding the areas of the villages, the more significant clustered pattern is as we enlarge the radius of density analysis. On the map, the core areas of village combination patterns are labeled in blue. Hence these areas are the reference locations to search for areas with enriched resources or locations for small-scale community-based day-care institutions. If the institutions are set at dark (core) areas, geographically it is closer to the intensive residential areas of persons with chronic psychosis, and allegedly the services can be provided to more potential target subjects.

If the expansion of institutions is limited, only set institutions with large radius of core areas, so the potential size of target subjects is larger and the time and financial costs of the users on transportation to day-care centers can be reduced. This can also be the example of location planning for other types of handicapped citizens. In addition, the density analysis of all kinds of main disabilities discussed previously can also be the references for setting the locations of institutions. As to comprehensive considerations, the choice of location for institutions is based on the density graph made of the distribution of handicapped citizens.

Spatial strategies can be applied to regional systems to prioritize and support areas lacking of services. In addition, the analysis of individual service area discussed in previous sections have pointed out the centrality of service locations, which may not meet the requirements based on the volume of service and the ideal location of the demand. To overcome these challenges, strategies such as extensively establishment of institutions, changing locations or expanding service items at adjacent points are all welcome. As to functional systems, the planning of new services to meet the demands on future community supportive services or the requirements of new laws are both needed to assure that the service system is able to respond to all demands completely.

References

1. Sun-Shen Yang S-TC, Fung-Wei Liau, Li-Hwa Hsiao. Assessing Quality of Life in Limb Disabled Elderly. Taipei City Medical Journal 2007; 4(3): pp 226–34.
2. Tsai Y-H. The Study of School Adjustment for Students with Intellectual Disability in Universities 2014.
3. Berridge V. Disability and Social Policy in Britain since 1750. The English Historical Review. 2007; 122(499):1432–3.
4. Dear MJ, Wolch JR. Landscapes of despair: From deinstitutionalization to homelessness: Princeton University Press; 2014.
5. Atherton I, Wiles J. Geographies of Care: Space, Place and the Voluntary Sector. 2004.
6. Exworthy M. Geographies of Care: Space, Place and the Voluntary Sector. Health & Social Care in the Community. 2002; 10(5):410–1.
7. Ding Tsu-Jen, Tsai Bo-Wen. A new generation of geographic information system Arc view 9X 2007.
8. Taiwan Ministry of Health and Welfare Address. 2015 December, 照 Long-Term Care-System completion in phases, Access: http://www.mohw.gov.tw/CHT/LTC/DM1_P.aspx?f_list_no=917&fod_list_no=0&doc_no=53040

A Practical Study on the Relation of Out-of-Hospital Cardiac Arrest and Chronic Diseases

Jui-Hung Kao, Ro-Ting Lin, Li-Min Hsu and Tse-Chun Wang

Abstract This study has several segments: reviewing and collecting the OHCA patients with Registry Data of New Taipei city from 2010 to 2011: analyzing and appling International Utestein formula standards for data extraction; using regression analysis to search for the relationship between the variations and OHCA 2 h survival rate; appliing OHCA modeling to identify risk factors coefficient; using Apriori to analize the association rules, and adapting decision tree to create a Risk Decision Model. The correlation chronic diseases: Heart Disease, Diabetes Disease, and Hypertensive Disease have a significant correction on OHCA 2 h survival risk factors. Meanwhile, senior citizens have had high risk could not be revive form standard emergency measure. OHCA Decision tree model helped EMDand rescue units to allocate pre-hospital medical resources and contributes to an accurate clinical quality assessments to assist researchers to analyze medical policy for different OHCA patients.

J.-H. Kao (✉)
Graduate Institute of Biomedical Electronics and Bioinformatics, College of Electrical Engineering and Computer Science, National Taiwan University, Taipei, Taiwan
e-mail: kao.jui.hung@gmail.com

J.-H. Kao · R.-T. Lin
Center for GIS, RCHSS, Research Center for Humanities and Social Sciences, Academia Sinica, Taipei, Taiwan

L.-M. Hsu
National Taiwan University Hospital, Taipei, Taiwan

L.-M. Hsu
Department of Social Welfare, Taipei City Government, Taipei, Taiwan

T.-C. Wang
Department of Information and Communications, Shih Hsin University, Taipei, Taiwan

© Springer Nature Singapore Pte Ltd. 2018

N.Y. Yen and J.C. Hung (eds.), *Frontier Computing*, Lecture Notes in Electrical Engineering 422, DOI 10.1007/978-981-10-3187-8_78

1 Introduction

Since the first ambulance was imported, the members of fire departments in Taiwan have been dedicating themselves to emergency service training. Though all the emergency rescue measures have been improved, the causes and problems of Out-of Hospital-cardiac Arrest (OHCA) caused are still actively being explored by the experts from emergency medicine. The main causes of Out-of-Hospital Cardiac Arrest are classified into two categories: medical causes and surgical causes, the latter mainly caused by car accidents or falls. The top three OHCA medical causes are heart disease, hypertension, and diabetes mellitus.

According to the National Fire Agency, Ministry of Interior, fire departments in Taiwan provide critical services, such as fire prevention, disaster prevention, carbon monoxide poisoning prevention, drowning prevention, disaster prevention and protection act. The Ministry of Interior composes eight strategies which involves many different types of people and agencies including fire services act, disaster management, disaster prevention promotion and protection, command communication system improvement, effective integrate civilian resources, emergency medical services, civilian disaster prevention education promotion, implement self-Management of fire safety to reach the goal of providing a safe environment for cilivians. OCHA patients are categorized under emergency medical services (EMS), and ambulances are the responsibilities of the fire departments.

Emergency medical services (EMS) are usually activated by immediate response to bystanders' emergency calls. Followed by dispatching relevant medical personnel: Emergency Medical Technicians (EMT), who deliver the initial treatments such as Advance Life Support (ALS) and Basic Life Support (BLS) based on the seriousness of individual case. Emergency Medical Technicians are also divided into three different levels, from basic levels to advanced levels: Emergency Medical Technician-1, Emergency Medical Technician-2, and Emergency Medical Technician-Paramedic. Each level operates different competences while providing emergency medical services. According to rule 18, Regulations of Emergency Medical Services (2013), there should be at least two EMTs in the ambulance while applying emergency medical services or sending patients to hospitals. Intensive ambulance requires two EMTs, one of whom should be a doctor, medical staff or in the position of EMT-2.

It was mentioned on the boolklet of Taipei City Fire Department, the common misleading idea in terms of EMS was Emergency Medical Technician-Paramedics (EMT-P) were supposed to be dispatched to every emergency medical case. With very limitedworking force, the idea was considered unrealistic. According to Cai's study (2009), the levels and numbers of Emergency Medical Technician would not affect the outcomes of ROSC (Return Of Spontaneous Circulation), since most ROSC care were integrated shorter onsite treatment time and early cardiopulmonary resuscitation (CPR). Sending EMT-P should have been regarding on the life threatening situations. Accurate medical technician dispatch not only saved life's but also limited medical resource waste. After Intelligent Computer Aided Dispatch

System was activated, ALS dispatch rate, ALS excessive usage rate, BLS dispatch rate were sharply raised. However, ALS idling rate raised sharply, as well. The results suggested further adjustments were still needed.

2 Related Works

According to Wu and Yus' [1] study, nationwide average annual incidence rate (cases per 100,000 population) of OHCA was 73 and male incidence rate was 2.5 times higher than female's. In Taipei City, there were 53 Out-of Hospital-cardiac Arrest arrests in every 1000,000 population. Average male annual incidence rate is twice higher than female's. The data above was not included the numbers without using emergency medical services or the cardiac sudden deaths. It was highly possible the actual number would be much higher. This has become drawn quite attention in the epidemiology area.

According to the data gathered from Ministry of Interior [2], the total population in Taiwan 2012 was 23,315,822, increased 90,910 in numbers in comparison with year 2011; Total number in death in 2012 was 155,239 (male 61%, female 39%), slightly increased 2,033 people in comparison with year 2011. Among all the seasons of the year, Out-of Hospital-cardiac Arrest accrued most frequently in winter time [3, 4]. The peak periods were on Mondays, Fridays and Saturdays, from 6 a.m. to 10 a.m. and 5 p.m. to 8 p.m. Particulate matter at high conservation was positively associated with Out-of Hospital-cardiac Arrest, particularly on the male patients aged between 40 and 74 [4].

Boller [5] stated 80% of adult Out-of Hospital-cardiac Arrest patients and 55% of In Hospital Cardiac Arrests (IHCA) adult patients failed return to spontaneous circulation (ROSC) after receiving CPR. This result probably was associated with the fact a lot of patients who experienced Out-of Hospital-cardiac Arrest had a pre-medical history of heart disease, hypertension, and diabetes. According to the data from the Ministry of Health and Welfare [2], heart disease was the top two, diabetes was the top four and hypertension was the top eight causes to fatality in New Taipei City in 2012. Further explanations will be written in the next paragraph.

3 Methodology

In the studies the related to cardiac arrest and CPR, in particular in the area of epidemiological studies, each difference of the relevant variables definition has caused the different report results, in which contained important prognostic indicators, such as recovery from circulation and survival rates.

ZP Wu and CY Yu [6] noted in their research the so-called post-diseased patients' prognosis referred to the expected results and the probability that might occur. In other words, the prognosis had some risk factors: patients might have sequelas, the possible progress or the possibilities of death.

Søholm H [7] published the research on OHCA patients' improving survival situations in higher comparative Copenhagen area hospitals and non-tertiary hospitals. The researcher pointed out that pre-hospital features such as age, initial rhythm, and period from the alarm to the emergency operations units' arrival, early defibrillation, witness cardiac arrest, and bystander cardiopulmonary resuscitation (CPR) were known to be important predictors of outcome factors. The results showed that entering higher hospital could decrease OHCA patients' mortality rate and the care levels for the recovery are likely to be important prognosis factors.

The OHCA data this study used was based on the international standard UteStein formula, which enabled us to evaluate and compare national, regional and hospital-based emergency medical services. In New Taipei City, from January 1, 2010 to December 31, 2011, a total number of 2,788 stroke cases; in 2011, a total of 2,778 cases, and 5,566 cases of the total number were gathered for this research. The researchers used international Utstein format classification, chose 4,681 underlying patients with potential heart disease as the study samples.

4 Discussion and Result

Through Preliminary exploration of linear regression analysis, the researches tried to understand these variations from the relationship between the risk factors and extraction of OHCA, modeling OHCA between 2 h survival risk factors, and logistic regression to identify risk factors from OHCA factors. After that, the researchers used decision tree classification algorithm, to analyze OHCA multivariate, to explain the survival rate with risk factors set forth under the joint effect of these variables in the model.

Linear Regression Analysis is a feature selection. Schuller et al. [8] pointed out the Linear Regression was affective selection with high overall accuracy. This study adopted Linear Regression Analysis to select eigenvalues and got fourteen significant factors related to OHCA 2 h surviving.

Applying Linear Regression Analysis was used to find out OHCA potential factors, and the purpose of using Logistic Regression Analysis was to identify risk factors. Zhou et al. [9] adopted Logistic Regression to identify postoperative complications and risk factors from the patients with gastric cancer resections. While using regression models to analyze the data; if variables as categorical variables, then Logistic Regression should be used to process the data.

Apriori algorithm was mainly for researchers to seek the association rules in the backup project through frequent patterns. Then they calculated the probability of each backup item appearance. After that, the researchers would measure the appropriateness of the backup items as association rules according to the set support and confidence. Based on the value of support, researchers would try to find the frequent item sets to conduct the associated analysis among diseases, to suggest the degree of support and reliability of patients suffering from diseases A and B concurrently, and then predict the further hidden risks and diseases as well. Apriori algorithm used frequent patterns to look for the relationship appearing in the data. Han et al. [10] measured the effectiveness and uncertainty through support and confidence.

While using CART data to analyze data, the data attributes would be classified into two different groups: continuous and discontinuous. Continuous data could be used to predict and estimate weight values into variables, and discontinuous data could be applied to the categories of identification. When processing continuous-type data, algorithms, Regression Tree was frequently being used; while checking categorical data, algorithms, and researchers attended to use Classification tree. Win et al. [11] adopted CART through vital signs and laboratory data to set up Classification and Regression Tree for patients with Acute Decompensated Heart Failure (ADHF) in order to monitor the different levels of hospital mortality risks. The Classification and Regression Tree provided an effective risk factor model for the clinicians.

From January 2010 until December 2011, a total of 5,566 out-of-hospital cardiac arrest patients were recorded onto the registration of New Taipei City Fire Department Emergency statistical summary table. The pre-hospital cardiac arrest, using Utstein Style of the total number of the resulting matrix was 5,566 cases, and none of the cases was abandoned by an attempt to aid. Among all the patients with heart disease in the previous medical history, there were 3,729 cases arrest witnessed, 947 cases arrest not witnessed; while the ambulance crew witnessed five cases from the initial rhythm was VF accounted for 335 cases. The initial rhythm of VT was accounted for 18 cases, and the initial rhythm of ASY as the majority, total 2,109 cases. Other rhythms were 1,267 cases: 65 cases of circulatory system recovered and 2,274 cases of circulatory system didn't recovered. 42 cases of patients who recovered from their circulatory system were sent to the intensive care unit and 54 cases were sent to the emergency room. 22 cases were sent to the intensive care unit and were alive for 24 h. In the end, only 20 cases were alive and were able to get out of the hospital.

In this study, linear regression was used to stepwise for regression variables. The result was found from frequency allocation: OHCA mostly happened to patients over 65 years old. Most of the OHCA cases were with internal diseases and most of the OHCA patients were sent to first stage emergency hospital without referral. ROSC rate was about 2.5 percent, off which happened in the most pre-hospital ROSC. EMT-P was sent most of the time. However, most of the CPR oxygen compressions ratio was incorrect. Most patients who had AED were given an electric shock appropriately. Their initial heart rate and heart rate coding before AED Systole cardiac arrest. The patients who had a previous medical history of great suffering chronic; the emergency service time from arriving to leaving the scene was between 6 and 12 min.

In this study, Logistic regression for survival OHCA occurrence of risk factors and a 2-h analysis were found that the risk of death of the OHCA patients age 65 increased to 84.4% comparing to the OHCA patients less then age 65. Sending the patients to a third stage emergency hospital, the survival rate was 15.75 times to sending the patients to non-emergency responsibility hospital; 2.4 times of sending the patients to a first stage emergency responsibility hospital emergency and 1.3 times to sending the patients to a second emergency responsibilities hospital. When Rhythm encoded without using AED defibrillation shock (VT/VF) increased 26.1% the risk of mortality. Surgical OHAC patients in comparison with medical OHA patients, the formers' death rate increased 67.7%. And the survival rate of receiving EMT-P emergence care is 1.25 times higher than receiving the emergency care of general EMTs.

The study found that 62.53% of the patients with a medical history of OHCA in the past suffering from one or more chronic diseases and had a significant impact on survival rate (P-value = 0.001). This study used Apriori Analysis to seek connections among the samples: All of the 2927 examples with at least one chronic disease. The researchers set the minimum Support as 20%; minimum, Confidence as 60%. They received seven connections: five issues connected with high blood pressure, 2 connected with diabetes. Through Utstein style, all the chronic diseases were related to Diabetes Disease and Hypertensive Disease. The model results of this stud in the study was conducted by Classification and Regression Tree algorithm from 4681 cases and logistic regression aiming at risk factor for the survival of the 2 h of occurrence of OHCA. The model results are shown as below.

R01:	IF{AED Initial Heartbeat Coding = 1 and First Aid Capabilities Hospital = 3 and Sending EMT − P = 1}THEN{Colon = 1}
R02:	IF{AED Initial Heartbeat Coding = 1 and First Aid Capabilities Hospital = 3 and Sending EMT − P = 0}THEN{Colon = 1}
R03:	IF{AED Initial Heartbeat Coding = 1 and First Aid Capabilities Hospital ≤ 2 and OHCA Type = 2}THEN{Colon = 1}
R04:	IF{AED Initial Heartbeat Coding = 1 and First Aid Capabilities Hospital ≤ 2 and OHCA Type = 1}THEN{Colon = 1}
R05:	IF{(AED Initial Heartbeat Coding = 0)and(Age ≤ 65) and(OHCA Type = 2)THEN{Colon = 1}
R06:	IF{(AED Initial Heartbeat Coding = 0)and(Age ≤ 65) and(OHCA Type = 1)}THEN{Colon = 1}
R07:	IF{(AED Initial Heartbeat Coding = 0)and(Age > 65) and(Sending EMT − P = 1)THEN{Colon = 1}
R08:	IF{(AED Initial Heartbeat Coding = 0)and(Age > 65) and(Sending EMT − P = 0)THEN{Colon = 1}
R09:	IF{(AED Initial Heartbeat Coding = 2)and(Age ≤ 65) and(Ventilation Ratios for CPR = 2)THEN{Colon = 1}
R10:	IF{(AED Initial Heartbeat Coding = 2)and(Age ≤ 65) and(Ventilation Ratios for CPR = 1)THEN{Colon = 1}
11:	IF{(AED Initial Heartbeat Coding = 2) and(Age > 65)THEN{Colon = 1}
R12:	IF{(AED Initial Heartbeat Coding = 2) and(Age > 65)Else{Colon = 0}

Through six risk factors, Classification and Regression Tree went through the resulting prediction mode and received eleven rules. Take Rule number one (R01) as an example, OHCA patients' AED Initial Heartbeat Coding was Asystole/PEA, and were sent to First-Rate Capabilities Hospital. After receiving ENT-P, the survival rate was 28.05% (Colon = 1); Rule number two (R02) indicated OHCA patients' AED Initial Heartbeat Coding Asystole/PEA and were sent to First-Rate Capabilities Hospital without ENT-P. Their survival rate was 20.9% (Colon = 1); Rule number three (R03) showed that OHCA surgical patients' AED Initial Heartbeat Coding was Asystole/PEA and were sent to no First-Rate Capabilities Hospital. Their survival rate was 12.1% (Colon = 1); Rule number four (R04) indicated OHCA medical patients' AED Initial Heartbeat Coding were Asystole/PEA, and patients were sent to no First-Rate Capabilities Hospital, their survival rate was 20.4% (Colon = 1); Rule number five(R05)demonstrated surgical OHCA patients aged under sixty–five had no AED Initial Heartbeat Coding, their survival rate was 22.0% (Colon = 1); Rule number six(R06)indicated OHCA medical patients aged under sixty-five had no AED Initial Heartbeat Coding, survival rate was 33.1% (Colon = 1); Rule number seven(R07)showed that OHCA patient aged over sixty-five and had no AED Initial Heartbeat Coding, received ENT-P emergency services,the survival rate was 26.9% (Colon = 1); Rule number eight (R08) proposed that OHCA patients aged over sixty-five and had no AED Initial Heartbeat Coding, received ENT-P emergency services and their survival rate was 18.3% (Colon = 1); Rule number nine (R09) included an appearance of OHCA patients aged under sixty-five and their AED Initial Heartbeat Coding was VT/VF. However, with the incorrect CPR circulation oxygen ratio, the survival rate was 35.9% (Colon = 1); Rule number ten (R10) explored aged under sixty-five OHCA patients' AED determine presence was VT/VF, with correct CPR circulation oxygen ratio, their survival rate was 50.7% (Colon = 1). Rule number eleven (R11) checked assumptions of aged over sixty-five OHCA patients' AED determine presence was VT/VF and their survival rate was 28.3% (Colon = 1).

5 Conclusion

Prognosis refers to the possibility of an expected result accruing once a patient has been diagnosed. In other words, prognosis refers to a patient's possible complications and percentage of recovery or death under certain risk factors. Risk factors are the variable that affect a patient having a good or bad prognosis. With the use of Classification and Regression Tree, four main risk factors that affect an OHCA patients' survival rate were yieled. First, age was a determining factor. Second, patients with surgical causes were exposed in higher risks than the patients suffered from medical causes. Third, survival rates would have been higher if the OHCA patient were delivered sent to mid or high stage emergency hospitals. Fourth, sending EMT-P, which would provide more rapid evacuation, more supports of the respiratory tract, ventilation maintaining, administered intravenous fluids, medications, and more effective CPR and other first aid medical care as professionals also increased the survival rates of OHCA patients (25.9% vs.21.3%.) The key elements of better survival rate from EMT-P are providing patients Logistic regression extraction risk factors used in the decision in the study are: Age, OHCA Type, First Aid Capabilities Hospital, Sending EMT-P, AED Initial Heartbeat Coding, Ventilating Ratios for CPR. OHCA patients can be layered for the three groups, which help first aid medical dispatchers (Emergency medical dispatcher, EMD) allocate OHCA patients in the pre-hospital medical resources. This simple model can be a practical to providing rescue units, and contribute to clinical quality assessments to assist researchers to analyze the medical policy of different OHCA patients and enhance the survival rate of the OHCA patients.

References

1. W. Zong-ping and Y. Chi-Yu, “A Study on Resuscitative Outcome of Out-of-Hospital Cardiac Arrest Patients-The Implementation Experience of the Fire Bureau of Taipei County,” *Journal of Crisis Management,* vol. 8, pp. 9–18, 2011.
2. M. o. H. a. Welfare. (2013). *Department of Medical Affairs-Emergency Medical Services*. Available: http://www.mohw.gov.tw/CHT/DOMA/DM1.aspx?f_list_no=935&fod_list_no=5772.
3. C. Ming-Tai, “An Evaluation of Emergency Medical Dispatch System in Taipei,” Master of Public Health Degree Program College of Public Health, National Taiwan University, Master Thesis, 2011.
4. Y. Zui-Shen, “Association between Particulate Matter and the Incidence of Out-of-Hospital Cardiac Arrest,” Institute of Epidemiology and Preventive Medicine, National Taiwan University, Doctoral Dissertation, 2012.
5. B. Manuel, “Will models of naturally occurring disease in animals reduce the bench-to-bedside gap in biomedical research?,” *Zhonghua wei zhong bing ji jiu yi xue,* vol. 25, pp. 5–7, 2013.
6. Z.P. Wu CYY. A Study on Resuscitative Outcome of Out-of-Hospital Cardiac Arrest Patients-The Implementation Experience of the Fire Bureau of Taipei County. Journal of Crisis Management. 2011; Vol. 8(No. 1): pp. 9–18.
7. Søholm H, Wachtell K, Nielsen SL, Bro-Jeppesen J, Pedersen F, Wanscher M, et al. Tertiary centres have improved survival compared to other hospitals in the Copenhagen area after out-of-hospital cardiac arrest. Resuscitation. 2012.
8. Schuller B, Friedmann F, Eyben F, editors. The Munich Biovoice Corpus: Effects of physical exercising, heart rate, and skin conductance on human speech production. Language Resources and Evaluation Conference; 2014.
9. Zhou J, Zhou Y, Cao S, Li S, Wang H, Niu Z, et al. Multivariate logistic regression analysis of postoperative complications and risk model establishment of gastrectomy for gastric cancer: A single-center cohort report. Scandinavian journal of gastroenterology. 2015; 51(1):8–15.
10. Han J, Kamber M, Pei J. Data mining: concepts and techniques: concepts and techniques: Elsevier; 2011.
11. Win S, Hussain I, Hebl V, Redfield MM. Mortality and Readmissions After Heart Failure Hospitalization in a Community Based Cohort: Estimating Risk Using the Acute Decompensated Heart Failure National Registry (ADHERE) Classification and Regression Tree (CART) Algorithm. Circulation. 2015; 132(Suppl 3):A12302-A.

Bring Independent Practice Technology to After-School: Preliminary Results from the Mixed Roles of Teachers and Parents Study to Explore Children Effects in Chinese Characters Learning

Chi Nung Chu

Abstract This paper discusses the efficacy of independent practice technology on Chinese characters learning for the children with parents absent after school. The self-generated practicing with the design of Chinese Characters Practicing Platform incorporates two modules in Learning Materials Editor and Multiple Chinese Input Methods Practicing Engine. The Chinese Characters Practicing Platform provides children with the opportunities of independent review and preview. This implementation of independent practice technology after school shows children have developed new skills and increased interest in Chinese characters learning, and increased engagement in school.

Keywords Chinese Characters Practicing Platform · Information technology · After-school learning

1 Introduction

Word identification is the essential skill to read that correlates reading comprehension [3, 8, 10]. It is well recognized that difficulties in automatic word recognition significantly affect a reader's ability to effectively comprehend what they are reading [9, 11]. Automatic reading involves the development of strong orthographic representations, which allows fast and accurate identification of whole words made up of specific letter patterns. It is important that prior to the stage where children read orthographically, they apply alphabetic strategies to analyze words [6]. However written Chinese is a logographic orthography that differs greatly from alphabetic writing systems [7]. Having the adequate skill to identify words does not

C.N. Chu (✉)
Department of Management of Information System, China University of Technology, No. 56, Sec. 3, Shinglung Rd., Wenshan Chiu, Taipei 116, Taiwan ROC
e-mail: nung@cute.edu.tw

© Springer Nature Singapore Pte Ltd. 2018

N.Y. Yen and J.C. Hung (eds.), *Frontier Computing*, Lecture Notes in Electrical Engineering 422, DOI 10.1007/978-981-10-3187-8_79

mean the one has the ability of reading comprehension, but lacking of adequate skill to identify words does mean the one will certainly be unable actually to complete the reading comprehension task.

The adoption of information technology to after-school learning creates a great opportunity to advance school works [2, 5, 12]. The digital learning is believed by employing technologies that can bring opportunities for more effective learning to scale [1, 4].

The shift means that students will be practicing by themselves. That means teachers and parents will be thinking about learning work product than test scores. These practices will result in better opportunity for low-income families, single-parent families and of families with two parents working outside the home that have grown dramatically in the last few decades [13]. There is a growing recognition of the good that after-school learning can do for children, but it is also a direct result of fundamental changes that have taken place in the way Taiwan families work and live.

2 System Architecture

The design of Chinese Characters Practicing Platform (CCPP) (Fig. 1) which provides the practices of letter-to-sound correspondences services with Chinese characters learning was developed by the Visual Basic 6.0. It consists of two modules in Learning Materials Editor and Multiple Chinese Input Methods Practicing Engine. The module of Learning Materials Editor provides the teachers/parents with an interface to build up the database of learning materials for children after-school practice (Fig. 2). With the functions of timer, the statistics in Chinese character entry rate and correctness rate, the CCPP could show users with the practicing results associated with the children (Fig. 3).

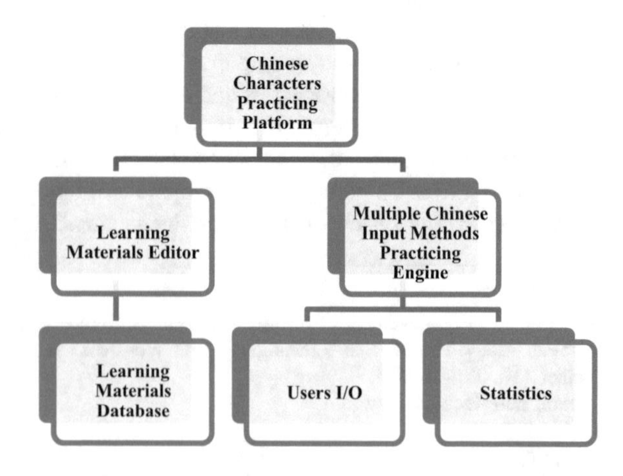

Fig. 1 Framework of Chinese Characters Practicing Platform

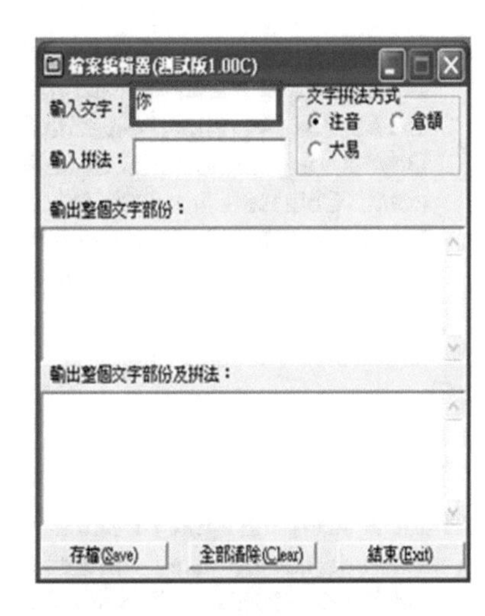

Fig. 2 Module of Learning Materials Editor

Fig. 3 Module of Multiple Chinese Input Methods Practicing Engines

3 Benefits Evaluation

This Chinese Characters Practicing Platform was designed as a support for children in the learning process of Chinese characters after school in Taipei. There are three anticipated after-school practicing effects in Chinese characters as the follows:

1. Encouraging the teachers and parents preparing personalized Chinese character learning materials. The manipulation interfaces of Multiple Chinese Input Methods Practicing Engine are simplified to reduce the learning curve for the expertise users.

2. Providing easier use of independent practicing environment. As using Multiple Chinese Input Methods Practicing Engine, no additional on-site teachers/parents are needed in comparison with traditional schooling.
3. Developing the potential of children with parents absent after school for practicing Chinese characters independently. The Chinese Characters Practicing Platform would facilitate children in the learning process of words, phrases, sentences and even paragraph.

4 Conclusion

Children need the kind of extras after school that advantaged children take for granted but that too many children don't get as learning alone. This study aimed at enhancing children's academic achievement in Chinese character learning with the design of CCPP. The results show that the effects of implementing information technology facilitate children independent practicing. Children do show increased interest and ability in Chinese character learning, improved school attendance and increased engagement in school. The CCPP works with an independent practicing environment to allow children to proceed to review or preview what they are engaged in the Chinese characters learning. A mentor-like CCPP in the after-school interval could help the children learn independently with parents absent.

References

1. Becker, H. J. (2000). Pedagogical Motivations for Student Computer Use That Lead to Student Engagement. Educational Technology, 40(5), 5–17.
2. Buabeng-Andoh, C. (2012). Factors influencing teachers' adoption and integration of information and communication technology into teaching: A review of the literature. International Journal of Education and Development using Information and Communication Technology, 8(1), 136.
3. Catts, H. W., Hogan, T. P., & Fey, M. E. (2003). Subgrouping poor readers on the basis of individual differences in reading-related abilities. Journal of Learning Disabilities, 36(2), 151–164.
4. Cristia, J., Ibarrarán, P., Cueto, S., Santiago, A., & Severín, E. (2012). Technology and child development: Evidence from the one laptop per child program.
5. Dias, L. B. (1999). Integrating technology: some things you should know. Learning & Leading with Technology, 27(3), 10–13.
6. Frith, U. (1985). Beneath the surface of developmental dyslexia. In Paterson, K.E., Marshall, J.C. and Coltheart, M. (Eds.), Surface dyslexia: neuropsychological and cognitive studies of phonological reading. Hillsdale, NJ: Lawrence Erlbaum Associates.
7. Julie Lo (1994). How to Enhance the Instruction in Chinese Character Recognition. Elementary Education, 35, 14–18.
8. Kieras, D. and Just, M. (1984) New Methods in reading comprehension research, Hillsdale, NJ: Erlbaum.
9. Lyon, G.R.: Towards a Definition of Dyslexia. Annals of Dyslexia, 45:3–27 (1995).

10. Mercer, C. D., & Mercer, A. R. (2001). Teaching students with learning problems. Upper Saddle River, NJ: Merrill Prentice Hall.
11. Torgesen, J.K., Rashotte, C.A., and Alexander, A.W.: Principles of Fluency Instruction in Reading: Relationships with Established Empirical Outcomes. In M. Wolf (Ed.), Dyslexia, Fuency, and the Brain. Timonium, MD: York Press (2001).
12. Mahoney, J. L., Lord, H., Carryl, E. (2005). An ecological analysis of after-school program participation and the development of academic performance and motivational attributes for disadvantaged children. Child Development, 76(4), 811–825.
13. Vadeboncoeur, J. A. (2006). Engaging young people: Learning in informal contexts. Review of Research in Education, 239–278.

A Web Accessibility Study Between the Degradation of Cognitive Perceptions and Physiology for the Aging People

Chi Nung Chu

Abstract The goal of this study is to aid the development of rich and dynamic mobile web sites. It collects the relevant engineering practices of cognitive and physiological decline with the aging people, promoting those that enable a better user experience and warning against those that are considered harmful. The results of the investigation could be considered as a possible measure that might be implemented towards the goal of providing as accessible an experience as possible on a mobile device. The specifications to promote the widespread deployment enhance the functionality and availability of the web sites.

Keywords Web accessibility · Cognitive perceptions · Aging people · Mobile devices

1 Introduction

The increasing growth of aged people is becoming more cautious about spending their time on the Internet efficiently. Recently, there has been an explosion of e-service and e-commerce from the web 2.0 social networks and government on the Internet [1, 5, 6, 10]. Many have transformed the way we share information across the web. As the technology and service of network are evolving constantly, the so-called Web 2.0 platform concept with the various types of demands has been developed in order to meet the users' needs. The growing challenge of access to the Internet posed by cognitive and physiological decline with the aging people has provoked critical interface design issues with the potential for cost-effectiveness and time-effectiveness [2, 3, 7–9]. In the meantime, although there are guidelines which could be followed according to the W3C Web Content Accessibility Guidelines 2.0 [11], the features of usability for the aging people with cognitive and physiological

C.N. Chu (✉)
China University of Technology, Department of Management of Information System, No. 56, Sec. 3, Shinglung Rd., Wenshan Chiu, Taipei, Taiwan 116, ROC
e-mail: nung@cute.edu.tw

© Springer Nature Singapore Pte Ltd. 2018

N.Y. Yen and J.C. Hung (eds.), *Frontier Computing*, Lecture Notes in Electrical Engineering 422, DOI 10.1007/978-981-10-3187-8_80

decline are limited in their lack of assistive technology as accessing to the Internet for them. Access to the Internet has thus emerged as an integral part of human society.

How to promote the availability and efficiency in accessing to the Internet for the aging people, therefore there are some specific requirements of accessibility and access efficiency faced with the mobile devices that they use. This paper investigated the development of accessible and efficient changes in the way web pages that could be made and used for the aging people with the degradation of cognitive perceptions and physiology.

2 Method and Results

There are two modes to be investigated for the interface design investigation: the system mode and the human mode [4]. The web accessibility evaluation between the degradation of cognitive perceptions and physiology for the aging people adopted cognitive engineering. In the process of evaluations, the participants' surfing behaviors were observed in two parts: user-familiar web sites and experiment-dedicated web sites. The research methods were implemented in an iterative cycle of investigation which was characterized by observations, an ideation phase, and rapid prototype and testing. Each iteration builds on the lessons learned from the previous cycle, and the process terminates either when the results are appropriate or when the allotted time has run out. The preliminary results are as follows.

(1) **Icon should be big enough to be touched**

As the accurate movability of the aging people is less controllable, the icon objects on the touchscreens can only be launched effectively if they are pointed accurately by the finger (Table 1).

Table 1 Successful rate of pointing icon with different age

Length/Width of icon	Actual icon display	Successful rate of pointing with age (%)	
		50–59	60–69
2 cm		95	75
1.3		70	50
0.9		60	40

(2) **On-screen keyboard should be manipulative**

The written Chinese is a logographic orthography that differs greatly from alphabetic writing systems. As spelling a Chinese character in a successive multiple keystrokes on the on-screen keyboard (Fig. 1), it's hard for the aging people with degraded vision and finger movability to the entry works (Table 2).

(3) **Any selection on the on-screen icon should be responsive**

As the sensitivity of the finger for the aging people is low, the alternative responses are needed to make sure the icon selection is activated (Table 3).

(4) **Previous interface layout option should be remained after new version operating system updated**

As the cognition of the aging people is kept to what they were familiar, the layout of manipulation environment would better be consistent in the way they are familiar with (Table 4).

Fig. 1 Chinese On-screen keyboard

Table 2 Spelling error rate with Chinese on-screen keyboard

Number of keystrokes	Testing examples	Error rate of spelling with age (%)	
		50–59	60–69
3	的：ㄉㄜ˙ 好：ㄏㄠˇ	42.9	57.1
4	經：ㄐㄧㄥ- 昨：ㄗㄨㄛˊ	43.8	56.3

Table 3 Awareness rate with icon selection

One-shot successful rate of icon selection with age	
50–59	60–69
82.6%	61.4%

Table 4 Confident rate in new version operating system manipulation

Confident rate of new version operating system manipulation with age	
50–59	60–69
52.6%	21.3%

3 Conclusion

This paper developed the accessible guidelines for the aging people with degradation of cognitive perceptions and physiology. The accessible guidelines do not represent a new version of the W3C Web Content Accessibility Guidelines at all, but merely continues to adopt the concepts to meet the aging people special needs. The rising demands for cognitive and physiological decline-driven web sites where much of the content must be rendered for ease and efficient access on the web pages have meant that the aging people without the assistive technology are given worse performance versus a site purely composed of general designs. Accessibility for the aging people with degradation of cognitive perceptions and physiology may also suffer in a web site designed with W3C Web Content Accessibility Guidelines. The satisfaction for the people is never full filled. In the future, the ongoing e-service and e-commerce will be expanded to become an indispensable environment in our daily life. The availability of a broad range of accessible and efficient online services and share information will be continually enhanced. The more we concern, the more quality of life the aging people with degradation of cognitive perceptions and physiology can get.

References

1. Lenhart, A., Purcell, K., Smith, A., & Zickuhr, K. (2010). Social Media & Mobile Internet Use among Teens and Young Adults. Millennials. Pew internet & American life project.
2. Magnusson, L., Hanson, E., & Borg, M. (2004). A literature review study of information and communication technology as a support for frail older people living at home and their family carers. Technology and Disability, 16(4), 223–235.
3. Marquié, J. C., Jourdan-Boddaert, L., & Huet, N. (2002). Do older adults underestimate their actual computer knowledge?. Behaviour & Information Technology, 21(4), 273–280.
4. Norman, D. A. (1986). Cognitive Engineering. User centered system design: New perspectives on human-computer interaction, 3161.
5. Norman, D. A., & Verganti, R. (2014). Incremental and radical innovation: Design research vs. technology and meaning change. Design Issues, 30(1), 78–96.
6. Reisenwitz, T., Iyer, R., Kuhlmeier, D. B., & Eastman, J. K. (2007). The elderly's internet usage: an updated look. Journal of Consumer Marketing, 24(7), 406–418.
7. Silva, S., Almeida, N., Pereira, C., Martins, A. I., Rosa, A. F., e Silva, M. O., & Teixeira, A. (2015, August). Design and Development of Multimodal Applications: A Vision on Key Issues and Methods. In International Conference on Universal Access in Human-Computer Interaction (pp. 109–120). Springer International Publishing.

8. Silva, S., Braga, D., & Teixeira, A. (2014, June). AgeCI: HCI and age diversity. In International Conference on Universal Access in Human-Computer Interaction (pp. 179–190). Springer International Publishing.
9. Wagner, N., Hassanein, K., & Head, M. (2010). Computer use by older adults: A multi-disciplinary review. Computers in human behavior, 26(5), 870–882.
10. Wetter-Edman, K. (2014). Design for Service: A framework for articulating designers' contribution as interpreter of users' experience.
11. World Wide Web Consortium. (2008). Web content accessibility guidelines (WCAG) 2.0.

The Study of Salient Object and BoF with SIFT for Image Retrieval

Yih-Chearng Shiue, Sheng-Hung Lo, Yi-Cheng Tian and Cheng-Wei Lin

Abstract Big data generated search difficult problems, thus effectively retrieve digital images, image retrieval has attracted much attention in recent years. First, this study presents detection technology is expected to identify significant object image through the significant items, and reduce the influence of back-ground noise of the object. Through the significant object detection processed image, be used in the scale-invariant feature transform (SIFT) features to capture an image, and then through the k-means clustering algorithm clustering feature vectors for all of the images to obtain a bag-of-features (BoF) vectors as the basic conditions for images retrieval. Finally, the system is expected to improve through the search pattern, as well as improve the accuracy of images search, images search system to make a real attempt to solve the big data, images search difficult problems arising.

Keywords Image retrieval · Content-based image retrieval · Scale-invariant feature transform · Bag-of-features model · K-means clustering

Y.-C. Shiue · S.-H. Lo (✉) · C.-W. Lin
Department of Information Management, National Central University, Taoyuan, Taiwan
e-mail: shenghung.lo@gmail.com

Y.-C. Shiue
e-mail: ycs@mgt.ncu.edu.tw

C.-W. Lin
e-mail: domo.lin@gmail.com

Y.-C. Tian
Center for General Education, Hsin Sheng College of Medical Care and Management, Taoyuan, Taiwan
e-mail: tyc@hsc.edu.tw

© Springer Nature Singapore Pte Ltd. 2018

N.Y. Yen and J.C. Hung (eds.), *Frontier Computing*, Lecture Notes in Electrical Engineering 422, DOI 10.1007/978-981-10-3187-8_81

1 Introduction

1.1 Background and Motivation

Introduced in the 1990s, content-based image retrieval (CBIR) differs from keyword-based image search solutions in that CBIR bases its search on the features that can be extracted from an image (e.g., color, texture, and shape). Such features are categorized as low-level features that can be transformed into numerical vectors. The feature vectors can subsequently be used to evaluate the similarity between images to retrieve the most visually similar images. At present, image searches are performed largely through such online search engines as Google Image, Microsoft Bing, TinEye, and GazoPa. Most CBIR systems perform retrieval on the basis of an input similar image/photos or user-drawn sketch as the query image. Users can input similar images/photos as the query image to conduct an online image search, through which similar or precise images are retrieved using search algorithms. This query-by-example method uses an entire image for the query, which differs from the object-based search solution proposed in this study. Querying with user-drawn sketches requires users to have sufficient drawing capability; however, this is often ineffective because not every user is skilled at drawing. Accordingly, this study proposes an image search system to address the limitations of CBIR systems. The proposed system performed image searches using only object images as the query image. This enables users to specify any object in an image as the query image without the inconvenience of using entire images or user-drawn sketches for the query.

1.2 Purpose

Search method
The proposed image search system provides various object images for multiobject query, enabling the user to select one or multiple object images as the query image to retrieve the target image.

Image retrieval precision
To improve its image retrieval precision, the BoF model based on the SIFT algorithm was integrated with salient object detection.

Construction of the image search system

- Rectangular salient images were derived through salient object detection. The SIFT algorithm was implemented to extract features from these images and the BoF model was adopted to establish BoF vectors of the images.
- The image search system can use a combination of two to three object images as the query image to retrieve images containing multiple objects.

- The system adopts Euclidean distance to estimate the similarity between the object images and target images stored in a database; the target images and their corresponding names are then presented in descending order according to their similarity to the object images.

2 Literature Review

2.1 Content-Based Image Retrieval

CBIR is an image search solution based on visual feature extraction and searches for images stored in a multimedia database that correspond with the contents of an image specified by the user. Figure 1 presents the image-matching process of a CBIR system.

In a CBIR system, image contents are described by common low-level visual features such as color, texture, and shape [1]. These features are extracted from images to estimate the similarity between the query image and images stored in a database, through which the most similar images are retrieved. However, the features do not accurately describe the semantic concepts of images, as indicated by the summary of strengths and weaknesses of the features. In addition to low-level visual features, mid-level features have received increasing attention in recent years. Such features can be detected using SIFT, which is a computerized algorithm that that can be implemented to detect and describe local features in images. In addition, it is translation-, scale-, and rotation-invariant [2]. The SIFT algorithm has been experimentally verified to be the most robust local feature descriptor in the presence of geometric transformation, as compared with other descriptors of the same type [3].

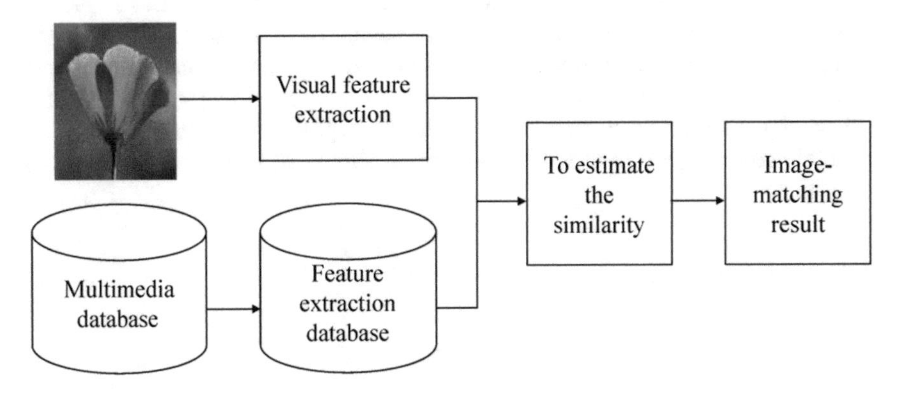

Fig. 1 Image-matching process of CBIR systems [8]

2.1.1 Bag-of-Features Model

The BoF model is based on the bag-of-words (BoW) model, a text-mining method that assumes that all words in a document are independent of each other, thereby disregarding the order of words, which are instead grouped into vocabularies, and lexical vectors are used as the representation of the document. Similarly, the BoF model treats each visual representation as an orderless collection of local features, and its effectiveness in image processing has been evidenced by previous studies [4]. This model represents images comprises two concepts: local features and codebooks [5, 6].

2.1.2 Scale-Invariant Feature Transform

The SIFT algorithm detects and describes the local features of an image. To identify scale-space extrema and extract extremum features that are invariant to location, scaling, and rotation, the key-points of the image are calculated through the following four stages: (1) scale-space extrama detection, (2) key-point localization, (3) orientation assignment, and (4) key-point descriptor generation.

2.2 *Salient Object Detection*

Immediately before the human eye engages in image recognition, it identifies the salient regions of an image with remarkable speed and precision. Thus, approximating human object recognition capability is a major topic in machine vision and has received extensive attention in this field. This study adopted the method for salient object detection proposed by Liu et al. [5]. Liu et al. [5] revealed a supervised approach to salient object detection, proposing a set of novel features to describe salient objects, and optimally combining three features through conditional random field learning. These features are multiscale contrast (MSC), center-surround histogram (CSH), and color spatial distribution (CSD). After the feature map is normalized, these three features are optimally combined through conditional random field learning to generate a binary label map that separates a salient object from the background. To identify regions of interest, the smallest rectangle containing at least 95% salient pixels should be identified in the binary label map.

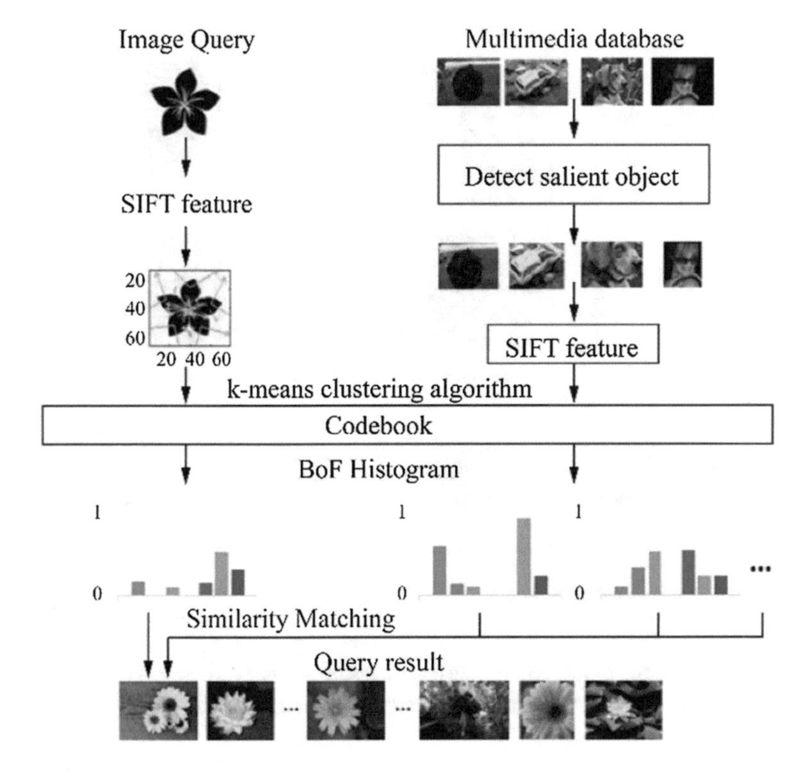

Fig. 2 Research framework

3 System Development

3.1 Research Framework

The framework of the proposed image search system integrates the BoF model based on SIFT with salient object detection. Based on this framework (Fig. 2), the image search system performs searches in three stages: (1) salient object detection, (2) BoF-based image transformation, and (3) similarity measurement.

3.1.1 Salient Object Detection

At this stage, the features of MSC, CSH, and CSD are defined as salient objects in images. During the experiment, all the images stored in the image database were resized to a minimum of 200 pixels in length or width (those smaller than 200×200 pixels retained their original size). However, the resized images contained distortions and slightly affected the precision of image search. After the

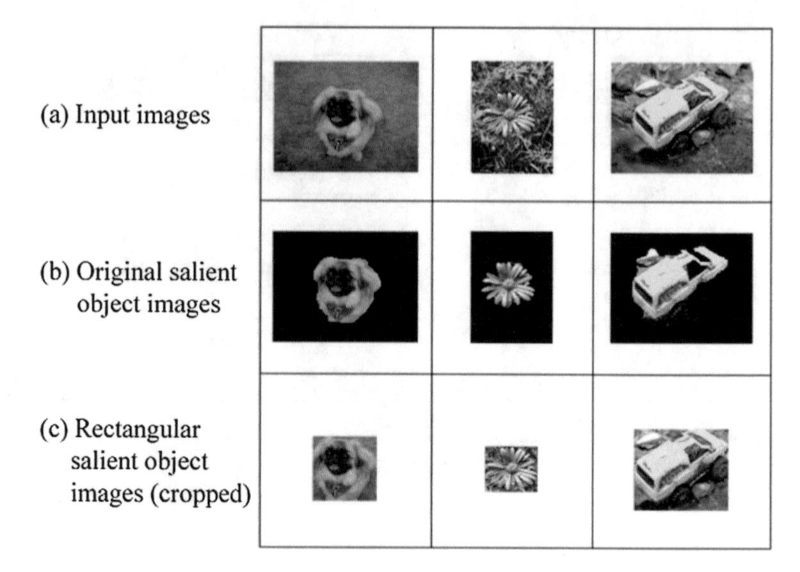

Fig. 3 Original versus cropped salient object images

salient objects were detected, the image backgrounds were filtered to retain the salient object images. The results of performing salient object detection on three images are shown by Fig. 3.

3.1.2 Image Transformation Based on Bag of Features

The BoF model transforms the query image into a set of visual words. The transformation comprises three stages: feature extraction and description, visual vocabulary creation, and feature quantization (Fig. 4) [7].

4 Experimental Design and Results

4.1 Experimental Environment

Establish a model

1. Image path: displays the link of a selected image data set folder.
2. Select folder: selects an image data set folder for image retrieval.
3. Indexing: performs salient object detection and BoF transformation for the selected image data set.

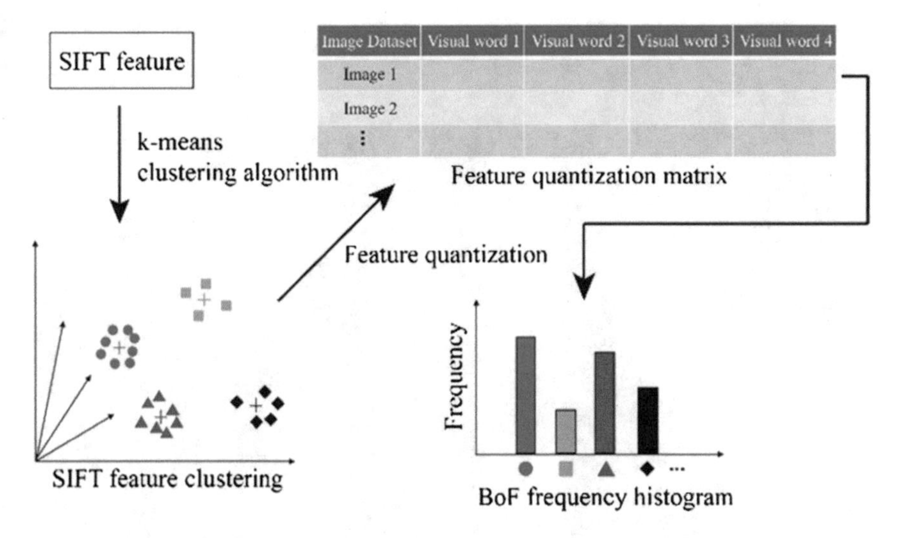

Fig. 4 Feature quantization

Object image
Toggle menu: The image search system provides the following ten types of object image: "bird," "ship," "butterfly," "car," "cat," "dog," "flower," "horse," "house," and "human". The default object image is "bird." Each type of object image comprises five relevant pictures. More object images can be incorporated into the system to retrieve different images containing different objects.

Query image
When an object picture listed on the Object Image panel is selected, the picture appears as the query image. A total of three object pictures can be selected as the query images. After the query images are determined, the user clicks the "search" button to begin searching for target images. Ten types of image were extracted from the MSRA-A image data set for the experiment [5], namely "bird," "ship," "butterfly," "car," "cat," "dog," "flower," "horse," "house," and "human". Each image type comprised 100 experimental images, totaling 1000 images.

Search results
By estimating the similarity between the query images and database images, the system lists the top 25 images according to similarity in the Search Results panel. If no target image is found, any picture from the search results that is deemed most similar to the target image can be selected as the query image to conduct further researches. Figure 5 presents the system search results obtained using the "butterfly" type object image as the query image. A retrieved image can be chosen as the query image to improve the similarity of search results according to the target image.

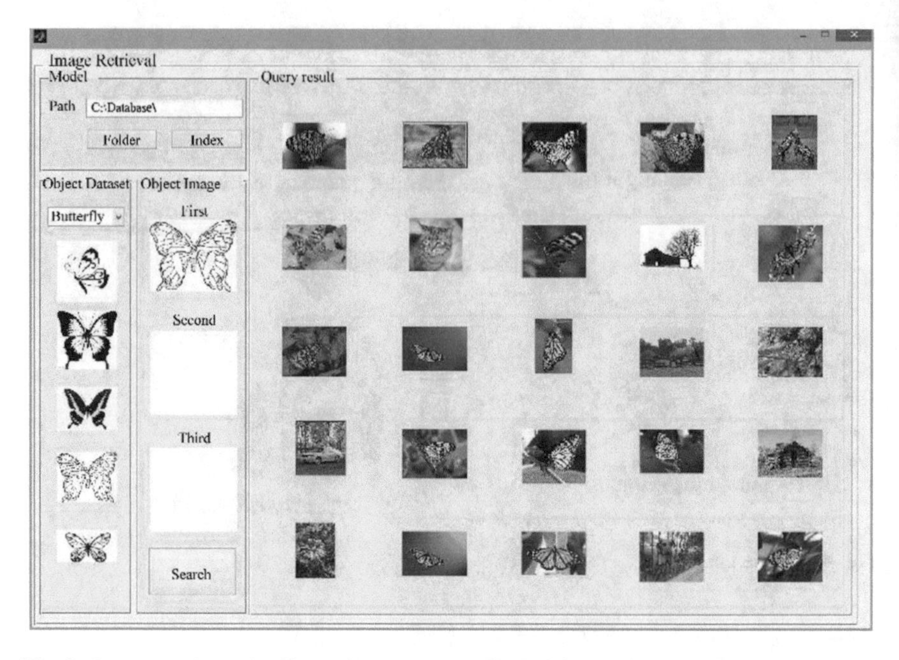

Fig. 5 Image search results (Query image source MSRA-A image data set [5])

4.2 Analysis of Experimental Results

4.2.1 Cropped Formats of Salient Object Images

After undergoing salient object detection, all the input images were cropped in two formats to obtain the salient object images. The percentage of images whose categories were completely identified was then estimated using both cropped formats of the salient object images (Table 1). For example, 70 salient object images of dogs were identified; thus, 70% of the salient object images of dogs were identifiable in terms of category (each salient object image comprised 100 pictures of the same category). As shown in Table 1, objects in some of the original salient image types ("ship," "cat," "horse," and "human") were less identifiable than in the corresponding rectangular ones, whereas those in other original salient images ("bird," "butterfly," "flower," and "house") were almost equally identifiable. Overall, objects in the rectangular salient object images were more identifiable than those in original ones.

4.2.2 Codebook Size

After undergoing salient object detection, all the salient object images were transformed into visual vectors through the BoF model. To create the visual

Table 1 Percentages (%) of objects identified in the original and rectangular salient object images

	Bird	Ship	Butterfly	Car	Cat	Dog	Flower	Horse	House	Human
Original salient object images	99	81	98	93	85	91	100	83	99	87
Rectangular salient object images (cropped)	100	92	100	100	96	97	100	96	100	94

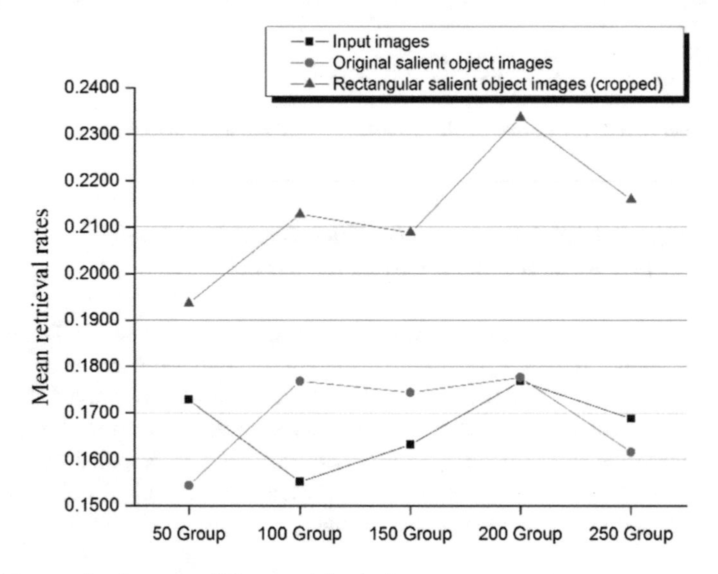

Fig. 6 Mean retrieval rates at different codebook sizes

vocabularies, the k-means clustering algorithm was implemented to establish a codebook. The size of a codebook is related to the results of an image search; in other words, cluster size affects image retrieval performance. Thus, in this Experiment, codebook size was configured as 50, 100, 150, 200, and 250 to determine the optimal size for image retrieval. Because objects in original salient images were less identifiable than those in rectangular salient images, both types of salient images were compared according to the image retrieval precision. In addition, this figure suggests that the mean retrieval rates for the rectangular salient images were noticeably higher than those for the original salient images. Thus, the cluster size was set to 200 in the subsequent experiments. These results are depicted in Fig. 6, which presents the mean image retrieval precision rates for the 10 object image types stratified by codebook size. In addition, this figure suggests that the mean retrieval rates for the rectangular salient images were noticeably higher than those for the original salient images.

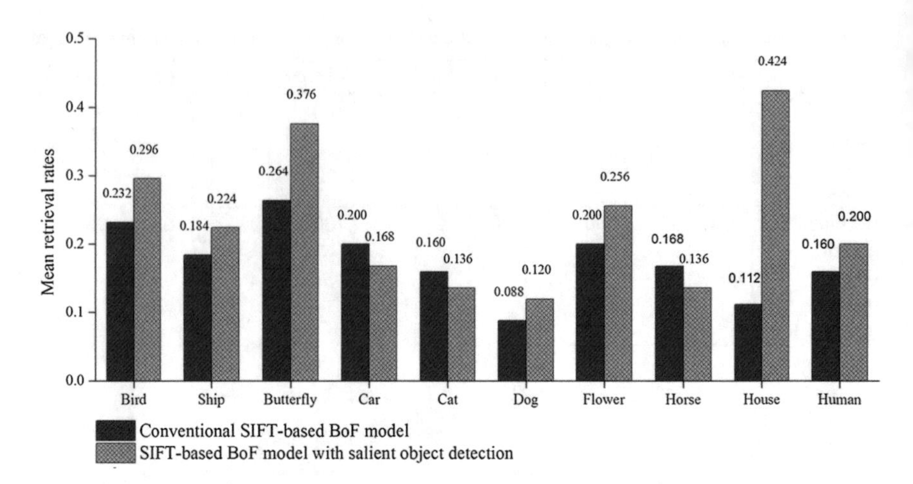

Fig. 7 Image retrieval performance the SIFT-based BoF model with and without salient object detection

4.2.3 Image Retrieval Performance: SIFT-Based BoF Models with and without Salient Object Detection

This experiment revealed that, compared with the conventional SIFT-based BoF model, the image retrieval precision of the SIFT-based BoF model with salient object detection improved. The mean retrieval precision rates improved for most of the original salient images that underwent salient object detection. Nevertheless, few of the original salient images, namely cars, cats, and horses, attained decreased precision rates (Fig. 7), likely because the contours of the salient objects of these images were partially removed during cropping.

5 Discussion and Conclusion

Current image retrieval technologies use either similar images/photos or sketches as the query images; however, both retrieval solutions have their limitations. Image querying based on a similar image/photo uses the entire image/photo, whereas image querying with sketches requires a sufficient impression of the target image and drawing capability of the user. To address the limitations of conventional image retrieval solutions, this study proposed using object images as query images. This object-based image retrieval solution enables users to conduct searches with any object(s) within a target image. After a conventional SIFT-based BoF model was integrated with salient object detection to improve image retrieval precision, an image search system was constructed on the basis of the proposed research framework. The findings are as follows:

(1) *Modified image retrieval solution*

An image retrieval precision analysis confirmed the feasibility of the proposed object-based image retrieval solution, which is more convenient and easier to use than conventional retrieval solutions based on images or sketches that resemble the target image.

(2) *Image retrieval precision*

After salient object detection was performed to identify salient objects in images, which filtered the image backgrounds and yielded rectangular salient object images, querying with the object images was less susceptible to interference from the background and other complex image features. In addition, the codebook size experiment revealed that the highest retrieval precision was attained with an image cluster size of 200. With the codebook size set to 200, the experiment on single-object image retrieval precision attained an average retrieval precision rate of 0.234 for all the object images (estimated by summing and averaging the mean retrieval precision rates of all these images). By contrast, the average retrieval precision rate for all the object images in the input images (without salient object detection) was 0.177, verifying that the retrieval precision of the SIFT-based BoF model can be improved through integrating the single-object querying approach with salient object detection.

The experiment on the double-object image retrieval yielded lower retrieval precision but a higher recall rate because the number of images containing two objects was limited. The experiment on triple-object image retrieval also yielded lower retrieval precision but a higher recall rate since relatively few objects contained three objects. The low retrieval precision in both experiments indicated that, because salient object detection retained only the salient objects in the images and filtered the image features, some images could not be retrieved on the basis of their content.

References

1. Ojala, T., Pietikäinen, M., Mäenpää, T.: Multiresolution gray-scale and rotation invariant texture classification with local binary patterns. Pattern Analysis and Machine Intelligence, IEEE Transactions on **24**(7), 971–987 (2002). doi:10.1109/Tpami.2002.1017623
2. Lowe, D.G.: Object recognition from local scale-invariant features. In: Computer vision, 1999. The proceedings of the seventh IEEE international conference on 1999, pp. 1150–1157. IEEE
3. Mikolajczyk, K., Schmid, C.: A performance evaluation of local descriptors. Pattern Analysis and Machine Intelligence, IEEE Transactions on **27**(10), 1615–1630 (2005). doi:10.1109/TPAMI.2005.188
4. Zhang, S., Tian, Q., Hua, G., Huang, Q., Gao, W.: Generating descriptive visual words and visual phrases for large-scale image applications. Image Processing, IEEE Transactions on **20**(9), 2664–2677 (2011). doi:10.1109/TIP.2011.2128333

5. Liu, T., Yuan, Z., Sun, J., Wang, J., Zheng, N., Tang, X., Shum, H.-Y.: Learning to detect a salient object. Pattern Analysis and Machine Intelligence, IEEE Transactions on **33**(2), 353–367 (2011). doi:10.1109/TPAMI.2010.70
6. Yuan, X., Yu, J., Qin, Z., Wan, T.: A SIFT-LBP image retrieval model based on bag of features. In: IEEE International Conference on Image Processing 2011
7. Lv, H., Huang, X., Yang, L., Liu, T., Wang, P.: A k-means clustering algorithm based on the distribution of SIFT. In: Information Science and Technology (ICIST), 2013 International Conference on 2013, pp. 1301–1304. IEEE
8. Khokher, A., Talwar, R.: Content-based Image Retrieval: Feature Extraction Techniques and Applications. In: Conference proceedings 2012

Social Information Sub-topic Clustering Using News Tag

Ping-I Chen and Fu-Jheng Jheng

Abstract Discussions in social networks change frequently as users express their different points of view. Most of the topic tracking systems designed to follow these changes use a set of keywords to filter articles in order to identify key points or user opinions. These systems group all of the articles together even though some are closely related or provide no new information. Therefore, it is difficult to ascertain how many distinct topics of discussion have occurred in a given time period without reading all the articles. This research describes a method for dividing a tracked topic that contains a set of articles into several sub-topics automatically. We provide data scientists with a dashboard that directly displays sub-topic categories and allows them to identify important related issues over a daily or weekly period.

1 Introduction

We created a system called iFeel to help data scientists analyze social media. iFeel uses a set of keywords to describe the main topic of observation and filters all the related articles into sub-topic categories to calculate statistics information and create dashboards. In this way, even data scientists without a computer science background can follow those dashboards and use their domain expertise to write reports and provide insight for their customers. Figure 1 shows our work flow.

P.-I. Chen (✉) · F.-J. Jheng
Innovative DigiTech-Enabled Application and Service Institute,
Institute for Information Industry, Taipei City, Taiwan
e-mail: be@iii.org.tw

F.-J. Jheng
e-mail: fjcheng@iii.org.tw

© Springer Nature Singapore Pte Ltd. 2018

N.Y. Yen and J.C. Hung (eds.), *Frontier Computing*, Lecture Notes
in Electrical Engineering 422, DOI 10.1007/978-981-10-3187-8_82

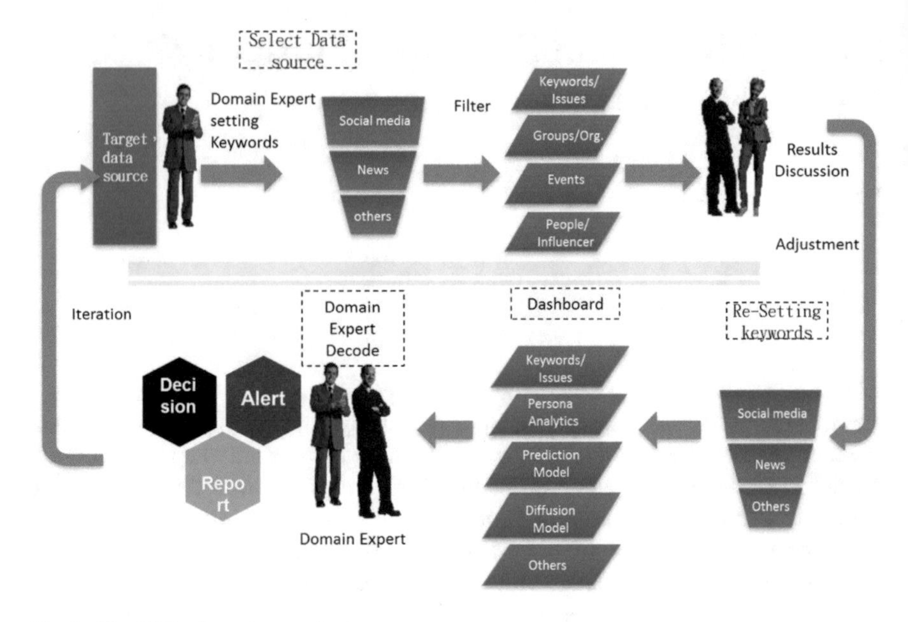

Fig. 1 The SOP of our data analysis process

iFeel uses a crawler to gather about 600,000 articles per day. Although the system has already used keyword sets to track highly related articles, each topic can contain nearly 2,000 articles on average and some popular discussion topics will generate more than 10,000 articles per day. It is impossible for data scientists to read all of those articles to find out whether the general tenor of the discussion has changed.

Initially, we used the Term Frequency-Inverse Document Frequency (TF-IDF) weights of terms in each article to extract some keywords to use as tags. Then, we randomly selected k articles and used the cosine similarity measure of each article's tags to allocate the remaining articles to these k clusters. Our goal was to treat each cluster as a sub-topic category but we found that because the articles were from the same tracked topic, their contents were very similar. Therefore, our auto-tagging procedure produced similar tags for, and allocated similar articles to, each cluster.

Some news websites in Taiwan allow reporters to tag their own articles. We decided to use this professionally generated content to extract sub-topics. We used a web crawler to crawl news websites and save the tags along with each article. Then, we used the association rule to find the largest frequent item sets to use as the head of each cluster. In this way, we divided articles in the same topic into several sub-topics successfully, allowing data scientists to view each cluster and determine if there is any new information or if the aspect of discussion that day has changed.

2 Related Works

A good article clustering/classification system depends on good feature selection method. How to select the representative feature from documents is a main problem? In the previous research, many feature selection methods were proposed. We will introduce three common used feature selection methods.

2.1 *TF-IDF*

The TF-IDF was proposed by Salton and Buckley [1], which is commonly used in information retrieval and text mining. It is a statistical measure used to evaluate the importance of the word according to the collected data. The importance increases proportionally to the number of times a word appears in the document but is offset by the frequency of the word in the corpus. The basic formula of TFIDF is as follows:

$$\mathrm{tf}_{i,j} = \frac{n_{i,j}}{\sum_k n_{k,j}} \qquad \mathrm{idf}_i = \log\frac{N}{n_i} \qquad w_{i,j} = \mathrm{tf}_{i,j} \times \mathrm{idf}_i$$

The $\mathrm{tf}_{i,j}$ is a measurement of the importance of the term ti within the particular document d_j; $n_{i,j}$ is the number of occurrences of term t_i in document d_j; $\sum_{\mathbf{k}} \mathbf{n}_{\mathbf{k},\mathbf{j}}$ is the sum of number of occurrences of all terms in document d_j. idf_i is a measure of the general importance of the term. N is the total number of documents in the corpus; n_i is the number of documents where the term t_i appears. The disadvantage is that the IDF need to be calculated based on pre-collected document corpus. It is not easy to apply in dynamic information filtering scenarios.

2.2 *Word Co-occurrence*

Word co-occurrence is another widely used feature selection method [2]. The main idea is that if the co-occurrence frequency of these words is high in many documents, there may be strong relations between these words, and these words may be more salient in feature space. The basic formula of Word co-occurrence is as follows:

$$r_{ij} = \frac{f(t_i \cap t_j)}{MAX(f(t_i), f(t_j))}$$

The attributes t_i and t_j are the term pair that we want to calculate their word co-occurrence relationship; $f(t_i)$ is the number of documents where the term t_i appears; $f(t_i \cap t_j)$ is the number of documents where both t_i and t_j appears. The disadvantage of this method is same as TF-IDF, the co-occurrence relationship also need to be calculated based on pre-collected document corpus.

3 System Design

Hashtags are labels or metadata tags used on social networks to make it easier to find messages with a similar theme or specific content. They help users determine the topic of an article quickly and make the article be speed rapidly. Over the past few years, Internet news sources have started to use hashtags in articles to make them more easily readable and to accelerate the virality of the news. Our system groups articles based on reporter-generated tags on Internet news sites. We can use these groups to understand the context of each tag, and subsequently to search for related topics. In this chapter, we discuss our data source and our method for grouping topics using tags.

3.1 Data Collection

We used Social Event Radar (SER), a set of systems that deal with community analysis, to collect data for our system. Since 2012, SER has used a web crawler to collect data from websites such as Facebook, blogs, forums, and PTT (the most well-known BBS in Taiwan), as shown in Fig. 2. SER has saved more than 50 Tb of data in our database and this amount increases by about 300 Gb of data per month. SER provides a complete framework for users to search for data they need for a variety of applications. SER has abundant information including data from 257 domestic and foreign news websites. For our dataset, we used tagged news articles from 143 Taiwanese news sites.

3.2 Sub-topic Clustering System

With the growth of the Internet community, many companies have started to use the Internet to market their products and investigate their public reputation. Furthermore, some companies observe social network discussions to plan crisis management strategies. Our Social Event Tracking System provides complete function for analyzing community and news data using user-defined keywords. A common problem users may encounter is the enormous number of collected articles retrieved

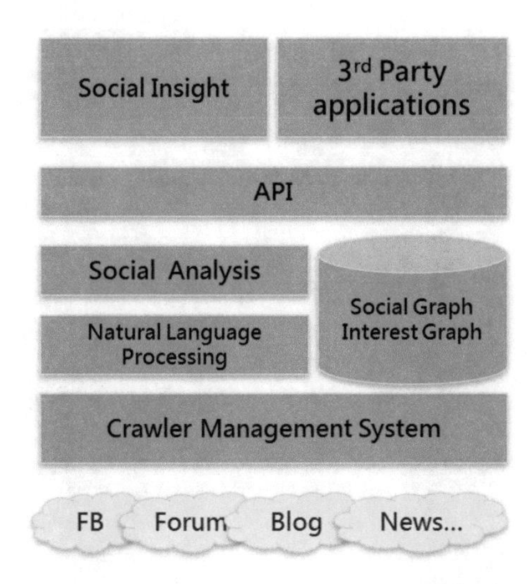

Fig. 2 SER architecture

by their keywords. When establishing keywords, it can be difficult for users to determine whether the content of the articles returned by their keywords matches their expectations. In particular, users may be unable to find the data they need because their keywords are too imprecise. Additionally, it is hard for users to identify and group the topics these keywords are discussing while also decoding the information in the article. Our Sub-topic Clustering System makes it easier for users to cluster the data from the articles retrieved, and therefore to find the information they need. This system also makes it easier for users to understand which topics pertaining to the keywords occurred within the allotted time and which incidents pertaining to the topic are most closely related.

In the SER news database, there are 257 news media data source, including parts of the Mandarin data (143 Taiwanese news sites that publish 45,000 news articles per day) that we used for this system. We used the keywords set up by users to filter the articles' titles and contents. Then we regard the tag added by the reporter of the article as the Item Set, and think of every news article as a transaction. We organize the transactions into a complete Data Set, and then input it into an FP-Growth algorithm. Users can adjust the minimum support threshold via a web page so that they can select the most relevant results. After doing FP-Growth, we consider the node of the FP-Tree to be the cluster center and compound every affiliate transaction appearing in the node into a group. Each group contains one or more tags. Groups may then be merged together. If all of a group's Items are complete and it is the only one contained by another group, we merge the smaller group with the larger one. Finally, we used Mike Bostock's Zoomable Circle Packing for data visualization.

Take Tsai Ing-Wen, the president-elect in 2016, as an example. In Fig. 3, we can see that the tags related to Tsai Ing-Wen over a 24 h period are, "客家, 國軍, 林全, 內閣人事, 南庄, 國宴菜, 台三線, 蔡英文, 國宴, 三軍". We can understand the relevance of the designated keyword during the allotted time via these tags.

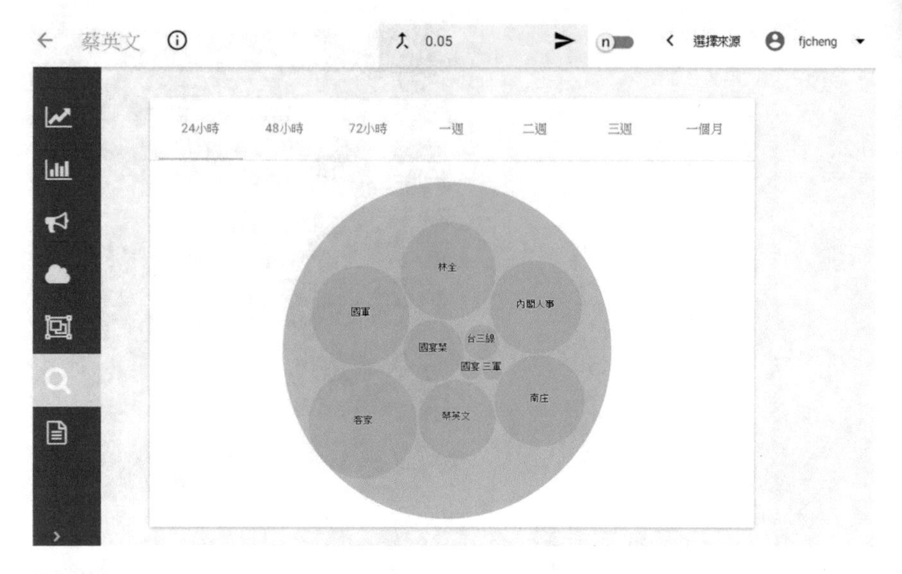

Fig. 3 Clustering result of the System

Furthermore, we can inspect these tag topics to determine which sub-topic tags relate to others. We often bring up the sub-topics when helping users understand the complete content of a topic. For example, “南庄” is a township in Miaoli County in Taiwan. We find that in Fig. 4, when “南庄” is mentioned, there are some related words (南庄, 產業, 客家, 台三線, 文化) that are also mentioned often. In fact, “南

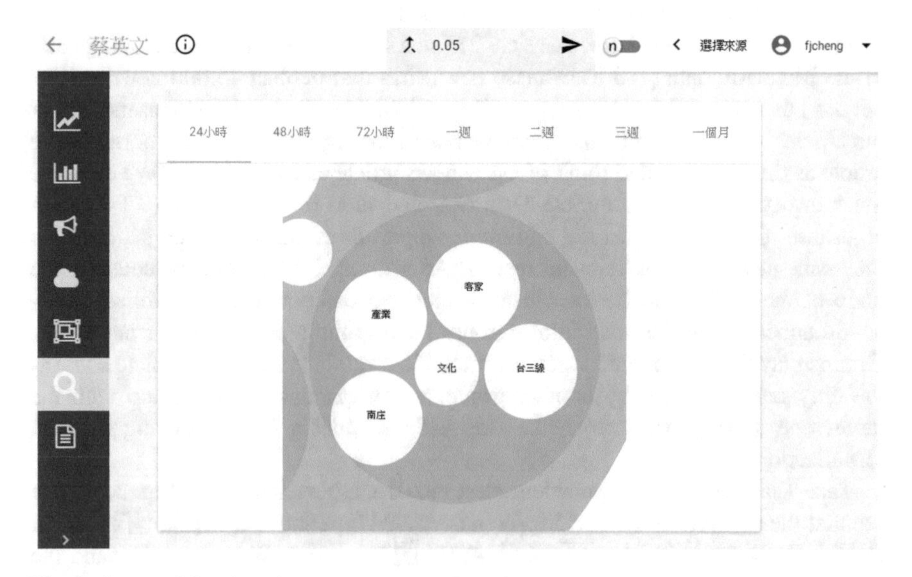

Fig. 4 Group of the clustering

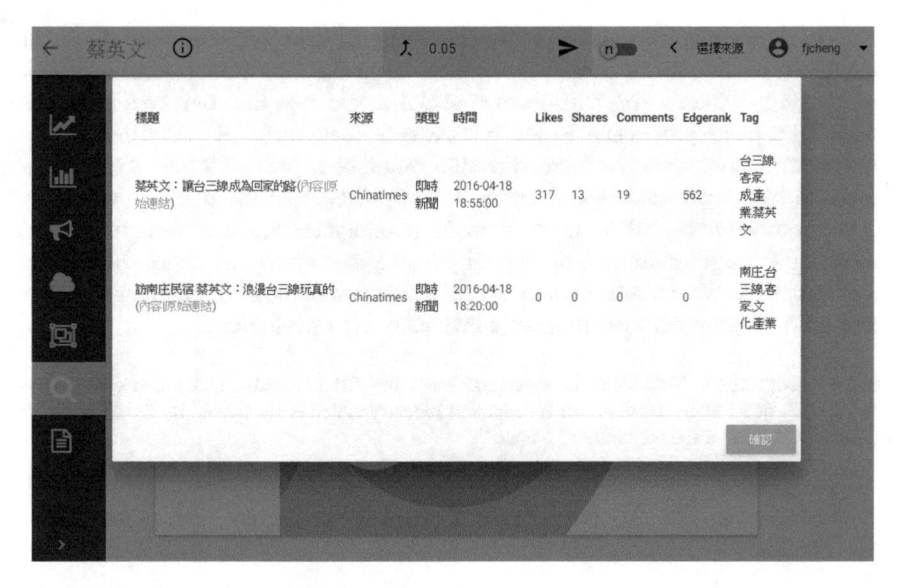

Fig. 5 The article lists after tag-clicking

庄" is a Hakka village on Provincial Highway 3 where Tsai Ing-Wen proposed policy to revitalize Hakka cultural industries during her presidential campaign.

In the 24 h after this picture emerges, Tsai Ing-Wen might release a related statement or plan that causes the media to report this incident. If we want to understand more of the content of the news, we can click on every tag to read the related articles and better understand the cause and effect. Figure 5 shows the article lists after tag-clicking. From these lists, it clear that the topic shows the statements about the Provincial Highway 3 cultural industries development after Tsai Ing-Wen visited "南庄". From the above example, we can see that users can classify large datasets in a short time and understand which topics are highly relevant using our Sub-topic Clustering System.

4 Conclusions

Many companies and governments are eager to solve the problems of quickly ascertaining what is happening in the world and discovering the relationship between two incidents. In this article, we introduced a Sub-topic Clustering System that allows users to explore events and quickly and understand them. We use reporter-generated tags on news articles and make use of the FP-Growth algorithm to find the relevance between each tag and then perform clustering. The results show that our system can assist users with understanding related sub-topics efficiently.

While we designed our Sub-topic Clustering System with the news media in mind, we would like to extend its application to the flourishing world of social media. In the future, we can integrate social media tags into our system or use natural language processing to incorporate comments made by social media followers. Another interesting future direction would be to provide different aspects to users to help them understand different viewpoints from the public towards an event. Furthermore, when users without a computer science background are adjusting the support threshold, they might not know which values are suitable for their purposes. We can add an auto-adjust feature to find the ideal minimum support threshold, making our system more convenient and user-friendly.

Acknowledgements This study is conducted under the "III Innovative and Prospective Technologies Project" of the Institute for Information Industry which is subsidized by the Ministry of Economy Affairs of the Republic of China.

References

1. Salton, G., & Buckley, C. (1988). Term-weighting approaches in automatic text retrieval. Inf. Process. Manage., 24(5), 513–523.
2. Liu, Y.C., Wang, X.L., & Liu, B.Q. (2004). A feature selection algorithm for document clustering based on word co-occurrence frequency. Proceedings of 2004 International Conference on Machine Learning and Cybernetics.

Evaluation of Shopping Website

Cheng-Yueh Tsai, Kang-Wei Chai and Wei-Ming Ou

Abstract The aim of this article is compare Yahoo mall to Ruten mall website. The study used questionnaire and experiment to find the user's satisfaction and operation performance of two websites. One questionnaire study, it's found user's feel is different in before and after experiments. On experiment study, we compare two times for user finding merchandise on yahoo mall and ruten mall. The two goods of user research are Hair dryer and I Phone 6S in study. In this study find: 1. The "word size" and "advertisement" are effected user's feel in website. 2. The "gender" and "category merchandise" are effected the performance of user operation on website.

Keywords Shopping website · Screen · Word size · Advertisement

1 Introduction

In recent decade, Internet is very important for our live, ex: communication, getting knowledge, shopping etc.; it is a part of our live. Because internet is so important, the interface of people and computer is more and more important too. Website is most popular interface of internet for people. However, website-design would effect the operation and feeling of people. Well website-design is one important factor for

C.-Y. Tsai (✉)
Department of Marketing Logistics Management, Hsin Sheng Junior College of Medical Care and Management, Taoyuan, Taiwan, ROC
e-mail: wwu@hsc.edu.tw

K.-W. Chai
Department of Marketing and Logistics Management, Far East University, Tainan, Taiwan, ROC
e-mail: frica@cc.feu.edu.tw

W.-M. Ou
Department of Marketing Management, Shih Chien, Kaohsiung, Taiwan, ROC
e-mail: wm515@yahoo.com

© Springer Nature Singapore Pte Ltd. 2018

N.Y. Yen and J.C. Hung (eds.), *Frontier Computing*, Lecture Notes in Electrical Engineering 422, DOI 10.1007/978-981-10-3187-8_83

getting order in business internet. In this research, we are compare two shopping website in Taiwan: Yahoo mall and Ruten mall. Use questionnaire and experiment to find out the factor that effect usability on website.

2 Related Work

When people are browsing website, they are like to read funny topic, more pictures. However, people shopping on website, the most important they care is finding something they want quickly. A well website-design can make this purpose [1]. Internet is part of a multi channel promotion strategy, however, it is more and more important [2]. If people feel well, they will be keep in website longer. It is a potential business. Many research study potential influence on customer and consumption on internet [3, 4]. Some people do not want to purchase due to the risk of their personal information [5, 6].

3 Methodology

Two experiments in this study, One is questionnaire for test user's feel before and after operating task. The other is recorded task performance when user found merchandise by website. Study has 30 volunteers.

3.1 Task 1: Questionnaire

First, user fills in the questionnaire before experiment. After experiment, user fills another questionnaire. All questionnaires are about user thinking which effect operation performance in website. The step is as Fig. 1. The questionnaire is divided into three parts. The first part content is layout of website, second part content is colors and advertisement of website, the last part content is user's personal data.

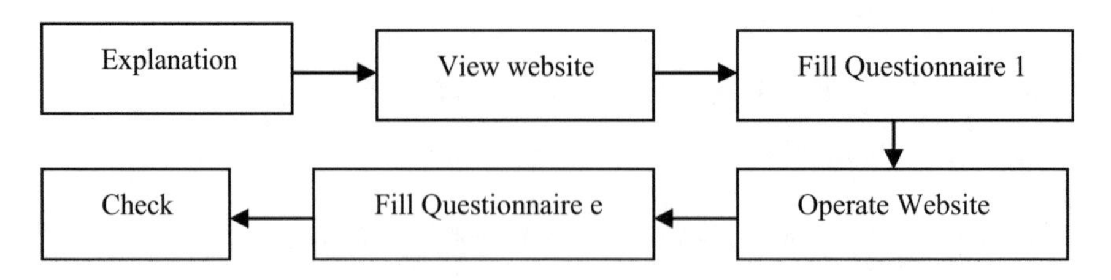

Fig. 1 Operation process in questionnaire

Fig. 2 Ruten mall (*left*) and Yahoo mall (*right*)

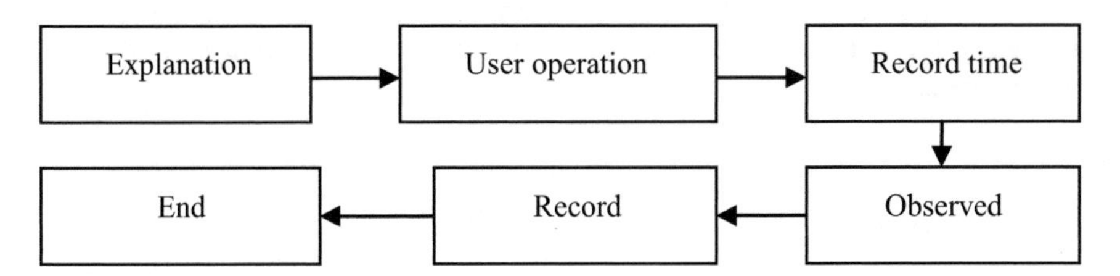

Fig. 3 Experiment process

3.2 *Task 2: Experiment*

The user is required find out I phone 6 s and Hair dryer in two websites: Yahoo mall and Ruten mall (in Fig. 2). In search process, It is record the time which user spend by clock and which user's condition by camera. The steps are like as Fig. 3.

4 Results and Discussion

In this research, twenty-five women and five men are in thirty volunteers. 13 volunteers' age is lower than 19 years old. 8 volunteers' age is in 20–26 years old. 4 volunteers' age is in 30–39 years old, 2 volunteers' age is in 40–39 years old, 3 volunteers' age is higher than 50 years old. Twenty-two volunteers have experiences in buy something on Yahoo mall and Ruten mall.

4.1 *Questionnaire Analysis*

Analysis the first questionnaire, user thinks four effects of website performance are "layout", "color", "size of the word" and "advertisement". However, second

Table 1 Spend time in yahoo mall versus ruten mall

Find targets	Yahoo mall	Ruten mall
Average second	122.24	100.04
I phone 6 s	52.83	90.66
Hair dryer	191.66	109.43

Table 2 Gender effect searching time

Gender	Mean	Stand error
Man	82.85	45.7295
Woman	116.92	97.9544

questionnaire show "size of the word" and "advertisement" are effects to performance of website. It finds that people's surface validity and content validity are different. In this research find that "size of the word" and "advertisement" is real effects factors.

4.2 *Experiment*

In this study, user been requested search I phone 6 s and Hair dryer in Yahoo mall and Retun mall. They be claimed must use website's rout and not keywords. The results are in Table 1.

In Table 1 shows as when user search I phone 6 s in Yahoo mall (52.83 s) is quickly than that in Ruten mall (90.66 s). In T test analysis, searching I phone 6 s and Hair dryer times has statistic signal. This result shows that the path design of Yahoo mall is better than that of Ruten mall for searching I phone 6 s. On the other hand, the path design of Ruten mall is better than that of Yahoo mall fro searching Hair dryer. For T test, gender is statistics signal for searching time, show as Table 2. The average time of man is 82.85 s and that of women is 116.92 s. This result tells us maybe man's computer ability is better than women. It means that man can use website fluently or man has better cognitive ability.

5 Conclusion

In this paper, we use questionnaire and implementation to study two shopping websites design. It found: 1. word size and advertisement are interfere use's operating in shopping website. 2. Yahoo mall and Ruten mall are different in layout. 3C product, like as I phone, maybe suitable for Yahoo mall. Appliances, like as hair dryer, maybe suitable for Ruten mall. 3. Gender is effect to use's searching time on website. Mans performance is better than that of women.

References

1. William C. M, Rachel C. W., & Charles O. K.: An examination of retail website design and conversion rate. Journal of Business Research, in press (2016)
2. Grewal, D., Gopalkrishnan, R. I., Levy, M.: Internet retailing: Enablers, limiters and marketing consequences. Journal of Business Research, 57(7), 703–713 (2004)
3. Richard, M., Chandra, R.: A model of consumer web navigational behavior: Conceptual development and application. Journal of Business Research, 58(8), 1019–1029 (2005)
4. Sicilia, M., Ruiz, S., Munuera, J. L.: Effects of interactivity in a website. Journal of Advertising, 34(3), 31–45 (2005)
5. Koufaris, M., Hampton-Sosa, W.: The development of initial trust in an online company by new customers. Information and Management, 41(3), 377–397 (2004)
6. Tarafdar, M., Zhang, J.: Determinants of reach and loyalty - A study of website performance and implications for website design. The Journal of Computer Information Systems, 48(2), 16–24 (2007)

Social Network and Consumer Behavior Analysis: A Case Study in the Shopping District

Pin-Liang Chen, Ping-Che Yang and Tsun Ku

Abstract The increasing popularity of social networking services and the mobile devices have changed people's life and brought the new business challenges. The goal of this study was to analyze the characteristics and purchase probability of different customers groups in the Shimen shopping district and extract business value. We extracted a point-earning app's user records about the Shimen shopping district. The dates of those user records were between June 11, 2015 and July 12, 2015. Furthermore, we collected the Facebook information of the users. All statistical procedures were performed with our Customer Behavior Analysis System. The main customers in Shimen shopping district were younger groups. Different groups of users had divergent favorites because of their age and gender. The popular trend nowadays in the younger people and the reward points the shop gave may affect the conversion rate of a shop. Many people shopped at several similar stores in one day and that might mean they shopped purposefully not blindly. After customizing the point-earning app's assignments and promotions by the analysis results, we got an up to 6% conversion rate improvement.

Keywords Social network analysis · Consumer behavior analysis · Target marketing

P.-L. Chen (✉) · P.-C. Yang · T. Ku
Institute for Information Industry, Taipei, Taiwan, ROC
e-mail: mileschen@iii.org.tw

P.-C. Yang
e-mail: maciaclark@iii.org.tw

T. Ku
e-mail: cujing@iii.org.tw

© Springer Nature Singapore Pte Ltd. 2018

N.Y. Yen and J.C. Hung (eds.), *Frontier Computing*, Lecture Notes in Electrical Engineering 422, DOI 10.1007/978-981-10-3187-8_84

1 Introduction

Traditional advertising and marketing used in the past focused on the non-specific customers. However, most of them have diverse interests and the effects are not significant. Target marketing makes the promotion, pricing and distribution of products or services more cost-effective. Analyzing the shopping logs to find the target customers is adopted for ages. Besides, the increasing popularity of social networking services and the mobile devices have changed people's life. Social network analysis has become more and more important. To overcome the new business challenges, service providers have to improve their marketing strategies, analyze the existing and potential customers' behavior, and develop the precision marketing methods. The goal is to offer the right products to the right customers at right time and right location.

In the past, there were several studies focused on the analysis of customers [1–6]. Liao et al. found the differences between heavy and non-heavy users in top video apps based on Chunghwa Telecom network connection records [5]. He et al. analyzed the trend of tweets numbers for the big three pizza chains by text mining [3]. Such analysis results can be used to design appropriate marketing plans to improve sales [7]. However, those studies used only the social network data or private customer data.

In this study, we combined two types of user records, the social network user information and the customer shopping records. We got the customer shopping records from the online to offline (O2O) corporation we cooperated with. It combined the social media marketing and the service of earning reward points. Customers could use its point-earning app and get reward points from completing its assignments, for example, spending over NT$100 in the designated store. Its clients included POYA, Carrefour, SK-II, etc. Customers can exchange the reward points for some gifts, for example, donuts, tart or gift certificates. The combination of social network and consumer behavior analysis provided the power to extract business value from the large social and customer shopping data. The analysis results can be used to find the target customers, customize the point-earning app's assignments and promotions, and raise the conversion rates. In this paper, we used the customer shopping records from Shimen shopping district, a shopping district sells foods, accessories and apparel.

The primary goal of this study was to analyze the characteristics and purchase probability of different customers groups in the Shimen shopping district. All statistical procedures were performed with the Customer Behavior Analysis System we developed before. The secondary goal was to find the target customers by the analysis results and customize the point-earning app's assignments and promotions to improve the conversion rates of its clients.

2 Subjects/Materials and Methods

2.1 Customer Behavior Analysis System

The Customer Behavior Analysis System was a social and customer behavior analysis tool we developed in Institute for Information Industry. It was written in Java, Python, PHP, JavaScript, Tableau and the statistical procedures were built with the Apache commons mathematics library. The Customer Behavior Analysis System contained several web crawlers [8–10] to obtain different heterogeneous social and EC customer behavior data, including Facebook, Taobao, Tmall, PTT, blogs, web forums, etc. Besides, it also collected the customer shopping data and beacon data of the corporations we cooperated with. The framework diagram of the Customer Behavior Analysis System was shown in Fig. 1.

The Customer Behavior Analysis System merged different heterogeneous data and provided several types of analysis modules: association analysis, time series analysis, sentiment classification, sleeping customer detection, etc. The association analysis module found the hidden association between different customer groups, customers' preferences, shopping logs and stores. The time series analysis module discovered the peak sales and the customer shopping periods on different types of products. The sentiment classification module extracted keywords from user opinions and classified those opinions into different categories. The sleeping

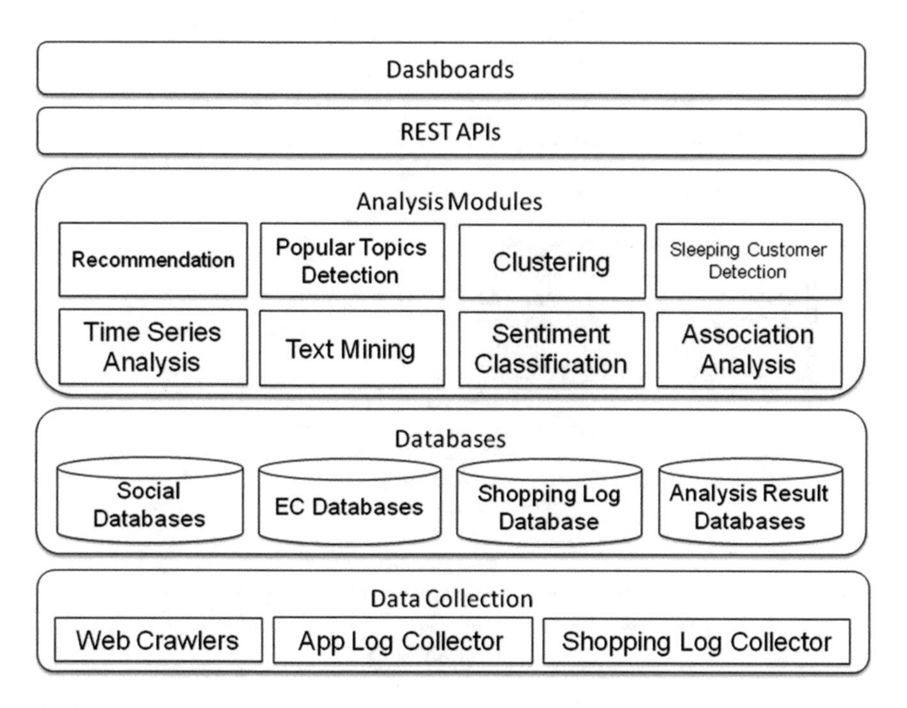

Fig. 1 The framework diagram of Customer Behavior Analysis System

customer detection module predicted the potential sleeping customers and sent alert to the corporations.

In this study, the architecture of modules used in the social network and customer behavior analysis was shown in Fig. 2. Those analysis modules were provided for the corporations we cooperated with by REST APIs or dashboards, helping them to improve their sales performance. Several analysis methods are developing and will be included in the Customer Behavior Analysis System in the future.

2.2 Data Source

In this section, we describe the data source we used in the paper. We collected the information of the users who installed the point-earning app. The point-earning app would give the users two types of assignments. One was to check into the store or scan the barcode of some products; another was to purchase some products in the store. When a user finished one assignment, he/she would be rewarded some reward points and the record would be saved in the database. Besides, the point-earning app needed the authorization to access user's information on Facebook. Therefore,

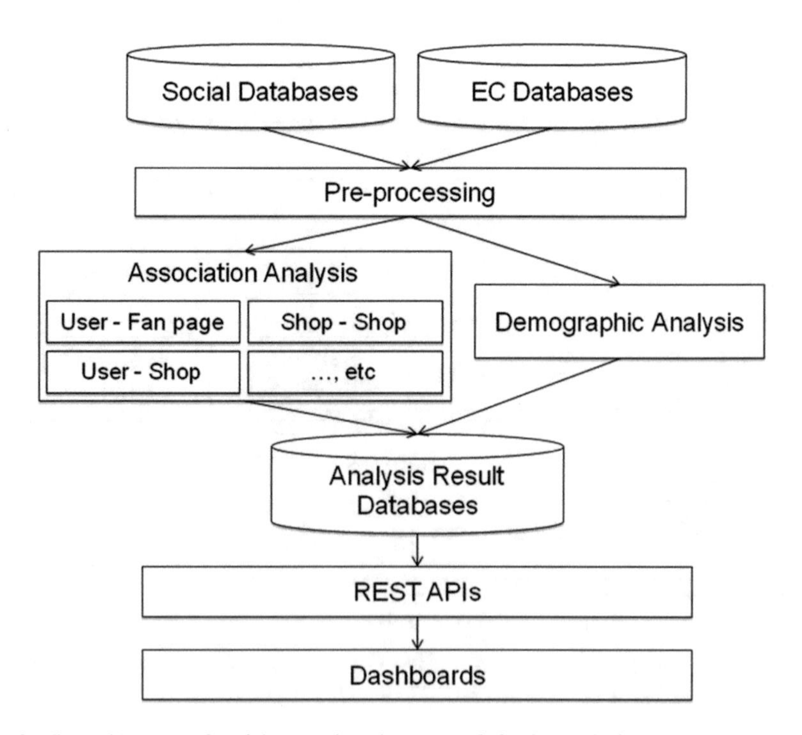

Fig. 2 The architecture of social network and customer behavior analysis

we had two types of information, the user records on the point-earning app and their information on Facebook.

In the data collection process, we collected the user records of the point-earning app from October 14, 2014 to August 12, 2015. The user records contained user Facebook ID, user email, invite code, shop name, assignment name, execution time, synchronization time, beacon ID, reward points, app platform, invoice information, invoice amount, and status. There were two types of records on the point-earning app according to the assignment name. One was the check-in assignment; another was the shopping assignment.

Furthermore, we collected the Facebook information of the point-earning app users by their Facebook ID. The user information on Facebook contained user Facebook ID, user email, gender, birthday, city he/she lived in, number of followers, the fan pages' IDs he/she liked, the fan pages' names he/she liked, and the fan pages' categories.

2.3 Study Samples

In the study, we extracted user records about the Shimen shopping district. The dates of those user records were between June 11, 2015 and July 12, 2015. Those records covered 98 stores in Shimen shopping district. There were 98 kinds of check-in assignments, included checking into the ASUS store, checking into the HTC store, checking into the KFC, etc. There were 92 kinds of shopping assignments, included spending any money in ASUS store, spending any money in HTC store, buying chicken in KFC, etc. To avoid the influence of the error records, we deleted the duplicate records or the records with incomplete status.

We separated the study samples into 8 groups, and those groups were mutually exclusive. The group of student included those were 24 years old and under. That was because most of Taiwan students found full-time jobs after they got Master's degree. The group of petite bourgeoisie included those were between 25 and 39 years old. Each of these groups was further separated into male and female groups. The groups of father and mother included those were 40 years old and above. Besides, the group of mother with babies included those were 25 years old and above, followed Facebook pages with keywords pregnancy, baby, infant, etc. Finally, the group of others included those with no information on Facebook.

2.4 Events

The goal of the point-earning app was to encourage the customers to purchase some products in the store. Therefore, the event was whether the user buy or not in the store. If one user finally bought something in the store, we put him/her into the event group. Otherwise, we put him/her into the non-event group if he/she only checked into the store or scanned barcode.

2.5 *Statistical Analysis*

All statistical procedures were performed with our Customer Behavior Analysis System. The top favorite fan pages in each user groups were calculated and ordered by the number of users followed those fan pages in each group. The conversion rate of one shop was defined as the number of users buying something in the shop divided by the number of check-in users. In the analysis of cross-store shopping, we calculated the number of users buying something in one shop and traced the following shops they checked into see if they buying something there or not.

3 Results

A total of 1078 point-earning app users who completed any assignments in Shimen shopping district were included in the case study cohort. Table 1 shows the demographics of point-earning app users who completed any assignments in Shimen. The female subjects (54.5%) were slightly more than the male subjects (43.4%). The main subjects were 18–39 years old (78.5%). The 7.7% subjects had one or more followers on Facebook. Nearly half of the subjects (44.7%) purchased some things in Shimen.

Table 2 shows the definition and number of subjects of each user group. Most of the subjects were petite bourgeoisie groups, petite bourgeoisie girl (21.6%) and petite bourgeoisie man (22.5%). The number of girl students (22.7%) was 1.5 times more than the number of boy students (13.3%). The main customers in Shimen shopping district were younger groups.

We calculated the favorite fan pages of each user group and listed all fan pages with more than 1% fans in each user group. However, the table was too large to be presented, so we only showed the top 1 favorite fan page of each user group in Table 3. The girl student group preferred the fan pages of Chinese singers (85.4%) and actors (84.3%), including May Day, Hebe, Jolin Tsai, etc. The boy student group preferred the fan pages of games (71.8%) and Chinese actors (69.2%), including League of Legends, Tetris Battle, Tower of Saviors, etc. The petite bourgeoisie girl group preferred the fan pages of online shopping (70.1%) and preferential information (65.8%), including Rakuten, Groupon, GOMAJI, etc. The petite bourgeoisie man group preferred the fan pages of games (62.6%) and news (49.6%), including Yahoo News, Apple Daily, ETtoday, etc. The mother group also preferred the fan pages of online shopping (43.3%). The father group preferred the fan pages of travel (56.7%) and online shopping (50%), including backpackers, i Walker, cola tour, etc. The mother with babies group preferred the fan pages of

Table 1 Demographics of point-earning app users who completed any assignments in Shimen*

Characteristics	Subjects (n = 1078)	
Age, median (IQR), y	27	(23–37)
<18	56	(5.2)
18–24	332	(30.8)
25–29	224	(20.8)
30–34	153	(14.2)
35–39	137	(12.7)
40–44	55	(5.1)
> = 45	71	(6.6)
No data	50	(4.6)
Gender		
Male	468	(43.4)
Female	587	(54.5)
No data	23	(2.1)
Follower		
No followers	995	(92.3)
One or more	83	(7.7)
Purchase		
Yes	482	(44.7)
No	596	(55.3)

Abbreviation: IQR, interquartile range
*Data are number (percentage) except where indicated

Table 2 Definition and number of subjects of each user group

User group	Subjects (%)		Description
Girl student	245	(22.7)	Female, 24 years old and under.
Boy student	143	(13.3)	Male, 24 years old and under.
Petite bourgeoisie girl	233	(21.6)	Female, between 25 and 39 years old
Petite bourgeoisie man	243	(22.5)	Male, between 25 and 39 years old
Mother	60	(5.6)	Female, 40 years old and above.
Father	59	(5.5)	Male, 40 years old and above.
Mother with babies	45	(4.2)	Female, 25 years old and above, follow Facebook pages with keywords pregnancy, baby, infant, etc.
Others	50	(4.6)	No information on Facebook.

occident singers (17.3%) and gender relationship (13.5%), including Lady Gaga, Avril Lavigne, Pairs, etc.

Table 4 shows the top 1 favorite shop category of each user group. The girl student group preferred the shops of accessories (39.1%), including Invisible

Table 3 The top 1 favorite fan page of each user group

User group	Fan page	Subjects (%)	
Girl student	Chinese singer	209	(85.4)
Boy student	Game	103	(71.8)
Petite bourgeoisie girl	Online shopping	163	(70.1)
Petite bourgeoisie man	Game	152	(62.6)
Mother	Online shopping	26	(43.3)
Father	Travel	33	(56.7)
Mother with babies	Occident singer	8	(17.3)

Table 4 The top 1 favorite shop category of each user group

User group	Shop category	Subjects (%)	
Girl student	Accessories	96	(39.1)
Boy student	Drinks	81	(56.6)
Petite bourgeoisie girl	Apparel	128	(54.9)
Petite bourgeoisie man	Sporting goods	109	(44.9)
Mother	Food	17	(28.3)
Father	Food	20	(33.9)
Mother with babies	Apparel	21	(46.7)

Shield, Moneyhold, WILD IDEA, etc. The boy student group preferred the beverages (56.6%), including COMBUY, CoCo, [15.25] Green Tea, etc. The petite bourgeoisie girl group preferred the apparel shops (54.9%), including San-Chuan (三川), Simple Clothing, COOL STYLE, etc. The petite bourgeoisie man group preferred the sporting goods (44.9%), including Momentum (摩曼頓), HOT-SHOES, etc. The mother and father group both preferred the food restaurants (28.3 and 33.9%), including KFC, PIZZA HUT, Victoria Restaurant, etc. The mother with babies group also preferred the apparel shops (46.7%).

Table 5 shows the top 10 conversion rates of shops in Shimen shopping district. KFC had the highest conversion rate (38.3%) and the highest number of check-in users (608). It also gave the highest reward points (250 points) and most of other shops gave 60–80 reward points. San-Chuan, which sells apparel, had the second highest conversion rate (35.9%). It had several branch stores (Room 26, 31, 28, etc.) and occupied 3 positions in the top 5 shops.

We analyzed the cross-store shopping behavior and listed all number of consumers in the original and next shop. However, the table was too large to be presented, so we only showed the sample of results in Table 6. There were 48 users shopping in the ASUS Phone Shop and 11 of these users also shopped in the HTC Phone Shop after (22.9%). Besides, there were 39 users shopping in the HTC Phone Shop and 20 of these users also shopped in the ASUS Phone Shop after (53.8%). There were 125 users shopping in one San-Chuan branch store (Room 28) and 64 of these users also shopped in another San-Chuan branch store (Room 31) after

(51.2%). Momentum sold sporting goods and there were 25 of 64 users shopping in two of its branch stores in one day. There were 20 users shopping in Fable Shoes and 7 of these users also shopped in J.T.S after (35.0%), and the two shops both sell shoes.

Table 7 shows the conversion rate of point-earning app users who completed any assignments in Shimen during different periods. 481 of 1078 (44.6%) subjects purchased some things in Shimen during June 11, 2015–July 12, 2015. According to the analysis results we provided, the O2O corporation customized the

Table 5 The top 10 conversion rates of shops in Shimen

Shop	Reward point	Consumers	Check-ins	Conversion rate (%)
KFC	250	233	608	38.3
San-Chuan (Room 26)	80	121	337	35.9
Paramount (百樂門)	80	137	413	33.2
San-Chuan (Room 31)	80	128	393	32.6
San-Chuan (Room 28)	80	125	392	31.9
Invisible shield	80	113	394	28.7
Moneyhold	80	109	385	28.3
[15.25] Green Tea	60	68	245	27.8
Momentum (峨嵋一)	60	81	334	24.3
Wild idea	80	82	361	22.7

Table 6 The sample of cross-store shopping analysis

Source Shop	Next Shop	Consumers in both shops	Consumers in source shop	Ratio (%)
ASUS Phone Shop	HTC Phone Shop	11	48	22.9
HTC Phone Shop	ASUS Phone Shop	20	39	51.3
San-Chuan (Room 28)	San-Chuan (Room 31)	64	125	51.2
Momentum (花花)	Momentum (峨嵋二)	25	64	39.1
Jolie cosmetics	Invisible shield	29	79	36.7
Fable shoes	J.T.S	7	20	35.0

Table 7 Conversion rate of point-earning app users who completed any assignments in Shimen during different periods

Period	Customers	All subjects	Conversion rate (%)
2015/06/11–2015/07/12	481	1078	44.6
2015/06/11–2015/08/12	690	1346	51.3

point-earning app's assignments and promotions. 690 of 1346 (51.3%) subjects purchased some things in Shimen during June 11, 2015–August 12, 2015. The conversion rate got an up to 6% improvement during one month.

4 Discussion

In this study, we analyzed the characteristics and purchase probability of the subjects in Shimen shopping district in different groups and got some interesting findings. For example, the main subjects were between 18 and 39 years old. Most of the subjects were petite bourgeoisie groups. The number of girl students was 1.5 times more than the number of boy students. The main customers in Shimen shopping district were younger groups. Finally, nearly half of the subjects (44.7%) purchased some things in Shimen.

We found that most of the girl students followed many stars' fan pages; whereas most of the boy students followed PC games and mobile games fan pages. Perhaps it was because the students had little economic pressure and focused on their interests. Interestingly, most of the petite bourgeoisie men followed fan pages about games too. Many petite bourgeoisie men also followed fan pages about news, and that meant they paid attention on the social events. Most of the petite bourgeoisie girls and mothers followed fan pages about online shopping and preferential information. That meant they had stable income and considered how to save money. Most of the fathers followed the fan pages about travel and online shopping. Perhaps it was because they had better financial ability and attached importance to quality of life. The mother with babies group followed occident singers and gender relationship. Perhaps it was because they were younger mothers and cared about their marital relationship. These results reflected the divergent favorites of different groups.

The favorite shop category of each user group also reflected their interests and financial ability. The girl student group liked accessories and the boy student group liked drinks, whereas the petite bourgeoisie girl liked apparel and the petite bourgeoisie man liked sporting goods which were more expensive. The father and mother groups both liked food restaurants. Perhaps it was because most shops sold young style goods which didn't fit in with their preference.

We also found that KFC had the highest conversion rate and the highest number of check-in users. Perhaps it was because it was the favorite fast food restaurant or it gave the highest reward points. San-Chua, which sells apparel, had the second highest conversion rate. It had several branch stores and occupied 3 positions in the top 5 shops. Other shops with high conversion rates sold accessories, sporting goods and mobile phone protection shield. That reflected the popular trend nowadays in the younger people.

We found several interesting things in the cross-store shopping analysis. The number of users buying something in ASUS Phone Shop after shopping in HTC Phone Shop was higher than the number of users buying something in HTC Phone

Shop after shopping in ASUS Phone Shop. It may reflect the favorite brand younger people liked. However, why those people shopping in different brands of mobile phone shops in one day was an interesting problem and worth further study. Besides, many people shopped at several similar stores in one day and that might mean they shopped purposefully not blindly.

The strengths of this study are that we cooperate with the information service providers. We combine the heterogeneous data from the social network and the consumer behavior data. We provide the analysis results for our customers and improve their conversion rates. There are also some limitations to this study. We collect the consumer behavior data from the information service providers' app. If someone is the customer in Shimen shopping district but not the information service providers' app user, we don't have his/her information.

In conclusion, we analyzed the characteristics and purchase probability of the subjects in Shimen shopping district in different groups to find the target customers. The main customers in Shimen shopping district were younger groups. Different groups of users had divergent favorites because of their age and gender. The popular trend nowadays in the younger people and the reward points the shop gave may affect the conversion rate of a shop. Most people shopped at several similar stores in one day and that might mean they shopped purposefully not blindly. Besides, we customized the point-earning app's assignments and promotions by the analysis results. The conversion rate of Shimen shopping district increased from 44.6 to 51.3% during one month, an up to 6% improvement.

Acknowledgements This study is conducted under the "System-of-systems driven emerging service business development project(1/4)" of the Institute for Information Industry which is subsidized by the Ministry of Economy Affairs of the Republic of China.

References

1. F. Bonchi, C. Castillo, A. Gionis, and A. Jaimes, "Social network analysis and mining for business applications," *ACM Transactions on Intelligent Systems and Technology,* vol. 2, 2011.
2. S. Dhaliwal, H. Jahangirli, R. Aljomai, A. Sarhan, W. Almansoori, and R. Alhajj, "Integrating social network analysis and data mining techniques into effective e-market framework," presented at the The 6th International Conference on Information Technology, 2013.
3. W. Hea, S. Zha, and L. Li, "Social media competitive analysis and text mining: a case study in the pizza industry," *International Journal of Information Management,* vol. 33, pp. 464–72, 2013.
4. H.-Y. Wu, K.-L. Liu, and C. Trappey, "Understanding customers using Facebook Pages: data mining users feedback using text analysis," in *Proceedings of the 2014 IEEE 18th International Conference on Computer Supported Cooperative Work in Design*, Hsinchu, 2014, pp. 346–50.
5. C.-H. Liao, Y.-H. Lei, K.-Y. Liou, J.-S. Lin, and H.-F. Yeh, "Using big data for profiling heavy users in top video apps," in *IEEE BigData Congress*, 2015.

6. G. Prassas, K. C. Pramataris, O. Papaemmanouil, and G. J. Doukidis, "A recommender system for online shopping based on past customer behaviour," presented at the In Proceedings of the 5th international conference on Intelligent user interfaces, 2001.
7. R. Bambini, P. Cremonesi, and R. Turrin, "A recommender system for an IPTV service provider: a real large-scale production environment," *Recommender Systems Handbook,* pp. 299–331, 2010.
8. C.-H. Tsai and P.-Y. Yang, "Social Event Radar (SER v1.0): design and implementation of a web crawlers based in social networks," presented at the The International Conference on Knowledge Community, 2014.
9. C.-H. Tsai, Y.-H. Cheng, T. Ku, and W.-F. Chien, "Object Architected Design and Efficient Dynamic Adjustment Mechanism of Distributed Web Crawlers," *International Journal of Interdisciplinary Telecommunications and Networking,* vol. 7, pp. 57–71, 2015.
10. T. Ku, C.-H. Tsai, P.-Y. Yang, and M.-J. Chen, "Social Event Radar (SER v2.0): efficient dynamic adjustment mechanism of distributed web crawlers in social networks," in *Proceedings of The World Congress in Computer Science, Computer Engineering, and Applied Computing*, 2014.

A Study of Identifying Online Shopping Behavior by using fsQCA Method

You-Shyang Chen, Chien-Ku Lin, Huan-Ming Chuang and Chyuan-Yuh Lin

Abstract The development of electronic commerce (e-commerce) has brought new business opportunities for shopping; thus, electronic stores (e-store) become new marketing channels. Good service quality is basic services and essentials for e-store. In order to understand impact of business relationship between the customer and the e-store, we use fuzzy set qualitative comparative analysis (fsQCA) method to analyze the framework of the study with empirical data and conclude three directions, as below: (1) The results of fsQCA reveal that situations combining promising positive reliability, responsiveness, assurance, environment quality, delivery quality and outcome quality can lead to a higher level of customer satisfaction and affective commitment. (2) The results exhibit that customers are more willing to purchase again if they experience positive service satisfaction or highly affective commitment. (3) Positive affective commitment supports customer advocacy intention.

Keywords E-store · Service quality · Fuzzy set qualitative comparative analysis

Y.-S. Chen
Department of Information Management, Hwa Hsia University of Technology,
111, Gong Jhuan Rd., Chung Ho District, New Taipei City 235, Taiwan, ROC
e-mail: ys_chen@cc.hwh.edu.tw

C.-K. Lin (✉)
Department of Business Administration, Feng Chia University,
100, Wenhwa Rd., Seatwen, Taichung 40724, Taiwan, ROC
e-mail: g9923808@yuntech.edu.tw

H.-M. Chuang · C.-Y. Lin
Department of Information Management, National Yunlin University
of Science and Technology, 123, University Road, Section 3, Douliou
Yunlin 64002, Taiwan, ROC
e-mail: chuanghm@yuntech.edu.tw

C.-Y. Lin
e-mail: g9923807@yuntech.org.tw

© Springer Nature Singapore Pte Ltd. 2018

N.Y. Yen and J.C. Hung (eds.), *Frontier Computing*, Lecture Notes in Electrical Engineering 422, DOI 10.1007/978-981-10-3187-8_85

1 Introduction

In a highly competitive and rapidly changing environment, Internet retailers increasingly emphasis service quality. By improving service quality, e-stores cannot only maintain a competitive advantage but also ensure survival in a competitive industry. Therefore, a finding better way to meet customer demand and attracts customers is an issue worth exploring. Among previous studies of service quality, including discussions of both traditional service quality and electronic service quality, many studies have assessed the overall service quality or dimensions of service quality. However, various dimensions of service quality still influence antecedents of customer loyalty and loyalty behavior differently. In this study, we select appropriate research methods based on research motivation and purpose to explore the influence of service quality dimensions on customers' antecedents of loyalty and behavior.

We use the fuzzy set Qualitative Comparative Analysis (fsQCA) method for advanced research of the experimental outcome of the combined cause-effect situations. Such finding may help explain some of macro-society phenomena and the different effects of the results based on the influence of causal relationships.

2 Literature Review

2.1 Service Quality Models

PZB SERVQUAL model. Since being proposed by Parasuraman, Zeithaml, and Berry in 1985 [1], SERVQUAL has been a good and widely used means for measuring service quality. Parasuraman et al. [1] conducted exploratory research and proposed five gaps of perceived service quality: consumer expectation—perception gap (Gap1), management perception—service quality specification gap (Gap2), service quality specification—service delivery gap (Gap3), service delivery—external communications gap (Gap4) and expected service—perceived service gap (Gap5), respectively. Later, in order to avoid overlaps between dimensions, the aforementioned dimensions were simplified into five dimensions: (1) reliability, (2) responsiveness, (3) assurance, (4) empathy, and (5) tangibles, in the SERVQUAL model also developed [2].

Quality of Electronic Services model (QES). QES differs from traditional service quality in two important ways. First, there is the self-service character of electronic services. Second, the service environment is largely created by specific design features of the graphic user interface [3].

2.2 Customer Loyalty and its Antecedents

Custome loyalty. Early approaches for assessing customer loyalty emphasized the behavioral aspect as repeat patronage or purchase frequency [4]. Nevertheless, behavior-oriented customer loyalty suffers a major shortcoming in that it does not tell the marketer why a brand was selected as a customer may be making a repeat purchase not because of true commitment but because of convenience, price, availability, or inertia due to habit [5]. The attitudinal customer loyalty approach proposes that loyalty involves much more than repeat purchase behavior and includes a favorable attitude reflecting a preference or commitment expressed over time [5].

Satisfaction. Many studies from a variety of fields have contributed to the discussion of satisfaction. This is because all commercial activities can generate customer satisfaction. Satisfaction is defined as an effective response to a purchase situation [6, 7].

Affective commitment. Within the marketing field, the concept of affective commitment is similar to that of organizational behavior, meaning that one party keeps a business relationship because they like their business partner and enjoy the partnership. They feel loyalty and a sense of belonging [8]. Affective commitment can also be considered as a psychological status.

2.3 Fuzzy Set Qualitative Comparative Analysis (fsQCA)

Coding work is the most basic research tool and skill in the field of social science. Coding sustains study of all causal conditions through the examination of each recipe and definition of activities or events within it through line-by-line coding. However, the weakness the QCA method is that coding work is highly professional and also time consuming. Thus, Ragin [9] developed fuzzy set qualitative comparative analysis (Fuzzy set QCA) which adopts the concept of QCA within its framework and process flow. Ragin [9] compared two categories of methods and explained that the differences between fsQCA and QCA are: (1) fsQCA selects and defines objects from literature reviews, not from QCA coding work. (2) fsQCA data collection is bound under specifically designed questionnaires rather than based on free interviews. (3) fsQCA uses setups of the fuzzy range scope for fuzzy membership function, calculation of membership scores, and the logic concept of Boolean algebra to convert and categorize membership scores into trust tables with multiple combinations and different levels of standards instead of grounded QCA.

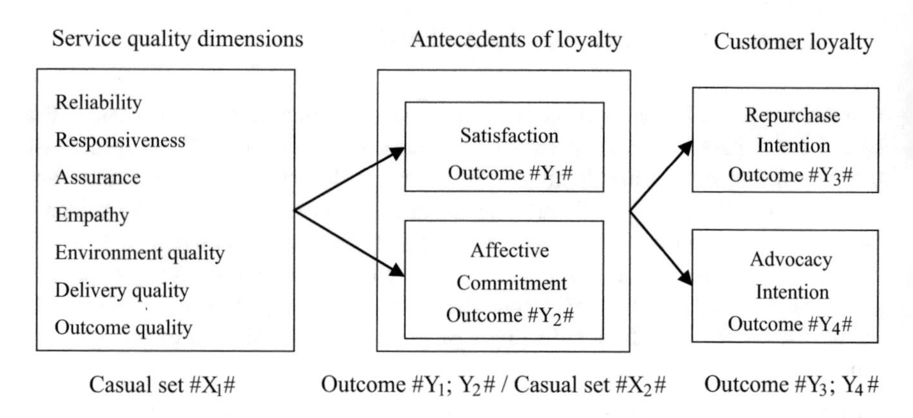

Fig. 1 Research framework of this study

3 Research Methodology

3.1 Research Framework

This section describes the use of fuzzy set qualitative comparative analysis (fsQCA) to find the results of the key elements and examine the logic of the object of analysis. The research framework of this study is shown in Fig. 1.

3.2 Data Collection

Through surveys, we target only consumers who have online shopping experiences and a certain quantity of samples is needed to reach the requirement of our study. This study expects to collect a wide range of information, so gender, age and educational background are not restricted in this study. Two questionnaires were delivered: first, convenience sampling was executed for the pilot study. Based on the results, the items were adjusted minimally and a formal questionnaire was developed. An online questionnaire was distributed for convenience sampling so as to collect the necessary amount of samples. The formal questionnaire was designed with Google Forms. The link of the questionnaire was posted in BBS and social networks so that it could be completed voluntarily. Last, invalid questionnaires were deleted when recovering the samples so as to increase the reliability and validity of the information.

3.3 Data Analysis and Results

This study conducts fsQCA analysis by using the software program fs/QCA 2.5. The program simplifies data patterns to identify potentially "sufficient" causal associations. Its ultimate products are a set of logical statements identifying factors or combinations of factors that appear as subsets of an outcome, along with mathematical measures with which to assess their utility. The framework of evolutionary fsQCA provides the methodology for exploring multiple relationships among service quality and customer loyalty. Not all of these combinations have exploratory power to explain the causal relationships. The combinations should enable the examination of the consistency value using threshold criteria that should be greater than 0.75 [10].

Linking service quality dimension to satisfaction. Following the above approach, we apply the frequency threshold of 20 and the consistency of 0.85. A fuzzy truth table algorithm is based on the logic concepts of Boolean algebra. The truth table is an analytic device that displays all logically possible combinations of causal conditions, indicates cases' distribution across these combinations and provides a simplified data map.

$$\text{rel} * \text{res} * \text{asu} * \text{enq} * \text{deq} * \text{ouq} \rightarrow \text{sat}$$

When one interprets the formula accordingly, it can thus be deduced that cases that combine a high level of reliability (rel) with a high level of responsiveness (res), high assurance, high environment quality (enq), high delivery quality, and a high level of outcome quality will have a high level of customer satisfaction (sat).

Linking service quality dimensions to affective commitment. Following the same approach, we apply the frequency threshold of 20 and the consistency of 0.85. Truth table rows directly correspond to the logical possibility of a particular causal combination. Only 2 of these combinations are actually represented empirically in our data. All other logically possible configurations are "remainders"—counterfactual configurations that lack empirical instances.

$$\text{rel} * \text{res} * \text{asu} * \text{enq} * \text{deq} * \text{ouq} \rightarrow \text{acm}$$

When one interprets the formula accordingly, it can thus be deduced that cases that combine a high level of reliability (rel) with a high level of responsiveness (res), high assurance, high environment quality (enq), high delivery quality, and a high level of outcome quality will have a high level of customer affective commitment (acm).

Linking antecedents of loyalty to repurchase intention. Following the above approach, we apply the frequency threshold of 20 and the consistency of 0.9. Truth table rows directly correspond to the logical possibility of a particular causal combination. Only 3 of these combinations are actually represented empirically in our data. All other logically possible configurations are "remainders"—counterfactual configurations that lack empirical instances.

$$sat + acm \rightarrow rep$$

When one interprets the formula accordingly, it can thus be deduced that cases that combine a high level of satisfaction (sat) or cases that combine high level of affective commitment (acm) will have a high level of customer repurchase intention (rep).

Linking antecedents of loyalty to advocacy intention. Following the same approach, we apply the frequency threshold of 20 and the consistency of 0.9. Truth table rows directly correspond to the logical possibility of a particular causal combination. Only 3 of these combinations are actually represented empirically in our data. All other logically possible configurations are "remainders"—counterfactual configurations that lack empirical instances.

$$acm \rightarrow adv$$

When one then interprets the formula accordingly, it can thus be deduced that cases that combine a high level of affective commitment (acm) will have a high level of customer advocacy intention (adv).

4 Conclusion

Based on our comparative case study of service quality dimensions, we draw three major conclusions, which we elaborate below. We also discuss the implications of these conclusions for e-store management. The first major conclusion of the present study suggest that in order to improve service quality and thus enhance customer satisfaction and affective commitment, e-store managers should be committed to improving service quality dimensions, including reliability, responsiveness, assurance, environment quality, delivery quality and outcome quality. Analysis results indicate that cases that combine a high level of reliability with a high level of responsiveness, high assurance, high environment quality, high delivery quality, and a high level of outcome quality will have a high level of customer satisfaction and affective commitment. The second conclusion of the present study has to do with promoting customer repurchase intention; e-store managers should be devoted to improving customer satisfaction or affective commitment. This result indicates that cases that combine a high level of satisfaction or cases that combine a high level of affective commitment will have a high level of customer repurchase intention. The third conclusion of the present study concerns encouraging customer advocacy intention; e-store managers should make efforts to improve customer affective commitment. This result indicates that cases that combine a high level of affective commitment will have a high level of customer advocacy intention. Final, the bringing together of complexity theory, fsQCA, and pattern research in service quality dominant logic is a fit-like-a-glove union for advancing theory, method, and practice in service quality research. Complexity theory is a useful lens for seeing that simple antecedent conditions relate to positive, negative and neutral outcome conditions—which of these three relationships occur depends on the observed complex antecedent conditions in which the simple antecedent conditions occur. Individual service facets in service performances with successful service outcomes do not all have to consist of positive ingredients in all possible complex but parsimonious patterns.

References

1. Parasuraman, A., Zeithaml, V. A., Berry, L. L.: A Conceptual Model of Service Quality and Its Implications for Future Research. Journal of Marketing. 49 (4), 41–50 (1985)
2. Parasuraman, A., Zeithaml, V. A., Berry, L. L.: SERVQUAL: A Multiple-Item Scale for Measuring Consumer Perceptions of Service Quality. Journal of Retailing. 64 (1), 12–40 (1988)
3. Fassnacht, M., Koese, I.: Quality of Electronic Services Conceptualizing and Testing A Hierarchical Model. Journal of Service Research. 9 (1), 19–37 (2006)
4. Anderson, R. E., Srinivasan, S. S.: E-Satisfaction and E-Loyalty: A Contingency Framework. Psychology & Marketing. 20 (2), 123–138 (2003)
5. Zikmund, W. G., McLeod, J. R., Gilbert, F. W.: Customer Relationship Management: Integrating Marketing Strategy and Information Technology. NY: Wiley, New York (2003)
6. Anderson, J. C., Narus, J. A.: A Model of Distributor Firm and Manufacturer Firm Working Partnerships. The Journal of Marketing. 54, 42–58 (1990)
7. Bennet, R., Hartel, C. E., McColl-Kennedy, J. R.: Experience As a Moderator of Involvement and Satisfaction on Brand Loyalty in a Business-to-Business Settings. Industrial Marketing Management. 34 (1), 97–107 (2005)
8. Geyskens, I., Steenkamp, J. B. E., Scheer, L. K., Kumar, N.: The Effects of Trust and Interdependence on Relationship Commitment: A Trans-Atlantic Study. International Journal of Research in Marketing. 13 (4), 303–317 (1996)
9. Ragin, C. C.: Redesigning Social Inquiry: Fuzzy Sets and Beyond. Wiley Online Library (2008)
10. Ragin, C. C., Rubinson, C., Schaefer, D., Anderson, S., Williams, E., Giesel, H.: User's Guide to Fuzzy-Set/Qualitative Comparative Analysis. University of Arizona. 87 (2008)

Investigating the Acceptance Intention of Wearable Symbiotic Devices by Fuzzy Set Qualitative Comparative Analysis—Based on the Case of Mi Band

Huan-Ming Chuang, Li-Chuan Wang, Chien-Ku Lin, You-Shyang Chen and Kuo-Chuan Lai

Abstract Wearable symbiotic devices are fully functional, self-contained computer systems that can be worn, carried, or attached to the body, letting the user access information anytime and anywhere. Though the acceptance of information systems have been widely discussed and verified, unique features of wearable devices, such as the relocation of human-computer interface from external devices onto the body itself, concerns about perceived comfort and privacy merits further investigations that motive this study. An Internet questionnaire survey was conducted based on the case of Xiaomi Mi Band. With 548 valid responses, the study applied fuzzy set qualitative comparative analysis (fsQCA) under set-theoretical approach to explore important configuration of attributes contributing to the acceptance of wearable devices. Major research findings include the configurations of attributes contributing to favorable attitude as well as acceptance intention of wearable symbiotic devices. Then practical guidance for promoting wearable devices is suggested.

H.-M. Chuang · K.-C. Lai
Department of Information Management, National Yunlin University of Science and Technology, 123, University Rd., Section 3, Douliou Yunlin 640, Taiwan, ROC
e-mail: chuanghm@yuntech.edu.tw

K.-C. Lai
e-mail: n10223002@yuntech.edu.tw

L.-C. Wang · C.-K. Lin (✉)
Department of Business Administration, Feng Chia University, 100, Wenhwa Rd., Seatwen, Taichung 40724, Taiwan, ROC
e-mail: g9923808@yuntech.edu.tw

L.-C. Wang
e-mail: wlj@changiron.com

Y.-S. Chen
Department of Information Management, Hwa Hsia University of Technology, 111, Gong Jhuan Rd., Chung Ho District, New Taipei City 235, Taiwan, ROC
e-mail: ys_chen@cc.hwh.edu.tw

© Springer Nature Singapore Pte Ltd. 2018

N.Y. Yen and J.C. Hung (eds.), *Frontier Computing*, Lecture Notes in Electrical Engineering 422, DOI 10.1007/978-981-10-3187-8_86

Keywords Wearable symbiotic device • Theory of planned behavior • Set-theoretical approach • Qualitative comparative analysis

1 Introduction

Wearable symbiotic devices are fully functional, self-contained computer systems that can be worn, carried, or attached to the body, letting the user access information anytime and anywhere [1, 2]. Comparing with portable devices, wearable ones remain attached to the body regardless of its orientation and activity, do not need to be removed from the body in order to be operated, and do not require muscular effort to be carried around [2].

Among extant wearable devices, Xiaomi Mi Band (hereafter, Mi Band) is a wearable fitness tracker produced by Xiaomi Tech and unveiled during a Xiaomi launch event on 22 July 2014. Mi Band resembles a bracelet in its design, and can be worn on either hand or on the neck. The band's location can be set using the official Mi Band app.

Though the acceptance of information systems have been widely discussed and verified, unique features of wearable devices, such as the relocation of human-computer interface from external devices onto the body itself, concerns about perceived comfort and privacy merits further investigations that motive this study.

Research objectives can be summarized as follows. First, identify configurations of critical factors contributing to favorable attitude toward wearable devices. Second, investigate important configurations of antecedents for acceptance of wearable devices. Last, suggest practical strategies and tactics for the promotion of wearable devices.

2 Literature Review

2.1 Theory of Planned Behavior

Theory of planned behavior (TPB) is proposed for explaining general individual behavior emphasizing that individual behavior is driven by behavior intentions which are further determined by individual's attitude toward behavior, subjective norms (or social influence), and perceived behavioral control (or facilitating conditions) [3].

2.2 *Innovation Diffusion of Theory*

Current knowledge on new product/service adoption largely relies on Rogers' [4] innovation diffusion theory (IDT) which posits that perceived attributes of the innovation are the major drivers of adoption [5]. These attributes play the role of components of market knowledge [6] representing gestalt of information about customer needs and preferences available to the markers of new product/service.

2.3 *Other Important Dimensions Relevant to Wearable Symbiotic Systems*

Apart from above-mentioned general dimensions relevant for the acceptance of wearable symbiotic systems, Spagnolli et al. [2] maintain two dimensions should also be taken into considerations. First, perceived comfort relates to physical factors of the device such as size, weight, and textile fibers that might impact users' experienced comfort. Besides wearing devices in public might be perceived as funny and embarrassing that cause discomfort. Second, since wearable symbiotic devices collect personal sensitive information, perceived privacy could raise privacy issues such as misinterpretation of data disregarding their original meaning [7, 8].

2.4 *Fuzzy Set Qualitative Comparative Analysis*

Coding work is the most basic research tool and skill in the field of social science. Coding sustains study on all causal conditions through the examination of each recipe and definition of activities or events within it through line-by-line coding. However, the weakness of QCA method is that coding work is highly professional and also time consuming. Thus, Ragin [9] develops fuzzy set qualitative comparative analysis (Fuzzy set QCA) which adopts the concept of QCA with framework and process flow in the further research.

Epstein et al. [10] indicate that fsQCA offers several advantages. First, it is better-suited than regression for exploring causal configurations—situation in which variables have an impact only in combination with a high or low degree of one or more other factors. Second, fsQCA allows to identify multiple pathways to an outcome. Third, whereas regression is useful for examining tendentious relationships—the general tendency of a particular factor to influence an outcome of interest fsQCA is helpful in exploring a different kind of relationship: causal sufficiency.

3 Method

3.1 Research Framework

Based on literature review, research framework of this study can be illustrated as Fig. 1.

3.2 Profile of Research Subjects

This study used internet questionnaire survey to collect data. 601 responses were collected, excluding 53 invalid ones, there were 548 valid samples accounting for 91.18% valid response rate.

From the profile, respondents distribute quite evenly in age coverage. Most of them have degree of bachelor and above. Out of 548 respondents, 168 respondents have experience with wearable devices which accounts for 30.66%. Among those with wearable devices experience most of them 77.98% adopt Mi Band.

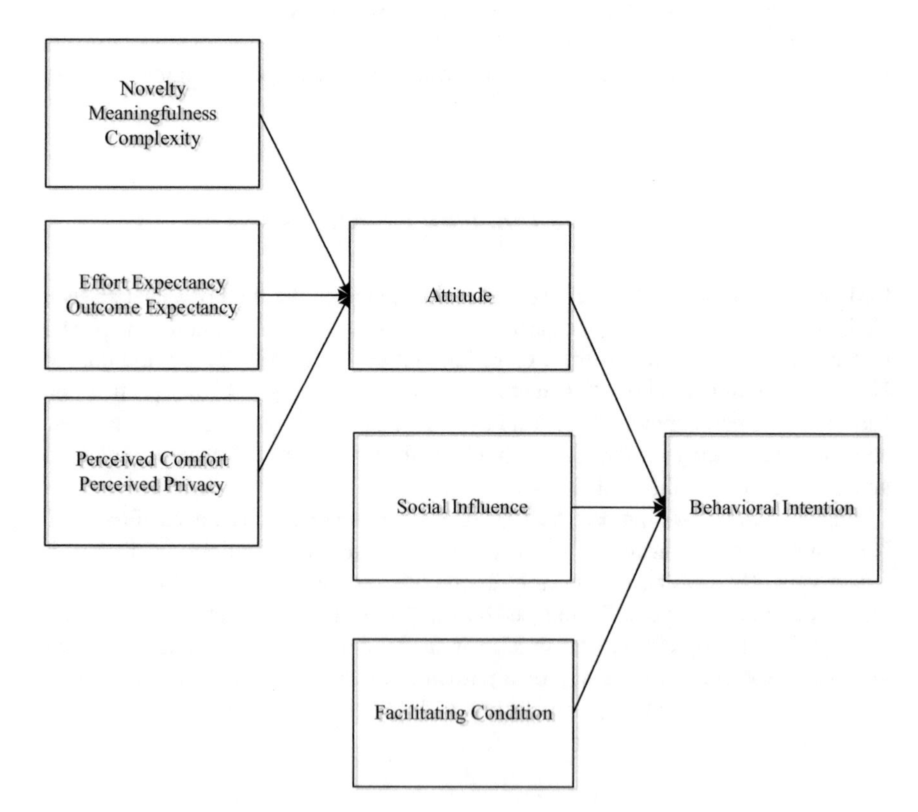

Fig. 1 Research framework

4 Data Analysis

4.1 *Calibration of Data*

Since collected data is based on Likert 7 scale, calibration was conducted to transform it into scale from 0 to 1.

4.2 *fsQCA Analysis Results*

Two models were analyzed by QCA as shown in Table 1. The results of two models can be summarized as Table 2.

5 Discussion

Based on the results of QCA analysis, in terms of favorable attitude toward wearable devices, it is found meaningfulness; outcome expectancy, effort expectancy, and perceived comfort are essential requirements. Then two distinctive customer segmentations can be identified. The first segmentation is composed of customers who prefer wearable devices with low complexity. In contrast, the other segmentation is formulated by customers who pursue novel products and be sensitive to their privacy.

Regarding the acceptance intention, two significant types of customers can also be identified. One has customers both favorable attitude toward wearable devices and well prepared facilitating condition. The other has customers who do not involve with wearable devices much and accept wearable devices just casually; therefore, it doesn't matter with favorable attitude, social influences, or facilitating condition.

Table 1 Models analyzed by QCA

Model	DV	IV
1	AT	NVE, MEAN, COM, EE, OE, PC, PP
2	BI	AT, SI, FC

Table 2 Results of QCA analysis

Model	Frequency cutoff	Consistency cutoff	Configurations	Solution coverage	Solution consistency
1	11	0.990	LOC * MEAN * OE * EE * PC NVE * MEAN * OE * EE * PC * PP	0.739	0.986
2	41	0.876	AT * FC ~AT * ~SI * ~FC	0.826	0.933

References

1. Buenaflor, C., Kim, H. C.: Six Human Factors to Acceptability of Wearable Computers. International Journal of Multimedia and Ubiquitous Engineering. 8 (3), 103–114 (2013)
2. Spagnolli, A., Guardigli, E., Orso, V., Varotto, A., Gamberini, L.: Measuring User Acceptance of Wearable Symbiotic Devices: Validation Study Across Application Scenarios. Symbiotic Interaction, Springer. 87–98 (2014)
3. Ajzen, I.: The Theory of Planned Behavior. Organizational Behavior and Human Decision Processes. 50 (2), 179–211 (1991)
4. Rogers, E. M.: Diffusion of Innovations. Simon and Schuster (2010)
5. Meuter, M. L., Bitner, M. J., Ostrom, A. L., Brown, S. W.: Choosing Among Alternative Service Delivery Modes: An Investigation of Customer Trial of Self-service Technologies. Journal of Marketing. 69 (2), 61–83 (2005)
6. Ordanini, A., Parasuraman, A., Rubera, G.: When the Recipe Is More Important Than the Ingredients: A Qualitative Comparative Analysis (QCA) of Service Innovation Configurations. Journal of Service Research. 17 (2), 134–149 (2014)
7. Britz, J. J.: Technology as a Threat to Privacy: Ethical Challenges and Guidelines for the Information Professionals. Microcomputers for Information Management. 13, 175–193 (1996)
8. Tavani, H. T.: Genomic Research and Data-Mining Technology: Implications for Personal Privacy and Informed Consent. Ethics and Information Technology. 6 (1), 15–28 (2004)
9. Ragin, C. C.: Redesigning Social Inquiry: Fuzzy Sets and Beyond. Wiley Online Library (2008)
10. Epstein, J., Duerr, D., Kenworthy, L., Ragin, C.: Comparative Employment Performance: A Fuzzy-Set Analysis. Method and Substance in Macrocomparative Analysis. 67–90 (2008)

A Study on the Acceptance Intention of "My Health Bank" from the Perspective of Health Belief Model

Li-Chuan Wang, Huan-Ming Chuang, You-Shyang Chen, Chien-Ku Lin and Yung-Shuen Wang

Abstract To help people effectively control healthcare-related experience and expenditure, Taiwan's Ministry of Health and Welfare launched the "My Health Bank" (MHB hereafter) system. This study focuses on important factors influencing acceptance intention of MHB, after extensive literature review, health belief model (HBM) combined with extended valence framework was adopted to establish a research framework. Then SmartPLS 2.0 (M3) was applied for structural equation modeling (SEM) analysis for 433 valid samples. Major conclusion can be summarized as follows: (1) perceived integrity and perceived competence have significant positive effect on users' trust toward MHB, (2) performance risk and time risk are two major concerns for users' risk perception toward MHB, (3) perceives severity and self-efficacy have significant and positive effect on users' perceived benefit toward MHB, and (4) perceived risk and perceived benefit have significant positive effect on users' acceptance intention of MHB. These findings can help suggesting strategies for effective MHB promotions.

L.-C. Wang · C.-K. Lin
Department of Business Administration, Feng Chia University, 100, Wenhwa Rd., Seatwen, Taichung 40724, Taiwan, ROC
e-mail: wlj@changiron.com

C.-K. Lin
e-mail: g9923808@yuntech.edu.tw

H.-M. Chuang · Y.-S. Wang
Department of Information Management, National Yunlin University of Science and Technology, 123, University Rd., Section 3, Douliou, Yunlin 640, Taiwan, ROC
e-mail: chuanghm@yuntech.edu.tw

Y.-S. Wang
e-mail: n10223302@yuntech.edu.tw

Y.-S. Chen (✉)
Department of Information Management, Hwa Hsia University of Technology, 111, Gong Jhuan Rd., Chung Ho District, New Taipei City 235, Taiwan, ROC
e-mail: ys_chen@cc.hwh.edu.tw

© Springer Nature Singapore Pte Ltd. 2018

N.Y. Yen and J.C. Hung (eds.), *Frontier Computing*, Lecture Notes in Electrical Engineering 422, DOI 10.1007/978-981-10-3187-8_87

Keywords My Health Bank • Health belief model • Extended valence framework • Self-efficacy • Acceptance intention

1 Introduction

To manage one's health, the general public usually relies on care from medical staff, while ignoring the importance of self-management. To help people effectively control healthcare-related experience and expenditure, Taiwan's Ministry of Health and Welfare launched the "My Health Bank" (MHB hereafter) system.

However, based on social-technical approach, the successful promotion of MHB needs not only technological side as system design and functionality, but also human factors. As a result, this study aims to investigate factors influencing the acceptance intention of MHB from a holistic perspective. Major purposes of this study can be summarized as: (1) understand the effects of perceived trust on perceived risk, perceived benefits, and intention to use MHB, (2) explore major factors influencing the acceptance of MHB, and (3) suggest practical strategies for effective promotion of MHB.

2 Literature Review

2.1 Health Belief Model

The health belief model (HBM) was established by social psychologists to explain preventive health behavior early in the 1950s [1]. The model posits that an individual's health behavior depends on the existence of certain beliefs toward a given condition [2].

2.2 Extended Valence Framework

The valence framework is developed from the economics and psychology literatures to understand consumer behaviors [3]. This framework identifies that perceived risk and perceived benefit are two fundamental aspects of consumers' purchasing behavior [4]. In general, consumers try to minimize undesired negative effects (i.e., perceived risks), meanwhile maximize positive effects (i.e., perceived benefits) of purchasing [3].

3 Research Method

3.1 Research Framework

Based on literature review, research framework of this study can be illustrated as Fig. 1.

3.2 Hypotheses Development

Hypotheses developed from research framework can be summarized as follows.

Hypotheses related to perceived trust.
H1a. Perceived integrity has significant positive effects on perceived trust
H1b. Perceived competence has significant positive effects on perceived trust
H1c. Perceived benevolence has significant positive effects on perceived trust
Hypotheses related to perceived risk.
H2a. Performance risk has significant positive effects on perceived risk
H2b. Psychology risk has significant positive effects on perceived risk
H2c. Time risk has significant positive effects on perceived risk
Hypotheses related to perceived benefits.
H3a. Perceived susceptibility has significant positive effects on perceived benefits
H3b. Perceived severity has significant positive effects on perceived benefits
H3c. Self-efficacy has significant positive effects on perceived benefits

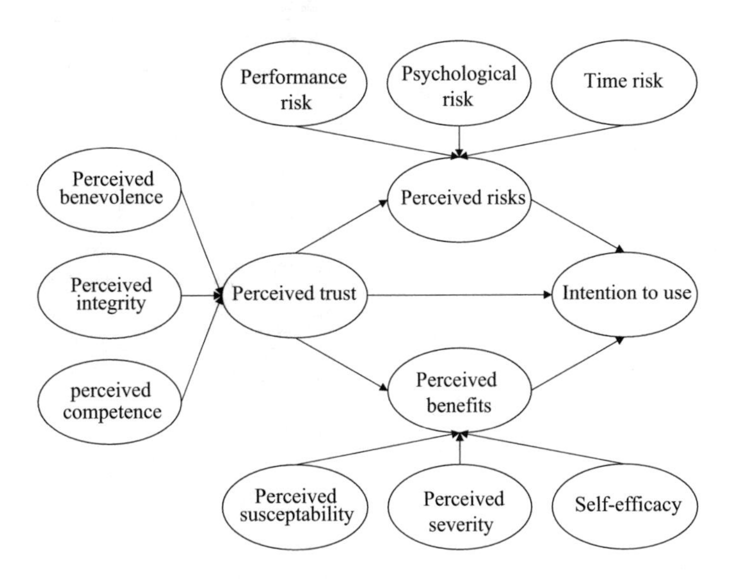

Fig. 1 Research framework of this study

Hypotheses related to intention to use.
H4a. Perceived trust has significant positive effects on perceived risk
H4b. Perceived trust has significant positive effects on perceived benefits
H4c. Perceived trust has significant positive effects on intention to use
H4d. Perceived risk has significant negative effects on perceived benefits
H4e. Perceived benefits has significant positive effects on perceived benefits

3.3 *Profile of Research Subjects*

The demographic profile of respondents for the research subjects is presented in Table 1.

Table 1 Demographic profile of respondents

Profile	Characteristic	N	%
Gender	Male	215	48.6
	Female	227	51.4
Age	20 and below	49	11.1
	21–30	219	49.5
	31–40	130	29.4
	41–50	34	7.7
	51 and above	10	2.3
Education	Elementary school and below	1	0.2
	Junior high school	2	0.4
	High school	23	5.2
	College	303	68.6
	Master and above	113	25.6
Occupation	Student	102	24.5
	Manufacturing	62	14.9
	Government employees	45	10.8
	IT industry	51	12.2
	Service industry	72	17.3
	Self-employed	9	2.2
	Professionals	41	9.8
	Unemployed	19	4.6
	Other	16	3.8

(continued)

Table 1 (continued)

Profile	Characteristic	N	%
Experiences with MHB usage per month	0	387	87.6
	1–5	55	12.4
	6–10	0	0
	11–15	0	0
	16 and above	0	0
The average number of medical treatment per month	0–5	437	98.9
	6–10	4	0.9
	11 and above	1	0.2

3.4 Data Analysis

This study used Structural Equation Modeling (SEM) for parameter assessment and hypothesis testing for the proposed casual model. Specifically, PLS algorithm and bootstrapping procedure were applied by SmartPLS software 2.0 (M3) [5].

4 Results

4.1 Assessment of Measurement Model

To examine the measurement model, outer loading, composite reliability (CR), average variance extracted (AVE) as convergent validity and discriminant validity were assessed [6].

All the outer loadings of the constructs are well above the minimum threshold value of 0.70. Besides, all the constructs have high level of internal consistency reliability. Furthermore, the AVE values are well above the threshold level of 0.5 indicating convergent validity.

4.2 Assessment of Structural Model

Once the measurement model has been confirmed as reliable and valid, next step is to evaluate the structural model results, which involves examining the model's predictive capabilities and the relationships between the constructs [6].

Assessing the significance and relevance of the structural model relationship was conducted by applying the PLS-SEM algorithm, which estimates the path coefficients to demonstrate the hypothesized relationships between the constructs. In addition to assessment of the size of the path coefficients, there significance was evaluated by bootstrapping option (5000 resample).

5 Discussion

The applications of extended valence framework and health belief model show satisfactory prediction on the intention to use of MHB that is aggressively promoted by Taiwan's Ministry of Health and Welfare. Major conclusion can be summarized as follows: (1) perceived integrity and perceived competence have significant positive effect on users' trust toward MHB, (2) performance risk and time risk are two major concerns for users' risk perception toward MHB, (3) Perceives severity and self-efficacy have significant and positive effect on users' perceived benefit toward MHB, and (4) perceived risk and perceived benefit have significant positive effect on users' acceptance intention of MHB.

Other than perceived benevolence, psychology risk, and perceived susceptibility, all constructs exert expected effects. Since MHB is promoted and supported by government, it is reasonable that perceived benevolence and psychology risk will not be major concerns. It should be noted that perceived susceptibility does not have significant effects on perceived benefits. This implies that motivation to use MHB does not be limited to people with greater health care needs. Consequently, the target users for the promotion of MHB could be anyone who is sensitive to their health conditions.

References

1. Rosenstock, I. M.: Historical Origins of the Health Belief Model. Health Education & Behavior. 2 (4), 328–335 (1974)
2. Chen, M. S., Land, K. C.: Testing the Health Belief Model: LISREL Analysis of Alternative Models of Causal Relationships between Health Beliefs and Preventive Dental Behavior. Social Psychology Quarterly. 49 (1), 45–60 (1986)
3. Kim, D. J., Ferrin, D. L., Rao, H. R.: Trust and Satisfaction, Two Stepping Stones for Successful E-Commerce Relationships: A Longitudinal Exploration. Information Systems Research. 20 (2), 237–257 (2009)
4. Peter, J. P., Tarpey Sr, L. X.: A Comparative Analysis of Three Consumer Decision Strategies. Journal of Consumer Research. 2 (1), 29–37 (1975)
5. Ringle, C. M., Wende, S., Will, S.: SmartPLS 2.0 (M3) Beta, Hamburg (2005)
6. Hair Jr, J. F., Hult, G. T. M., Ringle, C., Sarstedt, M.: A Primer on Partial Least Squares Structural Equation Modeling (PLS-SEM). Sage Publications (2013)

Application of Internet of Things: Seating Status Monitor System on Mobile Devices for Public Transportation

Cheng-Ming Chang and Sheng-Wen Hong

Abstract As we move from www (static pages web) to web2 (social networking web) to Web3 (ubiquitous computing web), the real time sensing anywhere of Internet of things (IoT) applies increasingly. Ubiquitous sensing enabled by Wireless Sensor Network (WSN) technologies covers many areas of modern day living. This offers the ability to collect day living indicators and monitor urban live. This paper presents a system which includes several sensor units to detect the change of seating status and send a flag to the receiver by wireless signal. The receiver transmits the flag to the cloud monitor system. The monitor system will update the real time seating status and send it to passenger's mobile devices such as smart phone, tablet PC and fixed devices such as metro platform's big screen as well, displaying seats layout map with number and location of "total seats", "be seated", "vacancy". Passenger will know the real time seating status before car arrives at the station through mobile devices such as mobile phone, tablet and through touch screen platform also. This system works for public transportation such as bus, metro, train etc.

Keywords Internet of things · Mobile device · Public transportation · Monitor system · Seating status

1 Introduction

According to the report of the APTA (American Public Transportation Association), Americans took 10.8 billion trips on public transportation in 2014 (the highest in 58 years). Also, public transit ridership is up 39% since 1995, outpacing pop-

C.-M. Chang (✉) · S.-W. Hong
Department of Information Management, Hwa Hsia University of Technology, New Taipei, Taiwan (ROC)
e-mail: cmchang@go.hwh.edu.tw

S.-W. Hong
e-mail: swh0205@gmail.com

© Springer Nature Singapore Pte Ltd. 2018

N.Y. Yen and J.C. Hung (eds.), *Frontier Computing*, Lecture Notes in Electrical Engineering 422, DOI 10.1007/978-981-10-3187-8_88

ulation growth, which is up 21% [1]. On the other hand, U.S. public transportation use saves 37 million metric tons of carbon emission every year. Also, Public transit use in the United States saves 4.7 billion gallons of gasoline annually, according to APTA [2]. In short, Public Transportation is used more frequently than decades ago and is friendly to the environment. This instigates the intention to build a system to improve the experience of taking public transportation.

The seating status monitor system includes several sensor units to detect the change of seating status and send a flag to the receiver by wireless signal. The receiver transmits the flag to the cloud monitor system. The monitor system will update the real time seating status and send it to passenger's mobile devices and platform's big screen, displaying seats layout map with number and location of "total seats", "be seated", "vacancy". Passenger will know the real time seating status before car arrives at the station through mobile devices such as mobile phone, tablet and through touch screen platform also. This system works for bus, metro, train etc.

The features include:

- Seats Layout map
- Vacancy or be seated real time status
- Mobile devices display
- Screen display on platform
- Vacancy finding before train arrives

2 System Description

The seating status monitor system structure be showed in Fig. 1. The system includes several sensor units to detect the change of seating status and send a flag to the receiver by wireless signal, the cloud monitor system to monitor the sensor units and manage the system, the display units to show the real time status.

- Sensor units:

Sensor unit includes three parts which are detecting units marking 113, 114, 115, 116 in Fig. 2, sending unit marking 111, 112 in Fig. 2, and receiving unit marking 12 in Fig. 1. The detecting unit, locate under the seat, will detect the seat status changes which are from vacancy to be seated and from be seated to vacancy through detecting the on and off change of electricity. The left graph of Fig. 2 shows vacant seat with off electricity. The right graph of Fig. 2 shows be seated seat with on electricity. The sending unit will send a flag of the seat status change to the cloud monitor system through wireless signal. Receiving unit will receive a flag of confirmation from cloud monitor system to make sure the system had received the flag.

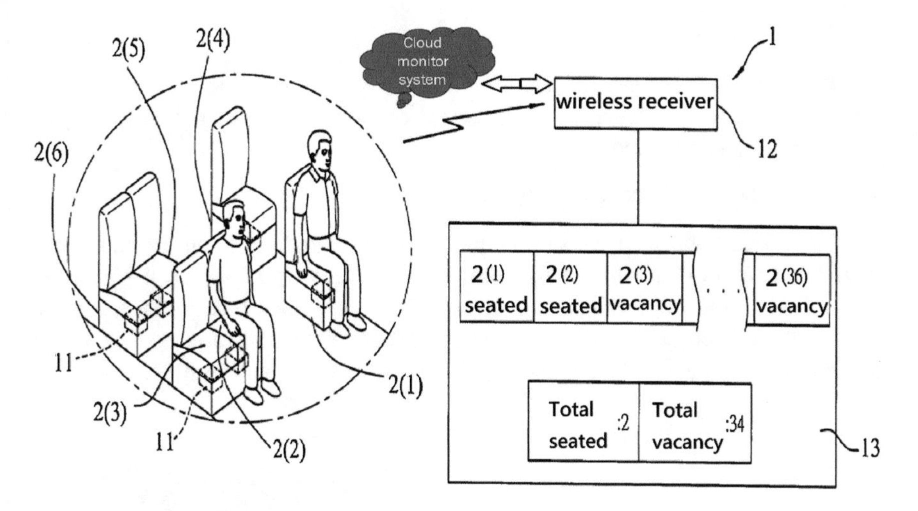

Fig. 1 Monitor system structure

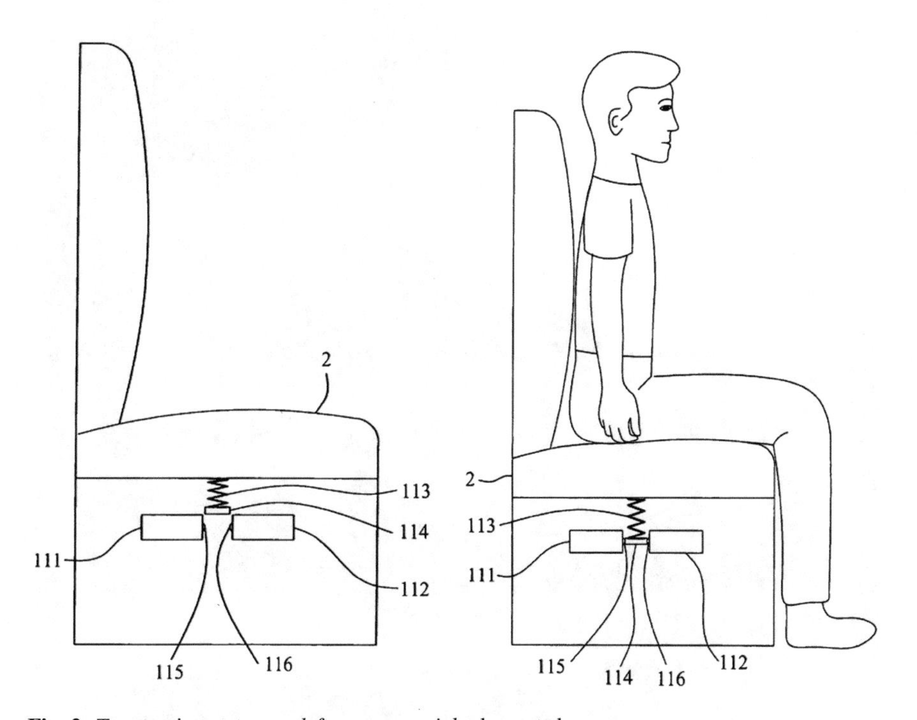

Fig. 2 Two seating statuses—*left*: vacancy, *right*: be seated

- Cloud monitor system:

The cloud monitor system, show in Fig. 1, will control and monitor the seat status change received from wireless receiver and update the seat layout map on the screen display simultaneously.

- Display on mobile device or big screen:

Figure 3 demonstrates the seat finding scenario which will help to find vacant seats before train or metro arriving station. Passengers will look up the total seat status first to find the carriage with vacant seats, then click that carriage, the seat layout map will pop out and display the number of total seats, number of be seated, number of vacancy. Passengers look up the real time seat status with mobile devices

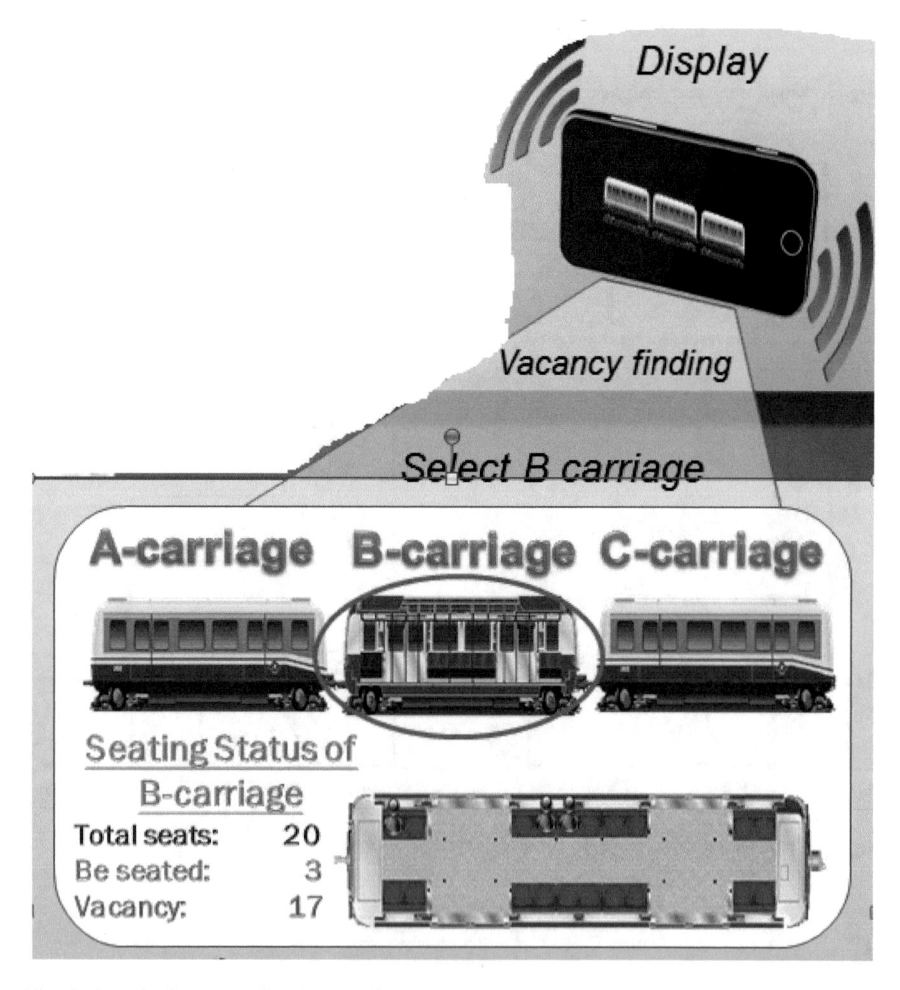

Fig. 3 Seat finding scenario with mobile device

or big touch screen at the station before public transportation arriving station. Then move to the destined position where carriage with vacant seat will arrive and hopefully they can find the exact vacant seat location very quickly.

3 Conclusion

The seating status monitor system features real time seats layout map, mobile devices display and big screen display as well. The system was invented to save the effort for passengers to find a vacant seat before the metro or train arrive the station. This empirical system aims to improve efficiency and comfortability for passengers in taking public transit.

Hopefully, this system can apply to any kinds of seat status finding application, like seat finding at theater, conference hall etc. and potential participants can check out the seat status by monitoring the real time seat status layout map through mobile devices and computer's screen too.

References

1. American Public Transportation Association, web site. http://www.apta.com/mediacenter/ptbenefits/Pages/default.aspx
2. Virginia Miller, "More than 35% of U.S. Public Transit Buses Use Alternative Fuels or Hybrid Technology- Public Transportation is Leading the Way in Green Vehicles", 4/22/2013 Transit News, http://www.apta.com/mediacenter/pressreleases/2013/Pages/130422_Earth-Day.aspx

To Explore Female Players Decision-Making Processes in MMORPGs: A Preliminary Model

Tung-Lin Chuang

Abstract The genre of massively multiplayer online role-playing games (MMORPGs) has become increasingly popular with the group of young. Online games have emerged as an important and interesting research area in a variety of academic and official settings. This study adopts ethnographic decision tree modelling (EDTM) which integrates qualitative and quantitative paradigms to model the decision-making process of whether or not to play online games by female undergraduate students. The model is built based on qualitative data gathered through in-depth interviews of 29 respondents, where 18 criteria for playing online games are elicited. The model will be tested using qualitative and quantitative data from interviews, and a structured questionnaire in the further study. Armed with the results, this study provides online game industries an understanding of the decision-making process of female players, and reveals theory standpoints for the researchers with similar interests.

Keywords Massively multiplayer online role-playing games • Ethnographic decision tree modelling

1 Introduction

With the increase in Internet usage, people from all over the world have already been able to participate together in on-line games, and this trend of playing on-line games continues. The emergence of on-line games, especially massively multiplayer online role-plating games (MMORPG), as a new culture on the Internet is apparent in the worldwide electronic game market and continuously dominated the entire game market within a few years of its emergence [3]. On-line games are an interactive entertainment industry where new business models, new markets,

T.-L. Chuang (✉)
Department of Advertising and Strategic Marketing, Ming Chuan University, Taipei, Taiwan
e-mail: tunglin@mail.mcu.edu.tw

© Springer Nature Singapore Pte Ltd. 2018

N.Y. Yen and J.C. Hung (eds.), *Frontier Computing*, Lecture Notes in Electrical Engineering 422, DOI 10.1007/978-981-10-3187-8_89

and new growth are being developed [21]. For these reasons, many researchers have used qualitative research methods to explore through the psychological perspective what expectations and factors contribute to players' intention to participate in on-line games [12, 13, 24]. However, it has been difficult for researchers in the traditional market research field to elicit and explicate exactly what customers wanted as they have used many of the common marketing research methods such as surveys, and have thus relied on verbal communication as the primary data-collection method. Thus, the main purposes of this study are twofold: First, to establish a model using ethnographic decision tree modeling (EDTM) to describe the decision-making process of Internet users. Second, the study desires to provide a framework and endeavors to gain an in-depth understanding of the decision-making process of whether or not to play on-line games.

2 Literature Review

Ethnography is defined as the science of describing a group of people and their culture (attributes or properties), and has a long tradition in anthropology [23]. The overarching characteristic of an ethnographic approach is its commitment to cultural interpretation [17]. By studying behavior in natural environments, ethnography is used to explore the culture and elicit knowledge of behavior and the contexts in which they occur [18]. In other words, ethnography describes a culture from the native's or insider's point of view and not from the researcher's or outsider's point of view [19]. This is important, as decision making needs to take into account the intrinsic viewpoint of the person making the decision [4].

Decision tree models are similar to ethnographic techniques or cognitive-science models like taxonomy, componential analysis, and plans or scripts [20, 22]. Thus, a real world decision should be made using decision criteria derived from interviews or participatory observations, and by analyzing the data to determine which factors the person considers to validate the decision criteria. Ethnographic decision tree modeling (EDTM) is a formal technique used to combine individual decision makers' criteria and rules or expert systems into a computer-programmed decision model for the group, which can be tested against actual choice data collected from a sample of decision makers in the group. A good model is a statement about a process and should provide a sense of process within it [11]. A good model is evaluated in terms of its ability to correctly predict other new facts, situations and events. While, in general, quantitative decision making uses complex mathematical models to predict a person's choice [28], it usually fails to capture the individual's emotions. To combat this, ethnographic methods use qualitative research paradigms and the cognitive ethnographic decision tree model methodology, while embodying quantitative research paradigms [4].

Decision trees had predicted, with a high degree of accuracy in a wide variety of choice contexts, a great number of decisions. These researches included that decisions made by Californian families' decisions regarding the sexual division of labor within the family for daily routine tasks [16], farmers' cropping decisions [1, 5] peasants' choice of treatment for illness in Pichatero, Mexico [26, 27], United States car buyers' choice of cars [6, 7]. In each case in which the method had been used, the predictability had been as high as 85–95% of the historical choice data used to test the model.

3 Research Design and Methodology

3.1 Research Procedure

This study was hybrid blending as proposed by Gladwin [4] who described the ethnographic method using qualitative research paradigms and the cognitive ethnographic decision tree models methodology embodying quantitative research paradigms. The immediate goal of the study is to seek and understand the process of decision making in MMORPGs by female players. This research process is thus carried out in two phases (show as Fig. 1). The first phase, models the concerns of female players who facing the choice of playing MMORPGs through the qualitative research method. In the second phase of verifying model, to use the quantitative research method tested the model until the predictability up to 85% success rate with the same population as first phase. However, the preliminary model will be verified in the further study (Fig. 2).

3.2 Data Samplings and Collection

The study assessed the content validity, three e-commerce scholars and four Internet users (two experienced and two unexperienced on-line games players) were invited to comment on the set of interview questions that derived from comprehensively searching relevant and up-to-date literature on on-line game and related topics, including online gamer demographic profiles. Furthermore, according to Market Intelligence Center (MIC) indicated that most of players whose educational

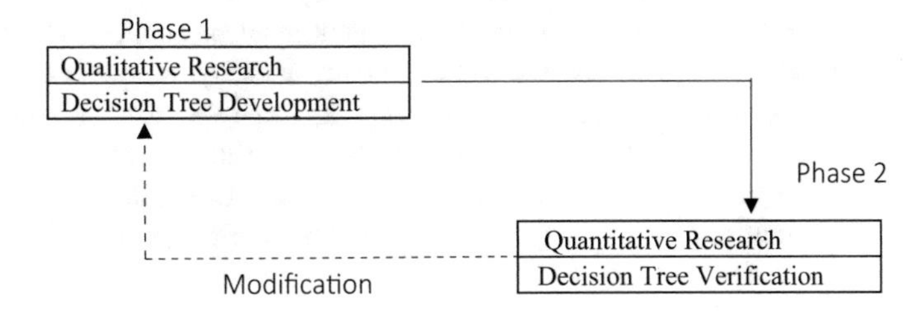

Fig. 1 The research design

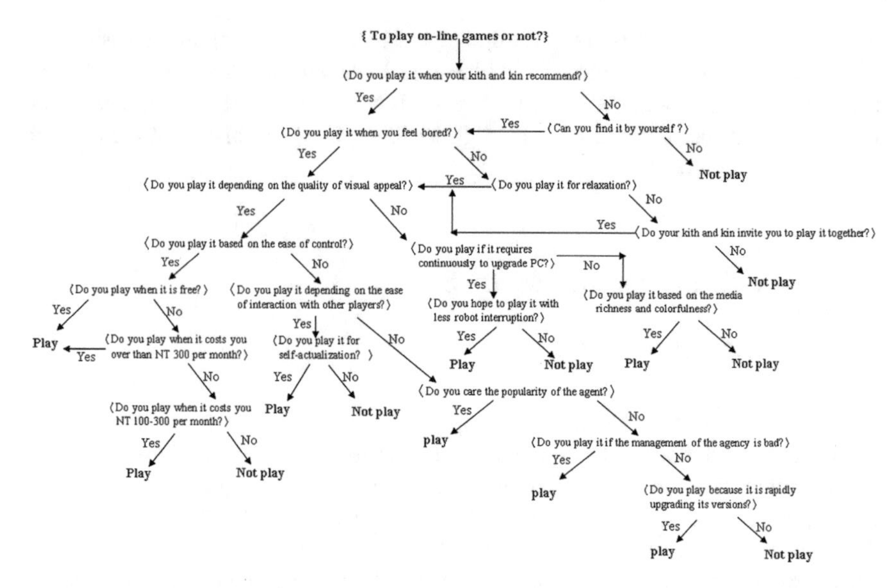

Fig. 2 A preliminary model of decision-making processes

background were undergraduate students in Taiwan. The average age was 22.8 years old in which most of part among 20–29 years old. All samplings in this study were corresponding to the statistics [14, 15]. In order to capture the unique criteria that Internet users had depended on, the ethnographic interview technique was adopted. A convenience sample of the 29 respondents, with years experienced of Internet access, were interviewed about 30–45 min until no new information emerged in the study. The average age in this sample is 18.96 which ranged from 19 to 25. The interviews adopted open-ended questions that addressed users' perceptions of on-line playing, reasons for playing or not playing on-line games, and influential factors in making a decision.

3.3 Data Analysis

In data analysis stage, the study adopted content analysis to elicit information from verbatim. Holsti offered a broad definition of content analysis as any technique for making inferences by objectively and systematically identifying specified characteristics of messages [8]. Circumstantially, content analysis was a systematic, replicable technique for compressing many words of text into fewer content categories based on explicit rules of coding [2, 10, 25]. Under Holsti definition [8], the technique of content analysis was not restricted to the domain of textual analysis, but might be applied to other areas. In order to allow for replication, however, the technique could only be applied to data that were durable in nature.

4 Research Finding

In the model building phase, the process includes an iterative process of asking a series of ethnographic questions, collecting ethnographic data, analyzing data, and discovering better questions that are designed to assist the researcher in identifying key decision criteria (show as Table 1). The outcome of 18 criteria used to build up the preliminary model and those criteria are phrased as questions, which incorporated the key words and concepts, identified from verbatim. Following the suggestions [4], the ordering of criterions in the decision tree can be developed through the use of frequency statistics and logic. The more frequencies of decision criteria, the higher located in the decision model. Each question on the model is represented one criterion. Moreover, each criterion is developed onto a singular question requiring a "yes" or "no" answer. A preliminary model of decision-making processes will be validated in the further study until 85% or more. If the model predicts less than 85% of decisions, it probably required modification again [4].

Table 1 The criteria of decision-making

No.	Decision criteria	Fre.
1	Do you play it for relaxation?	14
2	Do you play it when you feel bored?	13
3	Do you play it when your kith and kin recommend?	12
4	Do you play it depending on the quality of visual appeal?	11
5	Do you play it based on the ease of control?	11
6	Do you play when it is free?	9
7	Do you play it depending on the ease of interaction with other players?	8
8	Do your kith and kin invite you to play it together?	7
9	Do you play when it costs you over than NT300 per month?	4
10	Do you play when it costs you NT100-300 per month?	4
11	Do you hope to play it with less robot interruption?	2
12	Do you play it based on the media richness and colorfulness?	2
13	Do you care the popularity of it?	2
14	Do you play it for self-actualization?	2
15	Do you play if it requires continuously to upgrade PC?	2
16	Can you find it by yourself?	2
17	Do you play it if the management of the agency is bad?	1
18	Do you play because it is rapidly upgrading its versions?	1

5 Conclusions

This study adopts ethnographic decision tree modelling (EDTM) to model the decision-making process of whether or not to play online games by female undergraduate students. The model is built based on qualitative data gathered through in-depth interviews of 29 respondents, where 18 criteria for playing online games are elicited. Most of the players would like to play on-line games to kill time or for relaxlation. Moreover, they liked the on-line game because of the quality of visual appeal, no charge and ease of control. Contrary to [9] used technology acceptance model (TAM) to probe into why people play on-line games by quantitative paradigm. The result indicated that "perceived ease of use" effected "attitude toward playing an online game", and "attitude toward playing an online game" affected "intention to play an online game". The factor of "ease of use" playing an important part in both of researches were parallel to each other.

References

1. Barlett, P.: The Structure of Decision Making in Paso. American Ethnologist, 4(2), 285–307 (1977)
2. Berelson, B.: Content Analysis in Communication Research. Glencoe, Ill: Free Press (1952)
3. Chuang, Y. C.: Massively Multiplayer Online Role-Playing Game-Induced Seizures: A Neglected Health Problem in Internet Addiction. CyberPsychology & Behavior, 9(4), 451–456 (2006)
4. Gladwin, C. H.: Ethnographic Decision Modeling. Newbury Park. CA, Sage (1989)
5. Gladwin, C.: Contributions of Decision-Tree Methodology to a Farming Systems Program. Human Organization, 42(2), 146–157 (1983)
6. Gladwin, H., & Murtaugh, M.: Test of a Hierarchical Model of Auto Choice on Data from the National Transportation Survey. Human Organization, 43(3), 217–226 (1984)
7. Gladwin, H., & Murtaugh, M.: The Attentive Pre-attentive Distinction in Agricultural Decisions. In P. Barleet (Ed.), Agricultural Decision Making. New York: Academic Press, 115–136 (1980)
8. Holsti, O.R.: Content Analysis for the Social Sciences and Humanities. MA: Addison-Wesley (1969)
9. Hsu, C. L. & Lu, H. P.: Why do people play on-line games? An extended TAM with Social Influences and Flow Experience, Information & Management, 41(7), 853–868 (2004)
10. Krippendorff, K.: Content analysis: An introduction to its Methodology. Newbury Park, CA: Sage (1980)
11. Lave, C., & March, J.: An Introduction to Models in the Social Sciences. New York: Harper and Row (1975)
12. Liu, M. T.: Model Construction and Confirmation for On-line Game Players' Consensus. Journal of Management, 24(2), 191–209 (2007)
13. Magni, Taylor & Venkatesh.: "To Play or Not to Play": A Cross-Temporal Investigation Using Hedonic and Instrumental Perspectives to Explain User Intentions to Explore a Technology. International Journal of Human-Computer Studies, 68, 572–588 (2010)
14. Market Intelligence Center (MIC), http://mic.iii.org.tw/intelligence
15. Market Intelligence Center (MIC), http://mic.iii.org.tw/aisp/pressroom/press01.asp

16. Mukhopadhyay, C.: Testing a Decision Process Model of the Sexual Division of Labor within the Family. Human Organization, 43(3), 227–242 (1984)
17. Patton, M., & Westby, C.: Ethnography and Research: A Qualitative View. Topics in Language Disorders, 12(3), 1–14 (1992)
18. Punch, K. F.: Introduction to Social Research, Quantitative and Qualitative Approaches. London: Sage Publications (1998)
19. Spradley, J. P.: The Ethnographic Interview. New York: Holt, Rinehart & Winston (1979)
20. Schank, R., & Abelson, R. Scripts, plans, goals and understanding. New York: Wiley and sons (1979)
21. Sharp, C. E., & Rowe, M.: On-line Games and E-business: Architecture for Integrating Business Models and Services into On-line Games. IBM System Journal, 45(1), 161–179 (2006)
22. Werner, O., & Schoepfle, G. M.: Systematic Fieldwork, 1, Newbury Park, CA: Sage (1987)
23. Wallace, E.: Psychiatry and Its Nosology: A Historicao-Philosophical Overview. In Sadler, J., Wiggins, O., and Schwartz, M (eds.), Philosophical perspectives on psychiatric diagnostic categories. Baltimore: Johns Hopkins University Press, 16–86 (1994)
24. Wan, C. S., & Chiou, W. B.: Psychological Motives and Online Games Addiction: A Test of Flow Theory and Humanistic Needs Theory for Taiwanese Adolescents. CyberPsychology & Behavior, 9(3), 317–324 (2006)
25. Weber, R. P.: Basic Content Analysis, 2nd ed. Newbury Park, CA (1990)
26. Young, J. C.: A model of illness treatment decisions in Tarascan town. American Ethnologist 7(1), 106–131 (1980)
27. Young, J. C.: Medical Choice in a Mexican Village. New Brunswick, NJ: Rutegers University Press (1981)
28. Zhuang, C. H.: Decision-Making and Simulation Analysis for Capita Planning. Statistics Monthly, 81, 21–29 (1996)

The Research of the Cloud Service Adopted by the Corporations—A Perspective of Innovation Diffusion Theory

Ming-Kuen Chen, Yen-Ling Lin and Wei-Chih Huang

Abstract Current cloud service providers are devoted to the development of large-scale, systematic cloud service systems. However, because of related factors, large-scale, systematic firm acceptance of cloud services still cannot be achieved. Firm acceptance of cloud services is still uncertain. This study combines the Technology Acceptance Model and Innovation Diffusion Theory and probes the acceptance of cloud computing services introduced in firms through the use of pre-decision, in-decision and post-decision stages. According to the research findings, obtained using the process-oriented Innovation Diffusion Theory and the Technology Acceptance Model, cloud service cost, security and service continuity/completeness will influence the stages of cloud service introduction in the information technology industry.

Keywords Cloud service · Technology Acceptance Model · Innovation Diffusion Theory · Decision making

1 Introduction

With regard to IT innovations or the introduction of information systems, there are many factors in employees' technology acceptance. Past researchers constructed many technology acceptance behavior models, the Technology Acceptance Model (TAM) being the most complete and common of these [4]. According to past

M.-K. Chen · W.-C. Huang
Department of Business Management, National Taipei University of Technology, Taipei, Taiwan
e-mail: mkchen@ntut.edu.tw

W.-C. Huang
e-mail: raul3776@gmail.com

Y.-L. Lin (✉)
Department of Management, Fo Guang University, Yilan, Taiwan
e-mail: yllin@gm.fgu.edu.tw

© Springer Nature Singapore Pte Ltd. 2018

N.Y. Yen and J.C. Hung (eds.), *Frontier Computing*, Lecture Notes in Electrical Engineering 422, DOI 10.1007/978-981-10-3187-8_90

studies on the TAM, the model has three research dimensions: broad application to IT, the explanation of potential users' acceptance of IT innovations and the analysis and study of factors in users' acceptance of IT innovations. However, when studying IT, some researchers attempted to expand the TAM and included other possible factors [26]. Thus, based on Innovation Diffusion Theory (IDT), proposed by Roger [21], this study probes the factors in the acceptance of cloud services introduced in firms before the introduction of cloud computing services becomes prevalent. Using a literature review, the researcher discusses the characteristics and related challenges of cloud services and applies the TAM and IDT to the framework of decision stages. Using expert interviews, this study constructs the acceptance model of the introduction of cloud services in firms and uses the IT industry as an example to probe factors at different decision stages of cloud service introduction in firms.

2 Literature Review

2.1 Cloud Services

Currently, cloud computing service is the most popular topic in business circles and academia. Past studies have also suggested that firms can save costs using cloud services and improve flexibility and efficiency [5, 12, 16, 18, 23, 25]. However, cloud computing service is still developing, and before it becomes prevalent, many challenges and concerns will certainly emerge [12, 19, 24]. According to past research on the concerns and challenges of introducing cloud services, there are more questions and concerns regarding compatibility, trust, service continuity and security.

2.2 TAM

The TAM is the theoretical model that Davis [4] developed after revising rational action theory. The model is adopted to identify the provenance of IT innovations, explain and predict potential users' acceptance of IT innovations and analyze the decision-making factors used by users to accept IT innovations. The TAM is treated as the most complete and common theoretical model that explains and predicts potential users' acceptance of IT innovations, [9, 17, 30]. The TAM suggests that people's actual system use of an IT innovation is influenced by the behavioral intention to use. Behavioral intention to use is affected by attitude toward use, Perceived Usefulness (PU) and Perceived Ease of Use (PEOU). The model suggests that external factors in user acceptance of a new technology are based on two specific TAM beliefs [28]: PU and PEOU. The two critical moderators influence users' behavioral intention to use. Thus, in the original TAM, attitude factors include three basic relationships: PEOU influences PU, PEOU influences attitude, and PU influences attitude. The three relationships are highly consistent and precise in explaining new technology products and services.

2.3 IDT

IDT was first proposed by Rogers in [20], and Rogers later supplemented or revised related content. From a process-oriented perspective, the theory explains how an innovative product or service is accepted by customers or introduced. Roger divided the innovation decision into the following five stages: knowledge, persuasion, decision, implementation and confirmation.

Since Roger proposed IDT in 1962, the theory has been demonstrated by thousands of studies. According to a literature review, most of the current research on IDT adopts compatibility, as suggested by Rogers [22]. IDT is not only adopted to explore potential users' acceptance of IT innovations but also applied to firms' adoption of innovation technology [1, 7, 29]. Because it is considered appropriate to study potential users' acceptance of new IT using IDT and the TAM [11, 28, 30], this study uses these concepts to probe firms' introduction of cloud services.

3 Research Method

3.1 Research Structure

According to past research, when cloud services are introduced in firms, many problems and concerns materialize [12, 14, 15, 18, 24, 25]. Because cloud services are within the scope of operation of IT and information systems, to explore the acceptance of the introduction of cloud services in firms, this study refers to the TAM as the basic framework for the following reasons: (1) The TAM is adopted to recognize the prevalence of IT, to explain and predict potential users' acceptance of an IT innovation and to analyze the factors in users' acceptance of an IT innovation [3]; (2) The TAM is regarded as the most complete and common theoretical model explaining and predicting potential users' acceptance of an IT innovation attitude [9, 17, 30]; (3) The reliability of TAM questionnaires has been supported by many studies, experiments and modifications. There is no need to construct individual measurement tools in accordance with different types of IT [11, 26, 31].

3.2 Research Hypotheses

Based on the TAM and from the innovation diffusion perspective, this study uses three stages of IDT suggested by Yu and Tao [31] as the base of the research framework: pre-stage, in-stage and post-stage. Using a process-oriented view, the researcher attempts to identify how innovative products or services are accepted by firms and includes the characteristics of cloud services to establish the research framework of firms' acceptance of the introduction of cloud services. Based on the framework, the researcher identifies whether cost, security, service continuity/completeness, PU, PEOU and compatibility will influence different decision stages of firms' introduction of cloud services.

(1) Pre-decision stage

By conducting a literature review of the characteristics of cloud services and the concerns and challenges inherent in their introduction, the researcher attempts to identify whether firms will consider and plan costs, security and service continuity/completeness [12, 14, 15, 24, 25]. In a study on IDT, compatibility is the most frequently cited element of the five elements of innovation characteristics [2, 9, 27]. The study of the relationship between PU and PEOU is common in TAM research [2, 6, 27, 28].

Thus, in the pre-decision model, the researcher cites the IDT concepts cognition and persuasion to explore the process by which individuals recognize cloud services and actively search for related information. Therefore, in the pre-decision stage, the effects of factors on attitude and the effect of PEOU on PU are explored. The hypotheses are developed as follows:

H_1: The TAM has a significant influence on the firm's attitude toward the adoption of cloud services.

H_{1a}: Cost has a significant influence on attitude.
H_{1b}: Security has a significant influence on attitude.
H_{1c}: Service continuity/completeness has a significantly influence on attitude.
H_{1d}: PU has a significant influence on attitude.
H_{1e}: PEOU has a significant influence on attitude.
H_{1f}: Compatibility has a significant influence on attitude.

H_2: The firm's PEOU has a significant influence on PU.

(2) In-decision stage

In the in-decision model, the researcher cites the IDT decision stage to explore the process by which individuals recognize cloud service information, evaluate the advantages and disadvantages of the service and make an acceptance decision. Thus, during the in-decision stage, the researcher probes the relationship between the factors and the adoption decision and the effect of attitude on the adoption decision. In accordance with past research, the following hypotheses are developed for this study:

H_3: The TAM has a significant influence on the firm's adoption decision regarding cloud services.

H_{3a}: Cost has a significant influence on the adoption decision.
H_{3b}: Security has a significant influence on the adoption decision.
H_{3c}: Service continuity/completeness has a significant influence on the adoption decision.
H_{3d}: PU has a significant influence on the adoption decision.
H_{3e}: PEOU has a significant influence on the adoption decision.
H_{3f}: Compatibility has a significant influence on the adoption decision.

H_4: The firm's attitude has a significant influence on the firm's adoption decision regarding cloud services.

(3) Post-decision stage

In the post-decision model, the researcher cites the IDT concepts implementation and confirmation to explore individuals' decisions to adopt cloud services in accordance with different situations. This stage is regarded as a post-review stage and determines the continuity or rejection of adoption. Thus, during the post-decision stage, the researcher aims to identify the relationship between factors and the adoption decision and the effect of attitude on decision continuity. The following hypotheses are developed for this study:

H_5: The TAM has a significant influence on the firm's continuation decision regarding the adoption of cloud services.

H_{5a}: Cost has a significant influence on the continuation decision.
H_{5b}: Security has a significant influence on the continuation decision.
H_{5c}: Service continuity/completeness has a significant influence on the continuation decision.
H_{5d}: PU has a significant influence on the continuation decision.
H_{5e}: PEOU has a significant influence on the continuation decision.
H_{5f}: Compatibility has a significant influence on the continuation decision.

H_6: The firm's attitude has a significant influence on the firm's continuation decision regarding the adoption of cloud services.

3.3 Research Scope and Subjects

According to MIC in 2010, among global industries, the high-end technology industry has the highest rate of cloud computing service adoption, followed by financial services. High-end technology has been an important industry in Taiwan enhancing economic development. In Taiwan's 2010 ranking of international brand value, 6 of the top 10 brands are in the high-end technology category. Thus, high-end technology employees in Taiwan are the subjects of this study. In addition, the researcher focuses on departments that frequently use cloud services, such as the basic IT framework department and network management department, and on supervisors or employees in the investment planning department to explore the acceptance of the introduction of cloud services in firms. Regarding the sampling of questionnaires, for the sake of time, manpower and resources, in accordance with Hair et al.'s [10] suggestion that "the acceptable ratio would be a ten-to-one ratio of independent variables", this study aims for at least 70 valid questionnaires.

4 Research Results

4.1 Basic Sample Information

In this study, data are collected through the mailing of questionnaires. A total of 80 questionnaires are sent to high-end technology firms, resulting in 73 valid returns. The valid return rate is 91.25%. Because the subjects are high-end technology firms, 100% of the valid returns are from computer information and electronic industries. Regarding corporate size, 75.4% of the firms have more than TWD$51 hundred million in capital. Regarding number of employees, all firms have at least 100 employees, and more than 95.9% have at least 1,001 employees. Thus, the sample includes small and large companies. According to the returned samples, most of the respondents are male (65.8%), and regarding educational level, most of the participants graduated from university (91.8%) and 56.2% have a master's degree. Regarding working years, 26% have worked for at least 7 years, and most of the respondents have worked 1–3 years.

4.2 Reliability and Validity Analyses

According to Guielford [8], an α above 0.7 means that a variable has validity. In this study, the reliabilities of all variables are above 0.7. This result means that the scales have high reliability. Using a factor analysis, this study confirms the validity of the questionnaires. The absolute value of factor loadings must be 0.5, the Eigenvalue > 1 and the cumulative explained variance > 40%. These values indicate high validity [13]. In this study, the cumulative explained variance is above 40%; the dimension factor loadings are at least 0.5 (the lower limit). According to the above analysis, this study has construct validity. In addition, the researcher analyzes the relationship between dimensions and variables using Pearson correlation coefficient analysis. When the correlation coefficient is too high, the regression analysis may indicate collinearity. There is a positive correlation between the variables, and the Pearson correlation coefficients are below 0.678. Thus, there is no collinearity, and significance has been attained.

4.3 Research Results and Discussion

Consistent with the TAM and IDT, this study probes the acceptance of the introduction of cloud computing services in firms at different stages. Using a literature review and expert interviews, questionnaire scales are developed. The researcher explores the IT industry using a questionnaire survey. The validation results for the hypotheses on the pre-decision, in-decision and post-decision stages are shown in Table 1.

According to the analytical results of this study, at the pre-decision stage, the relationship among PU, PEOU and adoption attitude matches that in the TAM suggested by Davis [4]. However, at the in-decision and post-decision stages, only PU has a significant influence on the adoption decision at the in-decision stage. This result suggests that the TAM can only explain firms' attitudes toward the adoption of a new technology and cannot predict firms' adoption decisions and continuation decisions [31].

Furthermore, regarding the concerns and challenges associated with the introduction of cloud services indicated by the literature review, the results of this study suggest that cost, security and service continuity/completeness have a significant influence on the different stages of firm adoption of cloud services. At the pre-decision and post-decision stages, cost, security and service continuity/completeness all have a significant influence on firms' adoption of cloud services. However, at the in-decision stage, the researcher cites the IDT concept of decision making and explores individuals' acquisition of cloud service information, the measurement of the service and the determination of acceptance or rejection. At this stage, cost does not significantly influence firms' adoption decisions. This result should be attributed to the characteristics of cloud services. Because technological progress lowers the cost of the basic framework and enhances operational efficiency, it increases the services' economies of scale. Furthermore, most enterprises have adopted cloud services, which means that they have acquired the benefits of cloud services. At the in-decision stage, the factors in firms' adoption of cloud services include only security, service continuity/completeness, PU and compatibility. The post-decision stage is the process by which firms decide to either continuously use or reject the adoption. Currently, cloud service providers still make efforts to "make clouds". Although firms currently use cloud services, they still worry about the future of cloud service providers. Thus, security, cost and service continuity/completeness have significant effects.

Table 1 Validation results for hypotheses

Hypotheses	Content	Results
Pre-decision stage		
H_1	The TAM has a significant influence on the firm's attitude toward the adoption of cloud services	Supported
H_{1a}	Cost has a significant influence on attitude	Supported
H_{1b}	Security has a significant influence on attitude	Supported
H_{1c}	Service continuity/completeness has a significant influence on attitude	Supported
H_{1d}	PU has a significant influence on attitude	Supported
H_{1e}	PEOU has a significant influence on attitude	Supported
H_{1f}	Compatibility has a significant influence on attitude	Supported
H_2	The firm's PEOU has a significant influence on PU	Supported
In-decision stage		
H_3	The TAM has a significant influence on the firm's adoption decision regarding cloud services	Partially supported
H_{3a}	Cost has a significant influence on the adoption decision	Not supported
H_{3b}	Security has a significant influence on the adoption decision	Supported
H_{3c}	Service continuity/completeness has a significant influence on the adoption decision	Supported
H_{3d}	PU has a significant influence on the adoption decision	Supported
H_{3e}	PEOU has a significant influence on the adoption decision	Not supported
H_{3f}	Compatibility has a significant influence on the adoption decision	Supported
H_4	The firm's attitude has a significant influence on the firm's adoption decision regarding cloud services	Supported
Post-decision stage		
H_5	The TAM has a significant influence on the firm's continuation decision regarding the adoption of cloud services	Partially supported
H_{5a}	Cost has a significant influence on the continuation decision	Supported
H_{5b}	Security has a significant influence on the continuation decision	Supported
H_{5c}	Service continuity/completeness has a significant influence on the continuation decision	Supported
H_{5d}	PU has a significant influence on the continuation decision	Not supported
H_{5e}	PEOU has a significant influence on the continuation decision	Not supported
H_{5f}	Compatibility has a significant influence on the continuation decision	Not supported
H_6	The firm's attitude has a significant influence on the continuation decision regarding the adoption of cloud services	Supported

5 Conclusions

According to the research findings and the TAM and IDT, at the pre-decision stage, cost, security and service continuity/completeness of clouds significantly influence the firm's decision to adopt cloud services. However, at the in-decision and post-decision stages, only PU of the TAM significantly influences decisions. This result means that the TAM can only explain firms' adoption attitudes toward new technology and cannot predict firms' adoption decisions and continuation decisions. Cloud factors significantly influence these two stages. However, at the in-decision and post-decision stages, the effects of the variables differ. This result means that at the in-decision stage of adoption and at the post-decision stage of review, factors in the introduction of cloud services differ.

The acceptance model developed in this study on the introduction of cloud services in the IT industry can serve as a reference for future firms with investment intentions. In the academic field, few studies explore the acceptance of cloud services introduction in firms from the innovation diffusion perspective. Thus, future researchers can conduct case studies or model difference analyses on other industries.

References

1. Chau, P.Y.K. and Tam, K.Y.: Organizational adoption of open systems: a technology-push, technology-pull perspective. Information and Management, vol. 37 No. 2, pp. 229–239 (2000)
2. Chen, L. D., Gillenson M. L. and Sherrell, D. L.: Enticing online consumers: an extended technology acceptance perspective. Information and Management, vol. 39 No. 8, pp. 705–719 (2002)
3. Davis, F. D., Bagozzi, R. P. and Warshaw, P. R.: User Acceptance of Computer Technology: A Comparison of Two Theoretical Models. Management Science, vol. 35 No. 8, pp. 982–1002 (1989)
4. Davis, F.D.: Perceived usefulness, perceived ease of use, and user acceptance of information technologies. MIS Quarterly, vol. 13 No. 3, pp. 319–340 (1989)
5. Erdogmus, H.: Cloud computing: Does Nirvana hide behind the Nebula? IEEE Journal Software, vol. 26 No. 2, pp. 4–6 (2009)
6. Fang, X., Chan, S., Brzezinski, J., and Xu, S.: Moderating effects of task type on wireless technology acceptance. Journal of Management Information Systems, Vol. 22 No. 3, pp. 123–157 (2006)
7. Frambach, R.T. and Schillewaert, N.: Organizational innovation adoption: a multi-level framework of determinants and opportunities for future research. Journal of Business Research, vol. 55 No. 1, pp. 163–176 (2002)
8. Guielford, J. P.: Fundamental Statistics in Psychology and Education. New York: Mcgraw-Hill. (1965)
9. Gumussoy, C. A. and Calisir, F.: Understanding factors affecting e-reverse auction use: An integrative approach. Computers in Human Behavior, vol. 25, No. 4, pp. 975–988 (2009)
10. Hair, J. F., Anderson, R. E., Tatham, R. L. and Black, B. C.: Multivariate data analysis, 5th ed., Englewood Chiffs, NJ: Prentice-Hall International. (1998)

11. Karahanna E., Straub, D. W. and Chervany, N. L.: Information technology adoption across time: a cross-sectional comparison of pre-adoption and post-adoption beliefs. MIS Quart, vol. 23 No. 2, pp. 183–213 (1999)
12. Kaufman, L.M.: Can a Trusted Environment Provide Security?", *IEEE Journal Security and Privacy*, vol. 8 No. 1, pp. 50–52 (2010)
13. Kerlinger, F., and Lee, H.: Foundations of behavioral research. NY: Thomson Learning (2000)
14. Kim, W.: Cloud Computing: Today and Tomorrow. Journal of Object Technology, vol. 8 No. 1, pp. 65–72 (2009)
15. Kshetri, N.: Cloud Computing in Developing Economies. IEEE Journal Computer, vol. 43 No. 10, pp. 47–55 (2010)
16. Leavitt, N.: Is cloud computing really ready for prime time. IEEE Journal Computer, vol. 42 No. 1, pp. 15–22 (2009)
17. Legris, P., Ingham, J. and Collerette, P.: Why do People Use Information Technology? A Critical Review of the Technology Acceptance Model. Information and Management, vol. 40 No. 3, pp. 191–204 (2003)
18. Miller, H.G. and Veiga, J.: Cloud Computing: Will Commodity Services Benefit Users Long Term? IEEE Journal IT Professional, vol. 11 No. 6, pp. 57–59 (2009)
19. Mouratidis, H., Islam, S., Kalloniatis, C., & Gritzalis, S.: A framework to support selection of cloud providers based on security and privacy requirements. Journal of Systems and Software. (2013)
20. Rogers, E. M.: Diffusion of Innovations, New York: Free Press. (1962)
21. Rogers, E. M.: Diffusion of Innovations, 3rd ed., New York: Free Press. (1983)
22. Rogers, E. M.: Diffusion of Innovations, 5th ed., New York: Free Press. (2003)
23. Ryan, M. D.: Cloud computing security: The scientific challenge, and a survey of solutions. Journal of Systems and Software. (2013)
24. Sheth, A. and Ranabahu, A.: Semantic Modeling for Cloud Computing, Part 1, IEEE Journal Internet Computing, vol. 14 No. 3, pp. 81–83 (2010)
25. Sultan, N.: Cloud computing for education: A new dawn? International Journal of Information Management, vol. 30 No. 3–4, pp. 109-116 (2010)
26. Taylor, S. and Todd P. A.: Assessing IT Usage: The Role of Prior Experience, MIS Quarterly, vol. 19 No. 4, pp. 561–570 (1995)
27. Tunga, F. C., Changa S. C. and Chouc C. M.: An extension of trust and TAM model with IDT in the adoption of the electronic logistics information system in HIS in the medical industry. International Journal of Medical Informatics, vol. 47 No. 7, pp. 324–335 (2008)
28. Wu, J., and Wang, S.: What drives mobile commerce? An empirical evaluation of the revised technology acceptance model. Information and Management, vol. 42 No. 5, pp. 719–729 (2005)
29. Wu, Y., Cegielski, C. G., Hazen, B. T., & Hall, D. J.: Cloud Computing in Support of Supply Chain Information System Infrastructure: Understanding When to go to the Cloud. Journal of Supply Chain Management, vol. 49 No. 3, pp. 25–41 (2013)
30. Yi, M. Y., Jackson, J. D., Park, J. S. and Probst, J. C.: Understanding information technology acceptance by individual professional: Toward an integrative view. *I*nformation and Management, vol. 43 No. 3, pp. 350–363 (2006)
31. Yu, C. S. and Tao, Y. H.: Understanding Business-level technology adoption. Technovation, vol. 29 No. 2, pp. 92–109 (2009)

Examining the Learning Effect of Elderly Through Multimedia Assisted Instruction Design in Taiwan

Su Fen Chen

Abstract The elderly often fear technology products in Taiwan, because the complexity of technology and the lack of information create barriers for the elderly to use technology. The concurrent trends of an aging population and digital product advances means that new technology products can enrich elderly lives, if learning barriers are overcome. This study are utilized the quasi- experimental design, A multimedia-assisted teaching system (MAIS) is an intervention program. Seventy-eight older adults (over 65 years old) will participate in this study; 38 older adults belong to the comparison group, and the other 40 will be in the experimental group. Subjects' post-test score will be used for evaluating the learning outcome after implementing the intervention program. The study seeks to clarify how to design multimedia instruction to maximize learners' recall and understanding of the presented material.

Keywords Elderly · Multimedia · Cognitive load

1 Introduction

In almost every country around the globe, the proportion of people aged 65 and over is growing faster than any other age group, as a result of both longer life expectancies and a decline in fertility rates (U.S.C.B., 2010) [22]. In particular, according to the estimates by the Council for Economic Planning and Development of Taiwan's government, the senior population will continue to grow until 2051, when elderly people will account for 10.48–36.98% of the total population. Based

S.F. Chen (✉)
Department of Education, Keelung City Government, 8f, No. 164, Sec 2, Anle Rd., Anle Dist., Keelung City 20402, Taiwan (ROC)
e-mail: chensufen3@gmail.com

S.F. Chen
Social Service Program of Fordham University, Lincoln Center Campus, New York, NY 10023, USA

© Springer Nature Singapore Pte Ltd. 2018

N.Y. Yen and J.C. Hung (eds.), *Frontier Computing*, Lecture Notes in Electrical Engineering 422, DOI 10.1007/978-981-10-3187-8_91

on these issues, some have a negative image and misunderstanding or bias of older people's learning abilities, which affects the promotion of the opportunity for older adults to learn and develop their potential. In fact, Taiwanese caregivers for the elderly face not only the problem of elders' physical health, but also the need to encourage older people's willingness to learn actively, and to follow up the pulse of society to learn new skills and knowledge, which can be combined with their wealth of life experience to enjoy a happy life. Therefore, it is important to create a suitable operation for products' interface, as well as an effective multimedia teaching style to assist the elderly and make the e-learning environment accessible. Establishing a good learning environment and/or media for older adults will make it easier for them to adapt to digital technology, be happier in their lives, and re-engage in society.

1.1 Background and Significance

Due to recent medical, scientific, and technological progress, longevity has increased. A person's average life span in Taiwan is 79.18 years old; men have an average life expectancy of about 76.13 years, and women's averages 82.55 years [3]. However, longevity is not always a welcome development in individuals or societies, because general impression thought that the elderly have demonstrated functional decline and many suffer from dependency, depression, and dementia at individual level, and fear of bankruptcy the national budget (SSI), rising health care costs, and burden to younger generation at population level. The same situation in Taiwan, Taiwan's government recently presented a revised Law of Health Issuance in The Legislative Yuan, as an attempt to prevent financial deterioration due to the increasing aging population. To solve the problems facing aging societies, many studies have been done, but have focused on the limitations of older adults and primarily considered their functional decline and dependency. This research pointed out that older adults have more significant limitations and obstacles in learning new knowledge or new technologies when compared to young people [15].

In Taiwan, the elderly is higher barriers in participation the learning activities due to limited in physiological and physical function. Factors may include "physical and mental limitations and obstacles," "information processing limitations and obstacles," and so on [15, 16], because the elderly lack experience with complex operations of digital tools. Studies by Schieber and Baldwin [19] and Salthouse [20] pointed out that perception and cognitive abilities decline with age, and negative age-related cognitive results have been confirmed by numerous studies. Aging people's perceptive ability, strength, flexibility, mobility, memory, and learning ability are reduced [7]. However, Lamdin and Fugate [8] argued that many senior citizens are not only positive in physical health, but also willing to contribute to their family or community.

Paas, van Gerven, Tabbers [12] have pointed out that well-designed learning materials can compensate for the adverse effects of cognitive aging, enhance learning outcomes, and to shorten the gap between the performance of the elderly and the young. Many researchers have studied cognitive-oriented learning and found that the older adults do have the capability to learn [14]. However, the majority of related research always focuses on the adults' learning; the instructional design of the effectiveness of learning for the elderly is still less studied than other aspects. This is the reason why this proposal focuses on this topic.

Mayer's [9] study concluded that learning efficiency and training content will be affected by different media presentation. Especially, if future elderly citizens cannot rely on government assistance, it is important to design effective learning contents to help the elderly to learn new skills [18] therefore, other businesses and scholars recently also advocated the concept of designing a suitable assisted system for elderly learning depend on their physical and mental condition. For example, Mobilinux Inc. [12] gave developers who developed the Mobile Operating System Optimized for Portables a scalable platform that lets them create their own differentiated feature phones or smart phones for the elderly to use. Another developer named Troy Wolverton [21] also designed an operating system in smart phones for aging adults. It is likely that more electronic products with assisted instruction systems will be created, and the products using them will be marketed, because the technology is now simple and easy to access.

2 Literature Review

Humans' ways of receiving inputs from the external environment can be divided into five paths: vision, hearing, touch, smell, and taste. Therefore, multimedia information processing system works through using the multimedia presentation forms such as text, pictures, and sounds to transmit various messages [6]. This study hypothesizes that it easier to learn training content through additional multimedia text and images than through written instructions alone. Multimedia instruction will help to improve memory encoding. This study will attempt to test whether the performance of elderly learning will be enhanced by the multimedia instructional design.

In general, scholars usually discuss the relationships among intelligence, memory, learning, and age to understand the relationship between the cognitive factors and people's age. The "cognitive load" in this study refers to the burden of mental ability for older learners who watch the multimedia learning materials [10]. Sweller, van Merriënboer, and Paas [19] indicated that the cognitive load can be divided into intrinsic load and extraneous load according to their sources; it refers to the load imposed on the operation of memory during the task of information processing.

Park's et al. [16] assessment of the difference in cognitive performance between older adults and young people for the four aspects of the information processing are included Processing speed, working memory and sensory function. As for processing speed, the old and young people were able to complete the tests within the same timeframe, but the elderly group had more errors in their work. There were also response of speed/accuracy trade-offs, implying that the ability of the elderly to work effectively was affected by time pressure [13]. Therefore, the variable also will be included in my proposal to test the interaction of cognitive loading.

In addition, Working memory Craik and Byrd [2] describe the pattern of short-term memory "processing resource" as an instant information processor, temporary storage, and retrieval and conversion of messages; it means the mental capacity connected with instantaneous operations. Short-term memory encoding can be divided into auditory encoding and visual encoding [1]. Craik and Byrd [2] found that the elderly will have more restrictions in auditory coding, so they suggested making changes to the elderly environment, using simpler words and shorter sentences to reduce the demand for short-term memory. Therefore, this factor will be considered into designing the intervention program of this study.

Regarding Sensory function, the inquiry had found that physical function will be retrograded with age and that most people are plagued with visual and auditory deterioration. Most responses (93%) will show degradation in understanding capacity, knowledge, memory, and perceptual speed gradually decreasing with age, especially for the elderly over the age of 70 [8]. Mayer [9] pointed that learners would get different learning outcomes by using a diverse media system; specifically, she found that the learning outcomes with the text and graphic media are better than those using just text. New teaching software uses flash combined with animation, text, and audio-visual effects; thus, it is more narrative-based and rigorous for the learners. Meanwhile, the animation-based teaching resulted in the learners paying more attention, compared to static pictures. Levin, Anglin, and Carney [4] say that animation is an informational means of communication, with text, color, shape, still pictures, moving pictures, and combinations of sound mixed in a variety of forms, in order to achieve the effect of rapid learning. Multimedia teaching and research (Lindy and Zhang Jun Xiang) [4] found that senior citizens using animation-assisted learning systems performed significantly better than those using only static graphics. Above all, it is proven that the learning outcomes with mixing in a variety of forms media work better than those using just text or static graphics.

The learners who are older adults can memorize more messages for short data, while the long-form information is easier to forget [11]. There are obvious differences in how the elderly use new technology and how they function under a diversity environment in terms of learning [17]. Mayer [9] found the cognitive Models in Multimedia Learning as the following chart, showing how elderly learn or process information.

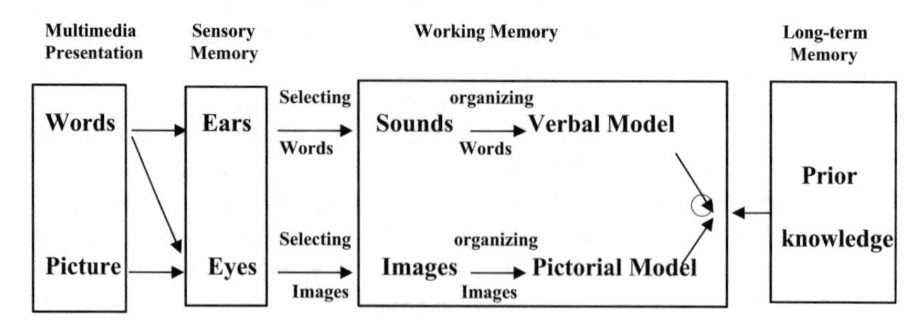

Additionally, socioeconomic status will affect the learning outcome; Tsai Ming-shun [21] indicated that the family's socioeconomic status and academic achievement is directly related. Scott [20] indicated that for families with a better socioeconomic status, it was easier for parents to be motivated for stronger off-spring. Because a higher socioeconomic status will allow people more resources, it is likely they also could have the more experiences with technology products than others. Caplan and Schooler [5] trained senior citizens to use graphics software, and found that older people took a longer time to complete their training, and showed a lower performance. Therefore, socioeconomic and age variables do act as moderators for learning outcomes, but are difficult to manipulate in this study.

2.1 Specific Aims

The study will focus on finding an interface to boost the efficiency of the elderly for learning outcomes. For example, when learning how to operate the functions of digital cameras, learners can control the display speed of multimedia in accordance with their cognitive loading. This study aims to explore the effects in learning outcomes and cognitive load of the elderly through a multimedia-assisted instruction system.

2.2 Research Questions

This study intends to examine the learning performance of the elderly using electronic products through presented modality and the method of speed-controlled multimedia. Therefore, the study will use the special functions of multimedia, and find ways to stimulate the learning outcomes of the elderly.

The research questions are summarized as follows, Does the multimedia-assisted instruction system (MAIS) affect the elderly learning outcomes? Does the presented speed of MAIS affect the elderly learning outcomes?

2.3 Hypotheses

By exploring these issues, the hypotheses in this study are as follows:

H1: Elderly who are in the treatment group will gain a higher learning test score via receiving the intervention of MAIS than the comparison group.
H2: Elderly who are in the treatment group will gain a higher learning test score through receiving the MAIS and playing speed-controlled instruction by themselves than those only receiving the MAIS.
H3: The playing of speed-controlled of MAIS will affect the elderly cognitive load.

3 Research Plan

3.1 Research Design

This study will use the quasi-experimental design method, pre-test and post-test with nonequivalent group. The study will examine the learning outcomes for the treatment group and the comparison group, the participants having been non-randomly assigned to each. The treatment group will receive the computer-assisted instruction system which uses a combined model of presentation and speed-controlled multimedia to teach the elderly the related photographic knowledge of the digital camera as the intervention content. The comparison group will not participate in the intervention program, but both groups will read the simplest instruction manual, which include the text and static graphics, in the limited time before the pre-test beginning, meaning that they will get the basic photographic knowledge. This intervention program is MAIS. Both of the treatments are MAIS, one played automatically by a computer, and the other and speed-controlled by learners, and so are the independent variables for this study. The learning outcome test is the dependent variable. This program will last for two weeks. The conceptual model can be seen in **Appendix 1**. There are 78 older adults (aged between 65 and 80 years old) are the participants who will be divided into two main groups depending on their willingness. The comparison group will be 38 people; the treatment group will be 40 people who are then divided into two sub-groups (meaning 20 subjects in the treatment group will receive the MAIS played automatically by computer, and the other 20 subjects will receive the MAIS as speed-controlled by themselves). The experiment requires the subjects to have their vision corrected to normal vision before testing. Because of the small samples recruited, the quasi-experimental design uses a within-subjects design. Although this study is designed to decrease threats to internal validity as much as possible, limitations remain. In particular, the elderly like to form the small groups according to their interests (not socioeconomic status or age). This experiment only can match similar demographics as much as possible in the comparison and treatment groups in order to reduce the between-subject differences.

Participants will be given a $20 incentive upon the completion of the pre-test and post-test study. The assisted-teaching contents are related to a digital camera's operation function and knowledge for photography.

4 Outcomes Measures

The learning outcome utilizes the paper and pencil test, which will contain three parts: one is a rote learning effectiveness test, the second is a meaningful effectiveness test, and the third part is a cognitive load test. The rote learning effectiveness test will consist of text-level problems to test the elderly learning content. The meaningful effectiveness test is a situation model level, to test the elderly's inferential understanding of learning materials, and how they integrate the original knowledge to new situations. The test requires learners to fill in blanks with the numbers shown on the computer screen during the playing MAIS. The aim is to examine the elderly cognitive load.

The learning testing utilizes the PISA (Programme for International Student Assessment) test scale. A PISA test is a logic test tool, to measure a group of students on information, mathematics, natural science, reading, and problem solving ability. PISA draws mainly from three levels to measure students' ability: the ability to access to the message, understand the message, and the ability to think and judge. This shows that the PISA test is to assess the learners' ability to access information, to obtain information, analyze information, evaluate information, integrate information, and express information, as well as personal ability to think independently. Therefore, it is suitable to examine the elderly learning outcome. The PISA test is used for the students of 28 OECD (Organization for Economic Cooperation and Development) countries and four non-OECD countries, and it is currently used broadly in Taiwan's schools; it implies good internal validity. There are six levels of scoring on this scale, worst = (F) under 50, bad = (E) 51–60, general = (D) 61–70, good = (C) 71–80, better = (B) 81–90, excellent = (A) 91–100. If the subjects in this study accept the whole intervention program, they will know how to take a photograph with a digital camera and related knowledge, thus scoring at least C, and even to A.

The paper and pencil test will have 20 questions, including 10 questions for the rote learning effectiveness test, and 10 questions for the meaningful effectiveness test. Additionally, there will be 3 numeric "fill in the blank" questions for cognitive load testing, which will be measured only for the experimental group.

4.1 *Sampling Plan*

The "purposive sampling" method will be used in this study, which is expected to recruit 78 older adults who are aged between 65 and 80 years old from the senior centers in Taipei, the said senior centers are in charge of Ministry of Education in Taiwan. The reason for this choice is that the researchers are the central government education officials; they can get effective assistance because of their geographical location and occupation. The study is limited to the elderly who are literate, and normal (or corrected to normal) hearing and vision. There are four senior centers in Jhongjheng District, Taipei city, and a total of 260 older adults are between 65 and 80 years old. According to the Ministry of Education of Taiwan previous study for happy satisfaction from senior center in 2010, only one-third of all the older adults will participate in the experiment. Therefore, the experiment for this study will be expected to use a nonrandom sample of one-third of all the elderly.

Due to the limitations of manpower and material resources, this study intends to use a "purposive sampling" method to get the required sampling base on the inclusion criteria. The majority of senior citizens are now retired. The experiment requires the subjects to have normal vision and hearing. In consideration for the elderly persons who may have hearing difficulties, the volume of sound will be adjustable to correct for the hearing ability of the subject before the experimental testing. In addition, all elderly subjects in this study for the content of training must give the consent and meet the following common criteria: who have (1) little or no contact with and/or practical use of the multimedia function; (2) have little or no experience with digital cameras; and (3) can identify and comprehend Chinese text and sound.

In addition, the sampling in this study are focus on elderly, so the instructor also will be recruit from the senior centers of Taiwan, he/she will be at least 5 years experience in teaching elderly, and would be best ever participated the related elderly research.

4.2 Internal Validity Threats

There are some threats to internal validity, discussed as follows:

4.2.1. All the subjects will be dispatched depending on what the subjects demand; so some older adults who can't operate the computer will participate in the treatment 2 group.
4.2.2. After the pre-testing, the elderly might learn the experimental material from their family or friends; and thus may get a different basic knowledge bias.
4.2.3. Because all the participates are elderly, there are some risks for their physical condition to complete the whole experimental process.
4.2.4. All the subjects have the different background and age, the level of MAIX maybe not suitable for the subjects.

4.3 Procedures

The form of Quasi-Experimental Designs is **Appendix 2**. The experimental procedure is divided into browsing the operation of photography functions and related knowledge of digital camera through the multimedia-assisted instruction system.

4.4 *The Limitation of the Internal Threats*

4.4.1. It uses non-random sampling to recruit eligible samples, as samples are not easily accessible.

4.4.2. This study is just for pilot testing. However, to decrease the bias in intervention, the assisted-teaching contents and learning outcome test will commit the National Taiwan University Professor of Industries Center to review the concept of delivery in the material, assess whether the statement is smooth and coherent, and verify the appropriate translation of the term, and reliability and validity. Following this learning material will be a flash movie, and pilot testing for five elderly to watch and understand the appropriateness in speed of the film's narration and/or animation presentation.
4.4.3. To decrease the limitation threats, all the subjects need to sign a protocol to verify they will obey the experimental principle, including don't collect other related knowledge and give up.

4.5 *Human Subject*

Before conducting the study, the research team will have a meeting with the leader and officials of senior centers of the potential eligible senior centers. During the meeting, the research team will provide a formal presentation to explicitly discuss the study, including the goals and purpose of the study, the procedure, risks and benefits, etc., and make sure they will support it completely.

All the potential elderly are recommended from the senior centers, and to further confirm their decision, the interviewees who are willing to participate in this research will be contacted (by phone when possible and if not, by person), where they will be informed about the research study and its purposes, as well as the voluntary nature of the study. Before the in-person interview, participants will be asked to sign an informed consent form that will be discussed with the interviewee. All the potential participants are guaranteed anonymity.

Confidentiality during the study is a little difficult to attain for potential elderly who are willing to participate the study, especially considering that elderly were identified through purposeful sampling of extreme cases with the help of senior center teachers.

5 Summary

The completion of this study will provide specific empirical data, and further form of the design principles of multimedia learning materials in order to address the elderly cognitive aging, and improve the quality of the elderly study. Because the object of concern is adult education literature which rarely explores the topics of learning styles, learning needs, learning preferences, and learning characteristics for elderly, studies have always been based on 25- to 64-year-old learners. The elderly index is rising in current society, and senior citizens' learning activities have been highly rich and diverse, but often lack the guidance of cognitive theory and empirical research results in the design concept of the learning materials. Therefore the activity design cannot reduce the cognitive load of the elderly to enhance their learning effectiveness. In this study, the greatest contribution will be the substantive contents that will promote teachers and designers of multimedia learning materials to address the phenomenon of elderly cognitive aging, and to understand the design principles of cognitive load theory-oriented learning materials, making and propagating suitable material for senior citizens to learn.

6 Future Direction

This study focuses on learning-related knowledge of multimedia and digital camera operation by the elderly, and cannot ensure the same solution will be occur in the operation of other digital products, and doesn't compare the performance between elderly and younger subjects. It is recommended that the related research can expand the age range of the subjects in the future.

In addition, learning outcomes assessment can be joined to record the learning process for the elderly, including eye movements and spoken dimensions, in order to understand whether the different learning materials will be lead to different cognitive processes and learning results. In general, the older adults may have individual differences in learning and familiarity for technology products because of different levels of education and previous experience, confusion arising from the text and icons, and difficulty in using the technology products. But this study will not explore the differences in the performance of learning for elderly who have more experience, good language skills, and higher education to use digital products. If the hypothesis in this study will be supported, then the intervention will be evidenced as effective for elderly learning outcomes, and the theoretical framework will be promoted to continue in the future; if it is opposite, the researchers need to find other treatment or intervention programs to assist the elderly in learning.

Appendix 1: Conceptual Model

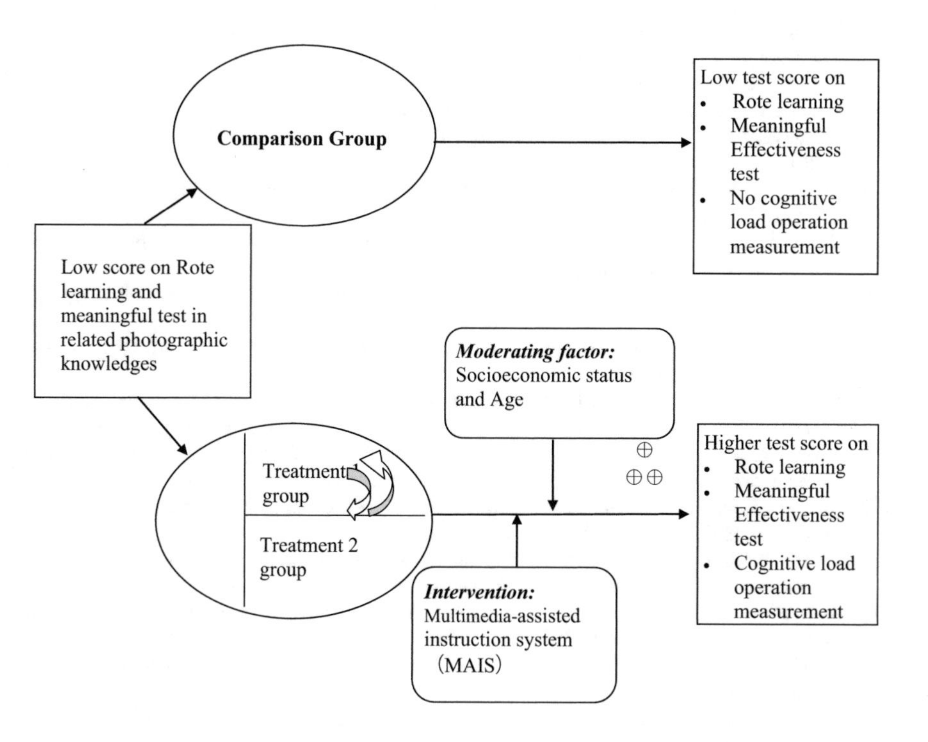

P.S. ⊕ it means increase the testing score.

⊕⊕ it means increase the testing score more than ⊕.

Appendix 2: Visual Notation for Research Design

- Untreated Control Group Design with Dependent Pretest and Posttest Samples Using a within-subjects design (Quasi-Experimental Design)

$$\frac{NR \quad C \quad \mathbf{O}_1 \qquad \mathbf{O}_2}{NR \quad E \quad \mathbf{O}_1 \quad \mathbf{X} \quad \mathbf{O}_1}.$$

References

1. Baddeley, A. D. (1992). Working memory. Science, 255, 556–559.
2. Craik, F. I. M., & Byrd, M. (1982). Aging and cognitive deficits: The role of attention resources. In F. I. M. Craik & S. Trehub (Eds.), Aging and Cognitive Processes, 191–211. New York: Plenum Press.
3. Council for Economic Planning and Development, 2010. Population Extrapolation Estimate for the Republic of China, Taiwan 97 Years To 145 Years. Email from http://www.cepd.gov.tw/m1.aspx?sNo=0000152.
4. Craik, F. I. M., Byrd, M. & Swanson, J. M. (1987). Patterns of memory loss in three elderly samples. Psychology and Aging, 2, 79–86.
5. Czaja, S. J., and Sharit, J. (1989). Age differences in the performance of computer-based work, Psychology and Aging, 8, 59–67.
6. Feld, S., & Fox A. A., (1994). Music and Language. Annual Review of Ant homology, 23, 25–35.
7. Fisk, A. D., Rogers, W. A. (1997). Handbook of Human Factors and the Older Adult, Academic, San Diego.
8. Lindenberger, U., & Baltes, P. B. (1997). Intellectual functioning in old and very old age: Cross-sectional results from the Berlin Aging Study. Psychology and Aging, 12, 410–432.
9. Mayer, R. E. (2003). The promise of multimedia learning: using the same instructional design methods across different media. Learning and Instruction, 13, 125–139.
10. Murphy, K. P. (1989). Is the relationship between cognitive ability and job performance stable over time? Human Performance, 2, 183–200.
11. Miller, G. A. (1956). The magical number seven, plus or minus two: some limits to our capacity for processing information. Psychological Review, 69, 81–97.
12. Montavista Software Inc.(2005). Mobile Operating System Optimized for Portables. Journal of Electronic Design, Vol. 53 Issue 16, p 38–39, ISSN: 00134872.
13. O'Neil, H. F. (Ed.). (1979a). Issues in Instructional Systems Development. New York: Academic Press.
14. O'Neil, H. F. (Ed.). (1979b). Procedures for Instructional Systems Development. New York: Academic Press.
15. Park, D. C. (1996). Psychological issues related to competence: Cognitive aging And Instrumental activities of daily living. In W. Schaie & S. Willis (Eds.), Social Structures and Aging, 66–82.
16. Mahwah, New Jersey: Erlbaum. Park, D. C., Smith, A. D., Lautenschlager, G., Earles, J., Frieske, D., Zwahr, M., & Gaines, C. (1996). Mediators of long-term memory performance across the life span. Psychology and Aging, 11, 621–637.
17. Park, D. C. (1999). Aging and the controlled and automatic processing of medical information and medical intentions. In D.C. Park, R.W. Morrell, & K. Shifren (Eds.), Processing of Medical Information in Aging Patients: Cognitive.
18. Park, D. C. (2000). Aging and the controlled and automatic processing of Medical information and medical intentions. In D.C. Park, R.W. Morrell, & K. Shifren (Eds.), Processing of Medical Information in Aging Patients: Cognitive Human Factors Perspectives. Mahwah, NJ: Erlbaum.
19. Schacter, D. L., Koutstaal, W. E., & Norman, K. A. (1998). False memories and Aging. Trends in Cognitive Sciences, 1, 229–236.
20. Salthouse, T. A. (1991). Theoretical Perspectives on Cognitive Aging. Hillsdale, NJ: Erlbaum.
21. Troy Wolverton, Dean Takahashi. San Jose Mercury News (CA) (2007). CES Notebook: Prepare to wow us: Silicon Valley's Palm and Reveal Parts of Strategies to Complete with their Rivals. Newspaper.
22. U.S. Census Bureau. (2010). THE NEXT FOUR DECADES The older population in the United States: 2010 to 2050. Retrieved September 25, 2010, from http://www.census.gov/prod/2010pubs/p25-1138.pdf.

Impact of eWoM on Consumers' Purchasing Intention in Social Commerce

Chun-Hui Wu, Ping-Lan Juan and Yu-Hei Liu

Abstract Social commerce is growing rapidly in e-commerce in recent years. However, few studies have investigated the relationships between eWOM and consumer purchasing intention in social media platforms. This study attempts to fill the void in the existing literature by building a theoretical model based on the social cognitive theory in order to explain the interactions among person cognition (belief towards the eWOM in social commerce), behavior (purchasing intention), and social environment (trusted social commerce environment) in social commerce. A total of 309 questionnaires were returned and 85% are valid ones, leaving a final sample of 262 used in the data analysis. The research model was tested using the component-based partial least squares structure equation modeling approach (PLS-SEM) with the smartPLS 3.0 software. All research hypotheses are accepted. This study makes a research contribution by providing additional insights into how trusted social environment may be utilized to leverage eWOM adoption in order to influence purchasing intention.

Keywords Social commerce · eWoM · Trust · Purchasing intention

1 Introduction

With the advancement of Web 2.0 technologies and applications, social media has found its way into nearly every aspect of our lives. It's how we communicate with each other, follow the news, and stay entertained. In recent years, companies established page through social media to encourage consumers sharing their experiences and interacting with each other and to increase sales and customer purchasing intention. On the other hand, consumers benefit from various social media platforms that support social interaction and user contributions in the

C.-H. Wu (✉) · P.-L. Juan · Y.-H. Liu
Department of Information Systems, National Formosa University,
Yunlin 632, Taiwan, ROC
e-mail: melody@nfu.edu.tw

© Springer Nature Singapore Pte Ltd. 2018

N.Y. Yen and J.C. Hung (eds.), *Frontier Computing*, Lecture Notes
in Electrical Engineering 422, DOI 10.1007/978-981-10-3187-8_92

acquisition of products and services [1, 2]. Nowadays, social media is becoming increasingly more influential on consumers because they learn from each other's shopping experience for understanding what they are purchasing and make more informed and accurate purchasing decisions. Along with the popularity and commercial success of various forms of social media, the term "social commerce" was coined in 2005 by David Beach, Product Manager on Yahoo! Shopping.

Social commerce was defined by [3] as an internet-based commercial application, using social media and web 2.0 technologies which assist social interaction and user-generated content in order to help consumers in their decision-making on purchases of products and services. In short, social commerce refers to e-commerce transactions mediated by social media such as Facebook, Twitter, Pinterest, and Instagram [4, 5]; therefore, it was considered as a subset of e-commerce [6]. Several studies have confirmed that social commerce results in significant financial and strategic benefits to companies [7]. Social commerce is growing rapidly in e-commerce in recent years, and its revenue has grown six times from a humble $5 billion in 2011 to eclipsing $30 billion in 2015 [8].

Social recommendations is a primary driver of the rise in social commerce [9]. According to Nielsen's survey [10], consumers have more confidence in their friends and other customers' reviews rather than in all other forms of advertising. Social recommendations can help consumers evaluate the quality of things by listening to the advice of their friends or other consumers. If they trust the recommendations, they generally act on the referral [11]. Supporting this, Nielsen's latest report [12] indicated "people don't trust advertising, at least not as much as they trust recommendations from friends and consumer opinions expressed online".

More specifically, social commerce is word of mouth (WOM) applied to e-commerce. WOM is defined as the act of exchanging information among consumers about a product or a service experience [13] and takes many forms online or off-line. Online WOM is called electronic WOM (eWOM) that is defined as "any positive or negative statement made by potential, actual, or former customers about a product or company, which is made available to a multitude of people and institutions via the Internet" [14]. An increasing number of consumers make purchase decisions according to electronic word-of-mouth (eWOM) on social media, and they often ask other friends, family members, or professionals for advices on purchases of products, services, and content. In today's digital era, eWOM has become increasingly essential in changing consumer attitudes and behavior towards products and services. Since the power of eWOM in marketing has grown dramatically, it has long been considered as an effective promotional instrument in social commerce [15]. Relevant studies have shown the impact of eWOM on consumer purchasing intention in blogs, discussion forums, review websites, and shopping websites.

However, few studies have investigated the relationships between eWOM and consumer purchasing intention in social media platforms [15]. This study attempts to fill the void in the existing literature by building a theoretical model based on the social cognitive theory in order to examine the influence of trusted environment and eWOM adoption on consumer purchasing intention in social commerce.

The remainder of this paper is structured as follows. Section 2 introduces about the related literatures and theory background. The research model and hypotheses are presented in Sect. 3, which is followed in Sect. 4 by data analysis results. Finally, conclusion are reported in Sect. 5.

2 Literature Review

2.1 Social Commerce

In the literature, there are various definitions by a variety of social media and ecommerce experts. Social commerce can be defined as both a broad and narrow concept. In a broad sense, social commerce is defined as "Social commerce involves the use of Internet-based media that allow people to participate in the marketing, selling, comparing, curating, buying, and sharing of products and services in both online and offline marketplaces, and in communities" [16]. In a narrower definition, social commerce is a form of e-commerce, and it involves using social media that supports social interaction and user contributions, to assist in the online buying and selling of products and services [17].

Social commerce is becoming increasingly more influential on consumers' purchasing decision so that it is gaining increasing attention in worldwide. According to the 2015 Internet Retailer's Social Media 500 report, the top 500 retailers gained $3.3 billion from social shopping in 2014, a 26% increase over 2013, and better than the average 16% growth rate for e-Commerce. It is predicted that the increase in 2016 may be even greater [18]. Companies are becoming intensely aware of the success of social commerce.

The success of social commerce depends mostly on its benefits to customers. Four major benefits for consumers list in the following [19]:

- Free information—The majority of consumers search online information about brands, products, and services before shopping. All this information and product ratings are for free.
- Saving time—Having lot of information, personalizing channels of information and product suppliers, receiving suggestions, using translation tools and other social and e-commerce tools help consumers save time in getting useful information and making purchasing process.
- Better product—Other consumer contributions result in product improvement, therefore it is beneficial for consumers to buy the new upgraded product and to avoid buying the previous product.
- Trusted environment—The trusted social commerce environment makes the consumer feel safe and supports them make buying decisions based on the advice of a network of friends, family members, and professionals whom they trust.

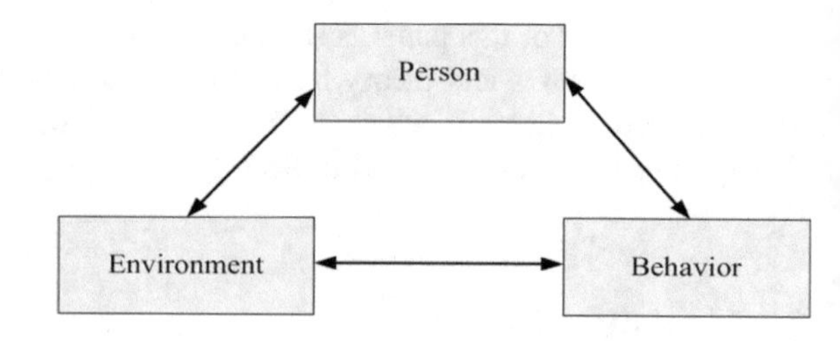

Fig. 1 A bi-directional interaction of social cognitive theory

2.2 *Social Cognitive Theory*

Social cognitive theory proposed by Bandura [20] is a well known learning model based on the cognition-motivation view. Social cognitive theory argues that a person's behavior is partially shaped and controlled by the influences of environment (i.e., social systems) and the person's cognition (e.g., expectations, beliefs). Social cognitive theory states a reciprocal relationship exiting between persons' cognitive properties, environment factors, and their behaviors.

A bi-directional interaction occurs between behavior and the environment. Bandura argued that an individual's behavior is both a product and a producer of his/her environment. Environment consists of social and physical environments. Social environment include family members, friends and colleagues. Physical environment is the size of a room, the ambient temperature or the availability of certain foods. Social cognitive theory states a reciprocal relationship exiting between persons' cognitive properties, environment factors, and their behaviors, shown as Fig. 1.

3 Research Methods

3.1 *Research Model and Hypotheses*

This study utilized the social cognitive theory to explain the interactions among person cognition (belief towards the eWOM in social commerce), behavior (purchasing intention), and social environment (trusted social commerce environment) in social commerce, shown as Fig. 2.

A theoretical model which consists of three research hypotheses was developed based on the social cognitive theory and previous studies. The theoretical model and research hypotheses were depicted as follows (Fig. 3).

H1: There is a positive relationship between trusted social environment and eWOM adoption in social commerce.

H2: There is a positive relationship between eWOM adoption and purchasing intention in social commerce.

H3: There is a positive relationship between trusted social environment and purchasing intention in social commerce.

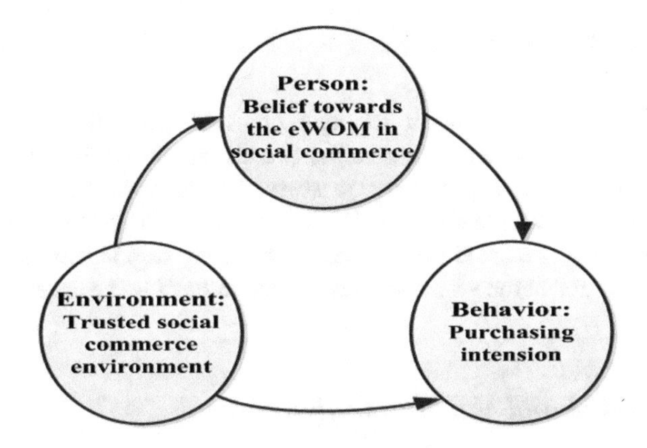

Fig. 2 Perspective of social cognitive theory in social commerce

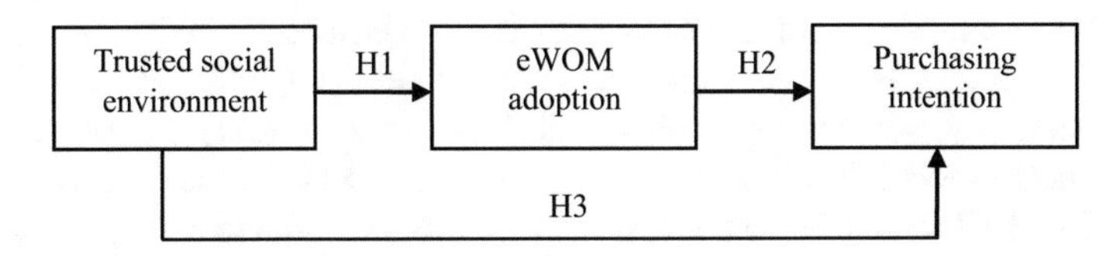

Fig. 3 Theoretical model

3.2 Survey Methodology

To test the hypothesis, survey method was employed using existing scales in the literature to measure the variables of trust, eWOM adoption, and purchasing intention. Questionnaire was adopted to collect research data from consumers of social commerce.

A total of 309 questionnaires were returned and 85% are valid ones, leaving a final sample of 262 used in the data analysis. All respondents have already experienced in social commerce, and over 80% of them had ever bought products/services on social media platforms.

4 Data Analysis and Results

The research model was tested using the component-based partial least squares structure equation modeling approach (PLS-SEM) with the smartPLS 3.0 software. The preliminary stage of data analysis involved assessment of construct reliability, and convergent and discriminant validity in this study. Then, the structure model was analyzed.

4.1 Non-response Bias Test

Non-response bias occurs when the opinions and perceptions of the survey respondents do not accurately represent the overall sample to whom the survey was sent. T-tests for non-response bias were assessed by comparing the answers of early versus late respondents to the survey [21]. T-test was employed to examine whether significant differences existed between early and lately respondents. No significant difference was found at the 95% confidence level, indicating the absence of the bias.

4.2 Reliability and Validity Analysis

Construct validity of the measures was tested by employing confirmatory factor analysis (CFA) which is a multivariate technique that facilitates testing of the psychometric properties of the scales used to measure variables. A series of empirical tests was used to examine the measurement properties of the indicators, namely reliability, convergent validity, and discriminant validity for each construct in the research model. Table 1 reports the correlation matrix, the average variance extracted (AVE), and the descriptive statistics of the principal constructs. Measurement reliability is assessed using composite reliability. It is suggested that a composite reliability of 0.70 or greater is considered acceptable for research. As in Table 1, the internal consistencies of all variables are considered acceptable since they exceed 0.70, signifying satisfactory reliability.

Discriminant validity is demonstrated when different instruments are used to measure the different construct, and the correlations between the measures of these different constructs are relatively weak. Discriminant validity is satisfied if both AVE estimates for every two construct pair are greater than their squared correlation. The results fully satisfy the requirement for discriminant validity.

Table 1 Descriptive statistics, correlations, and average variance extracted

Principal construct		Mean	Std.	CR	AVE	1	2	3
1	Trust	3.33	0.65	0.88	0.65	**0.80**		
2	eWOM adoption	3.69	0.59	0.80	0.58	0.52	**0.76**	
3	Purchasing intention	3.54	0.62	0.90	0.68	0.63	0.62	**0.83**

Note CR: Composite Reliability
The diagonal elements (in bold) represent the square root of AVE

4.3 Results of SmartPLS

The standardized path coefficients were used to test the research hypotheses, and the PLS path coefficients are shown in Fig. 4. Table 2 presents the results of the SEM model with the path coefficient and corresponding t-values.

Trusted social environment construct demonstrated statistically significant positive on both eWOM adoption and purchasing intention, which means hypotheses 1 and 3 can be considered to have empirical supports in data at hand. Also, eWOM adoption construct showed statistically significant positive and strong impact on purchasing intention, which means hypotheses 2 was supported. Our findings show that both trusted social environment and eWOM adoption play the influence on purchasing intention in social commerce. Trusted social environment has a positive impact on eWOM adoption, which in turn has a positive influence on purchasing intention. The total variance explained of purchasing intention in the examined model was 0.52.

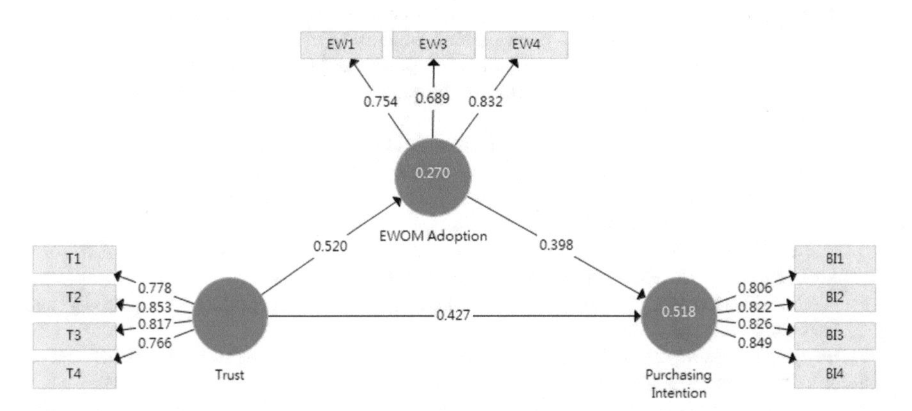

Fig. 4 PLS results

Table 2 Summary of hypothesis analysis

Hypothesis	Path coefficient	t-value	Result
H1: Trust social environment → eWOM	0.52	9.73***	Accepted
H2: eWOM → Purchasing intention	0.40	7.63***	Accepted
H3: Trust → Purchasing intention	0.43	8.46***	Accepted

Note *** indicates significant at $p = 0.001$ level

5 Conclusions

The results of this study support our hypotheses and make a research contribution by providing additional insights into how trusted social environment may be utilized to leverage eWOM adoption in order to influence purchasing intention. Success in social commerce is not simply a matter of utilizing certain channels, or even maintaining a high level of engagement, but of providing a trusted social environment in which consumers are willing to sharing their shopping experiences with others. Since a trusted social commerce environment makes the customer feel safe, therefore, the consumers will return to buy products/services in that trusted environment.

References

1. Kim, S., Park, H.: Effects of various characteristics of social commerce (s-commerce) on consumers' trust and trust performance. Int. J. of Info. Mgmt. 33(2), 318–332 (2013)
2. Xiang, L., Zheng, X., Lee, M. K., Zhao, D.: Exploring consumers' impulse buying behavior on social commerce platform: The role of parasocial interaction. Int. J. of Info. Mgmt. 36, 333–347 (2013)
3. Huang, Z., Yoon, S. Y., Benyoucef, M.: Adding social features to e-commerce. In Proceedings of the Conference on Info. Syst. Appl. Res. 1508 (2012)
4. Turban, E., King, D., Lee, J. K., Liang, T. P., Turban, D. C.: Electronic commerce: A managerial and social networks perspective. Springer. (2015)
5. Barnes, N. G.: EWOM Drives Social Commerce: A Survey of Millennials in US and Abroad. J. of Marketing Development and Competitiveness, 9(2), 36 (2015)
6. Aiello, G., Donvito, R. (eds.). Global Branding and Country of Origin: Creativity and Passion. Routledge (2016)
7. Leggatt, H.: Survey: Small Businesses Find Success with Social Networking (2010)
8. eMarketer report: Social Commerce Revenues Worldwide (2015)
9. Laudon, K., Laudon, J.: Management Information Systems: International Edition, 11/E. Pearson Higher Education (2014)
10. Nielsen Report, Word of Mouth Still Most Trusted Resource, http://digitalintelligencetoday.com/word-of-mouth-still-most-trusted-resource-says-nielsen-implications-for-social-commerce/ (2015)
11. Kotler, P., Ang, S. H., Tan, C. T.: Marketing and Management: An Asian Perspective. 609 (2003)
12. Nielsen Report, Global Trust in Advertising report, http://www.nielsen.com/us/en/insights/reports-downloads/2012/global-trust-in-advertising-and-brand-messages.html
13. Katz, E., Lazarsfeld, P.: Personal Influence. Free Press, New York (1955)
14. Hennig-Thurau, T., Gwinner, K. P., Walsh, G., Gremler, D. D.: Electronic word-of-mouth via consumer-opinion platforms: What motivates consumers to articulate themselves on the Internet?. J. of Interactive Marketing, 18(1), 38–52 (2004)
15. Erkan, I., Evans, C.: The influence of eWOM in social media on consumers' purchase intentions: An extended approach to information adoption. Computers in Human Behavior, 61, 47–55 (2016)
16. Zhou, L., Zhang, P., Zimmermann, H. D.: Social commerce research: An integrated view. Electronic Commerce Research and Applications, 12(2), 61–68 (2013)

17. Wang, C., & Zhang, P.: The evolution of social commerce: The people, management, technology, and information dimensions. Communications of the Association for Information Systems, 31(5), 1–23 (2012)
18. Hughes, B.: Small business trends, http://smallbiztrends.com/2015/12/social-commerce-2016-business-trends.html (2015)
19. Mustafi, S.: The Benefits of Social Commerce. University of Applied Sciences Northwestern Switzerland (2010)
20. Bandura, A.: Social foundations of thought and action: A social cognitive theory. Prentice-Hall, Inc. (1986)
21. Lambert, D. M., Harrington, T. C.: Measuring nonresponse bias in customer service mail surveys. J. of Busi. Logistics, 11(2), 5–25 (1990)

Predicting Earning Per Share with Financial Ratios

You-Shyang Chen, Jerome Chih-Lung Chou and Hsiu-Chen Huang

Abstract Predicting EPS (earning per share) with financial ratios is of high practical value. In this study, we applied two methods of data mining to predict EPS and compared their performance. We used three classifiers to find rules for predicting EPS from financial ratios, and compared the difference of prediction performance with and without data discretization. The experimental data set was extracted from the online financial database of Taiwan Economic Journal (TEJ) and collected from the 2009–2013 financial statements of six different industries. 26 condition attributes were selected from financial statements of listed public companies. The decision attribute, EPS, was classified into two and three classes. The result shows that there are three key determinants: Operating income per share, Times interest earned, and Total assets growth rate, and the method of with data discretization is of better prediction accuracy.

Keywords Data mining · Data discretization · Classification · Financial ratio · EPS

1 Introduction

Stock has been a popular tool of investment for people in Taiwan. There are over one thousand listed companies for investors to choose in Taiwan Stock Exchange. However, return on stock is risky, and without analysis, investing stock is like gambling. The analysis of stock has two threads: fundamental analysis and technical analysis. Analyzing financial reports belongs to fundamental analysis, assuming the market price of stock will center to the company's intrinsic value. We took this perspective and tried to apply techniques of data mining to analyze financial ratios to predict earning per share. High earning per share is the key reason for investors of fundamental analysis to hold the stock for a longer period of time.

Y.-S. Chen · J.C.-L. Chou (✉) · H.-C. Huang
Hwa Hsia University of Technology, Taipei, Taiwan
e-mail: Jerome@go.hwh.edu.tw

© Springer Nature Singapore Pte Ltd. 2018

N.Y. Yen and J.C. Hung (eds.), *Frontier Computing*, Lecture Notes in Electrical Engineering 422, DOI 10.1007/978-981-10-3187-8_93

In analyzing financial ratios, the techniques of data mining we applied here were data discretization and classification.

2 Literature Review

2.1 Financial Ratio

Some financial ratios are predictive to EPS. In this study we selected 21 financial ratios that are reported to be more relevant to EPS. They are Cash flow ratio, Net asset value of each share, Cash flow of each share, Per share sales, Per share operating income, Current ratio, Quick ratio, Debt ratio, Accounts receivable turnover ratio, Inventory turnover ratio, Fixed asset turnover ratio, Operating income margin, Return on net asset, Cash ratio, Times interest earned, Interest expense, Year-on-year percentage total assets, Total asset turnover ratio, Net worth turnover ratio, and Stock return ratio.

2.2 Stacking

Stacking classifier is training a learning algorithm to combine the predictions of several other learning algorithms. A logistic regression model is often used as the combiner. Stacking typically yields performance better than any single one of the trained models [1].

2.3 RBF Network

Radial basis function (RBF) network is an artificial neural network that uses radial basis functions as activation functions. Typical radial basis function networks have three layers: an input layer, a hidden layer with a non-linear RBF activation function and a linear output layer. RBF networks learn how to transform input data into a desired response, and are widely used for pattern classification [2–4].

2.4 Naive Bayes

Naive Bayes classifiers are simple probabilistic classifiers based on applying Bayes' theorem with strong independence assumptions between the features. Naive Bayes classifiers assume that the value of a particular feature is independent of the value of any other feature, given the class variable. Regardless of their naive design and oversimplified assumptions, naive Bayes classifiers have worked quite well in many complex real-world situations [5, 6].

3 The Proposed Procedure with an Empirical Study

Step 1 Interviewing experts and collecting data
We consulted experts in financial management and investment, and learned that the database of Taiwan Economic Journal is suitable for the objective of this study.
Step 2 Downloading data from Taiwan Economic Journal
We downloaded real data from TEJ. The time period of the data set is from 2009/1/1 to 2013/12/31.
Step 3 Preprocessing the data
We filtered six industries for analysis. We cleaned irrelevant columns, calculated and add new columns that were not in the data set, such as Times interest earned and Stock return ratio, combined columns of common stock and special stock into total capital, and verified the accuracy of TEJ data with official reports from TWSE (Taiwan Stock Exchange).
Step 4 Verifying features
The decision feature of this study is EPSEPS, and the condition features are 21 financial ratiosFinancial ratio plus Time, Season, Industry, and Total capital.
Step 5 Discretizing the dataset
We tried two ways of datamining for performance comparison: with and without data discretizationData discretization. Data discretization is done by applying filter in the software Weca. Value of EPS was split into 2 classes (P: positive and N: negative and zero) or 3 classes (A: $0 \geq$ EPS, B: $3.75 \geq$ EPS > 0, and C: EPS > 3.75).
Step 6 Applying classifiers
We tried three classifiers: Stacking, RBF network, and Naive Bayes and compare their performance under two and three classes of EPS.

4 Analytical Result and Discussion

Test of cross-validation and test of percentage split were conducted. The cross-validation test uses 90% of data as training subset, and the percentage split test uses 67%. The results are shown in Tables 1, 2, 3 and 4. Model B differs from Model A by using discretized condition features as input, and Model B's performance is better when using classifiers of RBF network and Naive Bayes, but of no

Table 1 Test of cross-validation with EPS of 2 classes

Model	A (%)	B (%)[a]
Stacking	72.2033	72.2033
RBF network	74.3301	90.0893
Naive Bayes	65.3126	88.9834

[a]With data discretization

Table 2 Test of percentage split with EPS of 2 classes

Model	A (%)	B (%)[a]
Stacking	72.6804	72.6804
RBF network	73.6469	91.3015
Naive Bayes	69.4588	90.6572

[a]With data discretization

Table 3 Test of cross-validation with EPS of 3 classes

Model	A (%)	B (%)[a]
Stacking	71.2675	71.2675
RBF network	77.3926	88.7495
Naive Bayes	66.9715	86.4738

[a]With data discretization

Table 4 Test of percentage split with EPS of 3 classes

Model	A (%)	B (%)[a]
Stacking	71.7784	71.7784
RBF network	78.6727	90.3995
Naive Bayes	70.6186	89.1108

[a]With data discretization

difference when classifier Stacking is applied. Using 2 classes rather than 3 classes of EPS as decision feature also yields better model performance.

In summarizing rules mined in this study, Operating income per share, Times interest earned, and Total assets growth rate are determinant factors in classifying EPS. In the model of data discretization, the classifier RBF network provides the best EPS prediction of 2 classes with the accuracy of 91.3015%.

5 Conclusion

Predicting EPS with financial ratios has practical value in financial management and investment. In this study, we identified RBF network is a relatively effective classifier in mining out companies of positive EPS from massive financial ratios.

Furthermore, among those ratios, we identified Operating income per share, Times interest earned, and Total assets growth rate are crucial in predicting the class (positive or negative) of EPS. In future research, we plan to test more classifiers to compare their performance.

References

1. Giacinto, G. Roli, F.: Dynamic classifier selection based on multiple classifier behaviour. Pattern Recognition. 34, 9, 1879–1881 (2001).
2. Govindarajan, M.: Performance Evaluation of Bagged RBF Classifier for Data Mining Applications. IJIEEB. 5, 5, 49–56 (2013).
3. Lampariello, F. Sciandrone, M.: Efficient training of RBF neural networks for pattern recognition. IEEE Trans. Neural Netw. 12, 5, 1235–1242 (2001).
4. Park, J. Sandberg, I.: Universal Approximation Using Radial-Basis-Function Networks. Neural Computation. 3, 2, 246–257 (1991).
5. Mizianty, M. et al.: Discretization as the enabling technique for the Naive Bayes and semi-Naive Bayes-based classification. The Knowledge Engineering Review. 25, 04, 421–449 (2010).
6. Webb, G. et al.: Not So Naive Bayes: Aggregating One-Dependence Estimators. Mach Learn. 58, 1, 5–24 (2005).

Performance Evaluation of Data Mining Technologies: An Example of ERP System Adoption

You-Shyang Chen, Chien-Ku Lin and Wen-Shan Chen

Abstract Traditionally, most studies highlight on exploring the critical success factors (CSFs) of ERP system implementation, and the lack of effective methods for identifying ERP system adoption. In addition, it is an important and interesting issue to help ERP system vendor selecting a suitable customer who will survey an appropriate decision to adopt ERP system. We compare the results of the decisional feature database constructed by two classification prediction models, Models 1 and 2, and find out the critical factors of industrial evaluation for ERP system summarized through the empirical results and hypothesis. The empirical results include Model 1: the accuracy of percentage split without feature selection can reach 89.7810% at maximum, with the minimum value of 57.6642%, and Model 2: the accuracy of percentage split with expert feature selection can reach 89.7810% at maximum, with the minimum value of 54.0146%.

Keywords Enterprise resource planning (ERP) system · Expert feature selection · Classification model

Y.-S. Chen (✉) · W.-S. Chen
Department of Information Management, Hwa Hsia University of Technology,
111, Gong Jhuan Rd., Chung Ho District, 235, New Taipei City, Taiwan, ROC
e-mail: ys_chen@cc.hwh.edu.tw

W.-S. Chen
e-mail: farcom1203@gmail.com

C.-K. Lin
Department of Business Administration, Feng Chia University,
100, Wenhwa Rd., Seatwen, Taichung 40724, Taiwan, ROC
e-mail: g9923808@yuntech.edu.tw

© Springer Nature Singapore Pte Ltd. 2018

N.Y. Yen and J.C. Hung (eds.), *Frontier Computing*, Lecture Notes
in Electrical Engineering 422, DOI 10.1007/978-981-10-3187-8_94

1 Introduction

It's been over a decade since the first wave of Enterprise Resource Planning (ERP) introduction. Most software and hardware are aging, depreciated or inadequate. Most of these enterprises that badly needed to change servers several years ago, delay the upgrade to now due to the economic recession and the growth of unemployment rate. Thus, the enterprises may have the intention of upgrading ERP now. When an enterprise is determined to introduce ERP system, how should it select the most suitable one among various commercial software markets and a lot of ERP system suppliers? Regardless of the price, all management systems at home and abroad expect to make the organizational operation more efficient by software upgrade or introduction. It shows the requirements for ERP software in the enterprises at home and abroad. How can the enterprise select the software that meets its requirements, so as to make ERP system feasible and maximize its performance with the shortest time, the lowest cost and the most efficient way? The motivation of the study is to validate the related operation procedure design and empirical analysis. The Management Information System (MIS) [1] is constructed to differentiate one enterprise in the competitive business environment. The study takes the historical data of more than 400 listed companies for data mining, and finds out the determinants quickly by validating the features of related data, finding out the decision-making factors and conducting empirical analysis. It can provide the business for ERP information vendor and assist the enterprise to introduce ERP system smoothly under the shortest time, the lowest time and the most efficient way, so as to maximize the ERP performance. It can also provide the enterprises with the ERP systems used by competitors as reference for constructing ERP system in the future. Moreover, it is expected to achieve the following objectives: (1) providing for the information vendor to predict the results of ERP system features more effectively and accurately; (2) finding out the important determinants for selecting ERP system.

2 Literature Review

2.1 Enterprise Resource Planning

Davis [2] defines the information technology as: "an integrated human-machine system that can provide information and support routine operation, management and decision-making activities of an organization. Such system can be applied on computer hardware and software, manual program, model and database." ERP [3] is an enterprise management theory proposed, a famous advisory company in America. Put it simple, ERP is a large, modularized and integrated process-oriented system. It integrates the information flow within an enterprise, such as finance, accounting, manufacturing, procurement, business and inventory. It provides the decision-making information quickly, so as to improve the operating performance and quick response ability of an enterprise. To survive and achieve stable profits under the impacts of financial crisis and low interest rate, the enterprise applies system integration, reforms the internal systems, and constructs ERP system, so as to increase the production efficiency and utilize the resources effectively. ERP [4] is a technology used for enterprise information integration. Its core database aggregates the data of various business activities and procedures in the enterprise, and also includes the functions of each department in the enterprise, which can be accessed remotely via network. In this way, it achieves resource sharing and supports module applications.

2.2 Data Mining

Data mining is used to discover the information hidden behind your data, so it is actually a part of knowledge discovery. Data mining uses a lot of statistical analysis and model methods to search for useful patterns and relationships in the database. The process of knowledge discovery shows critical influence on successful data mining application, which can ensure data mining to obtain meaningful results. Grupe and Mehdi Owrang [5] mention it is to discover new facts from the existing database that are still unknown to the experts. Data mining [6] takes advantage of statistical and machine learning algorithms to search for the hidden knowledge and laws with business value from large amount of data in a heuristic way, which can be applied by the automated business strategies.

2.3 Feature Selection

Feature selection [7] is a common term in data mining, which is also known as subset selection. It can be usually used in machine learning field, and is a tool and technique that integrates with learning algorithm and is used to reduce the input under manageable size for processing and analysis. Moreover, it is a process that selects the identifiable and valid items from the set of original features based on specific evaluation indicators, so as to determine the optimal feature subset and optimize the performance indicators.

2.4 Classification Algorithms

In this study, three types of classification algorithms are introduced, including Tree, Bayes, and Functions, as follows.

(1) **Tree**. Tree [8] is a powerful and popular classification and prediction tool. With the supervisory learning method, it firstly sets the decision tree in the instance set. Based on the tree diagram, the decision tree makes a tree structure for data classification by using the conditions and rules. The rules of the decision tree can be texts, which is easy to understand. Also it can be transformed into Structured Query Language (SQL) database for query. Therefore, it is also called semantic tree. The data records falling in specific class can be searched to form data tree composed of root, node and leaf just like a tree.
(2) **Bayes**. The core theory of Bayes classifier [9] is to calculate the possibility of classifying into each class based on Bayes corollary, which is only applicable to classification. Bayes corollary is an effective tool to make prediction under uncertainties. It infers the posterior distribution of the population mainly based on multiple possible prior distributions and empirical distribution, so as to calculate the possible values.
(3) **Functions**. The general function classifier [10] can be divided into two types: (1) SMO: Support vector machine of sequential optimization, and (2) Logistic: Logistic regression.

2.5 Evaluation of Classification Algorithms

The study uses percentage split to evaluate the classification algorithms to be used. Percentage split [11] extracts a portion of data based on a certain percentage for test. It conducts evaluation based on the prediction effect of the instances in the classifier. The data mining literatures generally use 67% (2/3) for training to get the model. After the model is generated, the rest 33% (1/3) is used to test the accuracy of the model. Since the training and test percentages can be only slightly adjusted, we use 2/3 for training to get model, and 1/3 to test and validate the accuracy of that model.

3 Research Methodology

3.1 Research Framework

This section mainly illustrates flowchart of the proposal hybrid model to be used in the study, including Models 1 and 2, which mainly includes 9 steps as detailed below. Figure 1 shows the flowchart.

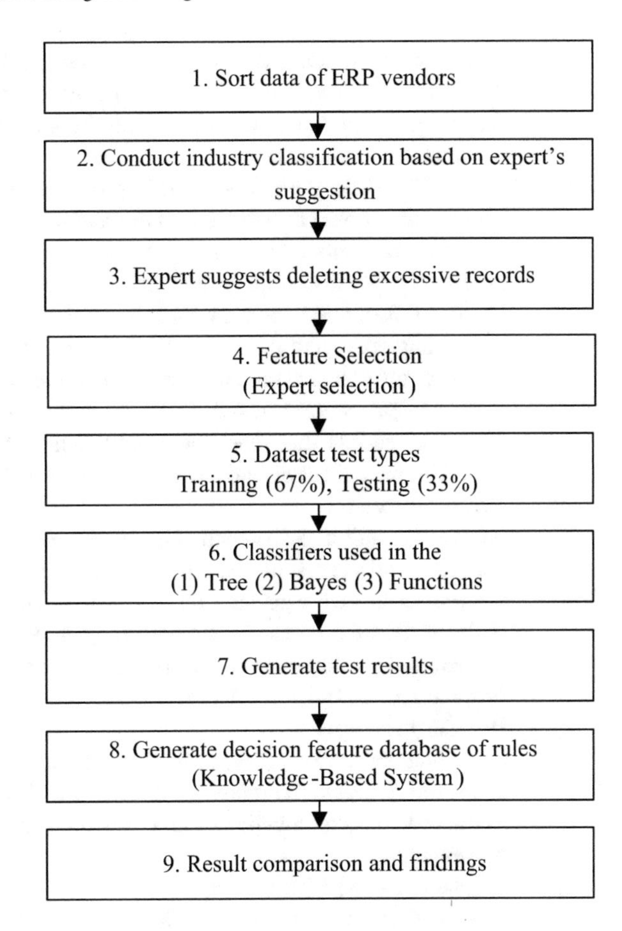
1. Sort data of ERP vendors
2. Conduct industry classification based on expert's suggestion
3. Expert suggests deleting excessive records
4. Feature Selection (Expert selection)
5. Dataset test types Training (67%), Testing (33%)
6. Classifiers used in the (1) Tree (2) Bayes (3) Functions
7. Generate test results
8. Generate decision feature database of rules (Knowledge-Based System)
9. Result comparison and findings

Fig. 1 Research framework of this study

3.2 Data Collection

The study data is the customer data provided by the domestic ERP vendors, including 708 data records in the period of 2006–2010. Moreover, the experts interviewed by the study have more than 20 years of business experience in Taiwan. Therefore, we follow the expert's suggestions in the steps of data pre-processing, data deletion, industry classification and feature selection.

3.3 Data Analysis and Results

Step 1 **Sort data of ERP vendors**
Among the 30 items of vendor's basic data in the initial planning, 29 strings are condition features, while 1 string is decision feature, which are assigned with domain name and string data type respectively. In the data

type, the data characteristics of each string are defined, each of which has fixed computation format. For example, the string can link up a phrase of meaningful texts, and the number can be used to perform basic or more complicated computation.

Step 2 **Conduct industry classification based on expert's suggestion**
The study conducts industry classification for the data based on the expert's suggestion: The industries with many data records are taken as the criteria for classification. The manufacturing industry, construction industry and information industry take the majority of Taiwan industries. The expert suggests taking the types of manufacturing industry, construction industry, information industry and communication industry as the subjects.

Step 3 **Data pre-processing based on the above 3 steps**
The data records that meet the requirements of the study are reduced to 416. The data finally obtained is the results of 11 strings.

Step 4 **Feature selection**
Feature selection by expert: The expert directly selects the important features and decision features among the data that goes through pre-processing as suggested by the expert. The expert suggests retaining such features as the establishment time, capital amount, region, reason to change, old customer level and software source, which include 5 condition features and 1 decision feature.

Step 5 **Dataset test types**
Percentage Split: 67% (2/3) of the samples are taken as the train set, which is used to generate the model. The rest 33% (1/3) of the samples are taken as the test set, which validates and tests the accuracy of our models after it is generated.

Step 6 **Classifiers used in the study**
The classifiers used in the study include 2 algorithms of Tree, 2 algorithms of Bayes and 2 algorithms of Functions.

Step 7 **Generate test results**
As suggested by the industrial experts, the study classifies the ERP software source into Domestic (D) and Foreign (F). They are explored based on two models respectively: (1) Model 1: accuracy of percentage split without feature selection and data discretization; and (2) Model 2: accuracy of percentage split with expert feature selection but without data discretization. The test objectives are: (1) to validate the difference between the single model and hybrid model; (2) to validate the difference of feature selection.

Step 8 **Get the decision tree**
It adopts the mining tool that is commonly used in the academia, and draws the tree of decision rules through the nodes actually generated by J48.

Step 9 **Results comparison and findings**
We compare the results of the decision feature database constructed by different classification prediction models, and find out the critical factors of

Table 1 Results comparison

Model	Classify	Classifier	Percentage split (67%)
Model 1	Tree	Simple Cart	89.7810%
		J48	89.0511%
	Bayes	BayesNet	85.4015%
		NaiveBayes	82.4818%
	Functions	SPegasos	83.2117%
		Vote Perceptron	57.6642%
Model 2	Tree	Simple cart	89.7810%
		J48	87.5912%
	Bayes	BayesNet	86.8613%
		NaiveBayes	82.4818%
	Functions	SPegasos	83.9416%
		Vote perceptron	54.0146%

industrial evaluation for ERP system summarized through the empirical results and hypothesis. Two empirical results are determined, including Model 1: the accuracy of percentage split without feature selection can reach 89.7810% at maximum, with the minimum value of 57.6642%, and Mode 2: the accuracy of percentage split with expert feature selection can reach 89.7810% at maximum, with the minimum value of 54.0146%. It is as shown in Table 1.

4 Conclusion

When exploring the ERP software selection, the most important thing is the capital amount of ERP system supplier, which should include the market share and the word-of-mouth of the service quality. In practice, it will set a threshold on the capital amount of the customer, and start to cut in from a certain capital amount based on the budget of its own products. Finally it will know whether it is a target customer based on the budget limit.

The technological advantage among the domestic enterprises can create exclusive industrial development means. It doesn't only have the cost advantage, but also acquire the technically outstanding and flexible production method, which can play a critical role globally. With the data analysis of scientific algorithms, the study hopes to provide the enterprises with available study data as follows: The two critical factors that influence ERP system selection are the budget limit and the capital amount of the ERP supplier, so it may make good use of these two condition features.

ERP system industry is intensively competitive, so it is very important to make use of the existing data for data mining and finding out the most favorable information. The study excludes other factors such as service, scalability and user interface convenience, but only explores finding out the pure factors of decision features by using data mining and through the hybrid model of the study. The industrial contribution is: The study results can provide reference for ERP system vendor, and make enterprise select the suitable ERP system more efficiently and conveniently. When selecting the suitable ERP software, the most important thing for enterprises is the capital amount of the ERP system vendor.

Acknowledgements The authors would like to express sincere appreciation to the Ministry of Science and Technology of the Republic of China, Taiwan, in financially supporting this research under Contract No. MOST 103-2221-E-146-003-MY2.

References

1. Pérez-Méndez, J. A., Machado-Cabezas, Á.: Relationship Between Management Information Systems and Corporate Performance. Revista de Contabilidad. 18 (1), 32–43 (2015)
2. Davis, G.: Management Information System: Conceptual Foundations, Structure and Development Network. Mc Graw-Hill. (1985)
3. Wylie, L.: ERP: A Vision of the Next-Generation MRP II. Computer Integrated Manufacturing. 300 (339.2), 1–5 (1990)
4. Shen, Y. C., Chen, P. S., Wang, C. H.: A Study of Enterprise Resource Planning (ERP) System Performance Measurement Using the Quantitative Balanced Scorecard Approach. Computers in Industry. 75, 127–139 (2016)
5. Grupe, F. H., Mehdi Owrang, M.: Data Base Mining Discovering New Knowledge and Competitive Advantage. Information System Management. 12 (4), 26–31 (1995)
6. Sohrabi, M. K., Akbari, S.: A Comprehensive Study on the Effects of Using Data Mining Techniques to Predict Tie Strength. Computers in Human Behavior. 60, 534–541 (2016)
7. Sayed, S. A. F., Nabil, E., Badr, A.: A Binary Clonal Flower Pollination Algorithm for Feature Selection. Pattern Recognition Letters. 77, 21–27 (2016)
8. Akkaş, E., Akin, L., Çubukçu, H. E., Artuner, H.: Application of Decision Tree Algorithm for Classification and Identification of Natural Minerals Using SEM–EDS. Computers & Geosciences. 80, 38–48 (2015)
9. Mujalli, R. O., López, G., Garach, L.: Bayes Classifiers for Imbalanced Traffic Accidents Datasets. Accident Analysis & Prevention. 88, 37–51 (2016)
10. Varando, G., Bielza, C., Larrañaga, P.: Decision Functions for Chain Classifiers Based on Bayesian Networks for Multi-Label Classification. International Journal of Approximate Reasoning. 68, 164–178 (2016)
11. Anuradha, C., Velmurugan, T.: A Comparative Analysis on the Evaluation of Classification Algorithms in the Prediction of Students Performance. Indian Journal of Science and Technology. 8 (15), 1–12 (2015)

Developing the Framework for Evaluating the Interaction Design of Mobile Augmented Reality Systems for Cultural Learning

Cheng-Wei Chiang, Li-Chieh Chen and Jing-Wei Liu

Abstract Mobile augmented reality (MAR) is an increasingly popular technology for enhancing how students interact with and learn about the cultural environment and cultural objects in the physical world. Sometime the learning effect is influenced by the delighted value, which is always subjective and learning—driven. In order to ensure successful launch the mobile augment reality of cultural interactive learning tool, it is extremely important to predict the delighted value of design alternatives systematically based on the common language understood by both students and designers. However, the framework for communicating and evaluating such value from interested perspective is not available in the literature. Therefore, the objective of this research is to extract key frameworks of delighted value from interested perspective and develop an effective algorithm to evaluate MAR cultural learning system. First, through literature review and the interview of participants, many scenarios of learning influence were collected. A focus group was invited to identify the essential elements that influence the delighted value of MAR cultural learning system. Followed by a large scale questionnaire survey and factor analysis, four frameworks were extracted. These frameworks, name as CARE framework in brief, included communication, association, reflection, and engagement. Second, the perception differences of MAR cultural learning paper prototypes were conducted to verify the validity of CARE framework for comparative studied. The findings of this study demonstrated that CARE framework was effective for solution designing in MAR cultural interactive learning tool.

C.-W. Chiang (✉) · L.-C. Chen
The Graduate Institute of Design Science, Tatung University, Taipei, Taiwan
e-mail: waylan@gmail.com

L.-C. Chen
e-mail: lcchen@ttu.edu.tw

J.-W. Liu
Department of Digital Game and Animation Design, Taipei College of Maritime Technology, Taipei, Taiwan
e-mail: jwliu99@gmail.com

© Springer Nature Singapore Pte Ltd. 2018

N.Y. Yen and J.C. Hung (eds.), *Frontier Computing*, Lecture Notes in Electrical Engineering 422, DOI 10.1007/978-981-10-3187-8_95

Keywords Mobile augmented reality design • Cultural learning • Frameworks of delighted value

1 Introduction

Today, more than a third of all adults and more than fifty percent of all college students in the U.S. have smart phone. Commercially available smart phones have enough processor and graphic engine power to run augment reality applications, and network connectivity fast enough to download high resolution videos. Because many students carry mobile devices everywhere they go, providing educational material through devices strikes us as a prime opportunity to reach the aspirational goal of "learning anytime, anywhere," which expands Weiser's vision [1] of ubiquitous computing (ubicomp) to everyday educational contexts.

Mobile Augmented Reality (MAR)—which involves the dynamic overlay of digital information in the user's view through mobile device-is an increasingly popular technology for enhancing how people interact with and learn about the environment and objects in the physical world [2]. Using MAR on smart phones, it explore ways to draw students' attention and delighted value to the physical environment they are immersed. Delighted value is related to the cultural learning interesting on cultural emotional ground [3]. Delighted value has been described as a relatively enduring predisposition to attend to certain objects and events and to engage in certain activities [e.g., 4–6].

The psychological state of delighted value can also be generated by specific environmental stimuli and is referred to as situational interest [7]. Sometime, students pay attention to learning culture was influenced by their delighted value and environments. However, interactive learning design practitioners often misunderstand what students really need. That is, they often encounter the problems of not being able to know students' preference and reactions. This is a barrier that every cultural education seeks to overcome during the process of cultural interactive learning development. If a common language between designers and students could be found, such as the interests of culture frameworks of the mobile augment reality of cultural interactive learning tool, the interesting gap between the two could be effectively narrowed.

To address this issue, the objectives of this research include identifying the elements of cultural environmental learning dimensions that influence MRT learning delighted value, extracting the cultural interactive learning frameworks, constructing an evaluation method for applying the cultural learning frameworks, and examining the validity of the cultural element dimension frameworks and evaluation method through case studies.

2 Building on Previous Work

In this section we review the types of MAR technology in physical environment and mobile learning for the cultural learning of our framework of evaluating the Interaction Design.

Technological advances revolutionized people's way of learning, working, and entertainment and lifestyle. MAR's extended mobility enables digital information to be displayed in more places away from the desktop, in the users' physical and personal world. This has opened up environmental opportunities for mobile learning [8]. In MAR systems, augmentable targets can be streets, buildings, natural areas and objects, even people and other moving targets. MAR enables the physical environment to be directly annotated and described in situ, guiding users to pay attention and get more interests to particular parts of objects in their environment [9, 10]. If designers have an effective evaluation method to estimate the extent of perception of students, then this would be a great benefit to the improvement of design process and design quality.

In this paper, there were two parts of literature review relevant to this research topic. The first part summarized experts' view points on factors that influence cultural environmental learning design and compiled students' cultural environmental learning concerns for mobile augment reality of cultural interactive learning tool design. The second part explored the current application of using quantitative analysis in the domain of cultural learning to evaluate design strategies and alternatives.

2.1 *Mobile Augment Reality*

Recently the technological properties of mobile devices have been harnessed in a combined way to employ a technique that has not been seen before on mobile devices: augment reality. It however remains unclear why augmented reality applications and mobile devices are such a good match and in what context it might be useful. To answer these questions, mobile augment Reality (MAR) extends the augment reality (AR) paradigm which is to display digital information in the user's view via head-mounted displays or other displays, so that objects in the physical world and digital world appear to spatially co-exist [11]. AR is often understood as part of the mixed reality continuum focusing on augmenting the real worlds [12]. In the following section, we discuss aspects of MAR that can contribute uniquely to students learning about objects or cultural environment.

2.2 MAR Design and Cultural Environmental Learning Dimension

Many researchers had developed methods for enhancing positive responses of cultural environmental learning on mobile devices or augment reality. In addition to mobile devices and augment reality, user experiences on the usage and the personal identification caused by learning device ownership are also important. Design is in fact a kind of communication [13]. Designer must have a deep understanding of users' environmental learning and communication with them. In fact, changing the user experiences and design content was the common approach in either design-driven innovation or user-oriented design [14].

To accomplish successful communication, the first and the most important issue is how to correctly measure learners' environmental responses. In order to address this issue, some measurement scales had been developed and applied to the filed on marketing and behavior analysis on the web, such as pleasure, arousal and dominance (PDA) scale, different environment scale (DES) [7], and self-assessment-manikin (SAM) scale [15]. Although these scales were effective for marketing and online shopping [16], they were not developed specifically for the purpose of physical MRT cultural environmental learning design and evaluation. In addition, they did not cover the issue of measuring environmental responses due to learner experiences on MRT usage.

2.3 MAR and Situated Learning

In contrast to traditional classroom learning, mobile learning has the opportunity to take the abstract and decontextualized knowledge (often offered in the classroom) and apply it back to the real world scenarios the students are immersed in. Mobile learning can help situate the learning in the learner's personal environment outside of the traditional classrooms. Situated cognition theory suggests that knowledge cannot be separated from the context in which it is embedded and learning results from students acting in apprentice-like situations in interaction with experts [17]; Brown et al. (989). One of the primary challenges in scaffolding novices to become experts is helping them first develop a "professional vision"—being able to see the world from an expert's perspective.

3 Development of the Framework and the Evaluation Method

To identify the elements that construct cultural environmental learning dimension, this study used literature analysis and structured interview with designers and students. The results of interview then served as the data for a focus group to encode, select, and determine the elements.

3.1 Exploring the Elements of Cultural Environment Learning Dimension

To identify the elements that construct cultural environmental learning dimension, this study used literature analysis and structured interview with designers and students. The results of interview then served as the data for a focus group to encode, select, and determine the elements.

As mentioned in previous, designers and students during the cultural environmental learning process usually do not have a common evaluation standard. To overcome this problem, this study first invited twelve participants with different levels of cultural interactive learning design or cultural working background. During the interview process, every participant needed to raise one to three related cases for explanation of their experiences of cultural environment learning and designing interactive with cultural environment.

Finally, the group classified the characteristics into thirteen elements. They are cultural of conservation, cognitive geography, cultural visual imagery, knowledge of aesthetics, fusion design, elements transformation, creative opportunities, internal and art and external shape layers, special meaning, cultural story, cultural emotion, information services, and technology integration. These were used in the next stage when data reduction was used to extract the major factors of the cultural environmental learning dimension.

3.2 Extracting the Major Factors of CARE Framework

Given the 13 elements obtained from the previous stage, the goal of this stage was to identify the structure among these elements through factor analysis. To help participant point out the factor that influences their cultural interactive learning efficiency.

The result of the pre-test were analyzed by factor analysis and the elements were analyzed, deleted, merged or corrected, and at the same time common elements were found and give names to extract the key cultural environmental learning dimension frameworks. After proceeding with the first factor analysis, according to the analysis result, the factor loading of "cultural of conservation" and "cognitive geography" did not exceed 0.50. Thus these two relatively inconspicuous elements were deleted. The data then underwent the second factor analysis. The principle component analysis of factor analysis was used to extract common factors, and by choosing common factors with eigenvalue larger than 1 as the selection principle, four major factors were chosen, which explained 51.087% of total variance. By then using varimax method, the select factors were rotated, resulting in more significant and easily explainable factors. Detailed results are show in Table 1.

Factor 1 consisted of three elements. They were cultural visual image, shape and form characteristics, and aesthetic exploration. These factor loadings were within the range of 0.501–0.628. The eigenvalue was 1.820, which explained 18.196% of total variance. Given the first four elements were related to characteristics directly expressed by cultural aesthetic shape and cultural image communication, Factor 1 was thus named as "Communication".

Factor 2 consisted of cultural story, cultural emotion, and heritage value, with factor loadings within the range of 0.56–0.715. The eigenvalue was 1.415, which explained 14.152% of total variance. Since these elements allowed learners to invoke wonderful feelings or touching stories, Factor 2 was thus named as "Association".

Factor 3 was main make up of Fusion of old and new elements and elements transformation, within the range 0.555–0.557. The eigenvalue was 1.092, which explained 10.915% of total variance. At the first glance, these two elements seemed unrelated, but in fact they had a common characteristic. That is, there were strong bond between learners' needs and design features. Learners benefited from these design features and understood the knowledge of culture. In such a case, it is very difficult for the learners to stop thinking the culture. Thus the authors borrowed a term from design connection and name the factor "Reflection".

Table 1 The result of factor analysis for the cultural environmental learning dimension

Elements	Factor 1	Factor 2	Factor 3	Factor 4
Cultural visual image	**0.628**	0.097	0.108	0.369
Shape and form characteristics	**0.617**	0.245	0.353	0.261
Aesthetic exploration	**0.501**	0.343	0.253	−0.199
Cultural story	0.217	**0.665**	0.016	−0.013
Cultural emotion	0.274	**0.715**	−0.072	−0.048
Heritage value	−0.225	**0.56**	−0.159	−0.477
Fusion of old and new elements	0.19	0.245	**0.555**	0.288
Cultural elements transformation	0.375	0.004	**0.702**	−0.404
Information service quality	−0.172	0.086	−0.053	**0.851**
Technology integration	0.163	−0.054	−0.013	**0.8**
Innovation	−0.156	0.109	−0.007	**0.757**
Variance (%)	18.196	14.152	12.727	10.915
Cumulative (%)	18.196	34.594	48.556	61.069

Note Bold-italic text indicates the absolute value of factor loading is >0.5

Factor 4 was mainly made up of information service quality, technology integration, and Innovation, with factor loading within the range of 0.757–0.815. The eigenvalue was 1.273, which explained 12.727% of total variance. Given that these elements were related cultural services and innovation complement, which were compatible with creations of need theory. Factor 3 was named as “Engagement”.

Based on the results of factor analysis, communication (C), association (A), reflection (R), engagement (E) were used construct the evaluation dimensions of CARE frameworks. This frameworks consisted of four dimensions and eleven evaluation criteria, as shown in Fig. 1.

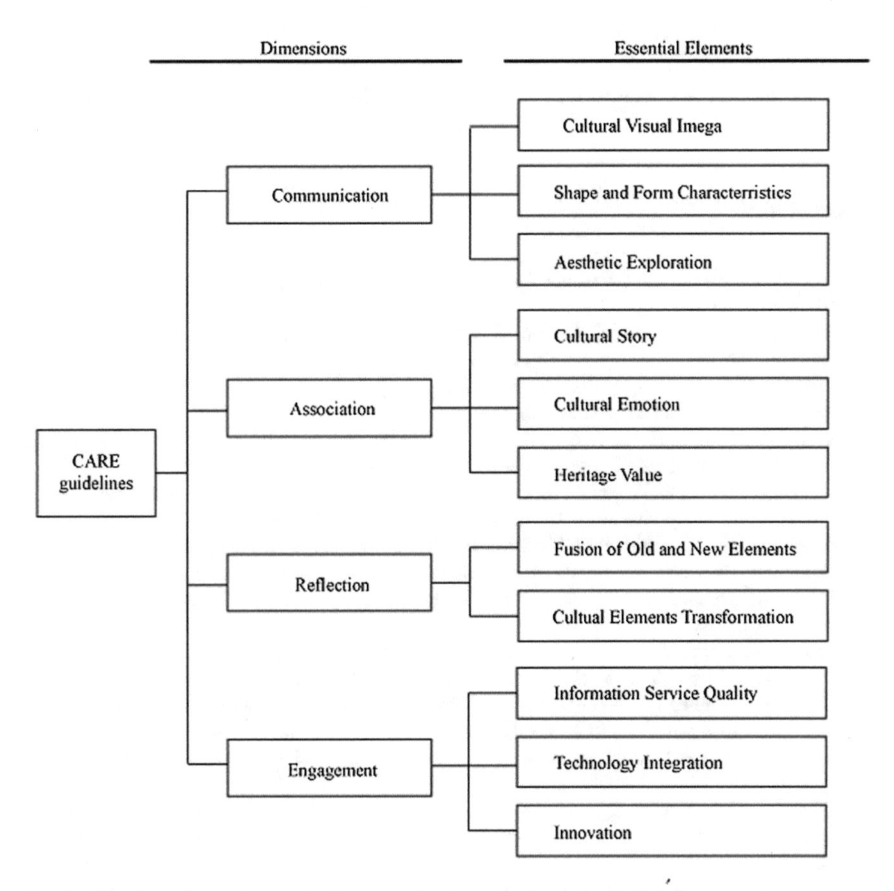

Fig. 1 The four dimensions and eleven evaluation criteria of the CARE frameworks

4 Recommendation for Future Research

To test whether the CARE frameworks is able to distinguish the difference among MAR systems in cultural environmental learning dimensions, the authors will collect data from 30 participants with different personalities, selected from different gender, age, and year from Taipei College of Maritime Technology.

On the other hand, after studying design literature and interviewing the scholars of temples, the four-person focus group with experts from academia research will select three well-know classic design mobile AR cultural learning system as experiment sample.

These systems introducing the important cultural elements would show up to provide augmented information and 3D models on the mobile screen. In the MAR systems, animated characters and culture elements in the cultural objects were re-built to deliver the story of the ancient. These contents help students to understand the history and traditional cultural objects in the physical world.

References

1. Weiser, M., 1991. The Computer for the Twenty-First Century, *Scientific American*, pp. 94–10, September 1991.
2. Bier, E., Stone, M., Pier, K., Buxton, W., DeRose, T.,1993. Toolglass and Magic Lenses: The See-Through Interface. *Proc. SIGGRAPH* '93, ACM Press (1993), 73–80.
3. Cristian, V.,2011. Entry points, interests and attitudes. An integrative approach of learning. *Procedia Socail and Behavioral Sciences, 11,* 77–81.
4. Krapp, A., Hidi, S. & Renninger, A., 1992. Interests, learning and development. *The role of interest in learning and development* (pp. 3–25). Hillsdale, NJ:Erlbaum.
5. Renninger, K. A., & Wozniak, R. H., 1985. Effect of interest on attention shift, recognition, and recall in young children. *Developmental Psychology, 21,* 624–632.
6. Renninger, K. A., 2000. Individual interest and its implications for understanding intrinsic motivation. In C. Sansone & J. M. Harackiewicz (Eds.), *Intrinsic and extrinsic motivation: The search for optimum motivation and performance* (pp. 373–404). New York: Academic Press.
7. Hidi, S., & Baird, W., 1988. Strategies for increasing text-based interest and students' recall of expository text. *Reading Research Quarterly, 23,* 465–483.
8. Gleue, T., Dähne, P, 2001. Design and implementation of a mobile device for outdoor augmented reality in the ARCHEOGUIDE project. In: *Proceedings of the 2001 conference on virtual reality, archeology, and cultural heritage*, Glyfada, Greece. ACM Press, New York, pp 161–168.
9. Höllerer T, & Feiner S, 2004. Mobile augmented reality. In: Karimi H, Hammad A (eds) *Telegeoinformatics: location-based computing and services*. Taylor & Francis Books, London
10. Wellner P., Mackay W. & Gold R, 1993. Back to the real world. *Comm ACM* 36(7):24–26.
11. Milgram P, Kishino F, 1994. A taxonomy of mixed reality visual display. *IEICE Trans Inf Syst*, E77-D(12):1321–1329.
12. Azuma, R.T., 1993. Tracking requirements for augment reality. *Communication of ACM*, 36 (7): 50–51.

13. Norman, D. A., 2004. Emotional design: Why we love (or hate) everyday things. NewYork: Basic Books
14. Veryzer, R. W., & Borja de Mozota, B., 2005. The impact of user-oriented design on new product development: An examination of fundamental relationships. *Journal of Product Innovation Management, 22(2), 128–143.* (2005).
15. Lang, P. J., 1985. The cognitive psychophysiology of emotion: Anxiety and the anxiety disorders. Hillsdale: Lawrence Erlbaum.
16. Huang, M. H., 2001. The theory of emotions in marketing. Journal of Business and Psychology, 16(2), 239–247.
17. Lave, J. & Wenger, E., 1991. *Situated Learning. Legitimate Peripheral Participation.* Cambridge University Press. Cambridge. (1991).

An Attraction Reservation System for Amusement Park Services

Pei-Chun Lee, Sheng-Shih Wang, Yu-Chen Shu, Zih-Ci Liou, Hao-Chien Cheng and Yi-Chen Lin

Abstract Many amusement parks have been operated with information and communications technologies to provide friendly and efficient tourist services. However, these systems leave various levels of decisions to tourists to make by themselves, which may incur pressure on tourists. To address this issue, we have developed a SAP system with smart tour suggestion. However, this SAP system would not be exhaustive without an attraction reservation scheme. Therefore, this paper proposes an attraction reservation system as an extension to the SAP system. The proposed system comprises three subsystems: The mobile app subsystem is an interface for tourists to make reservations for attractions, and stores the booking tickets for later verification. The central subsystem performs necessary computing and database management. The ticket-scanning subsystem scans the tourist's booking ticket for further verification. The demonstrations of the implemented prototyping system show that the proposed system does function.

Keywords Amusement park · Mobile app · Reservation system · Tourist service

---

P.-C. Lee · S.-S. Wang (✉) · Y.-C. Shu · Z.-C. Liou · H.-C. Cheng · Y.-C. Lin
Department of Information Management, Minghsin University of Science and Technology, Xinfeng, Hsinchu, Taiwan
e-mail: sswang@must.edu.tw

P.-C. Lee
e-mail: pjlee@must.edu.tw

Y.-C. Shu
e-mail: B01090011@std.must.edu.tw

Z.-C. Liou
e-mail: B03093101@std.must.edu.tw

H.-C. Cheng
e-mail: B02092111@std.must.edu.tw

Y.-C. Lin
e-mail: B01090009@std.must.edu.tw

© Springer Nature Singapore Pte Ltd. 2018

N.Y. Yen and J.C. Hung (eds.), *Frontier Computing*, Lecture Notes in Electrical Engineering 422, DOI 10.1007/978-981-10-3187-8_96

1 Introduction

According to the World Tourism Organization in a report of year 2000 edition, tourism has become the primary source of foreign exchange in many countries [1]. As a part of tourism industry, amusement parks have been playing an important role. Therefore, the development of amusement parks is one of the significant factors in the improvement of the hospitality and tourism industry worldwide. Several major issues regarding amusement parks include the visitors' satisfaction, loyalty and revisit intention. Information and communications technologies (ICT) can be good facilitation to these issues while the technologies have already become a major trend in the industry of amusement parks as observed.

Famous theme parks such as Walt Disney World [2], Universal Orlando Resort [3], Ocean Park Hong Kong [4], Janfusun Fancy World [5] and Leofoo Village Theme Park [6] in Taiwan have introduced ICT-based techniques into their park services, especially mobile apps [7–15] which get more and more popular in recent years. In addition to basic park and attraction information inquiry, these ICT-based park services have been designed to solve several existent problems which put tourists under pressure, such as tiresome long wait times for attractions. For example, Walt Disney World [16] and Universal Orlando Resort [17] have been carrying out FastPass+ or Express™ Pass schemes for highly popular attractions and providing tourists with the estimated wait time of each attraction. Janfusun Fancy World [5, 13] presents a solution to show three colors representing different wait time periods of attractions.

The solutions mentioned in the examples are all useful yet leave various levels of decisions to tourists to make by themselves, which may cause pressure on tourists and negatively impact tourists' experiences in the tour. What if the park service goes the extra mile to give tourists an even easier experience by automatically planning customized schedules for individual tourists? In response to this question, we have proposed a smart amusement park (SAP) system [18] to accommodate the tour pressure problem with an innovative solution called the GPS-based dynamic scheduling scheme to offer tourists the customized best plans of tour. In addition, the GPS-based dynamic map function is also provided as a supplementary feature. However, the SAP system would not be considered exhaustive without an attraction reservation scheme. Therefore, in this paper, we develop an attraction reservation (AR) system as an extension to the previously proposed SAP system, and this AR system can be integrated into the SAP system.

The rest of this paper is organized as follows. Section 2 gives an overview of the proposed attraction reservation (AR) system. Section 3 elaborates the AR system in detail. Section 4 presents the implementation and demonstration of the prototyping AR system. Finally, Sect. 5 concludes this paper.

2 Overview

The AR system consists of three subsystems: mobile app subsystem, ticket-scanning subsystem, and central subsystem, as shown in Fig. 1. Because the AR system is an extension to our previously proposed SAP system [18], Fig. 1 can also be viewed as a partial excerpt of the SAP system with extra functional modules.

The mobile app subsystem is the mobile app provided by the amusement park for tourists to download and install on their smartphones or tablets, and presents an integrated interface to tourists to take advantage of the attraction reservation (AR) services. As Fig. 1 shows, the subsystem has three modules, including Park Info Module, Tour Suggestion Module, and My Record Module, which will be elaborated in Sect. 3.

The ticket-scanning subsystem consists of two facilities, including the ticket scanner and the gate controller deployed at the reservation (RES) entrances of attractions. When the tourist who has booked the attraction comes to the RES entrance during the appointed period and shows the booking ticket, the ticket scanner will scan and decode the ticket information for ticket verification performed by the central subsystem. Only if the ticket is valid, the RES entrance gate will open up for the tourist to pass.

The central subsystem is responsible for attraction reservation management and database management, and may be located in the control room of the amusement park. The Attraction Reservation Management Module comprises two sub-procedures: Reservation Request Handling and Booking Ticket Verification, which will be described in Sect. 3. This subsystem communicates with the mobile app subsystem (app) and the ticket-scanning subsystem to process and respond the requests from these two subsystems.

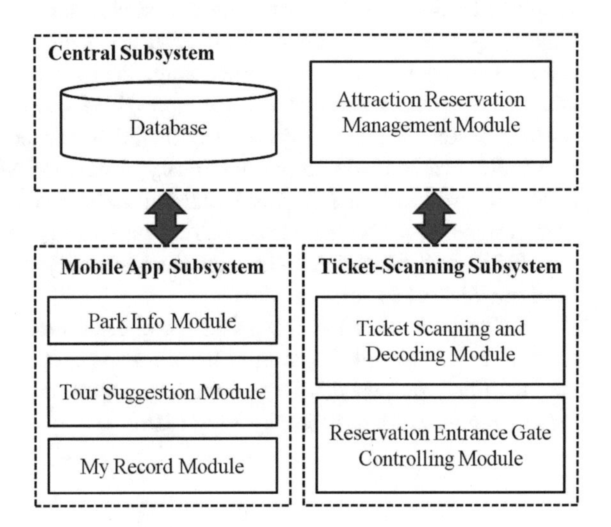

Fig. 1 Function blocks of the AR system

3 AR: Attraction Reservation System

This section describes the detailed design of the proposed AR system.

3.1 Mobile App Subsystem

The mobile app subsystem provides tourists with an integrated interface to take advantage of the AR services. Tourists need only to download and install the app into their smartphones or tablets and everything is on the go. Three main modules in this subsystem as in Fig. 1 are elaborated as follows.

Park Info Module. This module provides general information about the amusement park, such as News and Tour Guide. Inside Tour Guide, Attraction Info provides tourists with information about each attraction in the park, and when the tourists browse the attractions, they can add attractions into their personal My Play List. This list is to be used later in the Tour Suggestion Module, where the attraction reservation (AR) sub-module resides.

Tour Suggestion Module. In our previously proposed SAP system [18], this module consists of two features: GPS-based dynamic scheduling and GPS-based dynamic map. In our design, the AR sub-module resides in the GPS-based dynamic scheduling feature. To be more specific, the tourist can start to reserve the recommended attraction after the GPS-based dynamic scheduling scheme provides the recommended result. Nonetheless, the AR sub-module can also be designed independent of the GPS-based dynamic scheduling feature.

We elaborate a general scenario of attraction reservation and booking ticket verification as follows for clarity. When a tourist starts to reserve an attraction, app will show the tourist a list of bookable sessions of this attraction, each session with a current bookable quota. After the tourist chooses a bookable session and selects how many seats he/she wants to reserve, app will send a request message to the central subsystem for further processing. When receiving the confirmation reply message from the central subsystem, app will display the result and store a digital booking ticket for later verification at the RES entrance.

My Record Module. In our previously proposed SAP system, there are two lists in My Record Module: My Play Record and My Play List. Only the latter, My Play List, is relevant to the AR services. And necessarily, we add My Booking Tickets into this module for the AR services. Here, we only describe My Play List and My Booking Tickets as follows.

My Play List. This list comprises the chosen attractions selected by tourists on their own. It is also the operation basis for the GPS-based dynamic scheduling in the previously proposed SAP system. The attractions can be selected in Attraction Info or modified directly in My Play List.

My Booking Tickets. All the booking tickets are stored digitally in this list. When a tourist arrives at the RES entrance, he/she can draw out the corresponding ticket from this list and show the ticket to the ticket-scanning subsystem for further verification.

3.2 Ticket-Scanning Subsystem

This subsystem consists of a ticket scanner and a gate controller deployed at the RES entrance of each of the attractions. The ticket scanner can scan and decode the tourist's booking ticket and then request the central subsystem to verify whether the ticket is valid. Only if the ticket is valid, the gate controller will trigger the RES entrance gate to open up for the tourist to pass.

A simplified description considering only the successful case is elaborated as follows. When the tourist comes to the RES entrance, he/she can draw out the corresponding booking ticket from My Record Module/My Booking Tickets, and show the ticket to this subsystem. After this subsystem reads in and decodes the ticket information, it sends the information in *Ticket Verification Request* to the central subsystem for further verification. If the ticket-scanning subsystem receives a successful verification result, it will trigger the RES entrance gate to open up and allow the tourist to pass; otherwise, it will show the invalid result. A more detailed message flow chart including the failed cases between the ticket-scanning subsystem and the central subsystem is depicted in Fig. 2.

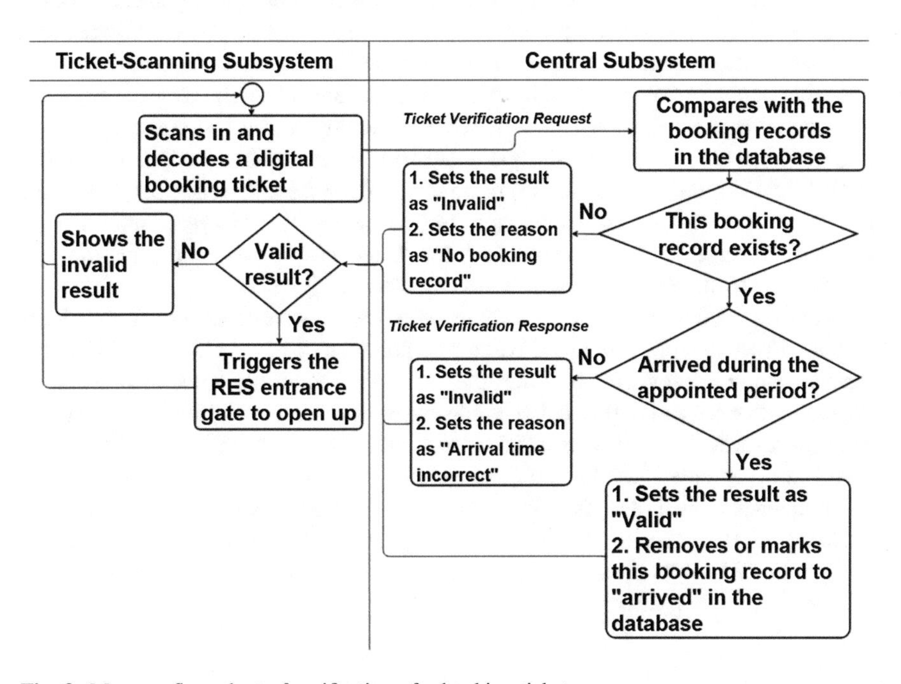

Fig. 2 Message flow chart of verification of a booking ticket

3.3 Central Subsystem

The central subsystem is responsible for attraction reservation management and database management. In addition to park information provision, this subsystem accepts requests from app, processes the requests accordingly, and replies app with corresponding results. Moreover, this subsystem receives the booking ticket information from the ticket-scanning subsystem and verifies the legality of the ticket.

In our previously proposed SAP system, the main procedure designed in the central subsystem is the Wait Time Estimation Algorithm. In the AR system, this algorithm actually still exists in the central subsystem, although this algorithm is not explicitly shown in Fig. 1 because it is not the main role of the AR system. This paper only focuses on the depiction of the Attraction Reservation Management Module.

The Attraction Reservation Management Module is kernel of our proposed AR system, and consists of two sub-procedures: Reservation Request Handling and Booking Ticket Verification. The former handles reservation requests from app; the latter verifies the validity of a booking ticket according to the information sent from the ticket-scanning subsystem.

Reservation Request Handling. A simplified description considering only the successful case is elaborated as follows. When a tourist starts to reserve an attraction, app will send a *Booking-Attraction Request* to the central subsystem. When receiving this request, the Attraction Reservation Management Module performs the Reservation Request Handling sub-procedure to process this request. The central subsystem will find out all bookable sessions, each with a current bookable quota, of the attraction from the database, and return the result to app via a *Booking-Attraction Response*. Later, when receiving a *Booking-Session-Amount Request* from app, the central subsystem will add a new booking record and update related fields in the database according to the booking information appointed by the tourist, and return a confirmation reply to app by *Booking Confirmation*. A more detailed message flow chart including the failed case between the mobile app subsystem and the central subsystem is depicted in Fig. 3.

Booking Ticket Verification. As the central subsystem receives a *Ticket Verification Request* from the ticket-scanning subsystem (see Fig. 2), the Attraction Reservation Management Module performs the Booking Ticket Verification sub-procedure to verify this ticket according to the received information and the booking records in the database. If the booking record exists and the tourist arrives during the appointed period, then the central subsystem recognizes this ticket is valid, updates related records in the database, and returns a valid result via a *Ticket Verification Response* to the ticket-scanning subsystem. Otherwise, the central subsystem will respond an invalid result via a *Ticket Verification Response* to the ticket-scanning subsystem.

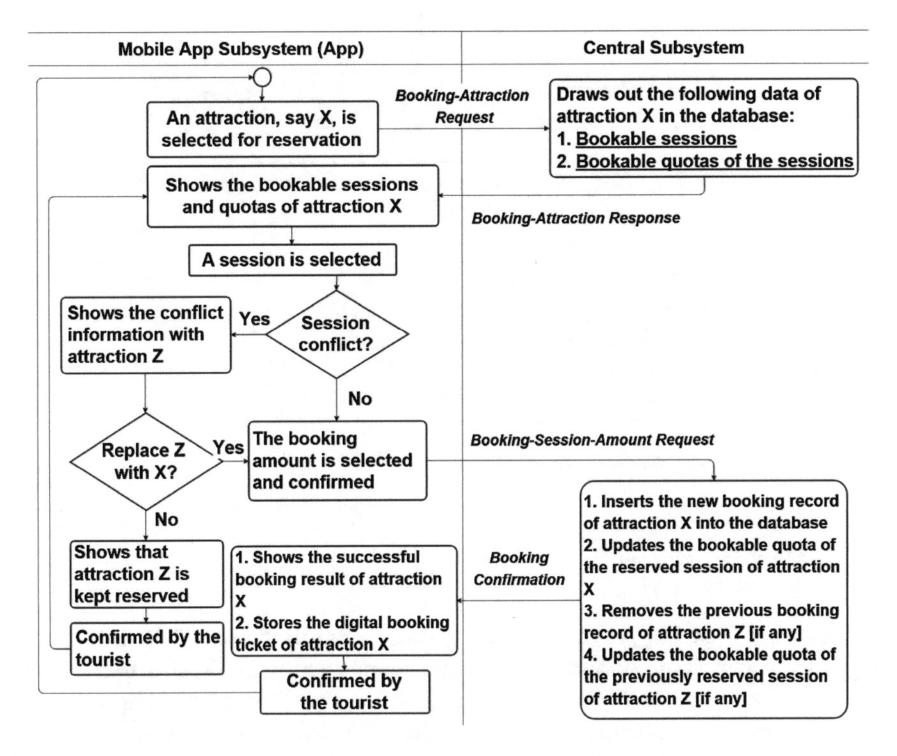

Fig. 3 Message flow chart of attraction reservation

4 System Implementation and Demonstration

4.1 Environment and Implementation Tools

The implementation environment of the prototyping AR system is shown in Fig. 4. The mobile app subsystem is developed in Android platform using Eclipse integrated development environment (IDE) with Android SDK. The booking ticket is generated in the form of QR codes. The central subsystem is implemented using Visual Studio C#, hosting on a desktop PC running Windows. Microsoft SQL Server serves as the system database. The ticket-scanning subsystem is implemented using Visual Studio C#, hosting on a notebook with an embedded webcam. The ticket scanner program is written as a QR code reader. The RES entrance gate is emulated by a program which exhibits a virtual gate on the notebook screen. As for the communication between the subsystems, the mobile app subsystem may communicate with the central subsystem via Wi-Fi or 3G/4G communications networks through the Internet, and the ticket-scanning subsystem may communicate with the central subsystem via Ethernet or Wi-Fi networks.

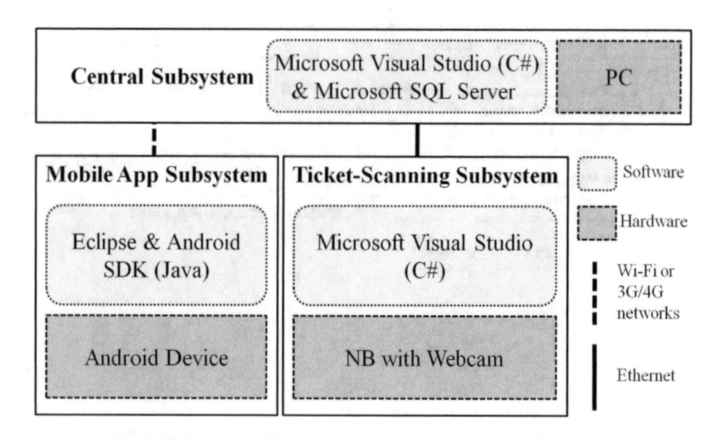

Fig. 4 Implementation environment of the prototyping AR system

4.2 Demonstration

The exhibition of the prototyping AR system is mostly presented in Traditional Chinese, where some necessary terms are commented with English translations besides. The testing field of the prototyping AR system is shown in Fig. 5. There are eight attractions located in the testing field. We will show some of our experimental results as follows. First, we test if the mobile app subsystem can access the database in the central subsystem correctly. We start the experiment from the Park Info menu of the mobile app subsystem. The successive screen captures are shown from Fig. 6a–e. Compared with the content in the database, we verify that the mobile app subsystem can access the database in the central subsystem and show the result correctly.

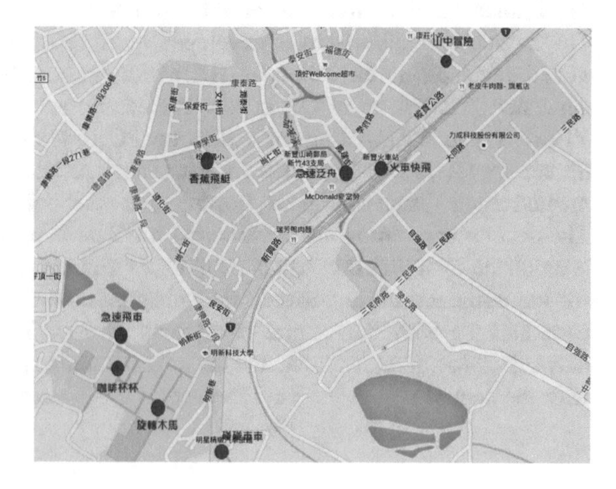

Fig. 5 Testing field of the prototyping AR system

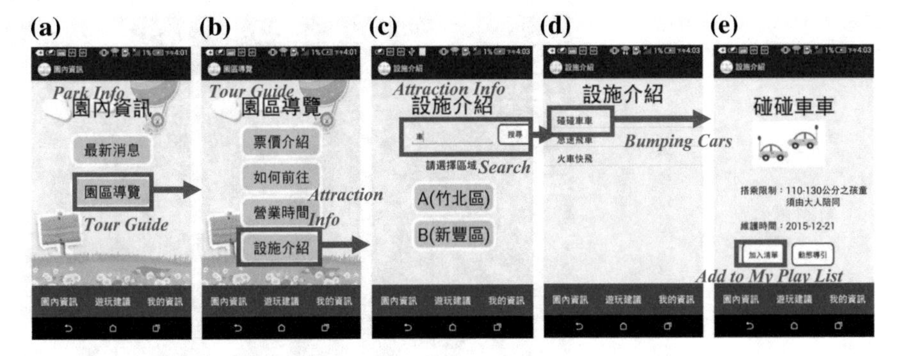

Fig. 6 Demonstration of Park Info

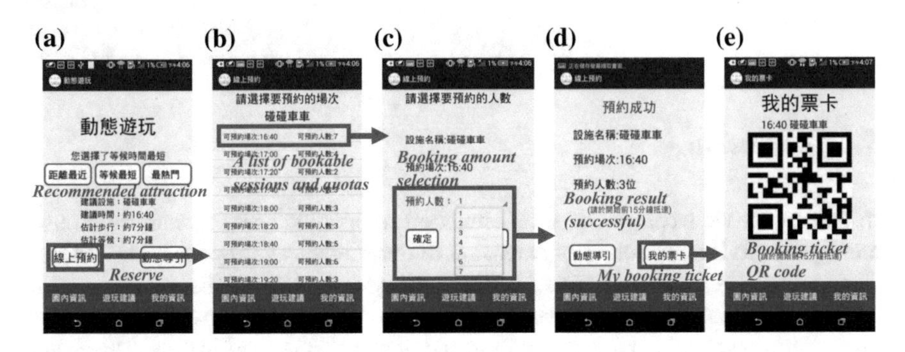

Fig. 7 Demonstration of attraction reservation

Second, the main features of the AR system, attraction reservation and booking ticket verification, are demonstrated as follows. We start the demonstration from the recommended attraction result (Fig. 7a) provided by the GPS-based dynamic scheduling function of the previously proposed SAP system. Then we start to reserve this recommended attraction. Figure 7b–d demonstrates the smartphone screen snapshots stepwise during the attraction reservation process. The generated booking ticket in the form of a QR code is shown in Fig. 7e. Figure 8 demonstrates how a user shows the QR code booking ticket (such as Fig. 7e) to the ticket-scanning subsystem (a webcam notebook) for ticket verification, where the picture was taken when we participated in a national contest with our prototyping system.

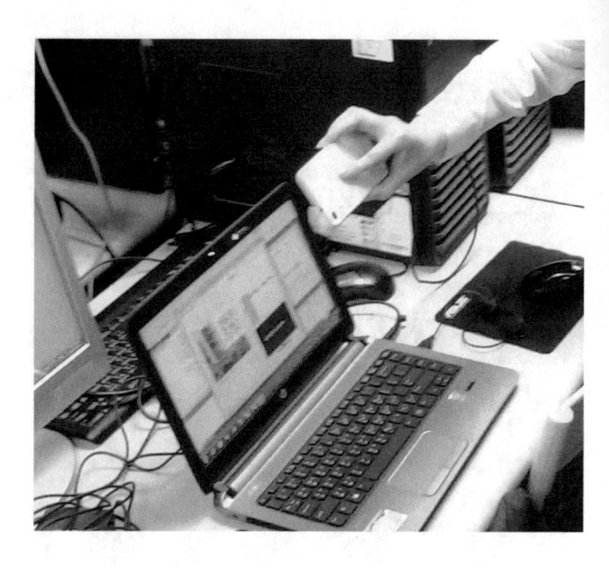

Fig. 8 Demonstration of booking ticket verification

5 Conclusion

In this paper, we have proposed an attraction reservation (AR) system for amusement park services, which is an extension to our previously proposed SAP system. The proposed AR system consists of three subsystems: the mobile app subsystem, the ticket-scanning subsystem, and the central subsystem, and provides two service procedures: reservation request handling and booking ticket verification. The design and implementation of the prototyping AR system are described in this paper, and the demonstrations show that the proposed AR system does function.

Acknowledgements This work is supported by the Ministry of Science and Technology of the Republic of China under Grant MOST 104-2622-E-159-008-CC3.

References

1. Tourism Bureau, M.O.T.C. Republic of China (Taiwan), http://admin.taiwan.net.tw/upload/contentfile/auser/b/wpage/chp1/1_1.1.htm
2. Walt Disney World®, https://disneyworld.disney.go.com/
3. Universal Orlando Resort, https://www.universalorlando.com/Home.aspx
4. Ocean Park Hong Kong, http://www.oceanpark.com.hk/html/en/home/
5. Janfusun Fancy World, http://fancyworld.janfusun.com.tw/en/
6. Leofoo Village Theme Park, http://www1.leofoo.com.tw/village/en/Info.aspx
7. My Disney Experience Mobile App on Google Play, https://play.google.com/store/apps/details?id=com.disney.mdx.wdw.google
8. My Disney Experience - Walt Disney World Mobile App on iTunes, https://itunes.apple.com/app/id547436543

9. Universal Orlando® Resort App on Google Play, https://play.google.com/store/apps/details?id=com.universalstudios.orlandoresort
10. The Official Universal Orlando® Resort App on iTunes, https://itunes.apple.com/tw/app/official-universal-orlando/id878217080
11. Ocean Park Hong Kong Mobile App on Google Play, https://play.google.com/store/apps/details?id=com.oceanpark.mobileapp
12. Ocean Park Hong Kong Mobile App on iTunes, https://itunes.apple.com/hk/app/ocean-park-hong-kong/id554122091
13. Janfusun Fancyworld Mobile App on Google Play, https://play.google.com/store/apps/details?id=tw.com.rainmaker.jfs
14. Leofoo Village Theme Park Mobile App on Google Play, https://play.google.com/store/apps/details?id=com.uni.leofoo
15. Leofoo Village Theme Park Mobile App on iTunes, https://itunes.apple.com/tw/app/liu-fu-cun-hao-hao-wan/id862930941?mt=8
16. FastPass+, https://disneyworld.disney.go.com/plan/my-disney-experience/fastpass-plus/
17. Universal Express™ Pass, https://www.universalorlando.com/Theme-Park-Tickets/Universal-Express/Express-Passes.aspx
18. Lee, P.C., Wang, S.S., Ku, P.H.: SAP: A Smart Amusement Park System for Tourist Services. In: ICITE 2015: 17th International Conference on Information Technology and Engineering, pp. 378–383. WASET (2015)

Digital Game Design—Brave Loop

Yen-Liang Wu, Shia Yu, Shih-Tav Kao, Chih-Yu Chen, Hao-Wen Liu, Po-Han Chen, Ci-Yuan Xu and Rong-Chi Chang

Abstract This game is a game full of action adventures. This game is designed to emphasize a 2.5D visual style. The game mainly relates to 3D roles and scene objects to create fantasies and gorgeous story and its scenarios. This game focuses on character interactive process, through switches of "white" and "shadow," players are able to understand the world of the story. Players need to think deep how creation and destruction power can influence the game world. This game provides teenagers with different philosophical experiences.

Keywords Digital game · Action-adventure · Fantasies

1 Introduction

One of people's important motives to play 'Games,' is that they hope to have "fun" from the games. This is also the public's cognition and visual impression of games. Buckingham [1] Computer Games published a book 'Computer Games.' In the book he made an observation and analysis on the games nowadays and the fun to play games. He proposed that a game has three aspects: Ludic, Representational, and Interaction.

Y.-L. Wu · S. Yu · S.-T. Kao · C.-Y. Chen · H.-W. Liu · P.-H. Chen · C.-Y. Xu
Department of Digital Media Design, Asia University, Taichung 413, Taiwan, ROC
e-mail: aw@asia.edu.tw

R.-C. Chang (✉)
Department of Technology Crime Investigation, Taiwan Police College, Taipei, Taiwan, ROC
e-mail: roger@mail.tpa.edu.tw

© Springer Nature Singapore Pte Ltd. 2018

N.Y. Yen and J.C. Hung (eds.), *Frontier Computing*, Lecture Notes in Electrical Engineering 422, DOI 10.1007/978-981-10-3187-8_97

In order to confront with the challenging targets, players are always indulged in thinking and high concentration such as dealing with new story messages, set strategies and solve problems, etc. While designing a game, designers merged themselves into the narration of the story, roles modeling and worldviews of the game, etc. In the course of playing and experience transmitting, players seem to have lost external senses such. For example, players are attracted by the dazzling and stunning sound effects. They will then be much indulged in the game world.

This game is designed to them on 'bravery' of adventure, expecting to create a positive sense. This game has created two roles of 'Creation' and 'Destruction' who are equipped with different powers by light and dark. Players can then obtain diversified information by different reaction generated from the interaction with the game roles. Levels of the game are continued in the form of beads and between the levels there is a rest stop. Players can decide to interact with the NPC or continue to go to the next level. And if they go to the next level, they will not able to go back to the previous one. While playing, players not only need to think how to solve problems and complete missions, there are also 'egg hunts' for the players to search and discover.

2 Game Development and Design

This game applies Unity engine. The visual effects adopt style of fantasies. In order to match the game roles, game interface adopts simplified style of black and white. Game interface is simple to provide players with clear information.

Player's skills are Creation and Destruction. Creative skills are road construction, vertical vine making and forming a large range of vine. Destruction forces are short-range damage, instant moving blasting and 10-s widespread damage. Energy consumption of these two forces is different.

The game has more than 30 virtual roles, each character has his (her) little story. Players can switch roles as Creation or Destruction and respectively dialogue with NPC to get diversified information. Figure 1 is the game architecture. Different roles of the game has his (her) own features and story. Table 1 lists the major roles and descriptions.

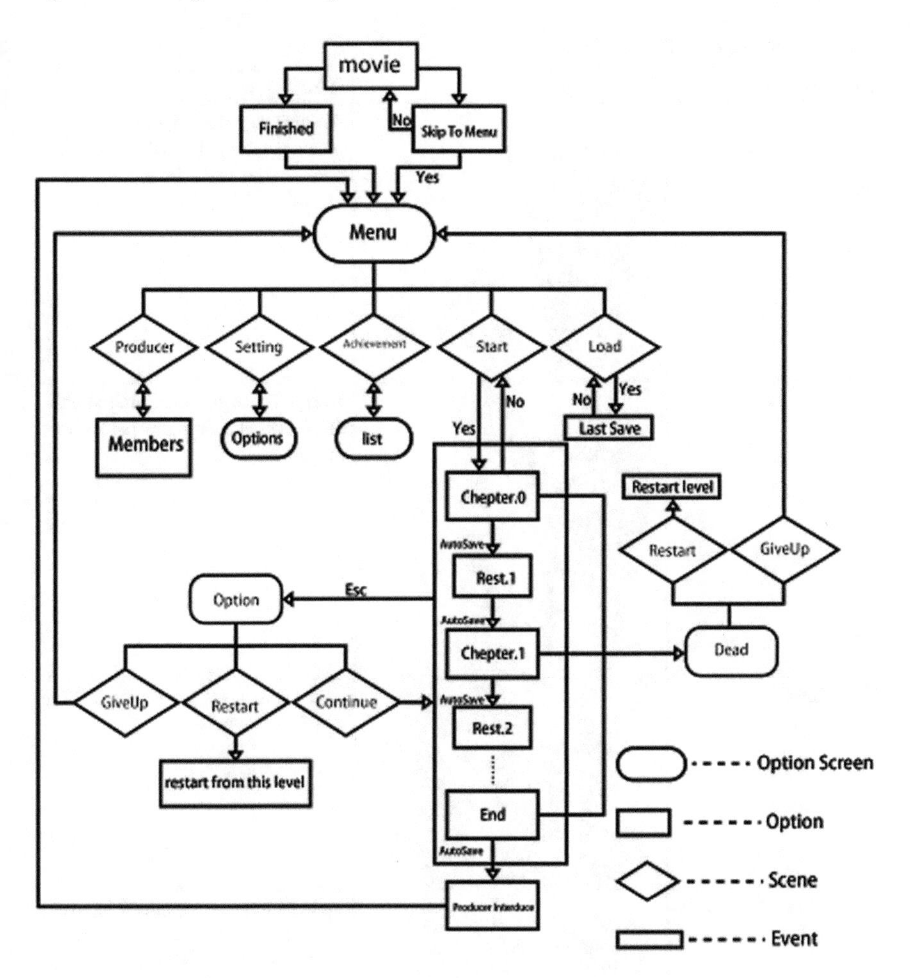

Fig. 1 Game architecture

Table 1 Introdcution of roles design

Name of roles	Design drawing	Description
Shiro		An ordinary person living on plants growing magic. She admires Braver who save the world in a legend
Brave/Yusha		Brave, white shadow, long silent, often asks his ordinary person not to go on
God/Kami		God in the game world; stays at the end of the world
Twinsister-ordinary and sage		Twin sisters are separated in two different worlds. They seem to know the truth of the world

3 Demonstrations

Game levels are forms in the horizontal with multiple routes. Players can choose a different route or way to move on. By applying switching power of Creation and Destructive to overcome the continual collapsing floors ahead and safely reach the end of each level. With level's moving on, each level will be gradually added some elements such as the player's new skills, new monsters, and new tricks. Besides the

Fig. 2 Game operation screen

(a) Opening of Level,plot is developed by dialogue between the two roles.

(b) Control arrow keys to move your role; floors behindwill be collapsing over time.

(c) When hollows appear, player uses Creation skills to pave the road to continue to move on.

(d) Destruction skill can be applied to eliminate obstacles ahead.

terrain obstacles player needs to break, he (she) will also encounter beasts and they will interrupt the player to move on or even attack the main role. The operation screen is shown in Fig. 2.

4 Conclusion

This game is designed to emphasize the role of visual styles and role features. The story is merged with fantasy and reality. Features of the natural environment are applied in scene design so that players can own a special experience. Players in the game need to think how to interactively switch between Creativity and Destructive. The game contains a little philosophical significance and is suitable for teenagers to experience.

Reference

1. Buckingham, D. 2006. Studying computer games. In *Computer Games: Text, Narrative, Player,* ed. D. Carr, D. Buckingham, A. Burn, and G. Schott, 1-13. Cambridge, U.K.: Polity Press.

An Adventure Game: Kido Senki Valhalla

Ch'eng-Han Hsu, Hsien-Che Ch'en, Cheng-Ru Jiang, Che-Hsien Lin, Wen-Yi Sung, Hsin-Ch'en Wang and Yi-Chun Liao

Abstract Kido Senki Valhalla that is a graphic adventure game with an original story designed by a student team 'Forward'. We focus on the graphic design of the mobile suits of game characters. A player will enjoy the game story with many CG illustrations. The end of game story will be created by the player. Soundtracks are also design by this student team. Game voice dubbing is also an interesting work for a student team to dub the dialogues in the game scene. All voice actors come from classmates. Over 2 h playing time is implemented by Unity game engine.

Keywords AVG · Kido Senki · Game storyboard

1 Introduction

"Adventure game" originated from the 1970s text computer game Colossal Cave Adventure, or as it was often referred to simply as Adventure [1], Adventure games have strong storylines with significant dialog, and sometimes make effective use of recorded dialog or narration from voice actors. This genre of game is known for representing dialog as a conversation tree. Players are able to engage a non-player character by choosing a line of pre-written dialog from a menu, which triggers a response from the game character. These conversations are often designed as a tree structure, with players deciding between each branch of dialog to pursue. However, there are always a finite number of branches to pursue, and some adventure games devolve into selecting each option one-by-one. Conversing with characters can reveal clues about how to solve puzzles, including hints about what that character

C.-H. Hsu · H.-C. Ch'en · C.-R. Jiang · C.-H. Lin · W.-Y. Sung · H.-C. Wang · Y.-C. Liao (✉)
Department of Digital Multimedia Design, China University of Technology, Taipei, Taiwan, ROC
e-mail: ech_liao@cute.edu.tw

C.-H. Hsu
e-mail: in61006@gmail.com

© Springer Nature Singapore Pte Ltd. 2018

N.Y. Yen and J.C. Hung (eds.), *Frontier Computing*, Lecture Notes in Electrical Engineering 422, DOI 10.1007/978-981-10-3187-8_98

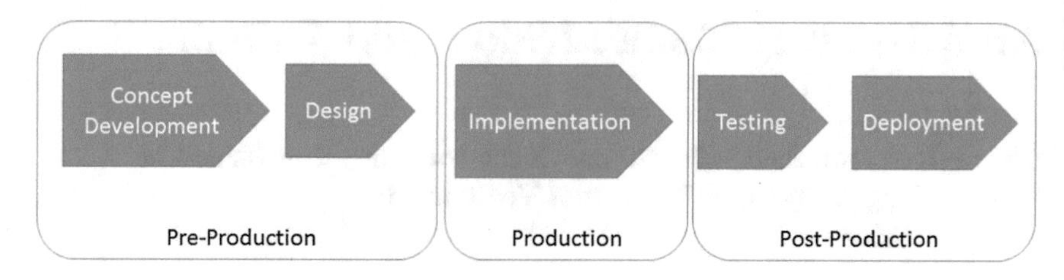

Fig. 1 Design steps

would want before they will cooperate with the player. Other conversations will have far-reaching consequences, deciding to disclose a valuable secret that has been entrusted to the player. Characters may also be convinced to reveal their own secrets, either through conversation or by giving them something that will benefit them [2].

A side-scrolling game, side-scroller or 2D is a video game in which the gameplay action is viewed from a side-view camera angle, and the onscreen characters generally move from the left side of the screen to the right (or less commonly, right to left) to meet an objective. These games make use of scrolling computer display technology. The move from single-screen or flip-screen graphics to scrolling graphics, during the golden age of video arcade games and during third-generation consoles, would prove to be a pivotal leap in game design, comparable to the move to 3D graphics during the fifth generation [3].

Sometimes 'Kido Senki' is known as 'Mobile Suit Gundam Wing'. Mobile suit is a feature of the robot for videogame players. In our game, mobile suits are designed in Japanese style. 'Kido Senki Valhalla' is an AGV game that combines of text adventure and 2D side-scrolling action game, players have to be a member of "Valhalla" to protect the homeland to fight! But a robot video game is not popular in Taiwan, especially for a student team. All designs made it special by adding CG illustrations, song lyrics, singing, voice, six members took one and a half years to complete the whole game. Game time is about 2 h, the quality and quantity of the game are provided in 100% in Fig. 1.

2 Story, Setting and Character

The design steps are three steps in which the segmentation of the steps is not clear in Fig. 2. For us, team members spent a lot of time to find the basic core mechanics. A big story was written down for 2 months. Finally, Unity game engine is used with a plugin package.

Besides, there three main components in our video game in Fig. 3. The unity game engine includes animation and sounds functions. But the dialogues function and the story of the game are not easy built in unity. The whole game story is over 2 h and a plugin asset is better to do this.

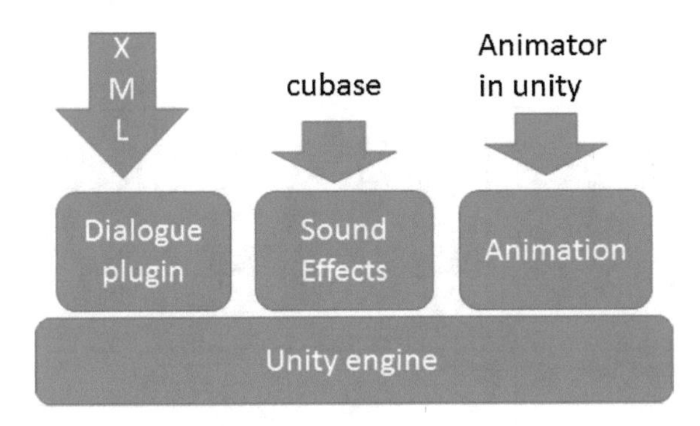

Fig. 2 Game components

Fig. 3 Components and animation of the mobile suit

This is an adventure game that combines side-scrolling and interactive action. Players will explore the whole story by a variety of main characters. Game World is a story which introduce main characters that manipulates the giant mobile suit to revolt enemies after the earth crisis that is invaded by numerous aliens.

Unfortunately, our game character, the female leader "Tiara" meets the murderer who makes her family bankrupt and the people dead in those years. Table 1 lists the major roles and descriptions. We designed these roles with everyone's ideas. after integrating, these differences are good for artists and can create special roles. To highlight the characters, Role's hair and clothing are exaggerated. Especially in the jersey unified design.

Table 1 Roles conception

Roles sketch-style1	Roles sketch-style2	Roles sketch-final
Name of roles	Design drawing	
Male leader—Asano Sho		
Female leader—Tiara. GLaria		
Anime Petit and Fix		

(continued)

Table 1 (continued)

Name of roles	Design drawing
Iris and Jodi	

2.1 *Mobile Suit Design*

The mechanical is our special design with personality and plot to set. Players will have the four mobile suits, the names are PA. Swats, Serket, Minerva and Ymir in Fig. 4. Other names come from the myth.

3 Demonstrations

A player drives the special mobile suit to attack the enemy, and the mission is to clear all enemies. In the last scene, there is a big boss that is waiting for players to challenge.

Game scene Features:

(1) Diversified enemies: find the fastest way to knock down every enemy! The attacking fun is not only one.
(2) Scoring: defeating more enemies fast can get the higher scores in a scene.
(3) Interactive story: the player can find out a game object to experience the interactive story.

The game screens are shown in Fig. 5.

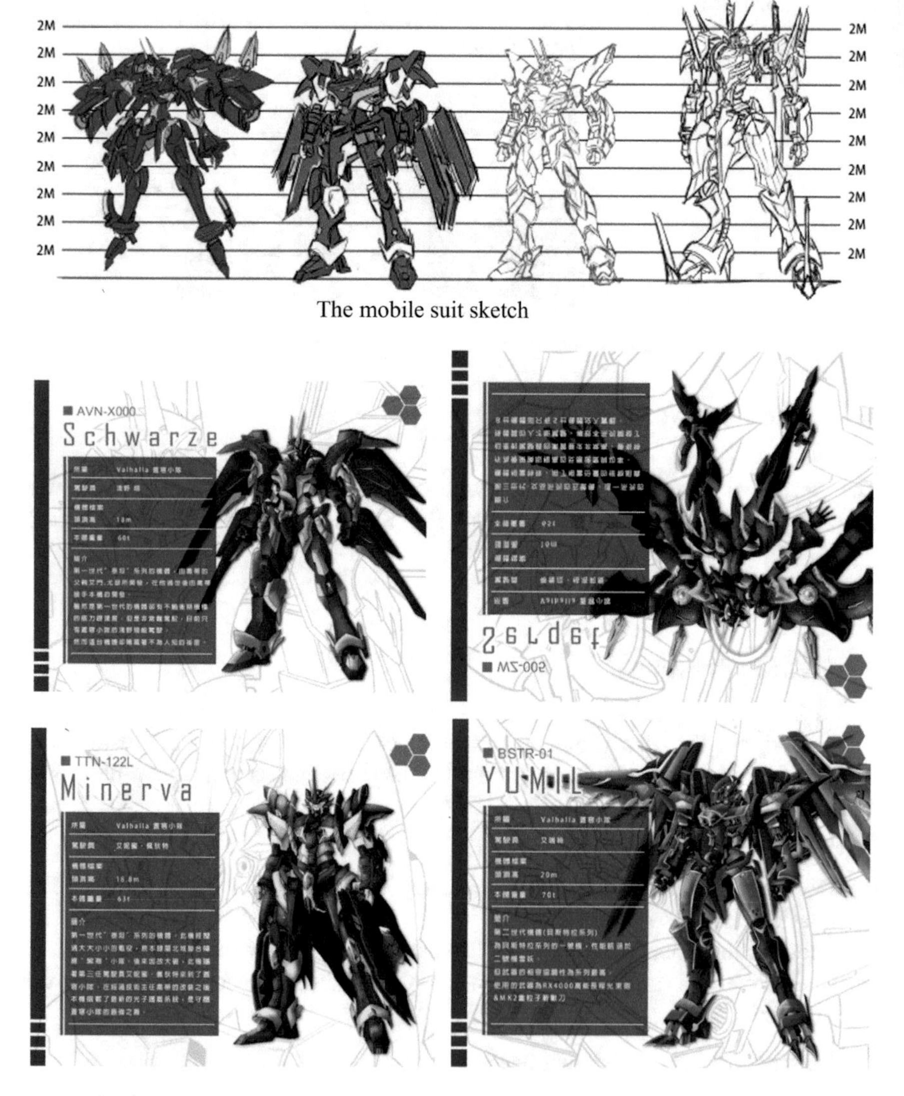

Fig. 4 The four mobile suits

(a) Drive a mobile suit

(b) Game dialogues.

(c) Game scenes

Fig. 5 Game screens

4 Conclusion

For the student team, they have learned the theoretical and practical foundations of video game production using the Unity 3D game engine. A great idea was achieved step by step that is very important. All practices will let we become a game designer, game artist, or game programmer.

The player can experience adventure and drama system in this game. The game features are:

(1) Interactive story: Player can decide story developing.
(2) Abundant sound effect: Rich sound effects and give the player the immersive experience.
(3) Illustration: a lot of the drama and collect CG artwork.
(4) The drama has a lot of picking. They can influence different ending.

References

1. Rollings, Andrew and Adams, Ernest (2003). Andrew Rollings and Ernest Adams on Game Design. New Riders. ISBN 1-59273-001-9.
2. https://en.wikipedia.org/wiki/Adventure_game.
3. Steven L. Kent. The Ultimate History of Video Games, ISBN 9780761536437.

Index

© Springer Nature Singapore Pte Ltd. 2018
N.Y. Yen and J.C. Hung (eds.), *Frontier Computing*, Lecture Notes in Electrical Engineering 422, DOI 10.1007/978-981-10-3187-8

Printed in the United States
By Bookmasters